Medicinal Chemistry

(Second Edition)

D. Sriram *and* **P. Yogeeswari**

BITS Pilani, Hyderabad Campus

Andhra Pradesh

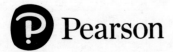

Associate Acquisitions Editor: Anita Yadav
Production Editor: Vamanan Namboodiri

Copyright © 2010 Dorling Kindersley (India) Pvt. Ltd

Copyright© 2007 Dorling Kindersley (India) Pvt. Ltd

This book is sold subject to the condition that it shall not, by way of trade or otherwise, be lent, resold, hired out, or otherwise circulated without the publisher's prior written consent in any form of binding or cover other than that in which it is published and without a similar condition including this condition being imposed on the subsequent purchaser and without limiting the rights under copyright reserved above, no part of this publication may be reproduced, stored in or introduced into a retrieval system, or transmitted in any form or by any means (electronic, mechanical, photocopying, recording or otherwise), without the prior written permission of both the copyright owner and the above-mentioned publisher of this book.

Although the author and publisher have made every effort to ensure that the information in this book was correct at the time of editing and printing, the author and publisher do not assume and hereby disclaim any liability to any party for any loss or damage arising out of the use of this book caused by errors or omissions, whether such errors or omissions result from negligence, accident or any other cause. Further, names, pictures, images, characters, businesses, places, events and incidents are either the products of the author's imagination or used in a fictitious manner. Any resemblance to actual persons, living or dead or actual events is purely coincidental and do not intend to hurt sentiments of any individual, community, sect or religion.

In case of binding mistake, misprints or missing pages etc., the publisher's entire liability and your exclusive remedy is replacement of this book within reasonable time of purchase by similar edition/reprint of the book.

ISBN 978-81-317-3144-4

First Impression, 2010

Published by Pearson India Education Services Pvt. Ltd, CIN: U72200TN2005PTC057128.

Head Office: 15th Floor, Tower-B, World Trade Tower, Plot No. 1, Block-C, Sector 16, Noida 201 301, Uttar Pradesh, India.

Registered Office: 7th Floor, SDB2, ODC 7, 8 & 9, Survey No.01 ELCOT IT/ ITES SEZ, Sholinganallur, Chennai – 600119, Tamil Nadu, India. Website: in.pearson.com,
Email: companysecretary.india@pearson.com

Digitally printed in India by Trinity Academy for Corporate Training Ltd, New Delhi in the year of 2021.

CONTENTS

Preface   xii

Chapter 1 Drug Discovery and Development   1
1.1   Introduction   2
1.2   How are the New Drugs Discovered?   2
1.3   Requirements for an Investigational Drug to be Tested in Human Volunteers   4
1.4   How are the Investigational Drugs Tested in Humans?   5
1.5   How Does Testing Continue after a New Drug is Approved?   6
1.6   Procedures Followed in Drug Design   6
      Further Readings   16
      Multiple Choice Questions   16
      Questions   17
      Solution to Multiple Choice Questions   18

Chapter 2 Principles of Drug Action   19
2.1   Introduction   20
2.2   Functional Group Review   20
2.3   Stereochemistry   27
2.4   Acid-Base Chemistry   29
2.5   Biologically Significant Nitrogen-Containing Compounds   34
2.6   Physicochemical Properties and Drug Action   36
2.7   Bioisosteric Replacement   39
2.8   Drug-Receptor Bonding   40
2.9   G Protein-Coupled Receptors   49
      Further Readings   50
      Multiple Choice Questions   50
      Questions   51
      Solution to Multiple Choice Questions   52

Chapter 3 Drug Metabolism and Prodrugs   53
3.1   Drug Metabolism   54
3.2   Prodrugs   55
      Further Readings   73
      Multiple Choice Questions   74
      Questions   74
      Solution to Multiple Choice Questions   74

Chapter 4 Computer-aided Drug Design   75
4.1   Introduction   76
4.2   Cadd Methodology   76
4.3   Small Molecule-Based Drug Design   76
4.4   Macromolecule-Based Drug Design   86
4.5   Substituent Constants Useful for Qsar Analysis   89
      Further Readings   90
      Multiple Choice Questions   90
      Questions   91

Solution to Multiple Choice Questions   93

Chapter 5 General Anaesthetics   94
- 5.1 Introduction—Anaesthesia and its Stages   95
- 5.2 Mechanism of Action   95
- 5.3 Classification   95
- 5.4 Inhalation Anaesthetics (Synthesis, Uses, and Dose)   96
- 5.5 Intravenous Anaesthetics (Synthesis, Uses, and Dose)   99
- 5.6 Newer Drugs   107
- *Further Readings*   108
- *Multiple Choice Questions*   108
- *Questions*   108
- *Solution to Multiple Choice Questions*   108

Chapter 6 Local Anaesthetics   109
- 6.1 Introduction   110
- 6.2 Ideal Properties of Local Anaesthetics   110
- 6.3 Mechanism of Action   110
- 6.4 Clinical Uses of Local Anaesthetics   111
- 6.5 Classification of Local Anaesthetics   111
- 6.6 Structures, Synthesis, and Structure-Activity Relationship (SAR) of Benzoic Acid Derivatives   113
- 6.7 Structures, Synthesis, and SAR of *P*-Aminobenzoic Acid Derivatives   116
- 6.8 Structures, Synthesis, and SAR of Anilides   119
- 6.9 Miscellaneous   122
- 6.10 Newer Drugs   126
- *Further Readings*   126
- *Multiple Choice Questions*   127
- *Questions*   128
- *Solution to Multiple Choice Questions*   128

Chapter 7 Sedatives, Hypnotics, and Anxiolytic Agents   129
- 7.1 Introduction   130
- 7.2 Classification of Sedative and Hypnotics   130
- 7.3 Barbiturates   133
- 7.4 Benzodiazepines   136
- 7.5 Benzodiazepine Antagonists   143
- 7.6 Miscellaneous   144
- 7.7 Anxiolytic Agents   148
- 7.8 Newer Drugs   150
- *Further Readings*   150
- *Multiple Choice Questions*   150
- *Questions*   151
- *Solution to Multiple Choice Questions*   152

Chapter 8 Anti-Epileptic Drugs   153
- 8.1 Introduction: Epilepsy and Types of Seizures   153
- 8.2 Classification   155
- 8.3 Structural Requirements for Anticonvulsant Agents and Their Structure-Activity Relationship   155
- 8.4 Hydantoins   158
- 8.5 Oxazolidinediones   159
- 8.6 Succinimides   159

8.7	Benzodiazepines	160
8.8	Gaba Analogues	161
8.9	Miscellaneous	163
8.10	Newer Anticonvulsants	164
	Further Readings	*170*
	Multiple Choice Questions	*171*
	Questions	*171*
	Solution to Multiple Choice Questions	*172*

Chapter 9 Antipsychotic Agents 173

9.1	Introduction	173
9.2	Typical or Classical Antipsychotic Agents: Synthesis, Mechanism of Action, and SAR	174
9.3	Atypical Antipsychotic Agents	185
9.4	Newer Drugs	191
	Further Readings	*193*
	Multiple Choice Questions	*193*
	Questions	*194*
	Solution to Multiple Choice Questions	*194*

Chapter 10 Antidepressants 195

10.1	Introduction	195
10.2	Classification	196
10.3	Mechanism of Action	196
10.4	Side-Effects of Classical Antidepressants	196
10.5	Monoamino Oxidase (MAO) Inhibitors	196
10.6	Tricyclic Antidepressants	198
10.7	Atypical Antidepressants	200
10.8	Miscellaneous Drugs	202
10.9	Newer Drugs	203

	Further Readings	*205*
	Multiple Choice Questions	*205*
	Questions	*205*
	Solution to Multiple Choice Questions	*206*

Chapter 11 Narcotic Analgesics 207

11.1	Introduction	207
11.2	Mechanism of Action	208
11.3	Pharmacophore Requirement	208
11.4	Clinical Uses	209
11.5	Classification	209
11.6	Morphine and its Derivatives	212
11.7	Phenyl (Ethyl) Piperidines	216
11.8	Diphenyl Heptanones	219
11.9	Benzazocin (Benzomorphan) Derivatives	220
11.10	Narcotic Antagonists	221
11.11	Newer Drugs	223
	Further Readings	*224*
	Multiple Choice Questions	*225*
	Questions	*226*
	Solution to Multiple Choice Questions	*226*

Chapter 12 Antipyretics and Non-Steroidal Anti-Inflammatory Drugs 227

12.1	Introduction	227
12.2	Classification	228
12.3	Biosynthesis of Eicosanoids	228
12.4	Mechanism of Anti-Inflammatory Action	229
12.5	Side-Effects	229

12.6	Salicylic Acid Derivatives	230		**Chapter 14**	**Antihistamines and Anti-Ulcer Agents**	**277**
12.7	P-Aminophenol Derivatives	231		14.1	Introduction	277
12.8	Pyrazolidinedione Derivatives	233		14.2	H_1 Receptor Antagonist	278
12.9	Anthranilic Acid Derivatives (Fenamates)	235		14.3	H_2 Receptor Antagonist	294
				14.4	Histamine H_3 Receptors	298
12.10	Aryl Alkanoic Acid Derivatives	236		14.5	Histamine H_4 Receptors	299
12.11	Oxicams	244		14.6	Proton Pump Inhibitors	299
12.12	SAR (Structure Activity Relationship of Oxicams)	244		14.7	Newer Drugs	302
					Further Readings	*302*
12.13	Miscellaneous Drugs	246			*Multiple Choice Questions*	*302*
12.14	Anti-Gout Drugs	247			*Questions*	*303*
12.15	Selective COX-2 Inhibitors	248			*Solution to Multiple Choice Questions*	*304*
12.16	Newer Drugs	250				
	Further Readings	*252*		**Chapter 15**	**Diuretics**	**305**
	Multiple Choice Questions	*252*		15.1	Introduction	305
	Questions	*253*		15.2	Uses	305
	Solution to Multiple Choice Questions	*254*		15.3	Classification	306
				15.4	Osmotic Diuretics	307
				15.5	Carbonic Anhydrase Inhibitors	309
Chapter 13	**Miscellaneous CNS Agents**	**255**		15.6	Thiazide Diuretics	311
				15.7	Potassium-Sparing Diuretics	314
13.1	Anti-Parkinsonism Agents	255		15.8	Loop Diuretics	318
13.2	Muscle Relaxants	260		15.9	Miscellaneous Drugs	320
13.3	Drugs for Alzheimer's Disease Treatment	266		15.10	Newer Drugs	324
13.4	CNS Stimulants	270			*Further Readings*	*325*
13.5	Newer Drugs	274			*Multiple Choice Questions*	*325*
	Further Readings	*275*			*Questions*	*327*
	Multiple Choice Questions	*275*			*Solution to Multiple Choice Questions*	*327*
	Questions	*276*				
	Solution to Multiple Choice Questions	*276*		**Chapter 16**	**Antihypertensive Agents**	**328**
				16.1	Introduction	328

16.2	Classification	330
16.3	Peripheral Anti-Adrenergic Drugs	330
16.4	Centrally Acting Antihypertensive Drugs	332
16.5	Direct Vasodilators	334
16.6	Ganglionic Blocking Agents	336
16.7	β-Adrenergic Blockers	336
16.8	Calcium Channel Blockers	339
16.9	Angiotensin-Converting Enzyme (ACE) Inhibitors	342
16.10	Angiotensin II Antagonists (AT$_2$ Antagonists)	345
16.11	Miscellaneous Drugs	347
16.12	Newer Drugs	347
	Further Readings	348
	Multiple Choice Questions	349
	Questions	350
	Solution to Multiple Choice Questions	350

Chapter 17 Antiarrhythmic Drugs 351

17.1	Introduction	351
17.2	Classification	352
17.3	Class II Agents	356
17.4	Class III Agents	356
	Further Readings	357
	Multiple Choice Questions	358
	Questions	358
	Solution to Multiple Choice Questions	358

Chapter 18 Antihyperlipidemic Agents 359

18.1	Introduction	359
18.2	Classification	361
18.3	HMG-COA Reductase Inhibitors (Statins)	361
18.4	Aryloxyisobutyric Acid Derivatives (Fibrates)	365
18.5	Miscellaneous Drugs	367
18.6	Newer Drugs	369
	Further Readings	371
	Multiple Choice Questions	371
	Questions	372
	Solution to Multiple Choice Questions	372

Chapter 19 Antianginal Drugs 373

19.1	Introduction	373
19.2	Classification	374
19.3	Nitrates/Nitrites	374
19.4	β-Adrenergic Blockers	377
19.5	Calcium Channel Blockers	377
19.6	Miscellaneous Drugs	377
19.7	Newer Drugs	378
	Further Readings	379
	Multiple Choice Questions	379
	Questions	380
	Solution to Multiple Choice Questions	380

Chapter 20 Insulin and Oral Hypoglycaemic Agents 381

20.1	Introduction	381
20.2	Insulin	382
20.3	Oral Hypoglycaemic Agents Classification	383
20.4	Strategy For Controlling Hyperglycaemia	385
20.5	Sulphonyl Ureas	386

20.6	Biguanides	387
20.7	Substituted Benzoic Acid Derivatives (Meglitinides)	388
20.8	Thiazolidinediones	389
20.9	Miscellaneous Drugs	390
20.10	Newer Drugs	391
	Further Readings	*392*
	Multiple Choice Questions	*392*
	Questions	*393*
	Solution to Multiple Choice Questions	*393*

Chapter 21 Oral Anticoagulants 394

21.1	Introduction	394
21.2	Classification	396
21.3	Coumarin Derivatives	397
21.4	Indanedione Derivatives	399
21.5	Newer Drugs	399
	Further Readings	*400*
	Multiple Choice Questions	*400*
	Questions	*400*
	Solution to Multiple Choice Questions	*400*

Chapter 22 Adrenergic Drugs 401

22.1	Introduction	401
22.2	Sympathomimetic Drugs	402
22.3	Sympathetic (Adrenergic) Blocking Agents (Sympatholytics)	408
22.4	Newer Drugs	411
	Further Readings	*411*
	Multiple Choice Questions	*412*
	Questions	*412*
	Solution to Multiple Choice Questions	*412*

Chapter 23 Cholinergic Drugs 413

23.1	Introduction	413
23.2	Cholinomimetics	415
23.3	Anticholinesterases	418
23.4	Cholinergic Blocking Agents (Antimuscarinic Drugs)	422
23.5	Newer Drugs	426
	Further Readings	*427*
	Multiple Choice Questions	*427*
	Questions	*428*
	Solution to Multiple Choice Questions	*428*

Chapter 24 Sulphonamides, Sulphones, and Dihydrofolate Reductase Inhibitors 429

24.1	Sulphonamides	429
24.2	Sulphones	435
24.3	Dihydrofolate Reductase (DHFR) Inhibitors	435
24.4	Newer Drug	437
	Further Readings	*437*
	Multiple Choice Questions	*437*
	Questions	*438*
	Solution to Multiple Choice Questions	*438*

Chapter 25 Quinolone Antibacterials 439

25.1	Introduction	439
25.2	Structures of Fluoroquinolones	439

25.3	Mechanism of Action	440
25.4	Uses	441
25.5	Classification	441
25.6	SAR of Quinolones	441
25.7	Adverse Effects	442
25.8	Synthesis of Fluoroquinolones	443
25.9	Newer Drugs	446
	Further Readings	446
	Multiple Choice Questions	446
	Questions	446
	Solution to Multiple Choice Questions	446

Chapter 26 Antibiotics 447

26.1	β-Lactam Antibiotics	448
26.2	Tetracycline Antibiotics	472
26.3	Macrolide Antibiotics	477
26.4	Aminoglycoside Antibiotics	482
26.5	Miscellaneous Antibiotics	482
26.6	Newer Drugs	484
	Further Readings	487
	Multiple Choice Questions	487
	Questions	488
	Solution to Multiple Choice Questions	488

Chapter 27 Antitubercular Agents 489

27.1	Introduction	489
27.2	Mechanisms of Action	489
	Further Readings	493
	Multiple Choice Questions	493
	Questions	494
	Solution to Multiple Choice Questions	494

Chapter 28 Antifungal Agents 495

28.1	Introduction	495
28.2	Classification	496
28.3	Antifungal Antibiotics	496
28.4	Azole Antifungals	497
28.5	Pyrimidine Derivatives	500
28.6	Miscellaneous Agents	500
28.7	Newer Drugs	502
	Further Readings	503
	Multiple Choice Questions	503
	Questions	504
	Solution to Multiple Choice Questions	504

Chapter 29 Antiviral Agents 505

29.1	Anti-HIV Agents	505
29.2	Anti-Herpes Simplex Virus (HSV) Agents	522
29.3	Miscellaneous Agents	527
29.4	Newer Drugs	528
	Further Readings	530
	Multiple Choice Questions	530
	Questions	531
	Solution to Multiple Choice Questions	532

Chapter 30 Antiprotozoal Agents 533

30.1	Introduction	533
30.2	Antiamoebic Agents	533
30.3	Antimalarial Agents	536
30.4	Anthelmintics	546
30.5	Miscellaneous Antiprotozoal Drugs	552
30.6	Newer Drug	554

	Further Readings	555
	Multiple Choice Questions	555
	Questions	556
	Solution to Multiple Choice Questions	556

Chapter 31 Anticancer Agents 557

31.1	Introduction	557
31.2	Limitations of Therapy	558
31.3	Classification	558
31.4	Alkylating Agents	558
31.5	Antimetabolites	568
31.6	Anticancer Antibiotics	574
31.7	Miscellaneous Drugs	575
31.8	Anticancer Plant Products	577
31.9	Hormones and Their Antagonists	579
31.10	Miscellaneous Drugs	580
31.11	Newer Drugs	582
	Further Readings	582
	Multiple Choice Questions	582
	Questions	583
	Solution to Multiple Choice Questions	584

Chapter 32 Prostaglandins 585

32.1	Introduction	585
32.2	Functions of Prostaglandins	586
32.3	Biosynthesis of Prostaglandins	586
32.4	Nomenclature	587
32.5	Structures of Therapeutically useful Prostaglandins	590
32.6	SAR of Prostaglandins	591
	Further Readings	591
	Multiple Choice Questions	591
	Questions	592
	Solution to Multiple Choice Questions	592

Chapter 33 Steroids 593

33.1	Introduction	593
33.2	Biosynthesis of Steroid Hormones	595
33.3	Overall Mechanism of Action of Steroid Hormones	597
33.4	Oestrogens	598
33.5	Progestins	603
33.6	Androgens	612
33.7	Glucocorticoids	619
	Further Readings	627
	Multiple Choice Questions	628
	Questions	628
	Solution to Multiple Choice Questions	628

Chapter 34 Miscellaneous Agents 629

34.1	Anti-Obesity Drugs	629
34.2	PDE-5 Inhibitors for Male Erectile Dysfunction (ED)	632
34.3	Anti-Migraine Drugs	636
34.4	Anti-Asthmatics	637
34.5	Thyroid and Anti-Thyroid Drugs	639
34.6	Anti-Platelet Drugs (Antithrombotics)	642
34.7	Immunopharmacological Drugs	644

34.8	Anti-Cough and Expectorants	645	**Chapter 35**	**Nomenclature of Medicinal Compounds**		**670**
34.9	Diagnostic Agents	646	35.1	Introduction		670
34.10	Antiseptics and Disinfectants	651	35.2	General Rules		670
			35.3	Heterocyclic Compounds		673
34.11	Some Important Antiseptics and Disinfectants	653	35.4	Examples		675
				Further Readings		*676*
34.12	Pharmaceutical Aid	655		*Questions*		*677*
34.13	Vitamins	658				
	Further Readings	*668*				
	Multiple Choice Questions	*668*	**Index**			**679**
	Questions	*669*				
	Solution to Multiple Choice Questions	*669*	**Color Plate Section**			

Preface

Medicinal Chemistry is a scientific discipline at the intersection of chemistry and pharmacy involved with the designing and developing of pharmaceutical drugs. Medicinal chemistry involves the identification, synthesis and development of new chemical entities suitable for therapeutic use. It also includes the study of existing drugs, their biological properties, and structure-activity relationships between the drugs. This book is an attempt to demystify medicinal chemistry, so that budding graduates can appreciate the power of drug design. The substantial scientific work that has been accomplished in recent years in the field of drug design has revolutionized the entire concept of medicinal chemistry.

The Post-GATT era and new international patenting pattern offer many challenges in the field of drug discovery. Keeping this view in mind, it is our endeavour to provide comprehensive textbook of medicinal chemistry covering synthesis, pharmacology, and structure--activity relationship of various classes of drugs. The book also emphasis on various new classes of drugs like anti-HIV, anti-obesity, PDE 5 inhibitors for male erectile dysfunction, anti-Alzheimer's drugs etc. In addition, each chapter has a reference section along with various multiple-choice questions and review questions that would help students to summarize the portions. This book is intended to provide knowledge and benefit to the students of degree and post-graduate students of pharmacy, as well as, personnel engaged in pharmaceutical industries.

In this second edition of the book, some new chapters in the emerging areas of medicinal chemistry have been incorporated. In all the chapters explanation on synthetic routes and pharmacological activity have been improved with additional information on new drugs in each class.

CHAPTER 1

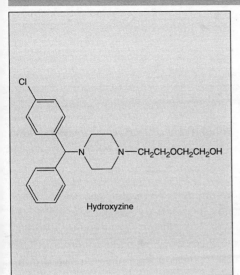

Hydroxyzine

Drug Discovery and Development

LEARNING OBJECTIVES

- Understand what is medicinal chemistry
- List of several stages of drug development
- Cost and duration of a drug to be brought from the bench to the market
- Identification of a key drug target
- Identification of lead chemical and how it can be developed as a drug
- Ethical issues in testing drugs on human beings
- Understand related terminologies like IND, NCE, NDA, and MAA
- Stages of different clinical studies
- Explain the different methods of drug discovery research
- Identification of drug-like compounds from nature
- Understand the importance of history behind drug discovery and how it can be further explored
- Explain the process of metabolism of drug in the biological system and its importance in drug discovery
- Identify the contribution of side-effects in further improvement of drug discovery
- Explain the importance of computer applications in the design of new drugs
- Understanding a real-life problem of treating HIV and the lessons learned from it
- Describe how a drug can be modified to improve their utility

1.1 INTRODUCTION

Medicinal chemistry involves the discovery, development, identification, and interpretation of the mode of action of biologically active compounds at the molecular level. Emphasis is put on drugs, but the significance of medicinal chemistry is also concerned with the study, identification, and synthesis of the metabolic products of drugs and related compounds. Medicinal chemistry covers three critical steps.

- **A discovery step,** involving the choice of the therapeutic target (receptor, enzyme, transport group, cellular, or *in vivo* model) and the identification (or discovery) and production of new active substances interacting with the selected target. Such compounds are usually called **lead compounds;** they can originate from synthetic organic chemistry, from natural sources, or from biotechnological processes. Drug design aims at the development of the drugs with high specificity and therapeutic index.

- **An optimization step,** which deals with the improvement of the lead structure. The optimization process takes primarily into account the increase in potency, selectivity, and toxicity. Its characteristics are the establishment and analysis of structure–activity relationships, in an ideal context to enable the understanding of the molecular mode of action. However, an assessment of the pharmacokinetic parameters such as absorption, distribution, metabolism, excretion, and oral bioavailability is almost systematically practiced at an early stage of the development in order to eliminate unsatisfactory candidates.

- **A development step,** whose purpose is the continuation of the improvement of the pharmacokinetic properties and the fine-tuning of the pharmaceutical properties (chemical formulation) of the active substances in order to render them suitable for clinical use. This chemical formulation process can consist in the preparation of better-absorbed compounds, of sustained release formulations, and of water-soluble derivatives, or in the elimination of properties related to the patient's compliance (causticity, irritation, painful injections, and undesirable organoleptic properties).

Discovering and bringing a new drug to the public typically costs a pharmaceutical company nearly $900 million (a study published in 2006 estimates that costs vary from around 500 million dollars to 2,000 million dollars depending on the therapy or the developing firm) and takes an average of 9 to 12 years (Fig. 1.1). In special circumstances, such as the search for effective drugs to treat AIDS and SARS, the Food and Drug Administration (FDA) has encouraged an abbreviated process for drug testing and approval called fast-tracking. The drug discovery and development process is designed to ensure that only those pharmaceutical products that are both safe and effective are brought to the market.

1.2 HOW ARE THE NEW DRUGS DISCOVERED?

New drugs begin in the laboratory with chemists, scientists, and pharmacologists who identify cellular and genetic factors that play a role in specific diseases. They search for chemical and biological substances that target these biological markers and are likely to have drug-like effects. Out of every 5,000 new compounds identified during the discovery process, only five could be considered safe for testing in human volunteers after pre-clinical evaluations. After three to six years of further clinical testing in

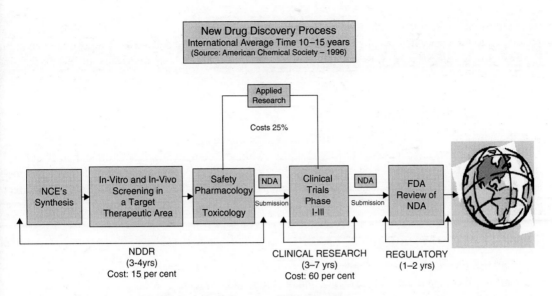

FIGURE 1.1 Various stages of drug discovery.

patients, only one of these compounds would ultimately be approved as a marketed drug for treatment. The following sequence of research activities begins the process that results in the development of new medicines. (*See Fig. 1.2 in the coloured set of pages.*)

1.2.1 Target Identification

Drugs usually act on either cellular or genetic regions in the body known as targets, which are believed to be associated with a particular disease condition. Rational drug design is an important concept in pharmaceutical research. The goal is to identify a key drug target based on a thorough understanding of regulatory networks and metabolic pathways, and to design a highly specific drug based on the known three-dimensional (3D) structure of that target. Scientists use a variety of techniques to identify and isolate a target, and learn more about its functions and how these influence the incidence of a disease. Compounds are then identified as being effective in the treatment of a specific disease, by exploring the various interactions that they exhibit with the drug targets. Allosteric effectors of haemoglobin and dihydrofolate reductase inhibitors, related to trimethoprim, were the very first biologically active molecules that were derived from protein 3D structures. The antihypertensive captopril was the first therapeutically used drug that resulted from a structure-based design. There are several other examples of rational drug design using targets with known 3D structure, including the HIV protease inhibitors amprenavir and nelfinavir and the influenza virus inhibitor zanaminivir.

1.2.2 Target Prioritization/Validation

To select targets most likely to be useful in the development of new treatments for diseases, researchers analyse and compare each drug target to others based on their association with a specific disease and their ability to regulate the activities of biological and chemical compounds in the body. Tests are conducted to confirm whether interactions with the drug target are associated with a desired change in the behaviour of diseased cells. Research scientists can then identify compounds that have an effect on the selected target.

1.2.3   Lead Identification

A lead compound or substance is one that is believed to have a potential to treat a particular disease. Laboratory scientists can compare known substances with new compounds to determine their likelihood of success. Leads are sometimes developed as collections, or libraries, of individual molecules that possess properties needed in a new drug. Testing is then done on each of these molecules to confirm its effect on the drug target.

1.2.4   Lead Optimization

Lead optimization compares the properties of various lead compounds and provides information to help pharmaceutical companies select the compound or compounds with the greatest potential to be developed into safe and effective medicines. Often during this same stage of development, lead prioritization studies are also conducted in living organisms (*in vivo*) and in cells in the test tube (*in vitro*) to compare the effects of various lead compounds and how they are metabolized in the body.

1.3   REQUIREMENTS FOR AN INVESTIGATIONAL DRUG TO BE TESTED IN HUMAN VOLUNTEERS

In the preclinical stage of drug development, an investigational drug must be tested extensively in the laboratory to ensure that it will be safe to administer to humans. Testing at this stage can take from one to five years and must provide information about the pharmaceutical composition of the drug, its safety, how the drug will be formulated and manufactured, and how it will be administered to the first human subjects.

Preclinical technology. During the preclinical development of a drug, laboratory tests document the effect of the investigational drug in living organisms (*in vivo*) and in cells in the test tube (*in vitro*).

Chemistry manufacturing and controls (CMC)/Pharmaceutics. The results of preclinical testing are used by experts in pharmaceutical methods to determine how to best formulate the drug for its intended clinical use. For example, a drug that is intended to act on the sinuses may be formulated as a timed-release capsule or as a nasal spray. Regulatory agencies require testing that documents the characteristics, chemical composition, purity, quality, and potency of the formulation's active ingredient and of the formulation itself.

Pharmacology/Toxicology. Pharmacological testing determines the effects of the candidate drug on the body. Toxicological studies are conducted to identify potential risks to humans.

The results of all the testing procedures must be provided to the FDA in the United States and/or other appropriate regulatory agencies in order to obtain permission to begin clinical testing in humans. Regulatory agencies review the specific tests and recommend further documentation that is required to proceed to the next stage of development.

1.4 HOW ARE THE INVESTIGATIONAL DRUGS TESTED IN HUMANS?

Testing of an investigational new drug begins with submission of information about the drug and application for permission to begin administration to healthy volunteers or patients.

1.4.1 Investigational New Drug (IND)/Clinical Trial Exception (CTX)/ Clinical Trial Authorization (CTA) Applications

INDs (in the United States), CTXs (in the United Kingdom), and CTAs (in Australia) are examples of requests submitted to appropriate regulatory authorities for permission to conduct investigational research. This research can include testing of a new dosage form or new use of a drug already in clinical use.

In addition to obtaining permission from appropriate regulatory authorities, an institutional or independent review board (IRB) or ethical advisory board must approve the protocol for testing. An IRB is an independent committee of physicians, community advocates, and others to ensure that a clinical trial is ethical and the rights of study participants are protected.

Clinical testing is usually described as consisting of Phase I, Phase II, and Phase III clinical studies. In each successive phase, increasing numbers of patients are tested (Fig. 1.3).

Phase I Clinical Studies

Phase I studies are designed to verify safety and tolerability of the candidate drug in humans, and typically take six to nine months. These are the first studies conducted in humans. A small number of subjects, usually from 20 to 100 healthy volunteers, take the investigational drug for short periods of time. Testing includes observation and careful documentation of how the drug acts in the body, how it is absorbed, distributed, metabolized, and excreted.

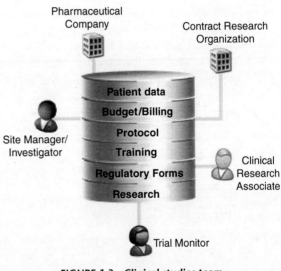

FIGURE 1.3 Clinical studies team.

Phase II Clinical Studies

Phase II studies are designed to determine effectiveness and to further study the safety of the candidate drug in humans. Depending upon the type of investigational drug and the condition it treats, this phase of development generally takes from six months to three years. Testing is conducted with up to several hundred patients suffering from the condition the investigational drug is designed to treat. This testing determines safety and effectiveness of the drug in treating the condition and establishes the minimum and maximum effective doses. Most Phase II clinical trials are randomized, or randomly divided into groups, one of which receives the investigational drug and another gets a placebo containing no medication, and sometimes a third receives a current standard treatment to which the new investigational drug will be compared. In addition, most Phase II studies are double-blinded, meaning that neither patients nor researchers evaluating the compound know who is receiving the investigational drug or placebo.

Phase III Clinical Studies

Phase III studies provide expanded testing of effectiveness and safety of an investigational drug, usually in randomized and blinded clinical trials. Depending upon the type of drug candidate and the condition it treats, this phase usually requires one to four years of testing. In Phase III, safety and efficacy testing are conducted with several hundred to thousands of volunteer patients suffering from the condition the investigational drug treats.

1.4.2 New Drug Application (NDA)/Marketing Authorization Application (MAA)

NDAs (in the United States) and MAAs (in the United Kingdom) are examples of applications to market a new drug. Such applications document safety and efficacy of the investigational drug and contain all the information collected during the drug development process. At the conclusion of successful preclinical and clinical testing, this series of documents is submitted to the FDA in the United States or to the applicable regulatory authorities in other countries. The application must present substantial evidence that the drug will have the effect it is represented to have when people use it or under the conditions for which it is prescribed, recommended, or suggested in the labelling. Obtaining approval to market a new drug frequently takes between six months and two years.

1.5 HOW DOES TESTING CONTINUE AFTER A NEW DRUG IS APPROVED?

After the FDA (or other regulatory agency for drugs marketed outside the United States) approves a new drug, pharmaceutical companies may conduct additional studies, including Phase IIIb and Phase IV studies. Late-stage drug development studies of approved, marketed drugs may continue for several months to several years.

1.5.1 Phase IIIb/IV Studies

Phase IIIb trials, which often begin before approval, may supplement or complete earlier trials by providing additional safety data or they may test the approved drug for additional conditions for which it may prove useful. Phase IV studies expand the testing of a proven drug to broader patient populations and compare the long-term effectiveness and/or cost of the drug with the other marketed drugs available to treat the same condition.

1.5.2 Post-Approval Studies

Post-approval studies test a marketed drug in new age groups or patient types. Some studies focus on previously unknown side effects or related risk factors. As with all stages of drug development testing, the purpose is to ensure the safety and effectiveness of marketed drugs.

1.6 PROCEDURES FOLLOWED IN DRUG DESIGN

1. The search for lead compound
2. Molecular modification of lead compounds

Medicinal Chemistry

1.6.1 The Search for Lead Compound

A lead compound in drug discovery is a chemical compound that has pharmacological or biological activity, and whose chemical structure is used as a starting point for chemical modifications in order to improve potency, selectivity, or pharmacokinetic parameters. Lead compounds were identified by the following ways:

Isolation of bioactive compounds from natural source

(i) The early 1920s saw the outbreak of a previously unrecognized disease of cattle in the northern United States and Canada. Cattle would die of uncontrollable bleeding from very minor injuries, or sometimes drop dead of internal haemorrhage with no external signs of injury. In 1921, Frank Schofield, a Canadian veterinarian, determined that the cattle were ingesting mouldy silage made from sweet clover that functioned as a potent anticoagulant. The identity of the anticoagulant substance in mouldy sweet clover remained a mystery until 1940, when Karl Paul Link of University of Wisconsin determined that it was the 4-hydroxycoumarin derivative dicoumarol.

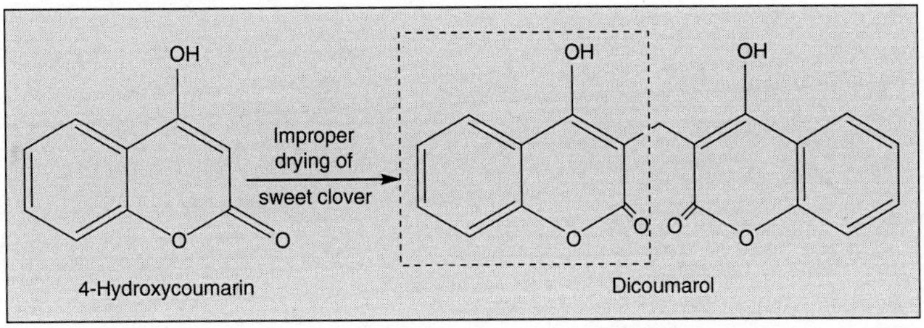

Structure-activity study reveals that for anticoagulant activity the following structural requirements were necessary:

- A 4-hydroxycoumarin ring
- A 3-substituent that may be alkyl or (sub)alkyl

From this lead, Link continued working on developing more potent coumarin-based anticoagulants, resulting in warfarin in 1948 and other drugs. (The name 'warfarin' stems from the acronym *WARF*, for *Wisconsin Alumni Research Foundation*, + the ending *-arin* indicating its link with coumarin.)

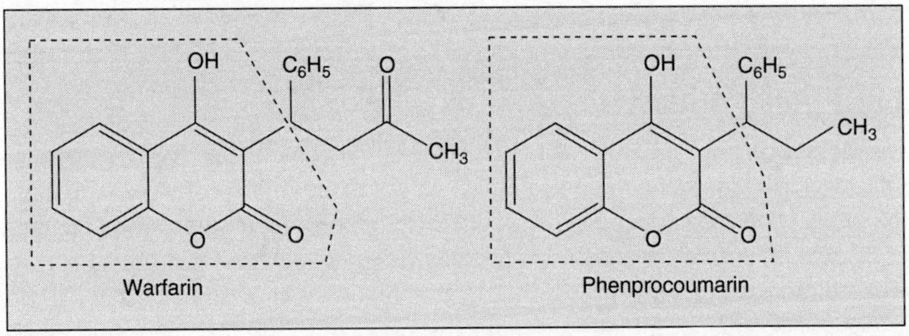

(ii) Similarly, all local anaesthetics were developed from the natural lead cocaine.

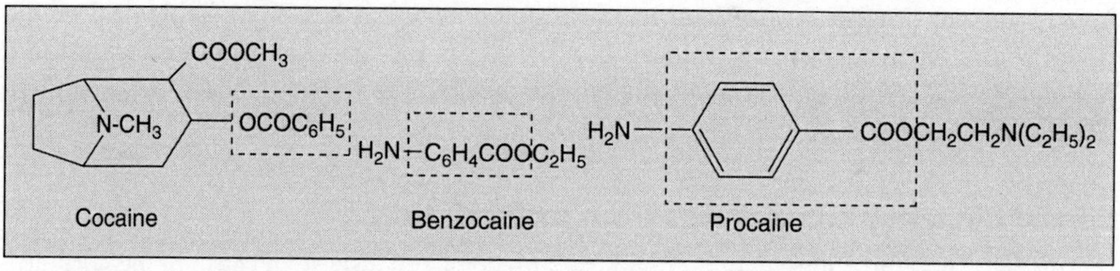

Cocaine, Benzocaine, Procaine

(iii) Anti-malarials were developed from quinine as a lead.

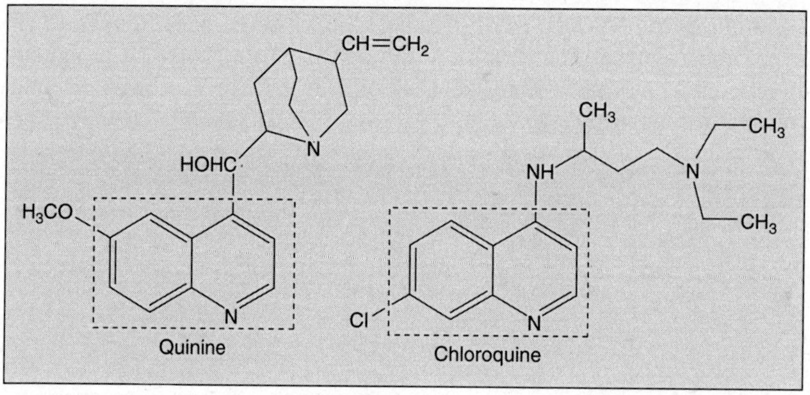

Quinine, Chloroquine

(iv) Narcotic analgesics were derived from morphine.

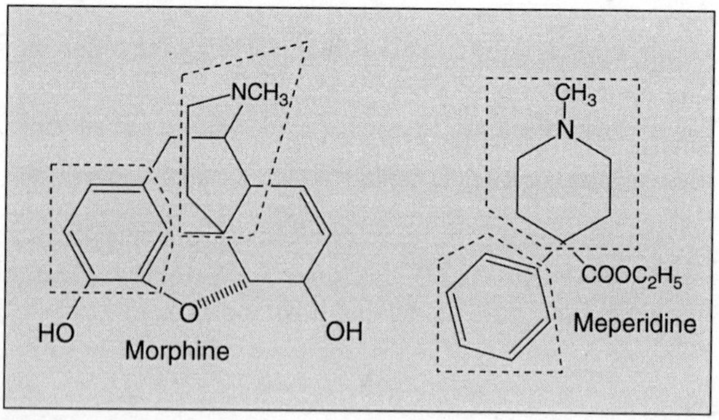

Morphine, Meperidine

Accidental drug/lead discovery

(i) Discovery of penicillin

In the late 1800s, bacteriologists and microbiologists set out to identify substances with therapeutic potential. One of the greatest problems faced by these scientists during their studies was the contamination of 'pure' cultures by invading micro-organisms, especially fungi or bacteria—a problem that still plagues the modern-day microbiologists. It is this problem of contamination which is most often identified as leading to the 'chance' observation that eventually led to the discovery of penicillin. These studies of contaminated cultures led to a series of observations by late 19th-century bacteriologists and microbiologists, describing the effect of mould on bacterial growth. These observations were as follows:

- In 1874, William Roberts observed that cultures of the mould *Penicillium glaucum* did not exhibit bacterial contamination.
- French scientists Louis Pasteur and Jules Francois Joubert observed that growth of the *anthrax bacilli* was inhibited when the cultures became contaminated with mould.
- The English surgeon Joseph Lister noted in 1871 that samples of urine contaminated with mould did not allow the growth of bacteria. Lister unsuccessfully attempted to identify the agent in the mould which inhibited bacterial growth.

In 1928, Alexander Fleming was researching the properties of the group of bacteria known as *Staphylococci;* his problem during this research was the frequent contamination of culture plates with airborne moulds. One day, he observed a contaminated culture plate and noted that the *Staphylococci* bacteria had burst in the area immediately surrounding an invading mould growth. He realized that something in the mould was inhibiting growth of the surrounding bacteria. Subsequently, Fleming isolated an extract from the mould and he named it penicillin.

(ii) Discovery of sildenafil [Viagra]

It would not become a drug for heart diseases, researchers at Pfizer sighed. For years they worked on sildenafil, an inhibitor of the PDE5 (*Phosphodiesterase* 5) enzyme, which they hoped would be effective in relaxing coronary arteries and relieving chest pain. Their hope was dashed by 1992. Gloomily, the researchers terminated the trial and asked participants to return the unused drug. Many men refused, clutching to the drug as if it was gold. Idiosyncrasies being present in all clinical tests, researchers gave the objection little thought until they heard rumours about the drug's side effects on sex life and, more important, read a paper on the role of PDE5 in the chemical pathway of erection. Gloom evaporated in the excitement that sildenafil may be a blockbuster after all. This time, their expectation was confirmed by new clinical tests on impotent men. They stumbled on an effective drug for erectile dysfunction which Pfizer would market as Viagra (Fig. 1.4).

FIGURE 1.4 Viagra tablets.

(iii) Discovery of insulin

In 1889, the Polish–German physician Oscar Minkowski in collaboration with Joseph von Mering removed the pancreas from a healthy dog to test its assumed role in digestion. Several days after the dog's pancreas was removed, Minkowski's animal keeper noticed a swarm of flies feeding on the dog's urine. On testing the urine, they found that there was sugar in the dog's urine, establishing for the first time a relationship between the pancreas and diabetes. Since that discovery in 1889, many scientists tried to extract the pancreas' secretion and use it as a remedy for diabetes. Some almost made it; others failed, but not necessarily because they were more ignorant. Then, Frederick Banting, Charles Best, and John Macleod succeeded to extract insulin and demonstrate its therapeutic efficacy in 1921.

Examination of metabolite

(i) Acetanilide was known for many years to be an analgesic (painkiller). It was subsequently found that acetanilide is metabolized to acetaminophen (Paracetamol) in the liver. Today, acetaminophen is one of the leading over-the-counter pain medications.

(ii) Hydroxyzine is a first-generation antihistamine, of the piperazine class that is an H_1 receptor antagonist. It has the major drawback of sedation. It is metabolized in the liver; the main metabolite through oxidation of the alcohol moiety to a carboxylic acid is cetirizine. The metabolite was isolated and screened, and was found that because of carboxylic acid group in its structure it was not able to cross the blood brain barrier of the central nervous system. Cetirizine is the second-generation antihistamine devoid of sedation.

(iii) Iproniazid is a monoamine oxidase inhibitor (MAOI) that was developed as the first antidepressant. It was originally intended to treat tuberculosis but it was then discovered that patients given iproniazid became 'inappropriately happy'; subsequently, further research on iproniazid revealed that it cleaved into isonicotinic acid (inhibits mycolic acid synthesis of *Mycobacterium tuberculosis*) and isopropylhydrazide (inhibits MAO) *in vivo*. Isopropylhydrazide was responsible for antidepressant property, and from that lead, other drugs like phenalzine and isocarboxazide were developed.

$$\text{Iproniazid} \xrightarrow{\textit{in vivo}} \text{Isonicotinic acid} + \text{Isopropylhydrazide}$$

Exploitation of side effects of drugs

Side effects refer to pharmacological activity demonstrated by a drug in addition to its desired bioactivity.

(i) Minoxidil is a vasodilator and was exclusively used as an oral drug to treat high blood pressure. It was, however, discovered to have the interesting side effects of hair growth and reversing baldness, and in the 1980s, Upjohn Corporation produced a topical solution that contained two per cent minoxidil to be used to treat baldness and hair loss.

(ii) During the clinical trial (Phase I) of cardiovascular drug sildenafil, it was observed that penile erections were a common side effect in the multiple dose of sildenafil. Furthermore, emerging data implicated nitric oxide (NO) as a key mediator of the neural and haemodynamic effects that lead to penile erection in men. In particular, in the early 1990s, Ignarro and others reported that NO is the neurotransmitter that is released from cavernous nerves during sexual stimulation. The NO diffuses into vascular smooth muscle cells of the penis, stimulating the production of cGMP and leading to corpus cavernosum smooth muscle relaxation, vasocongestion, veno-occlusion (by constriction of the venous outflow from the penis against the tunica albuginea), and ultimately, erection. Sildenafil is the first drug marketed for the treatment of male erectile dysfunction discovered by exploitation of side-effects approach.

Random screening

Random screening involves no intellectualization; all compounds are tested in the bioassay without regard to their structures. Prior to 1935, random screening was essentially the only approach; today, this method is used to a lesser degree. However, random screening programmes are still very important in order to discover drugs or leads that have unexpected and unusual structures for various targets.

Zidovudine (AZT) was the first drug approved for the treatment of AIDS. Jerome Horwitz of Barbara Ann Karmanos Cancer Institute and Wayne State University School of Medicine first synthesized AZT in 1964, under a US National Institutes of Health (NIH) grant. It was originally intended to treat cancer, but failed to show efficacy and had an unacceptably high side-effect profile. The drug then faded from view until February 1985, when Samuel Broder, Hiroaki Mitsuya, and Robert Yarchoan, three scientists in the National Cancer Institute (NCI), collaborating with Janet Rideout and several other scientists at Burroughs Wellcome, started working on it, and they have found that zidovudine inhibited HIV *in vitro* and marketed it as the first AIDS drug.

Rational drug design

(i) Receptor-based drug design

The receptor is the target (usually a protein) to which a drug molecule binds in order to cause a biological effect. Emil Fischer (in 1894) proposed the analogy that a drug molecule is like a key that fits into a lock. The specificity of a drug molecule for its target is like the specificity of a key for its lock; if you make similar keys, they may also 'fit' into the 'lock'. The term 'receptor' was coined by Ehrlich in 1909 to describe the target. But it was many decades later when X-ray crystallography finally let us see the three-dimensional structure of the receptor. Eventually, we also got to see crystal structures of ligands bound into the active sites of proteins.

Nowadays, it is not uncommon to identify the gene coding for a target protein; clone, express, and purify the protein; crystallize the protein; and then use the three-dimensional structure to guide the design of ligands. We will discuss one well-studied example: the HIV protease. HIV infects cells and directs the cellular machinery to make viral proteins and RNA. Several of the proteins are synthesized in one continuous chain (polyprotein). The polyprotein is cleaved into smaller chains that can then assemble to form new virus particles. The cleavage, which takes place at specific sites on the polyprotein, is carried out by the virus' protease enzyme. If you can block the activity of the protease, you can prevent the synthesis of new virus.

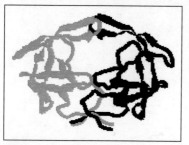

FIGURE 1.5 X-ray crystallographic structure of HIV-1 protease. The protein is a dimer; one subunit is shown in gray, and the other black. The large open space (top centre) is the active site.

In 1988 the role of the HIV protease was established, and its amino-acid sequence was determined. It was shown to have considerable homology to *Rous sarcoma* virus protease, the structure of which had been determined crystallographically. Thus, the first model of HIV protease was produced by homology modelling. The *sarcoma* virus protease was used as a starting point, and side chains were modified to match those of the HIV protease. This can be done because protein tertiary structure (3D-folding) is highly conserved (even more conserved than primary structure or sequence). In 1989 the structure of HIV protease was determined by X-ray crystallography, and the homology-modelled structure was shown to be essentially correct (Fig 1.5).

There are numerous proteases throughout all living organisms: digestive enzymes and blood-clotting factors are just two examples in humans. The catalytic mechanism provides a way to classify proteases into three large families. Aspartic proteases contain two aspartic acid side chains that participate in cleavage of the peptide bond. Serine proteases use a serine-histidine-aspartate triad. Metalloproteases use a metal ion (e.g., Zn^{++}) to help catalyse proteolysis.

Some proteases are fairly non-specific, and others have very tight specificity. The figure below shows, in schematic form, how the amino acid side chains may interact with 'specificity pockets' on the protease to establish the protease's specificity. S1, S2, S3, S1', S2', S3', etc., label the specificity pockets, while the side chains P1, P2, . . . are the protein specificity determinants.

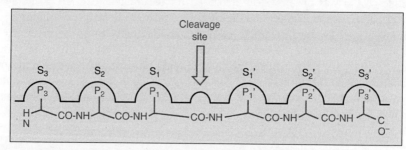

In HIV protease, S_3 and S_3' are essentially non-existent; the other sites are specific for large hydrophobic side chains such as phenylalanine and proline. Therefore, in designing an inhibitor, there should be substituents that look something like these side chains. Another consideration in designing an inhibitor is that the inhibitor should not be hydrolyzed by the protease, and so, amide bonds should be avoided. In fact, in the central part of the inhibitor, it is useful to mimic the transition state or intermediate in the hydrolysis reaction.

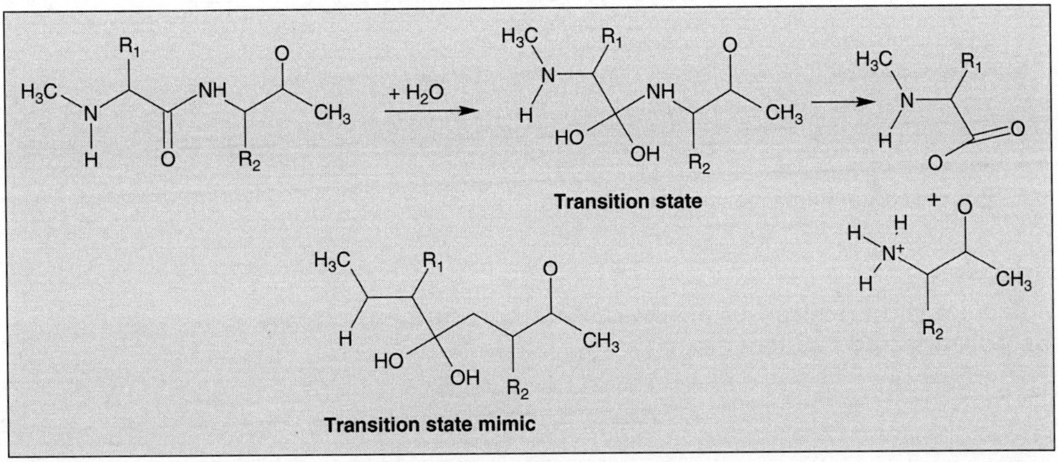

[Figure: reaction scheme showing substrate → transition state → products, with a "Transition state mimic" structure below]

This is because enzymes work by facilitating the development of the transition state, which is on the path from substrate to product. They do this in part by binding more favourably to the transition state than to either substrate or product.

It was known from work on renin, another aspartic protease, that replacement of the peptide C=O with a hydroxy (—OH) group makes a good transition-state mimic. But renin has different specificity sites than HIV protease; so, known renin inhibitors were not very good HIV protease inhibitors. It is just as well if HIV protease inhibitors also inhibited renin, they would have side effects involving hypotension. Several examples of successful protease inhibitors are illustrated below. The structures are shown here to illustrate the principles involved in designing a drug molecule to fit into a receptor site. These include hydrogen bonding interactions in the active site as well as hydrophobic interactions in the specificity sites. There is one particularly important feature that is routinely observed in crystal structures of HIV protease with inhibitors bound in the active site. This is the presence of a tightly bound water molecule near the top of the active site. It forms two hydrogen bonds from water hydrogen's to oxygen atoms on the inhibitor (in the natural function of the protease, hydrogen bonds to two peptide backbone C=O oxygen atoms). In turn, the water oxygen atom forms two hydrogen bonds to peptide backbone N—H atoms of the protease. (*See Fig. 1.6 in the coloured set of pages.*)

1.6.2 Other Examples of Designed Drugs

1. Cimetidine, the prototypical H_2-receptor antagonist from which the later members of the class were developed
2. Dorzolamide, a carbonic anhydrase inhibitor used to treat glaucoma
3. Many of the atypical antipsychotics
4. Selective COX-2 inhibitor NSAIDs
5. SSRIs (selective serotonin reuptake inhibitors), a class of antidepressants

6. Zanamivir, an antiviral drug
7. Enfuvirtide, a peptide HIV entry inhibitor

1.6.3 Molecular Modification of Lead Compounds

Molecular modification is undertaken to prepare compounds with

a. greater potencies than the lead compounds;
b. reduced toxicity;
c. greater specificity of drug action;
d. improved duration of action;
e. improved stability; and
f. reduced cost of production.

Examples

Molecular modification of antihistamine hetramine increases the potency and reduces toxicity

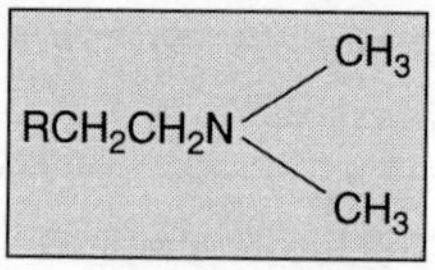

$$RCH_2CH_2N(CH_3)_2$$

R	ED_{50} (µM/Kg)[a]	LD_{50} (µM/Kg)[b]	T.I.[c] (ED_{50}/LD_{50})
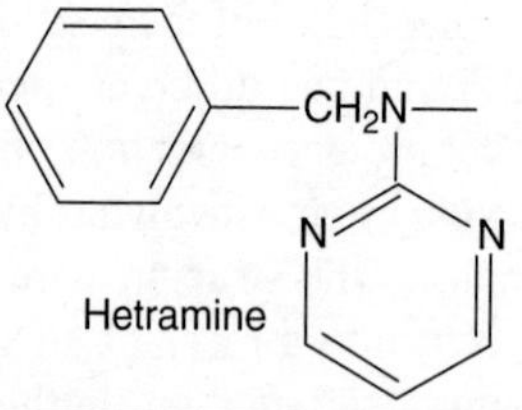 Hetramine	0.51	260	510
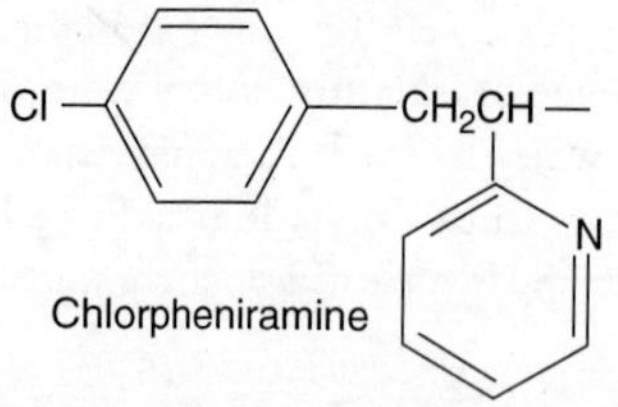 Chlorpheniramine	0.47	680	1447
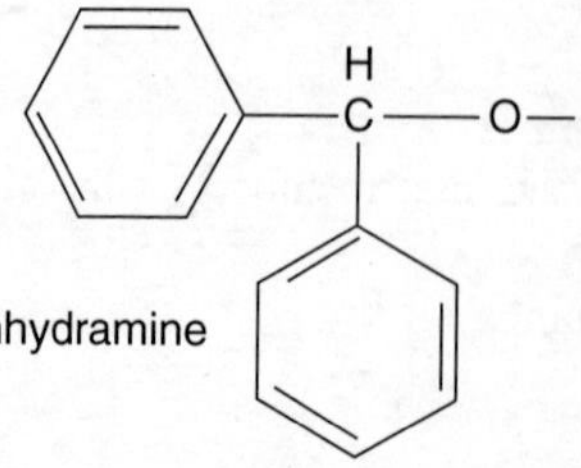 Diphenhydramine	0.21	480	2286

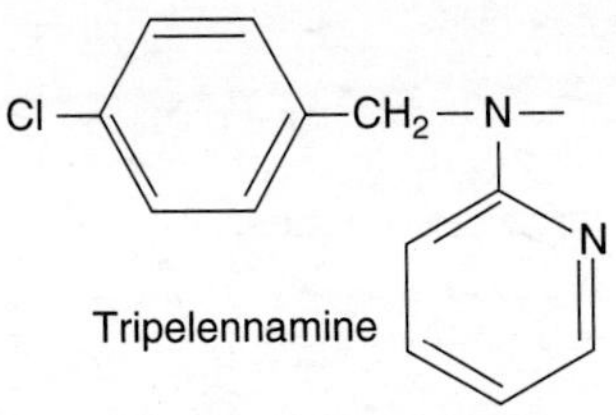

		0.012	230	19167
Tripelennamine (Cl-C6H4-CH2-N-pyridyl)				
Pyrilamine (H3CO-C6H4-CH2-N-pyridyl)		0.0037	250	67568

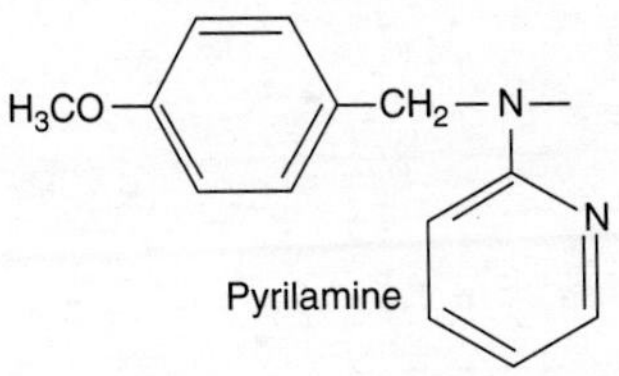

[a] 50% effective dose; [b] 50% lethal dose; [c] therapeutic index

Molecular modification produces orally active compound

Progesterone is poorly absorbed by oral ingestion because it undergoes extensive first pass metabolism at α, β-unsaturated ketone, and also at C_{20} keto function to form 3, 20 diol. Progesterone was modified by introducing hydroxyl group at C_{17} and methyl group at C_5 position, which inhibits metabolism and provides orally active compound Medroxyprogesterone.

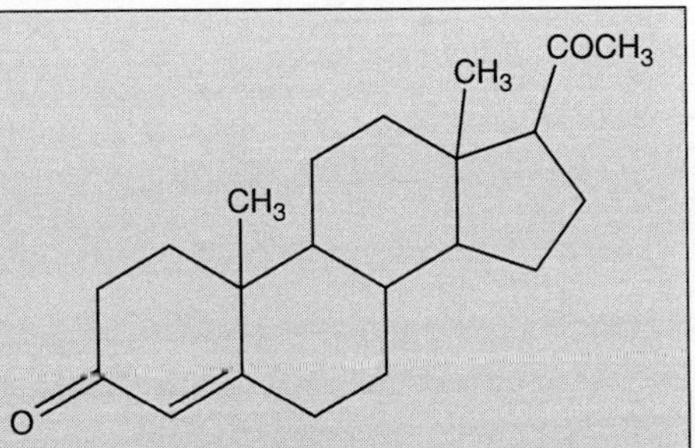

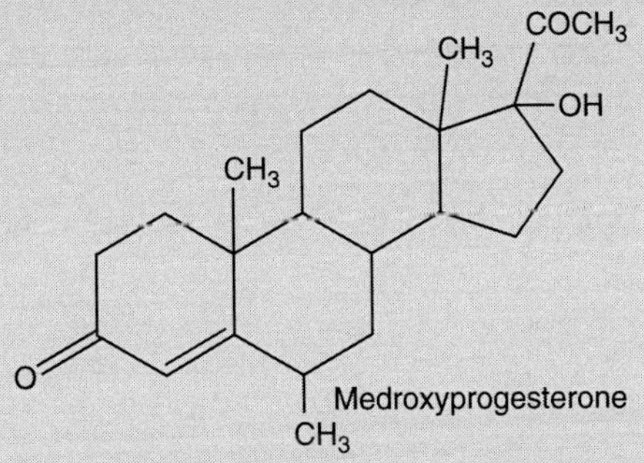

Molecular modification to give increased lipid solubility and affects duration of action

Pentobarbitone is the short duration of action (less than three hours) barbiturate. Replacement of the oxygen at C-2 by a sulphur atom leads to Thiopentone, which has quick onset and short duration of action—because of its high partition coefficient, it enters in/out of the brain fast.

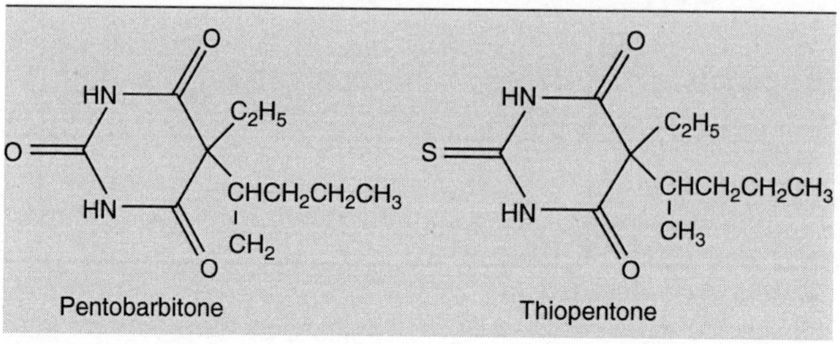

Molecular modification to alter metabolism

Procaine is the local anaesthetics and also used as an anti-arrhythmic drug. It is rapidly hydrolyzed *in vivo* by esterases. However, replacement of the ester function of procaine by an amide gave a much more metabolically stable analogue, Procainamide.

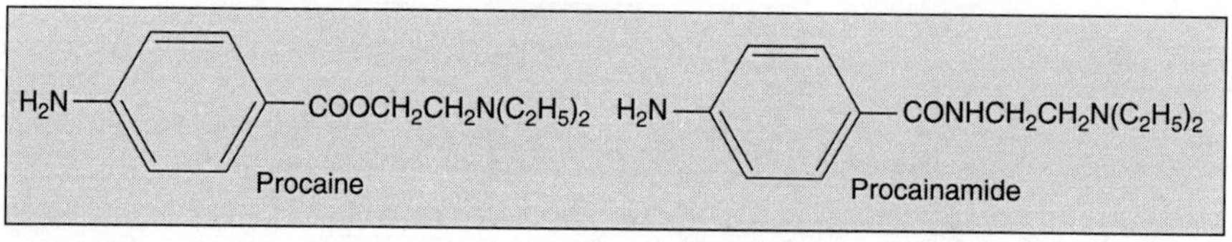

FURTHER READINGS

Burger's Medicinal Chemistry & Drug Discovery, Manfred E. Wolf (ed.), Wiley Interscience.

MULTIPLE-CHOICE QUESTIONS

1. What does FDA stand for?
   a. Federal Department of Drug Administration
   b. Food and Drug Act
   c. Food and Drug Administration
   d. Federal Drug Association

2. International average time for drug to be discovered involves
   a. 10 years
   b. 2 years
   c. 25 years
   d. 5 years

3. The first step in the drug discovery process is
   a. Lead Modification
   b. Lead Identification
   c. Lead Validation
   d. Lead Optimization

4. The requests submitted to appropriate regulatory authorities for permission to conduct investigational research include one of the following:
   a. CMC
   b. IND
   c. NCE
   d. IRB

5. The safety of the candidate drug in humans are studied in
   a. Phase I
   b. Phase II
   c. Phase III
   d. Phase IV

6. In which country is marketing authorization application (MAA) done?
   a. India
   b. USA
   c. UK
   d. South Africa

7. The studies that test a marketed drug in new age groups or patient types is done
   a. Pre-approval
   b. Post-approval
   c. During approval
   d. Such study never done

8. Which drug was discovered in the following case 'Cattle would die of uncontrollable bleeding from very minor injuries.'
   a. Warfarin
   b. Dicoumarol
   c. Quinine
   d. Morphine

9. Meperidine was derived from which of the following lead?
   a. Cocaine
   b. Morphine
   c. Coumarin
   d. Quinine

10. An example of drug among the following discovered by random screening is
    a. Morphine
    b. Penicillin
    c. Zidovudine
    d. Paracetamol

11. The drug identified by metabolite-based study is
    a. Isoniazid
    b. Warfarin
    c. Meperidine
    d. Insulin

12. Which of the following has a poor oral absorption?
    a. Progesterone
    b. Pentobarbitone
    c. Medroxyprogesterone
    d. Thiobarbital

QUESTIONS

1. Briefly explain the drug discovery process.
2. What is rational drug design? Cite its advantages over conventional design.
3. How is a lead identified? Explain with suitable examples.
4. Give a short note on lead identification via side-effect exploitation.
5. How is a drug approved for marketing?

6. What is structure-based drug design? Give some examples.
7. What are clinical trials? Why are these studies essential?
8. What is a lead and how is it modified?
9. How to reduce toxicity of a drug by chemical modification? Explain with a suitable example.
10. Give a note on accidental drug discovery.

SOLUTION TO MULTIPLE-CHOICE QUESTIONS

1. c;
2. a;
3. b;
4. b;
5. b;
6. c;
7. b;
8. b;
9. b;
10. c;
11. a;
12. a.

CHAPTER 2

Principles of Drug Action

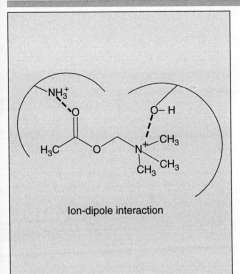

Ion-dipole interaction

LEARNING OBJECTIVES

- Description of the basic chemistry of drugs and their relation to their properties
- Define the hybridization, isomerism, chirality, stereochemistry, heteroatoms, and aromaticity, and their relation to the physicochemical properties of a drug
- Define the solubility, protein interaction, and hydrogen bonding related to functional groups present in a drug structure
- Define acid-base chemistry in relation to drug structure and properties like pK_a
- Define factors important to attain acidity or basicity
- Importance of heteroatoms in a drug structure, especially nitrogen
- Explain the drug action and importance of studying physicochemical properties
- Understand how drug encounters the biological environment
- Explain how the physicochemical properties can be determined or calculated
- Define Ferguson principle of determining drug action
- Define bioisomerism
- Explain how drug interacts with the target
- Define and explain drug-receptor theories
- Explain the working model of drug-receptor interactions
- Define agonists and antagonists
- Differentiate various antagonists
- Define and explain covalent and non-covalent bonding in drug-receptor interactions
- Define hydrogen bonding

- Explain hydrophobic effect and charge-transfer interaction
- Define G-protein coupled receptors

2.1 INTRODUCTION

Drugs are organic compounds, and as a result, their activity, their solubility in plasma, and their distribution to various tissues are dependent on their physicochemical properties. Even the interaction of a drug with a receptor or an enzyme is dependent on characteristics of a drug molecule, such as ionization, electron distribution, polarity, and electronegativity. To understand drug action, we must also understand the physicochemical parameters that make this action possible. The following sections are intended to explain the acid-base and physicochemical properties that determine drug action.

2.2 FUNCTIONAL GROUP REVIEW

2.2.1 Simple Hydrocarbons

As shown in the figure below, there are three electronic configurations of carbon which are regular components of drug molecules. Carbon has an atomic number of 6, and a molecular weight of 12.01, and has 4 electrons in its valence shell. It can exist in three distinct geometric forms based on three distinct hybrids orbital. The carbon species designated sp^3 is tetrahedral in shape, with bond angles of 109.5 degrees. Hydrocarbons containing sp^3 carbons are known as alkanes. In this form of carbon, chirality is possible, as will be discussed later. The carbon designated sp^2 is planar and trigonal in shape, due to the presence of a π orbital containing one electron. These hybrid carbons are present in the double bond, and the compounds containing this group are known as alkenes. The bond angles in this form of carbon are 120 degrees.

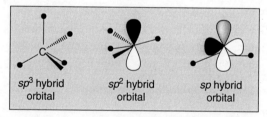

sp^3 hybrid orbital $\qquad$ sp^2 hybrid orbital $\qquad$ sp hybrid orbital

Carbon can also exist in a linear form known as sp hybrid, and compounds containing these carbons are known as alkynes. The bond angles in alkynes are 180 degrees, and they have two sets of π orbital, each containing one π electron.

Like carbon, many heteroatoms possess specific geometries that contribute to the overall shape of drug molecules. Nitrogen, an abundant heteroatom in drug molecules, has an atomic number of 7 and a molecular weight of 14.008, and has 5 electrons in its valence shell. It generally exists in a tetrahedral shape that is similar to carbon, except that one of the 4 bonds is paired to a lone pair of electrons. As we shall see later, this lone pair is crucial in determining the acid-base properties of drug molecules at physiological pH. Unlike carbon, the tetrahedral form of nitrogen is not chiral, since at ambient temperature it undergoes a rapid inversion known as Walden inversion, wherein the lone pair shifts from one side of the atom to the other and back again. As will be discussed later, nitrogen can also exist in a trigonal form that is analogous to an sp²-hybridized carbon.

Oxygen, which has an atomic number of 8 and a molecular weight of 16.00, has 6 electrons in the valence shell, and is most often found in the form shown here. In this form, oxygen has 2 pairs of π electrons, and a bond angle of 104.5 degrees. As you are aware, oxygen can also exist in a doubly-bonded form, such as is found in a carbonyl group. There are still two pairs of π electrons, but the geometry of this type of oxygen is obviously quite different.

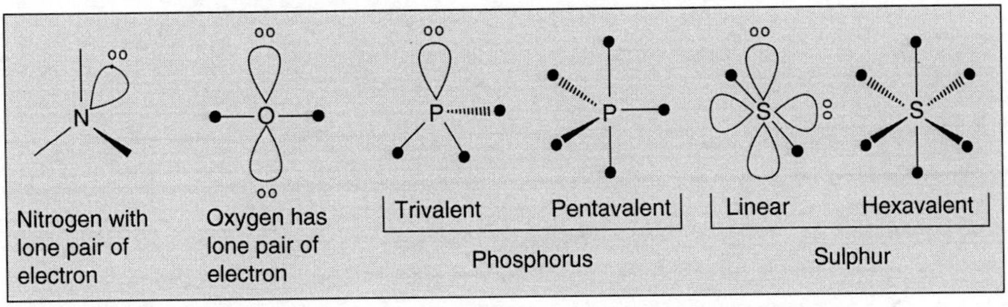

Phosphorus has an atomic number of 15, and a molecular weight of 30.97, with 5 electrons in the valence shell. In drug molecules, it exists in two major forms, the trivalent form (with one lone pair of electrons) and a pentavalent form (no lone pairs). In drug molecules, there are also two prevalent forms of sulphur, which has 6 electrons in its valence shell. One is a linear form with two lone pairs of electron, as shown above, and the other is a hexavalent form with no lone pairs.

The hydrocarbons known as alkanes have no electronegative groups, and cannot form hydrogen bonds. This is due to the absence of a dipole moment, in which electrons are pulled towards an electronegative atom. As such, alkanes are very insoluble in water, since formation of H-bonds with water is a prerequisite to water solubility. Alkanes are also chemically unreactive. When four different groups are bound to a single carbon, the possibility of isomers arises. As shown below, compounds with a single chiral centre can exist as enantiomers, which are isomeric forms that are mirror images.

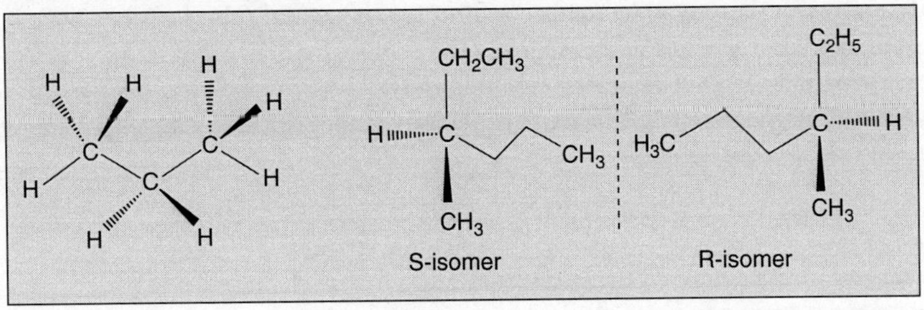

Alkenes are carbon compounds that contain sp²-hybridized carbons, as seen below. These analogues still have no electronegative groups, and since there is no dipole moment, they are unable to form hydrogen bond. This imparts low-water solubility to this series. Alkenes are hydrophobic, and somewhat chemically unreactive. In addition, alkenes exhibit geometric isomerism, and can exist in *cis* and *trans* forms. The *cis* form is known as the Z-isomer, derived from *zusammen* (the German word for 'together'), and the *trans* form is termed the E-isomer, from the word *entgegen* (the German word for 'opposite').

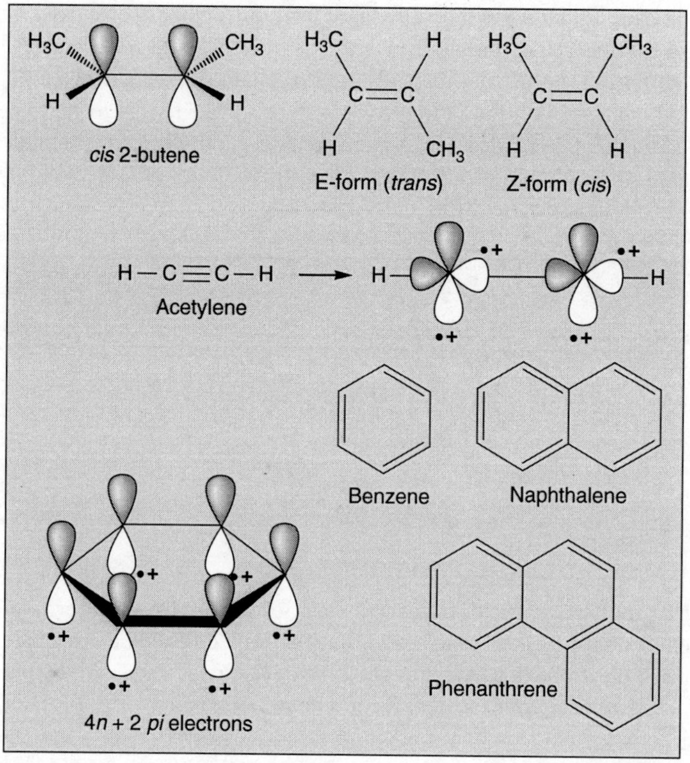

Chapter 2: Principles of Drug Action

cis 2-butene

E-form (*trans*)

Z-form (*cis*)

Acetylene

Benzene

Naphthalene

Phenanthrene

$4n + 2$ *pi* electrons

The third common form of carbon found in drug compounds is the alkyne containing sp-hybridized carbon, as shown above. There is a strong dipole in these compounds, such that the terminal hydrogen is acidic in strongly basic conditions. At physiological pH, these compounds are hydrophobic and exhibit poor water solubility.

A subset of the alkenes is the aromatic hydrocarbons. Aromatic hydrocarbons contain conjugate double bonds, and when these double bonds conform to the Huckul rule (i.e., they have $4n + 2$ π electrons), the resulting hydrocarbon is aromatic.

Hydrocarbons that contain halogens, such as butyl bromide (below), have a strong permanent dipole, due to the electronegativity of the halogen. However, they are still unable to form hydrogen bond and, as such, have poor water solubility.

Butyl bromide

A key point to remember is that the solubility of an organic compound in water is dictated by two factors: whether it can form hydrogen bonds with water, and/or whether it dissociates to form an ion. Compounds that lack these two traits will be generally water-insoluble.

2.2.2 Hydrocarbons Bonded to Heteroatoms

Heteroatoms such as oxygen, nitrogen, and sulphur give drug molecules the ability to form hydrogen bonds, and thus impart some degree of water solubility. However, the overall solubility of a given molecule also depends on the hydrophobicity of the alkyl group (i.e., octyl alcohol would be less soluble than ethyl alcohol).

As shown in the figure below, alcohols can exist as primary, secondary, or tertiary alcohols, depending on the number of groups appended to the carbon attached to the oxygen. Alcohols also possess a permanent dipole moment, and as such they can undergo hydrogen bonding themselves, and to water. In addition, alcohols can undergo biological oxidation, an important feature in the metabolism and excretion of many drugs. A primary alcohol such as ethanol can be oxidized *in vivo* to an aldehyde (acetaldehyde), and then to the corresponding carboxylic acid (acetic acid). Secondary alcohols are oxidized to the corresponding ketone (e.g., isopropanol is converted to acetone), and tertiary alcohols are stable to oxidation, since they do not possess the α hydrogen needed to participate in the reaction.

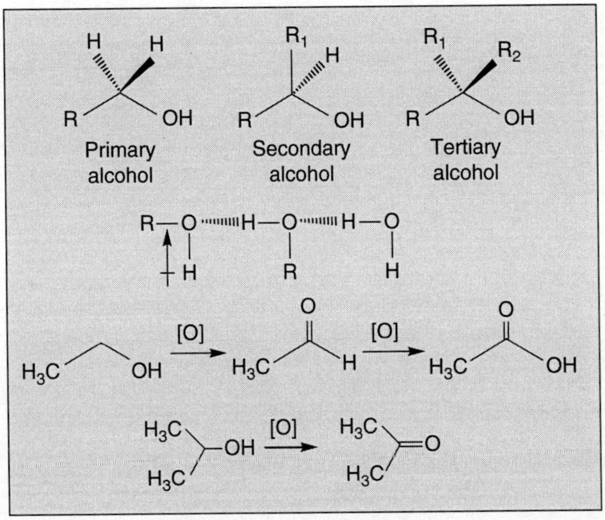

When a hydroxyl group is appended to an aromatic ring, the resulting alcohol is known as a phenol. The simplest example is the phenol derived from benzene, which is known as phenol. Phenols are weak acids, because they can dissociate in water to form the corresponding phenolate anion. This dissociation is more facile due to resonance stabilization of the phenolate, in which the negative charge delocalizes into the aromatic system. Phenol acidity is strongly affected by other substituents on the aromatic ring. As shown below, *p*-nitrophenol is more acidic than phenol, due to the electron-withdrawing group (EWG) (nitro), while *p*-ethylphenol is less acidic, owing to the ethyl electron-releasing group (ERG). It should also be pointed out that phenols, when treated with aqueous base, can form the corresponding salt form, and these entities can be isolated. This becomes important when a phenolic drug needs to be dissolved in an aqueous environment.

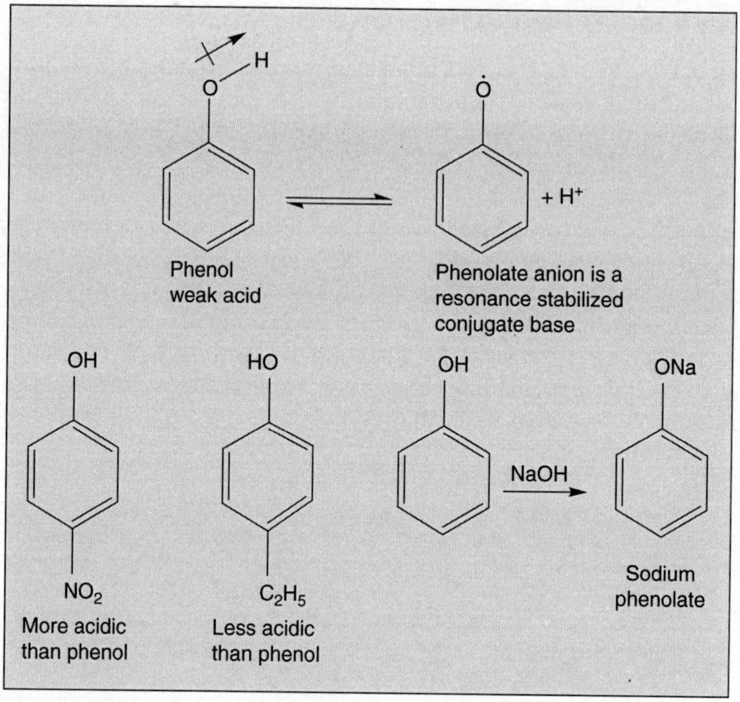

Ether is an oxygen containing a functional group wherein the oxygen is flanked by two alkyl groups. Although these compounds can H-bond weakly to water, they are not sufficiently polar to be water-soluble. They are also chemically inert unless exposed to a spark or flame. The figure below is the general structure of ether, showing the two lone pairs of electrons on the oxygen, where R1 and R2 are both alkyl and aryl. The structure on the right is, of course, the ether bunny.

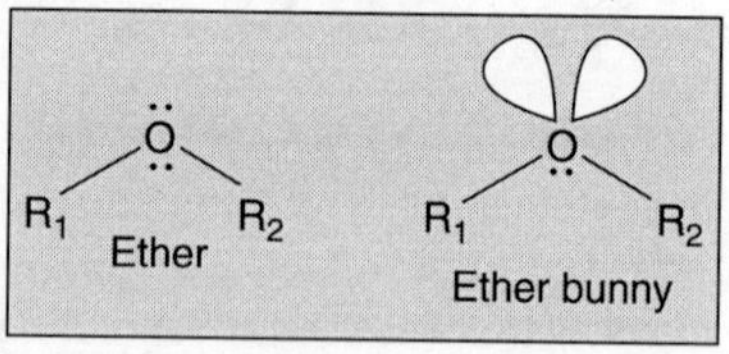

Aldehydes and ketones contain a carbonyl group, which is responsible for the properties of these molecules. As shown below, the electronegative oxygen pulls electrons and sets up a dipole moment. This confers a partial positive charge on the sp^2 carbonyl carbon, and a partial negative charge on the oxygen. As shown, aldehydes and ketones can form hydrogen bond with water, conferring some degree of water solubility. They also undergo a keto-enol tautomerism, as shown in the figure below.

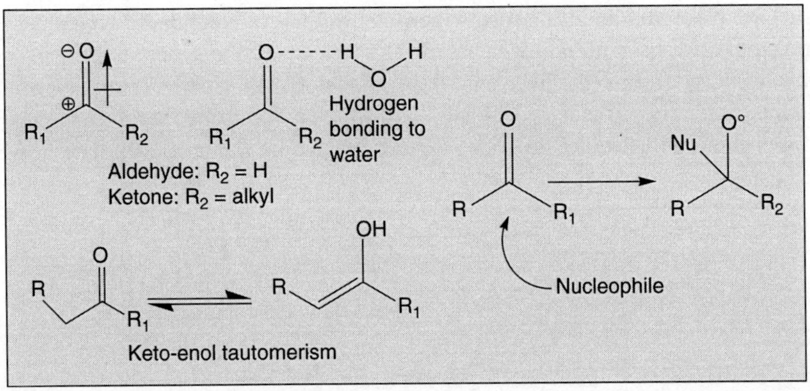

Because the carbonyl carbon has a partial positive charge, it becomes susceptible to nucleophilic attack. This is easily accomplished, since the carbonyl has the ability to accommodate the negative charge. In such a reaction, the carbonyl carbon is converted from sp^2 to sp^3, and then back again to sp^2 following the departure of a leaving group, and this process is known as nucleophilic substitution.

One of the most prevalent acid-base functional groups in drug molecules is the amine group. As seen below, these nitrogen-containing compounds can exist as primary, secondary, tertiary, and quaternary amines, depending on how many alkyl groups are appended to the nitrogen. Note that in the primary, secondary, and tertiary amine molecules, a lone pair of electrons is present, and as such these compounds are weak bases. When the lone pair is used to form a covalent bond with a fourth alkyl group, the resulting quaternary ammonium compound has a permanent positive charge. It does not have the ability to donate or accept a proton, and is thus a neutral compound. Amines are able to form hydrogen bond in two possible orientations, and as a result they have considerable water solubility.

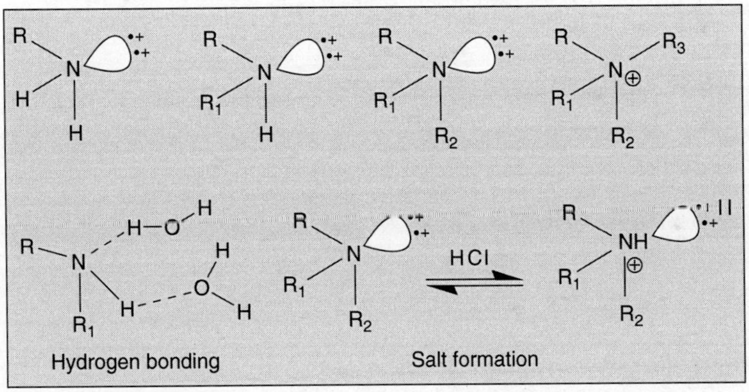

The key to determining the basicity of an amine is to determine the availability of the lone pair of electrons. If the lone pair is more accessible, the compound is more basic, and vice versa. The lone pair also enables amines to form salts; thus, if a basic amine is treated with HCl, the corresponding hydrochloride salt is formed.

The structure of the alkaloid drug morphine is shown below. Note that it contains a number of the functional groups that we have already discussed. The tertiary amine group and the phenolic hydroxyl are the only two of these groups that have acid-base characteristics *in vivo*. Treatment of morphine with HCl results in the formation of the hydrochloride salt, which is ionic and therefore highly water-soluble. Treatment of morphine with NaOH produces the corresponding sodium phenolate salt, which is also water-soluble.

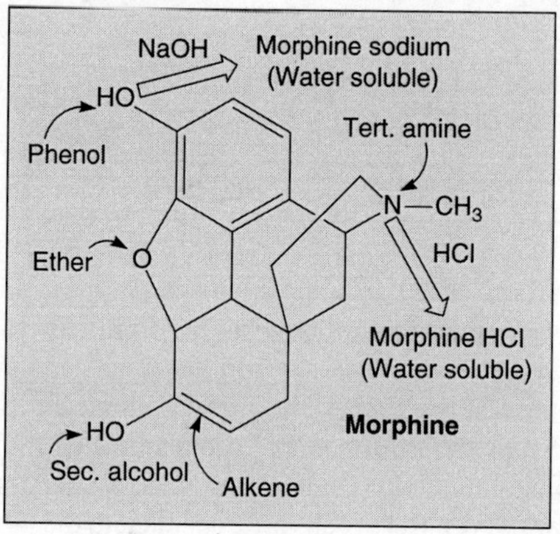

Carboxylic acids are very much water-soluble due to their ability to form hydrogen bonds with water and with themselves, as shown below. They also dissociate very easily because the conjugate base form is resonance-stabilized. When electron-withdrawing groups (EWG) are added to the α carbon, as shown, the acidity of the carboxylate is enhanced. Thus, fluoromethylacetic acid is more acidic than acetic acid, and trifluoroacetic acid is even more acidic. It should be pointed out that carboxylic acids, when treated with NaOH, produce the corresponding sodium salts.

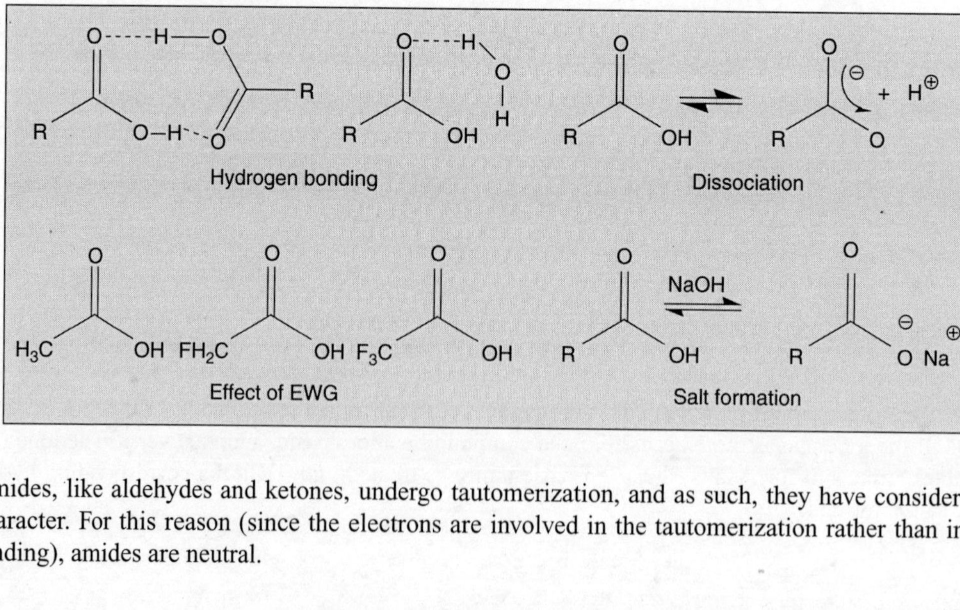

Amides, like aldehydes and ketones, undergo tautomerization, and as such, they have considerable sp² character. For this reason (since the electrons are involved in the tautomerization rather than in proton binding), amides are neutral.

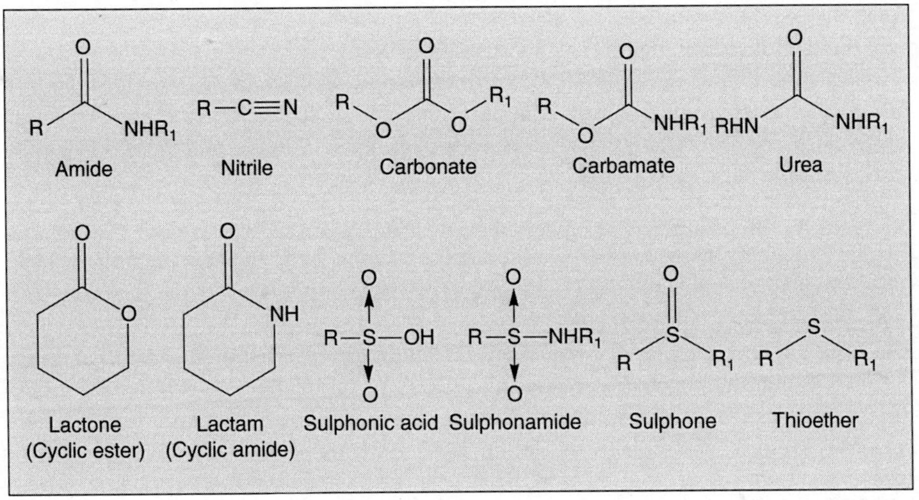

There are a number of derivatives of the carboxyl group which occur frequently in drug molecules, as shown above. The amide group has already been discussed. Carbonates and carbamates, like amides, are neutral in terms of acid-base properties. Ureas are also neutral, and lactones (cyclic esters) and lactams (cyclic amides), like their open-chain homologues, are also neutral. Other important functional groups include nitriles (neutral), sulphonic acids (acidic), sulphonamides (acidic), sulphones (neutral), and thioethers (neutral).

2.3 STEREOCHEMISTRY

Drug molecules must generally interact with biomolecules in a very specific way to elicit a pharmacological response. Because biomolecules are chiral, they often discriminate between isomers of a given drug molecule. In some cases, all isomers of a drug are equipotent; in some cases only one isomer is active; and it is even possible for one isomer to act as an antagonist to the action of the first. There are a number of ways to measure and denote chirality. Some of these are experimentally derived, while others depend on a representation of the three-dimensional structure.

When molecules with one or more chiral centres have the same empirical formula, and they are mirror images of one another, the isomers are known as enantiomers. Enantiomers have identical physical properties, and as such are very difficult to separate by conventional means such as chromatography or crystallization. There are three major ways to express the chirality of drug molecules:

1. *D* and *L* – these letters stand for *dextro* and *levo*, which are abbreviations of the Latin words for 'right' and 'left'. In order to use this method, the structure must be compared to the 3-carbon sugar D- or L- glyceraldehyde. This is done by placing the 'most oxidized group' at the bottom, and comparing the Fischer projections of the two molecules. Although this method is good for sugars, it cannot be used for large drug molecules, since in many cases it is quite ambiguous.
2. *d* and *l* – the lowercase letters *d* and *l* are used to express chirality, but in this case the values of *d* and *l* are determined experimentally. A solution containing the compound is placed in a polarimeter, and a beam of plane-polarized light is passed through the solution. The light will rotate either right (dextrorotatory or +) or left (levorotatory or -). Enantiomers that are pure produce equal and opposite

rotation (i.e., if the *d* form is +25 degrees, the *l* form will be –25 degrees). A racemic mixture (a 50 : 50 mixture of enantiomers) has a net rotation of zero.
3. *R* and *S* – these letters refer to the Latin words *rectus* (right) and *sinister* (left). Unlike *d* and *l*, *R* and *S* can be determined by examination of the structure. The letters *R* and *S* are assigned for a given chiral centre by placing the lowest priority group in the back, and then applying the following three priority rules:

Rule 1: The higher the atomic number, the higher the priority (e.g., O has priority over C).
Rule 2: If the two atoms being compared are the same, then move to the next atom in the chain.
Rule 3: In case of a tie, double bonds count double, and triple bonds count triple. Thus, C=C has priority over C—C, but not over C—C—N.

The assignment of 1-amino-2-hydroxybutane (figure below) is straightforward and easy, since the lowest priority group, hydrogen, is already in the back of the plane. Oxygen has priority over C—N, which has priority over C—C; thus, the molecule as shown is in the S-configuration. However, when the lowest priority group is NOT considered in the back, assignment of *R* and *S* can be tricky, unless you use Woster's Steering Wheel trick method, which can be as shown below for 1-amino-1-hydroxyetan. In this case, *H* is the lowest priority group, but it is not in the back of the plane. The simple solution is to interchange the group in the back with the lowest priority group. When two groups on a chiral centre are exchanged, the resulting molecule is in the other enantiomeric form. The priorities can then be assigned as usual, keeping in mind that you are assigning the enantiomer of the original compound. In our example, the priorities are O > N > C, and so the molecule is R. This means that the original molecule is in the S-configuration.

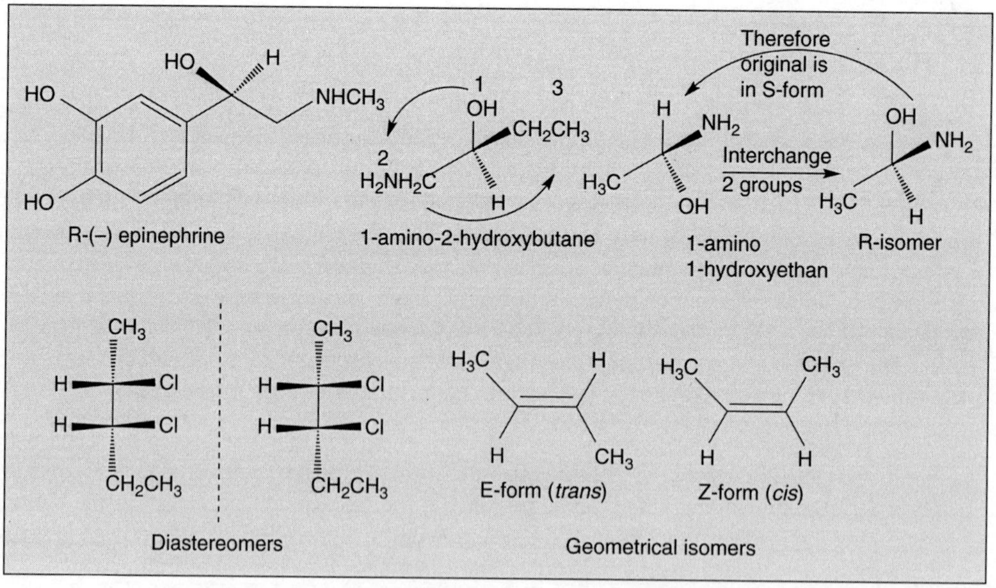

There are two additional forms of stereochemistry which are prevalent in drug molecules, as shown above. Recall that when a molecule has two or more chiral centres, it can exist as diastereomers. Diastereomers are isomers that are not mirror images, as shown above. Diastereomers have different physicochemical properties, and thus can be separated by chromatography, fractional crystallization, or other methods. Note that when a molecule has two chiral centres, there are four distinct isomeric forms. This gives rise to two pairs of enantiomers and two pairs of diastereomers. As discussed earlier, alkenes exhibit geometrical isomerism, and can exist in E (*trans*) and Z (*cis*) forms.

2.4 ACID-BASE CHEMISTRY

2.4.1 Introduction and Examples

You may recall that there are two prevailing theories that pertain to acid-base: the Lewis acid and base theory, and the Bronsted–Lowry theory. For weak organic acids and bases, only the Bronsted–Lowry theory is relevant. According to this theory:

- A B–L acid is a compound that acts as a proton donor, and
- A B–L base is a compound that acts as a proton acceptor.

It is critical to understand the chemistry of the functional groups that act as organic acids and bases, in order to be able to predict their behaviour in solution. Each acid-base equilibrium has an acid form (H:B) and a base form (B:), as shown in the general equation at the bottom of the figure below. Recall that the degree of dissociation of an organic acid or base is represented by the dissociation constant K_a, and that this is expressed as the inverse log value, pK_a. Thus, the pK_a value represents the overall reaction, and not the individual acid-base forms. Either the acid or the base form can be cationic, anionic, or neutral. Consider the examples below. Acetic acid dissociates in water to set up the acid-base equilibrium as shown in the figure. Note that the H:B form is neutral, while the conjugate base (B:) form is anionic. Conversely, methylamine is neutral in the B: form, while the conjugate acid H:B is cationic.

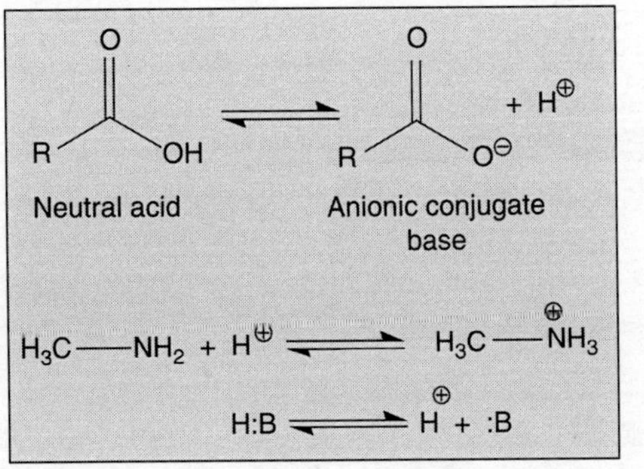

In the figure below, two acid-base equilibria are shown. In the first reaction, *m*-methylphenol, with a pK_a of 10.08, dissociates to form the corresponding phenolate. In this example, the H:B form is a relatively weaker acid, and the conjugate base is a relatively stronger base. This is reflected in the high pK_a value, and indicates that the equilibrium would lie to the left at neutral pH. In the second example, acetic acid, with a pK_a of 4.75, is a relatively stronger acid, and the conjugate base is a weaker base. This indicates that the compound would favour the B: form at neutral pH, and this is reflected by the lower pK_a value.

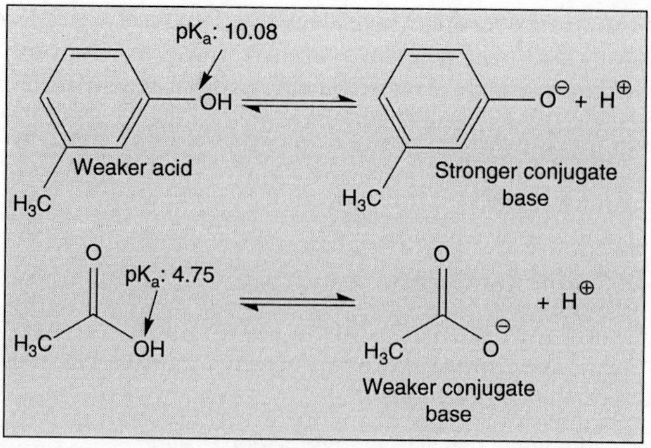

Chapter 2: Principles of Drug Action

[Figure: m-cresol (pK$_a$: 10.08) ionization to phenolate — Weaker acid ⇌ Stronger conjugate base + H$^+$; acetic acid (pK$_a$: 4.75) ionization to acetate — (stronger acid) ⇌ Weaker conjugate base + H$^+$]

In the next figure, two H:B forms are shown which have the same pK$_a$. Recall from our discussion above that phenols are weak acids and amines are weak bases; however, each compound has a B: and H:B form, as shown. *m*-Methylphenol is a molecular acid, meaning that it has a neutral charge. In this form, it is poorly water-soluble. If you treat this compound with aqueous base, it ionizes to the phenolate form. Since this form is anionic, it is now water-soluble. The second compound, N-methylpiperidine, is a cationic acid (i.e., it is positively charged), and in this form it is water-soluble. Treatment of this compound with aqueous base produces a molecular conjugate base, which is no longer water-soluble.

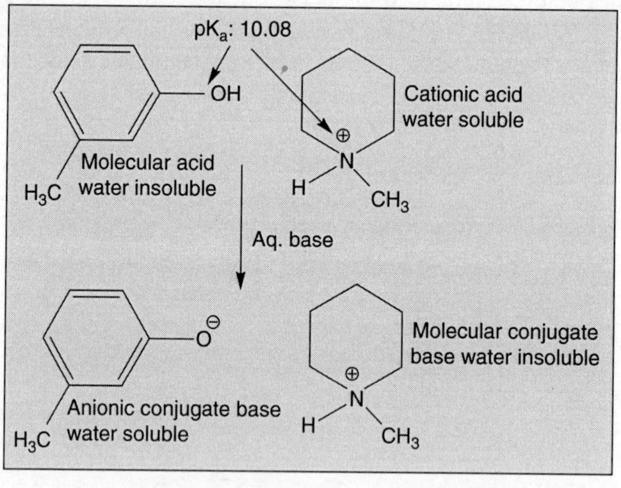

[Figure: m-methylphenol (Molecular acid, water insoluble, pK$_a$: 10.08) and N-methylpiperidinium (Cationic acid, water soluble) → Aq. base → m-methylphenolate (Anionic conjugate base, water soluble) and N-methylpiperidine (Molecular conjugate base, water insoluble)]

Acid-base reactions can be quantitated using the Henderson–Hasselbach equation, shown below. The equation is used to determine the percent ionization for a given acid-base pair.

$$pH = pK_a + \log \frac{[B:]}{[H:B]}$$

When calculating per cent ionized values, it is necessary to determine whether the B: or H:B form is the ionized species. This can be readily determined from the acid-base equation, where by convention the

H:B form is on the left and the B: form on the right. In the example below, methamphetamine, with a pK_a of 9.87, is dissolved in a solution at pH 7.87.

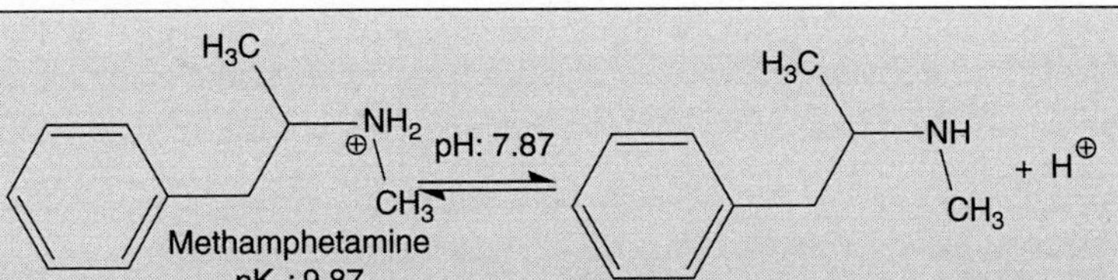

Methamphetamine
pK_a: 9.87

pH: 7.87

$$pH = pK_a + \log \frac{[B:]}{[H:B]}$$

Thus, for methamphetamine at pH 7.87

$$7.87 = 9.87 + \log \frac{[B:]}{[H:B]}$$

$$-2 = \log \frac{[B:]}{[H:B]}$$

In this example H:B is ionized

$$\frac{[B:]}{[H:B]} = \frac{1}{100}$$

$$\% \text{ ionized} = \frac{100}{101} = 99.01\%$$

In the second example, diethylbarbituric has an unionized H:B form, and an ionized B: form. It has a pK_a of 8.0, and is dissolved in fluid with a pH of 7.

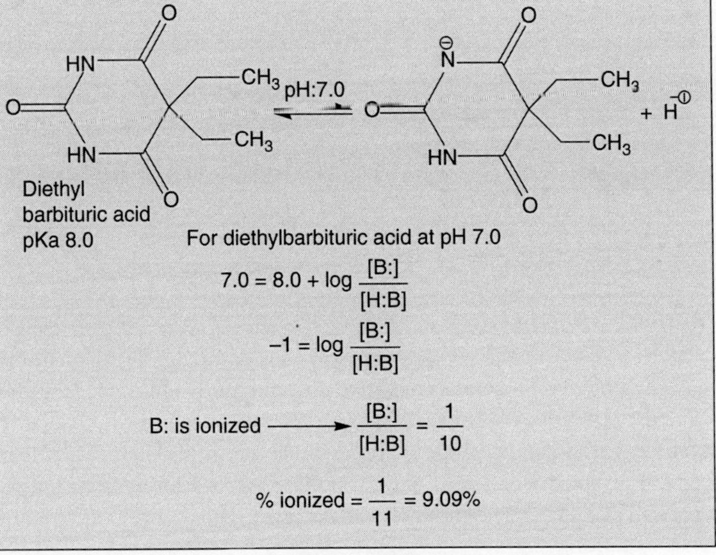

Diethyl barbituric acid
pKa 8.0

pH: 7.0

For diethylbarbituric acid at pH 7.0

$$7.0 = 8.0 + \log \frac{[B:]}{[H:B]}$$

$$-1 = \log \frac{[B:]}{[H:B]}$$

B: is ionized

$$\frac{[B:]}{[H:B]} = \frac{1}{10}$$

$$\% \text{ ionized} = \frac{1}{11} = 9.09\%$$

2.4.2 Factors Affecting Acidity and Basicity
Electronegativity

The acidity or basicity of a given functional group can be dramatically affected by the electronegativity of neighbouring groups or atoms. The term 'electronegativity' refers to the attraction of electrons by the nucleus of a neighbouring atom or group. There are two factors that affect the degree of electronegativity:

1. Electronegativity increases as the distance between the nucleus and the electron shell decreases (i.e., the atomic radius decreases).
2. Electronegativity increases as the number of protons in the nucleus increases.

Consider the upper righthand corner of the periodic table, as shown below. Electronegativity increases from left to right (increasing number of protons), and from bottom to top (decreasing electronic radius), making fluorine the most electronegative atom. It should be noted that electronegativity is not a constant value for a given atom; it depends on what the neighbouring atom is, and how well it pulls electrons. For example, Flourine atom next to carbon, as in ethyl fluoride, would be different than the value for fluorine next to another atom. Also, consider acetic acid, which has a pK_a of 4.75. The α carbon is more electronegative when attached to 2 chlorines, and thus the pK_a decreases to 1.29. Similarly, trichloroacetic acid has a pK_a of 0.65.

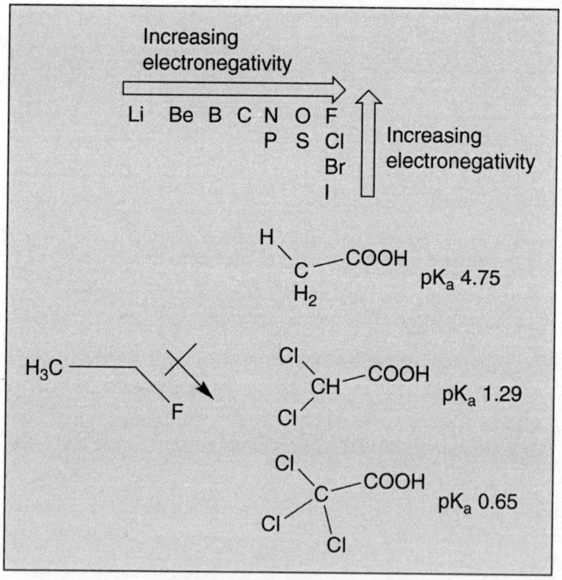

When a carbon is next to an electronegative atom, the carbons in neighbouring positions are subject to the inductive effect, as shown below. The electronegative chlorine pulls electrons from the adjacent carbon, giving the chlorine a partial negative charge and the α carbon a partial positive charge. The partial positive charge renders the α carbon somewhat electronegative, and it pulls electrons from the beta carbon, making it partially-partially positive. The β carbon pulls electrons away from the γ carbon, making it partially-partially-partially positive. The inductive effect, also known as chain induction, wears off after about 3–4 carbons. For this reason, the pK$_a$ of α-chloroacetic acid (2.84) is lower than β-chloroacetic acid (4.06), and even less than γ-chloroacetic acid (4.52). When a group is electronegative, it is referred to as -Is. -Is groups include the halogens (F, Cl, Br, I), ketones, oximes, and alkenes. Other strong -Is groups include nitro, which is more electronegative than F, OR, NR$_2$, and CR$_3$. Groups that are electropositive (i.e., electron releasing) include alkyl groups such as methyl, ethyl, and *t* -butyl.

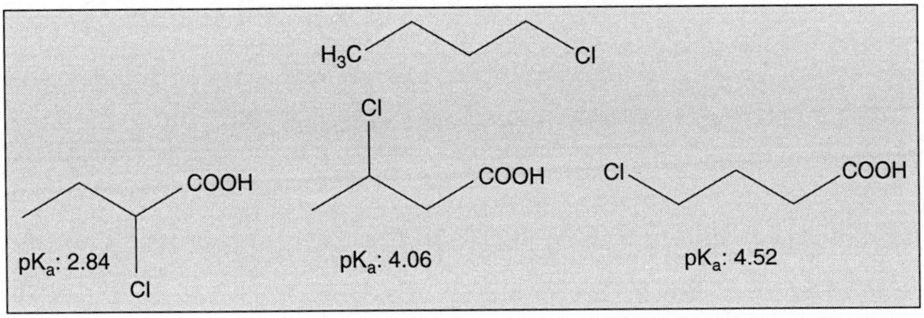

Resonance

This concept has already been discussed earlier. If the conjugate base of a weak acid can be resonance-stabilized, the acid form will be more acidic. We have already seen this effect in carboxylic acids; resonance stabilization also occurs with phenolic compounds, as seen below. Once the phenolate anion is formed, the negative charge can be accommodated in one of four resonance forms. Thus, the anion is stabilized, and the phenol is a stronger weak base.

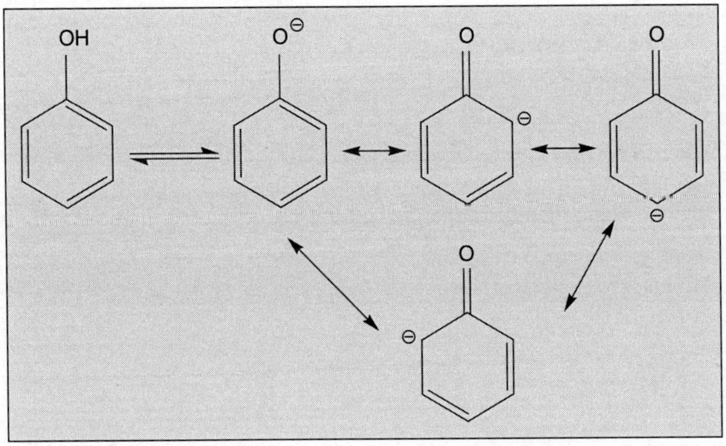

Do groups also have an effect on acidity and basicity?

For example, the compound *p*-nitrophenol is more acidic than phenol itself, due to the electron-withdrawing properties of the highly electronegative nitro group. Likewise, *p*-methylphenol would be less acidic than phenol, because of the electron-releasing properties of the alkyl group.

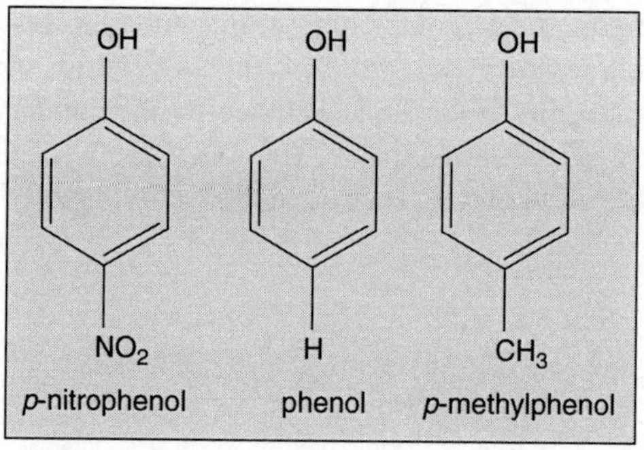

2.5 BIOLOGICALLY SIGNIFICANT NITROGEN-CONTAINING COMPOUNDS

As shown below, aliphatic amines are generally good weak bases in solution. As was mentioned earlier, their basicity depends on the availability of the lone pair of electrons on the nitrogen. Thus, as alkyl groups are added to the nitrogen, basicity increases, since the alkyl groups donate electrons to the system. In water, secondary amines are more basic than primary amines, as expected, but tertiary amines are less basic than both. This is because of a steric effect wherein the third alkyl group reduces the ability of the tertiary amine to H-bond. In organic solvents, tertiary are more basic than secondary, which are more basic than primary, as would be expected.

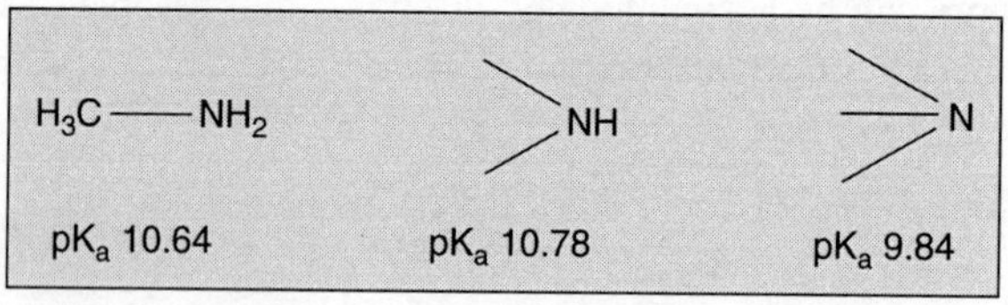

As seen below, aromatic amines can have unexpected acid-base properties. Consider aniline, shown in the figure below, which has a pK_a of 4.6. The nitrogen in aniline has an sp^3 configuration, and as such, the lone pair of electrons can interact with the aromatic electron cloud. This effect stabilizes the B: form of aniline, and thus it is less basic than the aliphatic counterpart, cyclohexylamine (pK_a 10.6).

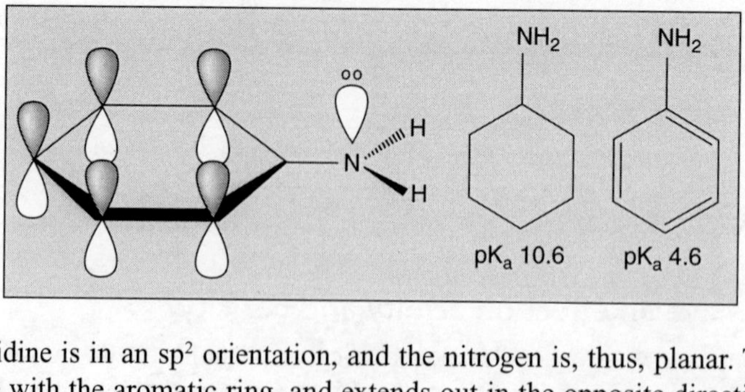

The nitrogen in pyridine is in an sp^2 orientation, and the nitrogen is, thus, planar. This means that the lone pair is in plane with the aromatic ring, and extends out in the opposite direction. Because it protrudes from the aromatic ring, it is available for bonding to hydrogen. However, the lone pair is also pulled in by the aromatic cloud, thus attenuating the basicity of pyridine. As a result, the molecule has a pK$_a$ of 5.2.

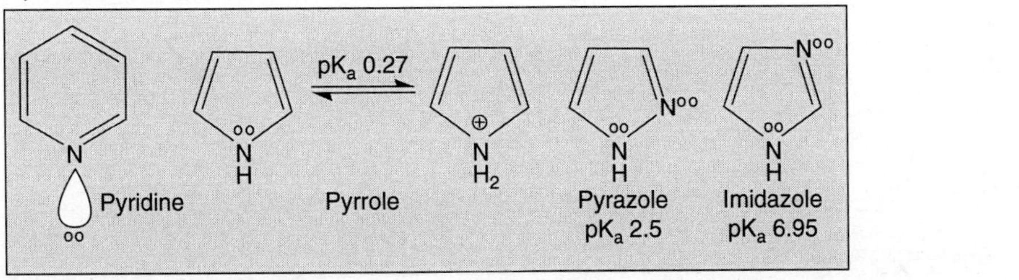

Pyrrole nitrogen is quite different from pyridine nitrogen. In these compounds, the lone pair is used to complete the aromatic cloud within the molecule. Thus, the 4 π electrons plus the two lone pair of electrons add up to 6π electrons, which is a Huckel number. In order to protonate pyrrole nitrogen, the aromaticity of the molecule would need to be destroyed. This is, of course, energetically unfavourable, and cannot be accomplished at physiological pH. Pyrrole nitrogen has a pK$_a$ of –0.27.

There are two common heterocycles that contain a pyridine- and pyrrole-type nitrogen, pyrazole and imidazole. The pyridine-type nitrogen in pyrazole has a pK$_a$ of 2.5. Because the pyridine-type nitrogen in imidazole is shielded from the pyrrole nitrogen by one carbon, it has a pK$_a$ of 6.95.

There are two biologically relevant carboxylic acid derivatives that should be mentioned, amides and amidines. As you will recall, amides (the type of bond present in peptides) undergo a keto-enol tautomerization, as shown below. Because the lone pair of electrons on the amide nitrogen is involved in this tautomerization, it is not available for bonding. As a result, amides are neutral compounds, and do not undergo acid-base reactions. By contrast, the related functional group known as an amidine is strongly basic, with a pK$_a$ in the range of 12.4. The basicity of amidines is due to the resonance stabilization of the conjugate acid form, which accepts a proton, and then delocalizes the positive charge as shown.

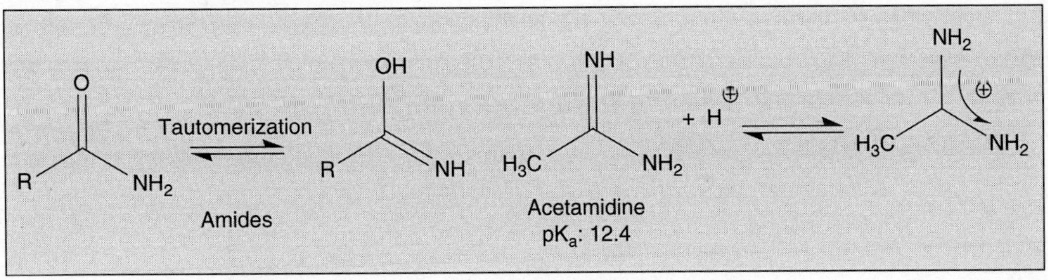

There are two common types of acidic nitrogen compounds, sulphonamides and imides, as shown below. Because of the electron-withdrawing character of the sulphone moiety in a sulphonamide, combined with the electron-withdrawing character of the phenyl ring, the bond between the nitrogen and the hydrogen is extremely weak. In fact, the bond is so weak that sulphonamides dissociate in water, and donate a proton like other weak acids. A similar situation is found in the case of nitrogen flanked by two carbonyls, a functional group known as an imide. The α carbonyls are EWG, and the resulting anion is resonance-stabilized as shown, with the negative charge being distributed over 5 atoms. Thus, imides act as weak organic acids.

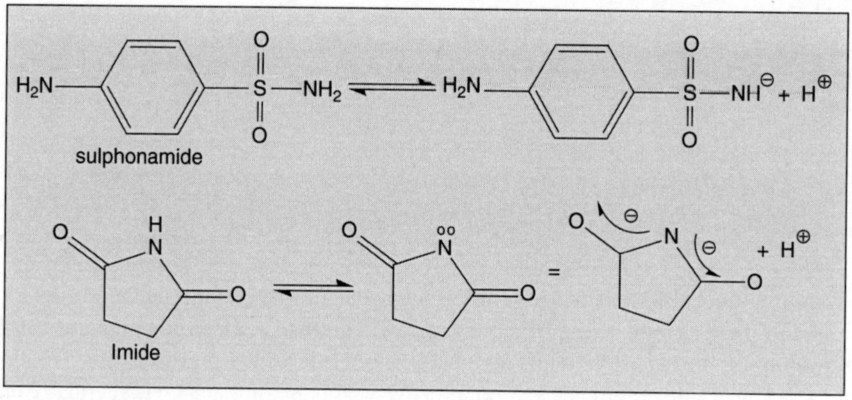

2.6 PHYSICOCHEMICAL PROPERTIES AND DRUG ACTION

In order to elicit a pharmacological effect, drugs must be sufficiently soluble in water to be absorbed and distributed throughout the body. They must also have sufficient lipophilicity to be able to pass through biological membranes. As was mentioned above, the ability of an organic molecule to dissolve in water is dictated by how well it can break into the lattice structure of water. As shown in the figure below, water has a dipole moment due to the 104.5-degree bond angle and the pull of electronegative oxygen on the attached hydrogen. This induced polarity gives water a higher boiling point and melting point than other hydrides (e.g., H-S-H, hydrogen sulphide, is gas at room temperature). This dipole also allows water to hydrogen bond, and in pure water, it H-bonds to itself, forming a lattice as shown below. Organic compounds that ionize are readily water-soluble, since they form an envelope of water molecules which increases the entropy of the system and decreases energy. Non-ionic, polar compounds such as those discussed above can also dissolve—they do not dissociate, but enter the water lattice by hydrogen bonding to water.

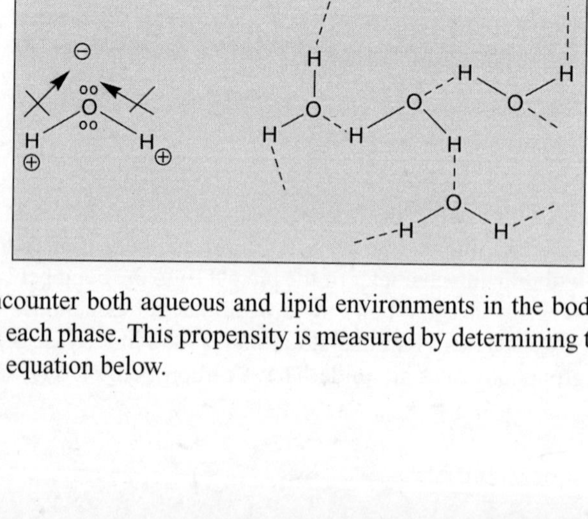

Because drugs must encounter both aqueous and lipid environments in the body, they must have some measure of solubility in each phase. This propensity is measured by determining the partition ratio, which is determined using the equation below.

$$P = \frac{[D]_{lipid}}{[D]_{water}}$$

The partition ratio is simply the ratio of the solubility of the drug in lipid (simulated by *n*-octanol) and its solubility in biological fluid (simulated by phosphate buffer at pH 7.4). The partition ratio of a given drug will determine its solubility in plasma, its ability to traverse cell membranes, and which tissues it will reach.

A number of theoretical representations of the relationship between physicochemical properties and drug action have been developed. One of the earliest of these is known as the Overton–Meyer Hypothesis. This theory was developed following the observation that neutral, lipid soluble substances have a depressant effect on neurons. The hypothesis states that, for these compounds, the higher the partition ratio P, the higher the pharmacological effect. This hypothesis was expanded upon by Ferguson, who extended the theory to include all drugs. The Ferguson Principle states that the concentration of a drug in plasma is directly proportional to its activity. This concentration can be measured, either as molarity or partial pressure. The Ferguson constant X is determined by measuring the molar concentration (or partial pressure) of a drug required for an effect, and dividing it by the molar solubility of the drug (or its partial pressure in the pure state). As seen below, if the value of X is between 0.1 and 1, the drug is said to have high thermodynamic activity. This means that the activity of the drug is based on its physicochemical properties only, such as in a gaseous anaesthetic. Such drugs are known as non-specific agents. When the value of X is less than 0.1, the drug is said to have low thermodynamic activity, meaning that the activity of the drug is based on its structure rather than physicochemical properties. Agents in this category are called specific agents, and their activity at low concentrations infers that they have a specific receptor.

The Ferguson principle

$$\frac{P_t}{P_o} = X \quad \text{or} \quad \frac{S_t}{S_o} = X$$

P_t is the partial pressure required for pharmacological activity
P_o is the partial pressure of the pure substance

S_t is the molar concentration required for pharmacological activity
S_t is the molar solubility of the compound

$X = 1$ to 0.1 means the drug has high thermodynamic activity
$X = < 0.1$ means the drug has low thermodynamic activity

The effect of substituents on the acidity and basicity of various functional groups has been discussed above. This effect can be quantified, and this was first done by Hammett, who measured the effect of various substituents on the acidity of benzoic acid. The derivation of the Hammett substituent coefficient is shown below. This coefficient (sigma, σ) is calculated by determining the dissociation constant K for a benzoate with substituent X, and dividing it by the K for benzoic acid (where the substituent is H). The log of this ratio is then the Hammett substituent coefficient. The values for the Hammett coefficient are available in tabular form, and are used in mathematical models of activity known as Quantitative Structure-Activity Relationships, or QSAR.

Deviation from Hammett substituent coefficient

$$\log \frac{K_X}{K_H} = \text{sigma}$$

In addition to contending with both lipid and aqueous environments, specific agents must also interact with cellular macromolecules such as receptors and enzymes. Because these macromolecules are chiral, it is not surprising that the stereochemistry of a drug can impact its ability to bind to its target. It is possible that both enantiomers of a given compound can have activity at the same receptor, but more often one isomer is active, while the other is inactive. It is also possible for the 'wrong' isomer to act as an antagonist, or it may have toxic effects not seen in the 'right' isomer. Consider the drugs shown below: levorphanol and dextromethorphan. Levorphanol is a powerful narcotic analgesic with a high addiction liability; its enantiomer is dextromethorphan, which is widely used in cold preparations. It has no analgesic activity or addiction liability, but retains the anti-tussive action seen in levorphanol.

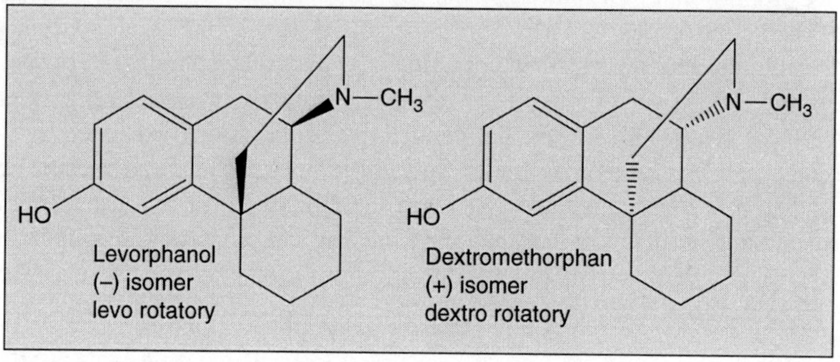

Levorphanol
(−) isomer
levo rotatory

Dextromethorphan
(+) isomer
dextro rotatory

The concept of isomeric potency can be generalized as outlined below. Consider a 'receptor' that has three binding areas (square, round, and hexagonal). When the 'right' isomer binds, it will fit precisely in all three of these sites, and this is termed a 'three-point attachment'. By contrast, the 'wrong' enantiomer can only produce a two-point attachment, and as such will be expected to be less active.

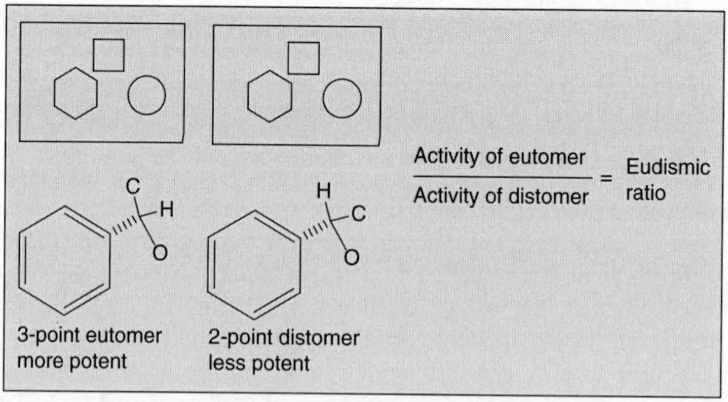

$$\frac{\text{Activity of eutomer}}{\text{Activity of distomer}} = \text{Eudismic ratio}$$

3-point eutomer
more potent

2-point distomer
less potent

According to this theory, the 'right' isomer is called the eutomer, and the 'wrong' isomer is called the distomer. The ratio of the activities of the eutomer and the distomer is called the eudismic ratio, as seen above, and converting the equation to log form affords the eudismic index EI.

2.7 BIOISOSTERIC REPLACEMENT

Bioisosteres are functional groups that have similar spatial and electronic character. In many cases, replacement of a group with a bioisoster results in a new compound that retains the activity of the parent. Thus, this approach is common in the pharmaceutical industry, since it allows them to generate marketable analogues of a known drug that has a patentable composition of matter. The figure below shows common isosteric replacements.

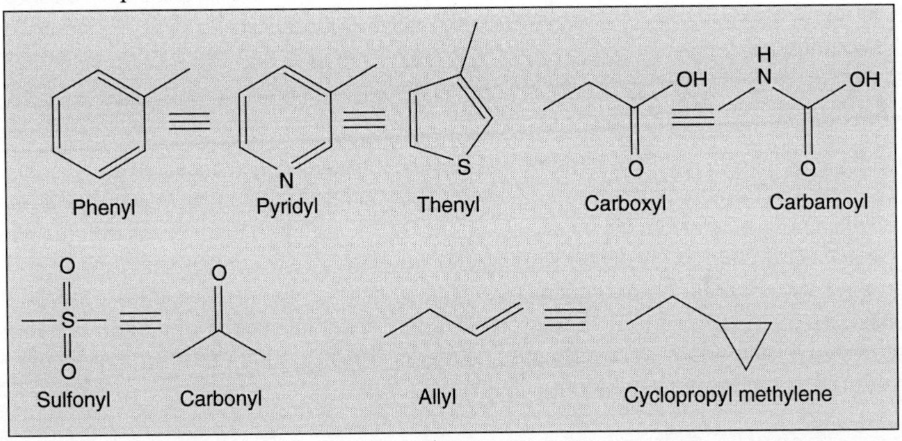

The requirement for bioisosteres to have similar spatial and electronic character is illustrated below. The phenothiazine ring system, on the left, is commonly found in anti-psychotic drugs such as chlorpromazine. The phenothiazine ring is planar due to the two aromatic rings and the intervening sulphur, which has 4 π electrons. If the sulphur is replaced by a double bond, the ring retains its π character, and the resulting dibenzazepine retains anti-psychotic activity. When the double bond is reduced, the ring is no longer planar, as shown below left. These compounds do not act as neuroleptics, and in fact have no CNS activity. They are mainly used as antihistamines.

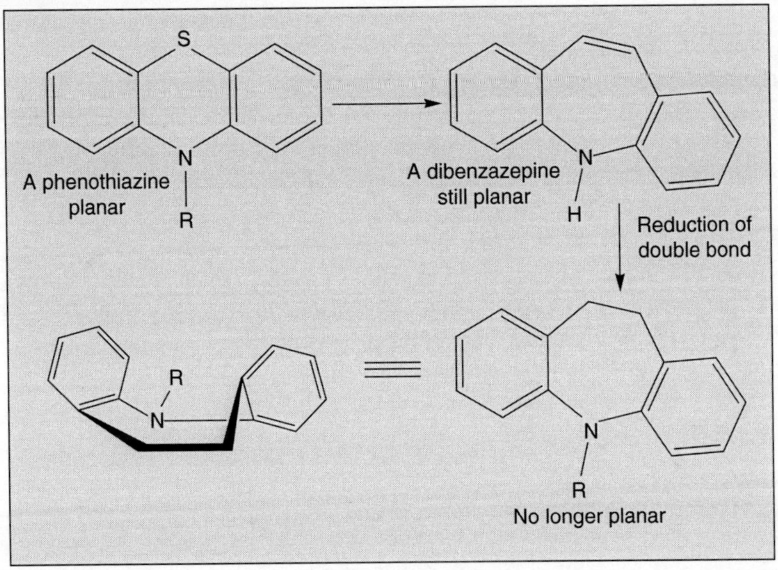

2.8 DRUG-RECEPTOR BONDING

Interaction forces between ligands and receptors contribute to the binding strength and biological efficacy. Receptor and enzymes are proteins of well-defined secondary and tertiary structure, and yields a 'unique' active site structure that is essential for the specific binding of ligands. Receptors are membrane-bound proteins and generally have helicies that span the membrane by total occupation of backbone H-bonding sites. The interior of the lipid membrane being hydrophobic, Extracellular region often glycosylated and intracellular region attached or bound to other proteins, such as kinases, G proteins, the receptors function as signal transmitters.

2.8.1 Receptor Theories

At the beginning of the 20th century, Langley and Ehrlich introduced the concept of a receptor that would mediate drug action. Clark was the first to quantify drug-induced biological responses and proposed a model to explain drug-mediated receptor activation. In 1901 Langley challenged the dominant hypothesis that drugs act at nerve endings, by demonstrating that nicotine acted at sympathetic ganglia even after the degeneration of the severed preganglionic nerve endings. In 1905, he introduced the concept of a receptive substance on the surface of skeletal muscle which mediated the action of a drug. It also postulated that these receptive substances were different in different species (citing the fact that nicotine-induced muscle paralysis in mammals was absent in crayfish).

Occupancy theory

One of the first models put forward by Clark to explain the activity of drugs at receptors quantified the relationship between drug concentration and observed effect—in particular, that the magnitude of the response is directly proportional to the amount of drug bound and that the maximum response would be elicited once all receptors were occupied. The response ceases when the drug dissociates from the receptor. The receptor–occupancy theory was subsequently modified by Ariens in 1954 and by Stephenson in 1956, to account for the intrinsic activity (efficacy) of a drug. According to them,

- A maximum effect can be produced by an agonist when occupying only a small proportion of the receptors termed as spare receptors.
- The response is not linearly related to the number of receptors occupied.
- Different drugs have varying capacity to initiate the response. Or, the occupancy of a receptor to the same extent by different compounds can produce responses of different magnitude. Also, different compounds that interact at the same site can have different efficacy.

Rate theory

Paton in 1961 explored a unique hypothesis to explain drug action in an attempt to provide a theoretical basis for some of his experimental findings. Paton postulated that intensity of pharmacological effect is proportional to the total number of encounters of the drug with the receptor per unit time. The proportion of receptors occupied at equilibrium (p) can be presented as

$$p = \frac{k_1 x}{k_2 + k_1(x)} = \frac{x}{x + k_2 l k_1}$$

where 'x' is the concentration of the drug in g/ml; k_1 and k_2 are the association and dissociation rate constants in $sec^{-1} g^{-1}$ ml and sec^{-1}, respectively.

Induced fit theory

This theory assumes that approach of a drug induces a conformational change of the active site. The induced-fit theory assumes that the substrate plays a role in determining the final shape of the enzyme and that the enzyme is partially flexible. This explains why certain compounds can bind to the enzyme but do not react because the enzyme has been distorted too much. Other molecules may be too small to induce the proper alignment and therefore cannot react. Only the proper substrate is capable of inducing the proper alignment of the active site. See the Fig. 2.1.

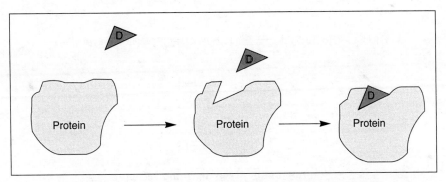

FIGURE 2.1   Drug-Protein binding and conformational changes.

Macromolecular perturbation theory

Receptors are highly flexible macromolecules. When an agonist interacts with the receptor, a specific conformational perturbation leads to a biological response, whereas a non-specific conformational perturbation due to antagonists binding leads to no response or an opposite response. If a drug contributes to both the possible perturbations, then it would result in partial agonism.

Activation–aggravation theory

This theory is considered to be an extension to the macromolecular perturbation theory, wherein a receptor exists in equilibrium between an activated state (bioactive) and an inactivated state (resting or bioinactive). According to this theory, even in the absence of a drug this dynamic equilibrium state would exist. Agonists bind to the active state and the antagonists bind to the inactivated state. A partial agonist would bind to both the states of the receptor.

The working model of drug–receptor interaction includes:

1. Ligand interacts with a binding site on the extracellular face of the receptor. Binding site may be located in the region of the receptor surrounded by membrane, and ligand (especially peptides) may enter the membrane prior to interacting.
2. Ligand binding induces a local conformational change.
3. Conformational change is propagated via the membrane-spanning helicies, or protein dimerization brings together functional regions inside the cell.
4. Leads to conformational changes on the intracellular face of the receptor and instigates cellular response (signal transduction) phosphorylation/dephosphorylation of G-proteins and cyclicAMP.

Alternative model: Protein dimerization or higher multimeric state

1. Ligand interacts with binding site on extracellular face of receptor (protein monomer).
2. Ligand induces protein dimerization.
3. Dimerized protein has functional groups adjacent to each other leading to cellular response.

2.8.2 Interaction of a Drug with Receptor

$$\text{DRUG} + \text{RECEPTOR} \underset{k_{off}}{\overset{k_{on}}{\rightleftharpoons}} \text{DRUG - RECEPTOR COMPLEX}$$

$$k_D = \frac{[Drug][Receptor]}{[Drug\text{-}Receptor]} = \frac{k_{off}}{k_{on}}$$

$$k_D = \frac{[Drug\text{-}Receptor]}{[Drug][Receptor]} = \frac{k_{off}}{k_{on}}$$

Equilibrium between drug and receptor

K_D (or K_I) is a dissociation constant note that the dissociated (unbound) species are in the numerator and the smaller the better (similar to IC_{50}). K_A (or K_{eq}) is an association constant. Note that the associated (bound) species are in the numerator and the dissociated species are in the denominator, hence the bigger the better. To maximize binding (small K_D, large K_A) k_{on} is initial by small as drug–receptor interactions are often diffusion-controlled; therefore, cannot be made larger (will be similar for a congeneric series of compounds) k_{off} small the stronger the drug-receptor interactions and the smaller k_{off} increase the drug–receptor interactions.

How strong? Can be determined from the vant Hoff equation ($K_A = 1/K_D$)

$$\Delta G^O = -RT \ln k_A$$

vant Hoff equation

Compound	K_D	ΔG(versus X)
X	100 mM	0.00
Y	1000 mM	1.36 kcal/mole
Z	10 mM	−1.36 kcal/mole

A change in the binding by a factor of 100 corresponds to a decrease (becoming more favourable) of the free energy by 2.7 kcal/mole; a 1,000-fold change is 4.1 kcal/mole. Thus, the change in free energy to impart specificity is relatively small.

2.8.3 Agonists and Antagonists

Efficacy can be defined as the amount of a biological response of a drug or intrinsic activity, and can be expressed in terms of percentages. A full agonist is thought to be a fully efficacious drug with 100 per cent biological response after interacting with the receptor. It is the opposite of an antagonist, which acts against and blocks an action. Agonists and antagonists are key agents in the chemistry of the human body. It is important to remember that potency and efficacy are different concepts and cannot be interchanged. If an agonist has high efficacy, it does not necessarily mean that it will display high potency, and vice versa. An agonist that produces the maximum response capable in that system is termed a full agonist, and anything producing a lower response is a partial agonist (Fig. 2.2).

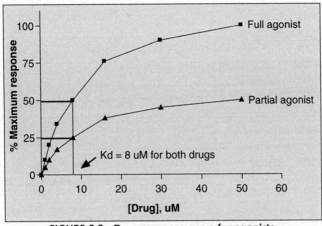

FIGURE 2.2 Dose-response curve for agonists.

One example of a drug that acts as a full agonist is isoproterenol, which mimics the action of adrenaline at β adrenoreceptors. A partial agonist exhibits less than 100 per cent biological response even with maximally effective concentration of agonist. Examples include buspirone and buprenorphine. They may also be considered as ligands that display both agonistic and antagonistic effects. When both a full agonist and a partial agonist are present, the partial agonist actually acts as a competitive antagonist, competing with the full agonist for receptor occupancy and producing a net decrease in the receptor activation observed with the full agonist alone. Receptors require an agonist-induced conformational change to be able to bind to G proteins or open a channel, with the partial agonist inducing a suboptimal conformational change in the receptor called the 'open door' concept. In another classic 2-state model, partial agonists are found to increase the proportion of receptors that are in the inactive (resting) state, compared to full agonists. This model is simple in assuming only two states of a receptor, without the many different conformations implied by the 'open door' concept described above. However, it does not always explain agonist pharmacology. More complex models (e.g., a 3-state model) have been proposed to explain agonist properties as well.

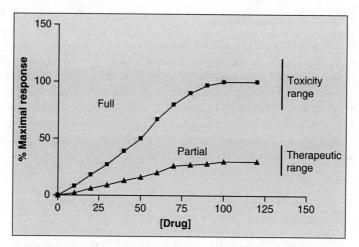

FIGURE 2.3 Dose-response curve depicting therapeutic index.

Partial agonist can be a good therapeutic agent when the safe dose range can be greatly extended (the maximum response only reaches the sub-100 per cent efficacy value and stays there as a plateau) (Fig. 2.3). An example of a clinically studied drug that binds to 100 per cent of its receptors but only has partial agonist characteristics is DMXB (GTS-21), a nicotinic channel receptor agonist.

Inverse agonists are those drugs that produce opposite biological response from that of agonists as shown in the figure below. Binding sites of inverse agonists are considered to be typically near to or the same as that of regular agonists and antagonists (Fig. 2.4). Receptors sensitive to inverse agonists include opiods, serotonin ($5HT_{2C}$, $5HT_{1A}$, $5HT_{1B}$, $5-HT_{1D}$), dopamine (D2, D3, D5), adrenergic (α, $\beta 1$, $\beta 2$), muscarinic (m1, m3, m5), histamine H2, calcitonin, bradykinin, and adenosine A1. Some classic receptor antagonists have been found to actually be inverse agonists, an example being atropine (muscarinic receptors).

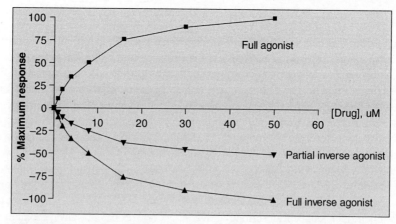

FIGURE 2.4 Dose-response curve depicting the action of inverse agonists.

The favourable binding of a drug to its target results in a decrease in energy, and can be the result of multiple bonding forces. For example, the binding of acetylcholine to its receptor involves 8 binding domains and 6 bonding modes.

The term 'antagonist' refers to any drug that will block, or partially block, a response. Antagonist can be competitive or non-competitive. A competitive antagonist binds reversibly to the same receptor as the agonist. A dose-response curve performed in the presence of a fixed concentration of antagonist will be shifted to the right, with the same maximum response and (generally) the same shape (Fig. 2.5). The EC_{50} is the concentration needed to obtain 50 per cent of the maximal response. Naloxone is a competitive antagonist to all opioid receptors.

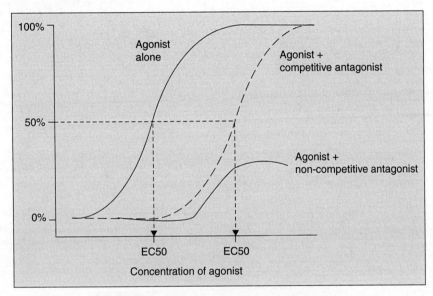

FIGURE 2.5 Dose-response curve of antagonists.

The actions of a non-competitive antagonist cannot be overcome by increasing the dose of agonist. This is because the agonist and antagonist binding sites are different; hence, the agonist will not displace the antagonist molecule (e.g., ketamine binds in the NMDA receptor channel pore, but the agonist, glutamate, binds to the extracellular surface of the receptor).

2.8.4 Covalent and Non-Covalent Bonding

Several chemical forces may result in a temporary binding of the drug to the receptor. Essentially, any bond could be involved with the drug–receptor interaction. Types of interactions contributing to ligand–receptor forces can be categorized into:

1. Covalent bonding
2. Non-covalent bonding

1. Covalent bonding: Approximately −100 kcal/mole formation of a covalent bond between a drug and a receptor will lead to irreversible binding. The covalent bonding is not usually preferred as it can produce a long-lasting effect and, hence, toxicity. Often, it is desirable for the drug effect to last only a limited time, and if prolonged can lead to toxic or lethal effects as in the case of CNS depressants. This type of bonding is beneficial when permanent cell destruction is desirable as in drug-DNA binding in designing chemotherapeutic agent, where a drug acts on a foreign organism or tumour cells. DNA as carrier of genetic information is a major target for drug interaction because of the ability to interfere with transcription (gene expression and protein synthesis) and DNA replication, a major step in cell growth and division. The latter is central for tumorigenesis and pathogenesis. Anticancer agents like nitrogen mustards and other alkylating agents form irreversible complexes with DNA by covalent bonding. Examples of drugs forming covalent bonds include nitrogen mustards (alkylating agents), anticholinestrase inhibitor (organophosphates, melathion), and α-blocker phenoxybenzamine. Drugs that bind to receptors using covalent bonding may cause a permanent bond, in which case the receptor or enzyme target is 'killed', or it may be transient. The electrophile involved in formation of the covalent bond is generally designed into the drug. For example, consider the nitrogen mustard shown below. The nitrogen lone pair displaces one of the chlorides, resulting in the formation of a highly reactive aziridine. The electrophilic aziridine then reacts with a nucleophile in the active site, forming a covalently bonded inhibitor.

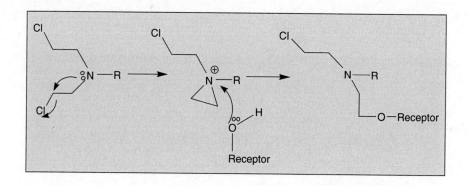

2. Non-covalent bonding: It is weaker compared to covalent bonding and preferred generally for drug effect to last for a limited duration. It does not involve sharing of electrons, but rather involves more dispersed variations of electromagnetic interactions. The energy released in the formation of non-covalent bonds is in the order of 1–5 kcal per mole.

(i) **Ionic bonding:** This is the strongest type of non-covalent bond that mainly occurs through attractions between opposite charges. This results from the attraction of ions with opposite charges. Given with q_1 and q_2, the charges of atoms on molecules 1 and 2, ε_o the dielectric constant and the distance between 1 and 2, the energy can be derived as

$$E = \frac{q_1 q_2}{4\pi\varepsilon_o r_{1,2}}$$

At physiologic pH, certain drugs ionize and interact with opposite charges in the active site pocket of their targets.

Certain amino acids like arginine and lysine get protonated in the physiological pH and provide a cationic environment, and certain other amino acids like aspartic acid and glutamic acid get deprotonated to provide an anionic group for ionic bonding to occur. This type of bonding is depicted below:

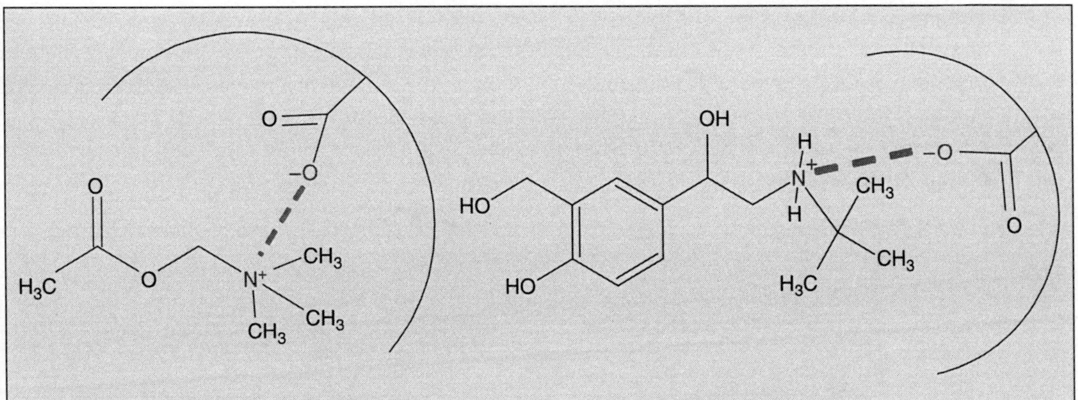

[Quaternary ammonium of acetylcholine]

[Protonated salbutamol]

(ii) **Ion-Dipole bonding:** This type of interaction results when there is an attraction between an ion and the partial charge of a dipole of opposite polarity. As a result of greater electronegativity of atoms like oxygen, sulphur, halogens, and nitrogen in a molecule, there will be an asymmetry in sharing of electrons to form electronic dipoles, as depicted below.

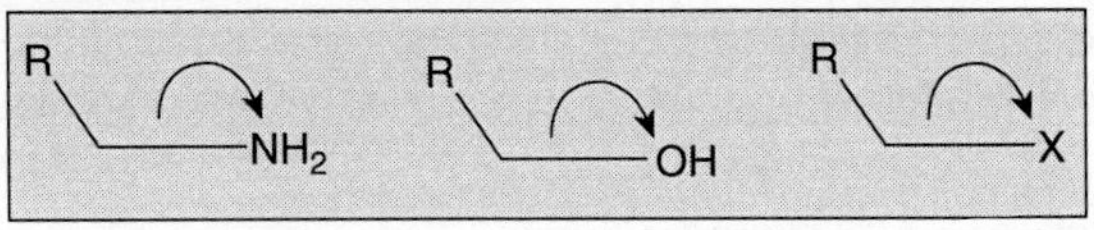

Because the charge of dipole is less than an ion, an ion-dipole bonding is weaker than an ion-ion interaction. In the receptor side, the amino acids with permanent dipoles include serine, threonine, aspargine, glutamine, tyrosine, tryptophan, cysteine, and histidine.

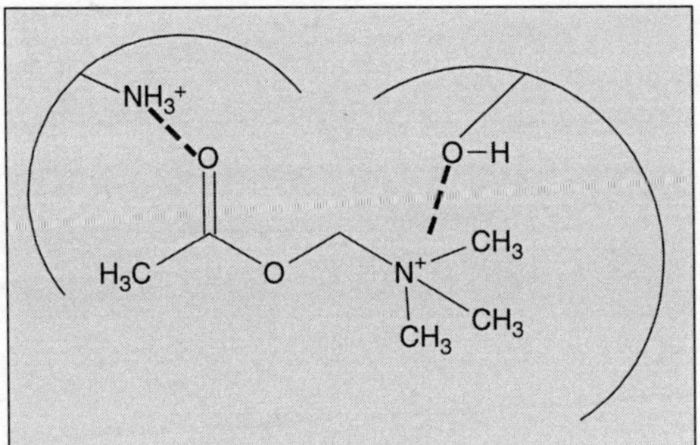

[Ion-dipole interaction]

(iii) **Dipole-Dipole:** Here, a partially positive atom in a dipole is attracted to a partially negative atom in another dipole. This type of bonding contributes to an energy calculated as

$$E = \frac{\mu_1 \mu_2}{4\pi\varepsilon_o r_{1,2}^3}$$

Wherein μ_1 and μ_2 are the dipole moments in molecule 1 and 2.

(iv) **Hydrogen bonding:** This is a dipole-dipole interaction where one of the constituents is hydrogen attached to a heteroatom. It could be considered as a special case of dipole-dipole interaction as it generally makes favourable contribution to the binding strength of ligands. The hydrogen bond donors can be a O-H, N-H, or S-H, and hydrogen bond acceptors could include N, O, or S, with a lone pair of electron C=O, OH, SH, C≡N, etc. In the figure below, intra- and inter-hydrogen bondings are depicted.

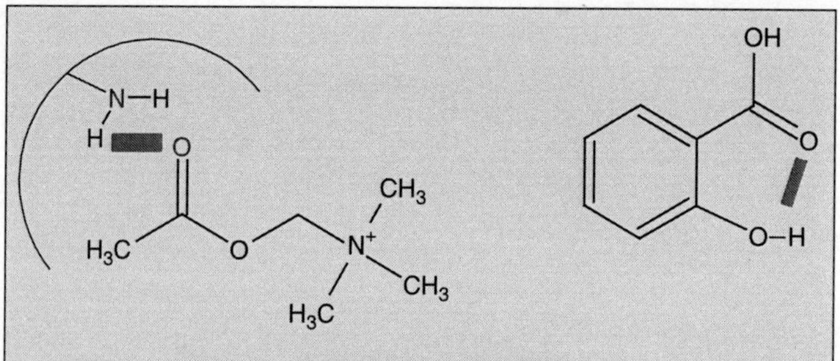

Inter-and Intra-Hydrogen bonding

(v) **The hydrophobic effect:** When two alkyl chains approach one another, water is extruded from the space in between them, resulting in an increase in entropy, and thus a decrease in energy. The pattern of hydrophobic interaction relative to ionic or hydrogen bonding can aid in specificity.

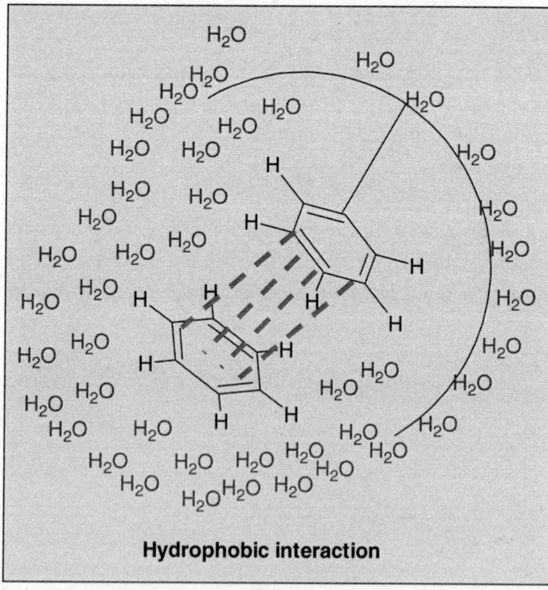

Hydrophobic interaction

(vi) **Charge-Transfer complexes:** When an electron-deficient region comes in closer contact with an electron-rich system, there is a dipole-induced dipole interaction, i.e., the electron-rich species transfers some charge to the electron-deficient species. A lone pair of electrons is 'shared' with a neighbouring group that has considerable π character. When the dipole moment of molecule 1 (μ_1), the polarizability of the adjacent molecule (α), and the distance r are known, then the energy of charge-transfer interaction can be derived as

$$E = \frac{1}{2}\frac{\mu_1^2 a_2}{r_{1,2}^6}$$

The following figure represents a charge-transfer interaction.

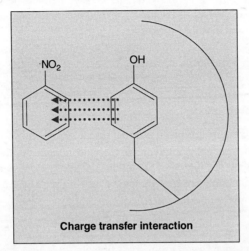

Charge transfer interaction

(vii) **Van der Waals forces:** One carbon in a chain approaches another carbon on a neighbouring chain, causing a perturbation known as an induced dipole. These opposite partial charges then attract one another. This type of force can also be termed as instantaneous dipole-induced dipole interactions, although the net average is zero. This will produce a net attraction even though the atoms interacting have no net charge. This type of interactions are dependent greatly on distance, and the energy varies inversely as the sixth power of the interatomic distances, as presented in the following formula:

$$E = \frac{3I_1 a_2^2}{4r_{1,2}^6}$$

Where I_1 is the ionization potential of molecule 1, α_2 is the polarizability of adjacent molecule 2, and $r_{1,2}$ is the distance between the interacting species 1 and 2.

2.9 G PROTEIN-COUPLED RECEPTORS

G protein-coupled receptors (GPCR), also known as seven transmembrane domain receptors, 7TM receptors, heptahelical receptors, serpentine receptors, and G protein-linked receptors (GPLR) comprise a large protein family of transmembrane receptors that sense molecules outside the cell and activate inside signal transduction pathways and, ultimately, cellular responses. The most familiar are the muscarinic acetylcholine receptors, the adrenergic receptors, the dopaminergic receptors, the opioids receptors, as well as many peptides receptors. A wide range of neurotransmitters, polypeptides, and inflammatory mediators transduce their signal into the interior of the cell by a specific interaction with receptors coupled to G proteins. G protein-coupled receptors are involved in many diseases, and are also the target of around half of all modern medicinal drugs.

GPCRs are involved in a wide variety of physiological processes, which include:

- **Behavioural and mood regulation:** Receptors in the mammalian brain bind several different neurotransmitters, including serotonin, dopamine, GABA, and glutamate.

- **Regulation of immune system activity and inflammation:** Chemokine receptors bind ligands that mediate intercellular communication between cells of the immune system; receptors such as histamine receptors bind inflammatory mediators and engage target cell types in the inflammatory response.
- **ANS transmission:** Both the sympathetic and parasympathetic nervous systems are regulated by GPCR pathways, responsible for control of many automatic functions of the body such as blood pressure, heart rate, and digestive processes.
- **Visual sense:** The opsins use a photoisomerization reaction to translate electromagnetic radiation into cellular signals. Rhodopsin, for example, uses the conversion of *11-cis*-retinal to *all-trans*-retinal for this purpose.
- **Smell sense:** Receptors of the olfactory epithelium bind odorants (olfactory receptors) and pheromones (vomeronasal receptors).

G protein-coupled receptors are activated by an external signal in the form of a ligand or other signal mediator. This creates a conformational change in the receptor, causing activation of a G protein. Further effect depends on the type of G protein. The targets of G proteins include enzymes such as adenylyl-cyclase, phospholipases C, and PI3K, channels that are specific for Ca^{++} or K^+. (*See Fig. 2.6 in the coloured set of pages.*)

FURTHER READINGS

Burger's Medicinal Chemistry and Drug Discovery, Abraham, D.J. (ed.), 6th Ed., 2003.

MULTIPLE-CHOICE QUESTIONS

1. The bond angles in sp carbon are usually
   a. 120
   b. 180
   c. 90
   d. 45

2. How many electrons are present in the valence shell of phosphorus?
   a. 2
   b. 3
   c. 4
   d. 5

3. Which of the following undergo tautomerization?
   a. Amides
   b. Alcohols
   c. Carboxylic acids
   d. Alkanes

4. Woster's Steering Wheel trick is used to assign for a molecule
   a. D/L-Form
   b. R/S-Form
   c. d/l-Form
   d. E/Z-Form

5. The concept that the concentration of a drug in plasma is directly related to its activity was given by
   a. Henderson–Hasselbach equation
   b. Overton–Meyer hypothesis
   c. Ferguson hypothesis
   d. Hammett's equation

6. Is-group does not include
   a. Halogens
   b. Oximes
   c. Alkanes
   d. Alkenes

7. Pyrrole nitrogen has a pK_a of
   a. 0.27
   b. –0.27
   c. 1.25
   d. –1.25

8. One of the following is bioisoster for carboxylic acid
   a. Tetrazole
   b. Carbonyl
   c. Amides
   d. Esters

9. The concept of nicotine that acted at sympathetic ganglia even after the degeneration of the severed preganglionic nerve endings was proposed by
   a. Langley
   b. Clark
   c. Ehlrich
   d. Paton
10. Which theory assumes that the substrate plays a role in determining the final shape of the enzyme?
    a. Rate theory
    b. Occupancy theory
    c. Induced-fit theory
    d. Activation–aggravation theory
11. One of the following receptors is sensitive to inverse agonist:
    a. $GABA_A$ receptor
    b. Seratonin $5-HT_3$ receptor
    c. Dopamine D_1 receptor
    d. Adrenergic α receptor
12. Which of the following can be a good therapeutic agent when the safe dose range can be greatly extended?
    a. Full agonist
    b. Partial agonist
    c. Inverse agonist
    d. Full antagonist
13. Which of the following is the strongest bonding between drug and receptor?
    a. Ionic bonding
    b. Hydrogen bonding
    c. Covalent bonding
    d. Van der Waals' bonding
14. Which interaction between a drug and a receptor would favour a permanent damage of killing of the cells?
    a. Inter-hydrogen bonding
    b. Intra-hydrogen bonding
    c. Covalent bonding
    d. Hydrophobic bonding
15. The charge of dipole compared to an ion is
    a. Lesser
    b. Same
    c. Greater
    d. Sometimes less and sometimes greater
16. In the biological pH, the amino acid with a permanent dipole is
    a. Alanine
    b. Phenylalanine
    c. Tyrosine
    d. Lysine

QUESTIONS

1. What are the factors that control solubility of a molecule in water?
2. How does electron withdrawing or releasing character of a substituent affect the properties of drugs?
3. What is pK_a? Why is it an important property for a drug? Explain with examples.
4. How does chirality influence the drug bioactivity? Give suitable examples.
5. Write a short note on Ferguson principle.
6. How is Hammett's constant derived and what is the importance of this constant in drug design?
7. What are bioisosters? Explain with examples.
8. What are the receptor theories?
9. Why is covalent bonding usually not preferred in drug design? When is it fruitful to have a covalent bonding between drug and the receptor?
10. For the following compound, draw an analogue of those compounds that will only be soluble in acidic conditions. Briefly explain why it is only soluble in acidic conditions.

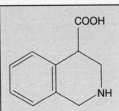

11. When applying homologation to enhance the efficacy of a drug candidate, it is often observed that the biological activity increases and then decreases. List the two reasons why this occurs, and explain how they make this occur.

12. Explain the difference between an agonist and an inverse agonist according to the 2-state model of a receptor.

13. The following figure depicts a type of interaction. Identify and explain.

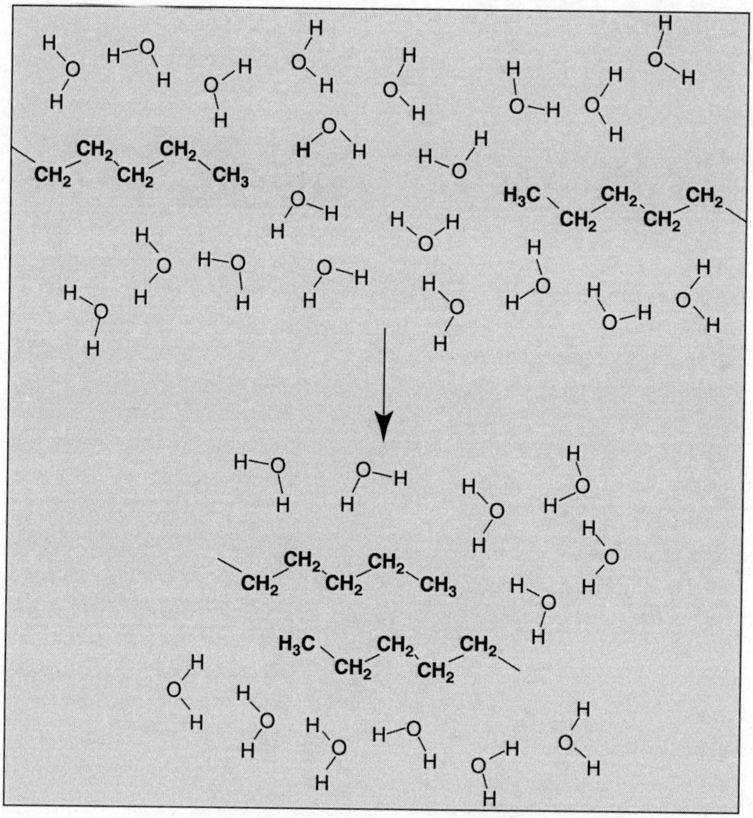

14. What effect will decreasing the pH from 7.4 to 5.4 have on the partition coefficient of ibuprofen?

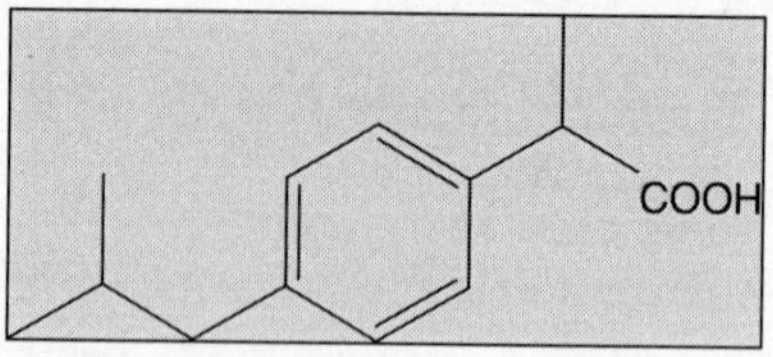

SOLUTION TO MULTIPLE-CHOICE QUESTIONS

1. b;
2. d;
3. a;
4. b;
5. c;
6. c;
7. b;
8. a;
9. a;
10. c;
11. d;
12. b;
13. c;
14. c;
15. a;
16. c.

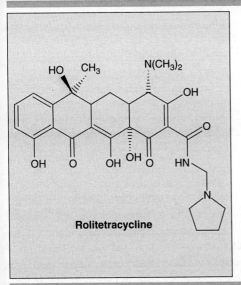
Rolitetracycline

CHAPTER 3

Drug Metabolism and Prodrugs

LEARNING OBJECTIVES

- Define metabolism
- Explain what happens to drug after action
- Various reactions that happen in the body
- Describe Phase I and Phase II reactions
- Define the concept of prodrugs
- Explain the need for prodrugs preparation
- Define the characteristics of prodrugs
- Explain the selective targeting with prodrugs
- List the classes of prodrugs
- Explain functional group-specific carrier linkages
- Describe how poor soluble drugs can be overcome
- Describe how bitter drugs can be made edible
- Describe how absorption and distribution of drug can be improved
- Describe how drugs can be improved upon on stability
- Describe how drugs can be made for prolonged release
- Describe how toxicity of drugs can be overcome
- Define mutual prodrugs
- Explain how bioprecursors can be prepared
- Explain the various reactions that are involved in the activation of bioprecursors

3.1 DRUG METABOLISM

Drug metabolism often converts lipophilic chemical compounds into more readily excreted polar products.

Quantitatively, the smooth endoplasmic reticulum of the liver cell is the principal organ of drug metabolism. Other sites of drug metabolism include epithelial cells of the gastrointestinal tract, lungs, kidneys, and the skin. These sites are usually responsible for localized toxicity reactions.

Drug metabolism involve two general types of reactions.

3.1.1 Phase I Reactions

These involve the biotransformation of a drug to a more polar metabolite by introducing or unmasking a functional group (e.g., —OH, —NH_2, —SH). In the case of pharmaceutical drugs, Phase I reactions can lead to either activation or inactivation of the drug. Phase I reactions (also termed non-synthetic reactions) may occur by oxidation, reduction, hydrolysis, cyclization, and decyclization reactions. Oxidation is the main reaction that involves the enzymatic addition of oxygen or removal of hydrogen, carried out by mixed-function oxidases, often in the liver. These oxidative reactions typically involve a cytochrome P450 haemoprotein, NADPH, and oxygen. The classes of pharmaceutical drugs that utilize this method for their metabolism include phenothiazines, paracetamol, and steroids.

Oxidation

The following enzymes are involved in oxidation:

- Cytochrome P450 mono-oxygenase system
- Flavin-containing mono-oxygenase system
- Alcohol dehydrogenase and aldehyde dehydrogenase
- Monoamine oxidase
- Co-oxidation by peroxidases

Examples of oxidative reactions with drugs

- Oxidation of aromatic moieties: Propranolol, Phenytoin, Phenobarbitol, Phenylbutazone, Diazepam
- Oxidation of olefins: Carbamazepine, Cyprohetadine, Diethylstilbestrol
- Oxidation of benzylic carbon atoms: Tolbutamide, Tolmetin
- Oxidation of allylic carbon atoms: Secobarbital
- Oxidation involving carbon–nitrogen heteroatom systems: Imipramine, Lignocaine, Diphenhydramine, Chlorpromazine
- Oxidation involving carbon–oxygen heteroatom systems: Phenacetin, Indomethacin
- Oxidation involving carbon–sulphur heteroatom systems: Thiopental, Thioridazine
- Oxidation of alcohols and aldehydes

Reduction

The following enzymes are involved in oxidation:

- NADPH-cytochrome P450 reductase
- Reduced (ferrous) cytochrome P450

Examples of reductive reactions with drugs

- Reduction of aldehyde and ketone: Chloral hydrate, Daunomycin, Naltrexone
- Reduction of nitro and azo compounds: Clonazepam, Metronidazole, Chloramphenicol, Sulfasalazine

Hydrolysis
The following enzymes are involved in oxidation:
- Esterases and amidases
- Epoxide hydrolase

Examples of hydrolytic reactions with drugs
- Hydrolysis of esters and amides: Aspirin, Clofibrate, Lignocaine, Indomethacin, Prazocin, Procainamide

If the metabolites of phase I reactions are sufficiently polar, they may be readily excreted at this point. However, many phase I products are not eliminated rapidly and undergo a subsequent reaction in which an endogenous substrate combines with the newly incorporated functional group to form a highly polar conjugate.

3.1.2 Phase II Reactions
Drugs or phase I metabolites that are not sufficiently polar to be excreted rapidly by the kidneys are made more hydrophilic by conjugation reactions (e.g., with glucuronic acid, sulphonates [commonly known as sulphation], glutathione, or amino acids), are usually detoxication in nature, and involve the interactions of the polar functional groups of phase I metabolites. Sites on drugs where conjugation reactions occur include carboxyl (—COOH), hydroxyl (—OH), amino (NH_2), and sulfhydryl (—SH) groups.

Methylation
- Methyl transferase: Alphamethyl dopa, Morphine

Sulphation
- Glutathione S-transferases: Isosorbide, Nitroglycerine
- Sulphotransferases: Salbutamol, Phenacetin

Acetylation
- N-acetyltransferases: Dapsone, Clonazepam, Sulphonamides, Hydralazine, Phenalzine
- Amino acid N-acyl transferases: Phenacemide, Isoniazid

Conjugation
- UDP-glucuronosyltransferases: Paracetamol, Propranolol, Meprobamate, Sulphisoxazole, Fenoprofen, Chloramphenicol
 (*See Fig. 3.1 in the coloured set of pages.*)

3.2 PRODRUGS

3.2.1 Introduction
All therapeutic agents possess various physicochemical and biological properties, which include desirable as well as undesirable ones. In general, researchers in the pharmaceutical world have been, and are,

concerned with minimizing the magnitude and number of undesirable properties of a drug while retaining, or at times enhancing, desirable therapeutic activity.

The chemical approach using drug derivatization offers perhaps the highest degree of flexibility in minimizing undesirable drug properties while retaining the desirable therapeutic activity (improving drug efficacy). Drug derivatization has long been recognized as an important means of producing better pharmaceuticals. Derivatives of drugs obtained by means of chemical modification can be broadly classified into two categories: namely, irreversible and reversible derivatives. As the name suggests, the irreversible derivative approach is generally used in the development of newer drug molecules to overcome the drawbacks, to improve the potency, or to incorporate some other desirable property not possessed by the existing parent drug molecule. The compounds, upon introduction to the appropriate biological environment, revert to the parent drug molecule by virtue of enzymatic and/or non-enzymatic means and elicit a response at the specific site.

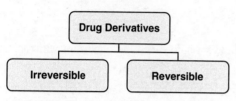

FIGURE 3.2 Types of drug derivatives.

A prodrug is a pharmacological substance (drug) that is administered in an inactive (or significantly less active) form. Once administered, the prodrug is metabolized *in vivo* into the active compound. It can also be defined as a process of purposely designing and synthesising a molecule that specifically requires 'bioactivation' to a pharmacologically active substance.

3.2.2 The Prodrug Concept

Albert and his coworkers were the first ones to suggest the concept of prodrug approach for increasing the efficiency of drugs in 1950. They described prodrugs as pharmacologically inactive chemical derivatives that could be used to alter the physicochemical properties of drugs in a temporary manner, to increase their usefulness and/or to decrease associated toxicity. Subsequently, such drug derivatives have also been called 'latentiated drugs', 'bioreversible derivatives', and 'congeners', but 'prodrug' is now the most commonly accepted term. Thus, prodrug can be defined as a drug derivative that undergoes biotransformation, enzymatically or non-enzymatically, inside the body before exhibiting its therapeutic effect. Ideally, the prodrug is converted to the original drug as soon as the derivative reaches the site of action, followed by rapid elimination of the released derivatizing group without causing side effects in the process. The definition of the prodrug indicates that the derivatizing group is covalently linked to the drug molecule. However, the term 'prodrug' has also been used for salts formed by the drug molecules.

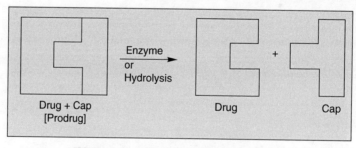

FIGURE 3.3 Concept of prodrug approach.

As shown in Fig. 3.3, a derivative of a known, active drug (D) can be capped to furnish a prodrug (PD). The PD enhances delivery characteristics and/or therapeutic value of the drug by transforming into the active drug via an enzymatic or a chemical process to remove the cap P at the site of action to regenerate D.

3.2.3 Why Use Prodrugs?

Prodrugs are designed for the following purposes:
1. Improve patient's acceptability (organoleptic properties or pain at injection point)
2. Alter and improve absorption
3. Alter biodistribution
4. Alter metabolism
5. Alter elimination
6. Reduce toxicity or adverse effects
7. Increase chemical stability
8. Prolong or shorten the biological action

3.2.4 Characteristics of Prodrug

In recent years, numerous prodrugs have been designed and developed to overcome barriers to drug utilization, such as low oral absorption properties, lack of site specificity, chemical instability, toxicity, bad taste, odour, and pain at application site. It has been suggested that the following characteristics of a prodrug must be improved for site-specific drug delivery.
1. The prodrug must be readily transported to the site of action as depicted in Fig. 3.4.

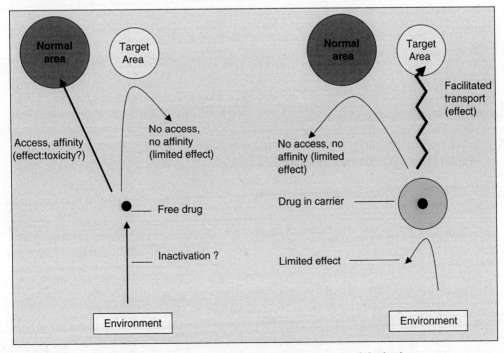

FIGURE 3.4 Prodrugs for targeting particular part of the body.

2. The prodrug must be selectively cleaved to the active drug, utilizing special enzymatic profile of the site.
3. Once the prodrug is selectively generated at the site of action, the tissue must retain the active drug without further degradation. *Fig. 3.5 (See coloured set of pages)* depicts the following properties of a prodrug.

3.2.5 Classification of Prodrugs

1. Carrier-linked prodrugs
2. Bioprecursors

Carrier-linked prodrugs

Carrier-linked prodrugs result from a temporary linkage of the active molecule with a transport moiety, which is mostly lipophilic in nature. A simple hydrolytic reaction cleaves this transport moiety. Such prodrugs are less active than parent compounds or even inactive. The transport moiety or promoiety (carrier group) will be chosen for its non-toxicity and its ability to ensure the release of the active principle with efficient kinetics.

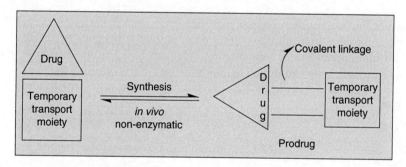

Classification Carrier-linked prodrugs can be further subdivided into:

- **Bipartite:** Composed of one carrier (group) attached to the drug
- **Tripartate:** Carrier group is attached via linker to drug
- **Mutual prodrugs:** Two drugs linked together

Carrier linkage for various functional groups
1. *Alcohols, phenols, and carboxylic acids*

    In general, alcohols, phenols, and carboxylic acids are converted to prodrugs that are esters. The reasons for preparing esters are:

(i) Esterase enzymes are ubiquitous, so metabolic regeneration of the drug is a facile process. Some types of esterases that mediate the hydrolytic processes include ester hydrolase, lipases, cholesterol esterases, acetylcholinesterases, carboxypeptidases, and cholinesterases.
(ii) It is possible to prepare ester derivative with any degree of hydrophilicity or lipophilicity. Example: Alcohol/phenol-containing drugs can be acylated with aliphatic (palmitate, propionate, etc.) or

aromatic (benzoate, etc.) carboxylic acids to increase lipophilicity, or with carboxylate-containing amino (glycine, etc.) or additional carboxylate group (succinate, etc.) to increase hydrophilicity (water solubility). Conversion to phosphate or sulphate esters also increases the water solubility.

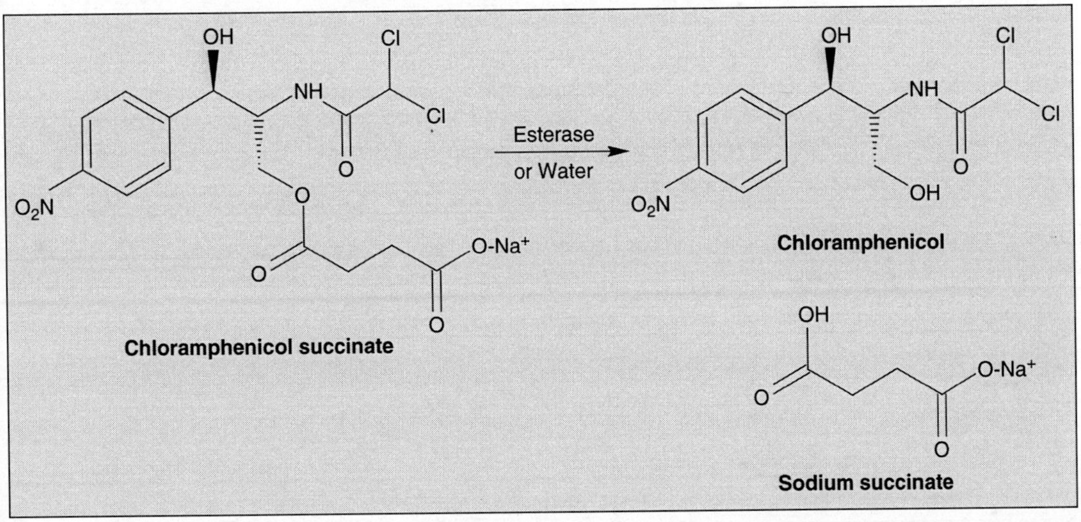

Chloramphenicol succinate → (Esterase or Water) → **Chloramphenicol** + **Sodium succinate**

(iii) Finally, a variety of stabilities of esters can be obtained by appropriate manipulation of electronic and steric factors. Erythromycin is a very bitter antibiotic easily destroyed by acidic pH. The lauryl sulphate salt of erythromycin (given in the following figure) was found to have lipophilicity and, hence, good absorption independent of food, and was also found to be less bitter and acid-stable.

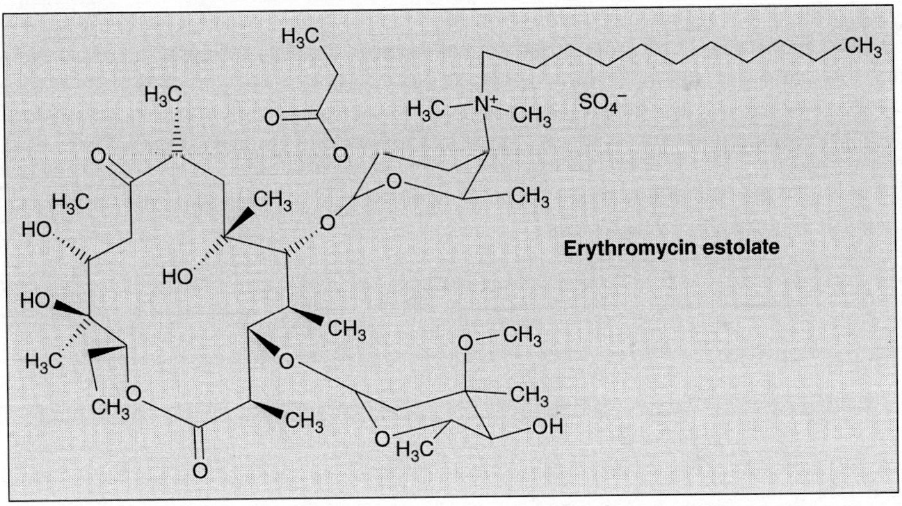

Erythromycin estolate

Example for electronic character:

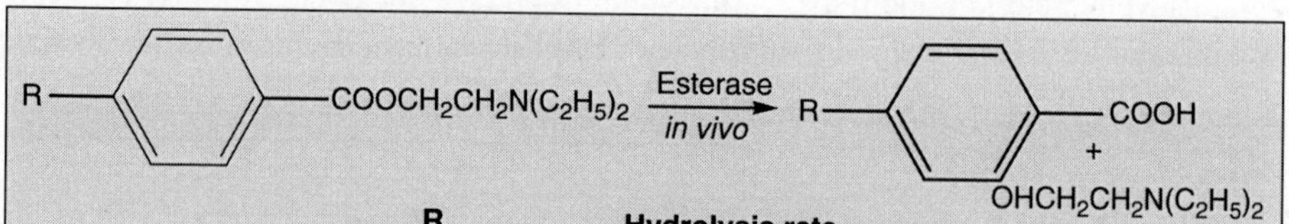

R—⟨C6H4⟩—COOCH$_2$CH$_2$N(C$_2$H$_5$)$_2$ →[Esterase, in vivo] R—⟨C6H4⟩—COOH + OHCH$_2$CH$_2$N(C$_2$H$_5$)$_2$

R	Hydrolysis rate microg/mL in serum/hr
F [Electron withdrawing]	3,000
NH$_2$ [Electron donating]	500

Example for steric character:

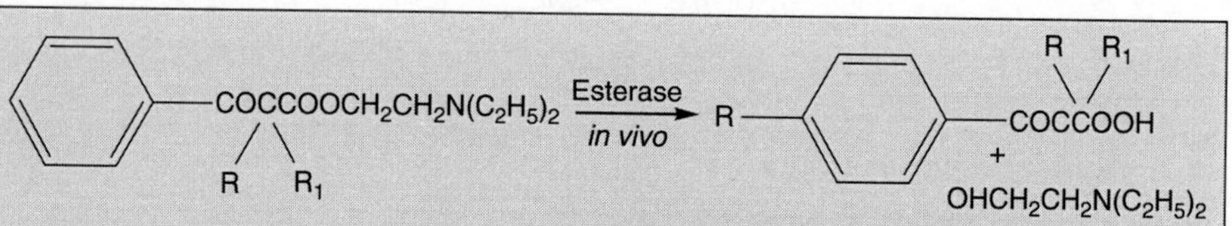

⟨C6H5⟩—COCCOOCH$_2$CH$_2$N(C$_2$H$_5$)$_2$ (with R, R$_1$ substituents) →[Esterase, in vivo] R—⟨C6H4⟩—COCCOOH (with R, R$_1$) + OHCH$_2$CH$_2$N(C$_2$H$_5$)$_2$

R	R$_1$	Hydrolysis rate microg/mL in serum/hr
H	H	500
CH$_3$	CH$_3$	30
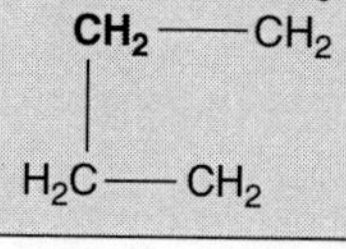 CH$_2$—CH$_2$ (cyclopropyl with H$_2$C—CH$_2$)		0 [Resistant to hydrolysis]

By using these approaches, a wide range of solubility can be achieved that will affect the absorption and distribution properties of the drug. These derivatives also can have an important effect on the dosage form—that is, whether used in tablet form or in aqueous solution.

1. *Alcohol/phen ol-containing drugs can be converted into the following derivatives:*

    - Esters of simple or functionalized aliphatic acid: R-O-OC-R$_1$ (Estradiol valerate)
    - Esters of amino acids: (Valacyclovir)

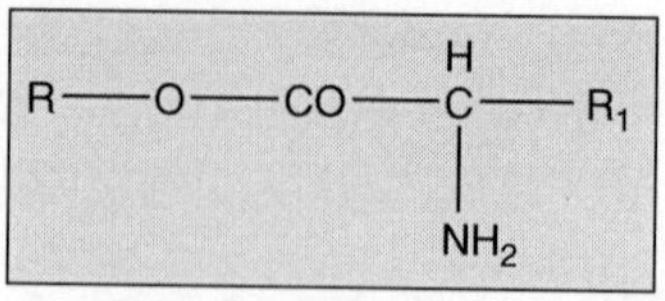

R—O—CO—CH(NH$_2$)—R$_1$

Medicinal Chemistry

- Esters of derivatized phosphoric acids: (Adefovir dipivoxil)

$$R-O-\underset{\underset{OR_2}{|}}{\overset{\overset{O}{\|}}{P}}-OR_1$$

- (Acyloxy)methyl/ethyl ethers: R—O—CH$_2$—O—CO—R$_1$
- (Alkoxycarbonyloxy)methyl/ethyl ethers: R—O—CH$_2$—O—CO—O—R$_1$
- Carbamic acid derivatives: R—O—CO—NH—R$_1$
- O-Glycosides with sugar molecules

2. *Acid-containing drugs can be converted into following derivatives*

- Esters of simple alcohol/phenol: R—CO—O—R$_1$
- Esters of alcohol containing an amino or amido function:
  R—CO—O—(CH$_2$)$_n$NR$_1$R$_2$; R—CO—O—(CH$_2$)$_n$NHR$_1$;
  R—CO—O—(CH$_2$)$_n$CONR$_1$R$_2$
- (Acyloxy)methyl/ethyl ethers: R—CO—O—CH$_2$O—CO—R
- N,N-Dialkylhydroxylamine derivatives: R—CO—O—NR$_1$R$_2$
- Pivaloyloxymethylester: R—CO—O—CH$_2$—CO—C(CH$_3$)$_3$ (Pivampicillin)
- Amides of amino acids:

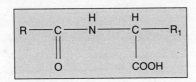

An example of a double ester is the prodrug of cefpodoxime. It is an oral third-generation cephalosporin antibiotic. To increase the absorption and to avoid acid-catalysed decomposition in GIT, a double ester prodrug was synthesised and successfully marketed. The structure and regeneration of the drug is presented below.

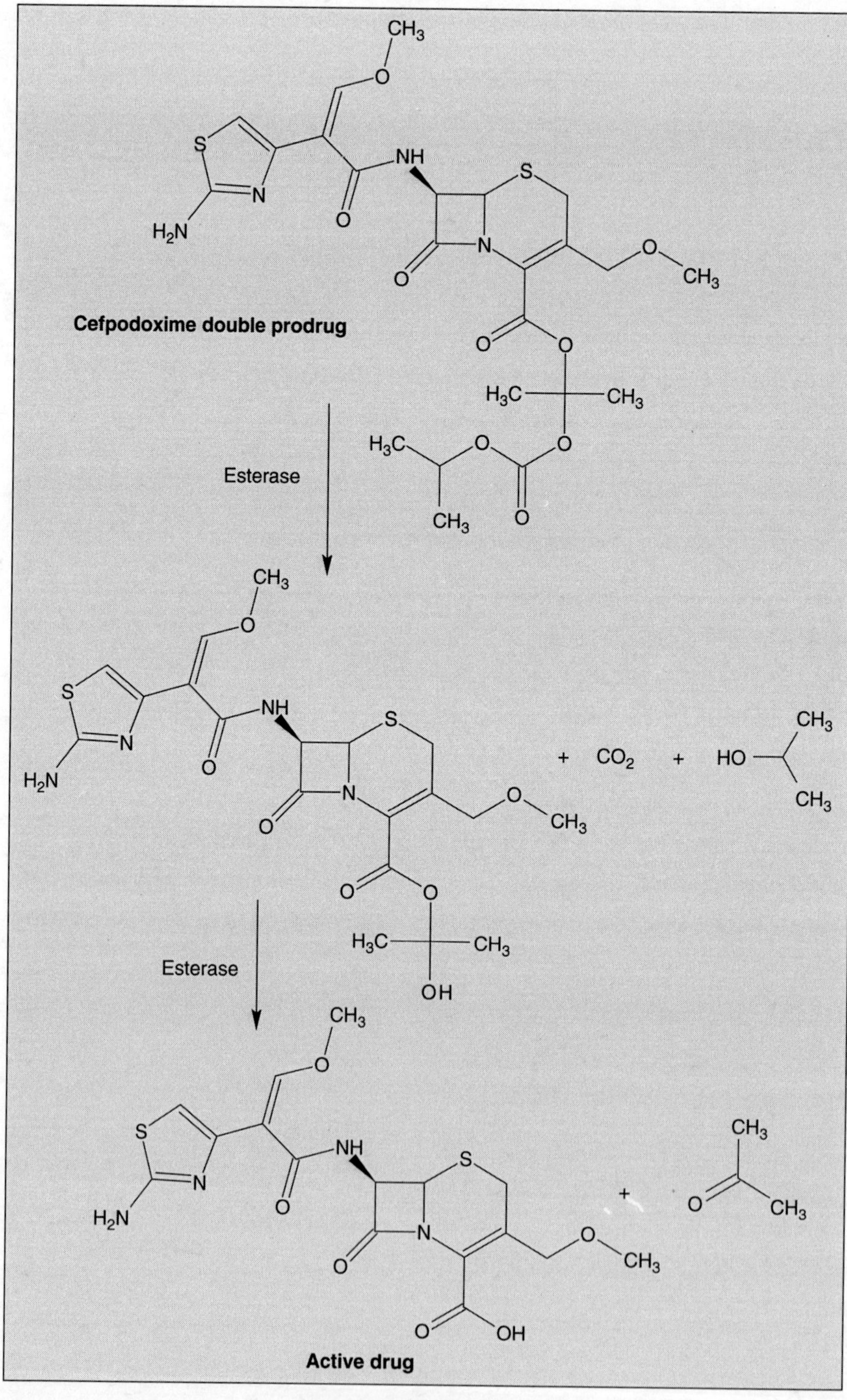

Chapter 3: Drug Metabolism and Prodrugs

Cefpodoxime double prodrug

↓ Esterase

→ + CO_2 + HO–CH(CH_3)$_2$

↓ Esterase

→ + CH_3COCH$_3$

Active drug

3. *Amines*

   Amino group-containing drugs can be modified as following derivatives:

   - Amides: R-NH-CO-R_1
   - Carbamates: R—NH—CO—O—R_1
   - Schiff base: R—N=CHR_1 or R—N=CR_1R_2 (example: Progabide)
   - Azo derivative: R—NH=NH—Ar (example: Sulphasalazine)
   - (Phosphoryloxy)methyl carbamate: R—NH—CO—O-CH_2—O—PO_3H_2

4. *Amides*

   Amide group-containing drugs can be modified as following derivatives:

   - Mannich bases: RCONH-CH_2-NR_1R_2 (example: Rolitetracycline, hetacillin)

Mannich bases are the most preferred prodrug form. Hetacillin is a prodrug of ampicillin, as shown below.

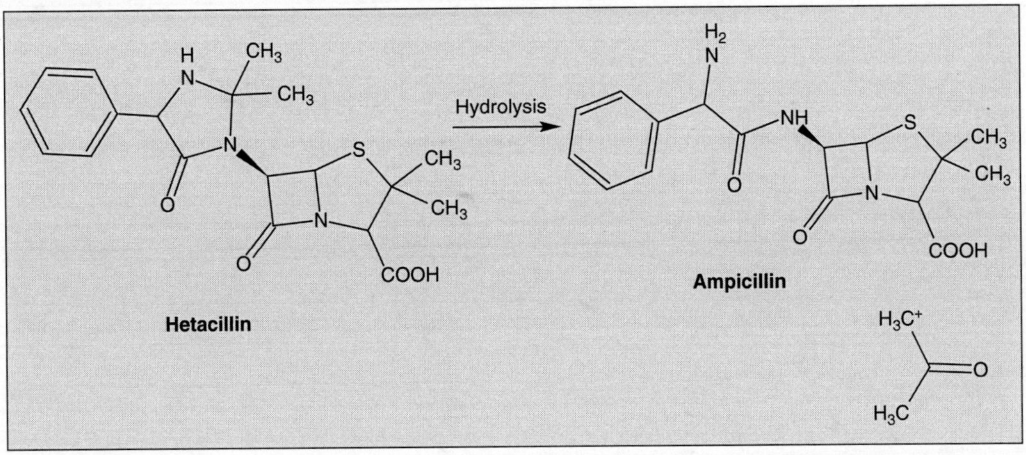

Similarly, rolitetracycline is also an example of mannich base prodrug of tetracycline with increased water solubility.

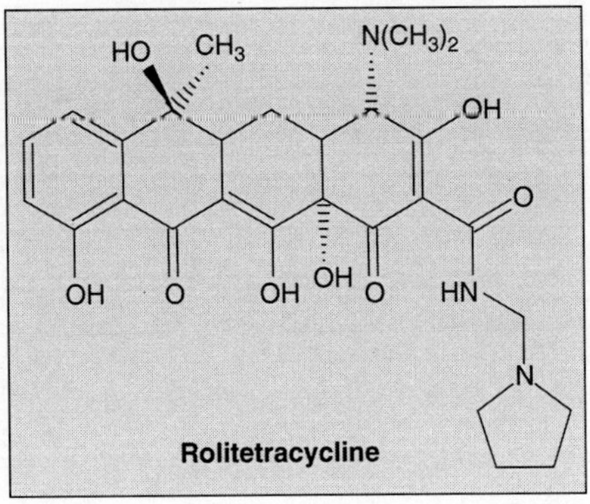

5. *Carbonyl*

   Carbonyl (aldehyde and ketone) group-containing drugs can be modified as Schiff bases, oximes, and acetal/ketals derivatives.

Examples of carrier-linked bipartite prodrugs

1. *Prodrugs for increased water solubility*

Methyl prednisolone (R=H) is a poorly water-soluble drug and cannot be administered by injection. Methyl prednisolone sodium succinate (R= —COCH$_2$CH$_2$COONa) is a water-soluble salt and can be given by injection.

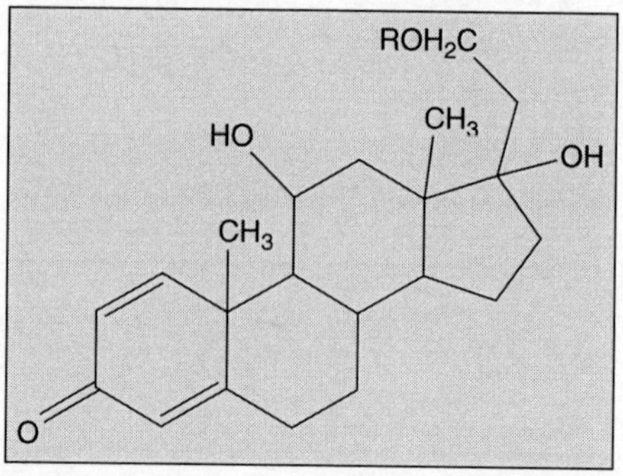

Fosphenytoin is a water-soluble phosphoric acid prodrug of phenytoin for administering the drug parenterally (intravenous injection). It is rapidly cleaved by phosphatases *in vivo* to form phenytoin.

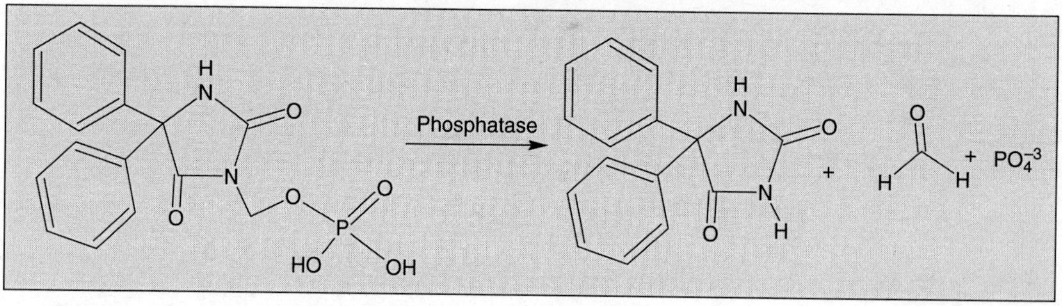

Artemisinin is an antimalarial drug that is insoluble in water. Their succinic acid ester derivative artesunate is water-soluble and can be used for parenteral use.

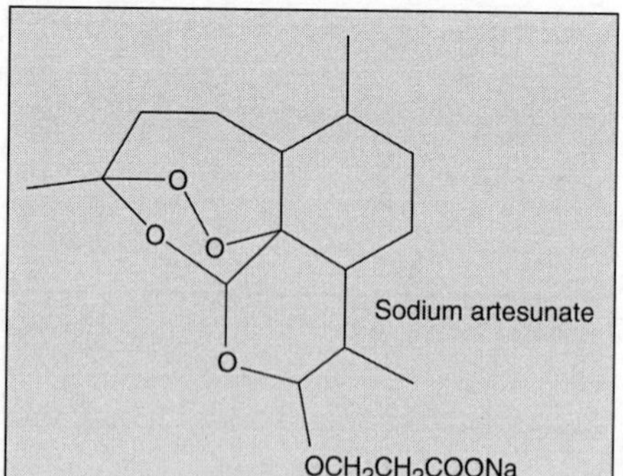

Sodium artesunate is converted to its active metabolite *in vivo*.

2. *Prodrugs for improved absorption and distribution*

Ampicillin is poorly absorbed from the gastro-intestinal tract (~30 per cent absorbed) after oral administration. To achieve the clinical efficiency and same blood level, one must give 2–3 times more ampicillin by mouth than by intramuscular injection. The non-absorbed part of the drug destroys the intestinal flora and causes diarrhoea. Modification of the polar COOH group of ampicillin with lipophilic and enzymatically labile esters gives prodrugs becampicillin and pivampicillin.

The absorption of these prodrugs is nearly 98–99 per cent, and they do not cause any side effect in GIT. The generation of parent ampicillin in the bloodstream takes place within 15minutes (0.8 gm of these prodrugs ≡ 2.0 gm of ampicillin).

3. *Prodrugs for site specificity*
In this approach of prodrug design, the site-specific drug delivery can be achieved by the tissue activation, which is the result of an enzyme unique to the tissue or present in higher concentration. For example, glycosidase enzymes are present in much higher concentration in bacteria associated with colon. This aspect can be utilized in the design of colon-specific drug delivery. Glycosides are hydrophilic in nature and are poorly absorbed in small intestine. Once they reach the colon, bacterial glycosides release the free drug to be absorbed in that region. Dexamethasone is corticosteroid used for anti-inflammatory properties and is hydrophobic in nature. It is absorbed efficiently in the intestinal tract and, as such, does not reach the colon area for treatment. However, when prodrug of dexamethasone—21-β-glucoside—is used, it is absorbed in colon more efficiently compared to its parent drug. The prodrug is hydrophilic in nature and, therefore, absorbed poorly in intestine. The glucosidase enzymes present in the bacteria located in colon release the parent hydrophobic drugs for absorption in the area.

4. *Prodrugs for stability*
Some prodrugs protect the drug from the first-pass metabolism. Propranolol on oral administration shows low bioavailability due to first-pass metabolism. The major metabolites are O-glucoronide, *p*-hydroxy

propranolol, and its gluocoronide. Preparing esters with succinate enhances the plasma levels of propranolol by eight times on oral administration.

Propranolol: OCH$_2$CH(OH)CH$_2$NHCH(CH$_3$)$_2$

Propranolol-O-succinate: OCH$_2$CHCH$_2$NHCH(CH$_3$)$_2$ | OCOCH$_2$CH$_2$COOH

5. *Prodrugs for prolonged release*
(i) Decreased secretion or lack of secretion of the enzyme insulin in pancreas leads to diabetes. Insulin is responsible for degradation of carbohydrate molecules to smaller units and is important in the catabolic process. Chronic diabetic patients take bovine insulin supplement through intravenous injections. Retention time of insulin in the blood is about six hours. So, patients need to administer required dose of insulin frequently. It is desirable to increase the retention characteristics of the enzyme to make it effective for prolonged periods. It was found that 9-fluorenylmethoxycarbonyl (Fmoc) protection of the hydroxy/amino groups of the enzyme makes it inactive and also increases its retention in blood for prolonged periods. The Fmoc group binds to the enzyme covalently, and in the process makes it inactive as well as reduces its rapid degradation by natural body process. However, at the pH of about 7.4 prevalent in the blood serum, the protected enzyme gets hydrolysed slowly and irreversibly back to the enzyme and Fmoc protecting group. The hydrolysis process was found to be slow and constant, which means that the release of enzyme is also slow and regulated. The hydrolysis rate can be fine-tuned by selecting derivatives of Fmoc protecting group or a number of Fmoc groups. It was shown that insulin having two Fmoc protecting groups was ten times more stable and more effective than the parent enzyme. It should be noted that the hydrolysis of the protecting group takes place in the blood without mediation from other enzymes.

Fmoc Insulin: —OC(=O)—INSULIN

(ii) Esterification of the 17-β hydroxyl group of testosterone with carboxylic acids decreases the polarity of the molecule, makes it more soluble in the lipid vehicles used for injection, and hence, slows the release of the injected steroid into the circulation. The longer the carbon chain in the ester, the more lipid soluble, and the steroid becomes more prolonged in action. Examples of prodrugs include testosterone propionate, testosterone enanthate, and testosterone cypionate.

6. *Prodrugs to minimize toxicity*
Derivatization of carboxylic acid group of indomethacin to amides selectively inhibits COX-2 enzyme and does not inhibit COX-1 enzyme, which is responsible for synthesis of cytoprotective prostaglandin in stomach. Indomethacin amides were devoid of gastric irritation.

Medicinal Chemistry 67

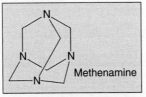
Indomethacin amides

7. Prodrugs to encourage patient compliance

Clindamycin HCl injection causes severe pain at the site of injection because of its low solubility in water (3mg/ml). By forming phosphate ester, solubility of the product rises to about 150mg/ml and this hydrolyses with $t_{1/2}$=10 minutes and does not produce any pain at the site of injection.

Chloramphenicol acetate is a bitter-taste compound, because this compound is soluble in saliva and interacts with taste receptor in mouth. Esterification with long-chain fatty acid makes drug less water-soluble and unable to dissolve in the saliva, and free from bitter taste.

8. Prodrugs to eliminate formulation problems

Formaldehyde (HCHO) is a flammable, colourless gas with pungent odour and is used as disinfectant. Solution of high concentration of HCHO is toxic. Consequently, it cannot be used directly in medicine. However, the reaction of 4 molecules of HCHO with 6 molecules of ammonia produces methenamine. In media of acidic pH (urine), methenamine (as enteric-coated tablet) hydrolyses to HCHO and ammonia, and is used as urinary antiseptic.

Methenamine

Tripartate prodrugs In this, the carrier is not connected directly to the drug, but to a linker arm that is attached to the drug.

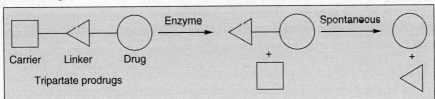
Tripartate prodrugs

Mutual prodrugs These are useful when two synergistic drugs need to be administered at the same site and at the same time. Mutual prodrug is bipartite or tripartate where a synergistic drug acts as the carrier.

Example 1: Sultamacillin For resistant bacteria, β-lactamase inhibitors are given in combination with antibiotics such as penicillanic acid sulphone (which inhibits the action of bacterial β-lactamases) and ampicillin. Simultaneous administration does not guarantee equivalent absorption or transportation to site of action. Sultamacillin is a tripartate mutual prodrug (double ester) of ampicillin and penicillanic acid sulphone.

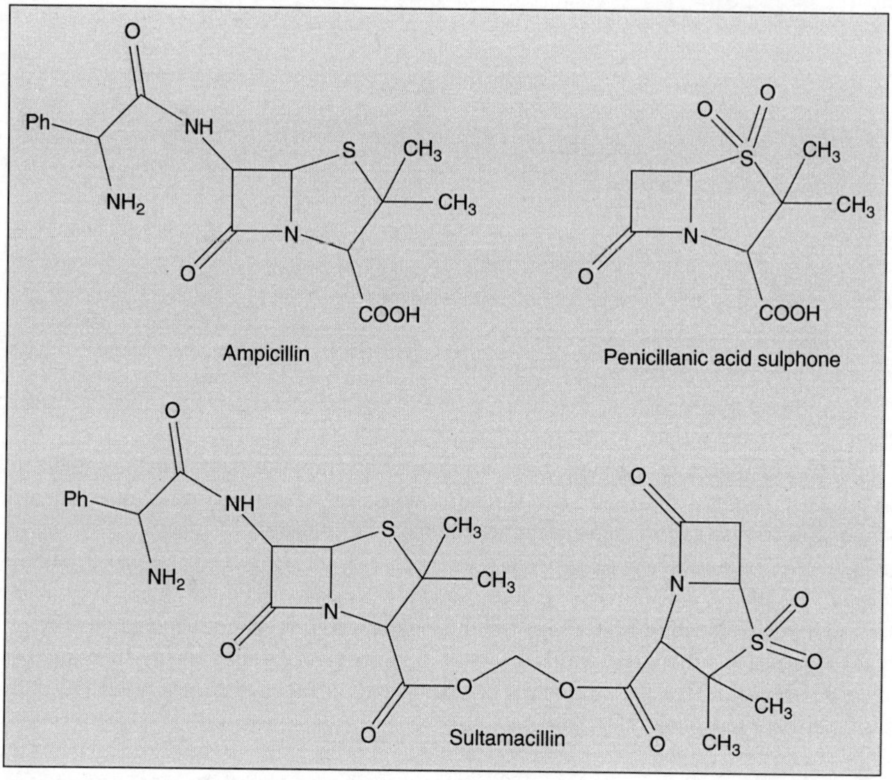

[Structures: Ampicillin, Penicillanic acid sulphone, Sultamacillin]

Example 2: *Estramustine* It is a chemotherapy agent used to treat metastatic carcinoma of prostate. It is a derivative of oestrogen (specifically, estradiol) with a nitrogen mustard-linked through carbamate moiety, which imparts effects similar to mechlorethamine. It is a bipartite mutual prodrug and is selectively taken up into oestrogen receptor positive cells, and then urethane linkage is hydrolysed.

- 17-α-estradiol slow prostate cell growth
- Nor nitrogen mustard is a weak alkylating agent

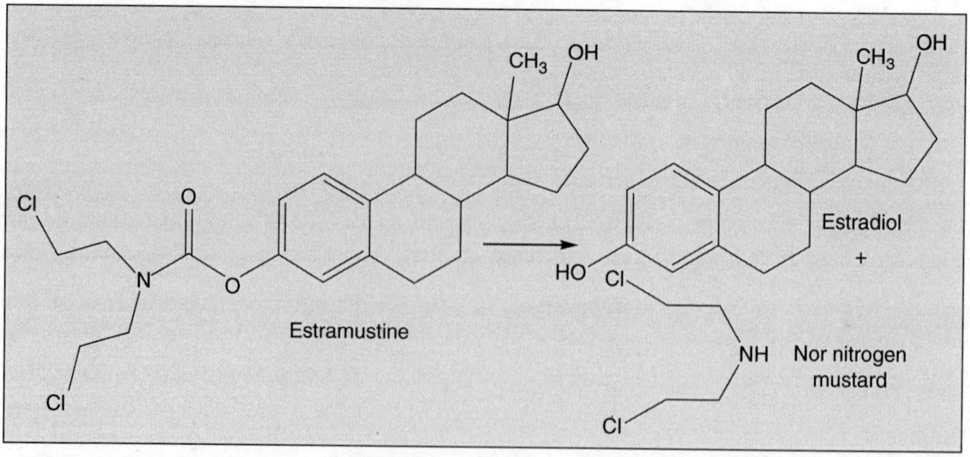

[Structures: Estramustine → Estradiol + Nor nitrogen mustard]

Example 3: *Sulphasalazine* Sulphasalazine is a bipartite mutual prodrug. In large intestines it gets activated to liberate 5-aminosalicylic acid, which in turn inhibits PG synthesis, and the sulphapyridine is useful for the treatment of infection. Hence, sulphasalazine is used in the treatment of ulcerative colitis.

3.2.6 Bioprecursor Prodrugs

A bioprecursor prodrug is a prodrug that does not imply the linkage to a carrier group, but results from a molecular modification of the active principle itself. This modification generates a new compound, able to be transformed metabolically or chemically, the resulting compound being the active principle. Metabolic activations of bioprecursor prodrugs take place by the following methods:

1. **Oxidative activation**
   - *N*-and *O*-dealkylation
   - Oxidative deamination
   - N-oxidation
   - Epoxidation

2. **Reductive activation**
   - Azo reduction
   - Sulphoxide reduction
   - Disulphide reduction

3. **Nucleotide activation**
4. **Phosphorylation activation**
5. **Cyclization**

N- **dealkylation:** The dealkylation of secondary and tertiary amines to yield primary and secondary amines, respectively, is one of the most important reactions in bioprecursor prodrug activation. Some of the N-substituents removed by oxidative dealkylation are methyl, ethyl, *n*-propyl, isopropyl, *n*-butyl, allyl, benzyl, and others having an α-hydrogen. Usually, dealkylation occurs with smaller alkyl group initially. Substituents that are more resistant to dealkylation include *tert*-butyl (no α-hydrogen) and the cyclopropylmethyl. In general, tertiary amines are dealkylated to secondary amines faster than secondary amines are dealkylated to primary amines.

Example: Open-ring analogue of anxiolytic drug alprazolam undergoes metabolic dealkylation followed by spontaneous cyclization to alprazolam.

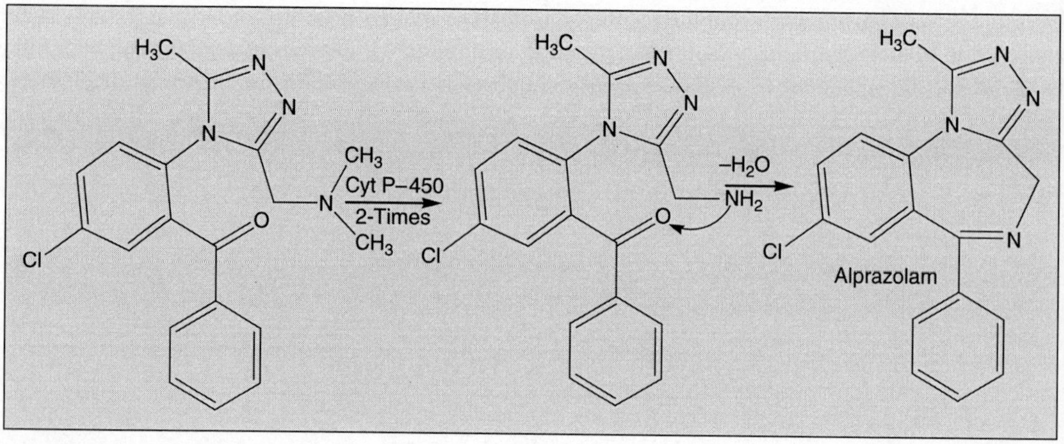

O-dealkylation: Oxidative *O*-dealkylation of ethers is a common metabolic reaction. The rate of *O*-dealkylation is a function of chain length, i.e., increasing chain length or branching and reducing the rate of dealkylation.

Example: Oxidative *O*-dealkylation of phenacetin results in *p*-acetaminophen (paracetamol).

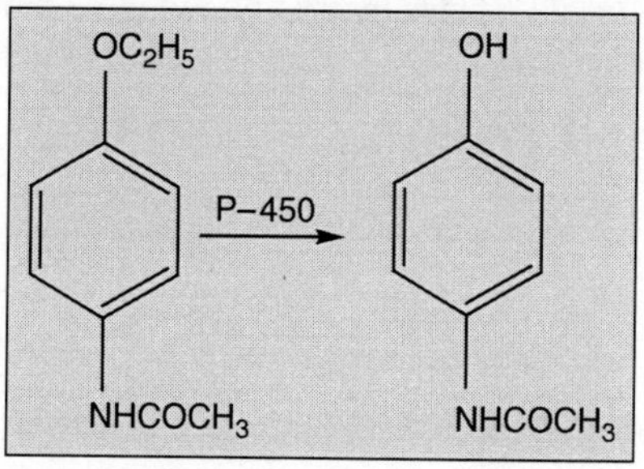

Oxidative deamination: Cyclophosphamide is an important drug for the treatment of a broad range of malignant diseases. Oxidative deamination leads to ring opening and release of the active phosphoramide mustard and parent nitrogen mustard, which alkylate DNA (N-7 position of guanine).

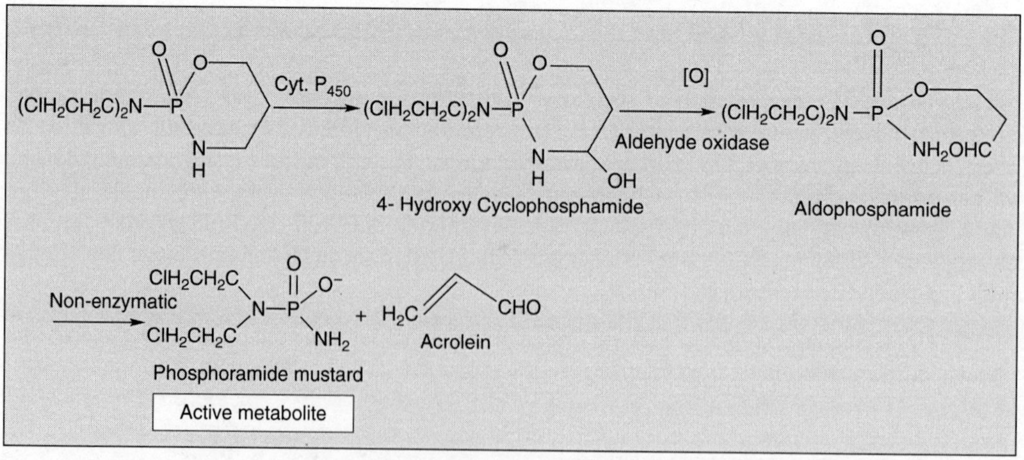

N-oxidation: Pralidoxime belongs to a family of compounds called oximes that bind to organophosphate-inactivated acetylcholinesterase. It is used to combat poisoning by organophosphates or acetylcholinesterase inhibitors (nerve gas). Because of poor lipid solubility, pralidoxime chloride cannot reactivate brain acetylcholinesterase although peripheral reactivation is effective. Bodor & coworkers' developed a reversible redox drug delivery system that allows blood-brain barrier permeability of the 5,6-dihydropyridine bioprecursor of pralidoxime. Once inside the brain, oxidation generates the parent drug.

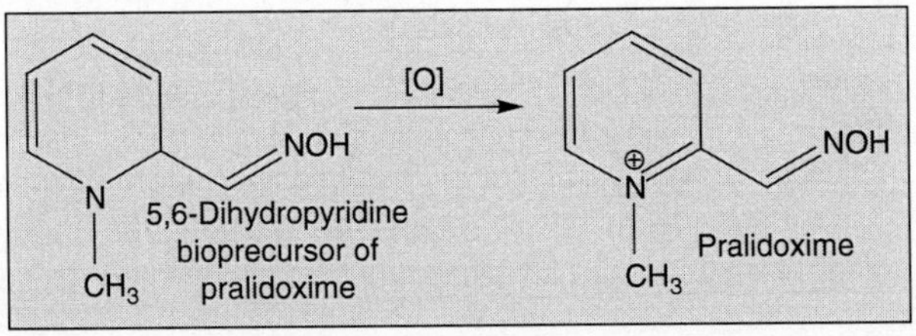

Epoxidation: Carbamazepine is an anticonvulsant and mood-stabilizing drug, used primarily in the treatment of epilepsy and bipolar disorder. Carbamazepine is metabolically transformed to the active epoxide derivative.

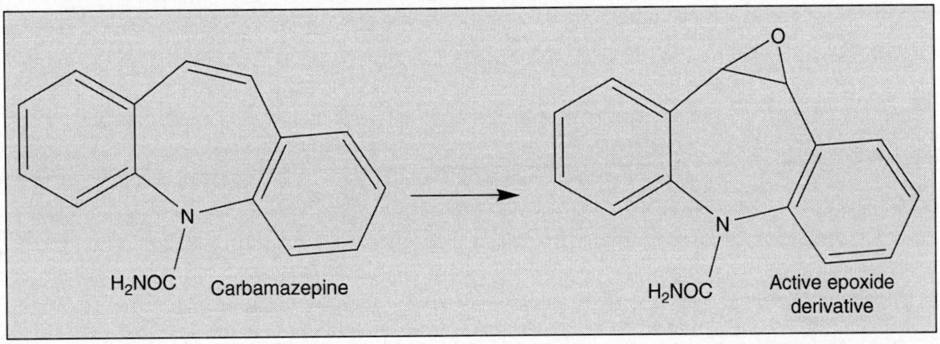

Azo reduction: Sulphasalazine is used in the treatment of inflammatory bowel disease (ulcerative colitis). Anaerobic bacteria in the lower bowel metabolically reduces azo group of sulphasalazine to the therapeutic agents 5-aminosalicylic acid (which act as analgesic) and sulphapyridine (which act as antibacterial).

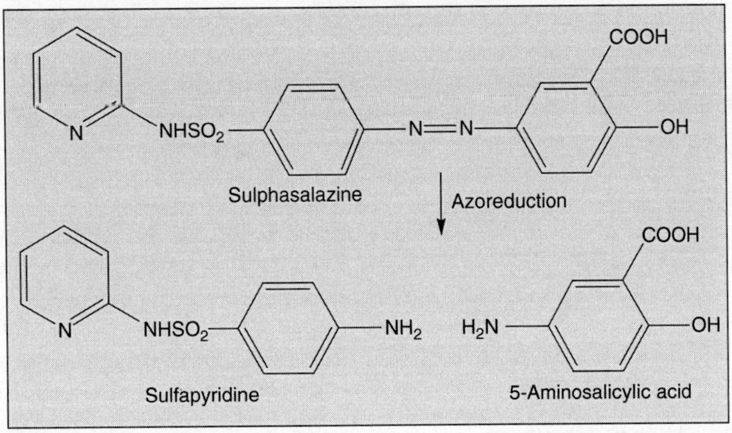

Sulphoxide reduction: Suldinac (sulphoxide) is inactive *in vitro* but highly active *in vivo*. It is reduced to the corresponding sulphide, which is formed by reduction, and is active both *in vitro* and *in vivo*.

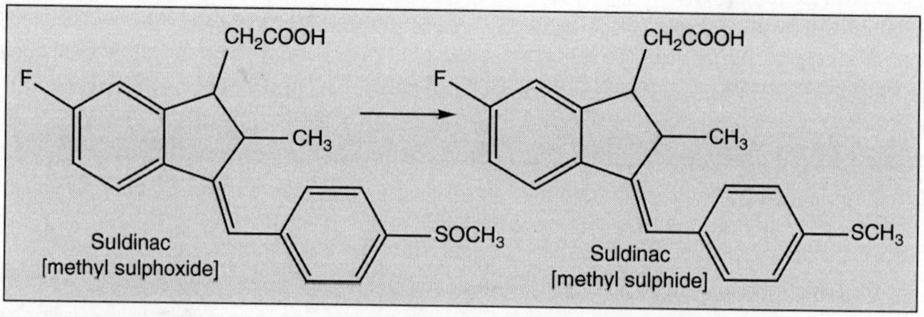

Disulphide reduction: Thiamin is poorly absorbed (quaternary ammonium) into the CNS from the GI tract. Brain abnormalities may arise from severe thiamin deficiency. The tetrahydrofurfuryl disulphide has enhanced lipid solubility and reacts with co-enzyme glutathione (GSH) to generate thiamin.

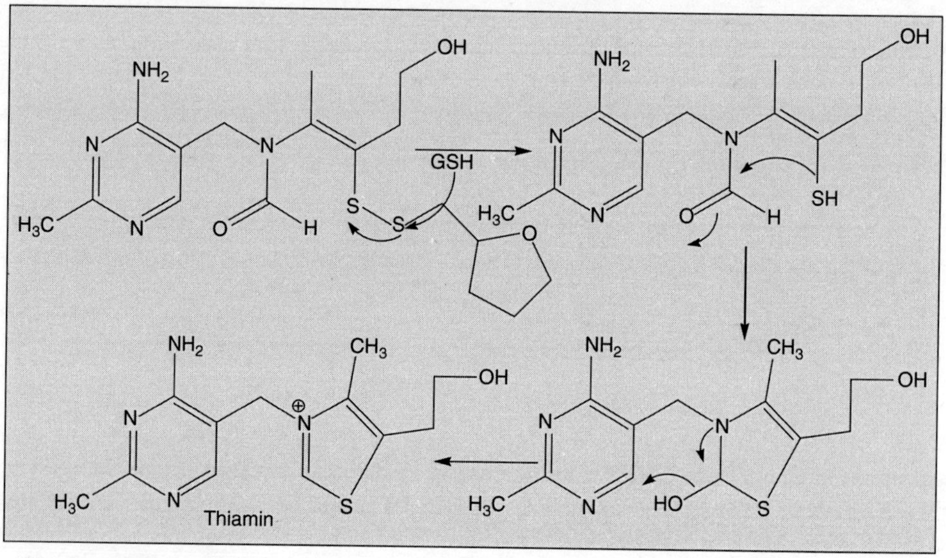

Nucleotide activation: 5-Fluorouracil is the prodrug and it is used in the management of the breast, colon, pancreas, and stomach carcinomas.

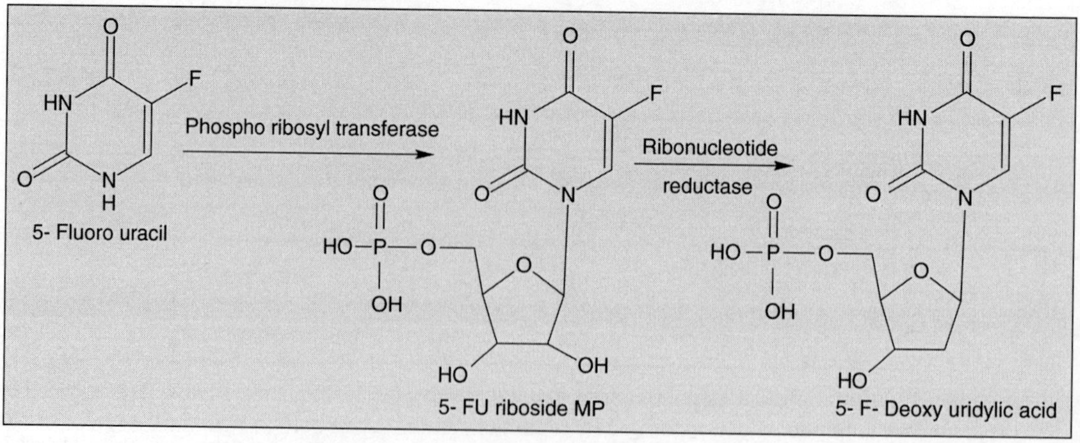

The active metabolite 5-fluoro deoxy uridylic acid is formed in two steps and competes with 2'-deoxy uridylate, and it reacts with enzyme Thymidylate synthase and co-factor 5, 10-methylene THF to form stable complex as a terminal product, and inhibits the thymidylate biosynthesis.

Phosphorylation activation: Acyclovir is highly effective against the genital herpes virus and resembles the structure of 2'-deoxyguanosine. A selective three-step phosphorylation is required for activity. Viral thymidine kinase generates the monophosphate (uninfected cells do not perform this phosphorylation); guanylate kinase generates the diphosphate; and a variety of enzymes (phoshpoglycerate kinase) generate the active triphosphate. Acyclovir triphosphate is a substrate for viral polymerase but not normal cellular polymerase. Incorporation of acyclovir triphosphate into viral DNA leads to an enzyme substrate complex that is not active (dead-end complex) after the subsequent or following deoxynucleotide triphosphate is incorporated. As a result, viral replication is disrupted.

```
Acyclovir (ACV) ──→ ACV-MP ──→ ACV-DP ──→ ACV-TP ──→ DNA
dGuo            ──→ dGMP   ──→ dGDP   ──→ dGTP   ──→ DNA
                    d Guo      GMP        NDP        Viral DNA
                    Kinase     Kinase     Kinase     Polymerase
```

Cyclization: Omeprazole is the proton pump inhibitor that contains a sulphinyl group in a bridge between substituted benzimidazole and pyridine rings. It reaches parietal cells from the blood and diffuses into the secretory canaliculi, where the drug becomes protonated and thereby trapped. The protonated agent rearranges to form a sulphenic acid and cyclized to form sulphonamide. The sulphonamide interacts covalently with sulfhydryl groups at critical sites in the extracellular domain of the H^+- K^+ ATPase and inhibits irreversibly, and thereby blocks gastric acid secretion.

Omeprazole → Sulphenic acid → Sulphenamide → Enzyme-Inhibitor complex

FURTHER READINGS

Burger's Medicinal Chemistry and Drug Discovery, Abraham, D.J. (ed.), 6th Ed., 2003.

MULTIPLE-CHOICE QUESTIONS

1. The advantage(s) of erythromycin estolate over erythromycin
   a. Less bitter taste
   b. Acid stable in stomach
   c. Good absorption profile
   d. All the above
2. Example for Mannich-base prodrug is
   a. Hetacillin
   b. Pivampicillin
   c. Rolitetracycline
   d. Valacyclovir
3. Example for Schiff-base prodrug is
   a. Fosphenytoin
   b. Progabide
   c. Vigabatrine
   d. Pivampicillin
4. Conversion of carboxylic acid group of indomethacin to its amide derivatives
   a. Selectively inhibits COX-2 enzyme
   b. Reduces gastric irritation
   c. Both (a) and (b)
   d. None of the above
5. Methenamine is the prodrug of
   a. Formaldehyde
   b. Mechlorethamine
   c. Metaprolol
   d. Mannitol
6. Example of mutual prodrug is
   a. Estramustine
   b. Sulphasalazine
   c. Sultamacillin
   d. All the above
7. Example of bioprecusor prodrug is
   a. Cyclophosphamide
   b. Pivampicillin
   c. Progabide
   d. Becampicillin
8. Primary metabolite of cyclophosphamide is
   a. Aldophosphamide
   b. Ketophosphamide
   c. Hydroxyphosphamide
   d. Carboxyphosphamide
9. Active metabolite of 5-fluorouracil is
   a. 5-fluoro-2'-deoxyuridylic acid monophosphate
   b. 5-fluoro-2',3'-dideoxyuridylic acid 5-fluoro-
   c. 5-fluorouracil monophosphate
   d. 5-fluorouracil triphosphate
10. Among the following, which is not a prodrug?
    a. Omeprazole
    b. Valacyclovir
    c. Alprazolam
    d. Propranolol

QUESTIONS

1. State true or false for the following statements. Justify your answer with appropriate reason(s).
   a. Omeprazole is a bioprecursor prodrug.
   b. Paracetamol is a prodrug.
   c. Cyclophosphamide is activated at the cell wall of cancer.
2. Write the various methods for preparing prodrug for a drug containing free primary amino group.
3. How do the electronic and steric factors affect the pharmacokinetic properties of prodrugs?
4. How are the following drugs cleaved into the active molecule *in vivo*? Write the mechanism of activation with structures.
   a. Becampicillin
   b. Fosphenytoin
   c. Cyclophosphamide
   d. Omeprazole
   e. Acyclovir

SOLUTION TO MULTIPLE-CHOICE QUESTIONS

1. d;
2. c;
3. b;
4. c;
5. a;
6. d;
7. a;
8. c;
9. a;
10. d.

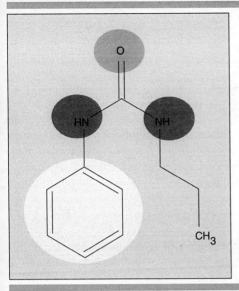

CHAPTER 4

Computer-aided Drug Design

LEARNING OBJECTIVES

- List the advantages of computer usage in drug discovery
- Explain the different methods employed in CADD
- Explain the importance of knowledge of structures of drug molecules
- Describe the methods to design new drugs from the known structures
- Describe the method of quantifying structural properties and biological properties, and their relation
- Define molecular properties
- Determine the various properties of functional group
- Define how to determine the molecular connectivity and their relation to drug activity
- List the methods of QSAR
- Explain the concept of manual interpretation of structure and biological activity
- Define a Topliss tree
- Define pharmacophore
- List the steps involved in pharmacophore model building
- Describe how 3D structural properties can help in the design of new drug molecules
- Define structure-based drug design
- Describe the methods of in-silico drug screening
- Define structural database
- Illustrate database screening strategy
- Explain how a drug structure can be built on the basis of target structure
- Illustrate the process of target-based ligand building

4.1 INTRODUCTION

Computer-aided drug design (CADD) is a strategy to meet the challenges faced in the drug discovery process. CADD is an organised guide to provide chemical insight into drug activity by helping in the drug discovery and predicting the drug properties. CADD works at the intersection of structural biology, biochemistry, medicinal chemistry, toxicology, pharmacology, biophysical chemistry, and information technology. Computational assessment of the binding affinity of enzyme inhibitors prior to synthesis is an important component of computer-aided drug design paradigms.

4.2 CADD METHODOLOGY

The basic or conventional drug discovery process starts with lead identification, modification by synthesis, biological testing, structure-activity studies, and further analogue design and synthesis. Methods of CADD are of two types depending upon whether the target protein structure is known/involved or not. In CADD, the ligands or drug is usually represented as small molecule and the protein or target is represented as macromolecule.

- Small molecule-based drug design
- Macromolecule-based drug design

4.3 SMALL MOLECULE-BASED DRUG DESIGN

Computational design of new ligands based on the structure of existing ligands is conventionally the more widely employed method. The method is often preferred due to the lack of information of protein or target-bound ligand structure. Structure-activity relationship studies play an important part of the ligand design. The methods involved in ligand-based design includes the following:

(i) Quantitative structure-activity relationship (QSAR)
(ii) Pharmacophore modelling
(iii) 3D QSAR

4.3.1 Quantitative Structure-Activity Relationships (QSAR)

We have alluded to the fact that drug-receptor interactions are dependent on physicochemical properties such as polarity, ionization, electron density, size, shape, and structure. A number of researchers have attempted to quantitate these parameters, and develop mathematical models for predicting the pharmacological activity of compounds that have not been made. This is a logical approach, since the pharmaceutical industry is able to market only one drug for every 10,000 compounds synthesised. The mathematical approaches developed to date are collectively known as quantitative structure-activity relationships (QSAR). QSAR attempts to relate or correlate the biological activity data with possible physicochemical parameters for a series of compounds from the same lead. Hence, QSAR method can be considered useful as a guide to predict bioactivity of new molecules before synthesis. It alleviates the need to determine molecular activity of hundreds of similar compounds that would take large amounts of resources to determine individually.

Molecular descriptors

There are a variety of microscopic or macroscopic properties of a molecule that influence its biological activity. Many properties of a molecule are hard to be determined experimentally. Due to the advancement in cheminformatics, today it is possible to directly compute the many properties of a molecule by atom-based or fragment-based contribution factor.

The most well-known and the most used descriptor in QSAR is log P (octanol/water partition coefficient), the hydrophobicity parameter. Log P has been very useful in correlating a wide range of activities due to its excellent modelling of the transport across various membranes in the body. Other descriptors include Hansch's π (hydrophobic) constant, Hammett's σ (electronic) constant, Taft's E_s (steric) constant, molar refractivity (MR), molecular weight, molar volume, parachor, chromatographic retention indices, quantum mechanics parameter like molecular connectivity, highest occupied molecular orbital (HOMO) and lowest unoccupied molecular orbital (LUMO) energies, net atomic charges, van der Waal's volume, UV-visible spectral shifts, polarizability index, hydrogen bond parameter, topological polar surface area (TPSA), and dihedral angle. CODESSA (comprehensive descriptors for structural and statistical analysis) is a comprehensive programme for developing quantitative structure-activity/property relationships (QSAR), which integrates all necessary mathematical and computational tools to calculate a large variety of molecular descriptors on the basis of the 3D geometrical and quantum chemical structural input of chemical compounds. Substituents constant for QSAR analysis are presented in a separate table at the end of this chapter.

Electronic parameters

Electronic properties of a molecule largely influence the chemical reactivity and hence the biological activity. Early works of Hammett contributed to the electronic property of substituents and their relation to chemical reactivity, and this is nowadays used more commonly in QSAR analysis. It is called the Hammett's σ constant and is derived for a substituent by determining the rate constant K_a of ionization of benzoic acid and substituted benzoic acid in water, as shown below.

$$\sigma_X = \log K_X - \log K_H$$

Electron-withdrawing groups are characterised by positive σ values, while electron-donating substituents have negative σ values. Substituents in aromatic ring had an orderly and quantitative effect on the dissociation rate. Ortho substitutions impart direct steric and polar effects, and are not generally correlated. An alternative to Hammett's constant, now pK_a is widely used in QSAR, particularly when transport phenomena is involved.

Another approach is the separation of sigma values into resonance (R) and inductive field (F) components. The sigma values are suggested to be a linear combination of these two effects as derived below:

$$\sigma_m = 0.60\ F = 0.27\ R$$
$$n = 42;\ r = 1;\ s = 0$$
$$\sigma_p = 0.56\ F + R$$
$$n = 42;\ r = 1;\ s = 0$$

The sign of F or R indicates the sign of the charge that the substituents place on the ring. Since nitro group makes the ring positive by both resonance and field effects, the R and F values are positive.

Other more known and used electronic parameters in QSAR are quantum chemical descriptors calculated by *ab initio* or semi-empirical techniques. Quantum chemical descriptors such as net atomic changes, highest occupied molecular orbital/lowest unoccupied molecular orbital (HOMO-LUMO) energies, frontier orbital electron densities, and superdelocalizabilities have been shown to correlate well with various biological activities.

Hydrophobicity descriptors

More than 100 years on, the most well-known hydrophobic descriptor in QSAR is log P. This is since the report of Overton and Meyer correlation of octanol/water partition coefficient and the narcotic potencies of some small molecules. In a landmark study, Hansch and his colleagues devised and used a multiparameter approach that included both electronic and hydrophobic terms, to establish a QSAR for a series of plant-growth regulators. This study laid the basis for the development of the QSAR paradigm and also firmly established the importance of lipophilicity in biosystems. Log P of any chemical compound can be determined experimentally by measuring the ratio of concentration of the compound in octanol to its concentration in water. The methods employed include shake flask, filter probe method, and centrifugal partition chromatographic technique.

$$P = \text{Concentration}_{octanol} / \text{Concentration}_{water}$$

Octanol is a suitable solvent for the measurement of partition coefficients for many reasons. It is cheap, relatively nontoxic, and chemically unreactive. The hydroxyl group has both hydrogen bond acceptor and hydrogen bond donor features capable of interacting with a large variety of polar groups. Despite its hydrophobic attributes, it is able to dissolve many more organic compounds. It is UV-transparent over a large range and has a vapour pressure low enough to allow for reproducible measurements. Hydrophobic solutes are not appreciably solvated by the octanol in the water phase unless their intrinsic log P is above 6.0. Octanol begins to absorb light below 220 nm and, thus, solute concentration determinations can be monitored by UV spectroscopy. More important, octanol acts as an excellent mimic for biomembranes because it shares the traits of phospholipids and proteins found in biological membranes due to long alkyl chain and polar hydroxyl group.

Chromatography provides an alternate tool for the estimation of hydrophobicity parameters. Thin-layer chromatographic (TLC) parameter R_f can be modified to R_m by the following formula, which is considered analogous to log P.

$$R_m = \log(1/R_f - 1)$$

The advantage of TLC over conventional log P determination is that minimum sample is utilised and it is easy to detect. The compounds under testing need not be pure, and even practically insoluble analogues can be analysed. The greatest advantages of TLC are that no quantification is required for concentration determination and that several compounds can be determined simultaneously. High-performance liquid chromatography (HPLC) is also a method of choice in measuring k', a hydrophobicity parameter analogous to log P. With t_r, the retention time of the compound, and t_o, the retention time of solute front, the k' can be calculated as:

$$k' = (t_r - t_o)/t_o$$

Reverse phase HPLC is considered a high-throughput hydrophobicity screening method for combinatorial libraries of compounds. A linear relationship was obtained between log P and k' as shown below:

$$\log P = 1.025(\pm 0.06) \log k' + 0.797$$
$$n = 33; r = 0.987; s = 0.127$$

Hansch derived the hydrophobicity contribution of substituents as π constant.

$$\pi_x = \log P_x - \log P_H$$

In this equation, P_X is the partition coefficient of a molecule with substituent X, and P_H is the partition coefficient of the unsubstituted molecule (i.e., $X = H$). A more positive number indicates a more lipophilic substituent. π_H is set to zero. The π_{nitro} is calculated from the log P of nitrobenzene and benzene.

$$\pi_{NO2} = \log P_{nitrobenzene} - \log P_{benzene}$$
$$= 1.85 - 2.13 = -0.28$$

Initially, the π-system was applied only to substitution on aromatic rings and when the hydrogen being replaced was of innocuous character, and apparently aliphatic fragments values were developed a few years later.

The molar refractivity (MR) is an additive constitutive property of a compound. Since MR is additive, each methylene group contributes a constant amount to the MR as with the case of log P. Hence, in a homologous series, log P and MR will be well correlated and the contributions of these two effects onto potency cannot be distinguished statistically. Parachor (PA) is another additive constitutive property of molecules which might be useful in QSAR. Molar volume, molar refractivity, and parachor are closely interrelated as presented below:

$$MR = MV \cdot n^2 - 1/n^2 + 2$$

'n' is the refractive index of the molecule and MV is the molar volume, which is measured with the following formula:

$$MV = \text{molecular weight}/\rho$$

'ρ' is the density of the compound. With molar volume and surface tension (γ) known, parachor can be easily determined from the formula below:

$$PA = MV \, \gamma^{1/4}$$

Steric descriptors

In 1950, Taft derived steric parameter by making an extension of the Hammett equation and named it as E_s. E_s is defined as follows:

$$E_s = \log (k_x/k_H)_A$$

where k_x and k_H represent the rate of acid hydrolysis of substituted methyl acetate and unsubstituted methyl acetate, respectively, as shown below:

Bulkier groups (X) like methyl block access to the carbonyl carbon by slowing the reaction and, hence, k_x will be less than k_H making E_s negative. Taft E_s values cannot be measured for many substituents because of the nature of the model reaction on which it is based. Charton, Kutter, and Hansch suggested that for spherically symmetrical groups one might use the radius as a measure of steric effects. Hence, an alternative to E_s parameter is the Charton's van der Waal's term as given below:

$$v_x = r_{vx} - r_{vH}$$
$$= r_{vx} - 1.20$$

The term r_v is the minimum van der Waal's radius of the substituent. There has been found to be a good correlation between E_s and van der Waal's radius as given below:

$$E_s = -2.062(\pm 0.86)v - 0.194(\pm 0.1)$$
$$n = 104; r = 0.978; s = 0.250$$

In 1976, Verloop and co-workers introduced a new set of steric substituent descriptors also termed as STERIMOL size parameters, which address the three-dimensional issues of a compound. Five parameters that define the size of a substituent include L, the length of the substituent along the axis of the bond between the substituent and the parent molecule. Perpendicular to this bond axis, four width parameters, B_1-B_4, are measured. B_1 is the minimal width; B_2, B_3, and B_4 are distances measured perpendicular to all other B values and ordered in increasing value. Thus, L, B_1-B_4 describe the positions, relative to the point of attachment and the bond axis, of five planes that closely surround the group. Verloop subsequently established the adequacy of just three parameters for QSAR analysis: a slightly modified length L, a minimum width B_1, and a maximum width B_5 that is orthogonal to L. (*Figure 4.1 depicts Verloop's sterimol parameters shown in the coloured set of pages.*)

The molar refractivity (MR) is an additive constitutive property of a compound and is one of the oldest steric parameters. Since MR is additive, each methylene group contributes a constant amount to the MR as with the case of log P. Hence, in a homologous series, log P and MR will be well correlated and the contributions of these two effects onto potency cannot be distinguished statistically. Parachor (PA) is another additive constitutive property of molecules which might be useful in QSAR. Molar volume, molar refractivity and parachors are closely interrelated as presented below,

$$MR = MV \cdot n^2 - 1 / n^2 + 2$$

n is the refractive index of the molecule and MV is the molar volume which is measured with the following formula,

$$MV = \text{molecular weight} / \rho$$

ρ is the density of the compound. With molar volume and surface tension (γ) known, parachor can be easily determined from the formula below

$$PA = MV \, \gamma^{1/4}$$

Indicator or dummy parameters

When deriving QSAR for a series of compounds, there could be a substitution or a group that is related or repeated. Such substitution can be understood by assigning an indication using indicator variables. These variables are arbitrarily assigned a value of one to indicate the presence of a particular group, and zero to indicate its absence so that the importance of the feature can be easily estimated from the regression equation. In no case an indicator variable is used for only one compound.

Molecular connectivity index (Randic branching index)

Milan Randic in 1975 proposed an algorithm to encode bond contributions to a molecular branching index. From this effort it has become possible to offer quantitative statements about the extent of branching in a molecule. This branching algorithm formed the basis of a structure-description paradigm called molecular connectivity, developed over the next decade (1976–86) by Kier and Hall. The molecule is regarded as a sum of the bonds connecting pairs of atoms. Each atom in a molecule is encoded by a cardinal number, δ^v, the count of all bonded atoms other than hydrogen calculated as:

$$\delta^v i = Z^v i - N_H / Z i - Z^v i - 1$$

In the above formula, $\delta^v i$ is the cardinal number of the ith atom in the molecule; $Z^v i$ is the valence number of the ith atom; N_H is the number of hydrogen attached to the atom; and $Z i$ is the atomic number of the ith atom. For example,

$$\delta^v_{CH3} = 4 - 3/6-4-1$$
$$\delta^v_{CH3} = 1$$
$$\text{Similarly, } \delta^v_{NH2} = 5-2/7-5-1$$
$$\delta^v_{NH2} = 3$$

The molecular connectivity index χ^v, the Chi index, is derived by combining these bond descriptors as given in the following formula:

$$\chi^v = \Sigma\ 1\ /\ \text{root of } \delta_i\ \delta_j$$

There are five general categories of molecular structure information intuitively described by the various Chi indices, which include the following: (1) degree of branching (emphasised in low-order Chi indices); (2) variable branching pattern (emphasised in high-order path Chi indices); (3) position and influence of heteroatoms (emphasised in the valence Chi indices); (4) patterns of adjacency (emphasised in the Chi cluster and path/cluster indices); and (5) degree of cyclicity (emphasised in the Chi chain indices). The statements are mainly based on the individual index.

Linear free energy relationship

The correlation of biological activity with physicochemical properties is often termed an extra thermodynamic relationship. The most widely used mathematical technique in QSAR is the linear free energy relationship, as it is very powerful in understanding the relationship of dependent (biological activity) and independent variables (physicochemical parameters or molecular descriptors). One of the first QSAR approaches to be developed was the Hansch linear free energy model. Hansch undertook a study for Robert Muir, a plant physiologist, in an attempt to understand how changes in chemical structure affected the potency of plant growth regulators. A simplest approach with Hammett's electronic parameter was worthless and, hence, emerged a multiparametric approach that is today known as multiple linear regression analysis, or QSAR. In this method, three parameters are measured and used in the Hansch equation. The first of these is the substituent hydrophobicity constant, π, which is calculated using the first equation below.

Hansch linear free energy model

$$\log 1/C = a\pi_x + bE_s + c\sigma + d$$

Free and Wilson method – the Fujita and Ban modification

$$\log 1/C = \sum a_i X_i + \mu_0$$

The second parameter is the Hammett coefficient, described above, and the third is the Taft constant E_S, which is a measure of steric bulk. These constants, which are either tabulated or calculated from tabulated data, are then plugged into the Hansch equation (the second equation above). The activity C is measured for 20–30 analogues, and after putting in the values for the three constants, regression analysis is used to determine a, b, c, and d. Using this equation, it is then possible to predict the activity of unmade analogues by inserting the appropriate constants for a given substituent and solving for c. The downside of this approach is that the data cannot be collected, and the values of the variable a–d cannot be determined, until a large number of analogues have been made. Also, the approach only works for substitutions at one position in a parent structure, usually an aromatic carbon.

Another approach to QSAR was developed by Free and Wilson, and later modified by Fujita and Ban. The Free-Wilson approach is truly a structure activity-based methodology because it incorporates the contributions made by various structural fragments to the overall biological activity. In this model, a_i is the contribution of the substituent to the activity of the analogue, and X_i is either 1 (substituent present) or 0 (no substituent present). Again, once a large number of analogues have been made, regression analysis is used to determine the variables a and μ, and the equation can be used to predict activity in as-yet unmade analogues. This method has the advantage of taking more than one substituent into account for a given molecule. Indicator variables are used to denote the presence or absence of a particular structure feature.

Fujita and Ban modified the approach in two important ways. The biological activity is expressed on a logarithmic scale, to bring it into line with the extra thermodynamic approach, as seen in the following equation:

$$\text{Log } X_C = \sum a_i X_i + \mu_0$$

Manual QSAR or Topliss approach

Another common method to predict activity is the Topliss scheme, as shown below. By this method, an unsubstituted aromatic ring within the parent is converted to the 4-chloro derivative. This results in a compound that is more active, less active, or equally active as the parent. Let us say, for the sake of an example, that the 4-chloro is less active. The Topliss scheme would then suggest making the 4-methoxy. If the 4 methoxy is less or equally active, the 3-chloro analogue is made. If the 4-methoxy is more active, the 4-(diethyl)amino analogue is made, and so on. Theoretically, this leads ultimately to the synthesis of the optimally active analogue.

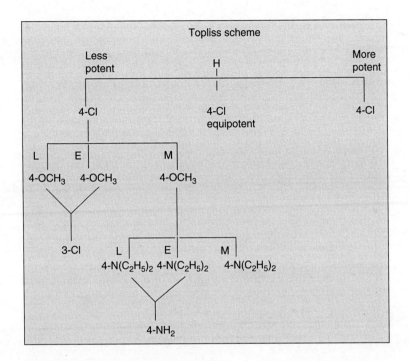

4.3.2 Pharmacophore-Based Drug Design

Pharmacophore is defined as a group of atoms in a molecule responsible for the bioactivity. Traditionally, a pharmacophore is defined as the specific 3D arrangement of functional groups within a molecular framework which are necessary to bind to a macromolecule and/or an enzyme active site. Pharmacophore modelling provides a useful framework for better understanding of bioactivity and can be used as a predictive tool in the design of compounds with improved potency, selectivity, and/or pharmacokinetic properties. Pharmacophore models are generated by analysing SAR and mapping common structural features of active analogues. Once a pharmacophore is established, the medicinal chemist has a host of 3D database search tools to retrieve novel compounds that fit the pharmacophore model. Numerous advances have been made in the computational perception and utilization of pharmacophores in drug discovery, database searching, and compound libraries.

Pharmacophore-based drug design involves three major steps that include: 1) data collection; 2) data analysis and model development; and 3) validation of the pharmacophore model and design of new entities. The crucial stage is the collection of data. Improper or insufficient data would end up in poor model development and, hence, poor drug design. Inactive molecule data can also be useful in certain cases to understand the receptor environment. The analysis stage starts with the distribution of collected data with respect to as many common features as possible. The analysis stage generates knowledge of the interlinkages between the common structural features and biological activity. The pharmacophore hypothesis developed should be validated with a test set of molecules, which can include a good number of inactives. The design phase involves integrating the pharmacophore template into a new molecule.

Chapter 4: Computer-aided Drug Design

The process of pharmacophore modelling can be understood in Figs. 4.2 and 4.3.

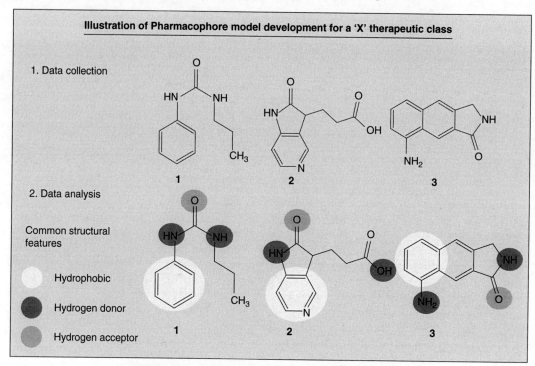

FIGURE 4.2 The process of pharmacophore modelling.

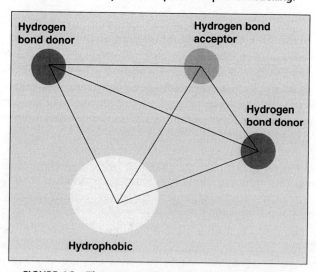

FIGURE 4.3 The process of pharmacophore modelling.

It is possible to derive pharmacophores in several ways: by analogy to a natural substrate or known ligand, by inference from a series of dissimilar biologically active molecules (the so-called active analogue approach), or by direct analysis of the structure of a target protein. Below are some examples of pharmacophore models extracted from various literatures. (*For Fig. 4.4 [b and d], see coloured set of pages.*)

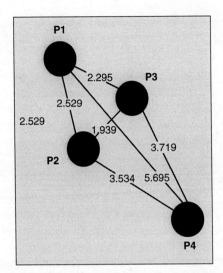

FIGURE 4.4a Point pharmacophore model for SARS CoV proteinase (P1 = Hydrogen bond acceptor, donor and hydrophobic centre; P2 = Hydrogen bond acceptor and hydrophobic centre; P3 = Hydrogen bond acceptor and donor; P4 = Hydrogen bond acceptor and donor)
Source: (Zhang et al., *European Journal of Medicinal Chemistry*, 2004.)

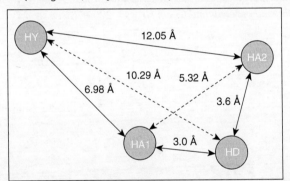

FIGURE 4.4c Point pharmacophore model for HIV-1 integrase inhibitors (HY = Hydrophobic; HA = Hydrogen bond acceptor; HD = Hydrogen bond donor 3D alignment of thrombin inhibitors based on a pharmacophore model)
Source: (Briggs et al., *Bioorganic Medicinal Chemistry Letters*, 14, 2004.)

4.3.3 Three-Dimensional QSAR (3D QSAR)

With the advent of molecular modelling, three-dimensional (3D) descriptors have replaced the traditional physicochemical and bi-dimensional descriptors. Researchers have created many forms of 3D descriptors containing shape and electrostatic information, and many methods of using such variables. Classical QSAR correlates biological activities of drugs with physicochemical properties and consider neither the 3D structure of the drugs nor their chirality. In 3D QSAR, steric, electrostatic, and hydrophobic interactions are probed at a large number of points in a 3D grid constructed around a set of molecules. (*See Fig. 4.5 in the coloured set of pages.*)

The assumptions made in 3D QSAR include the following:

- The biological activity produced is by the modelled compound and not its metabolite.
- The conformation studied is the bioactive one.

- The binding site is the same for all the modelled compounds.
- Pharmacokinetics, solvent effects, is not included.

Some of the common methods of 3D QSAR include comparative molecular field analysis (CoMFA), comparative molecular shape analysis (CoMSA), and comparative molecular similarity index analysis (CoMSIA). 3D QSAR is superior to 2D QSAR in the sense that even a diverse set of molecules can be modelled in the former. The output of 3D QSAR is not only a multiple linear equation, but also 3D graphics-based contour maps highlighting favourable and unfavourable interaction regions as shown in (*Fig. 4.6 in the coloured set of pages.*)

4.4 MACROMOLECULE-BASED DRUG DESIGN

In structure-based drug design, the three-dimensional structure of a drug target interacting with small molecules is used to guide drug discovery. The process starts with a detailed analysis of the binding site of the target protein, which is preferably complexed with a ligand. This complex unravels the binding mode and conformation of a ligand under investigation and indicates the essential aspects determining its binding affinity. It is then used to generate new ideas about ways of improving an existing ligand or of developing new alternative bonding skeletons. The success of structure-based drug design methodology has encouraged the development of various computational methods that can make use of structural information to suggest novel structures, which may either prove to be useful lead compounds or act as a stimulus to the creativity of designers. The following figure is a schematic representation of Cox-II inhibitor in the binding pocket residues of the enzyme (Source: Supuran et al., *Journal of Medicinal Chemistry*, 47, 2004). (*See Fig. 4.7 in the coloured set of pages.*)

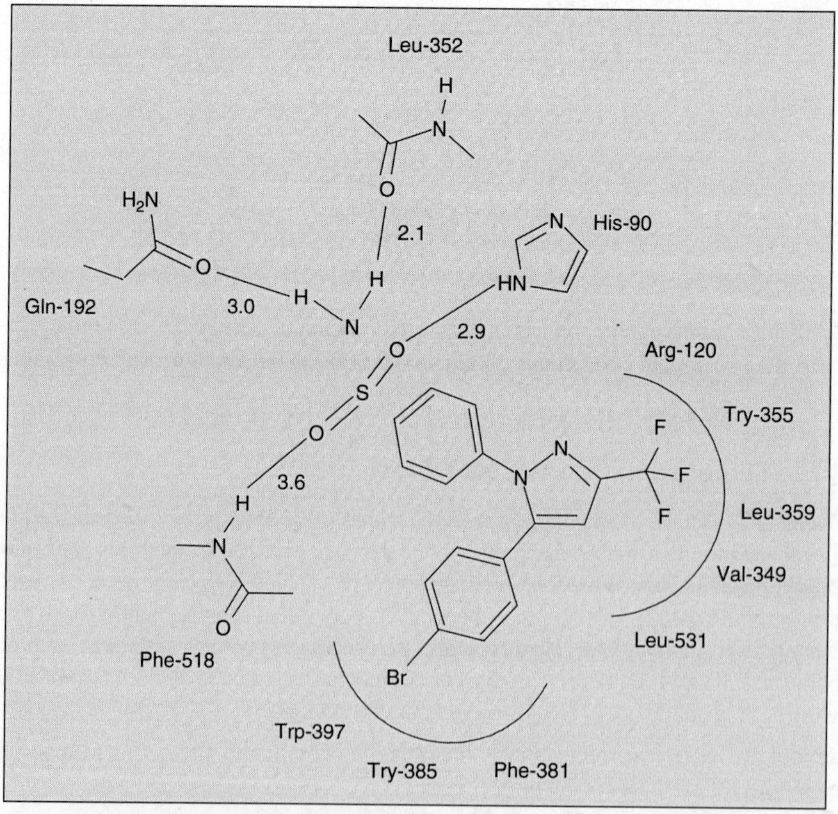

With the improvements in experimental techniques of X-ray crystallography and NMR, the amount of information concerning 3D structures of target proteins has increased a lot. Already in the mid-1990s, almost any (soluble) protein of pharmaceutical interest could be brought into suitable crystalline form and its structure determined. This greatly increased the number of repository of 3D structures of target proteins in the protein data bank (PDB). Cambridge structural database (CSD) is a repository for small-molecule (ligand) crystal structures. PDB and CSD together provide invaluable sources of new insights into non-bonded intermolecular interactions and molecular recognition as well as important aspects of molecular conformation and the roles of solvents, in particular water, in stabilizing molecular complexes.

There are two strategies employed in designing ligands based on target structure:

(i) Database searching method
(ii) Ligand building method

4.4.1 Database Searching Method

This method is about searching suitable ligands for a target protein in various small-molecule databases. A database is an organised body of related information of small molecules. In this case, a large number of molecules are screened to find those fitting the binding pocket of the receptor. Some researchers call this 'virtual screening' in analogy to the bioassay screening procedure employed in the traditional drug discovery. The key advantage of database searching is that it saves synthetic effort to obtain new lead compounds.

Database searching for ligands starts with a defined target structure and followed by active site analysis using various programmes like GRID. If the structure of target protein is unknown, a method called homology modelling is employed in which a known protein from the same family is selected as template and the structure of the unknown is derived using BLAST search software especially useful for this purpose. Various structure databases that can be employed in searching suitable ligands for the target protein include NCI (National Cancer Institute) database, CSD, Bell's organic database, and DEREK. The ligand structure database compounds will be docked on to the active site pocket model of the protein using various docking software that include DOCK, GOLD, Glide, and FlexX.

Three-dimensional database searching identifies existing molecules that match a hypothesis of the 3D requirements for bioactivity. Thus, it can be used to validate such pharmacophores and to suggest other existing compounds for testing to find a new lead. The current interest in 3D database searching was fuelled by the availability of tools for molecular modelling and pharmacophore mapping, and by the increasing numbers of 3D protein structures as targets for new drugs.

Three-dimensional searching programmes and their applications differ from each other in a number of respects. Types of database searching include:

- Geometric searches in which all 3D features are required to be present
- Pre-screening to reduce search time for geometric searches
- Searches in which not every query 3D feature need match
- Conformational flexibility searches
- 3D structure similarity searches

Some of the sources of 3D structure searching databases are CONCORD, AIMB, WIZRAD, and COBRA. Some of the successful molecules discovered by structure-based drug design using database searching methods include ligands for human protein FKBP, which is a protein from the immunophilin group, some T-type calcium channel blockers, glyceraldehyde phosphate dehydrogenase inhibitors, thrombin inhibitors, HIV enzyme inhibitors, aldose reductase inhibitors, and bacterial DNA gyrase inhibitors. (*Figure 4.8 is an illustration of database screening as shown in the coloured set of pages.*)

4.4.2 Ligand Building Method

This structure-based drug design method is about 'building' ligands. In this case, ligand molecules are built up within the constraints of the binding pocket by assembling small pieces in a stepwise manner. These pieces can be either atoms or fragments. The key advantage of this method is that novel structures, not contained in any database, can be generated.

The process of structure-based design involving ligand-building method starts with the target protein structure analysis. The protein databank structure can be downloaded easily. It is preferable to employ a structure with a known substrate or inhibitor-bound protein structure. A key step in receptor-based design strategies is to model the binding site as accurately as possible. This can be achieved in various ways, starting with an atomic resolution structure of the active site. The target protein is subjected to active site analysis using programmes like GRID or MCSS (multiple copy simultaneous simulation). Programmes like UCSF DOCK define the volume available to a ligand by filling the active site with spheres. Further constraints follow, using positions of H-bond acceptors and donors. Once favourable positions of atoms are indicated or specific functional groups or fragments are pre-docked into the binding site, the step of complete ligand assembly follows.

The active site residues are isolated and using a variety of structure-based design software, complimentary fragments at user-selected active sites are placed and connected randomly, and combinations of ligands are generated, docked, and scored.

The programme attempts to place atoms or fragments in the site, and assesses the fit. Since many possible solutions arise (hits), each has to be judged to decide which is most promising. This is called scoring. Programmes such as LEGEND, LUDI, Leap-Frog, SPROUT, HOOK, and PRO-LIGAND attempt this using different scoring techniques. These scoring functions vary from simple steric constraints and H-bond placement to explicit force fields and empirical or knowledge-based scoring methods. Programmes like GRID and LigBuilder set up a grid in the binding site and then assess interaction energies by placing probe atoms or fragments at each grid point. Scoring functions guide the growth and optimization of structures by assigning fitness values to the sampled space.

If the structure of the target is unknown, structures of known ligands can be used as an alternative strategy. The topological or three-dimensional structure of the ligand is used as a basis for a kind of homology modelling. One way to use this information is the ligand-pharmacophore model. This can be used to develop a pseudo-receptor model that can be adapted for use in receptor-based design programmes.

Protein flexibility, mediation of ligand binding through water molecules, and ligand specificity remain poorly understood aspects of drug design.

(*Figure 4.9 illustrates the process of ligand design as shown in the coloured set of pages.*)

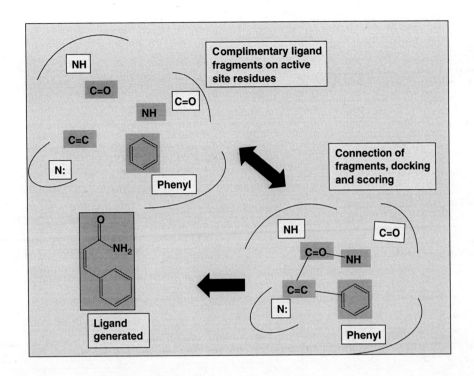

4.5 SUBSTITUENT CONSTANTS USEFUL FOR QSAR ANALYSIS

Substituent	π	σ_m	σ_p	MR	E_s	L	B_1	B_5	F	R
H	0.00	0.00	0.00	0.10	1.24				0.00	0.00
Br	0.86	0.39	0.23	0.88	0.08	3.82	1.95	1.95	0.44	-0.17
Cl	0.71	0.37	0.23	0.60	0.27	3.52	1.80	1.80	0.41	-0.15
F	0.14	0.34	0.06	0.09	0.78	2.65	1.35	1.35	0.43	-0.34
I	1.12	0.35	0.18	1.39	-0.16	4.23	2.15	2.15	0.40	-0.19
NO_2	-0.28	0.71	0.78	0.74	-1.28				0.67	0.16
OH	-0.67	0.12	-0.37	0.28	0.69	2.74	1.35	1.93	0.29	-0.64
NH_2	-1.23	-0.16	-0.66	0.54	0.63	2.78	1.35	1.97	0.02	-0.68
SH	0.39	0.25	0.15	0.92	0.17	3.47	1.70	2.33	0.28	-0.11
SO_2NH_2	-1.82	0.46	0.57	1.23		4.02	2.04	3.05	0.41	0.19
SO_2CH_3	-1.63	0.60	0.72	1.35		4.11	2.03	3.17	0.54	0.22
CN	-0.57	0.56	0.66	0.63	0.73	4.23	1.60	1.60	0.51	0.19
NCO		0.27	0.19	0.88					0.29	-0.08
NCS	1.15	0.48	0.38	1.72		4.29	1.50	4.24	0.51	-0.09
CHO	-0.65	0.35	0.42	0.69		3.53	1.60	2.36	0.31	0.13
COOH	-0.32	0.37	0.45	0.69		3.91	1.60	2.66	0.33	0.15
$COCH_3$	-0.55	0.38	0.53	1.12		4.06	1.60	3.13	0.32	0.20

(Continued)

Substituent	π	σ_m	σ_p	MR	E_s	L	B_1	B_5	F	R
OCOCH$_3$	−0.64	0.39	0.31	1.25		4.74	1.35	3.67	0.41	−0.07
CO(OCH$_3$)	−0.01	0.37	0.45	1.29		4.73	1.64	3.36	0.33	0.15
CONH$_2$	−1.49	0.28	0.36	0.98		4.06	1.50	3.07	0.24	0.14
CONHCH$_3$	−1.27	0.35	0.36	1.46		5.00	1.54	3.16	0.34	0.05
CH$_3$	0.56	−0.07	−0.17	0.57	0.00	2.87	1.52	2.04	−0.04	−0.13
CH$_2$CH$_3$	1.02	−0.07	−0.15	1.03	−0.07	4.11	1.52	3.17	−0.05	−0.10
OCH$_3$	−0.02	0.12	−0.27	0.79	0.69	3.98	1.35	3.07	0.26	−0.51
OCH$_2$CH$_3$	0.38	0.10	−0.24	1.25		4.80	1.35	3.36	0.22	−0.44
SCH$_2$CH$_3$	1.07	0.18	0.03	1.84		5.16	1.70	3.97	0.23	−0.18
CH$_2$OH	−1.03	0.00	0.00	0.72	0.03	3.97	1.52	2.70	−0.01	0.01
NHCH$_3$	−0.47	−0.30	−0.84	1.03		3.53	1.35	3.08	−0.11	−0.74
NHC$_2$H$_5$	0.08	−0.24	−0.61	1.50					0.10	−0.92
NHCOCH$_3$	−0.97	0.21	0.00	1.49		5.09	1.35	3.61	0.28	−0.26
NHCSCH$_3$	−0.42	0.24	0.12	2.34		5.09	1.45	4.38	0.27	−0.13
N(CH$_3$)$_2$	0.18	−0.15	−0.83	1.56		3.53	1.35	3.08	0.10	−0.92
CH$_2$N(CH$_3$)$_2$	−0.15	0.00	0.01	1.87		4.83	1.52	4.08		
C$_3$H$_7$	1.55	−0.07	−0.13	1.50	−0.36	4.92	1.52	3.49	−0.06	−0.08
CH(CH$_3$)$_2$	1.53	−0.07	−0.15	1.50	−0.47	4.11	1.90	3.17	−0.05	−0.10
OCH(CH$_3$)$_2$	0.36	0.10	−0.45	1.71					0.30	−0.72
OC$_3$H$_7$	1.05	0.10	−0.25	1.71		6.05	1.35	4.42	0.22	−0.45
Cyclopropyl	1.14	0.35	−0.30	1.35		4.14	1.55	3.24		
Cyclobutyl	1.51	0.37	−0.15	1.79	−0.06	4.77	1.77	3.82		
C(CH$_3$)$_3$	1.98	−0.10	−0.20	1.96	−1.54	4.11	2.60	3.17	−0.07	−0.13
CH(CH$_3$)C$_2$H$_5$	2.04	−0.08	−0.12	1.96	−1.13	4.92	1.90	3.49		
CH2CH(CH3)2	2.13	−0.06	−0.12	1.96	−0.93	6.17	1.52	4.54		

FURTHER READINGS

1. C. Hansch, T. Fujita, *Journal of American Chemical Society*, 86, 1616-1626, 1964.
2. M. E. Wolff (ed.), 'Burger's Medicinal Chemistry', Vol. I: *Principles and Practice*, Wiley, New York, 5th edn, 1995.

MULTIPLE-CHOICE QUESTIONS

1. QSAR method involves
   a. Target structure
   b. Target properties
   c. Ligand x-ray structure
   d. Ligand properties
2. One of the following is not used in QSAR:
   a. Molecular connectivity index
   b. Molecular similarity index
   c. Topological polar surface area
   d. Partition coefficient
3. One of the following is a quantum chemical parameter:
   a. STERIMOL
   b. Taft constant
   c. Highest occupied molecular orbital
   d. Hammett's constant

4. Which of the following is analogous to σ constant?
   a. log P
   b. R_f
   c. pK_a
   d. E_s
5. Which of the following is analogous to π constant?
   a. pK_a
   b. k'
   c. E_s
   d. MW
6. Which of the following is a QSAR technique performed manually?
   a. Hansch approach
   b. Fujita Ban approach
   c. Free Wilson approach
   d. Topliss approach
7. In 3D QSAR, blue regions indicate favourable points for
   a. Bulky groups
   b. Smaller groups
   c. Electron-rich groups
   d. Electron-deficient groups
8. In 3D QSAR, green regions indicate favourable points for
   a. Bulky groups
   b. Smaller groups
   c. Electron-rich groups
   d. Electron-deficient groups
9. In 3D QSAR, red regions indicate favourable points for
   a. Bulky groups
   b. Smaller groups
   c. Electron-rich groups
   d. Electron-deficient groups
10. 10. In 3D QSAR, yellow regions indicate favourable points for
    a. Bulky groups
    b. Smaller groups
    c. Electron-rich groups
    d. Electron-deficient groups

QUESTIONS

1. Isolate the pharmacophores for the following structures and explain.

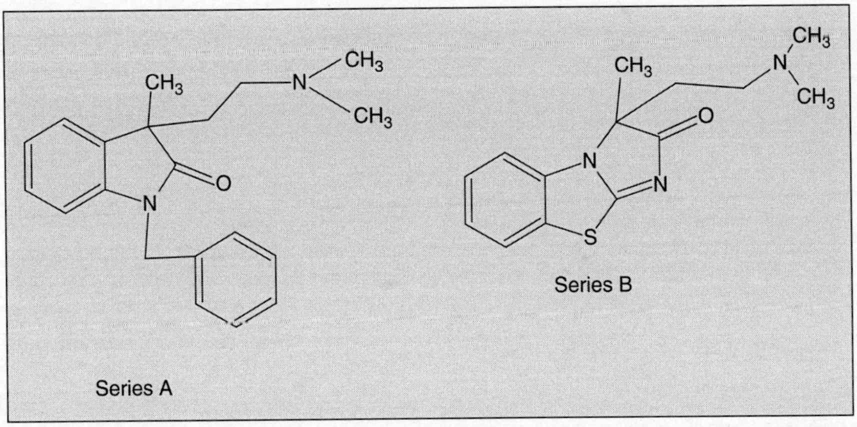

   Series A

   Series B

2. For each term on the right-hand side of the Hansch equation, indicate its meaning including a description of the relationship between the physical properties and the biological activity. Also, suggest one substituent that should enhance activity and justify your selection.

$$BA = -0.34\pi^2 + 0.95\pi - 2.01\sigma + 1.56 E_s + 6.77$$

3. Discuss the state of art in computational methods for lead discovery.
4. Which of these would be a good use for QSAR and why?
   (i) Direct ligand design
   (ii) Structure optimization
   (iii) Adding functional groups
5. Discuss analogue-based drug design with example.
6. How is a drug designed based on the structure of receptor?
7. How is TPSA and drug transport properties related? Explain using the graph given here.

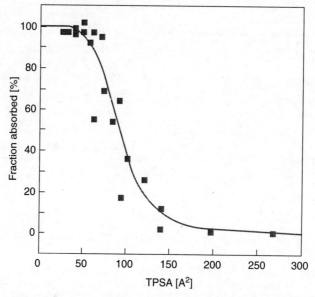

8. In the search for a new drug, a pharmacophore often serves as a template for the desired ligand. Discuss briefly.
9. For the following set of ACE inhibitors structures, identify pharmacophores and explain.

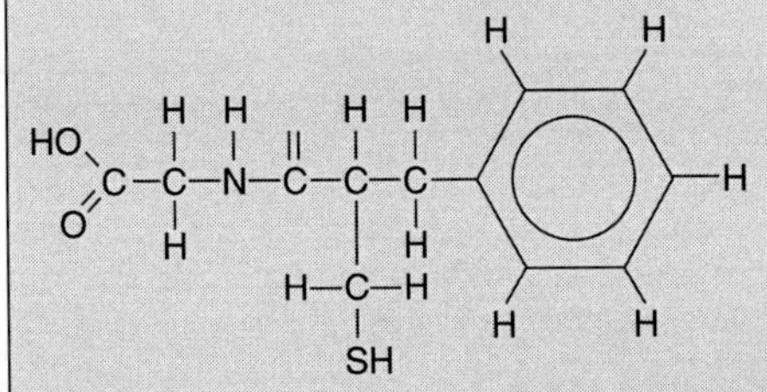

Medicinal Chemistry

mol3

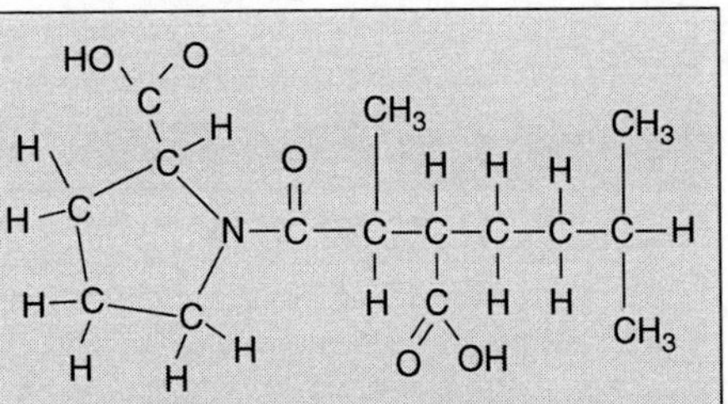

mol4

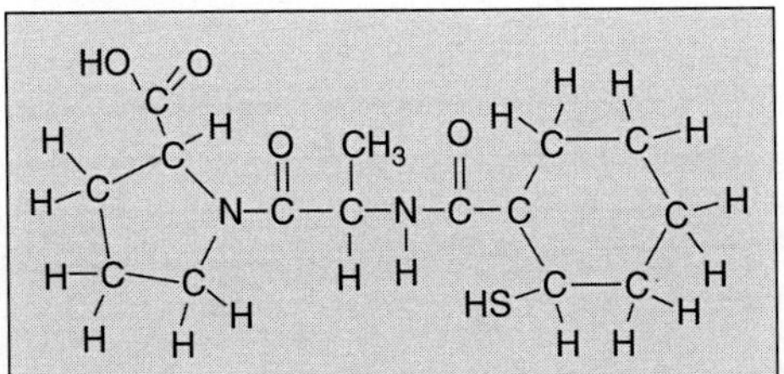

mol5

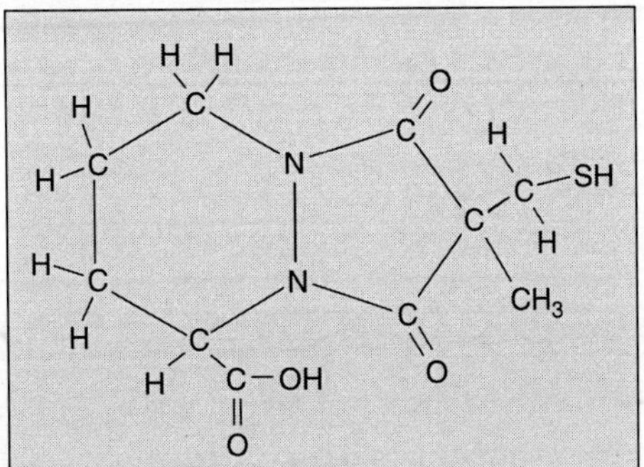

mol6

SOLUTION TO MULTIPLE-CHOICE QUESTIONS

1. d;
2. b;
3. c;
4. c;
5. b;
6. d;
7. d;
8. a;
9. c;
10. b.

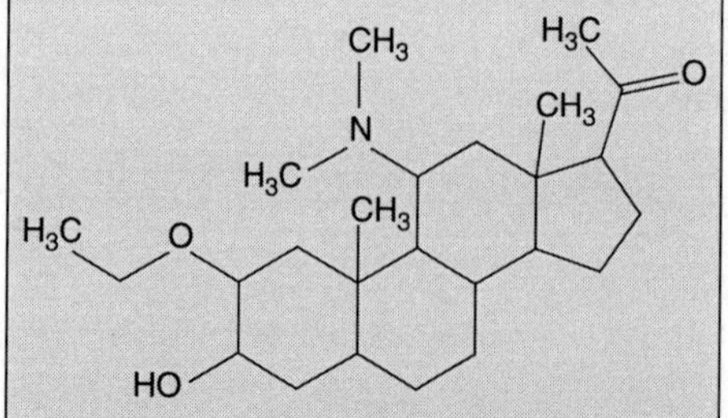

CHAPTER 5

General Anaesthetics

LEARNING OBJECTIVES

- Define the concept of anaesthesia
- List the different stages of attaining anaesthesia while conducting surgeries
- Explain the overall mechanisms of action of anaesthetics
- List the various classes of general anaesthetics
- Describe the liquid anaesthetics with examples
- Describe the synthetic routes of each anaesthetics
- Define the utility and doses of each anaesthetics
- Define the toxicity and side-effects of each anaesthetics
- List the anaesthetics useful in dentistry
- Define inhalation anaesthetics
- Describe the gaseous anaesthetics with examples
- Describe the utility of laughing gas and why it is called so
- Define some rapid-acting barbiturates
- Define rapid-acting non-barbiturates
- Describe benzodiazepine anaesthetic and their advantage
- List the narcotic analgesics used for general anaesthesia
- List other classes of anaesthetics that do not fall in the conventional class
- Describe some latest drugs in the class of general anaesthetics

5.1 INTRODUCTION—ANAESTHESIA AND ITS STAGES

General anaesthetics are CNS depressants that produce anaesthesia, which extends to the entire body, and are characterized by a state of unconsciousness, analgesia, and amnesia with skeletal muscle relaxation and loss of reflexes.

General anaesthetics are employed for surgical operations, and four stages of anaesthesia may be recognized as:

Stage I (Analgesia): The patient is conscious and experiences sensations of warmth, remoteness, drifting, falling, and giddiness. There is a marked reduction in the perception of painful stimuli. This stage is used often in obstetrics and minor surgery.

Stage II (Delirium): This stage begins with the loss of consciousness. Depression of higher centres produces a variety of effects including excitement, involuntary activity, and increased skeletal muscle tone and respiration.

Stage III (Surgical anaesthesia): This is the stage of unconsciousness and paralysis of reflexes. Respiration is regular and blood pressure is maintained. All surgical procedures are carried out in this stage.

Stage IV (Medullary paralysis): Respiratory and circulatory failures occur as depression of the vital centres of the medulla and brain stem occur.

5.2 MECHANISM OF ACTION

The wide variation in structures led to several theories of anaesthetic action. The mechanism by which inhalation anaesthetics manifest their effect is not exactly known. Since they do not belong to one chemical class of compounds, the correlations between structure and activity are also not known. There are a number of hypotheses that have been advanced to explain the action of general anaesthetics; however, none of them can adequately describe the entire spectrum of effects caused by general anaesthetics.

The action of general anaesthetics can be explained as a blockage of ion channels, or as specific changes in mechanisms of the release of neurotransmitters. Three of the proposed mechanisms are discussed here.

1. *Hydrate hypothesis:* Anaesthetic molecules can form hydrates with structured water, which can stop brain function in corresponding areas. However, the correlation between the ability to form hydrates and the activity of inhalation anaesthetics is not known.
2. *Ion channel hypothesis:* Anaesthetics block ion channels by interacting with cellular membranes and reducing the flow of Na^+ ions and increasing the flow of K^+ ions into the cell, which leads to the development of anaesthesia.
3. *Fluid membrane hypothesis:* Anaesthetics stabilize, or rather immobilize, the cell membrane, hampering membrane fluidity, which produces changes in the ion channel action.

5.3 CLASSIFICATION

1. Inhalation Anaesthetics
    (i) Liquid: enflurane, isoflurane, halothane, methoxyflurane, ether
    (ii) Gas: cyclopropane, nitrous oxide
2. Intravenous Anaesthetics
    (i) Ultrashort acting barbiturates: thiopentone, thiamylal, methohexital
    (ii) Arylcyclohexylamine: ketamine
    (iii) Benzodiazepine: midazolam

(iv) Narcotic analgesics: alfentanil, fentanyl
(v) Miscellaneous: etomidate, propofol

3. Newer Drugs: desflurane, sevoflurane, minaxolone

5.4 INHALATION ANAESTHETICS (SYNTHESIS, USES, AND DOSE)

5.4.1 Liquid Anaesthetics

The ideal liquid anaesthetics, yet to be discovered, should have a high margin of safety, produce surgical anaesthesia, have rapid and pleasant induction and recovery, be easily controlled and regulated, have no side-effects or toxicity, should not depress the cardiovascular and respiratory systems, be non-flammable and non-explosive, provide good analgesia and muscle relaxation, and have low cost.

Unfortunately, all available agents exhibit toxic properties that tend to limit their usefulness. Halothane sensitizes the myocardium to sympathoadrenal discharges and adrenaline. Consequently, serious and sometimes fatal cardiac arrhythmias may occur to the patients. Both enflurane and isoflurane are much less likely to sensitize the heart to adrenaline and sympathomimetic discharges. Except for ether, which is not hepatotoxic, all halogenated liquid anaesthetics are capable of producing liver damage. Methoxyflurane is used infrequently because of its low induction and renal toxicity.

Enflurane

2-Chloro-1,1,2-trifluoroethyldifluoromethyl ether

Synthesis

$$CHClF-CF_2-OCH_3 \xrightarrow[h\nu]{Cl_2} CHClF-CF_2-OCHCl_2 \xrightarrow{SbF_3} CHClF-CF_2-OCHF_2$$

2-chloro-1,1,2-trifluoro ethyl methyl ether

Enflurane is synthesised by chlorinating in light 2-chloro-1,1,2-trifluoroethylmethyl ether to give 2-chloro-1,1,2-trifluoroethyldichloromethyl ether, followed by substitution of chlorine atoms by fluorine on the dichloromethyl group using antimony (III) fluoride.

Uses: Enflurane is a pleasant-smelling, non-flammable, halogenated ether anaesthetic that provides rapid induction with little or no excitement. It provides better analgesia and muscular relaxation than halothane, but high concentrations may cause CVS depression and CNS stimulation.

Dose: Induction: 2.0%–4.5% in oxygen or with oxygen-nitrous oxide mixtures. Induction usually requires 7–10 minutes. Maintenance usually is accomplished with 0.5%–3% concentrations.

Isoflurane

1-Chloro-2,2,2-trifluoroethyl difluoromethyl ether

$$CF_3CH_2OH \xrightarrow[DMS]{(CH_3)_2SO} CF_3CH_2OCH_3 \xrightarrow[h\nu]{Cl_2} CF_3CH_2-O-CHCl_2$$

Trifluoroethanol   Methylation

$$\downarrow HF/SbCl_5, Cl_2$$

$$CF_3CHCl-O-CHF_2$$

Synthesis: Isoflurane is synthesised from 2,2,2-trifluoroethanol by methylating with dimethylsulphate. The resulting methyl ether undergoes chlorination by molecular chlorine to give 2-(dichloromethoxy)-1,1,1-trifluoroethane. In the subsequent interaction with hydrogen fluoride in the presence of antimony (V) chloride, chlorine atoms are ultimately replaced by fluorine atoms. The resulting ether again undergoes chlorination by molecular chlorine to give isoflurane.

Uses: Isoflurane, an isomer of enflurane, is a non-flammable inhalation anaesthetic for induction and maintenance of general anaesthesia. Induction of and recovery from isoflurane anaesthesia is rapid. Isoflurane is said to offer advantages over all available inhalation anaesthetics, especially in its lack of any important toxicity.

Dose: Induction: 1.5%–3.0% usually produce surgical anaesthesia in 7–10 minutes. Surgical levels of anaesthesia can be sustained with 1.0%–2.5% concentrations when nitrous oxide is used concomitantly.

Methoxyflurane

2,2-Dichloro-1,1-difluoroethylmethylether

Synthesis

$$CCl_3CHF_2 \xrightarrow{KOH} \underset{Cl}{\overset{Cl}{C}}=\underset{F}{\overset{F}{C}} \xrightarrow[KOH]{CH_3OH} CHCl_2CF_2-OCH_3$$

Methoxyflurane is synthesised from 1,1-difluoro-2,2,2-trichloroethane, which undergoes dehydrochlorination by potassium hydroxide to give 1,1-dichloro-2,2-difluoroethylene, to which methanol is added in the presence of potassium hydroxide.

Uses: Methoxyflurane is a potent liquid, volatile anaesthetic agent. A concentration of only 0.1%–2.0% in the inspired mixture will maintain surgical anaesthesia. It provides adequate analgesia and can be used alone in dentistry and obstetrics.

Dose: For analgesia: 0.3%–0.8% in air. For induction: 1.5%–3.0% vaporized by a 1:1 mixture of nitrous oxide and oxygen. For maintenance: 0.1%–2.0%.

Halothane

2-Bromo-2-chloro-1,1,1- trifluoroethane

Synthesis

$$\underset{\text{Trichloroethylene}}{CCl_2=CHCl} \xrightarrow[\underset{SbCl_3}{130°C}]{3HF} \underset{\text{2-chloro-1,1,1-trifluoroethane}}{CF_3-CH_2Cl} \xrightarrow{Br_2} CF_3CHClBr$$

Halothane is made by the addition of hydrogen fluoride to tricholoroethylene and simultaneous substitution of chlorine atoms in the presence of antimony (III) chloride at 130°C. The resulting 2-chloro-1,1,1-trifluorethane undergoes further bromination at 450 °C to form halothane.

Uses: Halothane is a potent, relatively safe, frequently employed general inhalation anaesthetic. Induction with halothane is smooth and rapid with little or no excitement. It is not a potent analgesic and skeletal

muscle relaxant. Therefore, it is used frequently in conjunction with nitrous oxide and with succinylcholine, tubocurarine, or gallamine.
Dose: For induction: 1.0%–4.0% vaporized by a flow of oxygen or nitrous oxide-oxygen mixture. For maintenance: 0.5%–1.5%.

Ether
Diethyl ether
Synthesis

(a) $C_2H_5OH \xrightarrow{H_2SO_4} C_2H_5OC_2H_5$

(b) $C_2H_5OH \xrightarrow{H_2SO_4} C_2H_5HSO_4 \xrightarrow{C_2H_5OH} C_2H_5OC_2H_5 + H_2SO_4$

Ethylsulfuric acid

Ether is prepared by intermolecular dehydration of alcohol. This direct reaction requires drastic conditions (heating to 140 degrees Celsius and an acid catalyst, usually concentrated sulphuric acid).
Use: Ether is an obsolete anaesthetic with a pungent, irritant odour. It is flammable and explosive at concentrations necessary for anaesthesia.

5.4.2 Gaseous Anaesthetics

Cyclopropane
Synthesis

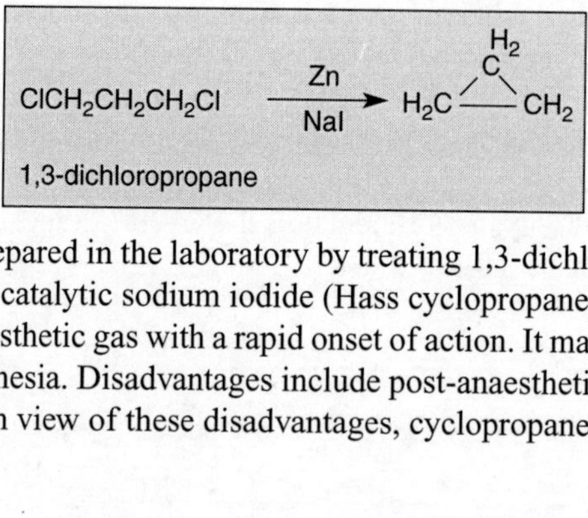

$ClCH_2CH_2CH_2Cl \xrightarrow[NaI]{Zn}$ cyclopropane

1,3-dichloropropane

Cyclopropane can be prepared in the laboratory by treating 1,3-dichloropropane, zinc dust in aqueous alcohol in the presence of catalytic sodium iodide (Hass cyclopropane process).
Uses: Cyclopropane is an anaesthetic gas with a rapid onset of action. It may be used for analgesia, induction, or maintenance of anaesthesia. Disadvantages include post-anaesthetic nausea, vomiting, headache, and malignant hypertension. In view of these disadvantages, cyclopropane is rarely used.

Nitrous oxide
Synthesis

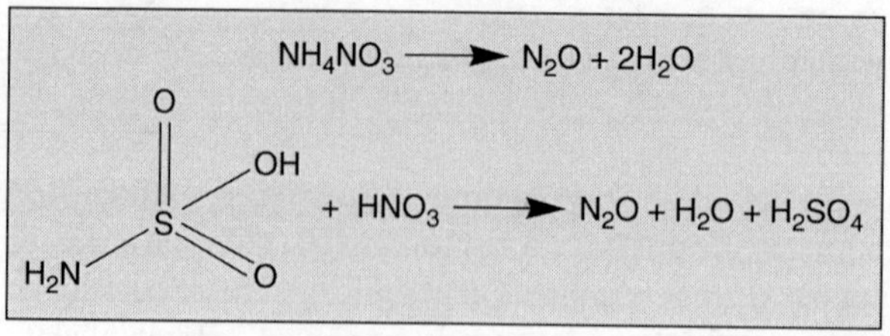

$NH_4NO_3 \longrightarrow N_2O + 2H_2O$

$H_2N-S(=O)_2-OH + HNO_3 \longrightarrow N_2O + H_2O + H_2SO_4$

Nitrous oxide is synthesised either by the thermal decomposition of ammonium nitrate, or by the oxidation of sulphamic acid by nitric acid.

Uses: Nitrous oxide is a weak anaesthetic with good analgesic properties and relatively no skeletal muscle-relaxant properties. Therefore, nitrous oxide is used in conjunction with other liquid anaesthetics. During its administration some patients become hysterical, and because of these characteristics it is often called 'laughing gas'. Nitrous oxide is used in dental surgery because of the rapid recovery that it allows, and it is employed in obstetrics to produce analgesia.

Dose: Analgesia: 25%–50%; maintenance: 30%–70%. Administered with at least 25%–30% oxygen.

5.5 INTRAVENOUS ANAESTHETICS (SYNTHESIS, USES, AND DOSE)

5.5.1 Ultrashort Acting Barbiturates

Rapid acting barbiturates injected most commonly are administered intravenously to induce or sustain surgical anaesthesia. Intravenous anaesthetics are suited best for the induction of anaesthesia and for short procedures, such as orthopaedic manipulations and operations, genito-urinary procedures, obstetric repair, and dilatation and curettage.

Thiopentone sodium

Sodium salt of 5-ethyl-5-(1-methylbutyl)-2-thiobarbiturate

Synthesis

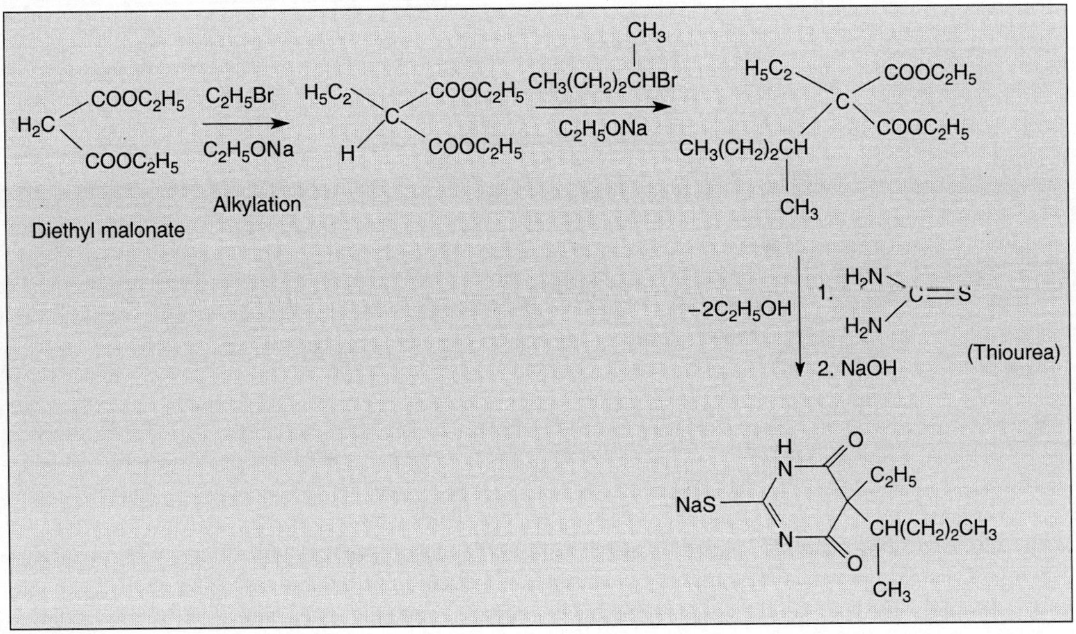

Thiopentone is synthesised from diethylmalonate by alkylating with ethyl bromide in presence of sodium ethoxide to form ethylmalonic ester, which further alkylated with 2-bromopentane in the presence of sodium ethoxide. The product ethyl-(1-methylbutyl)malonic ester undergoes heterocyclization with thiourea, using sodium ethoxide as a base.

Uses: It is the most commonly employed rapidly acting depressant of the CNS, which induces hypnosis and anaesthesia, but not analgesia. It produces anaesthesia within 30–40 seconds after I.V. injection. Recovery after small dose is rapid and it is indicated as the sole anaesthetic agent for brief (15 minutes) procedures, for induction of anaesthesia prior to administration of other anaesthetic agents.

Untoward reactions include respiratory depression, myocardial depression, cardiac arrhythmias, prolonged somnolence and recovery, sneezing, coughing, bronchospasm, larynchospasm, and shivering.

Dose: I.V. induction: 2 ml-3 ml of a 2.5% solution at intervals of 30–60 seconds; maintenance: 0.5 ml–2 ml as required.

Methohexital sodium

Sodium salt of 5-allyl-1-methyl-5-(1-methyl 2-pentynyl) barbiturate
Synthesis

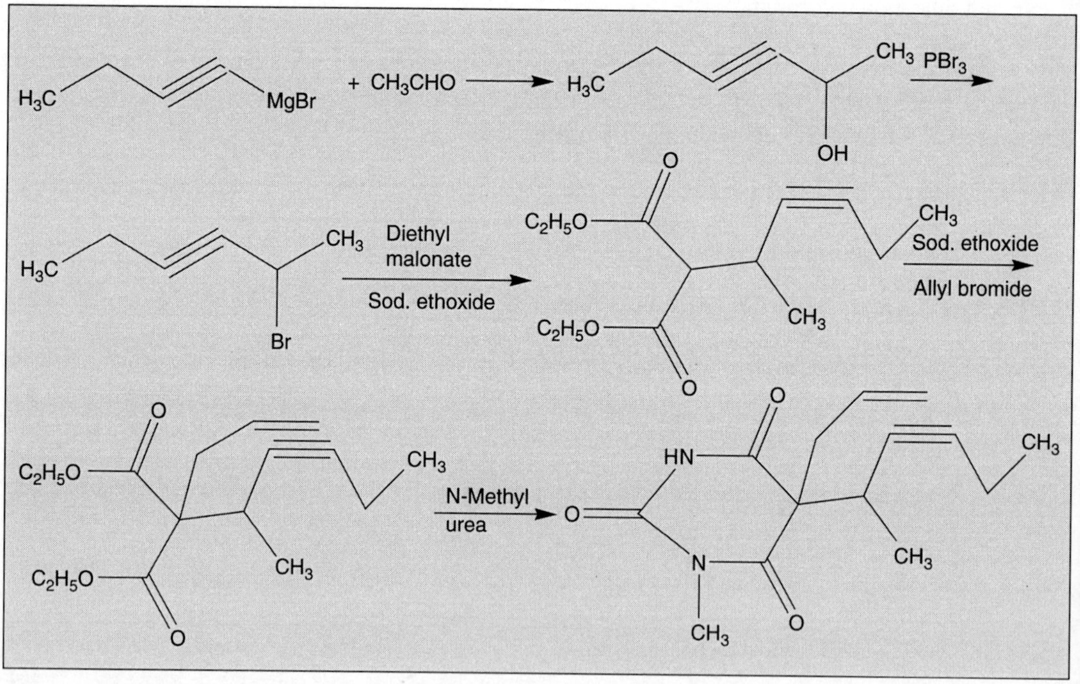

Methohexital is synthesised in the classic manner of making barbituric acid derivatives, in particular by the reaction of malonic ester derivatives with derivatives of urea. The resulting allyl-(1-methyl-2-pentynyl) malonic ester is synthesised by subsequent alkylation of the malonic ester itself, beginning with 2-bromo-3-hexyne, which gives (1-methyl-2-pentynyl) malonic ester, and then by allylbromide.

2-Bromo-3-hexyne is, in turn, synthesised from Normant's reagent, which is synthesised from 1-butyne and ethylmagnesium bromide and it is subsequent reaction with acetaldehyde followed by bromination of the resulting carbinol using phosphorous tribromide. Interaction of obtained dialkyl malonic ester prepared with N-methylurea gives desired methohexital.
Uses: Uses and untoward effects similar to those of thiopentone.
Dose: I.V. induction: 5 ml–12 ml of a 1% solution at the rate of 1 ml every second; maintenance: 2 ml–4 ml every 4–7 minutes as required.

Thiamylal sodium
Sodium salt of 5-allyl-5-(1-methylbutyl)-2-thiobarbiturate
Synthesis

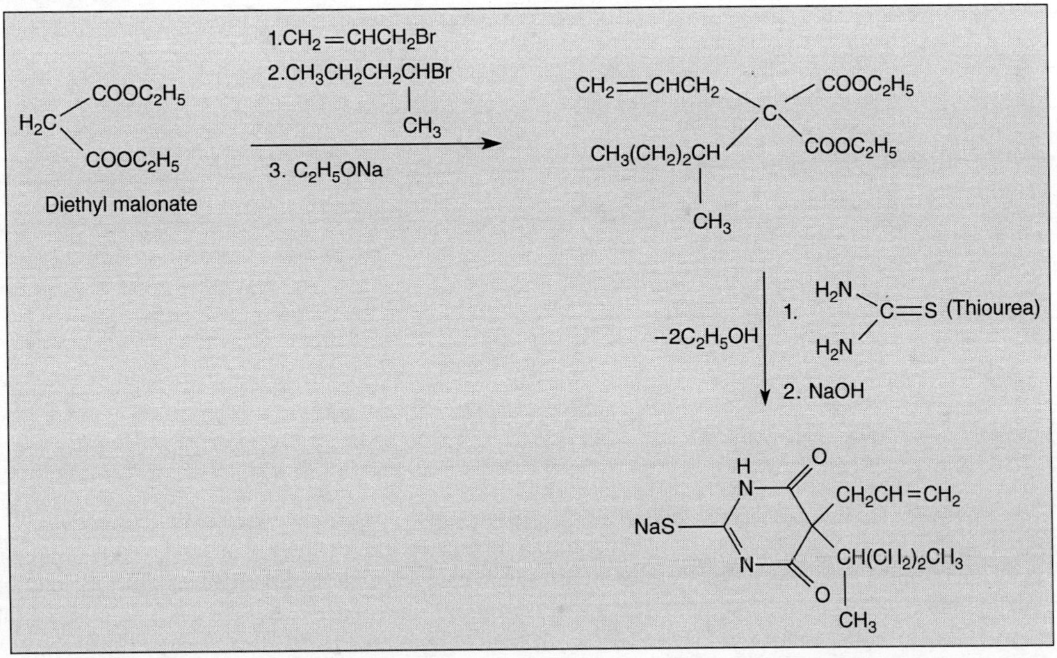

Thiamylal is synthesised similar to thiopentone sodium, using allyl bromide instead of ethyl bromide in the first step.
Uses: Uses and untoward effects similar to those of thiopentone.
Dose: Induction: 3 ml–6 ml of a 2.5% solution at the rate of 1 ml every 5 seconds; maintenance: 0.5 ml–1 ml of 0.3% solution by continuous drip as required.

5.5.2 Arylcyclohexylamine

Ketamine
2-(2'-Chlorophenyl)-2-(methylamino) cyclohexanone

Synthesis

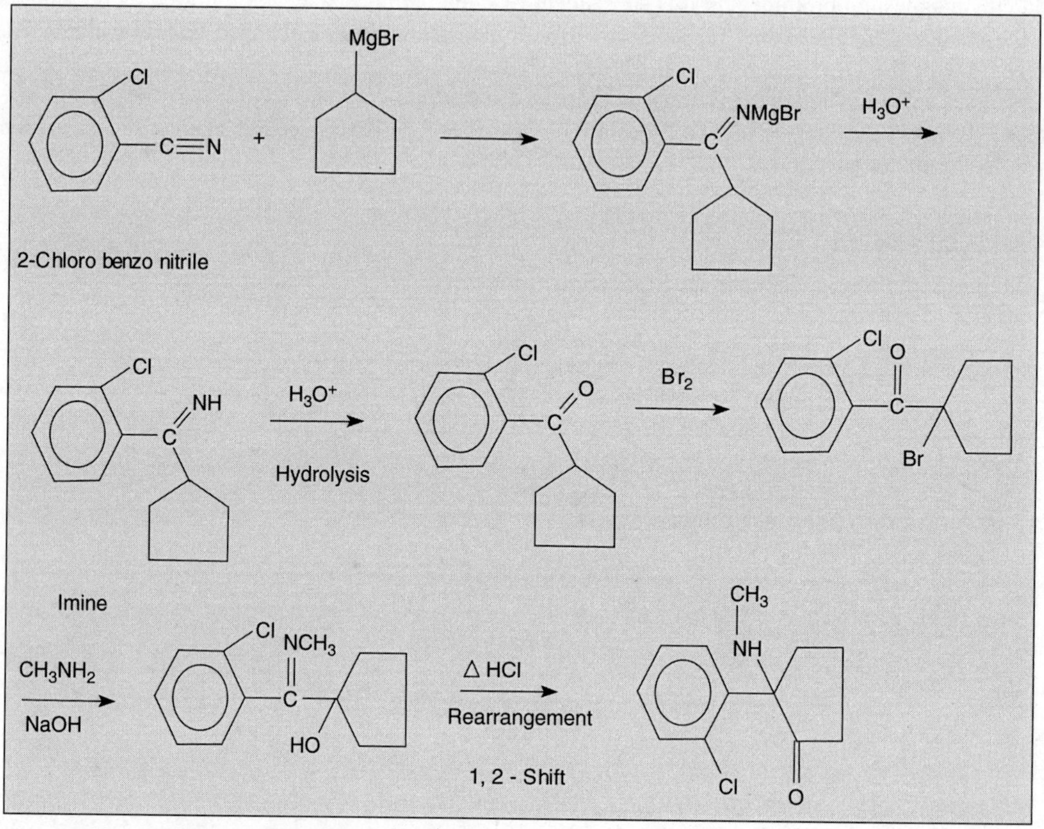

Ketamine is synthesised from 2-chlorobenzonitrile, which reacts with cyclopentylmagnesium bromide to give 1-(2-chlorobenzoyl)cyclopentane through two step intermediate. The next step is bromination, using bromine to the corresponding bromoketone, which upon interaction with an aqueous solution of methylimino forms the methylimino derivative. During this reaction, a simultaneous hydrolysis of the tertiary bromine atom occurs. On further heating the reaction product in decaline, a ring expansion rearrangement occurs, causing formation of ketamine.

Uses: A rapidly acting non-barbiturate general anaesthetics that produce anaesthesia characterized by profound analgesia. Intravenous doses (2 mg/Kg) produce surgical anaesthesia within 30 seconds and lasts about 10 minutes; intramuscularly (9–13 mg/Kg) produce surgical anaesthesia in 3–4 minutes and last from 12–25 minutes. The clinical anaesthetic state induced by ketamine is termed 'dissociative anaesthesia' since the patient may appear awake but is dissociated from the environment and does not respond to pain. Adverse reactions include elevated blood pressure and pulse rate.

5.5.3 Benzodiazepine

Midazolam

8-Chloro-6 (2'- fluorophenyl) - 2-methyl - imidazo benzodiazepine

Medicinal Chemistry 103

Synthesis

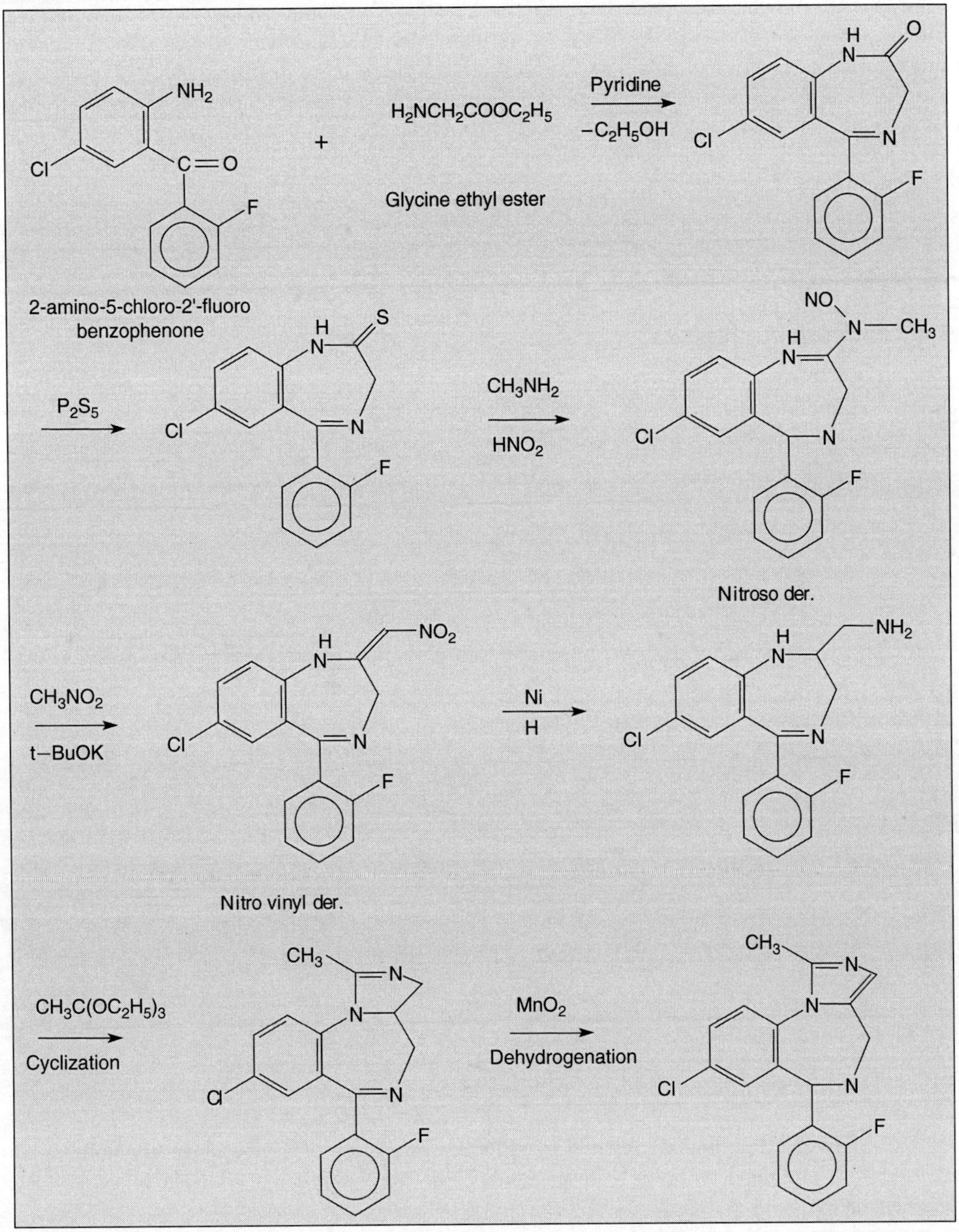

[Scheme: Synthesis of Midazolam]

Step 1: 2-amino-5-chloro-2'-fluoro benzophenone + H₂NCH₂COOC₂H₅ (Glycine ethyl ester), Pyridine, −C₂H₅OH → benzodiazepinone

Step 2: P₂S₅ → thione derivative

Step 3: CH₃NH₂ / HNO₂ → Nitroso der.

Step 4: CH₃NO₂, t-BuOK → Nitro vinyl der.

Step 5: Ni / H → amine derivative

Step 6: CH₃C(OC₂H₅)₃, Cyclization → imidazo-fused intermediate

Step 7: MnO₂, Dehydrogenation → Midazolam

Midazolam is prepared from 2-amino-5-chloro-2'-fluoro benzophenone, which undergoes cyclization with ethyl ester of glycine in presence of pyridine to form benzodiazepinone. Amide is converted to thioamide (which is much reactive) by treatment with phosphorouspentasulphide. Reaction of the thioamide with methylamine proceeds to give the amidine; this compound is transformed into a good leaving group by conversion to the N-nitroso derivative by treatment with nitrous acid. Condensation of this intermediate with the carbanion from nitro methane leads to displacement of N-nitroso group by methyl nitro derivative; the double bond shifts into conjugation with the nitro group to afford nitro vinyl derivative. Reduction with Raney nickel followed by reaction with methyl orthoacetate leads to fused imidazoline ring. Dehydrogenation with manganese dioxide converts it into an imidazole to give midazolam.

Uses: Midazolam has been used adjunctively with gaseous anaesthetics. The onset of its CNS effects is slower than that of thiopental, and it has a longer duration of action. Cases of severe post-operative respiratory depression have occurred.

5.5.4 Narcotic Analgesics

Fentanyl and alfentanil are used with other CNS depressants (nitrous oxide, benzodiazepines) in certain high-risk patients who may not survive for a full general anaesthetic.

Fentanyl citrate

N- (1- Phenylethyl - 4- piperidinyl) propionanilide citrate
Synthesis

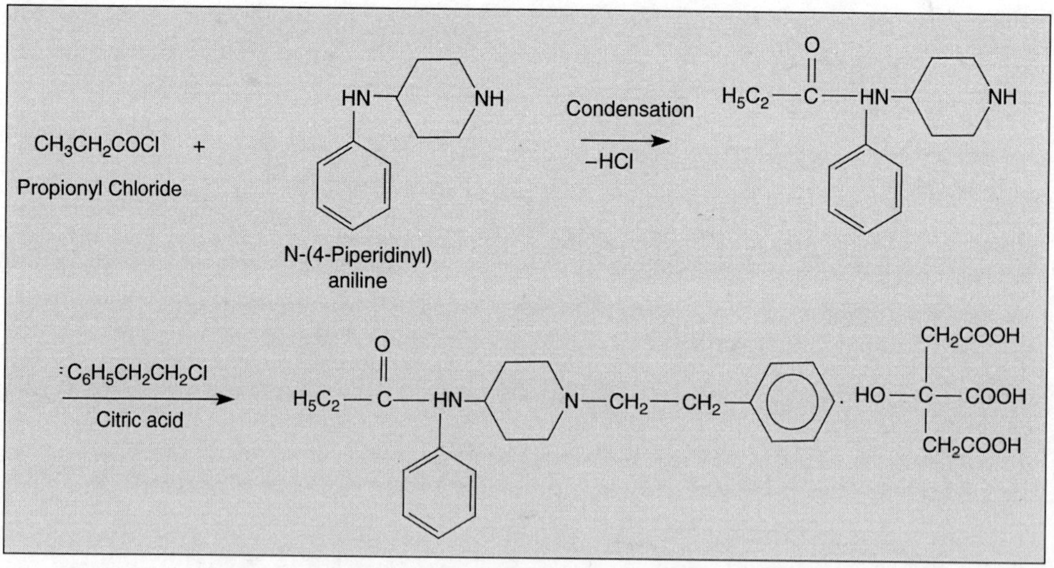

N-(4-Piperidinyl) aniline is prepared by reductive amination of 4-piperidone and aniline, which condensed with propionyl chloride to form amide. This on N-alkylation with phenyl ethyl chloride affords fentanyl.

Use: Fentanyl citrate relieves moderate to severe breakthrough pain.
Dose: For prompt analgesia during induction, I.V., 0.05 mg–0.1 mg, repeated at 2 to 3 minute intervals until desired effect is achieved; for maintenance I.V., 0.025 mg to 0.05 mg.

Alfentanil

N- [1- [2- (4- Ethyl- 4,5- dihydro - 5- oxo - 1H- tetrazol - 1-yl) ethyl] –4-(methoxymethyl)-4- piperidinyl] - N- phenyl propionamide

Synthesis

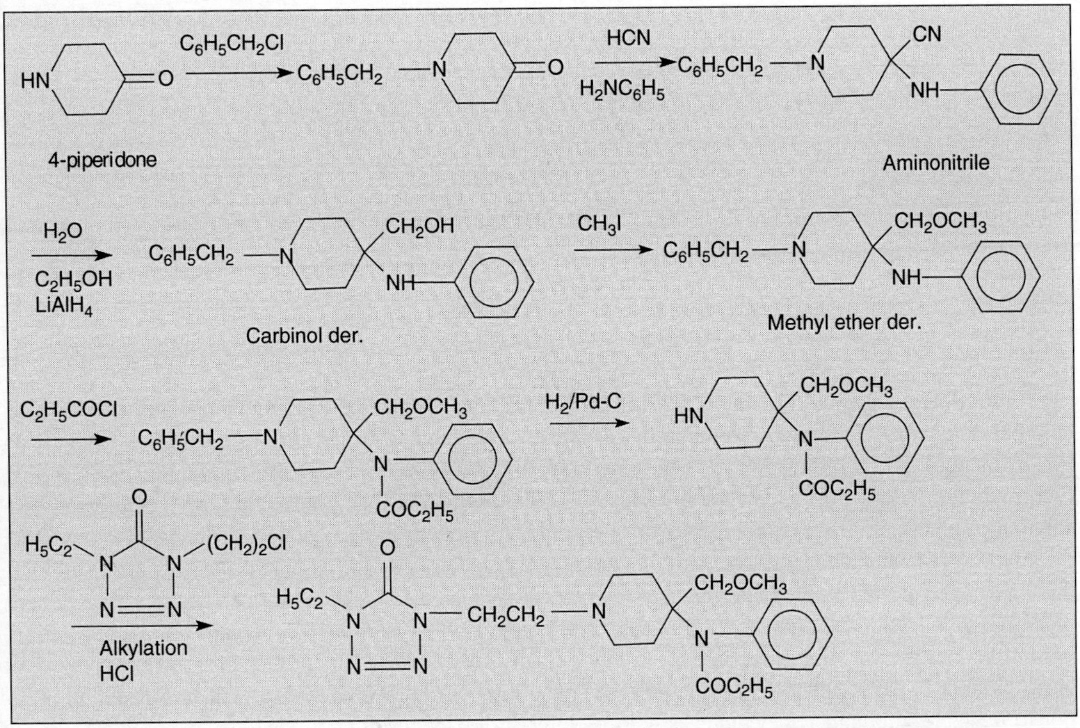

Alfentanil is synthesised from 1-benzylpiperidine-4-one by means of condensation with aniline in the presence of hydrogen cyanide. The resulting 4-anilino-4-cyano-1-benzylpiperidine undergoes ethanolysis, forming 4-anilino-4-carboethoxy-1-benzylpiperidine, which is reduced by lithium aluminium hydride into 4-anilino-4-hydroxymethyl-1-benzylpiperidine, which is methylated by methyl iodide to give 4-anilino-4-methoxymethyl-1-benzylpiperidine. The resulting product is acylated using propionyl chloride to give 1-benzyl-4-methoxymethyl-4-N-propionyl-anilinopiperidine, which undergoes debenzylation by hydrogen using a palladium on carbon catalyst to give 4-methoxymethyl-4-N-propionylanilinopiperidine, which on reaction with 1-(4-ethyl-4,5-dihydro-5-oxy-1H-tetrazol-1-yl)ethyl-2-chloride gives alfentanil.

Use: It relieves moderate to severe breakthrough pain.
Dose: For 30 minutes anaesthesia, induction: 8–20 µg/Kg; maintenance: 3–5 µg/Kg.

5.5.5 Miscellaneous Drugs

Etomidate

Ethyl - 1- (α- methylbenzyl) imidazole – 5-carboxylate

Synthesis

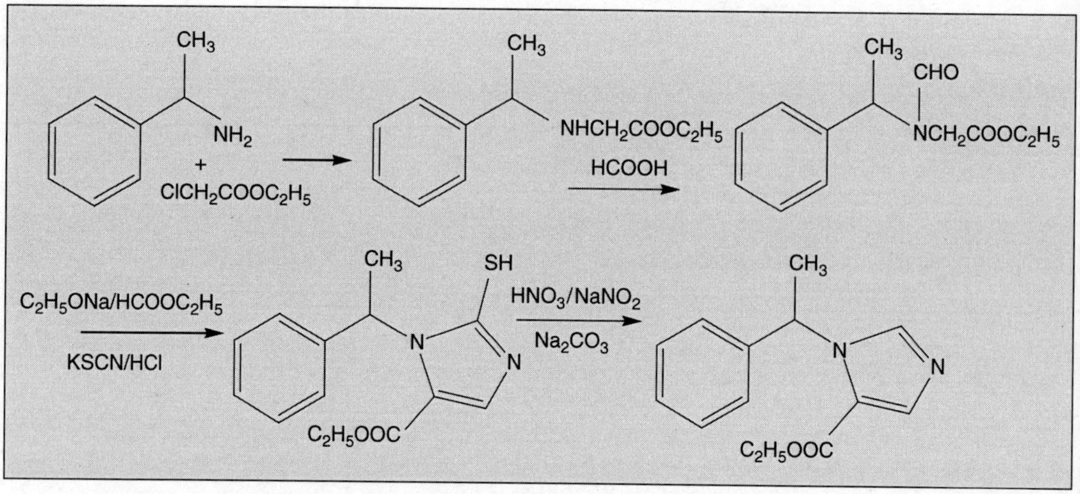

Etomidate is prepared by the following procedure. The reaction of α-methylbenzylamine with ethyl chloroacetate gives *N*-ethoxycarbonylmethyl-*N*-1-phenylethylamine, which undergoes further formylation by formic acid. The resulting *N*-ethoxycarbonylmethyl-*N*-formyl-*N*-1-phenylethylamine undergoes further C-formylation by ethylformate in the presence of sodium ethoxide. The product is further processed by a solution of potassium thiocyanate in hydrochloric acid. As a result of the reaction of thiocyanate ions with the amino group which occurs as a result of acidic hydrolysis of the *N*-formamide protecting group and further interaction of the obtained intermediate with the newly inserted aldehyde group, a Marckwald reaction-type heterocyclization takes place, resulting in formation of 5-ethoxycarbonyl-2-mercapto-1-(1-phenylethyl) imidazole. Finally, the thiol group is removed by oxidative dethionation upon interaction with a mixture of nitric and nitrous acids (nitric acid in the presence of sodium nitrite), which evidently occurs through formation of unstable sulphinic acid, which easily loses sulphur dioxide resulting in the desired etomidate.

Uses: Intravenous etomidate (0.2–0.6 mg/Kg) produces a rapid induction of anaesthesia with minimal cardiovascular and respiratory changes and without analgesic activity.

Propofol
2,6- Disopropyl phenol
Synthesis

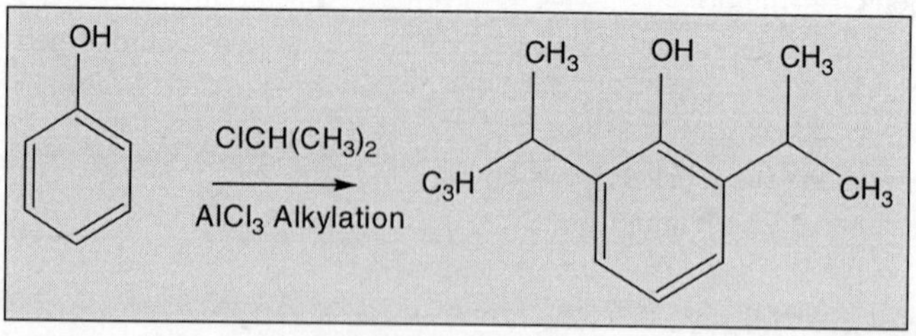

Propofol is prepared by bis-alkylating phenol with isopropyl chloride in the presence of Lewis acid.
Uses: Propofol is similar to the intravenous barbiturates in its rate of onset and duration of anaesthesia. It is used as an induction agent and for short anaesthetic procedures.

5.6 NEWER DRUGS

Desflurane
2-(Difluoromethoxy)-1,1,1,2-tetrafluoro-ethane

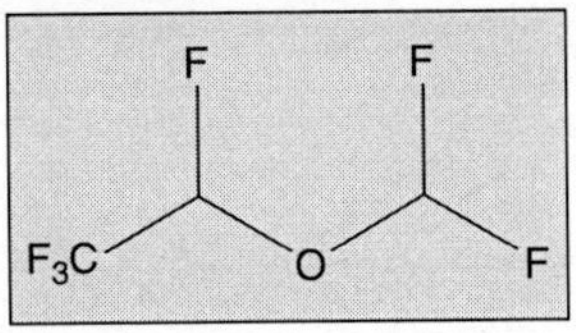

It is a highly fluorinated methyl ethyl ether used for maintenance of general anaesthesia. Together with sevoflurane, it is gradually replacing isoflurane for human use. It has the most rapid onset and offset of the volatile anaesthetic drugs used for general anaesthesia due to its low solubility in blood.

Sevoflurane
2,2,2-trifluoro-1-[trifluoromethyl]ethyl fluoromethyl ether

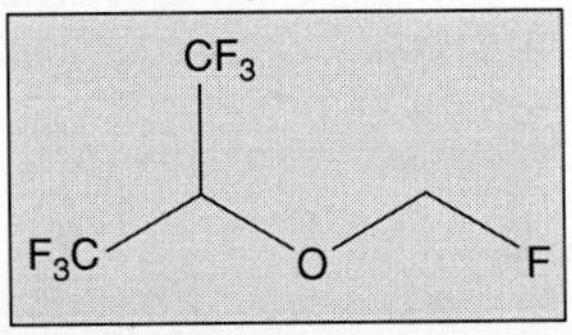

It is a sweet-smelling, non-flammable, highly fluorinated methyl isopropyl ether used for induction and maintenance of general anaesthesia. Together with desflurane, it is replacing isoflurane and halothane in modern anaesthesiology. It is often administered in a mixture of nitrous oxide and oxygen. Although desflurane has the lowest blood/gas coefficient of the currently used volatile anaesthetics, sevoflurane is the preferred agent for mask induction due to its lesser irritation to mucous membranes.

Minaxolone
1-(11-(dimethylamino)-2-ethoxy-hexadecahydro-3-hydroxy-10,13-dimethyl-1H-cyclopenta[a]phenanthren-17-yl)ethanone

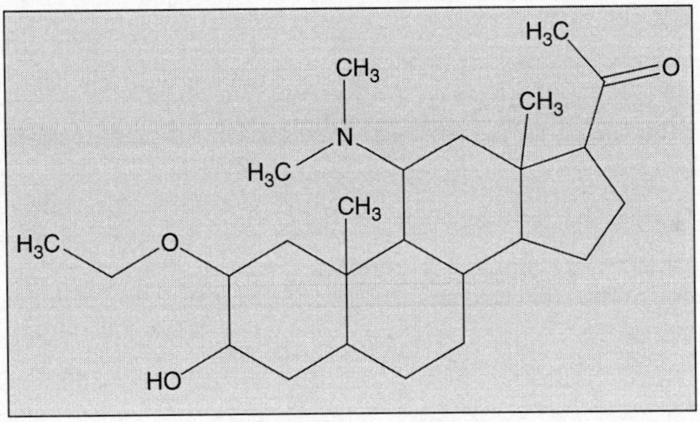

It is a new water-soluble steroid anaesthetic, and it appears to be a safe and effective intravenous anaesthetic with impressive recovery characteristics. Its only drawback would seem to be its high incidence of excitatory movements and hypertonus. It appears to be a promising intravenous anaesthetic agent worthy of further clinical investigation.

FURTHER READINGS

Burger's Medicinal Chemistry and Drug Discovery, Abraham, D.J. (ed.), 6th Ed., 2003.

MULTIPLE-CHOICE QUESTIONS

1. Which of the following agents has the slowest induction time?
   a. Nitrous oxide
   b. Enflurane
   c. Halothane
   d. Methoxyflurane
2. The major route of elimination of the volatile general anaesthetics is via
   a. kidneys
   b. skin
   c. lungs
   d. liver
3. The ideal general anaesthetic agent has
   a. high blood solubility
   b. a rapid recovery period
   c. long duration of action
   d. all of the above
4. Spinal anaesthesia is a form of general anaesthesia
   a. True
   b. False
5. The stage of general anaesthesia in which the patient may move about and mumble incoherently is stage _____.
   a. I
   b. II
   c. III
   d. IV
6. All of the following are characteristics of an ideal general anaesthetic except
   a. They should require uncomplicated equipment to administer.
   b. They should have a wide margin of safety.
   c. They should rapidly metabolize to avoid cumulative toxicity.
   d. They should provide rapid and uncomplicated induction and emergence.
7. Methohexital
   a. is a short-acting benzodiazepine, which can be reversed with flumazanil
   b. causes hypnosis
   c. has unpredictable absorption when given by the rectal route
   d. has no effect on blood pressure.

QUESTIONS

1. Write in detail on the different stages of anaesthesia.
2. Classify general anaesthetics with two examples for each class.
3. When are narcotic analgesics used along with anaesthetics, and how do they act?
4. Differentiate between topical and general anaesthetics.
5. Give the complete synthetic protocol for ketamine and etomidate.
6. Mention the structural requirements of anaesthetic barbiturates.
7. Why is nitrous oxide also called 'laughing gas'?
8. What do you mean by dissociative anaesthesia?
9. Compare the chemical and biological properties of phenobarbitone and thiopentone.

SOLUTION TO MULTIPLE-CHOICE QUESTIONS

1. d;
2. c;
3. b;
4. b;
5. a;
6. c;
7. b.

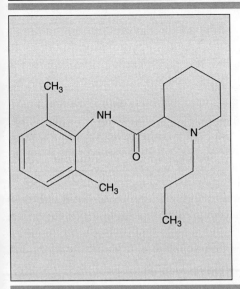

CHAPTER 6

Local Anaesthetics

LEARNING OBJECTIVES

- Compare general anaesthetics and local anaesthetics
- List the ideal properties of a local anaesthetics
- Explain the overall mechanisms of action of anaesthetics
- Describe the clinical utility of local anaesthetics
- Categorise the various modes of administration of local anaesthetics
- Learn the discovery pathway of present-day local anaesthetics that originated from natural drug cocaine
- List the various classes of local anaesthetics and their chemical structures
- Describe the synthesis, uses, and doses of benzoic acid class of local anaesthetics
- Describe the mechanism of action of benzocaine and how it is different from other anaesthetics
- Describe the relationship between structure of benzoic acid derivatives and their biological activity
- Learn which substitution is useful in improving the local anaesthetic property of benzoic acid derivatives
- Describe the most stable class of local anaesthetics
- Describe the relationship between structure of anilides and their biological activity
- List other classes of anaesthetics that do not fall in the conventional class
- Describe some latest drugs in the class of local anaesthetics

6.1 INTRODUCTION

Local anaesthetics are medications used for the purpose of temporary and reversible elimination of painful feelings in specific areas of the body by blocking transmission of nerve fibre impulses. These drugs, unlike general anaesthetics, cause a loss of feeling in specific areas while keeping the patient conscious.

Local anaesthetics are used for pain relief, soreness, itching, and irritation associated with disturbance of the integrity of the skin and mucous membranes (cuts, bites, wounds, rashes, allergic conditions, fungal infections, skin sores, and cracking). They are used during opthalmological procedures such as tonometry and gonioscopy, removal of foreign bodies, and minor surgical interventions. Local anaesthetics are widely used in surgery, gynaecology, and dentistry. In certain cases, local anaesthetics (lidocaine, procainamide) can be used as anti-arrhythmic drugs.

6.2 IDEAL PROPERTIES OF LOCAL ANAESTHETICS

1. Non-irritating to tissues and not causing any permanent damage
2. Low systemic toxicity
3. Effective whether injected into the tissue or applied locally to skin or mucous membranes
4. Rapid onset of anaesthesia and short duration of action

6.3 MECHANISM OF ACTION

A mechanism of local anaesthetic action in which they serve as **sodium channel blockers** has been proposed. According to this mechanism, the molecular targets of local anaesthetic action are the voltage-requiring sodium channels, which are present in all the neurons. The process of local anaesthesia by respective drugs can be represented in the following manner.

In a resting condition, there is a specific rest potential between the axoplasm and the inner parts of the cell. This rest potential is maintained by relative concentration of sodium and potassium ions along the membrane of the nerve. During nerve stimulation, the membrane is depolarised and sodium channels in that area are opened, allowing sodium ions to rush into the cell. At the peak of depolarization, potassium channels are opened. The last ones leave the cell and the cell is repolarised.

This process lasts 1–2 msec, after which the nerve cell, having transmitted the necessary impulse, restores its ion gradient.

It is believed that after introduction of local anaesthetic into the organism in the form of a water-soluble salt, equilibrium is established between the neutral and cationic forms of the used drug depending on the pK_a of the drug and the pH of the interstitial fluid. It is also believed that only the uncharged (neutral) drug form can pass through – it passes through connective tissue surrounding the nerve fibre and through the phospholipid plasma membrane into the axoplasm. In the axoplasm, the base is once again ionised until it reaches an appropriate value determined by intracellular pH.

It is suspected that **these drugs selectively bind with the intracellular surface of sodium channels and block the entrance of sodium ions into the cell. This leads to stoppage of the depolarization process**, which is necessary for the diffusion of action potentials, elevation of the threshold of electric nerve stimulation, and thus the elimination of pain. Since the binding process of anaesthetics to ion channels is reversible, the drug diffuses into the vascular system where it is metabolised, and nerve cell function is completely restored.

6.4 CLINICAL USES OF LOCAL ANAESTHETICS

Local anaesthesia is the loss of sensation in the body part without the loss of consciousness or the impairment of central control of vital functions, used for minor surgical procedures.

Local anaesthetics are categorised by the method of administration.

1. **Topical anaesthesia:** Anaesthesia of mucous membranes of the nose, throat, tracheobronchial tree, oesophagus, and genito-urinary tract can be produced by direct application of aqueous solutions of salts of many local anaesthetics or by suspension of the poorly soluble local anaesthetics. For prolonged duration of action, vasoconstriction can be achieved by the addition of a low concentration of a vasoconstrictor such as phenylephrine (0.005%): e.g., lignocaine (2–10%), cocaine (1–4%), and tetracaine (2%).
2. **Infiltration anaesthesia:** In this, local anaesthetics are injected directly into the tissue, which may be superficial tissue of the skin or deeper structures including intra-abdominal organs. The duration of infiltration anaesthesia can be doubled by the addition of epinephrine (5 µg/ml) to the injection solution: e.g., lignocaine (0.5–1.0%), procaine (0.5–1.0%), and bupivacaine (0.125–0.25%).
3. **Field block anaesthesia:** It is produced by subcutaneous injection of a local anaesthetics solution in such a manner as to anaesthetise the region distal to the injection. The advantage of field block anaesthesia is that less drug can be used to provide a greater area of anaesthesia than when infiltration anaesthesia is used: e.g., lignocaine (0.5–1.0%), procaine (0.5–1.0%), and bupivacaine (0.125–0.25%).
4. **Nerve block anaesthesia:** This involves injection of a solution of local anaesthetics into or about individual peripheral nerves or nerve plexus: e.g., lignocaine (1.0–1.5%), mepivacaine (up to 7mg/kg of 1.0–2.0%), and bupivacaine (2–3 mg/kg of 0.25–0.375%). Addition of 5 µg/ml epinephrine prolongs duration.
5. **Intravenous regional anaesthesia:** This technique relies on using the vasculature to bring the local anaesthetics solution to the nerve trunks and endings. It is used most often for the forearm and hand, but can also be adapted for the foot and distal leg: e.g., lignocaine (0.5%) and procaine (0.5%). (*See Fig. 6.1 in the coloured set of pages.*)
6. **Spinal anaesthesia:** It follows the injection of local anaesthetics into the cerebrospinal fluid in the lumbar space: e.g., lignocaine, tetracaine, and bupivacaine.
7. **Epidural anaesthesia**: In this, local anaesthetics is injected into epidural space – the space bounded by the ligamentum flavum posteriorly, the spina periosteum laterally, and the dura anteriorly: e.g., bupivacaine (0.5–0.75%), etidocaine (1.0–1.5%), lignocaine (2%), and chloroprocaine (2–3%). (*See Fig. 6.2 in the coloured set of pages.*)

6.5 CLASSIFICATION OF LOCAL ANAESTHETICS

1. Benzoic acid derivatives

Chapter 6: Local Anaesthetics

	R	R'
Cocaine	H	H₃COOC group with H₃C-N tropane ring (methyl ester of ecgonine tropane)
Hexylcaine	H	H₃C-CH(CH₃)-CH₂-NH-cyclohexyl
Meprylcaine	H	—H₂C—C(CH₃)₂—NH—CH₂CH₃
Isobucaine	H	—H₂C—C(CH₃)₂—NH—CH(CH₃)₂
Cyclomethycaine	cyclohexyl-O—	—H₂CH₂CH₂C—N(2-methylpiperidine)
Piperocaine	H	—H₂CH₂CH₂C—N(2-methylpiperidine)

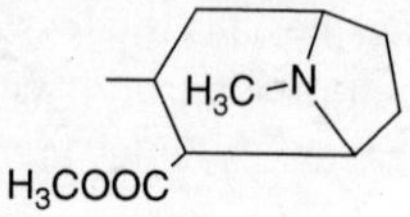

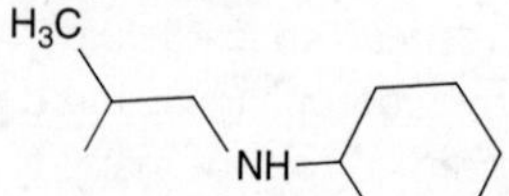

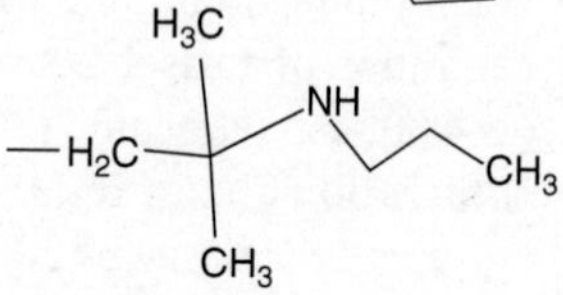

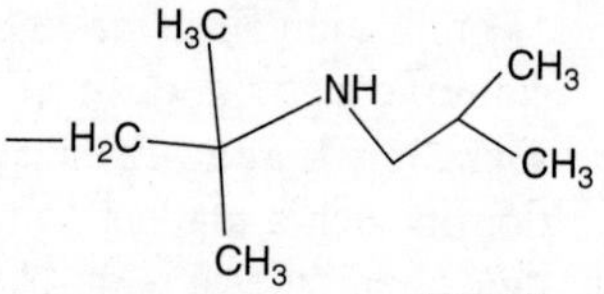

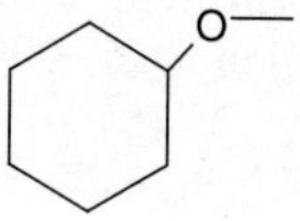

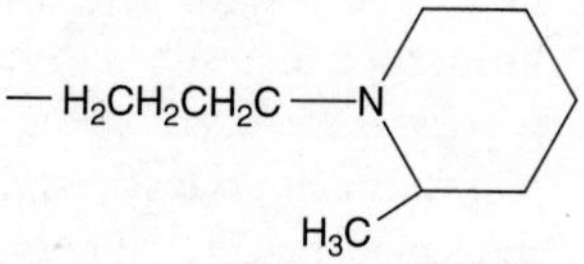

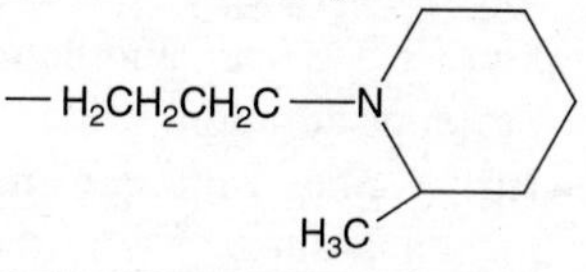

2. *p*-Aminobenzoic acid derivatives

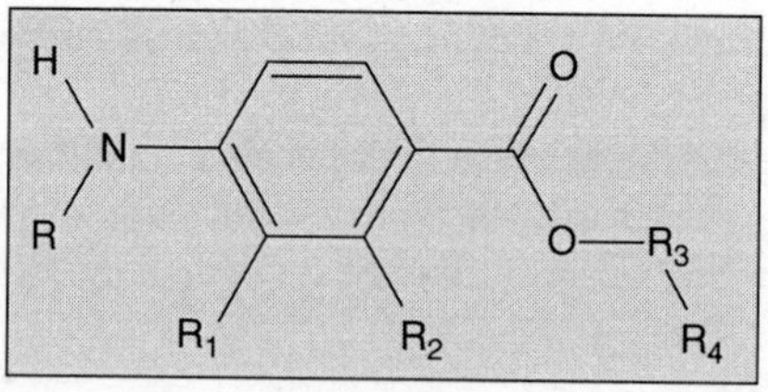

Core structure: R-NH-(phenyl with R₁, R₂ substituents)-C(=O)-O-R₃-R₄

	R	R₁	R₂	R₃	R₄
Benzocaine	H	H	H	—CH₂CH₃	—
Butamben	H	H	H	—(CH₂)₃CH₃	—
Procaine	H	H	H	—CH₂CH₂—	—N(C₂H₅)₂
Chloroprocaine	H	H	Cl	—CH₂CH₂—	—N(C₂H₅)₂
Tetracaine	Butyl	H	H	—CH₂CH₂—	—N(C₂H₅)₂
Butacaine	H	H	H	—CH₂CH₂—	—N(C₄H₉(n))₂
Benoxinate	H	H	butoxy	—CH₂CH₂—	—N(C₂H₅)₂
Propoxycaine	H	H	propyloxy	—CH₂CH₂—	—N(C₂H₅)₂

3. Anilides

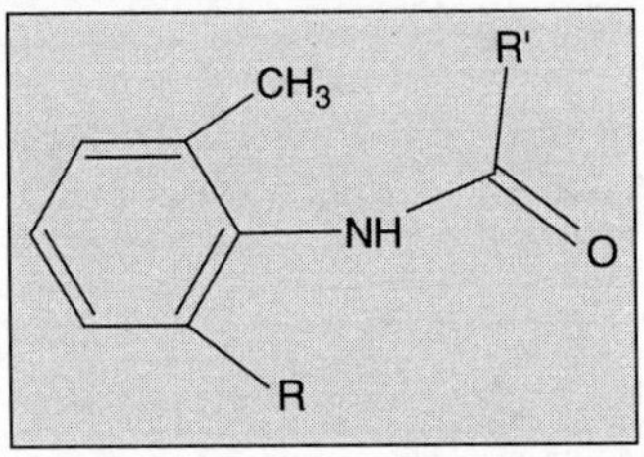

	R	R'
Lignocaine	CH₃	—CH₂N(C₂H₅)₂
Mepivacaine	CH₃	N-methyl-2-methylpiperidinyl
Bupivacaine	CH₃	N-butyl-2-methylpiperidinyl
Etidocaine	CH₃	—HC(C₂H₅)—N(C₂H₅)(C₃H₇)
Prilocaine	H	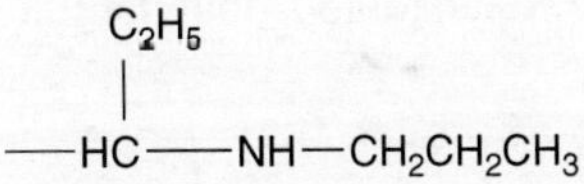 —HC(C₂H₅)—NH—CH₂CH₂CH₃

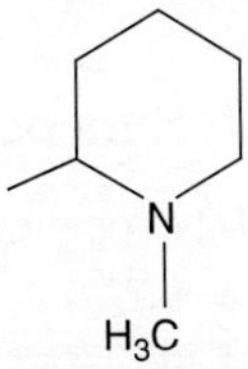

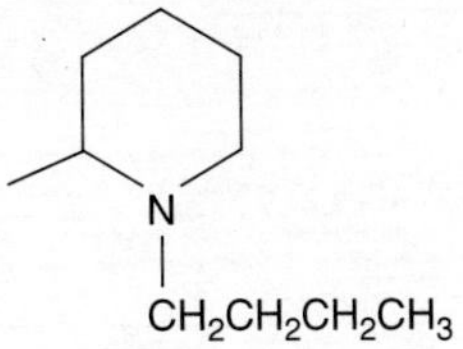

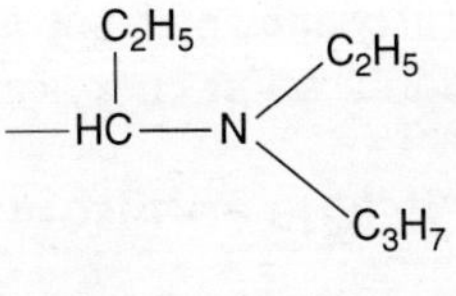

4. Miscellaneous: Phenacaine, Diperodon, Dimethisoquin, Pramoxine, Dyclonine, Dibucaine
5. Newer drugs: Ropivacaine, Levobupivacaine

6.6 STRUCTURES, SYNTHESIS, AND STRUCTURE-ACTIVITY RELATIONSHIP (SAR) OF BENZOIC ACID DERIVATIVES

6.6.1 Cocaine

(-)3-(Benzoyloxy)-8-methyl-8-azabicyclooctane-2-carboxylic acid methyl ester

Synthesis

Commercial production involves total extraction of cocaine and related alkaloidal bases from the leaves of *Erythroxylon coca*, followed by acid hydrolysis of the ester alkaloid to obtain the total content of (−) ecgonine. After purification of ecgonine, cocaine is synthesised by esterification with methanol and benzoic acid.

6.6.2 Hexylcaine Hydrochloride

1-(Cyclohexylamino)-2-propanol benzoate
Synthesis

Reductive alkylation of amino alcohol with cyclohexanone affords the secondary amine. Acylation with benzoyl chloride affords hexylcaine.

6.6.3 Meprylcaine and Isobucaine Hydrochloride

2-Methyl-2-(propylamino/isobutylamino)-1-propanol benzoate
Synthesis

Meprylcaine - $CH(CH_3)_2$
Isobucaine - $CH_2CH(CH_3)_2$

(Acid catalyzed N- to O- acyl migration)

Reductive alkylation of amino alcohol with acetone/isobutyraldehyde affords the corresponding amines. Acylation of the amine with benzoyl chloride probably goes initially to the amide. The acid catalysis used in the reaction leads to an N- to O-acyl migration to afford corresponding compounds.

6.6.4 Cyclomethycaine Sulphate

3-(2-Methylpiperidino)propyl-p-(cyclohexyloxy) benzoate sulphate
Synthesis

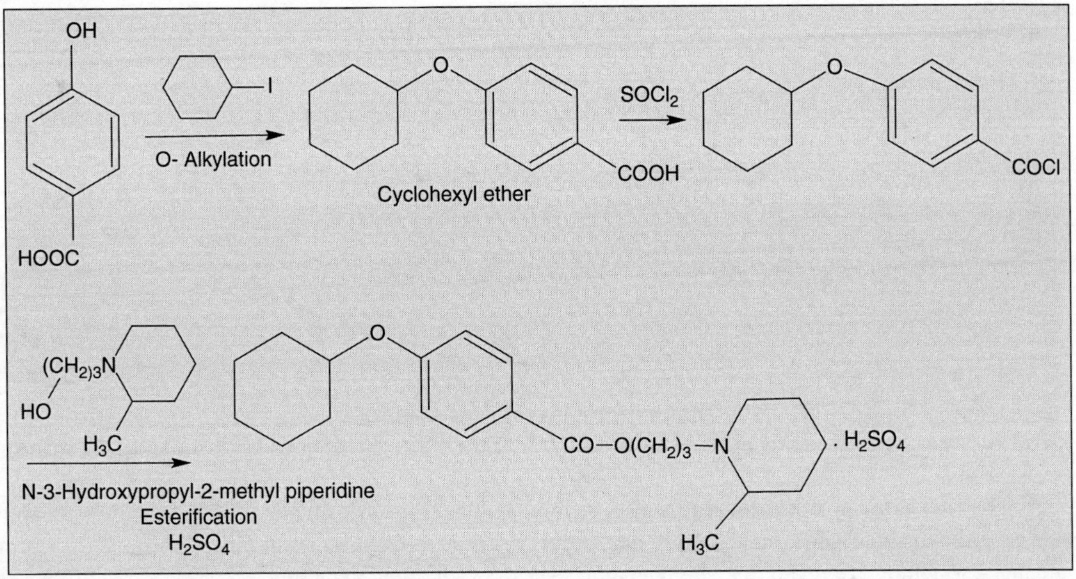

O-alkylation of *p*-hydroxy benzoic acid with cyclohexyl iodide affords the cyclohexyl ether under alkaline reaction condition. The acid group is activated with thionyl chloride and the acylation of the acid chloride with alcohol affords cyclomethycaine.

6.6.5 Piperocaine Hydrochloride

3-(2-Methylpiperidino)propyl benzoate
Synthesis

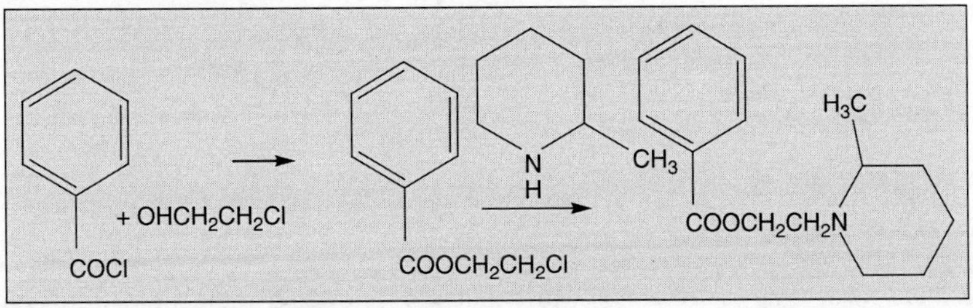

Benzoylation of 3-chloropropan-1-ol affords the haloester. Condensation of this intermediate with the reduction product of α-picoline affords piperocaine.

6.7 STRUCTURES, SYNTHESIS, AND SAR OF P-AMINOBENZOIC ACID DERIVATIVES

6.7.1 Benzocaine

Ethyl-*p*-aminobenzoate
Synthesis

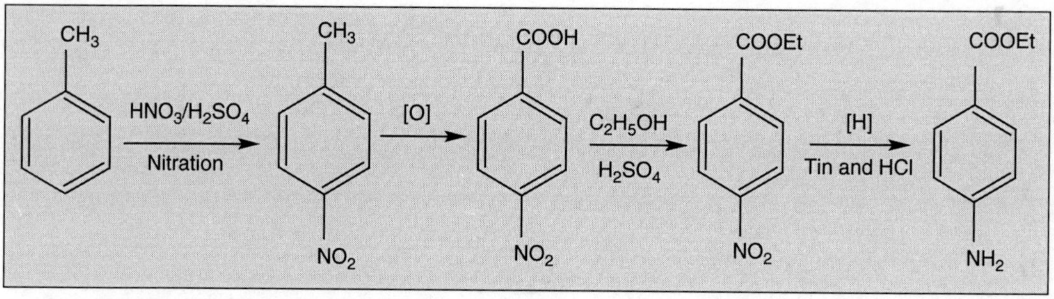

Toluene on nitration with nitrating mixture affords 4-nitro toluene, which on oxidation followed by esterification yields ethyl ester derivative. Nitro group is reduced with tin and HCl affords benzocaine.

The **mechanism of benzocaine** action differs slightly from that mentioned above. It presumably acts by diffusing across the phospholipid membrane and then stretching it out. This deforms the sodium channels, which in turn—and in a unique manner—lowers sodium conduction.

6.7.2 Butamben

Butyl-*p*-aminobenzoate
Synthesis

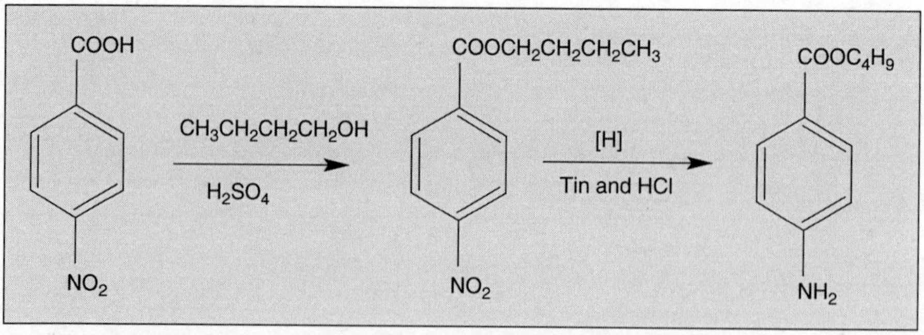

4-Nitro benzoic acid on esterification with *n*-butanol followed by reduction affords butamben.

6.7.3 Procaine Hydrochloride

2-(Diethylamino)ethyl-*p*-aminobenzoate
Synthesis

HO—CH$_2$CH$_2$Cl + HN(C$_2$H$_5$)(C$_2$H$_5$) [Ethylene chlorohydrin + Diethylamine] $\xrightarrow{-HCl}$ HO—CH$_2$CH$_2$N(C$_2$H$_5$)$_2$ [2-(Diethylamino) ethanol] $\xrightarrow{O_2N-C_6H_4-COCl \text{ (p-Nitrobenzoyl-chloride)}}$ *p*-O$_2$N—C$_6$H$_4$—COOCH$_2$CH$_2$N(C$_2$H$_5$)$_2$ $\xrightarrow[\text{Tin and HCl}]{[H]}$ *p*-H$_2$N—C$_6$H$_4$—COOCH$_2$CH$_2$N(C$_2$H$_5$)$_2$

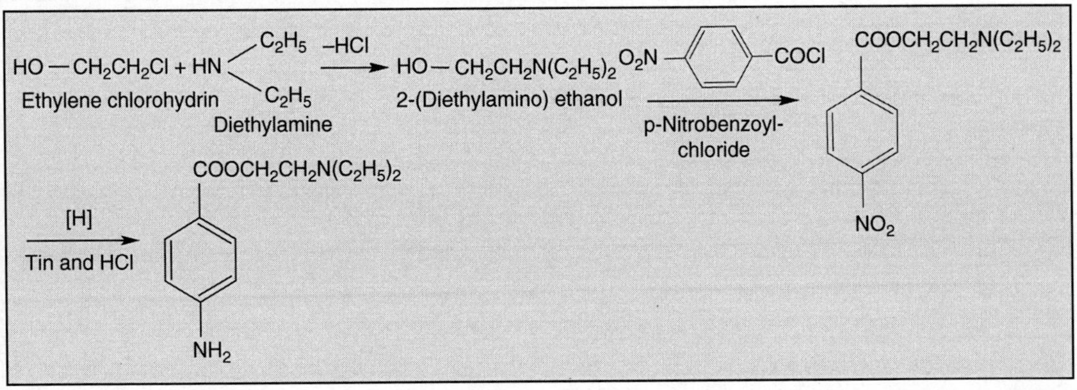

Ethylene on reaction with hypochlorous acid yields ethylene chlorohydrin, which on reaction with diethylamine affords 2-(diethylamino) ethanol. Benzoylation of alcohol followed by reduction affords procaine.

6.7.4 Chloroprocaine

Chloroprocaine differs structurally from procaine in having a chlorine substituent in the 2-position of the aromatic ring. The electron-withdrawing chlorine atom destabilizes the ester group to hydrolysis. Chloroprocaine is hydrolysed by plasma more than four times faster than procaine. It is more rapid in onset of action and more potent than procaine. Chloroprocaine is used in situations requiring fast-acting pain relief.

6.7.5 Tetracaine

2-(Dimethylamino)ethyl-*p*-(butylamino)benzoate
Synthesis

p-H$_2$N—C$_6$H$_4$—COOC$_2$H$_5$ $\xrightarrow[\text{N-Butylation}]{CH_3(CH_2)_3Br, \ Na_2CO_3}$ *p*-CH$_3$(CH$_2$)$_3$NH—C$_6$H$_4$—COOC$_2$H$_5$ $\xrightarrow[\text{Transesterification}]{HOCH_2CH_2N(C_2H_5)_2, \ C_2H_5ONa}$ *p*-CH$_3$(CH$_2$)$_3$NH—C$_6$H$_4$—COOCH$_2$CH$_2$N(C$_2$H$_5$)$_2$

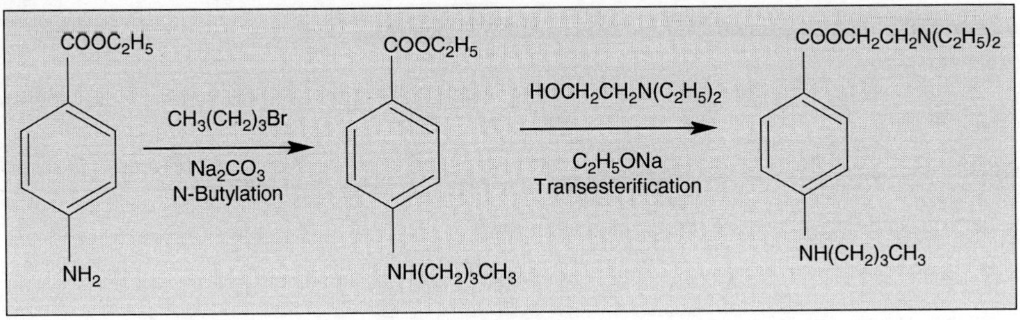

Benzocaine undergoes alkylation with butyl bromide in presence of base; this is followed by trans-esterification reaction with 2-(diethylamino) ethanol, which affords tetracaine.

6.7.6 Butacaine

3-(Di-n-butylamino)-1-propanol-p-amino benzoate
Synthesis
It is similar to that of procaine, but uses n-dibutylamine instead of diethylamine.

6.7.7 Benoxinate

2-(Diethylamino)ethyl-4-amino-2-butoxy benzoate
Synthesis

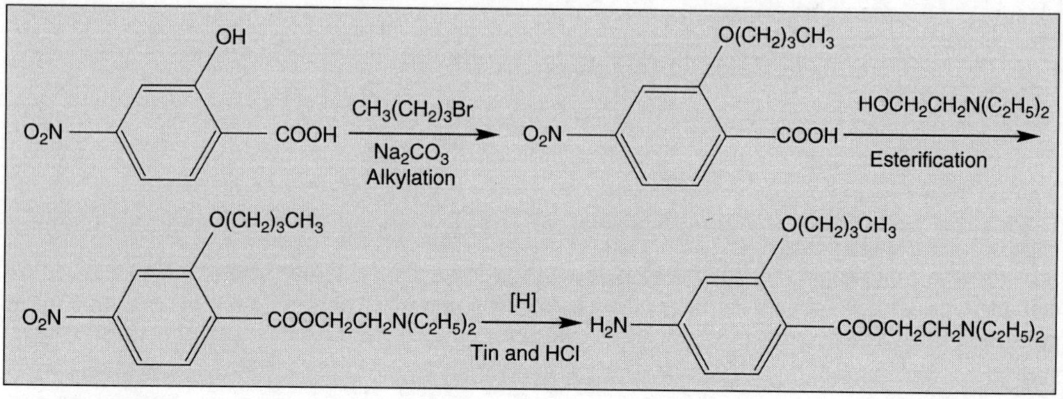

$$O_2N-C_6H_3(OH)-COOH \xrightarrow[\text{Alkylation}]{CH_3(CH_2)_3Br, Na_2CO_3} O_2N-C_6H_3(O(CH_2)_3CH_3)-COOH \xrightarrow[\text{Esterification}]{HOCH_2CH_2N(C_2H_5)_2}$$

$$O_2N-C_6H_3(O(CH_2)_3CH_3)-COOCH_2CH_2N(C_2H_5)_2 \xrightarrow[\text{Tin and HCl}]{[H]} H_2N-C_6H_3(O(CH_2)_3CH_3)-COOCH_2CH_2N(C_2H_5)_2$$

Benoxinate is prepared from 4-nitro salicylic acid by a three-step synthesis: first, by o-alkylation with n-butyl bromide; secondly, esterification with 2-(diethylamino) ethanol; and finally, reduction of nitro group affords benoxinate.

6.7.8 Propoxycaine

2-(Diethylamino)ethyl-4-amino-2-propoxy benzoate hydrochloride
Synthesis
It is similar to that of benoxinate, but alkylation is with n-propylbromide.

6.7.9 Proparacaine Hydrochloride

2-(Diethylamino)ethyl-3-amino-4-propoxy benzoate hydrochloride
Synthesis
It is similar to that of benoxinate, but starting material is 3-nitro salicylic acid and alkylation with propylbromide.

Structure-activity relationships (SAR) of benzoic acid derivatives

The benzoic acid derivatives are represented as follows:

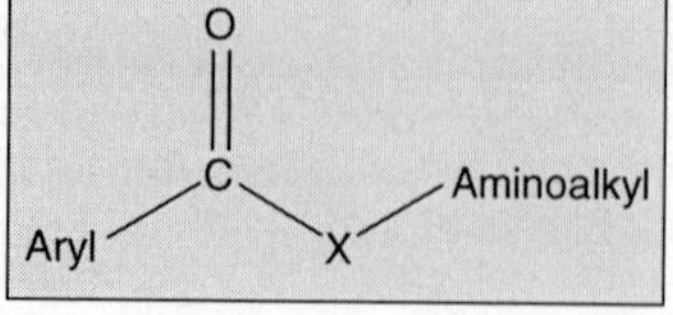

$$Aryl-C(=O)-X-Aminoalkyl$$

Aryl group

- The clinically useful local anaesthetics of this series possess an aryl radical attached directly to the carbonyl group.
- Substitution of aryl group with substituents that increase the electron density of the carbonyl oxygen enhances activity.
- Favourable substituents in aryl ring include (electron-donating groups) alkoxy (propoxycaine), amino (procaine), and alkylamino (tetracaine) groups in the *para* or *ortho* positions. This homologous series increases partition coefficients with increasing number of methylene group ($-CH_2-$). Local anaesthetics activity peaked with the C4-, C5-, or C6-homologous: e.g., tetracaine, cyclomethycaine.
- Aryl aliphatic radicals that contain a methylene group between the aryl radical and the carbonyl group result in compounds that have not found clinical use.

Bridge X

- The bridge X may be carbon, oxygen, nitrogen, or sulphur.
- In an isosteric procaine series, anaesthetic potency decreased in the following order: sulphur, oxygen, carbon, nitrogen.
- These modifications also affect duration of action and toxicity. In general, amides (X=N) are more resistant to metabolic hydrolysis than esters (X=O). Thioesters (X=S) may cause dermatitis.
- In procaine-like analogues, branching (especially at the alpha carbon) will increase duration of action. This effect is not seen in the lidocaine series.
- Increasing the chain length will increase potency but will also increase toxicity.

Aminoalkyl group

- The aminoalkyl group is not necessary for local anaesthetic activity, but it is used to form water-soluble salts (HCl salts).
- Tertiary amines result in more useful agents. The secondary amines appear to be of longer activity, but they are more irritating; primary amines are not very active and cause irritation.
- The tertiary amino group may be diethylamino, piperidine, or pyrrolidino, leading to the products that exhibit essentially the same degree of activity.
- The more hydrophilic morpholino group usually leads to diminished potency.
- Some analogues have no amino group at all, such as benzocaine. They are active but have poor water solubility.

6.8 STRUCTURES, SYNTHESIS, AND SAR OF ANILIDES

Agents of this class are more stable to hydrolysis. They are more potent, have a lower frequency of side-effects, and induce less irritation than benzoic acid derivatives.

6.8.1 Lidocaine Hydrochloride

Lignocaine, Xylocaine, 2-(Diethylamino)-N-(2,6-dimethylphenyl) acetamide

Synthesis

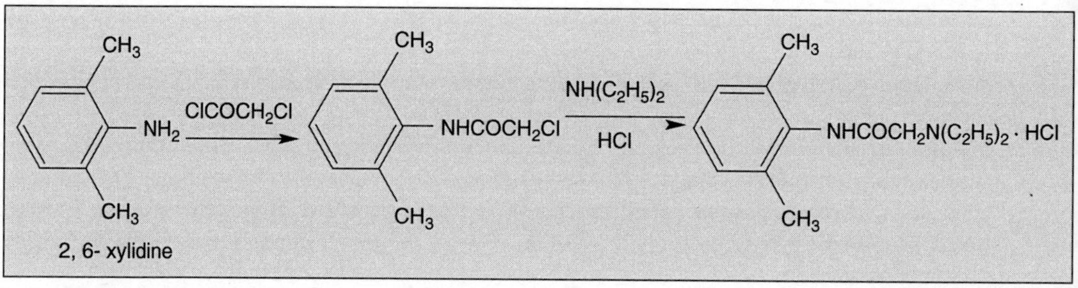

[Scheme: 2,6-xylidine + ClCOCH₂Cl → α-chloro-2,6-dimethylacetanilide (NHCOCH₂Cl), then + NH(C₂H₅)₂ / HCl → NHCOCH₂N(C₂H₅)₂ · HCl]

Lidocaine is synthesised from 2,6-dimethylaniline upon reaction with chloroacetyl chloride, which gives α-chloro-2,6-dimethylacetanilide, and its subsequent reaction with diethylamine affords lidocaine.

It is a widely employed amide-type local anaesthetics extremely resistant to metabolic hydrolysis. In addition to the relative stability of the amide bond, the 2,6-dimethyl substituents provide steric hindrance to attack of the carbonyl. Lidocaine is about twice as potent as procaine. It is also used as an effective cardiac depressant and may be administered intravenously in cardiac surgery and life-threatening arrhythmias.

6.8.2 Prilocaine Hydrochloride

N-(2-methylphenyl)-2-(propylamino) propanamide
Synthesis

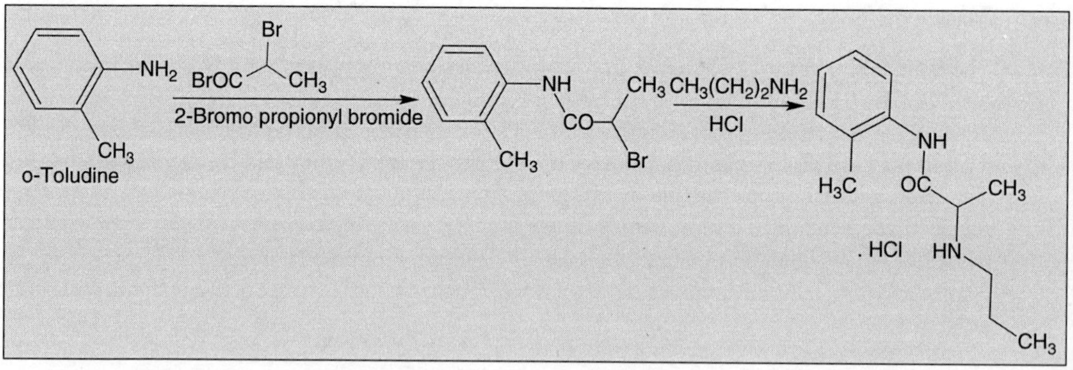

[Scheme: o-Toluidine + 2-Bromo propionyl bromide → intermediate (NHCO-CHBr-CH₃), then + CH₃(CH₂)₂NH₂ / HCl → Prilocaine · HCl]

Prilocaine is synthesised from 2-methylaniline upon reaction with 2-bromopropionyl chloride, followed by reaction with *n*-propylamine.

6.8.3 Mepivacaine and Bupivacaine Hydrochloride

N-(2,6-Dimethylphenyl)-1-methyl/ butyl-2-piperidine carboxamide

Synthesis

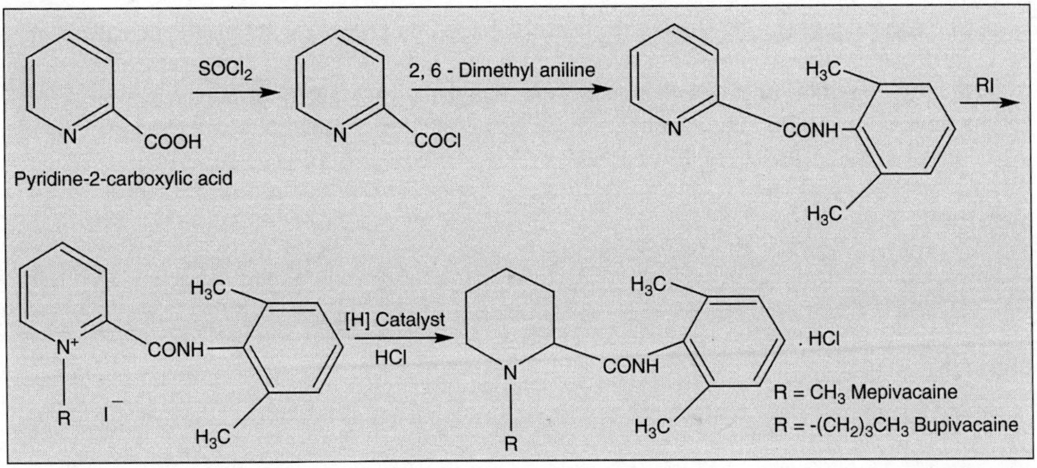

Reaction of 2,6-dimethylaniline with the acid chloride of pyridine carboxylic acid first gives the 2,6-xylidide of α-picolinic acid. The resulting 2,6-xylidide α-picolinic acid is alkylated with the corresponding alkyl halide, followed by reduction with hydrogen in the presence of platinum on carbon catalyst, and affords mepivacaine and bupivacaine.

6.8.4 Etidocaine Hydrochloride

2-(Ethylpropylamino)-2',6'-butyroxylidide
Synthesis

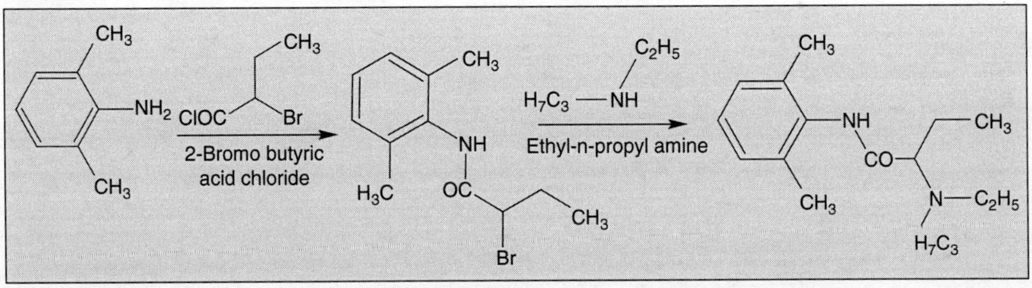

In the first stage of synthesis, 2,6-dimethylaniline is reacted with α-bromobutyric acid chloride to give the bromoanilide, which on amination with ethyl propyl amine affords etidocaine.

SAR of anilides

General structures of anilides are represented as follows:

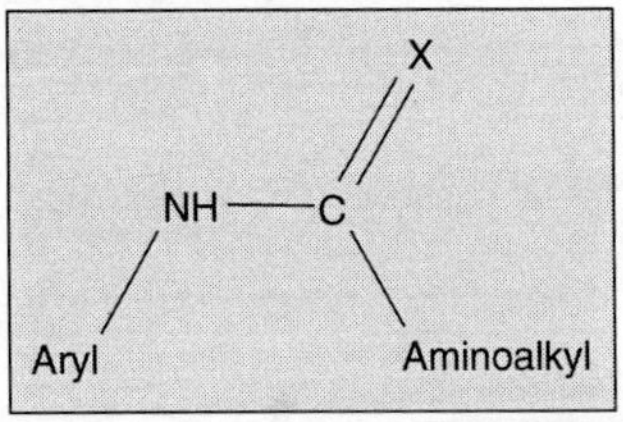

Aryl group

- The clinically useful local anaesthetics of this type possess a phenyl group attached to the sp^2 carbon atom through a nitrogen bridge.
- Substitution of the phenyl with a methyl group in the 2- or 2- and 6- position enhances activity. In addition, the methyl substituent(s) provide steric hindrance to hydrolysis of the amide bond and enhances the coefficient of distribution.

Substituent X

- X may be carbon, oxygen, or nitrogen. Among them Lidocaine series (X=O) has provided more useful products.

Aminoalkyl group

- The amino function has the capacity for salt formation and is considered the hydrophilic portion of the molecules.
- Tertiary amines (diethylamine, piperidines) are more useful because the primary and secondary amines are more irritating to tissue.

6.9　MISCELLANEOUS

6.9.1　Phenacaine Hydrochloride

N, N'-Bis (4-ethoxyphenyl)-ethanimidamide
Synthesis

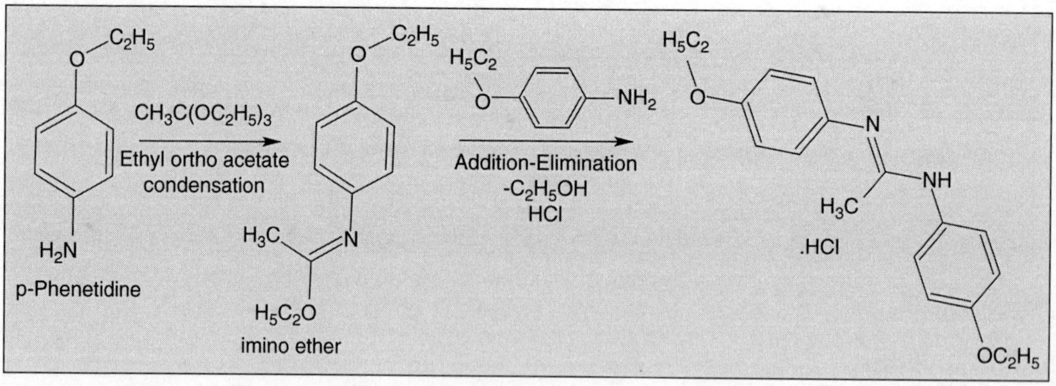

Condensation of 4-ethoxyaniline takes place with ethyl orthoacetate to afford the imino ether. Reaction of that intermediate with a second mole of 4-ethoxyaniline results in a net displacement of ethanol, probably by an addition-elimination scheme, and affords phenacaine.

Phenacaine is structurally related to the anilides in that an aromatic ring is attached to an sp^2 carbon through a nitrogen bridge. But it lacks the traditional ester or amide function and terminal aliphatic nitrogen.

6.9.2 Diperodon

3-Piperidino-1,2-propanediol dicarbanilate
Synthesis

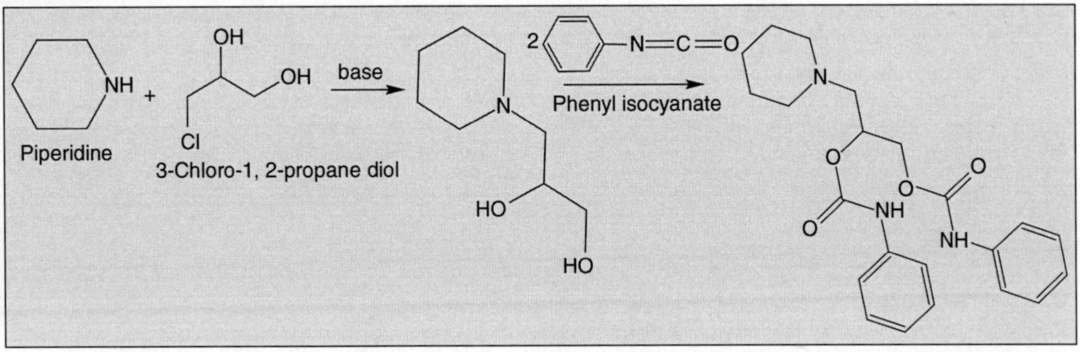

Alkylation of piperidine with 3-chloro-1,2-propane diol, followed by reaction with two moles of phenyl isocyanate, affords the *bis*-carbamate diperodon.

Structurally, it is related to the anilides in that an aromatic ring is attached to an sp^2 carbon by a nitrogen bridge.

6.9.3 Pramoxine Hydrochloride

4-[3-(4-Butoxyphenoxy)propyl]morpholine
Synthesis

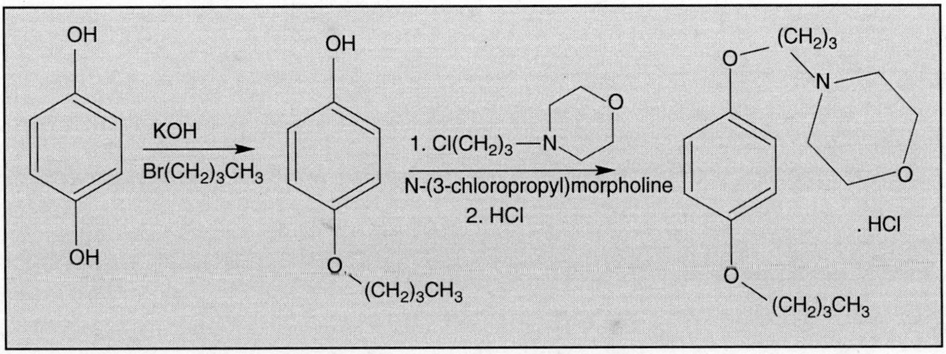

Alkylation of the mono potassium salt of hydroquinone with butyl bromide affords the ether, and alkylation of this with N-(3-chloropropyl)morpholine affords pramoxine.

Structurally, it is unrelated to either ester- or amide-type agents; simple ether linkage fulfils this function that exhibits local anaesthetic activity.

6.9.4 Dyclonine

4'-Butoxy-3-piperidinopropiophenone

Synthesis

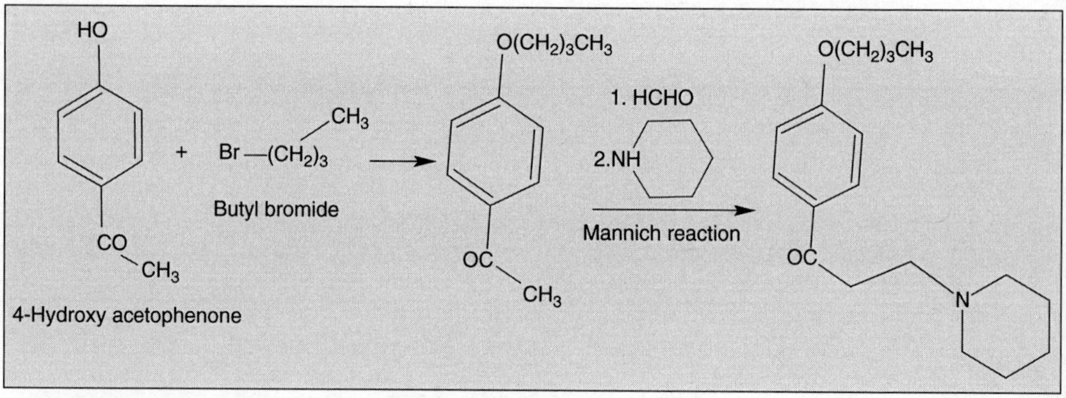

4-Hydroxy acetophenone undergoes O-alkylation with butyl bromide, followed by Mannich reaction with formaldehyde and piperidine, and affords dyclonine.

6.9.5 Dibucaine

2-Butoxy-N-(2-(diethylamino)ethyl)cinchoninamide (Cinchocaine)
Synthesis

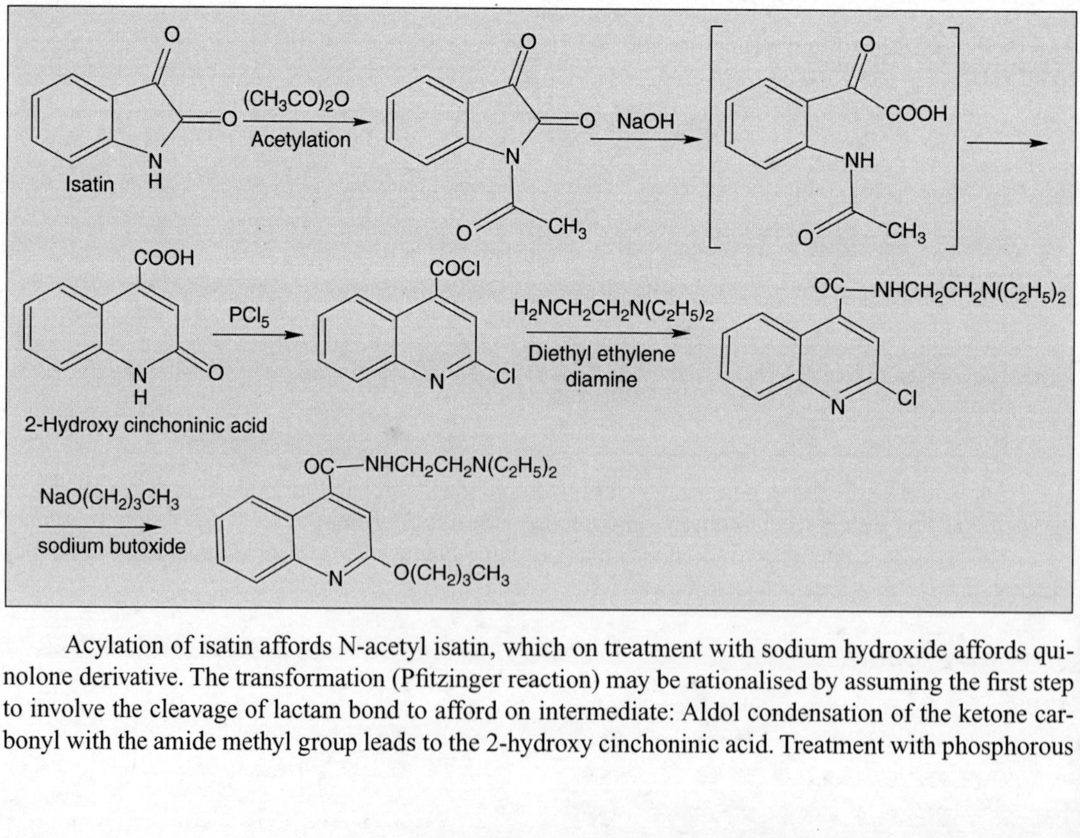

Acylation of isatin affords N-acetyl isatin, which on treatment with sodium hydroxide affords quinolone derivative. The transformation (Pfitzinger reaction) may be rationalised by assuming the first step to involve the cleavage of lactam bond to afford on intermediate: Aldol condensation of the ketone carbonyl with the amide methyl group leads to the 2-hydroxy cinchoninic acid. Treatment with phosphorous pentachloride serves both to form the acid chloride and to introduce the nuclear halogen. Condensation of the acid chloride with N,N-diethyl ethylenediamine followed by replacement of ring halogen with sodium butoxide affords dibucaine.

Local anaesthetics property of dibucaine (quinoline derivative) was discovered accidentally while research actually aimed for the preparation of antimalarial agents related to quinine. It is the most potent, most toxic, and longest-acting local anaesthetics.

6.9.6 Dimethisoquin Hydrochloride

3-Butyl-1-[2-(dimethylamino)ethoxy] isoquinoline
Synthesis

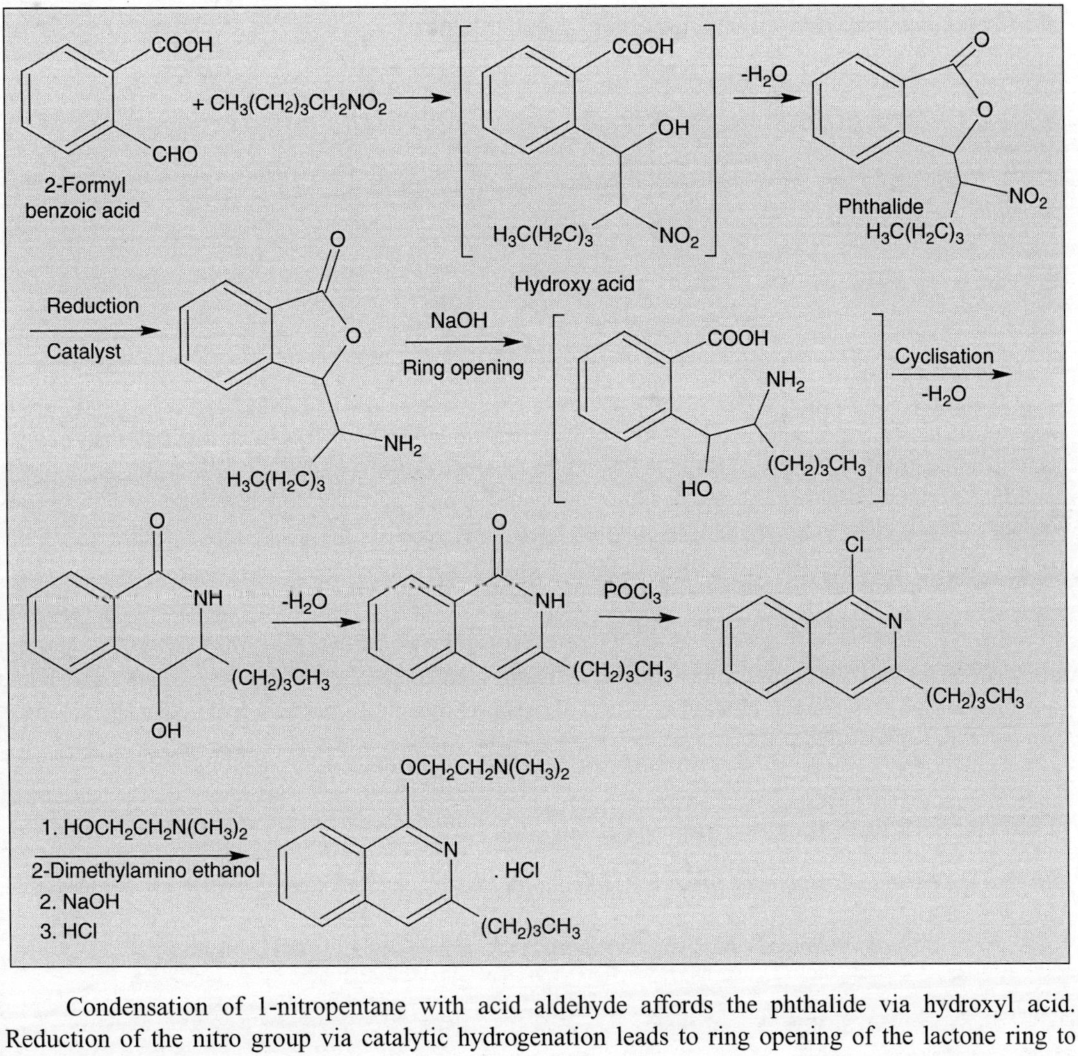

Condensation of 1-nitropentane with acid aldehyde affords the phthalide via hydroxyl acid. Reduction of the nitro group via catalytic hydrogenation leads to ring opening of the lactone ring to the intermediate amino acid. This cyclizes spontaneously to the isoquinoline derivative. Dehydration by means of strong acid followed by treatment with phosphorous oxychloride converts the oxygen function to the corresponding chloride via the enol form. Displacement of halogen with the sodium salt of 2-(diethylamino) ethanol affords dimethisoquin.

6.10 NEWER DRUGS

Newer local anaesthetics were introduced with the goal of reducing local tissue irritation, minimizing systemic cardiac and central nervous system (CNS) toxicity, and achieving a faster onset and longer duration of action.

6.10.1 Ropivacaine

(S)-N-(2,6-Dimethylphenyl)-1-propylpiperidine-2-carboxamide

It was developed after bupivacaine was noted to be associated with cardiac arrest in 0.5–0.75% of cases, particularly in pregnant women. Ropivacaine was found to have less cardiotoxicity than bupivacaine in animal models. Ropivacaine is indicated for local anaesthesia including infiltration, nerve block, epidural, and intrathecal anaesthesia in adults and children over 12 years. It is also indicated for peripheral nerve block and caudal epidural in children aged 1–12 years for surgical pain. It is also sometimes used for infiltration anaesthesia for surgical pain in children.

6.10.2 Levobupivacaine

It is the S-enantiomer of bupivacaine. Compared to bupivacaine, levobupivacaine is associated with less vasodilation and has a longer duration of action. It is approximately 13 per cent less potent (by molarity) than racemic bupivacaine.

FURTHER READINGS

Wilson and Gisvold's Textbook of Organic Medicinal & Pharmaceutical Chemistry, Block, J.H. & Beale, J.M., Jr. (eds), 11th Ed., 2004.
Foye's Principles of Medicinal Chemistry, Williams, D.A. & Lemke, T.L. (eds), 5th Ed., 2002.

MULTIPLE-CHOICE QUESTIONS

1. Which of the following local anaesthetics is least associated with dermatitis and asthma?

   a. [structure with CH3, NH, N-C2H5, C2H5, CH3]
   b. [structure with O, O-C2H5, H2N]
   c. [structure with O, O, CH3, N-CH3, H9C4-NH]

2. Which of the following statements is a characteristic of an ideal local anaesthetic?
   a. It should be reversible in action.
   b. It should be effective when used both systematically and topically.
   c. It should have maximum irritation either topically or at the site of injection.
   d. All of the above.

3. The starting material for the synthesis of dibucaine is
   a. 4-Chloroacetophenone
   b. Isatin
   c. Indole
   d. 2-Formyl benzoic acid

4. One of the following is true of anilide as local anaesthetics:
   a. Phenyl group is attached to the sp³ carbon atom.
   b. The methyl substituent provide steric hindrance to hydrolysis of the amide bond.
   c. Substitution of the phenyl with a methyl group reduces activity.
   d. All of the above.

5. Anaesthesia of mucous membranes is called
   a. Infiltration anaesthesia
   b. Field block anaesthesia
   c. Nerve block anaesthesia
   d. Topical analgesia

6. One of the following drugs does not belong to the benzoic acid class of anaesthetics:
   a. Benzocaine
   b. Cocaine
   c. Piperocaine
   d. Hexylcaine

7. One of the following does not contain the piperidine moiety in its structure:
   a. Mepivacaine
   b. Cyclomethycaine
   c. Lidocaine
   d. Bupivacaine

8. The drug that was discovered accidentally while research actually aimed for the preparation of antimalarial agents related to quinine was
   a. Quinacrine
   b. Dibucaine
   c. Dimethisoquin
   d. Benoxinate

9. The following drug synthesis involves Mannich reaction in one of the steps:
   a. Dibucaine
   b. Diclonine
   c. Lidocaine
   d. Diperodone

10. The local anaesthetic with a bicyclic ring is
    a. Lidocaine
    b. Diclonine
    c. Bupivacaine
    d. Cocaine

QUESTIONS

1. What is the prototype compound from which other local anaesthetics were designed? Give the structure.
2. Discuss in detail the SAR of benzoic acid derivatives on local anaesthetic activity.
3. Write down the synthetic protocol for the following:
    a. Lidocaine
    b. Hexylcaine
    c. Procaine
4. Classify local anaesthetics with structural examples for each class.
5. Discuss in detail the various categories of anaesthetics.

SOLUTION TO MULTIPLE-CHOICE QUESTIONS

1. a;
2. d;
3. b;
4. b;
5. d;
6. a;
7. c;
8. b;
9. b;
10. d.

CHAPTER 7

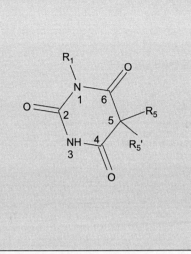

Sedatives, Hypnotics, and Anxiolytic Agents

LEARNING OBJECTIVES

- Define insomnia and various causes of insomnia
- Define sedation and hypnotics
- List the classes of sedatives and hypnotics
- Categorize barbiturates based on the duration of action
- Describe various benzodiapine classes of drugs useful as sedatives and hypnotics
- Describe the discovery of barbiturates
- Describe how barbiturates act on the CNS to deliver sedation and hypnosis
- Describe the methods of preparation of all barbiturates
- Describe the relationship between structure of barbiturates and their biological activity
- Learn the discovery pathway of benzodiazepines
- Define the mechanism behind the action of benzodiazepines as sedatives
- Describe the utility of Friedel-Crafts acylation in the synthesis of benzodiazepines
- Learn the differences between diazepam and oxazepam synthesis
- Describe the synthesis of benzodiazepine by an alternate method as in chlorazepate
- Describe the synthesis and uses of triazolobenzodiazepines
- Define the utility of imidobenzodiazepines in sedation other than general anaesthesia
- Derive a synthetic protocol for loprazolam
- Discuss the structure-activity relationship on benzodiazepines

Chapter 7: Sedatives, Hypnotics, and Anxiolytic Agents

- Describe the classes of benzodiazepine antagonists and their uses
- List other classes of sedatives and hypnotics that do not fall in the standard classification
- Define anxiolytics and their advantage over barbiturates
- Classify various anxiolytics based on their chemical structure
- Describe the new class of anxiolytics that are azaspirones
- Describe some newer drugs in the classes of sedatives, hypnotics, and anxiolytics

7.1 INTRODUCTION

Insomnia is a symptom, and its proper treatment depends on finding the cause of sleeplessness and treating the underlying etiology. The most common type of insomnia is transient insomnia due to acute situational factors. The typical factor is stress. Chronic insomnia is most commonly caused by psychiatric disorders. Numerous medical disorders can cause insomnia. Many drugs have been implicated as causing insomnia: alcohol, antihypertensives, antineoplastics, β-blockers, caffeine, corticosteroids, levodopa, nicotine, oral contraceptives, phenytoin, protriptyline, selective serotonin re-uptake inhibitors, stimulants, theophylline, and thyroid hormones. The underlying cause or causes of insomnia should be treated whenever possible. The primary indication for use of hypnotic agents in patients with insomnia is transient sleep disruption caused by acute stress.

Sedation is an intermediate degree of CNS depression, while hypnosis is a degree of CNS depression similar to natural sleep. Drugs of both classes are primarily CNS depressants, and a few of their effects, if not all, are evidently linked to action on the GABA-receptor complex.

Sedatives are CNS depressant drugs that reduce excitement and tension, and produce calmness and relaxation. **Hypnotics** are drugs that produce sleep similar to that of natural sleep. Both sedative and hypnotic actions may reside in the same drug; a small dose may act as sedative, whereas a large dose of the same drug may act as a hypnotic.

7.2 CLASSIFICATION OF SEDATIVE AND HYPNOTICS

1. **Barbiturates**

(i) Long duration of action (six or more hours)

	R	R_1	R_2
Phenobarbital	H	C_2H_5-	C_6H_5-
Mephobarbital	CH_3-	C_2H_5-	C_6H_5-
Metharbital	CH_3-	C_2H_5-	C_2H_5-

(ii) Intermediate duration of action (3–6 hours)

	R	R_1	R_2
Butabarbital	H	C_2H_5-	$CH_3CH_2CH(CH_3)-$
Amobarbital	H	C_2H_5-	$(CH_3)_2CHCH_2CH_2-$
Aprobarbital	H	$CH_2=CH-CH_2-$	$(CH_3)_2CH-$
Talbutal	H	$CH_2=CH\text{-}CH_2-$	$CH_3CH_2CH(CH_3)-$
Butalbital	H	$CH_2=CH-CH_2-$	$(CH_3)_2CHCH_2-$
Hexobarbital	CH_3-	CH_3-	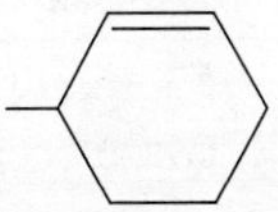 cyclohexenyl

(iii) Short duration of action (less than three hours)

	R	R_1	R_2
Pentobarbital	H	C_2H_5-	$CH_3CH_2CH_2CH(CH_3)-$
Secobarbital	H	$CH_2=CH\text{-}CH_2-$	$CH_3CH_2CH_2CH(CH_3)-$

2. Benzodiazepines

(i) 1,4-benzodiazepines

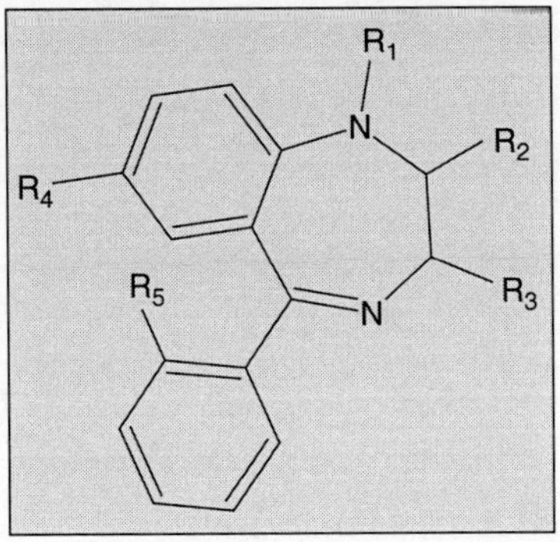

	R_1	R_2	R_3	R_4	R_5
Diazepam	CH$_3$	=O	H	Cl	H
Oxazepam	H	=O	OH	Cl	H
Clorazepate	H	=O	COOH	Cl	H
Prazepam	—H$_2$C—△	=O	H	Cl	H
Lorazepam	H	=O	OH	Cl	Cl
Halozepam	—CH$_2$CF$_3$	=O	H	Cl	H
Temazepam	CH$_3$	=O	OH	Cl	H
Flurazepam	—CH$_2$CH$_2$N(C$_2$H$_5$)$_2$	=O	H	Cl	F
Clonazepam	H	=O	H	Cl	NO$_2$
Nitrazepam	H	=O	H	NO$_2$	H
Quazepam	—CH$_2$CF$_2$	=S	H	Cl	F
Doxefazepam	—CH$_2$OH	=O	OH	Cl	F

(ii) Triazolobenzodiazepines

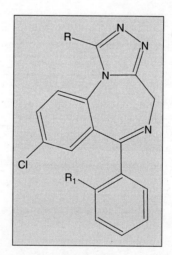

	R	R_1
Alprazolam	CH$_3$	H
Triazolam	CH$_3$	Cl
Estazolam	H	H

(iii) Imidazobenzodiazepines: Midazolam, Loprazolam
(iv) Miscellaneous: Bromazepam, Chlordiazepoxide, Brotizolam

3. **Miscellaneous:** Ethchlorvynol, ethinamate, glutethimide, meprobamate, zolpidem, zopiclone, zaleplon

Medicinal Chemistry 133

7.3 BARBITURATES

> Barbituric acid was first synthesised on December 6, 1864, by German researcher Adolf Von Baeyer. While barbituric acid itself does not have any effect on the central nervous system [you know the reason why?], to date, chemists have derived over 2,500 compounds that do possess pharmacologically active qualities.

Barbiturate action on the CNS is expressed in very diverse ways, ranging from small changes in patient behaviour to the onset of more obvious effects such as sedation, sleep, or general anaesthesia, depending, as a rule, on the administered dosage.

Barbiturates are used for brief periods of time for treating insomnia, since regular use of barbiturates (on average around three weeks) can lead to tolerance. Barbiturates are also used for controlling severe convulsive conditions and for treating various forms of epilepsy. They are used for pre- and post-operational sedation as well as in daytime sedation, for relieving patient anxiety, nervousness, and tension. Barbiturates are also used for treating catatonic and maniacal reactions, and as agents used in psychoanalysis (narcoanalysis and narcotherapy). Ultrashort acting barbiturates are used in anaesthesia. They are weak acids that form salts. As a rule, in order to exhibit central depressive action, barbiturates should contain two substituents on C5 of the hydrogenated pyrimidine ring.

Mechanism of action: Barbiturates bind to the allosteric site of GABA-A receptor, and activate and alter the transmission of ions by:

- Altering the structure of the ion channel, making it easier to open
- Increasing the affinity of GABA for its receptor by altering the receptor structure, so that GABA binds easily
- Prolonging the length of time the ion channel remains open by a magnitude of 4–5 folds

In high dose, barbiturate can open chloride ion channels in the absence of GABA.

$GABA_A$ receptor activation increases the membrane permeability to chloride ions that produces hyperpolarization in the cell body. This, in turn, will inhibit neuronal activity. (*See Fig. 7.1 in the coloured set of pages.*)

General method of preparation of 5,5-dialkyl barbiturates: This method is applicable for the synthesis of all barbiturates except phenobarbital and hexobarbital.

Chapter 7: Sedatives, Hypnotics, and Anxiolytic Agents

Diethyl malonate reacts with sodium ethoxide to form mono sodium salt of diethyl malonate, which in turn reacts with first alkyl halide to form mono alkyl diethyl malonate. This further reacts with second alkyl halide in presence of sodium ethoxide and gives dialkyl diethyl malonate. Dialkyl diethyl malonate on condensation with urea undergoes cyclization to form appropriate barbiturate.

7.3.1 Phenobarbital

5-Ethyl-5-phenyl barbituric acid

Synthesis

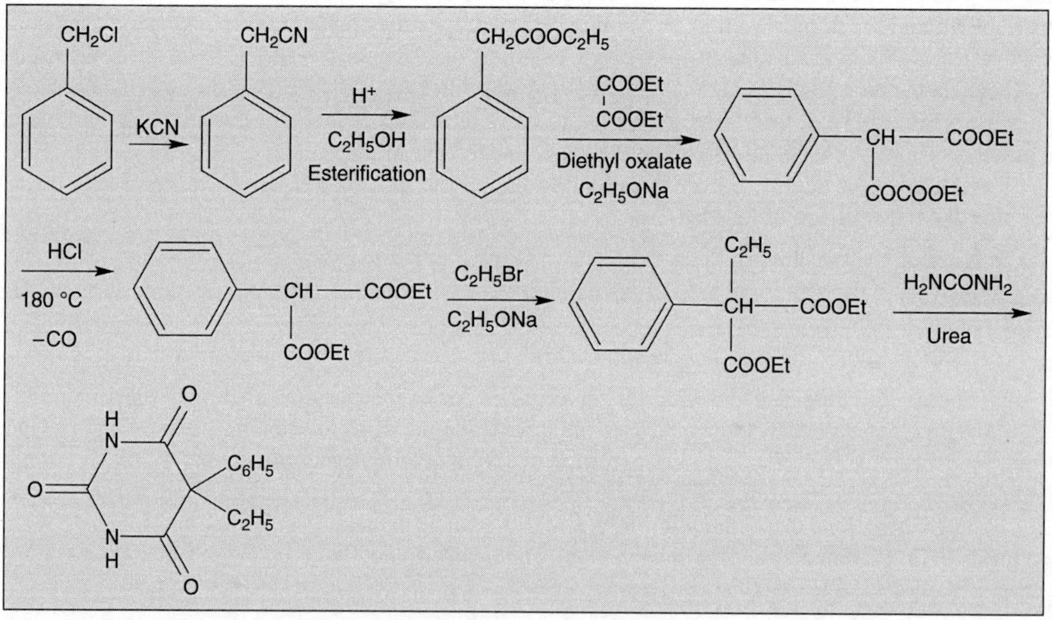

Benzyl chloride on reaction with potassium cyanide affords benzyl cyanide. Ethanolysis of benzyl cyanide in the presence of acid gives phenylacetic acid ethyl ester, the methylene group of which undergoes acylation using the diethyloxalate, giving diethyl ester of phenyloxobutandioic acid, which upon heating easily loses carbon oxide and turns into phenylmalonic ester. Alkylation of the obtained product using ethylbromide in the presence of sodium ethoxide leads to the formation of α-phenyl-α-ethylmalonic ester, the condensation of which with urea gives phenobarbital.

7.3.2 Hexobarbital

5-(1-Cyclohexen-1-yl)-1,5-dimethyl barbituric acid

Synthesis

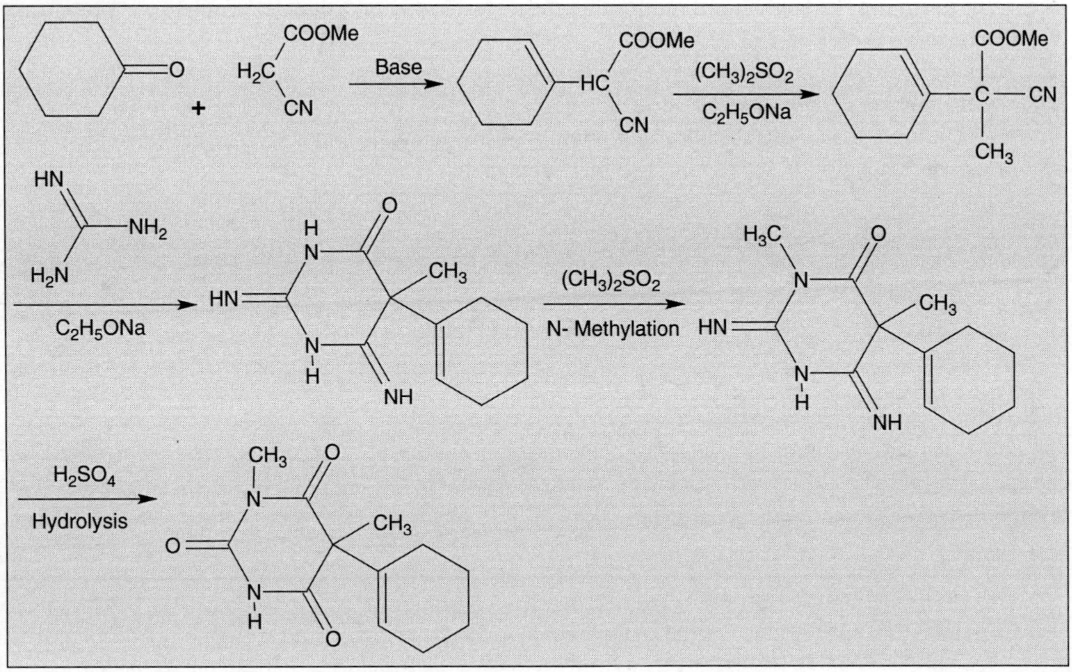

Knoevenagal condensation of cyclohexanone and methyl cyanoacetate gives alkylidene compounds, which undergo double-bond shift to form cyclohexenyl intermediate. Active hydrogen is methylated with dimethyl sulphate; this, followed by condensation with guanidine in the presence of sodium ethoxide, gives the diimino analogue of barbiturate. N-Methylation followed by hydrolysis affords hexobarbital.

SAR of barbiturates:

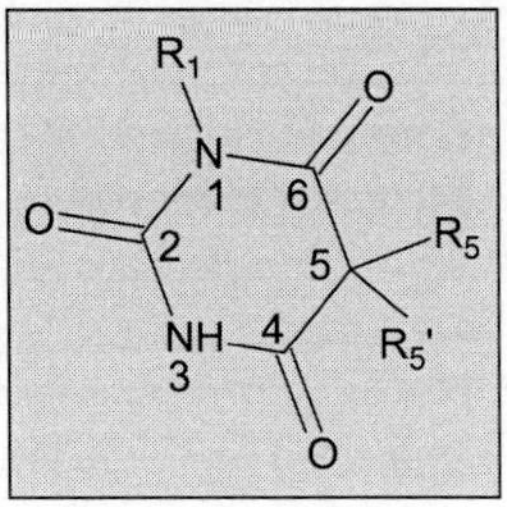

- Sedative and hypnotic activity increases with lipid solubility until the total number of carbon atoms of both substituents at C-5 is between 6 and 10. Further increase in the sum of the carbon atoms decreases activity in spite of further increase in lipid solubility, indicating that lipid solubility must remain within certain limits.
- Within the same series, the branched chain isomer has greater lipid solubility and activity, and shorter duration of action than the straight chain isomer.

- Within the same series, the unsaturated allyl (talbutal), alkynyl (methohexital), and cycloalkenyl (hexobarbital) derivatives are more active than corresponding saturated analogues with the same number of carbon atoms.
- Alicyclic or aromatic substituents at C-5 are more potent than the aliphatic substituents with the same number of carbon atoms.
- Conversion of a 5,5-disubstituted barbituric acid by methylation to a 1,5,5-trisubstituted barbituric acid does not change activity in a significant manner.
- Introduction of a polar substituent (NH_2, RNH, OH, COOH, SO_3H) into the aromatic group at C-5 decreases lipid solubility and potency.
- Replacement of the oxygen at C-2 by a sulphur atom increases quick onset and shortens duration of action, but replacement of additional carbonyl oxygen by sulphur decreases activity again, indicating that lipid solubility cannot be increased beyond limits.

7.4 BENZODIAZEPINES

The core chemical structure of 'classical' benzodiazepine drugs is a fusion between the benzene and diazepine ring systems. Many of these drugs contain the 5-phenyl-1,3-dihydro-1,4-benzodiazepine-2-one substructure.

> The first benzodiazepine, chlordiazepoxide (Librium), was discovered serendipitously in 1954 by the Austrian scientist Leo Sternbach (1908–2005), working for the pharmaceutical company Hoffmann-La Roche. Chlordiazepoxide was synthesised from work on a chemical dye, quinazolone-3-oxides.

Uses: Benzodiazepines are indicated for the symptomatic relief of tension and anxiety, acute alcohol withdrawal, adjunct therapy in skeletal muscle spasm, neurosis, pschychoneurosis, and management of status epilepticus.

The long-term use of benzodiazapines can cause physical dependence. The use of benzodiazepines should, therefore, commence only after medical consultation and it should be prescribed in the smallest dosage possible to provide an acceptable level of symptom relief.

Mechanism of action: Benzodiazepine receptors are present in the brain and they form part of a $GABA_A$ receptor-chloride ion channel macromolecular complex. Binding of benzodiazepines to these receptors activates $GABA_A$ receptor and increases chloride conductance by increasing the frequency of opening chloride ion channel. These, in turn, inhibit neuronal activity by hyper polarization and depolarization block. (*See Fig. 7.2 in the coloured set of pages.*)

Benzodiazepines bind at the interface of α and γ subunits on the $GABA_A$ receptor. Benzodiazepine binding also requires that alpha subunits contain a histidine amino acid residue (*i.e.*, α_1, α_2, α_3, and α_5 containing $GABA_A$ receptors). For this reason, benzodiazepines show no affinity for α_4 and α_6 subunits containing $GABA_A$ receptors, which contain an arginine instead of a histidine residue. Other sites on the $GABA_A$ receptor also bind neurosteroids, barbiturates, and certain anaesthetics.

In order for $GABA_A$ receptors to be sensitive to the action of benzodiazepines, they need to contain both α and γ subunits, where the benzodiazepine binds at the interface. Once bound, the benzodiazepine locks the $GABA_A$ receptor into a conformation where the neurotransmitter GABA has much higher affinity for the $GABA_A$ receptor, increasing the frequency of opening of the associated chloride ion channel and hyper polarizing the membrane. This potentiates the inhibitory effect of the available GABA, leading to sedatory and anxiolytic effects. As mentioned above, different benzodiazepines can have different affinities for $GABA_A$ receptors made up of different collections of subunits. For instance, benzodiazepines with high activity at the α_1 are associated with sedation, whereas those with higher affinity for $GABA_A$ receptors containing α_2 and/or α_3 subunits have good anti-anxiety activity.

Diazepam 7-Chloro-1,3-dihydro-1-methyl-5-phenyl-1,4-benzodiazepine-2-one

Synthesis

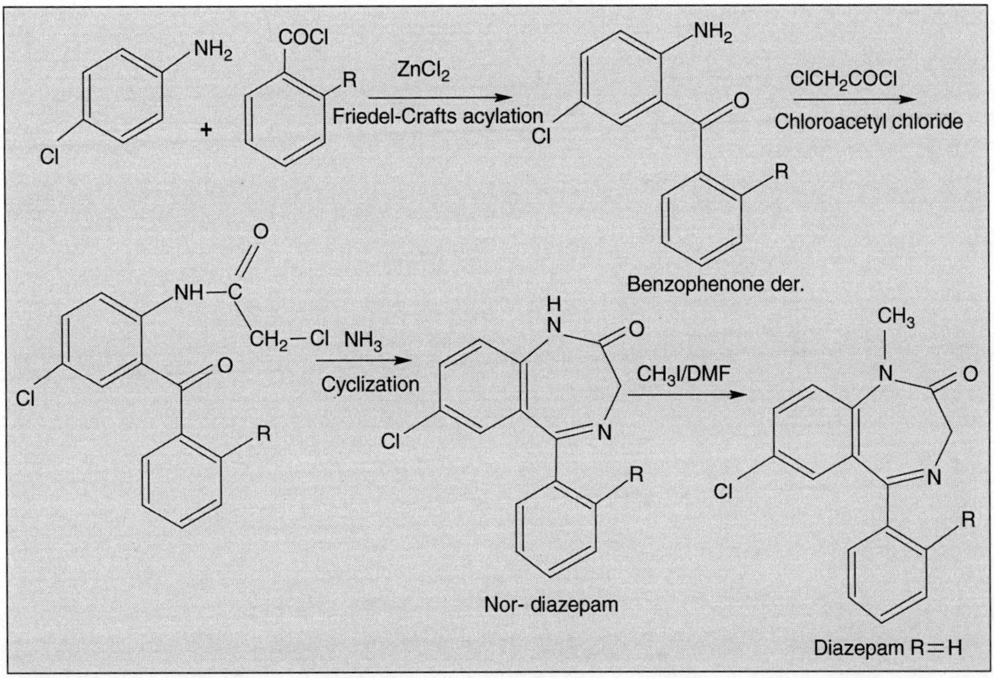

Friedel-Crafts acylation of 4-chloro aniline with corresponding benzoyl chloride in presence of Lewis acid affords benzophenone derivative. Acetylation of an amino group with chloroacetyl chloride gives the chloro acetamide. Heating with ammonia undergoes cyclization reaction to form nor-diazepam; N-methylation methyl iodide affords diazepam.

Prazepam Synthesised by alkylating nordiazepam with cyclopropylmethylbromide.

Halozepam Synthesised by alkylating nordiazepam with 1,1,1-trifluoroethylbromide [$BrCH_2CF_3$].

Flurazepam Synthesised by alkylating nordiazepam with diethylaminoethylchloride [$ClCH_2CH_2N(C_2H_5)_2$] and R=F.

Quazepam (R=F) Alkylation with $BrCH_2CF_3$ followed by treatment with P_2S_5 to convert C=S.

Clonazepam (R=NO_2) Obtained from nordiazepam.

Oxazepam 7-Chloro-1,3-dihydro-3-hydroxy-5-phenyl-1,4-benzodiazepine-2-one.

Synthesis

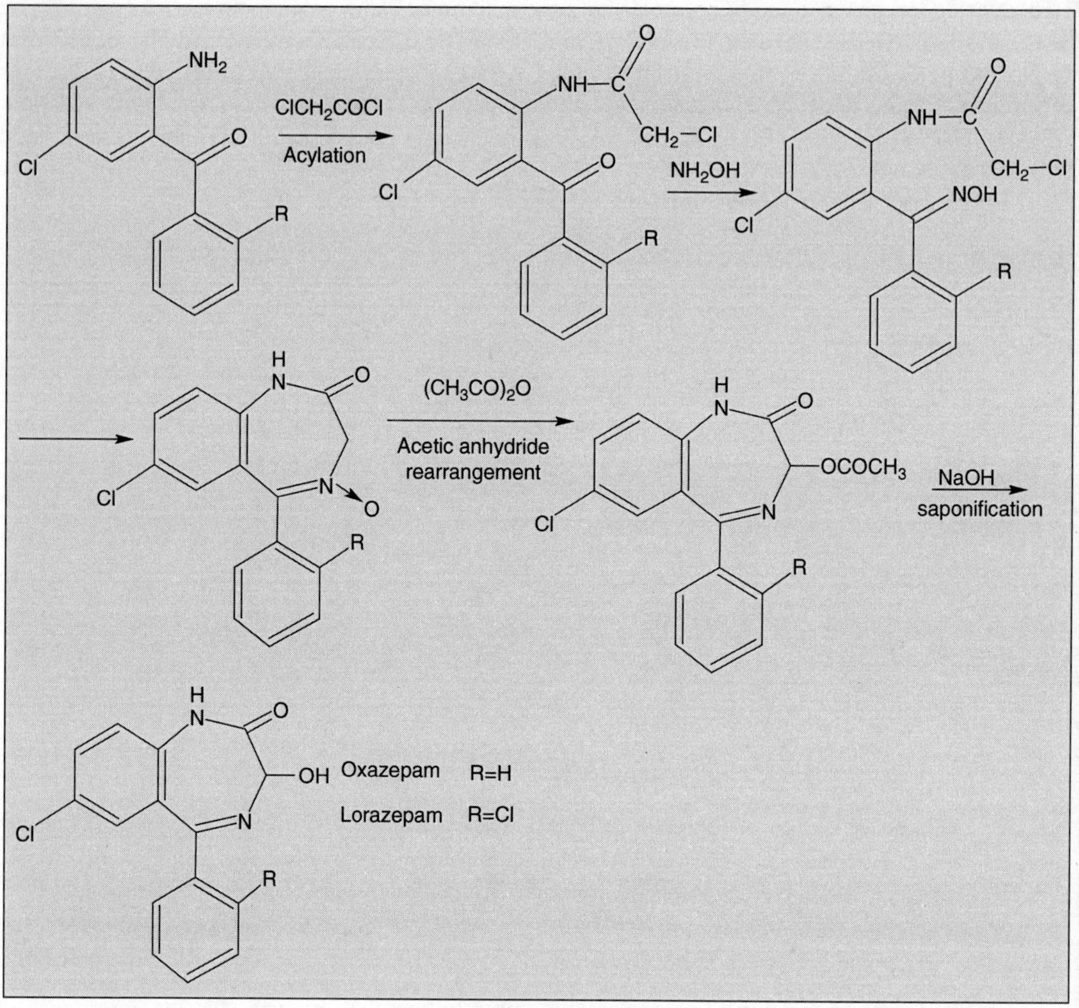

Acetylation of an amino group of benzophenone derivative with chloroacetyl chloride gives chloro acetamide. Heating the amide with hydroxylamine forms oxime, which undergoes cyclization to form 1,4-benzodiazepine-4-oxide. Reaction of N-oxide with acetic anhydride leads to the formation of O-acetate via the typical Polonovski rearrangement. Saponification with mild base gives oxazepam.

Clorazepate 7-chloro-2,3-dihydro-2-oxo-5-phenyl-1,4-benzodiazepine-3-carboxylic acid

Synthesis

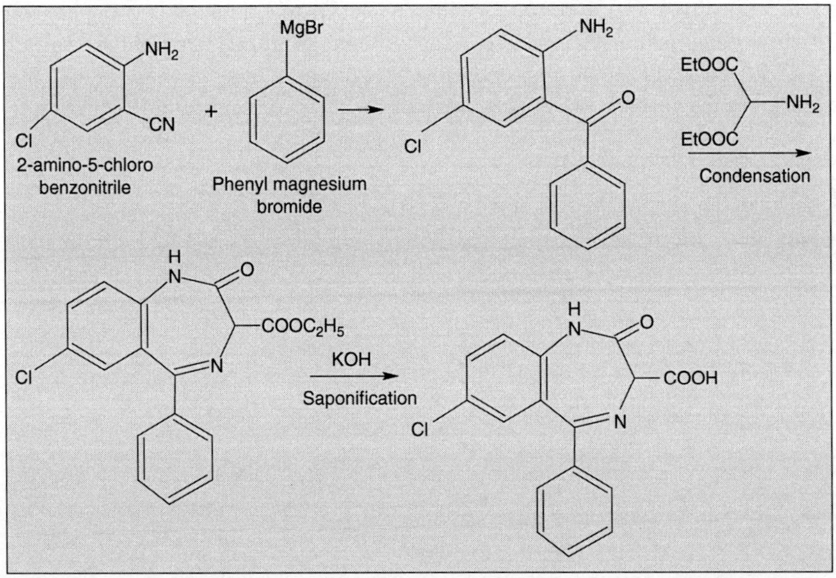

Benzophenone derivative is prepared by another method by reacting aryl nitrile with aryl magnesium bromide. Benzophenone derivative on condensation with amino diethyl malonate undergoes cyclization to form benzodiazepine derivative, which on saponification affords clorazepate.

7.4.1 Triazolobenzodiazepines

Synthesis

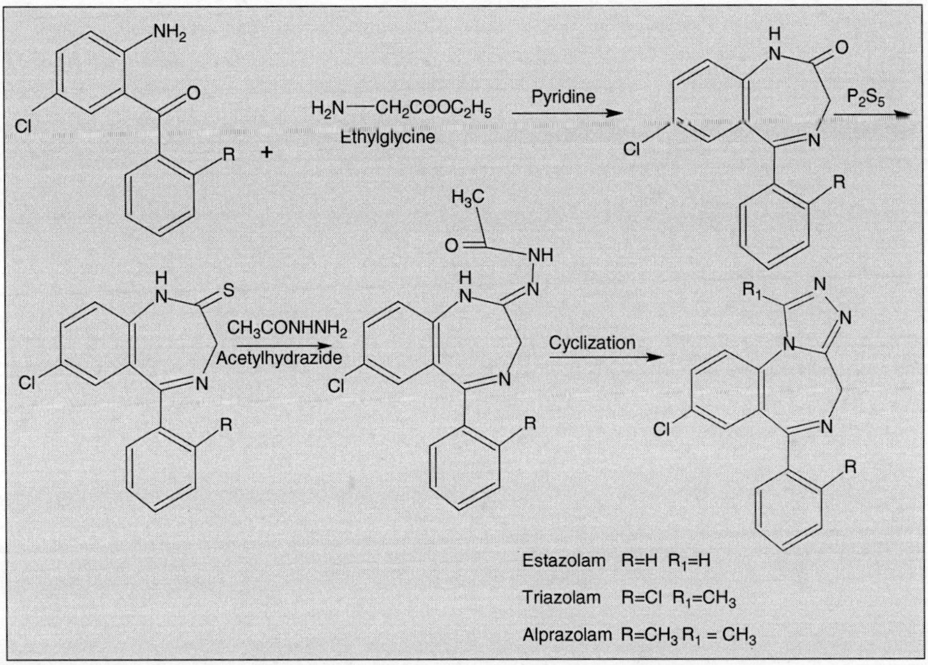

Estazolam R=H R_1=H
Triazolam R=Cl R_1=CH_3
Alprazolam R=CH_3 R_1 = CH_3

The reaction of 2-amino-2',5-dichlorobenzophenone with glycine ethyl ester gives 7-chloro-5-(2-chlorophenyl)-2,3-dihydro-1-H-1,4-benzodiazepine-2-one. By interacting this with phosphorus pentasulphide, the carbonyl group is transformed into a thiocarbonyl group, giving 7-chloro-5-(2-chlorophenyl)-2,3-dihydro-1-H-1,4-benzodiazepine-2-thione. The resulting cyclic thioamide on interaction with acetylhydrazine gives the corresponding acetylhydrazone, which upon heating cyclizes into triazolam.

7.4.2 Imidazobenzodiazepines

Midazolam Refer to the chapter 5 on General Anaesthetics

Loprazolam 6-(o-Chlorophenyl)-2,4-dihydro-2-[(4-methyl-1-piperazinyl)methylene]-8-nitro-1H-imidazo[1,2-a][1,4]benzodiazepine-1-one

Synthesis

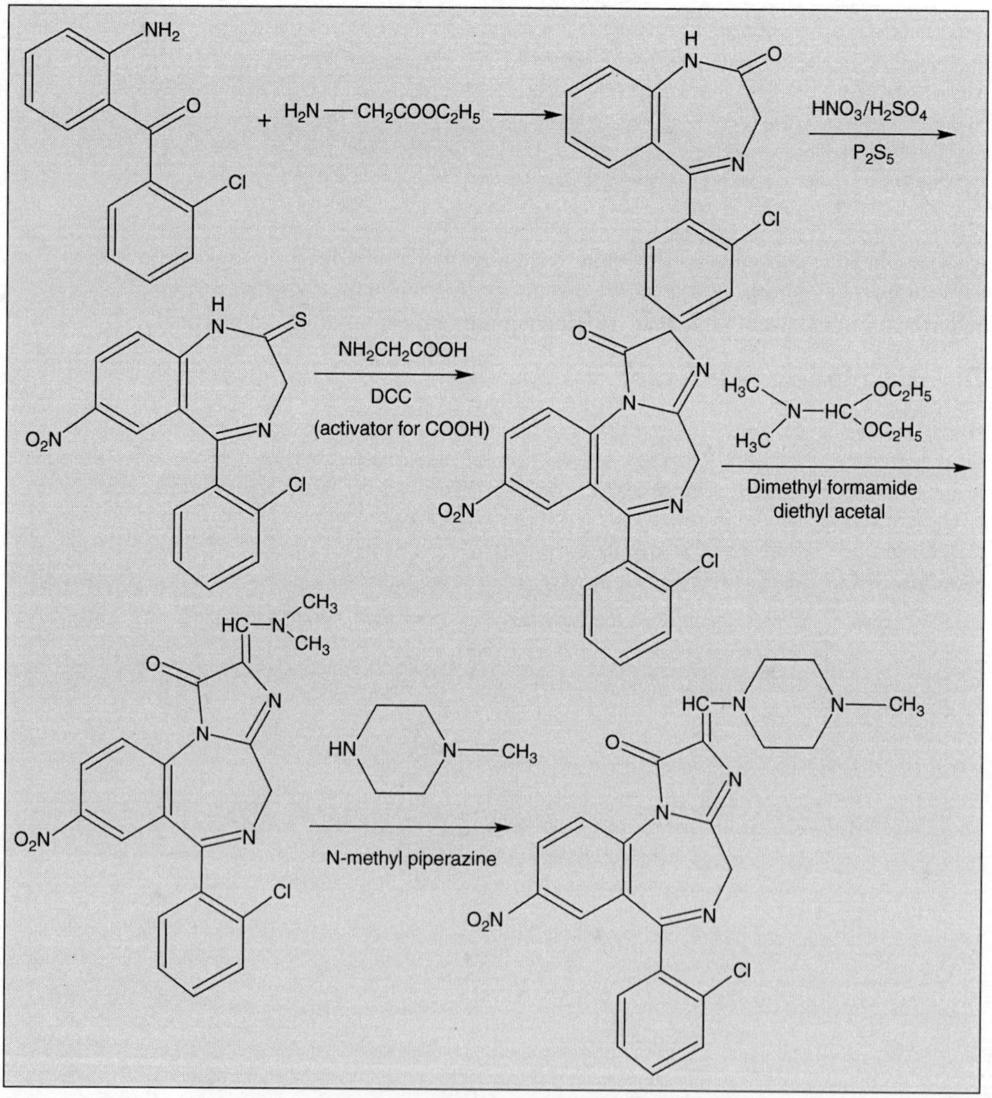

7.4.3 Miscellaneous Benzodiazepines

Bromazepam 7-Bromo-1,3-dihydro-5-pyrid-2'-yl-1,4-benzodiazepine-2-one

Synthesis

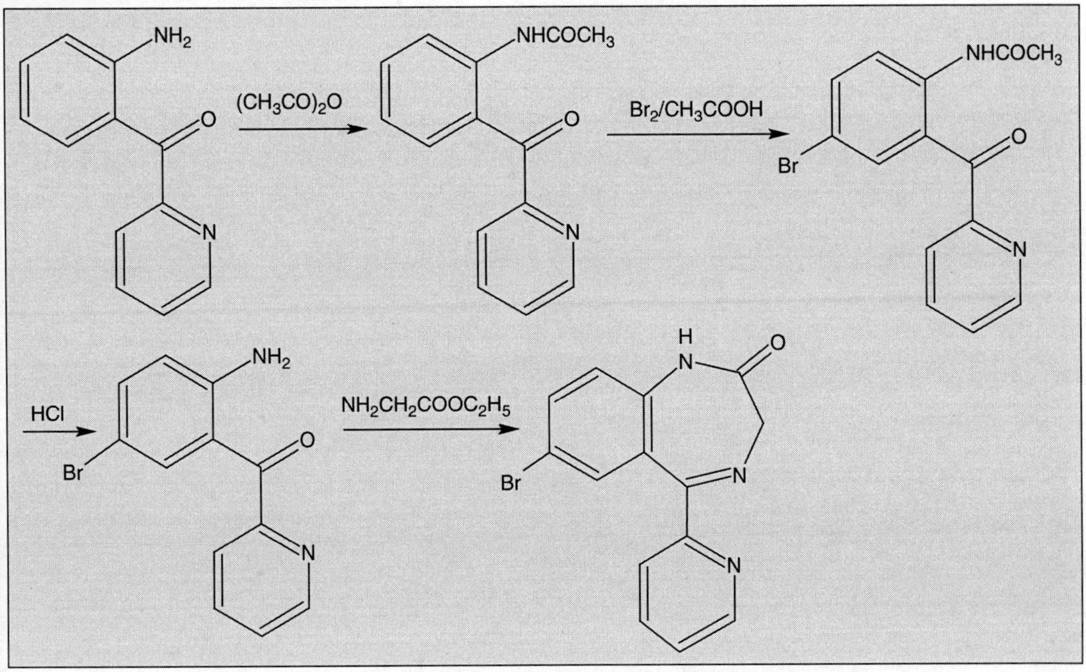

Chlordiazepoxide: 7-Chloro-2-(methylamino)-5-phenyl-1,4-benzodiazepine-4- oxide

Synthesis

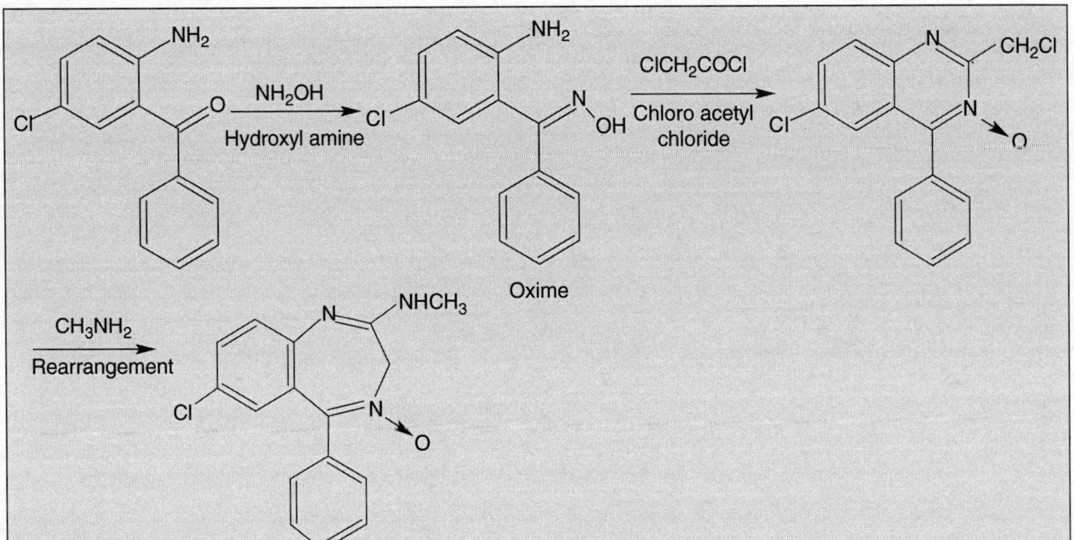

Reaction of the oxime from aminobenzophenone with chloroacetyl chloride gives the chloroacylamide; this compound is converted to the corresponding quinazoline *N*-oxide on treatment with hydrogen chloride. The *N*-oxide on treatment with methylamine undergoes rearrangement to afford chlordiazepoxide.

Brotizolam 2-Bromo-4-(2-chlorophenyl)-9-methyl-6H-thieno[3,2-f][1,2,4]-triazolo[4,3-a]-1,4-diazepine

Synthesis

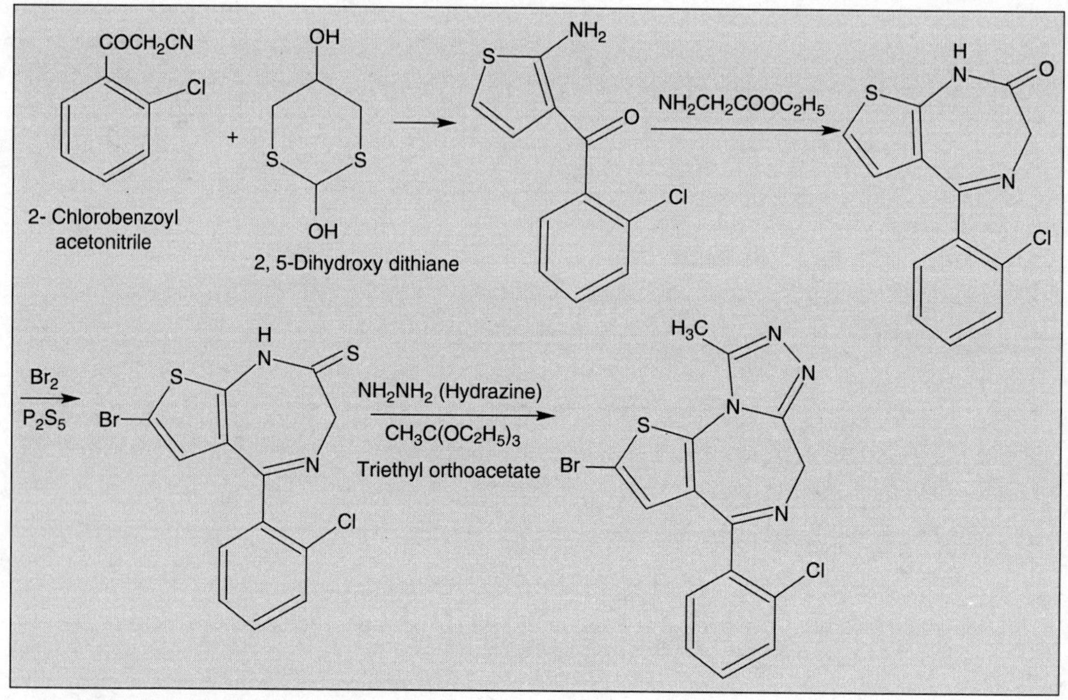

Thiophenophenone is prepared by treating 2-chlorobenzoyl acetonitrile and 2,5-dihydroxy dithiane. Construction of fused triazole ring follows the methods used for triazolam.

7.4.4 SAR of Benzodiazepines

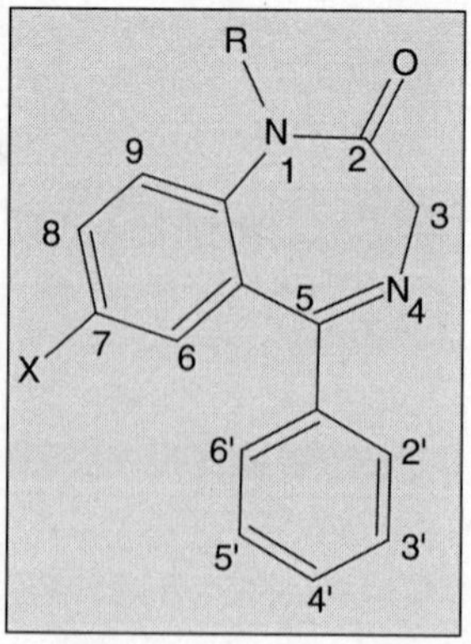

- The presence of electron-attracting substituents (Cl, F, Br, NO_2) at position 7 is required for activity, and the more electron attracting leads to more activity.
- Positions 6, 8, and 9 should not be substituted.
- A phenyl or pyridyl (bromazepam) at position 5 promotes activity. If phenyl ring is substituted with electron-attracting groups at 2', or 2', 6'-, then activity is increased. On the other hand, substituents in 3', 4', and 5' decrease activity greatly.
- Saturation of 4, 5 double bond or shift of double bond to the 3, 4 position decreases the activity.
- Alkyl substituents at 3-position decrease the activity, whereas substitution of the 3-position with polar hydroxyl and carboxyl group retains activity and have short half-lives.
- The N-substituent at 1 position should be small.
- Reduction of carbonyl function at 2-position to -CH_2- gives less potent compounds.
- Derivatives with additional rings (triazole or imidazole) joining the diazepine nucleus at the 1 and 2 positions are generally active.
- Replacement of the benzene ring by heteroaromatic (thiophene) ring resulted in enhanced anxiolytic properties.

7.5 BENZODIAZEPINE ANTAGONISTS

Flumazenil Ethyl 8-fluoro-5,6-dihydro-5-methyl-6-oxo-4*H*-imidazo[1,5-α][1,4]benzodiazepine-3-carboxylate

Synthesis

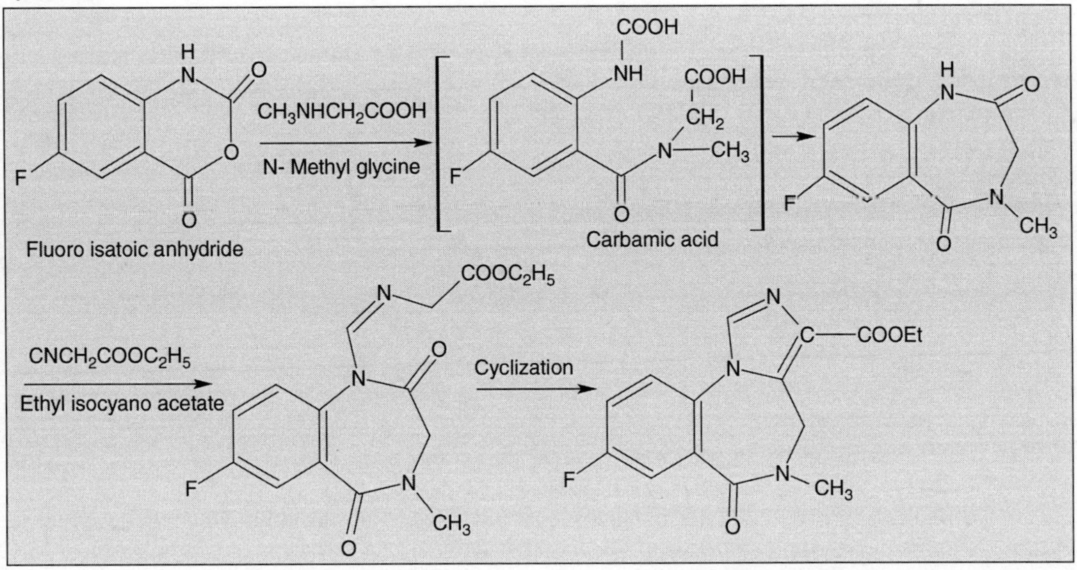

The benzodiazepinedione nucleus is obtained from the condensation of the fluoro isatoic anhydride with *N*-methylglycine. The first step involves acylation of the amino acid nitrogen by the activated anhydride carbonyl group. Loss of carbon dioxide from the resulting carbamic acid will lead to the amide.

This compound then cyclizes to the benzodiazepinedione. Reaction of this with ethyl isocyanoacetate then leads to addition of the only free amide nitrogen to the isocyanide function, to afford an intermediate such as the amidine. The doubly activated acetate methylene group then condenses with the ring carbonyl group to form an imidazole, affording flumazenil.

Flumazenil is the benzodiazepines analogue in which the benzene ring at the 5-position is omitted - which shows benzodiazepine antagonist activity.

Uses: Management of suspected benzodiazepines overdose and the reversal of sedative effect produced by benzodiazepines.

7.6 MISCELLANEOUS

Ethchlorvynol 1-Chloro-3-ethyl-1-penten-4-yn-3-ol

Synthesis

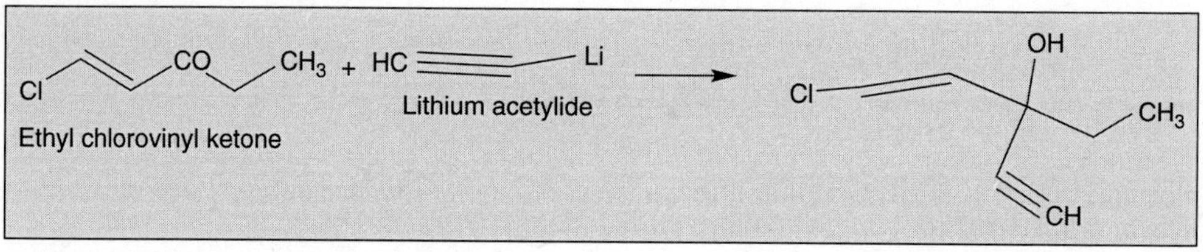

Ethchlorvynol is synthesised by the condensation of lithiated acetylene with 1-chloro-1-penten-3-one in liquid ammonia.

Use: Mild hypnotic at a dose of 500 mg induces sleep within 15 minutes to one hour, and has duration of action of approximately 5 hours.

Ethinamate: 1-Ethynyl-1-cyclohexyl carbamate

Synthesis

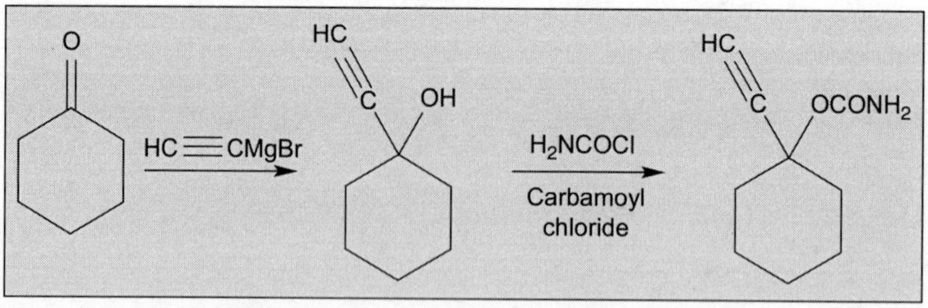

Ethinamate is synthesised by the condensation of acetylene magnesium bromide with cyclohexanone and the subsequent transformation of the resulting carbinol into carbamate by the subsequent reaction with phosgene, and later with ammonia.

Use: Mild sedative at a dose of 500 mg induces sleep within 20 minutes to 30 minutes.
Glutethimide 3-Ethyl-3-phenyl-2,6-piperidinedione

Synthesis

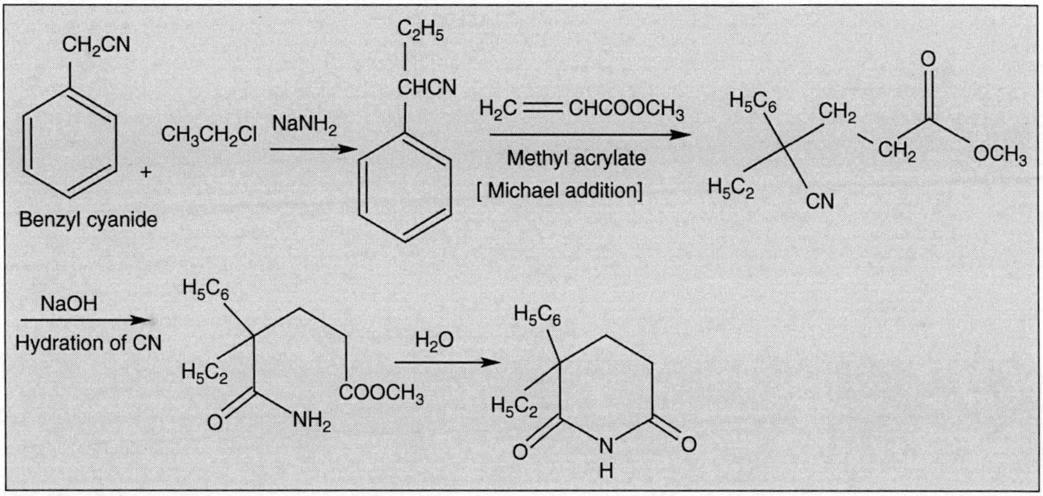

[Reaction scheme: Benzyl cyanide (CH_2CN on benzene) + CH_3CH_2Cl, $NaNH_2$ → 2-phenylbutyronitrile (CHCN with C_2H_5) → with $H_2C=CHCOOCH_3$ (Methyl acrylate) [Michael addition] → intermediate with H_5C_6, H_5C_2, CN, $CH_2CH_2COOCH_3$ → NaOH (Hydration of CN) → amide intermediate with H_5C_6, H_5C_2, $CONH_2$, $COOCH_3$ → H_2O → glutethimide (piperidine-2,6-dione with H_5C_6 and H_5C_2)]

Glutethimide is synthesised by addition of 2-phenylbutyronitrile (which is prepared by alkylating benzyl cyanide with ethyl chloride in presence of base) to the methylacrylate (Michael reaction), and the subsequent alkaline hydrolysis of the nitrile group in the obtained compound into an amide group, and the subsequent acidic cyclization of the product into the desired glutethimide.

Meprobamate 2-Methyl-2-Propyl-1,3-propanediol dicarbamate

Synthesis

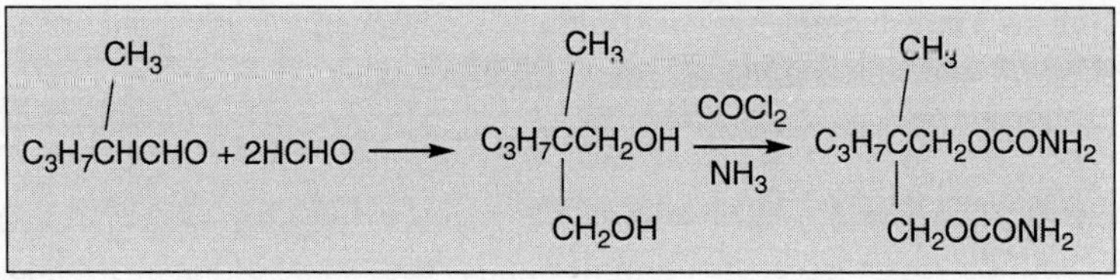

$$C_3H_7CHCHO + 2HCHO \longrightarrow C_3H_7C(CH_3)(CH_2OH)CH_2OH \xrightarrow[NH_3]{COCl_2} C_3H_7C(CH_3)(CH_2OCONH_2)CH_2OCONH_2$$

Meprobamate is synthesised by the reaction of 2-methylvaleraldehyde with two molecules of formaldehyde and the subsequent transformation of the resulting 2-methyl-2-propylpropan-1,3-diol into the dicarbamate via successive reactions with phosgene and ammonia.

Use: Management of anxiety disorders.

Zopiclone [8-(5-chloropyridin-2-yl)- 7-oxo-2,5,8-triazabicyclo [4.3.0]nona-1,3,5-trien-9-yl] 4- methylpiperazine-1-carboxylate

Synthesis

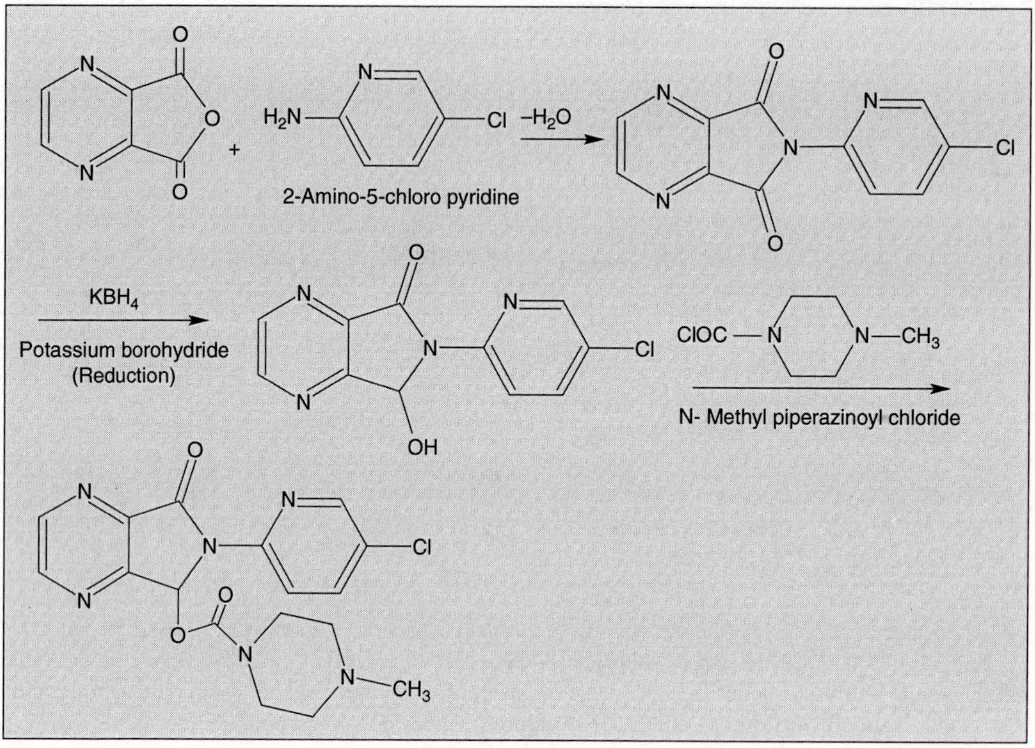

Amination of furopyrazine-5,7-dione with 2-amino-5-chloro pyridine, followed by reduction with potassium borohydride, gives alcohol. This, in turn, react with N-methyl piperazinoyl chloride affords zopiclone.

Use: Zopiclone is a new hypnotic agent structurally unrelated to barbiturates and benzodiazepines. It binds to the GABA-A-benzodiazepines receptor complex.

Zaleplon   N-[3-(7-Cyano-1,5,9-triazabicyclo[4.3.0]nona-2,4,6,8-tetraen-2-yl)phenyl]-N-ethyl-acetamide

Zaleplon is pyrazolo pyrimidine derivative and have properties of an ultrashort acting sleep inducer. Zaleplon interacts with the GABA receptor complex and shares some of the pharmacological properties of the benzodiazepines.

Synthesis

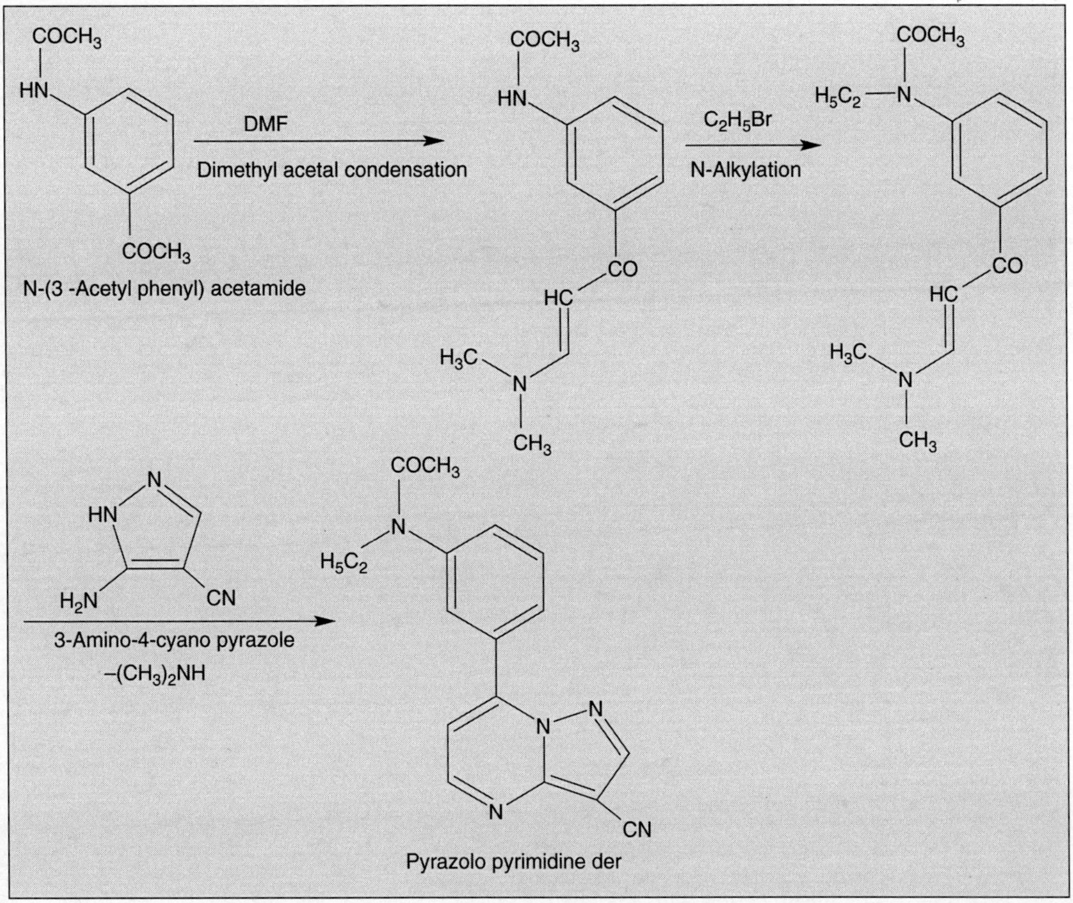

[Scheme: Synthesis of Zaleplon]

- N-(3-Acetyl phenyl) acetamide → (DMF, Dimethyl acetal condensation) → enamide intermediate → (C_2H_5Br, N-Alkylation) → N-ethyl enamide intermediate
- 3-Amino-4-cyano pyrazole + $-(CH_3)_2NH$ → Pyrazolo pyrimidine derivative (Zaleplon)

Condensation of the substituted acetophenone with dimethylformamide acetal gives the eneamide. The anilide nitrogen is then alkylated by means of sodium hydride and ethyl iodide to give intermediate, which on condensation with aminopyrazole derivative leads to addition-elimination sequence on the enamide function to give transient intermediate which on cyclization, leads to zaleplon.

Zolpidem N,N,6-Trimethyl-2-(4-methylphenyl)-imidazo(1,2-a)pyridine-3-acetamide

It is a short-acting non-benzodiazepine hypnotic that potentiates GABA, an inhibitory neurotransmitter, by binding to ω1 receptor subtype of the benzodiazepine receptors that are located on the GABA receptors.

Synthesis

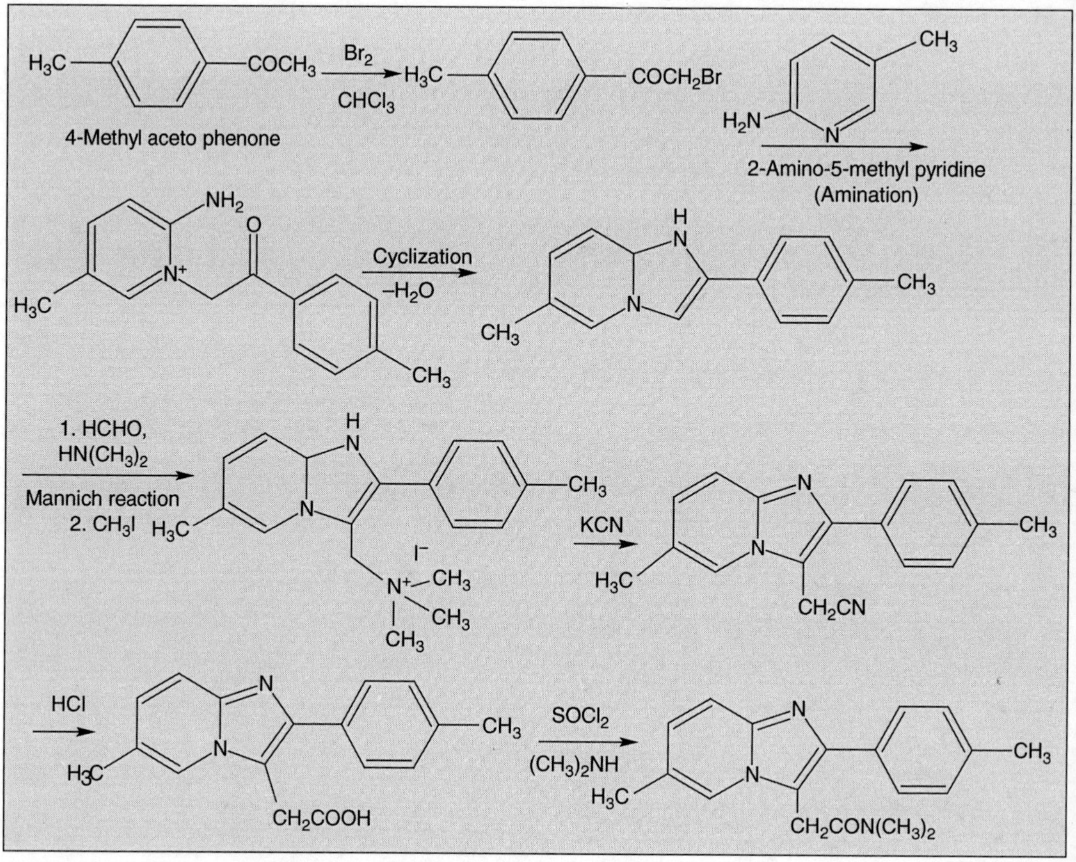

Reaction of 2-amino picoline with 4-methylphenacyl bromide leads directly to the imidazopyridine. The overall transformation can be rationalized by assuming initial alkylation on ring nitrogen; imine formation followed by bond reorganization then forms the imidazole. Treatment of imidazopyridine with formaldehyde and dimethylamine leads to the Mannich base. The dimethylamino group is then activated toward displacement by conversion to the quaternary salt through alkylation with methyl iodide. Reaction with potassium cyanide leads to the acetonitrile. The nitrile is then hydrolysed to the corresponding acid and this is then converted to dimethylamide.

7.7 ANXIOLYTIC AGENTS

Many illnesses are accompanied by anxiety, a worried state during which a syndrome characterized by feelings of helplessness, despair, dark premonitions, and asthenia begins to develop. It can be accompanied by headaches, increased perspiration, nausea, tachycardia, dry mouth, etc. A state of anxiety can originate from neurological reasons, and can also be of a somatopsychic nature, which is associated with pathological development in diseases of the cardiovascular system, neoplasms, hypertonia, and diseases of the gastrointestinal tract. Drugs used for relieving anxiety, stress, worry, and fear that do not detract attention from or affect psychomotor activity of the patient are called anxiolytics or minor tranquillizers. Most of them have sedative and hypnotic action, and in high doses their effects are in many ways similar to barbiturate action. However, the primary advantage of this group over barbiturates lies in their significantly increased value in terms of the ratio of sedative/hypnotic effects. In other words, the ratio between doses that reduce stress and doses that cause sleep is significantly higher in anxiolytics than in barbiturates.

Classification

1. Benzodiazepines: Chlordiazepoxide, Diazepam, Oxazepam, Clorazepate, Lorazepam, Prazepam, and Alprazolam
2. 1,3-Propanediol: Meprobamate
3. Azaspirodecanedione: Buspirone

Buspirone 8-[4-[4-(Pyrimidinyl)-1-piperazinyl]-butyl]azaspiro[4,5]decane-7,9-dione

Buspirone is an extremely specific drug that could possibly represent a new chemical class of anxiolytics - azaspirones. As an anxiolytic, its activity is equal to that of benzodiazepines; however, it is devoid of anticonvulsant and muscle relaxant properties, which are characteristic of benzodiazepines. It does not cause dependence or addiction. The mechanism of its action is not conclusively known. It does not act on the GABA receptors, which occur in benzodiazepine use; however, it has a high affinity for seratonin (5-HT) receptors and a moderate affinity for dopamine (D_2) receptors. Buspirone is effective as an anxiolytic. A few side effects of buspirone include dizziness, drowsiness, headaches, nervousness, fatigue, and weakness.

Synthesis

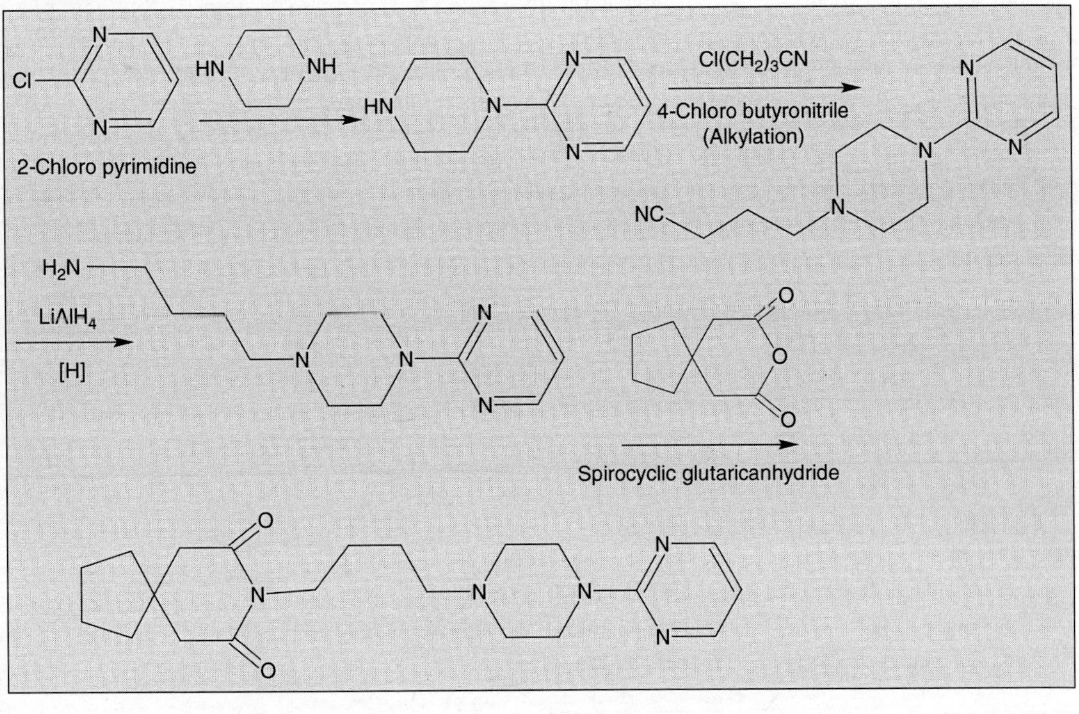

Buspirone is synthesised by the reaction of 1-(2-pyrimidyl)-4-(4-aminobutyl)piperazine with 8-oxaspiro[4,5]decan-7,9-dione. In turn, 1-(2-pyrimidyl)-4-(4-aminobutyl)piperazine is synthesised by the reaction of 1-(2-pyrimidyl)piperazine with 4-chlorobutyronitrile, giving 4-(2-pyrimidyl)-1-(3-cyanopropyl)piperazine, which is hydrogenated with Raney nickel into buspirone.

7.8 NEWER DRUGS

Eszopiclone It is the active stereoisomer of zopiclone, and belongs to the class of drugs known as cyclopyrrones. Eszopiclone is a hypnotic with a chemical structure unrelated to benzodiazepines or other drugs with known hypnotic properties. It is an agonist for the same family of receptors as benzodiazepines (primarily GABA), but selectively binds to the receptor subtype known as omega. It is believed that this selectivity is responsible for its strong hypnotic effects and lack of significant anxiolytic properties.

Clomethiazole (Chlormethiazole) 5-(2-Chloroethyl)-4-methyl-1,3-thiazole

It is a sedative and hypnotic that is widely used in treating and preventing symptoms of acute alcohol withdrawal. It is a drug that is structurally related to thiamine (vitamin B_1) but acts like a sedative, hypnotic, muscle relaxant, and anticonvulsant. Chlormethiazole acts on GABA receptors, which cause the release of the neurotransmitter GABA. GABA is a major inhibitory chemical in the brain involved in causing sleepiness and controlling anxiety and panic attacks. It acts by increasing the activity of GABA, thereby reducing the functioning of certain areas of the brain. This results in sleepiness, a decrease in anxiety, and relaxation of muscles. It also inhibits the enzyme alcohol dehydrogenase, which is responsible for breaking down alcohol in the body. This slows the rate of elimination of alcohol from the body, which helps to relieve the sudden effects of alcohol withdrawal in alcoholics.

FURTHER READINGS

Smith and Williams' Introduction to the Principles of Drug Design and Action, Smith, H.J. (ed.), 3rd ed., 1998.

MULTIPLE-CHOICE QUESTIONS

1. Which one of the following belongs to long-acting barbiturate?
    a. Pentobarbital
    b. Thiopental
    c. Phenobarbital
    d. Hexabutal

2. Replacement of the oxygen at C-2 of barbituric acid by a sulphur atom
   a. Has no change on activity
   b. Increases activity
   c. Decreases activity
   d. Shows anxiolytic activity
3. Clorazepate is
   a. 7-Chloro-1,3-dihydro-3-hydroxy-5-phenyl-1,4-benzodiazepine-2-one
   b. 7-Chloro-1,3-dihydro-1-methyl-5-phenyl-1,4-benzodiazepine-2-one
   c. 7-Chloro-2,3-dihydro-2-oxo-5-phenyl-1,4-benzodiazepine-3-carboxylic acid
   d. 7-Chloro-1,3-dihydro-2-oxo-5-phenyl-1,4-benzodiazepine-3-carboxylic acid
4. One of the following is not a triazolobenzodiazepine derivative:
   a. Alprazolam
   b. Triazolam
   c. Midazolam
   d. Estazolam
5. The drug that does not act on GABA receptor is
   a. Zopiclone
   b. Pentobarbitone
   c. Buspirone
   d. Brotizolam
6. One of the following is 'false' about benzodiazepines:
   a. Alkyl substituents at 3-position decreases the activity.
   b. The N-substituent at 1 position should be small.
   c. A phenyl or pyridyl at the 5-position decreases activity.
   d. The presence of electron-attracting substituents (Cl, F, Br, NO_2) at position 7 is required for activity.
7. Anxiolytic drug with no drowsiness is
   a. Diazepam
   b. Meprobamate
   c. Buspirone
   d. Alprazolam
8. Benzodiazepines in which benzene ring at the 5-position when omitted
   a. Increases the affinity towards the receptor
   b. Acts as antagonist to the receptor
   c. Acts at serotonin receptor
   d. Shows reduced sedative and hypnotic activity
9. Nordiazepam when alkylated with trifluoromethylbromide yields
   a. Diazepam
   b. Halozepam
   c. Oxazepam
   d. Clonazepam

QUESTIONS

1. Classify barbiturates on the basis of the duration of action with structural examples for each class.
2. Give the synthetic protocol of nordiazepam.
3. Discuss in detail the SAR of benzodiazepines.
4. Explain how the barbiturates act as sedative and hypnotics.
5. Give the general method of synthesis of barbiturates.
6. Differentiate between a sedative and a hypnotic.
7. Drugs such as benzodiazepines and alcohol that are generally depressant on behaviour often produce excitatory effects, particularly at low doses. How would you explain these apparently paradoxical effects?
8. How do benzodiazepines affect the action of GABA?
9. What are some of the current clinical uses of benzodiazepines?
10. What are the behavioural effects of sedative-hypnotic drugs in humans?

SOLUTION TO MULTIPLE-CHOICE QUESTIONS

1. c;
2. b;
3. c;
4. c;
5. c;
6. c;
7. c;
8. b;
9. b;

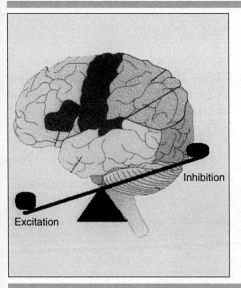

CHAPTER 8

Anti-Epileptic Drugs

LEARNING OBJECTIVES

- Define epilepsy and the causes of the disorder
- Categorize various forms of the disorder
- Categorize the anticonvulsant drugs based on the chemical class
- Learn the basic structural composition for a molecule to act as anticonvulsants
- Define pharmacophore model and how it can be derived
- Categorize anticonvulsants on the basis of treatment of different seizure types
- Define GABA and the importance of developing its analogues
- Describe some newer anticonvulsants

8.1 INTRODUCTION: EPILEPSY AND TYPES OF SEIZURES

Epilepsy is a brain disorder in which clusters of nerve cells, or neurons, in the brain sometimes signal abnormally. In epilepsy, the normal pattern of neuronal activity becomes disturbed, causing strange sensations, emotions, and behaviour, or sometimes convulsions, muscle spasms, and loss of consciousness. Epilepsy is a chronic disease that is characterised by paroxysmal attacks caused by pathologic excitation of cerebral neurons. Epilepsy is accompanied by various degrees of disturbance of consciousness. There are both convulsive and non-convulsive forms of epileptic attacks, each of which is characterised by distinctive clinical features. Moreover, there are specific changes in the electro-encephalogram for practically all varieties of epilepsy. Seizures are generated in the epileptogenic centre of the brain and can be nothing more than shaking of the extremities. If the convulsive discharge begins to spread and the excitation encompasses both hemispheres of the brain, seizures begin. Discharges induce major epileptic convulsive seizures (*grand mal*) and minor epileptic attacks (*petit mal*). Generally speaking, seizures are involuntary muscle contractions that can take place as a result of pathologic processes both inside and outside the brain. They can occur in response to toxins, trauma, hyperthermia, or medicinal overdose, or upon discontinuation of medication. Several mechanisms that underlie epilepsy have been postulated, including an imbalance between excitatory and inhibitory neurotransmission.

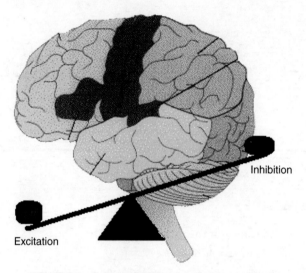

FIGURE 8.1 Representation of epileptic condition.

Epileptic seizures are categorized into four classes:

1. **Partial seizures** are those in which excessive neuronal discharges remain localized within a focal area of the brain. These are characterised by convulsions confined to a single limb, muscle group, or specific localized sensory disturbances usually without impairment of consciousness.
2. **Generalized seizures** impart unconsciousness. This includes those seizures known as *grand mal* and *petit mal*. Tonic-clonic (*grand mal*) seizures are characterised by major convulsions. Usually, there is a sequential tonic spasm of all the muscles in the body, followed by synchronous clonic jerking and a prolonged depression of all central functions. *Petit mal* (anti-absence) seizures are characterised by a brief and abrupt loss of consciousness, usually with some symmetrical clonic motor activity ranging from eyelid blinking to jerking of the entire body.
3. **Undetermined seizures** include neonatal seizures characterised by severe myoclonic epilepsy, and epilepsy with continuous spike-waves during slow-wave sleep.
4. **Special syndromes** or situation-related seizures occur only when there is an acute metabolic or toxic event due to alcohol, drugs, eclampsia, or non-ketonic hyperglycemia and febrile convulsions.

The mechanism of anti-epileptic drugs is not sufficiently clear, as the etiology of epilepsy is not yet completely understood. Some drugs block sodium channels, while others act on the GABA system. They enhance the GABA-dependent CNS inhibition. They also change the intracellular ratio of calcium and potassium ion concentrations, and block the *N*-methyl-D-aspartate (NMDA) receptor responsible for high-frequency discharges that appear during epilepsy.

8.2 CLASSIFICATION

1. Barbiturates: Phenobarbital, Mephobarbital, Metharbital (Refer to chapter 7 on 'Sedatives, Hypnotics, and Anxiolytic Agents')
2. Hydantoin (Imidazolidine-2,4-dione) derivative

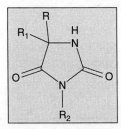

	R	R_1	R_2
Phenytoin	C_6H_5	C_6H_5	H
Phenylethyl hydantoin	C_6H_5	C_2H_5	H
Mephenytoin	C_6H_5	C_2H_5	CH_3
Ethotoin	C_6H_5	H	C_2H_5

3. Oxazolidinedione derivative

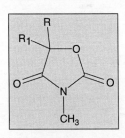

	R	R_1
Trimethadione	CH_3	CH_3
Paramethadione	CH_3	C_2H_5
Aloxidone	CH_3	$CH_2=CHCH_2$

4. Succinimides

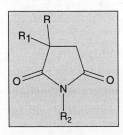

	R	R_1	R_2
Phensuximide	C_6H_5	H	CH_3
Methsuximide	C_6H_5	CH_3	CH_3
Ethosuximide	C_2H_5	CH_3	H

5. Benzodiazepines: Clobazam, clonazepam, diazepam, clorazepate (Refer to chapter 7 on 'Sedatives, Hypnotics, and Anxiolytic Agents')
6. Gamma-amino butyric acid (GABA) analogues: Progabide, tiagabin, vigabatrin, gabapentin
7. Miscellaneous: Carbamazepine, valproate, phenacemide, primidone
8. Newer anticonvulsants: Denzimol, dezinamide, fosphenytoin, lamotrigine, nafimidone, ralitoline, topiramate, zonisamide

8.3 STRUCTURAL REQUIREMENTS FOR ANTICONVULSANT AGENTS AND THEIR STRUCTURE–ACTIVITY RELATIONSHIP

It was stated that for a compound to act as anticonvulsants, the molecule should contain at least one aryl/lipophilic unit (**A**), one or two hydrogen acceptor–donor atoms (**HAD**), and an electron-donor atom (**D**) in a special spatial arrangement to be recommended for anticonvulsant activity. The well-known and structurally different compounds with anticonvulsant activity—carbamazepine, gabapentin, lamotrigine, mephobarbital, phenytoin, progabide, ralitoline, and zonisamide—are represented with their structural elements as follows.

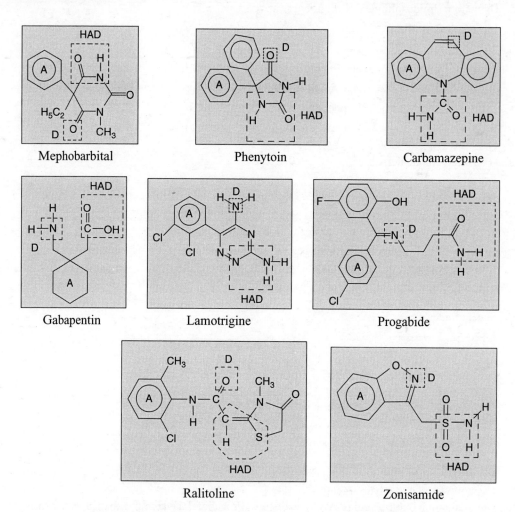

Mephobarbital   Phenytoin   Carbamazepine

Gabapentin   Lamotrigine   Progabide

Ralitoline   Zonisamide

A → Hydrophobic unit
D → Electron donor group
HAD → Hydrogen bond acceptor/donor unit

The distances between these structural elements should be optimal in the ranges depicted in the following figure. These distances are calculated using various computational tools for the 3D structures of the drugs.

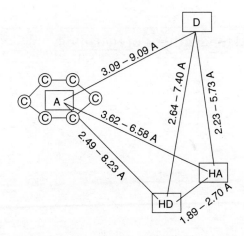

FIGURE 8.2 Distance ranges between the structural elements required for anticonvulsant activity.

Agents used for treating epileptic disorders		
Disorder	*Primary Agent(S)*	*Adjunctive Agent(S)*
Partial seizures		
Simple, complex, partial with secondarily generalized	Carbamazepine, Phenytoin, Phenobarbital, Primidone, Valproate	Lamotrigine, Gabapentin, Tiagabin, Topiramate
Generalized seizures		
Absence seizures	Ethosuximide, Valproate, Clonazepam	Lamotrigine
Myoclonic seizures	Valproate, Clonazepam	
Atonic seizures	Valproate	Lamotrigine
Tonic-clonic seizures (not preceded by partial seizure)	Carbamazepine, Phenytoin, Phenobarbital, Primidone, Valproate	Gabapentin, Lamotrigine
Status epilepticus	Diazepam, Lorazepam, followed by Phenytoin, Fosphenytoin	

8.4 HYDANTOINS
8.4.1 General Method for Synthesis of Hydantoin

Hydantoins are synthesised by the treatment of appropriate carbonyl compound with sodium cyanide in the presence of excess ammonium carbonate. The first step in this complex sequence can be visualized as addition of the elements of ammonia and hydrogen cyanide to give an α-aminonitrile. Addition of ammonia to the cyano group would then lead to an amidine. Carbon dioxide or carbonate ion present in the reaction mixture can then add to the basic amidine to afford carbamic acid-like intermediate; attack by the adjacent amino group will then close the ring and afford the imino derivative. This compound is then hydrolysed to a hydantoin by treatment with acid.

Phenytoin Using benzophenone as the starting material

> Phenytoin (diphenylhydantoin) was first synthesised by German chemist Heinrich Biltz in 1908. Biltz sold his discovery to Parke-Davis, which did not find an immediate use for it. In 1938, outside scientists including H. Houston Merritt and Tracy Putnam discovered phenytoin's usefulness for controlling seizures.

Phenylethyl hydantoin Using propiophenone as the starting material
Mephenytoin Alkylation of phenylethyl hydantoin with methyl iodide in the presence of base
Ethotoin Prepared from benzaldehyde and alkylation of hydantoin with ethyl iodide

Mechanism of action: Phenytoin and other hydantoins block the voltage-gated sodium channels in the brain. Voltage-gated sodium channels are responsible for the generation of action potentials of nerve fibres through selective transport of sodium ions across the cell membrane, leading to the rapid depolarization of the cell network and thereon to electrical excitability. Hydantoin inhibits the influx of sodium ions, prevents depolarization, and decreases electrical excitability.

Use: *Grand mal* seizures (phenytoin is the drug of choice) and complex partial seizures

8.5 OXAZOLIDINEDIONES
8.5.1 General Method for Synthesis of Oxazolidinedione Derivatives

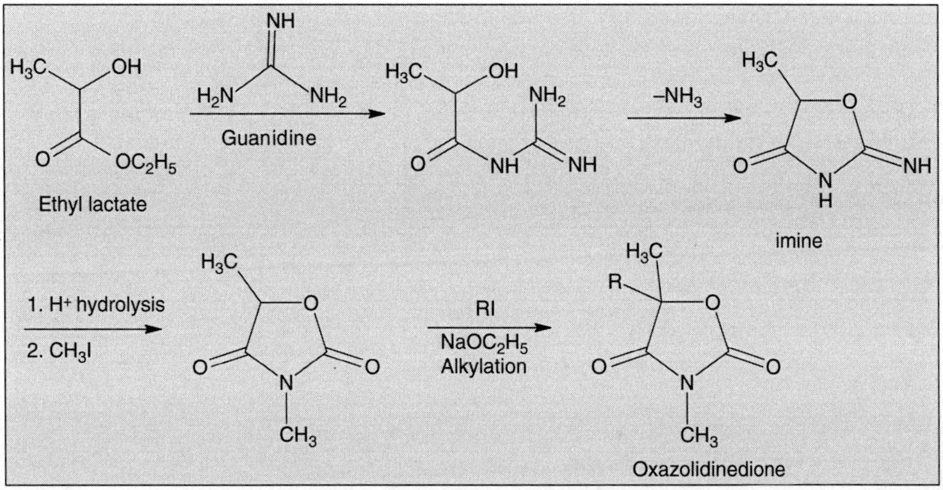

The reaction between ethyl lactate with guanidine proceeds probably via interchange of the ester to an acylated guanidine derivative. Addition of alkoxide to the imine followed by loss of ammonia leads to the formation of the iminooxazolidone. Further alkylation of this intermediate leads to various titled drugs.

Trimethadione Alkylation with CH_3I
Paramethadione Alkylation with C_2H_5I
Aloxidone Alkylation with $CH_2=CH-CH_2I$

Mechanism of action: Oxazolidinediones block T-type, voltage-dependent calcium channels in thalamic neurons and block the influx of calcium ions, thereby preventing the depolarization of the membrane. This decreases the electrical excitability of the neurons.

Use: *Petit mal* seizures

8.6 SUCCINIMIDES

Succinimides are a group of drugs that are derived from the amide of succinic acid. They are used in minor forms of epilepsy in which attacks are not observed.

8.6.1 General Method for the Synthesis of Succinimide

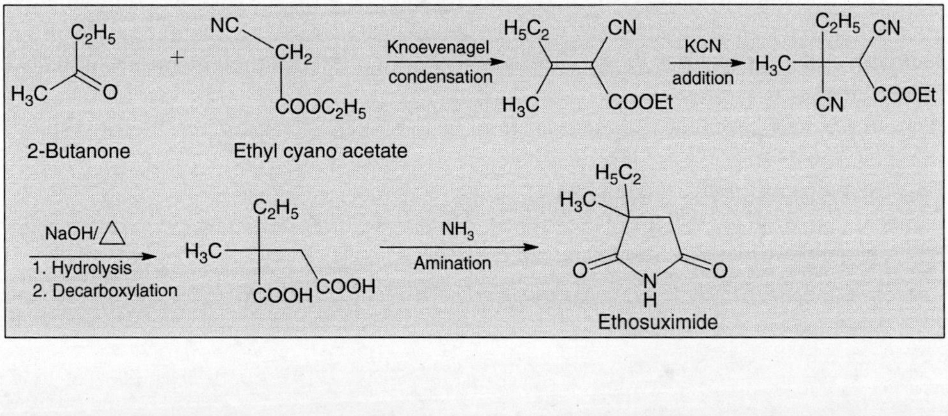

Ethosuximide (3-ethyl-3-methypyrrolidine-2,5-dione) is synthesised from methylethylketone and cyanoacetic ester, which are condensed in Knoevenagel reaction conditions. Then potassium cyanide is added to the resulting product. After acidic hydrolysis and decarboxylation of synthesised dinitrile, 2-methyl-2-ethylsuccinic acid is formed. Reacting this product with ammonia gives ethosuximide.

Phensuximide Synthesis starts with benzaldehyde (C_6H_5CHO) and in the last-step amination with CH_3NH_2.

Methsuximide Synthesis starts with acetophenone ($C_6H_5COCH_3$) and the amine is CH_3NH_2.

Mechanism of action: It inhibits the T-type, voltage-dependent calcium channels in thalamic neurons.

Use: Ethosuximide is the drug of choice for *petit mal* seizures.

8.7 BENZODIAZEPINES

Clobazam 1,5-Benzodiazepine derivative

Synthesis

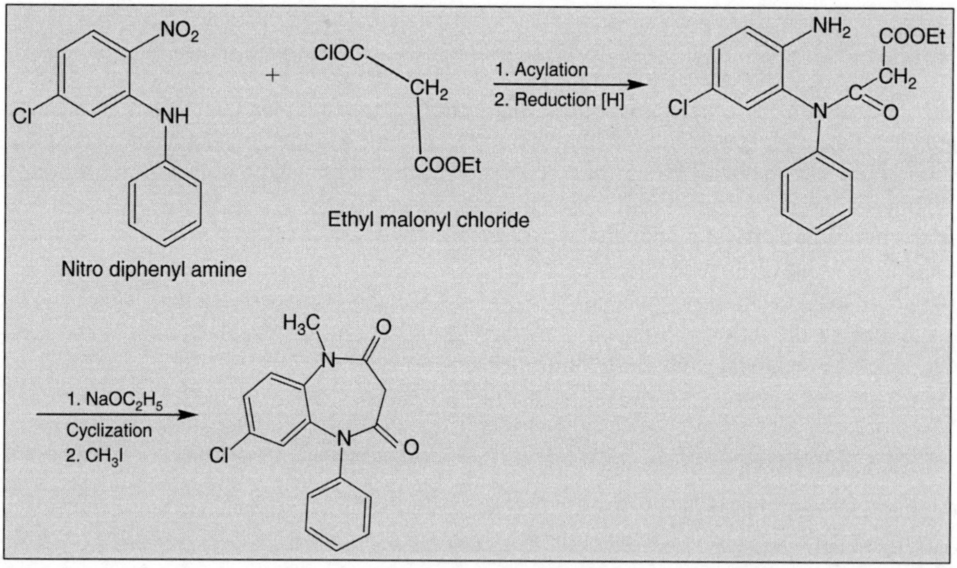

Acylation of the nitrodiphenylamine with ethyl malonyl chloride gives the corresponding amidoester; the nitro group is then reduced to the amine by catalytic hydrogenation to give the amino amido intermediate. Treatment with strong base closes the 1,5-benzodiazepine ring. Methylation by means of methyl iodide affords clobazam.

(For other benzodiazepines and mechanism of action, refer to chapter 7 on 'Sedatives, Hypnotics, and Anxiolytic Agents'.)

Uses: Myoclonic epilepsy (Clonazepam)
Partial and generalized epilepsy (Clobazam)
Status epilepticus or severe *grand mal* epilepsy (Diazepam)

8.8 GABA ANALOGUES

GABA is an inhibitory neurotransmitter. It cannot cross the blood-brain barrier. This problem is overcome by enhancing the lipid solubility by formation of Schiff's base of gabamide.

> In vertebrates, GABA acts at inhibitory synapses in the brain by binding to specific transmembrane receptors in the plasma membrane of both pre- and post-synaptic neuronal processes. This binding causes the opening of ion channels to allow the flow of either negatively charged chloride ions into the cell or positively charged potassium ions out of the cell. This action results in a negative change in the transmembrane potential, usually causing hyper polarization.

Progabide 4-[[(4-Chlorophenyl)(5-fluoro-2-hydroxyphenyl) methylene] amino] butanamide

Progabide cross the BBB readily and is biotransformed into its active acid metabolite. It acts as agonists for $GABA_A$ and $GABA_B$ receptors.

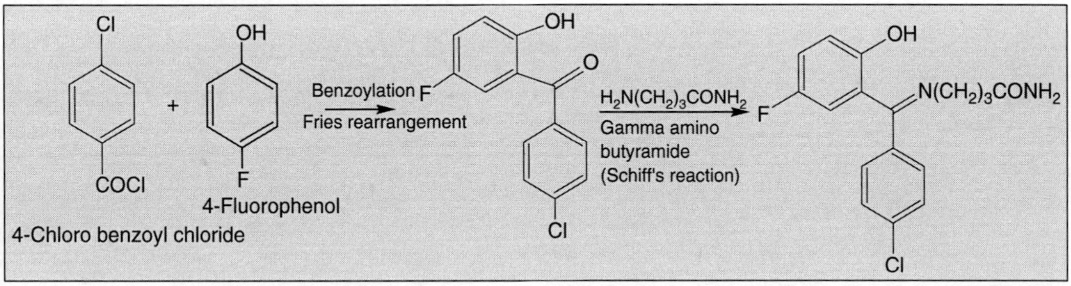

Acylation of 4-chlorophenol with 4-chlorobenzoyl chloride gives, after Fries rearrangement, benzophenone. Formation of the imines of this ketone with the amide of GABA gives progabide.

Progabide is the GABA agonist.

Uses: Simple and complex partial seizures, generalized tonic-clonic seizures, and myoclonic seizures

Vigabatrin γ-Vinyl GABA; 4-amino-5-hexenoic acid

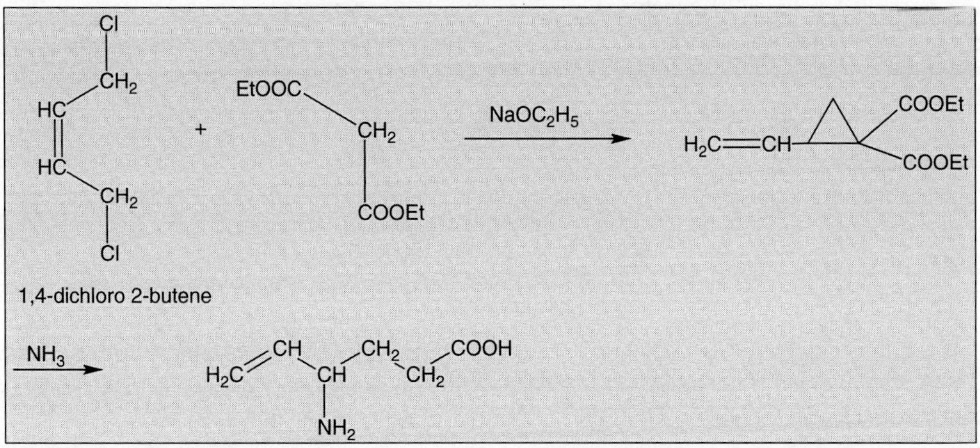

Diethyl malonate undergoes alkylation with 1,4-dichloro-2-butene in the presence of sodium ethoxide to form cyclic intermediate, which in turn reacts with ammonia to form vinyl GABA.

Mechanism of action: Vigabatrin readily crosses the BBB and raises brain GABA levels by virtue of GABA-transaminase enzyme inhibition. GABA-T is responsible for metabolism of GABA. The elevation of GABA in brain is most important in its anticonvulsant mechanism of action.

Tiagabin (R)1-[4,4-Bis(3-methyl-2-thienyl)-3-butenyl] nipecotic acid

Bis-thiophenyl ketone is prepared by the addition of thiophene-lithium reagent to thiophene carboxylate. The ketone is allowed to react with cyclopropylmagnesium bromide to give the tertiary alcohol. This readily forms carbocation with treatment with hydrogen bromide; that reactive species then undergoes the cyclopropylcarbynyl-homoallyl rearrangement to give the unsaturated bromide. Treatment of the bromide with ethyl nipecotate affords tiagabin.

Mechanism of action: Tiagabin is a GABA reuptake inhibitor that increases extracellular concentration of GABA, which in turn opens the chloride ion channels. These inhibit neuronal activity by hyper polarization and depolarization block.

Use: Partial seizures.

Gabapentin

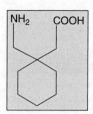

It is a lipid-soluble GABA analogue. It does not bind with $GABA_A$ receptor, causes no inhibition on GABA reuptake, and is not a GABA-T (GABA aminotransferase enzyme that metabolises GABA to succinic semi-aldehyde) inhibitor. Thus, the mechanism of action is unknown.

Medicinal Chemistry

8.9 MISCELLANEOUS

Carbamazepine Dibenzazepine 5-carboxamide

Carbamazepine, a urea derivative, is a broad-spectrum anti-seizure agent, but it is toxic.

Synthesis

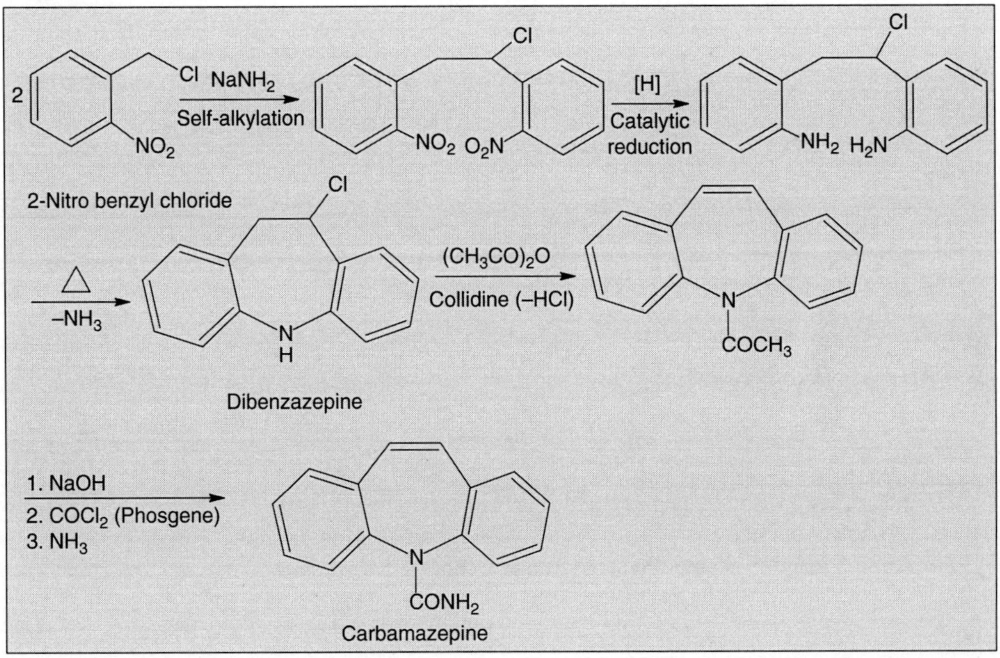

[Scheme: 2 equivalents of 2-nitrobenzyl chloride → NaNH$_2$ / Self-alkylation → dinitro diphenylmethane chloride intermediate → [H] Catalytic reduction → diamino intermediate → Δ, –NH$_3$ → Dibenzazepine (with CHCl group) → (CH$_3$CO)$_2$O, Collidine (–HCl) → N-acetyl dibenzazepine (COCH$_3$) → 1. NaOH, 2. COCl$_2$ (Phosgene), 3. NH$_3$ → Carbamazepine (CONH$_2$)]

Two molecules of 2-nitro benzaldehyde undergo self-alkylation in the presence of base; this followed by catalytic reduction gives diamino intermediate. Heating leads to formation of dibenzazepine nucleus, which on treatment with collidine followed by treatment with phosgene and ammonia affords carbamazepine.

Uses: Like phenytoin, carbamazepine inhibits voltage-dependent sodium channels and is used to treat partial seizures and *grand mal* seizures. It is also useful in the treatment of pain associated with trigeminal neuralgia.

Sodium Valproate Sodium salt of 2-propyl pentanoate

Synthesis

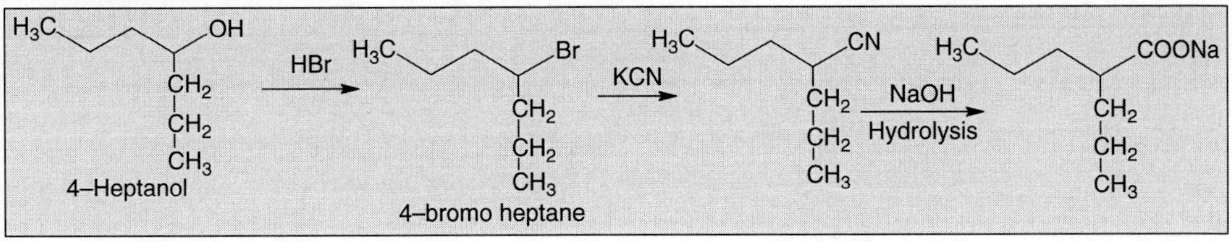

[Scheme: 4-Heptanol → HBr → 4-bromo heptane → KCN → nitrile (CN) → NaOH Hydrolysis → sodium 2-propyl pentanoate (COONa)]

4-Heptanol on treatment with hydrogen bromide affords 4-bromo heptane, which on treatment with potassium cyanide followed by alkaline hydrolysis affords sodium valproate.

Mechanism of action: Valproate exerts its anticonvulsant activity by blocking voltage-gated sodium ion channels and increases the GABA level in brain. The increase in GABA level is due to the inhibition of several enzymes involved in GABA metabolism and increases the activity of glutamic acid decarboxylate (GAD), the enzyme responsible for GABA synthesis.

Uses: Valproate is employed for the management of myoclonic and generalized tonic-clonic seizures.

Phenacemide Phenylacetyl urea

Synthesis

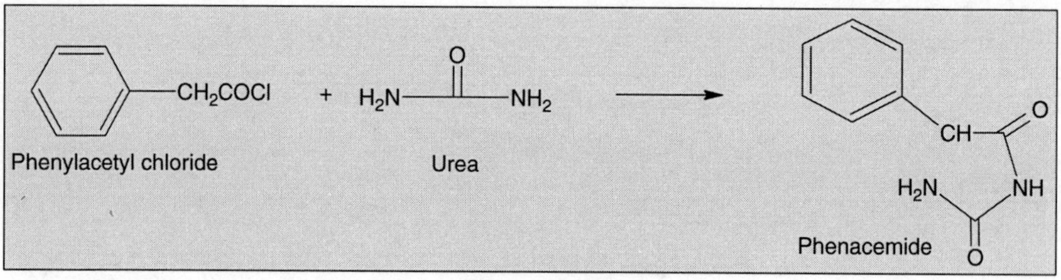

Acylation of urea with phenylacetyl chloride gives the phenacemide.
Use: It is a ring-opened analogue of 5-phenylhydantoin used for complex partial seizures.
Primidone 5-Ethyl-5-phenyldihydropyrimidine-4,6(1H,5H)-dione

Synthesis

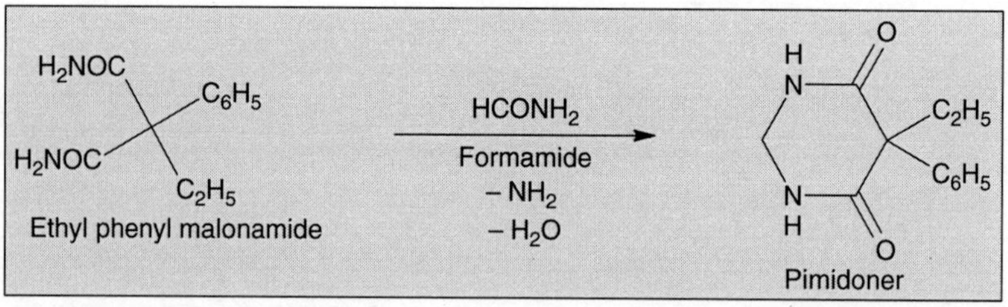

Ethyl phenyl malonamide condenses with formamide and affords primidone.
Uses: It is a 2-deoxy phenobarbital useful for the management of *grand mal*, complex partial, and focal epileptic seizures.

8.10 NEWER ANTICONVULSANTS

Over the past five years, a handful of new agents have been approved for use as anti-seizure medications.

Denzimol N-[β-[4-(β-Phenylethyl)phenyl]- β-hydroxyethyl]imidazole

Synthesis

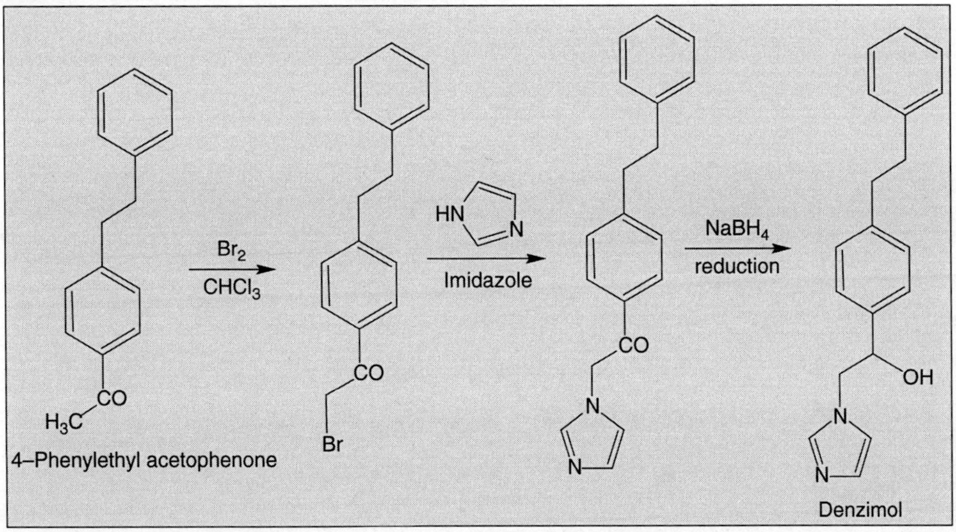

4-Phenylethyl acetophenone undergoes bromination in chloroform to give bromo acetophenone derivative, which on further reaction with imidazole undergoes dehydrohalogenation to form ketone intermediate. Ketone group is further reduced with sodium borohydride to afford alcohol denzimol.

Use: Complex partial seizures.

Dezinamide 3-[(3-Trifluoromethyl)phenoxy]-1-azetidine carboxamide

Synthesis

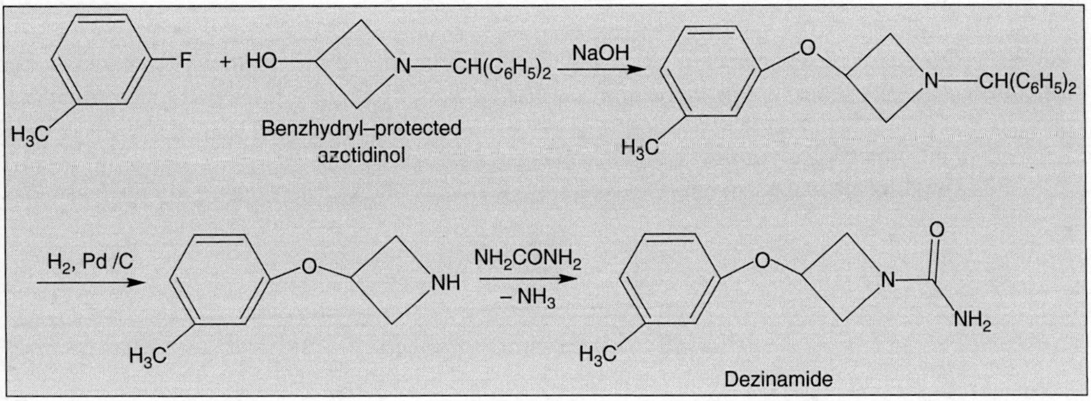

Replacement of fluorine in 3-fluorotoluene by oxygen in the benzhydryl-protected azetidinol in the presence of base leads to the ether. The protecting group is then removed by catalytic hydrogenation to afford the secondary amine. Amine exchange with urea proceeds to form the new urea and, thus, dezinamide.

Uses: Partial and *petit mal* seizures.

Fosphenytoin (2,5-Dioxo-4,4-diphenyl-imidazolidin-1-yl)methoxyphosphonic acid

It is a water-soluble phosphoric acid prodrug of phenytoin for administering the drug parenterally. Fosphenytoin has minimal activity before it is cleaved, but it is a water-soluble drug form that is suitable for intravenous injection. This preparation will be used to avoid the venous irritation produced when using parenteral phenytoin for status epilepticus.

Synthesis

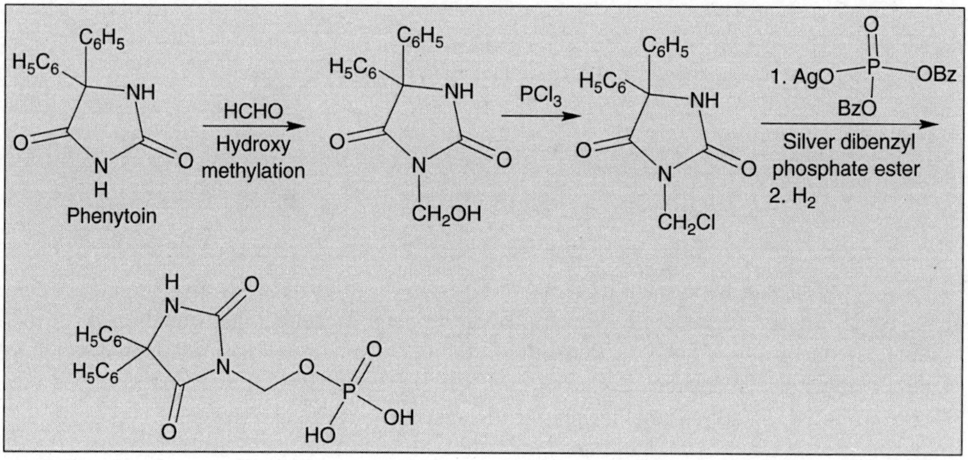

The first step in the preparation starts with the formation of the carbinolamine from addition of formaldehyde to phenytoin. The hydroxyl group is then replaced by chlorine through reaction with phosphorous trichloride. Displacement of halogen using silver dibenzylphosphate gives the phosphate ester. Removal of the protecting groups by hydrogenolysis gives fosphenytoin.

It is a prodrug form of phenytoin, and is rapidly cleaved by phosphatases *in vivo* to form phenytoin, as shown below.

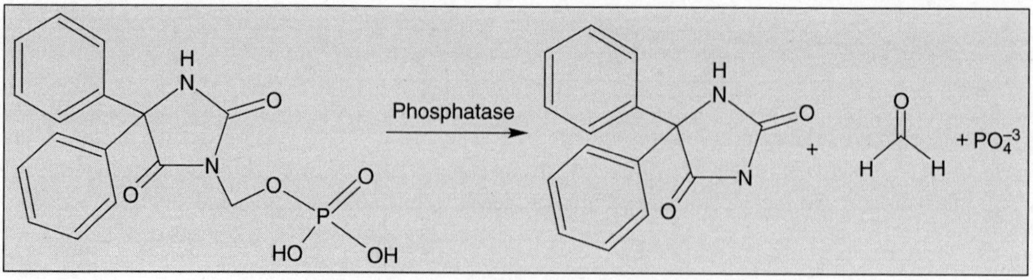

Lamotrigine 3,5-Diamino-6-(2,3-dichlorophenyl)-1,2,4-triazine

Synthesis

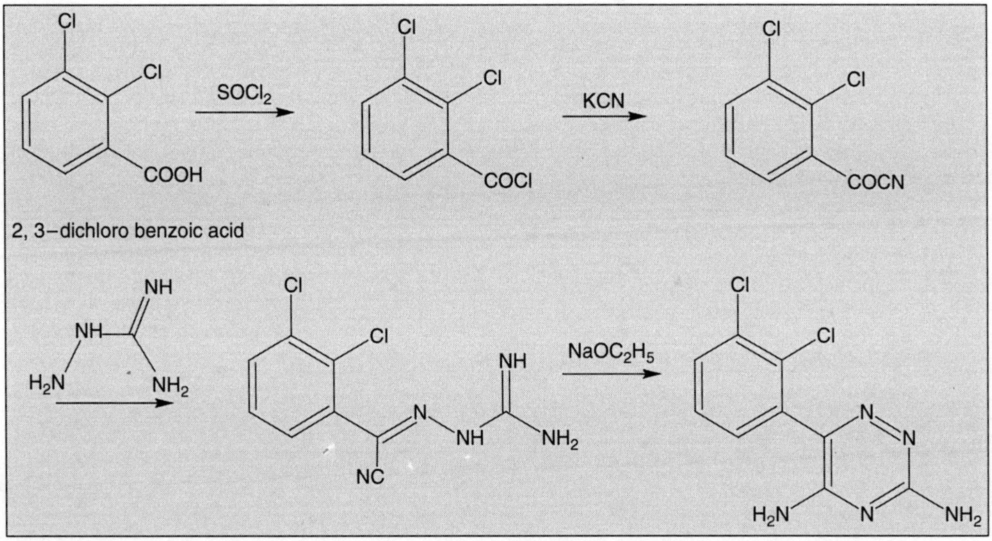

Acyl cyanide is prepared from acid by treatment with thionyl chloride and potassium cyanide. Condensation of acyl cyanide with aminoguanidine gives the imine as the initial product. Treatment of imine with base leads to addition of guanidino anion to the nitrile and, followed by bond reorganization, affords lamotrigine.

Uses: It is a sodium channel blocker used for the management of generalized tonic-clonic, *petit mal*, myoclonic seizures and myoclonic jerks in children.

Nafimidone 1-(2-Naphthoylmethyl)imidazole

Synthesis

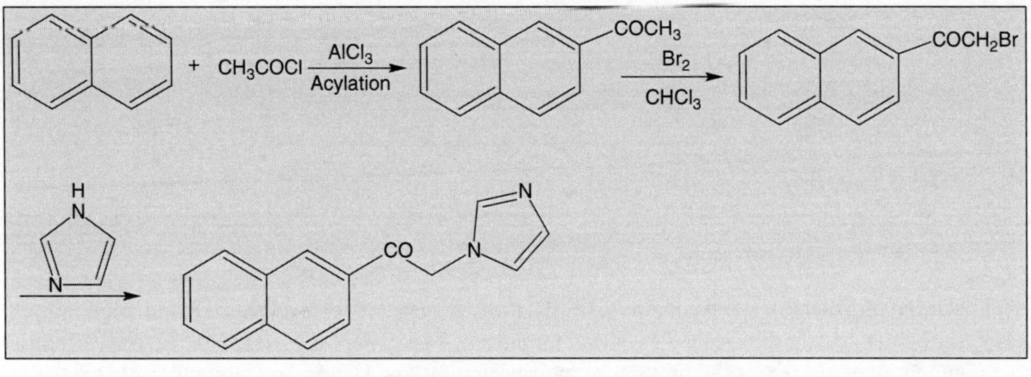

Nafimidone is prepared from naphthalene by Friedel–Crafts acylation with acetyl chloride, followed by bromination and finally reaction with imidazole.

Use: Tonic clonic and partial seizures.

Ralitoline N-(2-Chloro-6-methylphenyl)-(3-methyl-4-oxothiazolidine-2-ylidene)acetamide

Synthesis

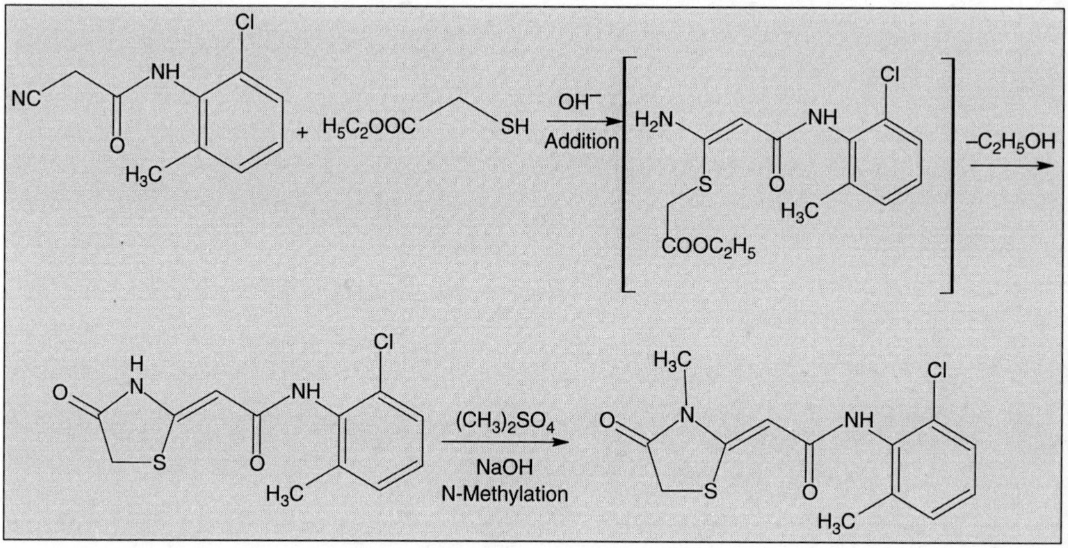

The synthesis starts with the base-catalysed addition of ethyl thioglycolate to the cyanoacetamide derivative. The mercapto group adds to the nitrile to give an unstable intermediate. The amine next displaces ethoxide from the adjacent ester under the reaction conditions to form thiazolone. Methylation of amide nitrogen with dimethyl sulphate and base then affords ralitoline.

Use: It is a sodium channel blocker used for the management of generalized seizures.

Topiramate 2,3:4,5-Bis-O-(1-methylethylidene)- β-D-fructopyranose sulphamate

Synthesis

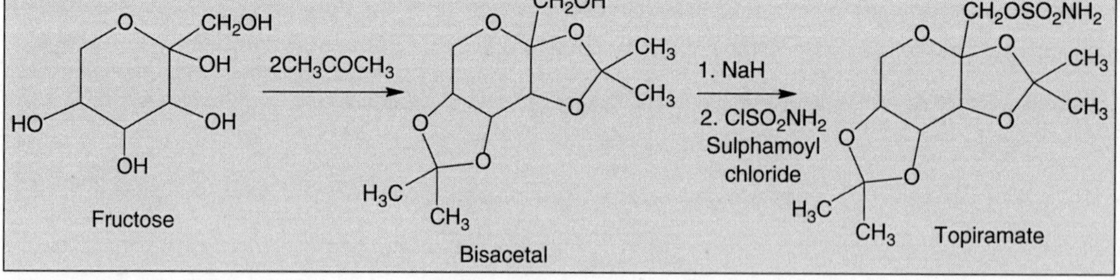

Reaction of fructose with acetone affords the bisacetal. The reactivity of the remaining free hydroxyl group is greatly diminished by steric hindrance. The group, thus, needs to be converted to the corresponding alkoxide by means of sodium hydride for further elaboration. Reaction of that alkoxide with sulphamoyl chloride leads to topiramate.

Use: Partial seizures.

Zonisamide 1,2-Benzisoxazole-3-methane sulphonamide.

Synthesis

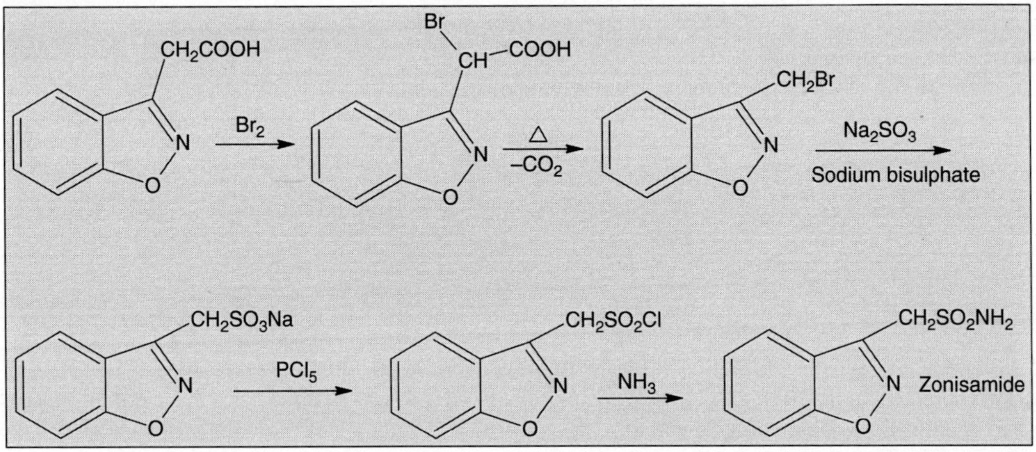

1,2-Benzisoxazole-3-acetic acid undergoes α-bromination on treatment with bromine; bromo derivative, on heating, undergoes decarboxylation to form bromomethyl intermediate. This in turn reacts with sodium bisulphate, phosphorous pentachloride, and ammonia, and converts to methane sulphonamide.

Use: Sodium channel blocker prevents tonic seizures.

Stiripentol 4,4-Dimethyl-1-[3,4(methylenedioxy)-phenyl]-1-penten-3-ol

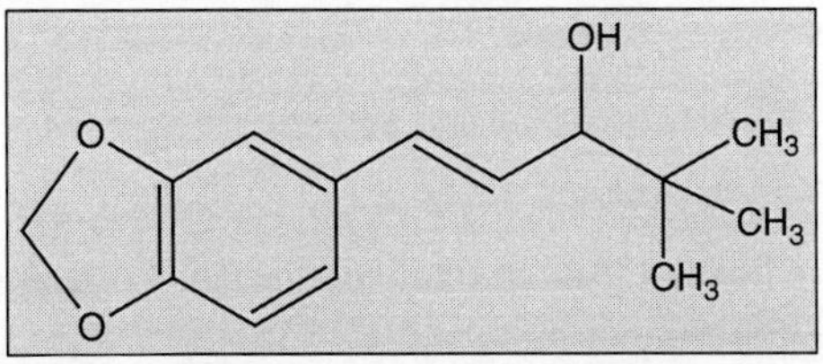

It is an anticonvulsant drug used in the treatment of epilepsy. It is unrelated to other anticonvulsants and belongs to the group of aromatic allylic alcohols. It can prevent the reuptake of GABA and inhibit its metabolism. It is indicated as an adjunctive therapy with sodium valproate and clobazam for treating severe myoclonic epilepsy in infancy.

Oxcarbazepine 10,11-Dihydro-10-oxo-5 H -dibenz(b,f)azepine-5-carboxamide

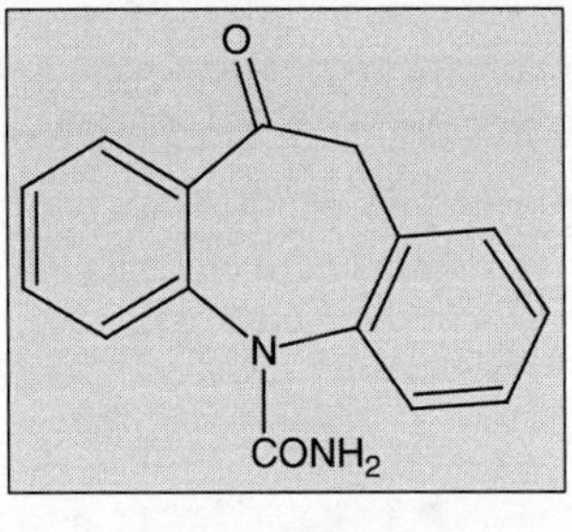

It is structurally a derivative of carbamazepine, adding an extra oxygen atom on the dibenzazepine ring. This difference helps reduce the impact on the liver of metabolising the drug, and also prevents the serious forms of anaemia or agranulocytosis occasionally associated with carbamazepine. Aside from this reduction in side-effects, it is thought to have the same mechanism as carbamazepine—that is, sodium channel inhibition.

Pregabalin (S)-3-(Aminomethyl)-5-methylhexanoic acid

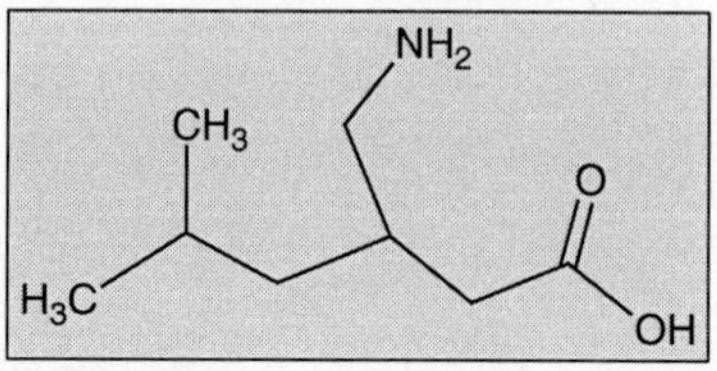

It is an anticonvulsant drug used for neuropathic pain, as an adjunct therapy for partial seizures. It has also been found effective for generalized anxiety disorder. It was designed as a more potent successor to gabapentin. Like gabapentin, pregabalin binds to the α2δ subunit of the voltage-dependent calcium channel in the central nervous system.

Levetiracetam 2-(2-Oxopyrrolidin-1-yl) butanamide

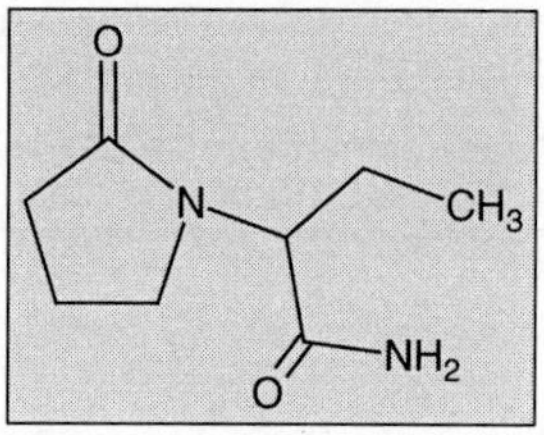

It is an anticonvulsant medication used to treat epilepsy. It is S- enantiomer of etiracetam, structurally similar to the prototypical nootropic drug piracetam.

FURTHER READINGS

1. Bauer, J. et al., 'Medical treatment of epilepsy', *Expert Opinion on Emerging Drugs*, 8, 457–467, 2003.
2. Beyenburg, S. et al., 'New drug for the treatment of epilepsy', *Postgraduate Medical Journal*, 80, 581–587, 2004.
3. P. Yogeeswari, J. Vaigunda Raghavendran, R. Thirumurugan, A. Saxena, and D. Sriram, 'Ion channels as important targets for antiepileptic drug design', *Curr. Drug Targets*, 5, 589–602, 2004.
4. P. Yogeeswari, D. Sriram, J.V. Raghavendran, and R. Thirumurugan, 'The GABA shunt: An attractive and potential therapeutic target in the treatment of epileptic disorders', *Curr. Drug Metabolism*, 6, 127–140, 2005.

MULTIPLE-CHOICE QUESTIONS

1. One of the following belongs to imidazolidine-2,4-dione class:
   a. Phenytoin
   b. Trimethadione
   c. Phensuximide
   d. Paramethadione
2. .... is the drug of choice for *grand mal* epilepsy.
   a. Diazepam
   b. Carbamazepine
   c. Phenytoin
   d. Lamotrigine
3. Oxazolidone acts by
   a. Acting as agonist for $GABA_A$ receptor
   b. Blocking the sodium ion channels
   c. Blocking the calcium ion channels
   d. Blocking potassium channels
4. Mechanism of action of carbamazepine is
   a. Allosterically enhances GABAergic inhibition
   b. Reduces L-glutamate-mediated excitation
   c. Blocks sustained repetitive neuronal firing
   d. Reduces calcium influx into neurons
5. Class of carbamazepine is
   a. Benzodiazepines
   b. Succinimides
   c. Barbiturates
   d. Iminostilbenes
6. Ethosuximide is indicated for
   a. Delirium tremor
   b. Generalized absence seizures
   c. *Petit mal* seizures
   d. Partial motor seizures
7. Mechanism of action of phenytoin is
   a. Blocks sustained repetitive neuronal firing
   b. Reduces L-glutamate-mediated excitation
   c. Enhances pre-synaptic release of GABA
   d. Allosterically enhances GABAergic inhibition
8. Indication of clonazepam is
   a. Myoclonic epilepsy
   b. Partial motor seizures
   c. Status epilepticus
   d. Partial complex seizures
9. Lamotrigine is effective in
   a. Partial seizures
   b. Generalized seizures
   c. Absence seizures
   d. All of the above
10. An enzyme inducer, this anticonvulsant is one of the metabolites of primidone (also an anticonvulsant).
    a. Phenytoin
    b. Phenobarbitone
    c. Carbamazepine
    d. Diazepam
11. This anticonvulsant is a prodrug derived from carbamazepine, with the benefits of fewer side-effects and fewer drug interactions.
    a. Lamotrigine
    b. Oxcarbazepine
    c. Procarbazepine
    d. Fosphenytoin

QUESTIONS

1. Explain, with structures, why oxcarbazepine would be expected to possess fewer side-effects than CBZ.
2. Why agents active against partial and secondary generalized tonic-clonic seizures are thought to be effective?
3. What are the advantages of fosphenytoin compared to phenytoin?
4. Which of the anti-seizure drugs is derived from fructose?
5. Explain the structural requirements for a compound to exhibit anticonvulsant properties.

6. Describe the differences between partial and generalized seizures, and between simple and complex partial seizures.
7. Name the heterocyclic ring present in the following drugs:
   a. Phenytoin
   b. Ethosuximide
   c. Carbamazepine
   d. Trimethadione
   e. Lamotrigine
   f. Phenobarbitone
8. Clonazepam, a benzodiazepine-type anticonvulsant, enhances GABAergic transmission by a mechanism other than direct $GABA_A$ receptor interaction. Explain how this is possible.

SOLUTION TO MULTIPLE-CHOICE QUESTIONS

1. a;
2. c;
3. c;
4. c;
5. d;
6. c;
7. a;
8. a;
9. d;
10. b;
11. b.

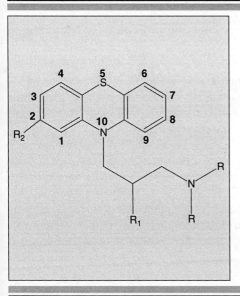

CHAPTER 9

Antipsychotic Agents

LEARNING OBJECTIVES

- Define antipsychotics and their utility in various conditions
- Categorize antipsychotics on the basis of the target at which they act
- Describe the synthetic routes of each antipsychotics
- Define the utility and doses of each antipsychotics
- Describe the mechanism by which antipsychotics act, and their implications
- Define the discovery and functional utility of phenothiazines
- Discuss the structure-activity relationship on phenothiazines
- Describe the neuroleptic SAR of butyrophenone
- Define atypical antipsychotics and the need for design
- Describe some newer drugs for psychosis treatment

9.1 INTRODUCTION

Antipsychotic drugs are used to treat psychosis, a severe mental illness in which people lose touch with reality (schizophrenia). Schizophrenia represents a group of psychiatric disorders characterized by disordered thought and behaviour. The symptoms of schizophrenia can be divided into two types: positive and negative. Positive symptoms are those not normally found in healthy individuals, including hallucinations, delusions, and thought disorder. Negative symptoms represent the loss of qualities normally present in healthy individuals, including impoverishment of thought, blunted emotion, attention deficit, and lack of motivation or initiative. Schizophrenia is a chronic disorder, typically arising in early adulthood and progressing throughout the rest of the individual's life. Although schizophrenia afflicts a relatively small percentage of the population, approximately one per cent over a lifetime, the impact is significant in disruption of families, loss of productivity, and a high incidence of suicide and premature death.

People with psychosis may hear voices, see things that are not really there, and have strange or untrue thoughts, such as believing that other people can hear their thoughts or are trying to harm them. They may also neglect their appearances and may stop talking or talk only 'nonsense'. Antipsychotic drugs are often known as **major tranquillizers, neuroleptics,** or **psychotropics.**

The specific etiology of psychotic disorders has not currently been sufficiently investigated. It is believed, however, that the initial cause of psychotic behaviour may originate from an imbalance of dopaminergic functions in the central nervous system (CNS). Many researchers adhere to the opinion that a large increase of dopamine activity in specific regions of the CNS is the cause of abnormal behaviour.

Antipsychotics are drugs that have a specific sedative effect, and improve the attitude and calm the behaviour of psychotic patients. These drugs cause emotional calmness and are extremely effective for treating patients with severe and chronic symptoms. A clear distinction should be made between antipsychotics used for treating severe and chronic psychosis, and anxiolytics intended for treating anxiety and stress associated with psychoneurotic or psychosomatic disorders. Antipsychotic drugs have a significantly stronger effect on the central nervous system, but they are not CNS depressants, and as a rule they are more toxic. However, even in long-term use they do not cause dependence and addiction, which is a very serious problem that originates from long-term use of anxiolytics. (*See Fig. 9.1 in the coloured set of pages.*)

Antipsychotic agents are classified into two types:

- Typical or classical antipsychotic agents
- Atypical antipsychotic agents

9.2 TYPICAL OR CLASSICAL ANTIPSYCHOTIC AGENTS: SYNTHESIS, MECHANISM OF ACTION, AND SAR

Typical or classical antipsychotic agents are subdivided into the following categories:

1. Phenothiazine derivatives

(i) Propyl dialkylamino side chain

	R_1	R
Promazine	—(CH$_2$)$_3$—N—(CH$_3$)$_2$	H
Chlorpromazine	—(CH$_2$)$_3$—N—(CH$_3$)$_2$	Cl
Triflupromazine	—(CH$_2$)$_3$—N—(CH$_3$)$_2$	CF$_3$
Piperacetazine	—(H$_2$C)$_3$—N⟨piperidine⟩—CH$_2$CH$_2$OH	COCH$_3$

(ii) Ethyl piperidyl side chain

	R	R_1
Thioridazine	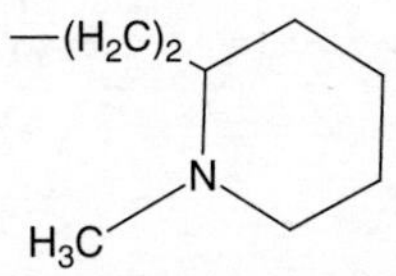 —(H$_2$C)$_2$—[1-methylpiperidin-2-yl]	SCH$_3$
Mesoridazine	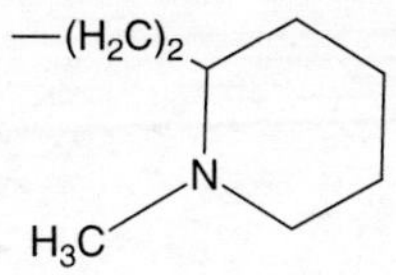 —(H$_2$C)$_2$—[1-methylpiperidin-2-yl]	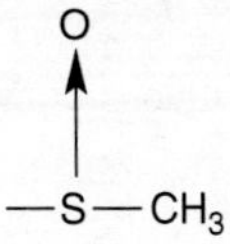 —S(=O)—CH$_3$

(iii) Propyl piperazine side chain

	R	R_1
Prochlorperazine	—(H$_2$C)$_3$—N(piperazine)N—CH$_3$	Cl
Trifluperazine	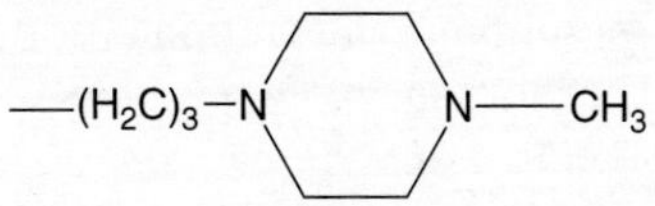 —(H$_2$C)$_3$—N(piperazine)N—CH$_3$	CF$_3$
Thiethylperazine	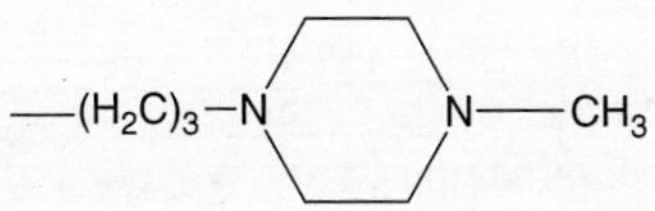 —(H$_2$C)$_3$—N(piperazine)N—CH$_3$	SCH$_2$CH$_3$
Perphenazine	—(H$_2$C)$_3$—N(piperazine)N—CH$_2$CH$_2$—OH	Cl
Fluphenazine	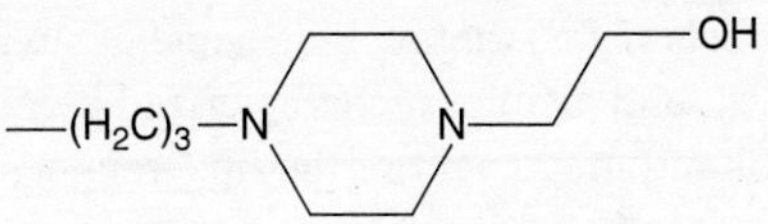 —(H$_2$C)$_3$—N(piperazine)N—CH$_2$CH$_2$—OH	CF$_3$
Acetophenazine	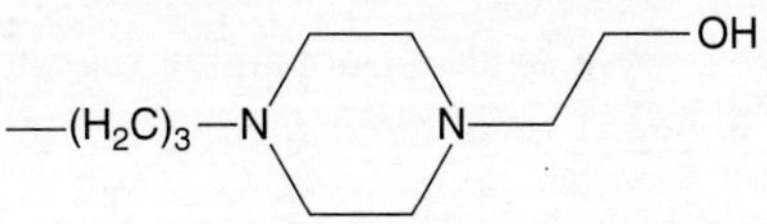 —(H$_2$C)$_3$—N(piperazine)N—CH$_2$CH$_2$—OH	COCH$_3$
Carphenazine	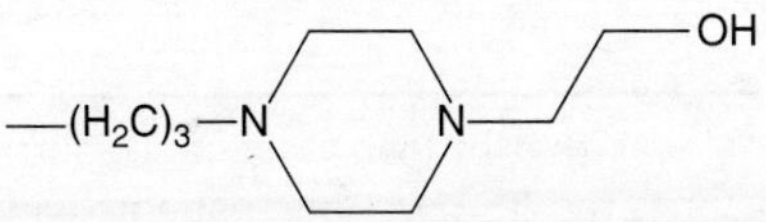 —(H$_2$C)$_3$—N(piperazine)N—CH$_2$CH$_2$—OH	COCH$_2$CH$_3$

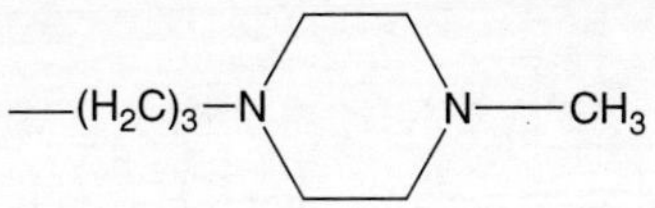

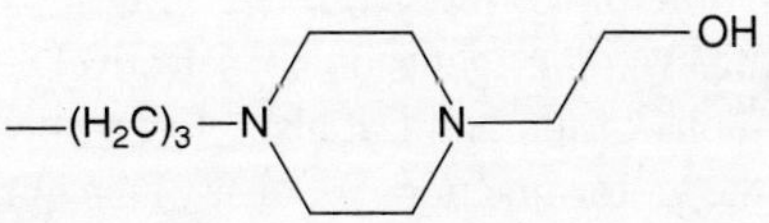

2. Thiothixene derivatives

	R	R_1
Chlorprothixene	=CHCH$_2$CH$_2$N(CH$_3$)$_2$	Cl
Thiothixene	=HC-CH$_2$-N(piperazine)N-CH$_3$	SO$_2$N(CH$_3$)$_2$

3. Butyrophenone derivatives: Haloperidol, Droperidol
4. Diphenylbutyl piperidines: Pimozide
5. Dibenzoxazepine: Loxapine
6. Dihydroindole: Molidone
7. Newer drugs

Mechanism of action: All typical antipsychotic agents currently employed clinically block post-synaptic dopaminergic D$_2$ receptors in the mesolimbic and prefrontal cortex regions of the brain, and act as a competitive antagonist of dopamine. The blockade of D$_2$ receptors is thought to be responsible for decreasing the positive symptoms of schizophrenia.

Central dopamine receptors are subdivided into D$_1$, D$_2$, and, according to some sources, D$_3$ receptors. These receptors have a high affinity for dopamine, but they differ in sensitivity to neuroleptics of various chemical classes. For example, drugs of the phenothiazine series are non-selective, competitive D$_1$ and D$_2$ antagonists. Unlike phenothiazines, antipsychotics of the butyrophenone series such as haloperidol display selective action only on D$_2$ receptors

Side-effects: Classical antipsychotic agents do not selectively block D$_2$ receptors present in the mesolimbic and prefrontal cortex regions of the brain; it also blocks dopamine D$_2$ receptors in the nigrostriatal region of brain, which controls the motor function. Because of the inhibition of D$_2$ striatal receptor, it causes:

1. Extra-pyramidal side-effects (EPS) that include acute dystonic reaction characterized by spasm of the tongue, face, and neck, and akathesia characterized by restlessness
2. Tardive dyskinesia, involving movement of the lips, tongue, and mouth, and purposeless motions of the extremities
3. Increase in prolactin levels by a block of dopamine receptors in the hypothalamus and leads to galactorrhea

9.2.1 Phenothiazines

The first published clinical trial was that of Jean Delay and Pierre Deniker at the Hospital Ste. Anne in Paris in 1952, in which they treated 38 psychotic patients with daily injections of chlorpromazine without the use of other sedating agents.

Phenothiazine derivatives are non-selective, competitive D_1 and D_2 antagonists that block dopamine activity on corresponding receptor sites. In addition, their action is expressed by blocking α-adrenoreceptors, serotonine, cholinergic, nicotinic, and muscarinic receptors.

Phenothiazines exhibit a complex pharmacological range of action on the CNS and the peripheral nervous system. In addition, they act on the endocrine system.

Every compound of this series differs to a certain degree from the other in their qualitative, yet primarily quantitative, characteristics. They all act on the CNS by causing moderate sedative and anti-emetic effects, affecting thermoregulatory processes, skeletal muscle, and endocrine system, and potentiating action of analgesics.

Phenothiazines have a diverse use in medicine. They are primarily used as antipsychotics. Despite the fact that they do not cure the disease, they reduce psychotic symptoms to a point where the patient is provided with a better sense of reality. Phenothiazines are sometimes used for relieving severe anxiety, especially in panic attacks caused by dependence on amphetamines or lysergic acid diethylamide (LSD). Phenothiazines are used for alleviating behavioural problems in children which do not respond to treatment of other agents. Phenothiazines are sometimes used during the pre-operational period because they relieve anxiety, control nausea, hiccups, and diarrhoea, and also cause muscle relaxation.

Synthesis of propyl dialkylamino side-chain derivatives

	R	
Promazine	H	
Chlorpromazine	Cl	
Triflupromazine	CF_3	

The first step in the synthesis of phenothiazine nucleus consists of displacement of the activated chlorine in the nitrobenzene by the salt from o-bromothiophenol to give the thioether. The nitro group is then reduced to form the aniline. Heating in a solvent such as DMF leads to displacement of bromine by amino nitrogen and formation of the phenothiazine. Alkylation of the anion from this intermediate with 3-dimethylaminopropylchloride in the presence of sodium amide affords titled compounds.

Synthesis of ethyl piperidyl side-chain derivatives

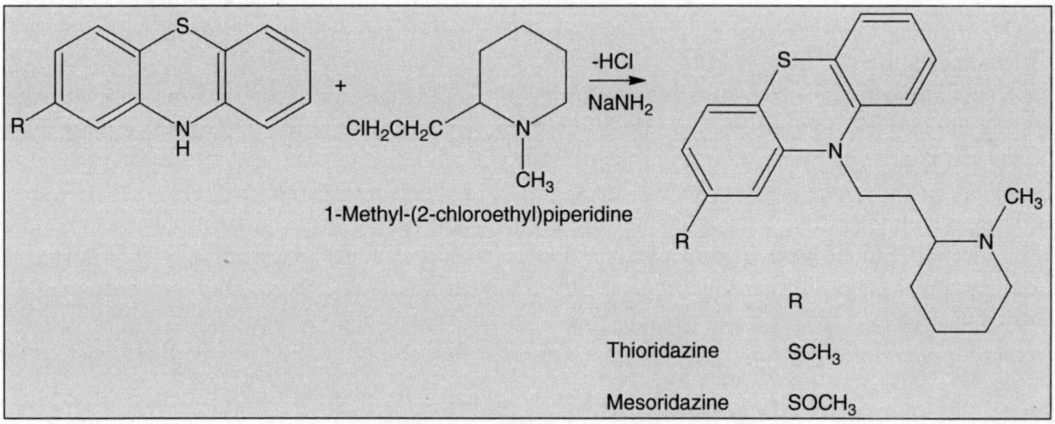

	R	
Thioridazine	SCH$_3$	
Mesoridazine	SOCH$_3$	

Appropriate phenothiazine derivative is alkylated by 2-(2-chloroethyl)-1-methylpiperidine in the presence of sodium amide, affording the desired mesoridazine/thioridazine.

Synthesis of propyl piperazine side-chain derivatives

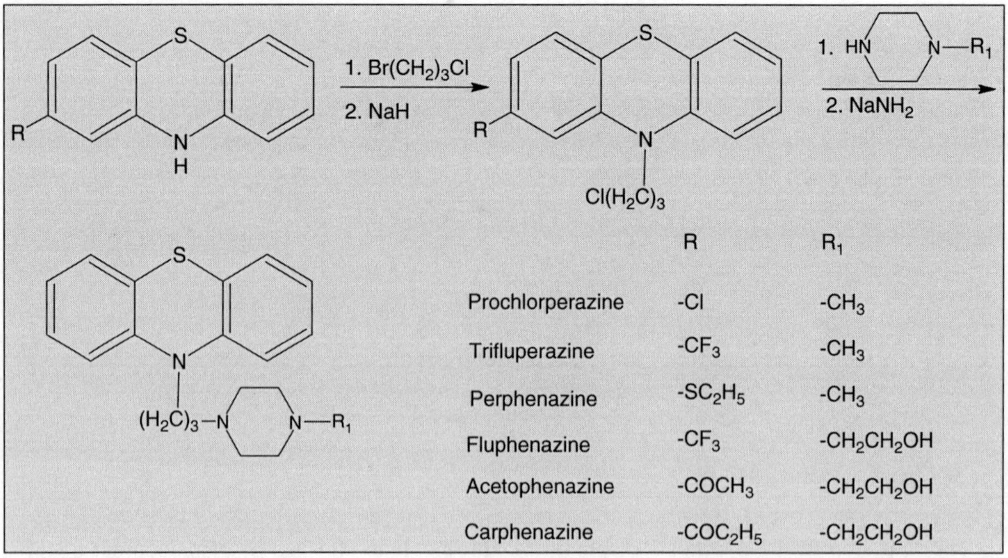

	R	R$_1$
Prochlorperazine	-Cl	-CH$_3$
Trifluperazine	-CF$_3$	-CH$_3$
Perphenazine	-SC$_2$H$_5$	-CH$_3$
Fluphenazine	-CF$_3$	-CH$_2$CH$_2$OH
Acetophenazine	-COCH$_3$	-CH$_2$CH$_2$OH
Carphenazine	-COC$_2$H$_5$	-CH$_2$CH$_2$OH

Prochlorperazine is synthesised by the alkylation of 2-chlorophenothizine using (4-methyl-1-piperazinyl)propyl-3-chloride in the presence of sodium amide, or alkylation of 2-chloro-10-[(3-chloropropyl)] phenothiazine using 1-methylpiperazine.

Structure–activity relationship of phenothiazines

1. **Alkyl side-chain (N-10)**

   - Maximum antipsychotic potency observed when the nitrogen of the phenothiazine ring and the side-chain nitrogen are connected by a three-carbon side chain.
   - Branching at the β–position of the side chain with small methyl groups ($R_1=CH_3$) decreases antipsychotic potency but enhances antihistaminic activity.
   - Maximum antipsychotic potency is observed in aminoalkylated phenothiazines having a tertiary amino group.
   - In general, alkylation of the basic amino group with groups larger than methyl decreases activity.
   - The piperidinyl and piperazinyl propyl derivatives are both somewhat less potent than the corresponding dimethylamino derivatives. But this leads to lower incidence of EPS, possibly due to increased central anti-muscarinic activity.
   - Among the piperazinyl propyl derivatives, hydroxy ethyl substitution substantially enhances the potency.

2. **Phenothiazine ring**

   - Potency increases in the following order of position of ring substitution: 1<4<3<2.
   - 2-Substitution in the phenothiazine nucleus increases neuroleptic potency in the following order: OH< H< CH_3< n-C_3H_7CO< C_2H_5CO< CH_3CO< SCH_3< Cl< CF_3.
   - In general, disubstitution has little effect or is detrimental to potency.

9.2.2 Thioxanthenes

Thioxanthenes differ structurally from phenothiazine in that the nitrogen atom of the central ring of the tricyclic system is replaced by carbon, which is joined to a side chain with a double bond.

Chlorprothixene 2-Chloro-9-(3'-dimethylaminopropylidene) thioxanthene

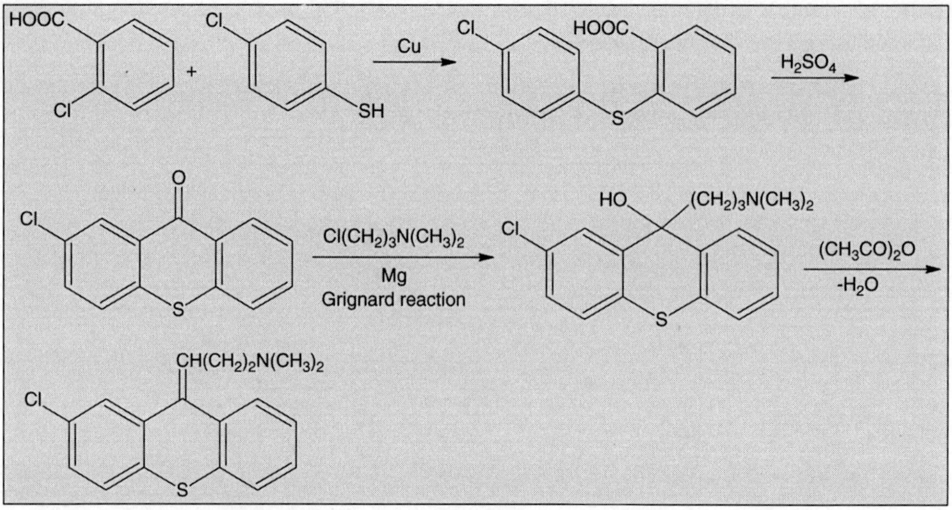

Condensation of *p*-chlorothiophenol with chlorobenzoic acid gives the thioether. This compound is then cyclized to the thioxanthone by means of sulphuric acid. Addition of the Grignard reagent from 1-chloro-3-dimethylaminopropane to the carbonyl group serves to add the required basic side chain. Dehydration of intermediate by means of acetic anhydride gives the olefin.

Thiothixene N,N-Dimethyl-9-[3-(4-methyl-1-piperazinyl)propylidene thioxanthene-2-sulphonamide

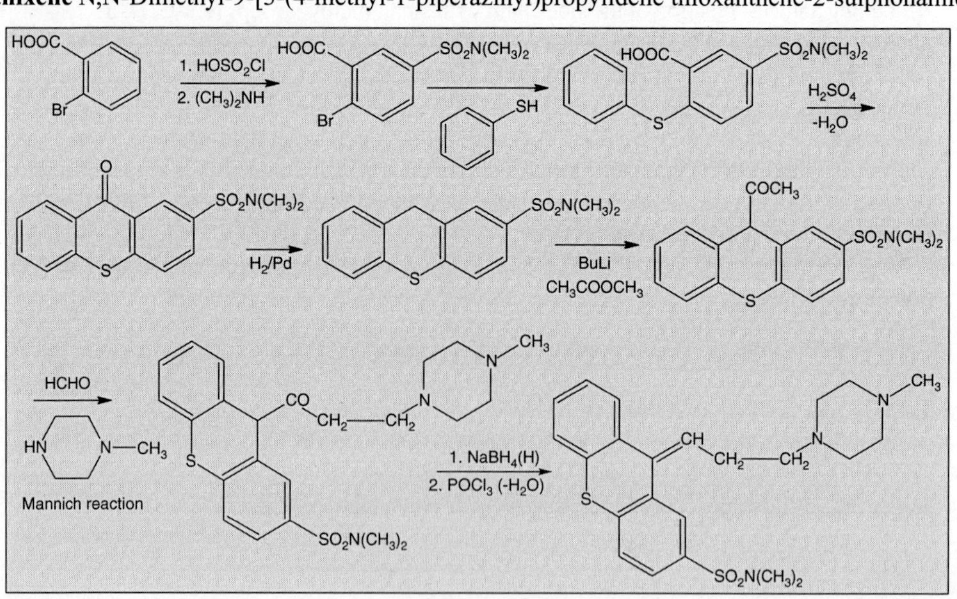

Reaction of the bromobenzoic acid with chlorosulphonic acid leads to the corresponding sulphonyl chloride; this compound affords the sulphonamide on treatment with dimethylamine. This product is then converted to a thioxanthone as shown above. The carbonyl group of thioxanthone is reduced to methylene using catalytic hydrogenation over palladium; this on treatment with butyllithium is followed by methyl acetate to afford methyl ketone. Aminomethylation of the resulting product with N-methylpiperazine and formaldehyde gives Mannich bases. The carbonyl group of the Mannich product is reduced to a secondary hydroxyl group using sodium borohydride, followed by the dehydration of the product with the help of phosphorous oxychloride to give the desired thiothoxene.

9.2.3 Butyrophenone Derivatives

A number of different compounds of the piperidine and piperazine series with p-fluorobutyrophenone group substitutions at the nitrogen atom display significant neuroleptic activity. There is a considerable interest in butyrophenone derivatives as antipsychotic agents as well as in anaesthesiology. They exhibit pharmacological effects and a mechanism of action very similar to that of phenothiazines and thioxanthenes in that they block dopaminergic receptors. However, they are more selective with respect to D_2 receptors.
Haloperidol 4-[4-(p-Chlorophenyl)-4-hydroxypiperidino]-4'-fluorobutyrophenone

In 1967, a butyrophenone named haloperidol was developed as the first therapeutic alternative to the phenothiazines that may be used to treat individuals whose psychoses do not respond to the phenothiazines. Its pharmacological effects are very similar to the phenothiazines. Haloperidol is an effective neuroleptic and also possesses anti-emetic properties; it has a marked tendency to provoke extrapyramidal effects and has relatively weak alpha-adrenolytic properties. It may also exhibit hypothermic and anorexiant effects and potentiate the action of barbiturates, general anaesthetics, and other CNS-depressant drugs.

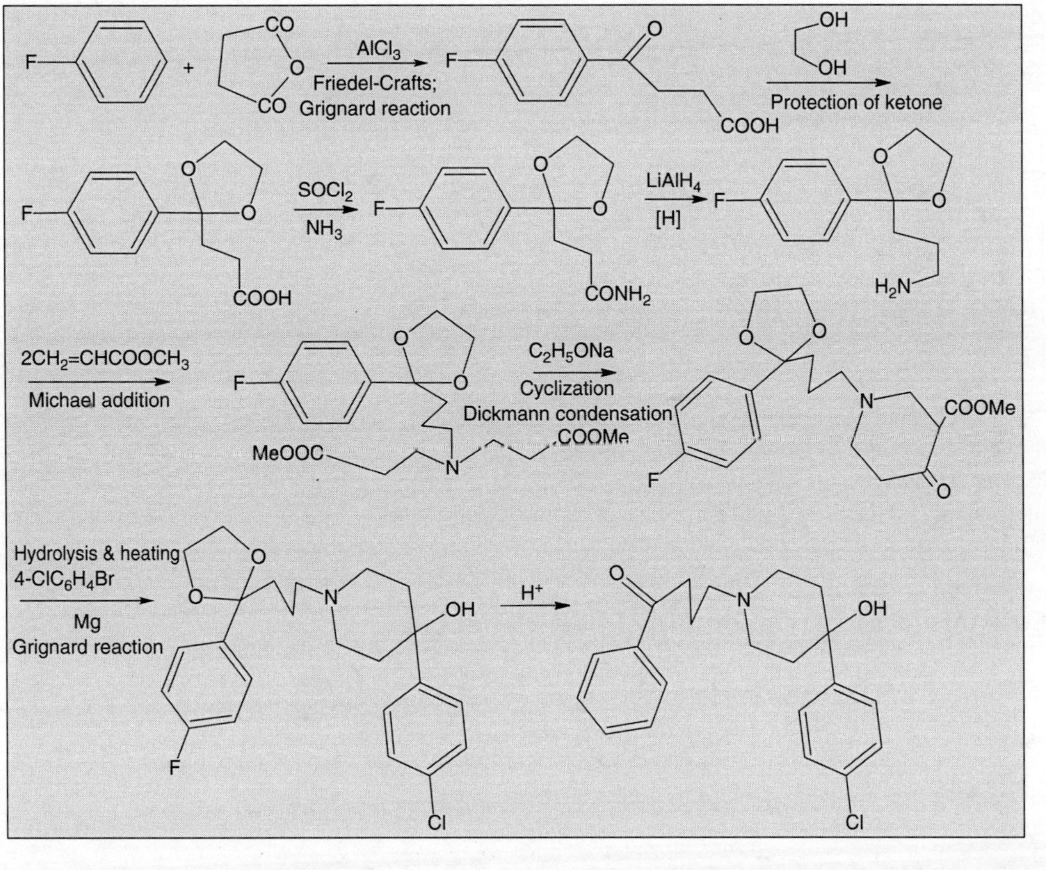

Chapter 9: Antipsychotic Agents

Friedel-Crafts acylation of 4-fluorobenzene with succinic anhydride gives keto acid. Successive protection of ketone as its acetal by reaction with ethylene glycol, conversion of the carboxyl group to acid chloride, and finally ammonolysis gives the amide. This compound is then reduced with lithium aluminium hydride to the primary amine. Conjugate addition of methyl acrylate to the primary amino group gives the diester. Dickmann condensation of diester leads to cyclization, which in turn condenses with Grignard reagent from *p*-chlorobenzene and gives the tertiary alcohol. Hydrolysis with aqueous acid leads to removal of the acetal-protecting group and affords haloperidol.

Droperidol 1-{1-[3-(*p*-Fluorobenzoyl)propyl]-1,2,3,6-tetrahydro-4-pyridyl}–2-benzimidazolinone

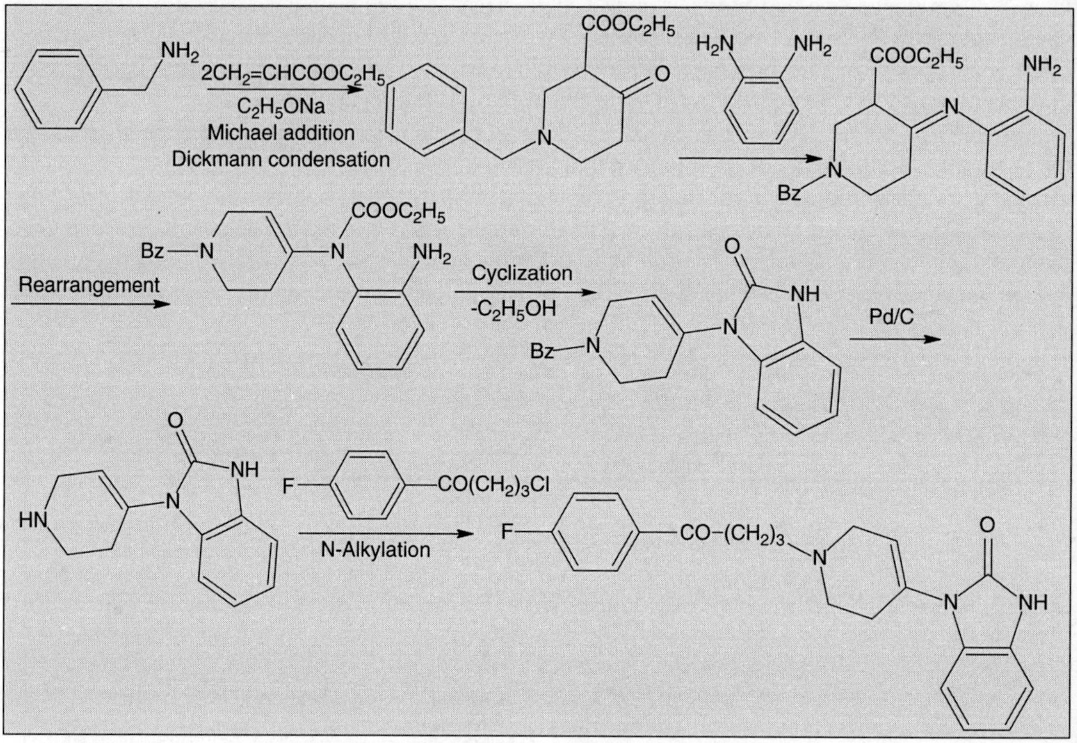

Droperidol is synthesised from 1-benzyl-3-carbethoxypiperidin-4-one, which is reacted with *o*-phenylendiamine. Evidently, the first derivative that is formed under the reaction conditions, 1,5-benzdiazepine, rearranges into 1-(1-benzyl-1,2,3,6-tetrahydro-4-piridyl)-2-benzymidazolone. Debenzylation of the resulting product with hydrogen over a palladium catalyst into 1-(1,2,3,6-tetrahydro-4-piridyl)-2-benzimidazolon and subsequent alkylation of this using 4'-chloro-4-fluorobutyrophenone yields droperidol.

Structure–activity relationship of butyrophenones

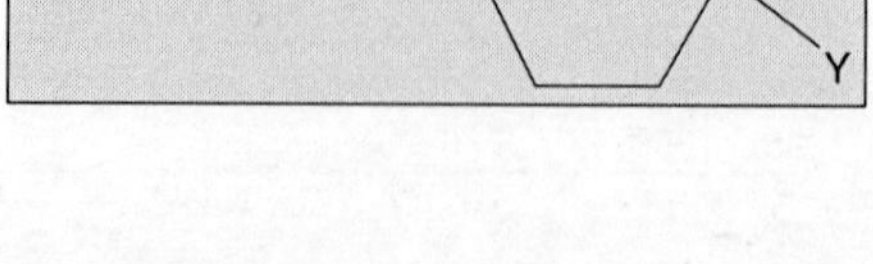

- Antipsychotic activity is seen when Ar is an aromatic system, in which fluoro substituents at *para* position aids activity.
- When X=carbonyl (C=O), optimal activity is seen, although other groups C(H)OH, C(H)aryl(Pimozide) also gives good activity.
- When n=3, activity is optimal, longer or shorter chains decrease the activity.
- The aliphatic amino nitrogen is required and highest activity is seen when it is incorporated into a cyclic form.
- Ar1 is an aromatic and is needed. It should be attached directly to the 4th position or occasionally separated from it by one intervening atom.
- The Y group can vary and assist activity.

9.2.4 Diphenylbutyl Piperidine

The principle distinctive feature of this series of drugs is their prolonged action. The mechanism of their action is not completely known; however, it is clear that they block dopaminergic activity.

Pimozide 1-[1-[4,4-Bis(4-fluorophenyl)butyl]-4-piperidinyl]-1,3-dihydrobenzimidazol-2-one

It is an analogue of droperidol, in which carbonyl compound was replaced with C(H)aryl group. Because of their high hydrophobic character, the compound was inherently long-acting.

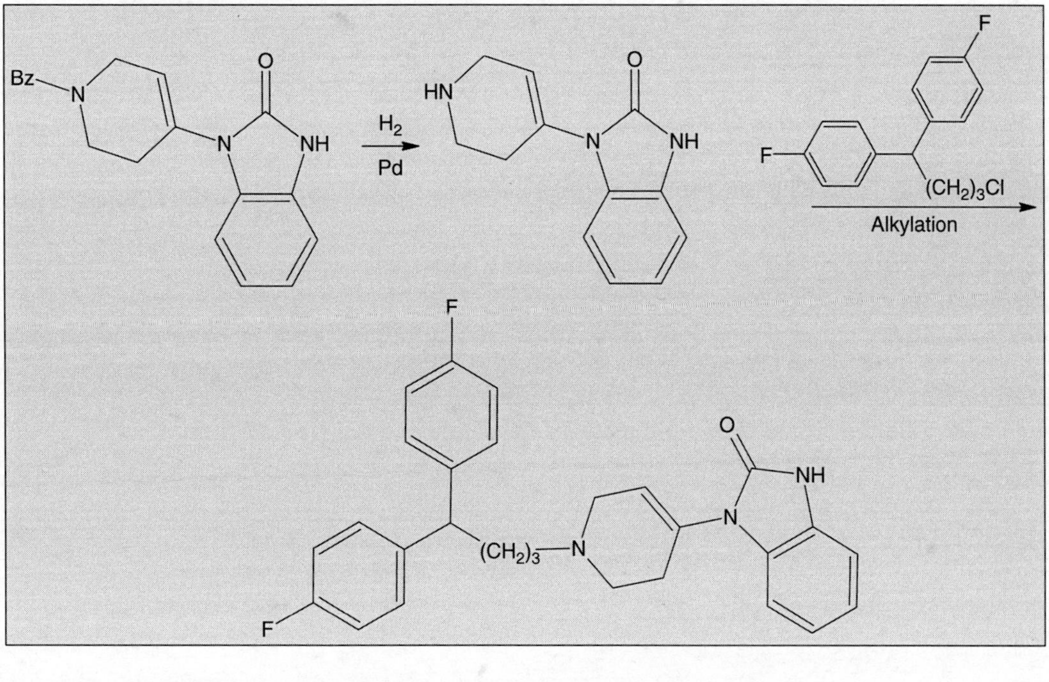

Alkylation of 1-(1,2,3,6-tetrahydro-4-piridyl)-2-benzimidazolon using 1,1-*bis*-(4-fluorophenyl) butyl chloride yields pimozide.

9.2.5 Dibenzoxazepine

Loxapine 2-Chloro-11-(4-methyl-1-piperazinyl)dibenz(1,4)-oxazepine

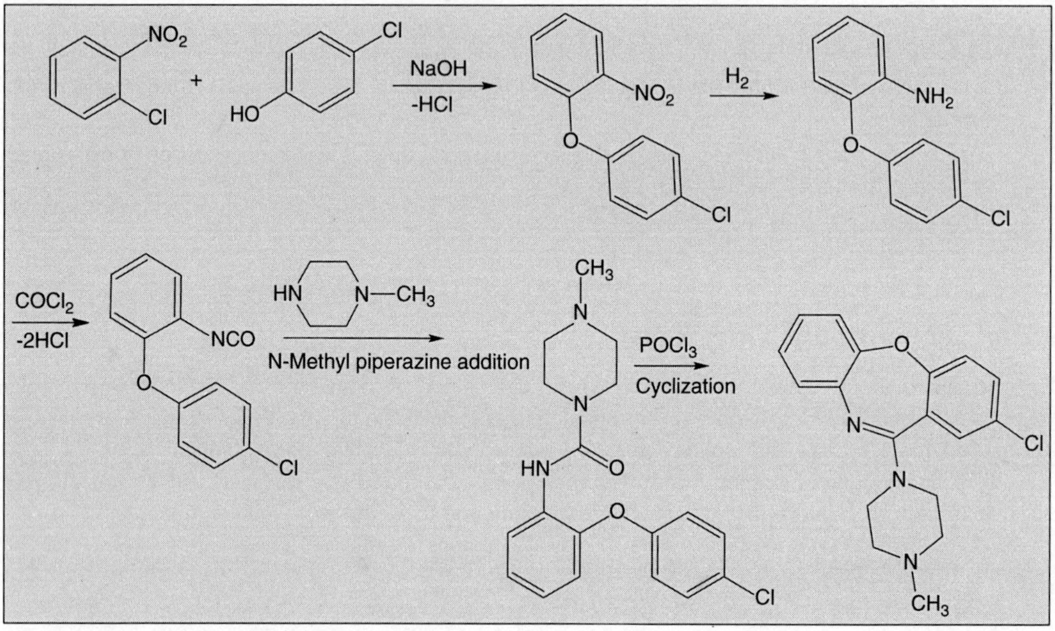

Displacement of chlorine on *o*-chloronitrobenzene by *p*-chlorophenoxide leads to the diphenylether derivative. Reduction of nitro group leads to the aniline, which on reaction with phosgene in the presence of base gives the isocyanate. Addition of *N*-methylpiperazine to the isocyanate function leads to a urea. Treatment of urea with phosphorous oxychloride leads to a cyclodehydration reaction via the imino chloride and affords loxapine.

9.2.6 Dihydroindole

Dihydroindole derivatives do not structurally belong to any of the classes of drugs examined above. However, their mechanism of action, indications of use and side-effects are very similar to phenothiazine derivatives.

Molindone 3-Ethyl-1,5,6,7-tetrahydro-2-methyl-5-(4-morpholinomethyl)-indol-4-one

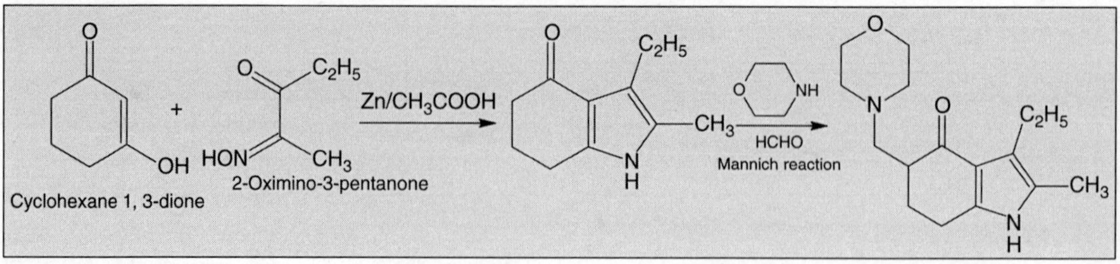

Molindone is synthesised by the nitrozation of diethylketone using nitric acid or methylnitrite into nitrozodiethylketone. Reduction of this product with zinc in acetic acid into 2-aminodiethylketone in the presence of cyclohexan-1,3-dion gives 3-ethyl-2-methyl-4,5,6,7-tetrahydroindol-4-one. Aminomethylation of this product using morpholine and formaldehyde gives molindone.

9.2.7 Miscellaneous Drugs

Oxypertine 5,6-Dimethoxy-2-methyl-3-[2-(4-phenylpiperazin-1-yl)ethyl]-1H-indole

The reaction between 5,6-methylenedioxyindole and oxalyl chloride gives ketoacid chloride intermediate, which on reaction with phenyl piperazine gives keto amide. This on reduction with lithium aluminium hydride reduces both carbonyls to oxypertine. It is a derivative of tryptophan.

9.3 ATYPICAL ANTIPSYCHOTIC AGENTS

Schizophrenia is now being treated with new medications that are commonly called **'atypical antipsychotics'**. These are new-generation neuroleptic agents useful for the treatment of schizophrenia and mania without EPS side-effects. Compared to the older 'conventional' antipsychotics, these medications appear to be equally effective for helping reduce the positive symptoms like hallucinations and delusions—but may be better than the older medications at relieving the negative symptoms of the illness, such as withdrawal, thinking problems, and lack of energy. (*See Fig. 9.2 in the coloured set of pages.*)

1. **D$_4$ receptor antagonist**
   The discrete localization of D$_4$ receptors in cortical and mesolimbic regions of the brain (regions believed to be involved in thoughts and emotions) and their relative absence from nigrostriatal region controlling motor function suggest dopamine D$_4$ receptor as a good target for the treatment of schizophrenia. An example of drugs acting through D$_4$ receptor antagonist is clozapine.
2. **Mixed D$_2$ and 5HT$_{2A}$ antagonist**
   Serotonin antagonism at 5HT$_{2A}$ receptors has been reported to improve the negative symptoms (additional impairment, flattened effect) and to reduce the occurrences of EPS effects. Examples of drugs that block both D$_2$ and 5HT$_{2A}$ receptors are risperidone, quetiapine, ziprasidone, olanzapine, iloperidone, and seretindole.

Clozapine 8-Chloro-11-(4-methyl-1-piperazinyl)-5*H*-dibenzo[b,e] [1,4]diazepine

Clozapine has been beneficial in the treatment of so-called **'treatment-resistant'** cases – the 30% to 60% of cases where traditional neuroleptics have not been successful. It is effective in individuals suffering from disorganization – for example, loose associations, inappropriate affect, incoherence and reduction in rational thought processes – and also relieves many of the negative symptoms, while lacking many of the extra-pyramidal side-effects associated with the phenothiazines.

Synthesis

Ullman coupling of anthranilic acid with 2-bromo-5-chloronitrobenzene gives the substituted anthranilate. The nitro group is then reduced to amine by means of catalytic hydrogenation, which is followed by cyclization and gives dibenzodiazepine nucleus. Reductive amination with N-methylpiperazine affords clozapine.

Quetiapine 2-(2-(4-Dibenzo[b,f][1,4]thiazepine- 11-yl-1-piperazinyl)ethoxy)ethanol

Quetiapine is used orally to treat psychotic disorders and symptoms such as hallucinations, delusions, and hostility. Quetiapine also has an antagonistic effect on the histamine H_1 receptor. This is thought to be responsible for the sedative effect of the drug.

Synthesis

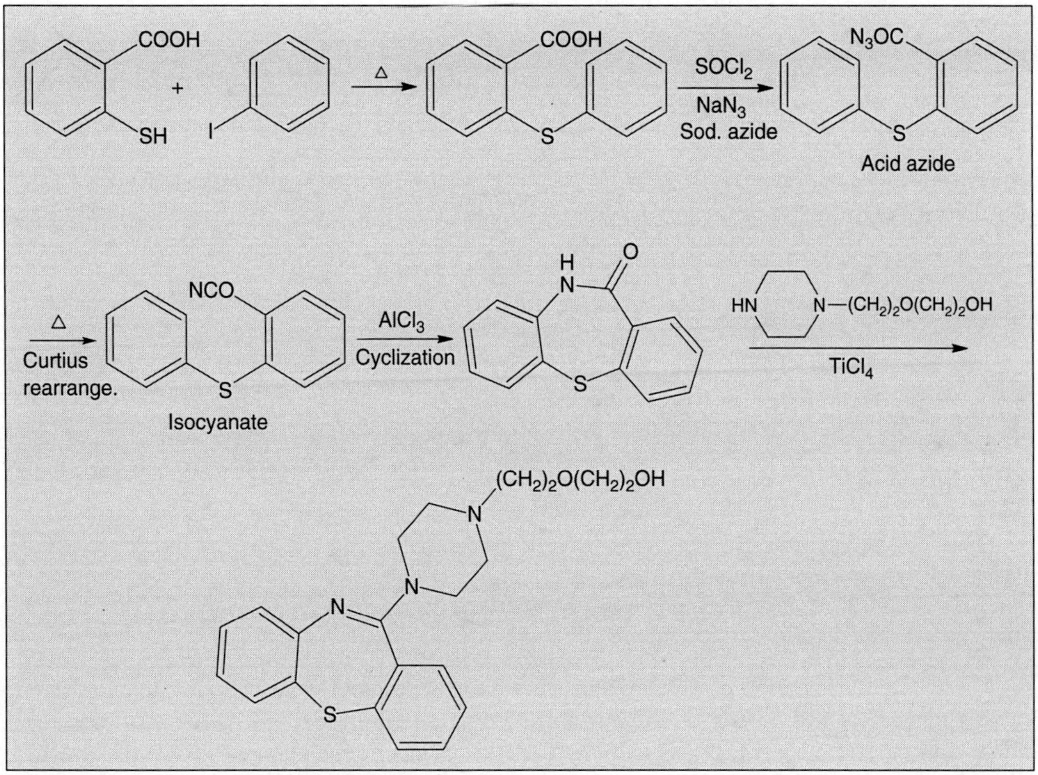

Diphenylthioether derivative is prepared by reacting 2-mercaptobenzoic acid and iodobenzene. The acid group is converted to acid azide by reacting with thionyl chloride, followed by sodium azide; this on heating loses nitrogen to form isocyanate. Isocyanate on reaction with aluminium chloride undergoes cyclization and affords dibenzo thiazepine nucleus. Side chain is attached by reductive amination with 1-(piperazinyl)ethoxy cthanol.

Risperidone 4-[2-[4-(6-Fluorobenzo[*d*]isoxazol-3-yl)-1-piperidyl]ethyl]-3-methyl-2,6-diazabicyclo[4.4.0] deca-1,3-dien-5-one

It is a uniquely balanced serotonin/dopamine antagonist (SDA) that offers considerable advantages in the first-line treatment of schizophrenia.

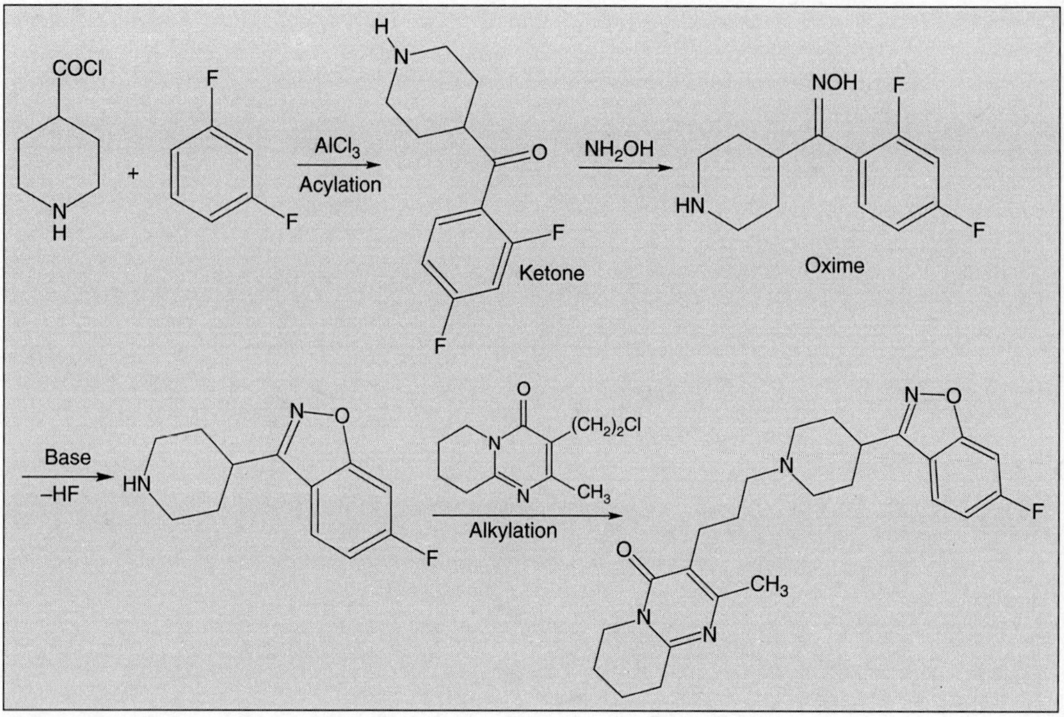

Friedel-Crafts acylation of difluorobenzene with isonipecotoyl chloride gives ketone derivative. Ketone group reacts with hydroxyl amine to form oxime, which cyclizes in presence of base to form 6-fluoro-3-(4-piperidyl)benzo[*d*] isoxazole. *N*-Alkylation of piperidinyl nitrogen affords risperidone

Ziprasidone   5-[2-[4-(1,2-Benzisothiazol-3-yl)-1-piperazinyl]ethyl]-6-chloro-1,3-dihydro-2*H*-indol-2-one

Ziprasidone has a high affinity for dopamine, serotonin, and alpha-adrenergic receptors, and a medium affinity for histaminic receptors. Antagonism at histaminic and alpha adrenergic receptors likely explains some of the side-effects of ziprasidone, such as sedation and orthostasis.

Synthesis

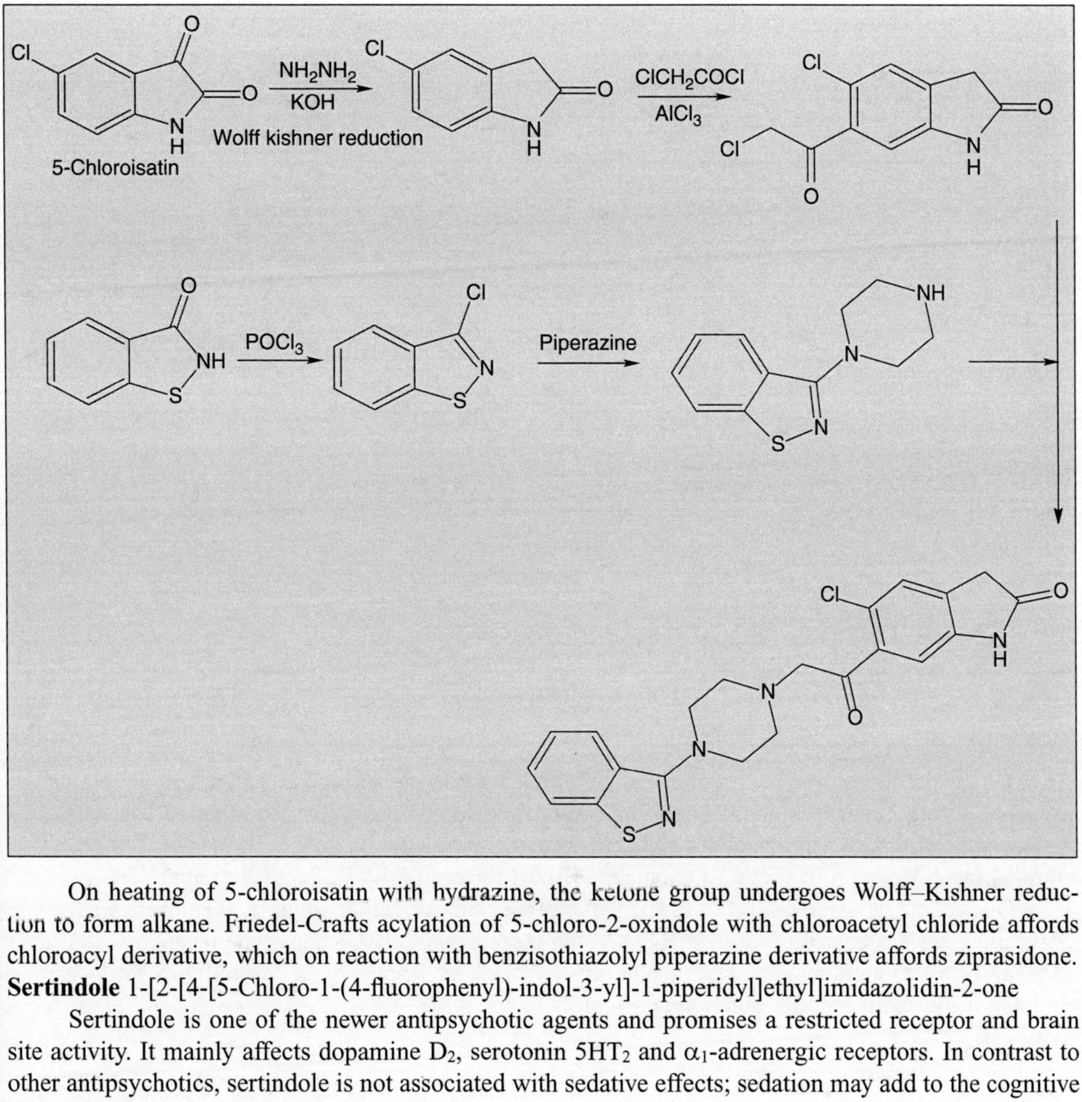

On heating of 5-chloroisatin with hydrazine, the ketone group undergoes Wolff–Kishner reduction to form alkane. Friedel-Crafts acylation of 5-chloro-2-oxindole with chloroacetyl chloride affords chloroacyl derivative, which on reaction with benzisothiazolyl piperazine derivative affords ziprasidone.

Sertindole 1-[2-[4-[5-Chloro-1-(4-fluorophenyl)-indol-3-yl]-1-piperidyl]ethyl]imidazolidin-2-one

Sertindole is one of the newer antipsychotic agents and promises a restricted receptor and brain site activity. It mainly affects dopamine D_2, serotonin $5HT_2$ and α_1-adrenergic receptors. In contrast to other antipsychotics, sertindole is not associated with sedative effects; sedation may add to the cognitive problems inherent in schizophrenia.

Synthesis

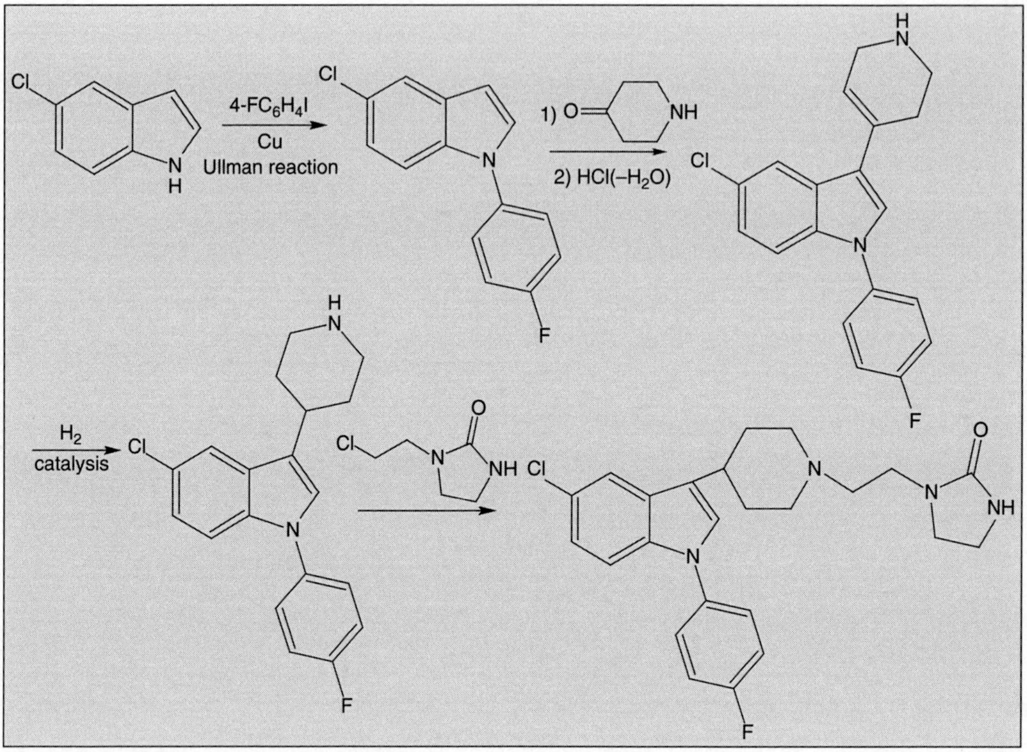

Ullmann condensation of 5-chloroindole with 4-fluoro iodobenzene gives 1-phenylindole derivative; this in turn condenses with 4-piperidone and, followed by reduction, gives piperidine derivative. N-Alkylation with 2-oxo imidazolidin ethyl chloride affords sertindole.

Olanzapine 2-Methyl-4-(4-methyl-1-piperazinyl)-10H-thieno[2,3-b][1,5]benzodiazepine

Olanzapine is structurally similar to clozapine, and is classified as a thienobenzodiazepine. Olanzapine has a higher affinity for 5HT$_2$ serotonin receptors than D$_2$ dopamine receptors. Like most atypical antipsychotics, compared to the older typical ones, olanzapine has a lower affinity for histamine, cholinergic muscarinic, and α-adrenergic receptors.

Synthesis

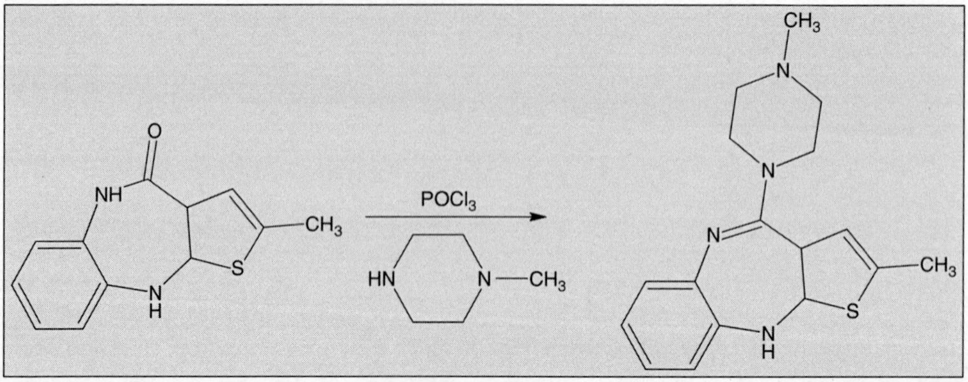

Medicinal Chemistry

2-Methyl-5,10-dihydro-4H-benzo[b]thieno[2,3-e][1,4]diazepin-4-one on reaction with phosphorous oxychloride forms enol and halo form, which on reaction with N-methylpiperazine affords olanzapine.

Iloperidone 1-[4-[3-[4-(6-Fluoro-1,2-benzisoxazol-3-yl)-1-piperidinyl]propoxy]-3-methoxyphenyl]ethanone

Iloperidone is a monoamine directed towards acting upon and antagonizing specific neurotransmitters. It is considered as an 'atypical' antipsychotic that is less likely to cause movement disorders in patients when compared to traditional methods of psychotic treatment. Iloperidone acts on both dopamine and serotonin receptors, making it a favourable choice against the competing drugs clozapine and olanzapine.

Synthesis

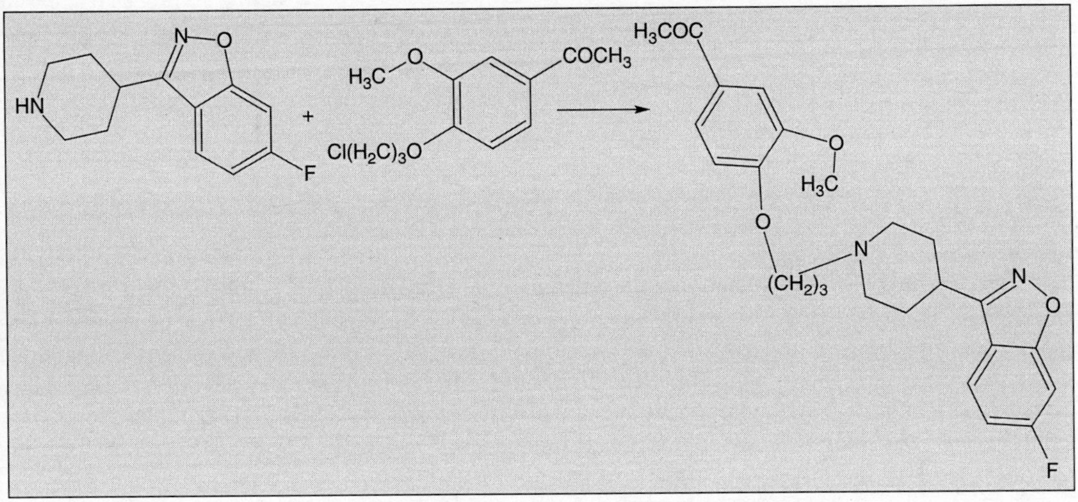

N-Alkylation of 6-fluoro-3-(4-piperidyl)benzo[d]isoxazole (for synthesis, refer to risperidone) with 1-[4-(3-chloropropoxy)-3-methoxyphenyl]-1-ethanone affords iloperidone.

9.4 NEWER DRUGS

Amisulpride 4-Amino-N-[(1 ethylpyrrolidin-2-yl)methyl]-5-ethylsulphonyl-2-methoxy-benzamide

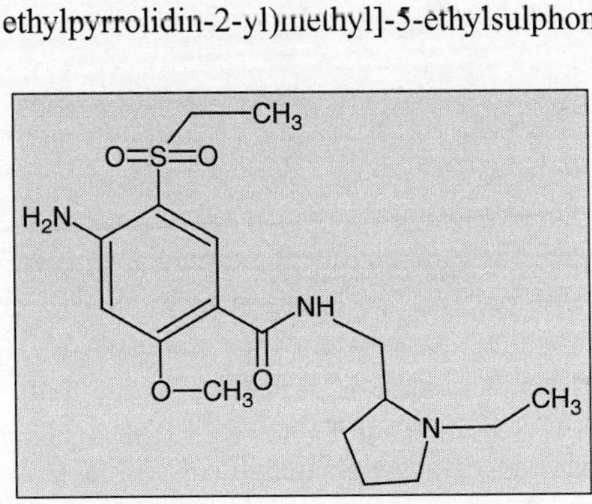

Amisulpride is a selective dopamine antagonist. It has a high affinity for D_2 and D_3 dopaminergic receptors. Lower doses (less than 50 mg) preferentially block D_2 autoreceptors that control the synthesis and release of dopamine. This results in an increase in dopaminergic transmission. This dopamine increase is hypothesized to cause a reduction in both depressive and negative symptoms. Higher doses of the drug block the post-synaptic dopamine receptors, resulting in an improvement in psychoses.

Paliperidone 3-[2-[4-(6-Fluorobenzo[d]isoxazol-3-yl)-1-piperidyl]ethyl]-7-hydroxy-4-methyl-1,5-diazabicyclo[4.4.0]deca-3,5-dien-2-one

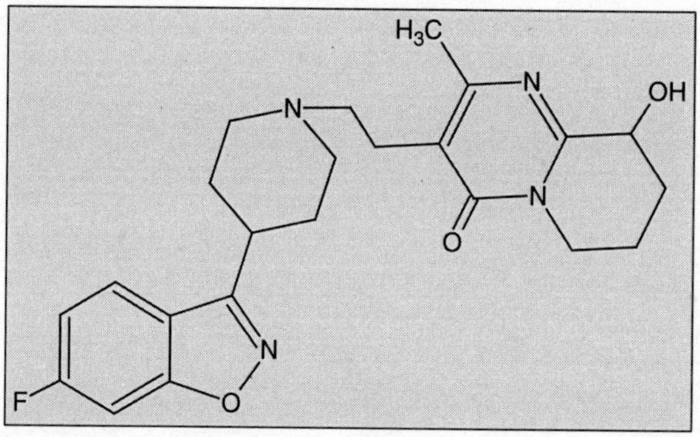

It is a primary active metabolite of the older antipsychotic risperidone (paliperidone is 9-hydroxy-risperidone, i.e., risperidone with an extra hydroxyl group). The therapeutic effect may be due to a combination of central dopamine Type 2 (D_2) and serotonin Type 2 ($5HT_{2a}$) receptor antagonism. Paliperidone also antagonizes at α_1 and α_2 adrenergic receptors and H_1 histaminergic receptors.

Aripiprazole 7-[4-[4-(2,3-Dichlorophenyl)piperazin-1-yl]butoxy]-3,4-dihydro-1H-quinolin-2-one

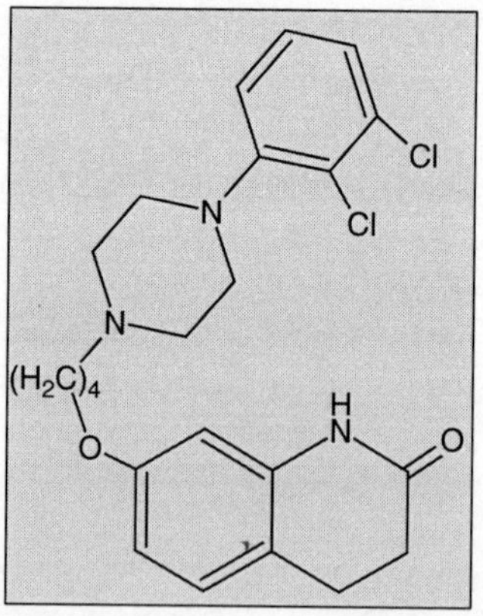

Its mechanism of action is different from the other atypical antipsychotics. Aripiprazole appears to mediate its antipsychotic effects primarily by partial agonism at the D_2 receptor, which has been shown to modulate dopaminergic activity in areas where dopamine activity may be increased or diminished, such as in the mesolimbic and mesocortical areas of the schizophrenic brain, respectively. In addition to its partial agonist activity at the D_2 receptor, aripiprazole is also a partial agonist at the $5HT_{1A}$ receptor, and like the other atypical antipsychotics, displays an antagonist profile at the $5HT_{2A}$ receptor. Aripiprazole has moderate affinity for histamine and alpha-adrenergic receptors, and no appreciable affinity for cholinergic muscarinic receptors.

FURTHER READINGS

1. Anand, J. et al., 2004. 'Drug therapy in schizophrenia', *Current Pharmaceutical Design*, 10, pp. 2205–2217.
2. Baldessarini, R.J., 1990. 'Drugs and the treatment of psychiatric disorders', in Gilman, Rall, Nies, et al., (ed.), *The Pharmacological Basis of Therapeutics* (8th ed.), New York: Pergamon, pp. 383–435.

MULTIPLE-CHOICE QUESTIONS

1. One of the following phenothiazines does not possess CF_3 group at C_2 position:
   a. Triflupromazine
   b. Trifluperazine
   c. Fluphenazine
   d. Prochlorperazine

2. Extra-pyramidal symptoms are a common side-effect of which of the following medications?
   a. Haloperidol
   b. Olanzapine
   c. Quetiapine
   d. Ziprasidone

3. Loxapine belongs to
   a. Dihydroindole
   b. Phenothiazine
   c. Dibenzoxazepine
   d. Diphenylbutyl piperidines

4. Which of the following statements correctly represents the SAR of phenothiazine?
   a. Maximum antipsychotic potency is observed in aminoalkylated derivatives.
   b. The aliphatic amino nitrogen is required and highest activity is seen when it is incorporated into a cyclic form.
   c. Potency increases in the following order of position of ring substitution: 1<3<4<2.<2.
   d. Disubstitution increases neuroleptic activity.

5. Chlorprothixene is
   a. 4-[4-(*p*-Chlorophenyl)-4-hydroxypiperidino]-4'-fluorobutyrophenone
   b. N,N-Dimethyl-9-[3-(4-methyl-1-piperazinyl) propylidene-thioxanthene-2-sulphonamide
   c. 2-Chloro-9-(3'-dimethylaminopropylidene) thioxanthene
   d. 2-Chloro-9-(3'-methylaminopropylidene) thioxanthene sulphonamide

6. The first therapeutic alternative to the phenothiazines is
   a. Butyrophenones
   b. Dihydroindole
   c. Dibenzoxazepine
   d. Iminostilbenes

7. Clozapine is synthesised from
   a. Anthranilimide
   b. Anthranilic acid
   c. Benzoic acid
   d. 2-Chlorobenzoic acid

8. The drug used orally to treat psychotic disorders with no extra-pyramidal side-effects is
   a. Haloperidol
   b. Prochlorperazine
   c. Quetiapine
   d. Molindone

9. In phenothiazines the modification that leads to lower incidence of EPS, possibly due to increased central anti-muscarinic activity, is
   a. Aminoalkylated phenothiazines
   b. Piperidinyl propyl derivatives
   c. Branching at the β–position of the side chain with small methyl groups
   d. None of the above

10. One of the following is not a mixed D_2 and $5HT_{2A}$ antagonist:
    a. Risperidone
    b. Olanzapine
    c. Iloperidone
    d. Clozapine

QUESTIONS

1. Why does the antipsychotic drug risperidone demonstrate low incidence and low severity of extra-pyramidal side-effects?
2. Which of the antipsychotic drugs are expected to demonstrate the highest incidence and severity of extra-pyramidal side-effects? Mention reasons for the same.
3. How could EPS be treated?
4. What are the characteristics of the positive and negative symptoms of schizophrenia?
5. What is the primary mechanism of action responsible for the effects of antipsychotic drugs?
6. Clozapine is generally referred to as an 'atypical' neuroleptic. What is the difference between 'atypical' and 'typical' neuroleptics (such as haloperidol)?

SOLUTION TO MULTIPLE-CHOICE QUESTIONS

1. d;
2. a;
3. c;
4. a;
5. c;
6. a;
7. b;
8. c;
9. b;
10. d.

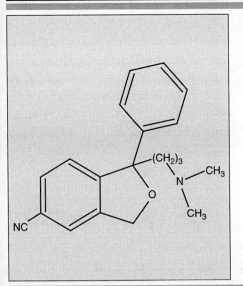

CHAPTER 10

Antidepressants

LEARNING OBJECTIVES

- Define antidepressants and their utility
- Categorize antidepressants on the basis of mode of action and chemical class
- Describe the mode of action of antidepressants and their side-effects
- Define monoamine oxide and their inhibition as an important strategy to treat depression
- Define atypical antidepressants and their significance
- List the other drugs that do not fall in the general classification
- Describe some latest drugs in the class of antidepressants

10.1 INTRODUCTION

Antidepressants are drugs that enhance alertness and may result in an increased output of behaviour. In other words, antidepressants are capable of removing or alleviating a number of disorders in the psycho-emotional realm referred to as 'depressive syndrome' in psycho-neurological practice.

In turn, conditions characterised by the term 'depression' include affective disorders, which are frequently accompanied by a number of other disturbances including unmotivated sorrow, sleep disorders, changes in appetite, various psychomotor disturbances, loss of interest in things once pleasurable, feelings of worthlessness, and often suicidal thoughts.

Despite the fact that the initial biochemical abnormalities responsible for depression and manic-depressive conditions have not been completely discovered, some facts suggest that depressive conditions may be caused by a lack of norepinephrine (noradrenaline) and serotonin. The majority of drugs used in treatment of such illnesses act by affecting the system of biogenic amines of the brain, thus leading to action of a mechanism that is capable of increasing their contents in respective parts of the brain.

They are used for the relief of symptoms of moderate and severe depression. Antidepressants are typically taken for at least 4 to 6 months.

Antidepressants are used for:

- Moderate to severe depressive illness
- Severe anxiety and panic attacks
- Obsessive compulsive disorders
- Chronic pain
- Eating disorders
- Post-traumatic stress disorder

10.2 CLASSIFICATION

1. Monoamine oxidase inhibitors (MAOIs): Isocarboxazide, Phenelzine, Tranylcypromine, and Pargyline
2. Tricyclic compounds

    a. Dibenzazepines: Imipramine, Cloimipramine, Trimipramine, and Desipramine
    b. Dibenzocycloheptanes: Amitriptyline, Nortriptyline, and Protriptyline
    c. Dibenzoxepines: Doxepine
    d. Dibenzoxazepines: Amoxapine and Loxapine

3. Atypical antidepressants: Bupropion and Trazodone
4. Miscellaneous: Fluoxetine and Sertraline
5. Newer drugs

10.3 MECHANISM OF ACTION

In depression, some of the neurotransmitter systems, particularly those of serotonin and noradrenaline, do not seem to work properly. Antidepressants potentiate central noradrenergic function, although they act through different mechanisms. MAOIs block the intracellular metabolism of biogenic amines; this results in increased amine concentrations in the nerve terminal, whereas tricyclic compounds and others inhibit the reuptake of norepinephrine and serotonin by nerve terminals, which in turn facilitates adrenergic neurotransmission and produces an antidepressant action.

10.4 SIDE-EFFECTS OF CLASSICAL ANTIDEPRESSANTS

MAOIs produce hypertensive crisis and is characterised by headache, palpitation, nausea, and vomiting. This reaction may be induced by ingestion of cheese, yeast extracts, chicken livers, pickles and chocolates.

Tricyclic compounds induce dryness of mouth, excessive perspiration, constipation, blurred vision, hypotension, drowsiness, and weight gain.

10.5 MONOAMINO OXIDASE (MAO) INHIBITORS

Monoamino oxidase is a complex enzymatic system that is present in practically every organ that catalyses deamination or inactivation of various natural, biogenic amines, in particular norepinephrine (noradrenaline), epinephrine (adrenaline), and serotonin. Inhibition of MAO increases the quantity of these biogenic amines in nerve endings. MAO inhibitors increase the intercellular concentration of endogenous amines by inhibiting their deamination, which seems to be the cause of their antidepressant action.

These drugs, which form stable complexes with MAO and thereby inhibit its action, have long been used in medicine as antidepressants, and are referred to as MAO inhibitors. It is possible that MAO inhibitors act not by complexation with the enzyme but by forming covalent bonds—that is, by irreversibly inactivating the enzyme.

Isocarboxazide 5-Methyl-3-isoxazole carboxylic acid-2-benzyl hydrazide

Synthesis

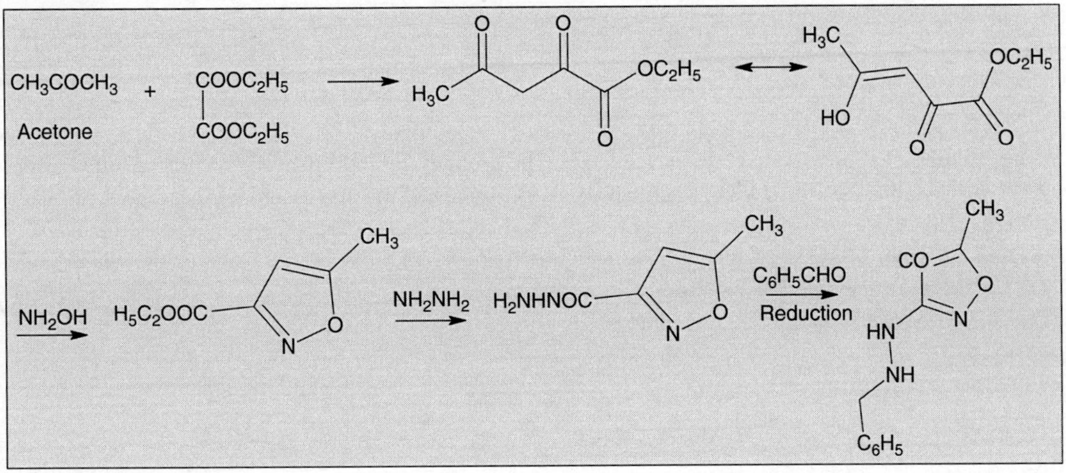

Condensation of acetone and diethyl oxalate in the presence of base gives diketone ester; this, in turn, condenses with hydroxylamine to form isoxazole ring system. The ester is then converted to the acid hydrazide by reaction with hydrazine hydrate. Condensation with benzaldehyde affords corresponding acid hydrazones, which on catalytic reduction affords isocarboxazide.

Phenelzine 2-(Phenylethyl) hydrazine

Synthesis

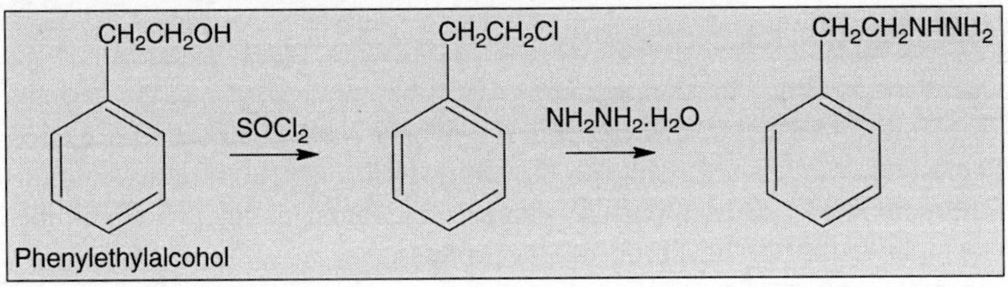

Phenelzine is synthesised by reacting 2-phenylethylchloride with hydrazine.

Tranylcypromine (+)-*trans*-2-Phenyl cyclopropylamine
Synthesis

It is synthesised from the ethyl ester of 2-phenylcyclopropancarboxylic acid, which is synthesised by the reaction of styrene with ethyl diazoacetate. The 2-phenylcyclopropancarboxylic acid ethyl ester is hydrolysed by alkali to 2-phenylcyclopropancarboxylic acid, and the *trans*-isomer is separated for further reactions. The reaction of the *trans*-isomer with thionyl chloride gives *trans*-2-phenylcyclopropancarboxylic acid chloride, which upon reaction with sodium azide gives the respective acid azide, which undergoes Curtius rearrangement to the transcyclopropylamine.

Pargyline N-Methyl-N-propynylbenzylamine
Synthesis

Pargyline is synthesised by alkylating *N*-methyl benzylamine with propynyl bromide.

10.6 TRICYCLIC ANTIDEPRESSANTS

The most frequently used drugs for depression are tricyclic antidepressants, the systems consisting of two benzene rings joined to a central 7-membered ring with a dialkylaminoalkyl group connected to the central ring. Depending on the substituents on the terminal nitrogen atom in the amine-containing side chain, they in turn are subdivided into tertiary (imipramine, amitriptyline, trimipramine, doxepin) and secondary (desipramine, nortriptyline, protriptyline) amines.

Tricyclic antidepressants elevate mood, increase physical activity, normalize appetite and sleep patterns, and reduce morbid preoccupation in 60%–70% of patients with major depression.

Tricyclic antidepressants inhibit the reuptake of the neurotransmitters serotonin and norepinephrine into their respective nerve terminals. Reuptake is the first step in the process of deactivating these neurotransmitters in the brain. After serotonin and norepinephrine are released from neurons, they are removed from the extracellular space by transporters (also known as reuptake sites) located on the cell membrane. The tricyclic antidepressants block these transporters. By inhibiting reuptake, the drugs allow serotonin and norepinephrine to remain active in the synapse longer, thereby correcting a presumed deficit in the activity of these neurotransmitters.

Synthesis of dibenzazepines:

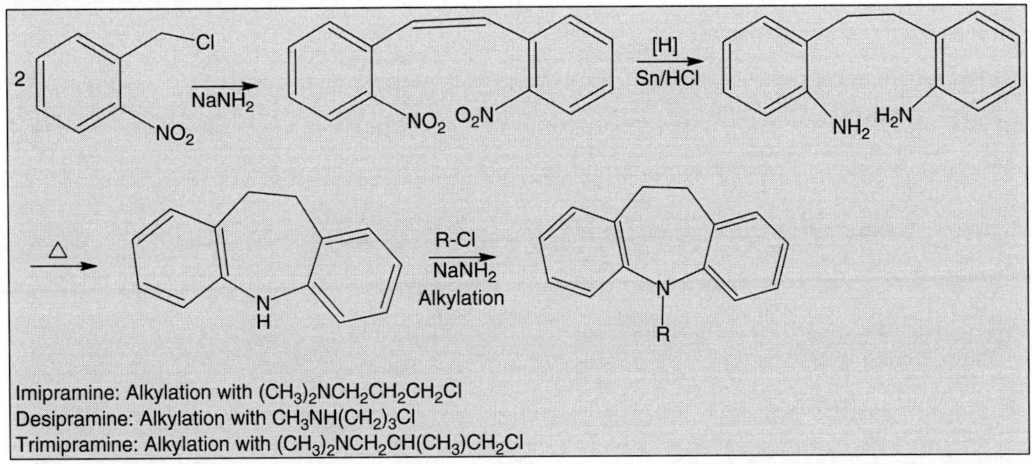
2
Cl
NO2
NaNH2
NO2 O2N
[H]
Sn/HCl
NH2 H2N
Δ
N
H
R-Cl
NaNH2
Alkylation
N
R
Imipramine: Alkylation with (CH3)2NCH2CH2CH2Cl
Desipramine: Alkylation with CH3NH(CH2)3Cl
Trimipramine: Alkylation with (CH3)2NCH2CH(CH3)CH2Cl

Dibenzazepines was synthesised by the alkylation of 10,11-dihydro-5*H*-dibenz[b,f]azepine (synthesis of this intermediate refer carbamazepine) using corresponding dimethylamino alkyl chloride in the presence of sodium amide.

Synthesis of dibenzocycloheptanes:

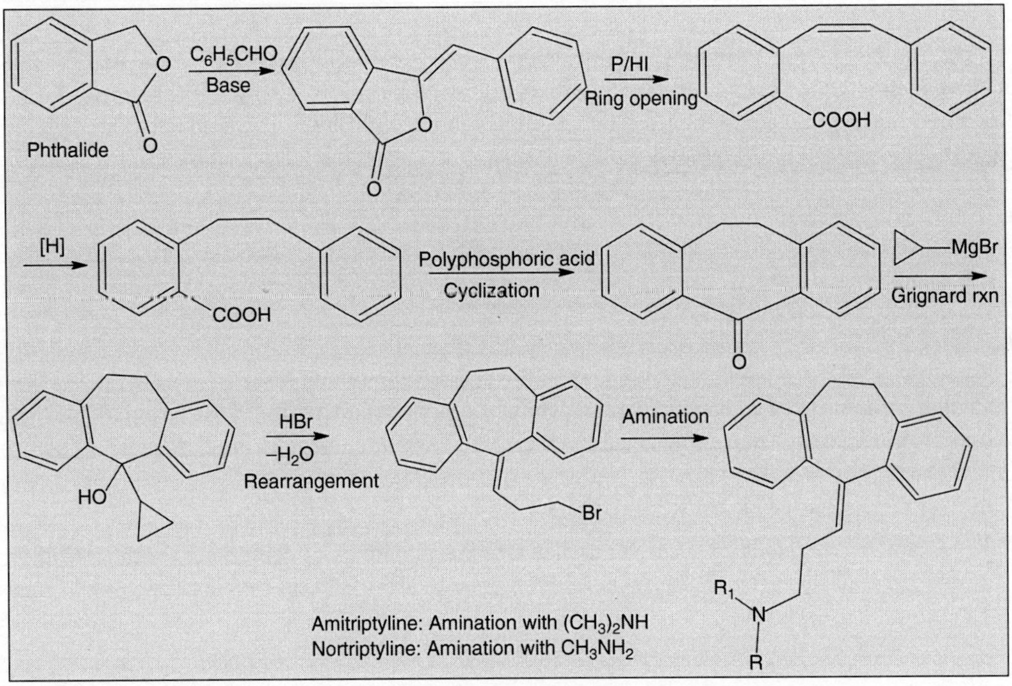
O
O
Phthalide
C6H5CHO
Base
O
O
P/HI
Ring opening
COOH
[H]
COOH
Polyphosphoric acid
Cyclization
O
MgBr
Grignard rxn
HO
HBr
–H2O
Rearrangement
Br
Amination
R1
N
R
Amitriptyline: Amination with (CH3)2NH
Nortriptyline: Amination with CH3NH2

Condensation of phthalide with benzaldehyde gives the benzal derivative, which is in effect an internal enol ester. Reaction with phosphorous and hydroiodic acid serves to reduce the carbonyl group; thus the stilbene carboxylic acid is obtained. Catalytic hydrogenation of the double bond followed by cyclization with polyphosphoric acid leads to dibenzocycloheptanone. The ketone group reacts with cyclopropylmagnesium bromide to give alcohol, which in presence of hydrogen bromide leads to rearrangement of the cyclopropyl carbinol to the homoallyl bromide. Displacement of halogen with appropriate amine affords titled compounds.

Doxepine 3-(Dibenzoxepine-11-ylidene) -N, N-dimethyl-1-propanamine

Synthesis

The initial 6,11-dihydrodibenz[b,e]oxepin-11-one is synthesised from the ethyl ester of 2-phenoxymethyl benzoic acid, which is easily synthesised by reacting ethyl 2-bromomethylbenzoate with phenol in the presence of a base. The resulting ester is hydrolysed into 2-phenoxymethylbenzoic acid, which is cyclized to 6,11-dihydrodibenz[b,e]oxepin-11-one by polyphosphoric acid. Doxepine is synthesised in an analogous manner by reacting 6,11-dihydrodibenz-[b,e]oxepin-11-one with 3-dimethylaminopropylmagnesium bromide and the subsequent dehydration of the resulting tertiary alcohol by hydrochloric acid.

Doxepine and amoxapine usually have less sedative effect and sometimes less muscarinic blocking action than conventional tricyclics.

Amoxapine 2-Chloro-11-(1-piperazinyl)dibenz [1,4] oxazepine

Amoxapine is N-desmethyl analogue of antipsychotic compound loxapine. The synthesis is similar to that of loxapine [Refer chapter 9 Antipsychotic agents], but differs in the use of piperazine instead of N-methyl piperazine. Amoxapine acts by inhibiting MAO and antagonizing dopamine receptor.

10.7 ATYPICAL ANTIDEPRESSANTS

Antidepressants are frequently prescribed to alleviate the symptoms of depression for those with bipolar disorder (manic depression). Atypical antidepressants are a subclass of antidepressants that are

Medicinal Chemistry

chemically unrelated to other antidepressants. These are new-generation compounds with relative safety, tolerability and less side-effects (anticholinergic and cardiac) when compared to other conventional derivatives.

Bupropion 1-(3-Chlorophenyl)-2-[(1,1-dimethylethyl)amino]-1-propanone

Bupropion is a novel, non-tricyclic antidepressant with a primary pharmacological action of monoamine uptake inhibition.

Synthesis

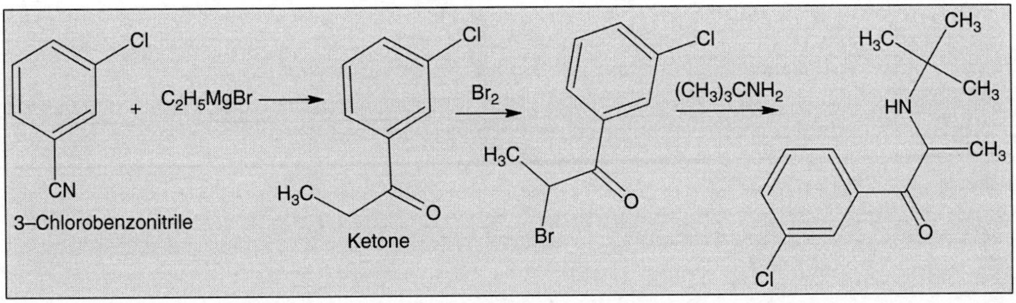

The synthesis of bupropion begins with the reaction of 3-chlorobenzonitrile with ethylmagnesium bromide to give 3-chloropropiophenone. Brominating this with bromine gives 3-chloro-α-bromopropiophenone, which on reaction with *tert*-butylamine gives bupropion.

Uses: Bupropion is used to treat depression. It is also used to treat bipolar depression and attention-deficit disorder. Bupropion is used to help people stop smoking as well.

Trazodone 2-[3-[4-(3-Chlorophenyl)-1-piperazinyl]propyl]-1,2,4-triazolo[4,3] pyridine-3-one

Synthesis

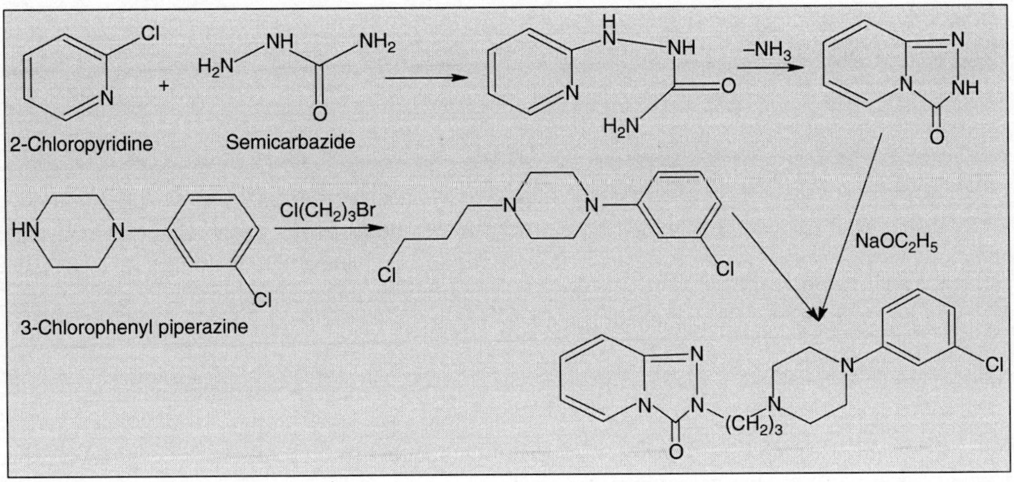

Trazodone is synthesised from 2-chloropyridine, the reaction of which with semicarbazide gives s-triazolo-3-one[4,3-a]pyridine. Alkylation of this product using 1-(3-chloropropyl)-4-(3-chlorophenyl) piperazine gives trazodone.

Antidepressant activity of trazodone is believed to be produced by blocking the reuptake of serotonin at the pre-synaptic neuronal membrane. Trazodone has no influence on the reuptake of norepinephrine or dopamine within the CNS. It has a sedative effect, which is believed to be produced by the alpha-adrenergic blocking action and modest histamine blockade.

10.8 MISCELLANEOUS DRUGS

Fluoxetine 3-[p-(Trifluoromethyl)-phenoxy]-N-methyl-3-phenylpropylamine

Fluoxetine belongs to a group of antidepressants known as the selective serotonin reuptake inhibitors (SSRIs). It acts as a reuptake inhibitor but is highly selective to 5-HT.

Synthesis

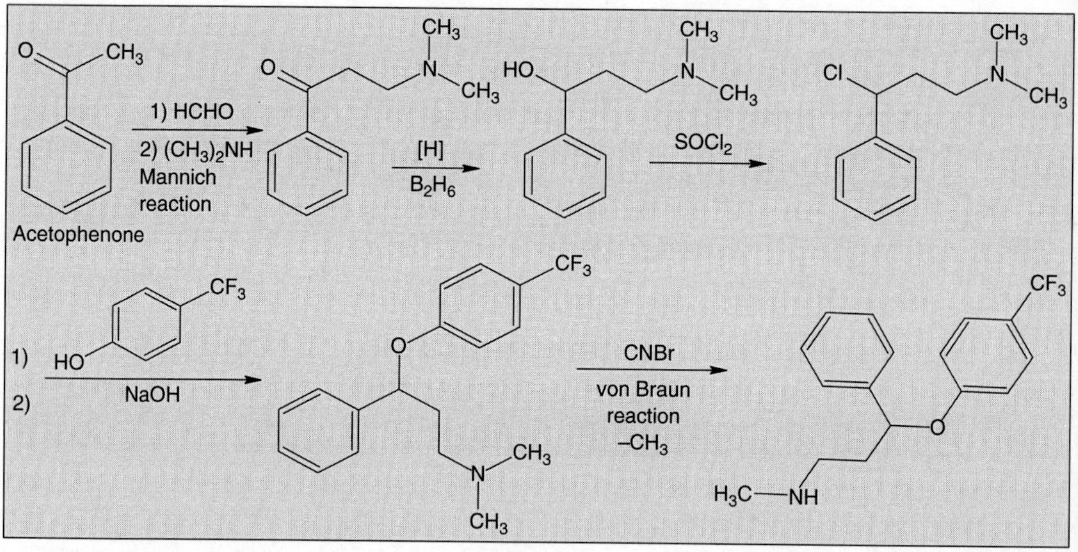

Synthesis starts with the Mannich base from the reaction of acetophenone, formaldehyde, and dimethylamine. The ketone is then reduced to an alcohol; this is converted to the chloride. Displacement of chloride with the phenoxide from treatment of 4-trifluoromethylphenol leads to the O-alkyl ether. One of the methyl groups on nitrogen is then removed by treatment of the intermediate with cyanogens bromide (von Braun reaction).

Sertraline (1S)-cis-4-(3,4-Dichlorophenyl)-1,2,3,4-tetrahydro-N-methyl-1-naphthalenamine

Sertraline is an antidepressant of the selective serotonin reuptake inhibitor (SSRI) class. Sertraline is primarily used to treat clinical depression in adult outpatients as well as obsessive-compulsive, panic, and social anxiety disorders in both adults and children.

Synthesis

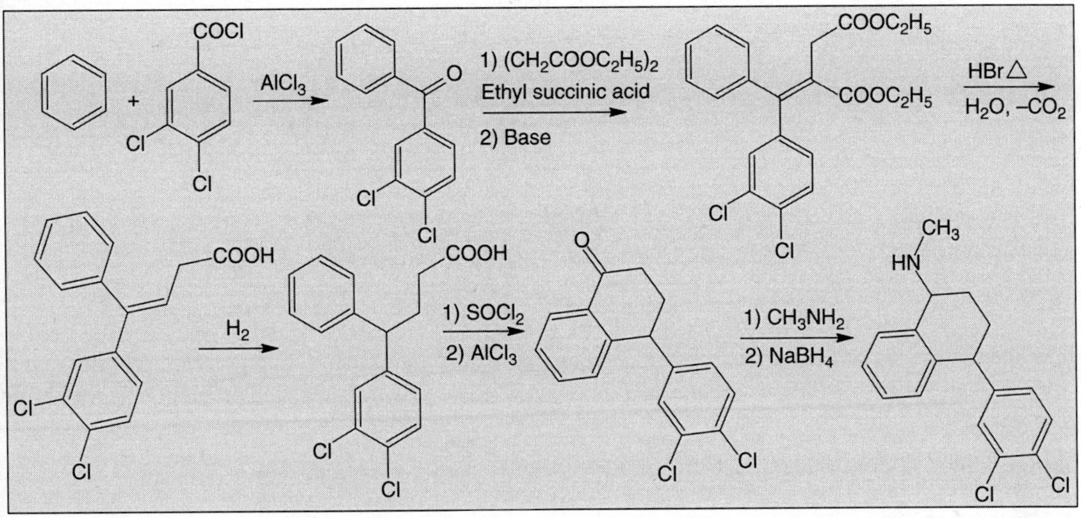

Friedel-Crafts benzoylation of benzene with 3,4-dichlorobenzoyl chloride gives benzophenone derivative. The Stobbe condensation of benzophenone with ethyl succinate gives the chain extended half-ester. Treatment of this product with hydrobromic acid leads to hydrolysis of the remaining ester group with concomitant decarboxylation, to give the acrylic acid; this is then hydrogenated to the diarylbutyric acid. Friedel-Crafts cyclization proceeds by attack on the more electron-rich of the benzene ring to afford teralone. This is converted to methylamino derivative by reductive amination and affords sertraline.

10.9 NEWER DRUGS

Escitalopram S-(+)-1-[3-(Dimethylamino)propyl]-1-(p-fluorophenyl)-5-phthalancarbonitrile

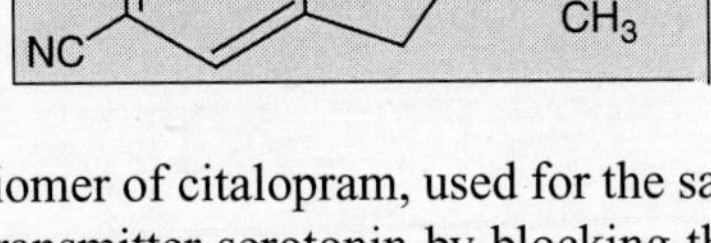

Escitalopram is an (S) enantiomer of citalopram, used for the same indication. It acts by increasing intra-synaptic levels of the neurotransmitter serotonin by blocking the reuptake of the neurotransmitter into the neuron. Of the SSRIs currently on the market, escitalopram has the highest affinity for the human serotonin transporter (SERT). It is approved for the treatment of major depressive disorder and generalized anxiety disorder; other indications include social anxiety disorder, panic disorder, and obsessive-compulsive disorder.

Paroxetine (*3S-trans*)-3-((1,3-Benzodioxol-5-yloxy)methyl)-4-(4-fluorophenyl)- piperidine

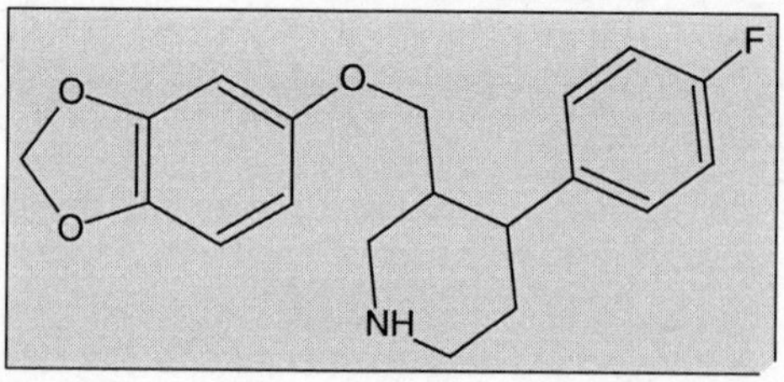

It is a selective serotonin reuptake inhibitor (SSRI) antidepressant. Paroxetine is primarily used to treat the symptoms of depression, obsessive-compulsive disorder, post-traumatic stress disorder, panic disorder, generalized anxiety disorder, social phobia/social anxiety disorder, and premenstrual dysphoric disorder.

Venlafaxine 1-[2-dimethylamino-1- (4-methoxyphenyl)- ethyl]cyclohexan-1-ol

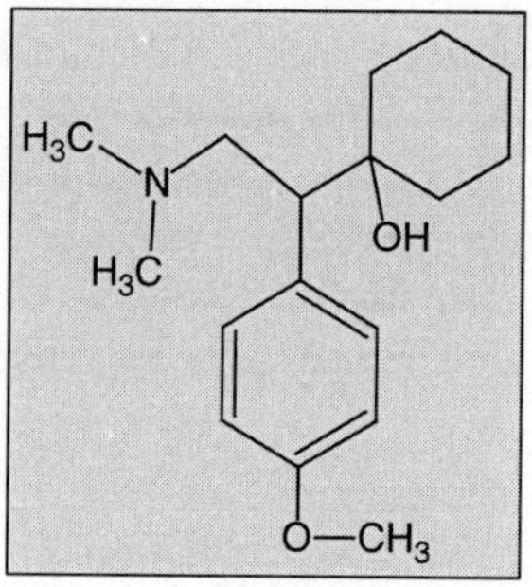

It is an antidepressant of the serotonin-norepinephrine reuptake inhibitor. Due to the pronounced side-effects and suspicions that venlafaxine may significantly increase the risk of suicide, it is not recommended as a first-line treatment of depression.

Duloxetine (+)-(*S*)-*N*-Methyl-3-(naphthalen-1-yloxy)-3-(thiophen-2-yl)propan-1-amine

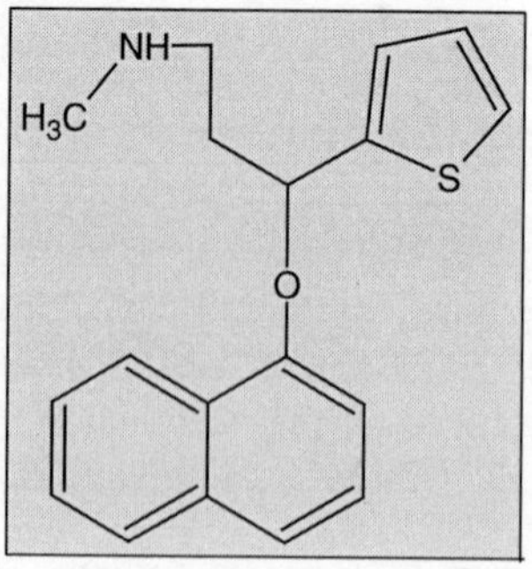

It is a drug that primarily targets major depressive disorder, generalized anxiety disorder (GAD), pain related to diabetic peripheral neuropathy, and, in some countries, stress urinary incontinence. When serotonin and norepinephrine are released from nerve cells (neurons) in the brain, they are reabsorbed into the nerve cells through reuptake. Duloxetine works by preventing serotonin, norepinephrine, and to a lesser extent dopamine from being reabsorbed into the nerve cells in the brain, specifically on the 5HT, NE, and D_2 receptors, respectively.

FURTHER READINGS

1. Leonard, Brian E. and David Healy, 1999. *Differential Effects of Antidepressants*, Taylor & Francis.
2. Foye's Principles of Medicinal Chemistry, Williams, D.A. & Lemke, T.L. (ed.), 5th Ed., 2002.

MULTIPLE-CHOICE QUESTIONS

1. Mechanism of action of nortriptyline is
   a. Dopamine antagonist
   b. Serotonin reuptake inhibitor
   c. Muscarinic antagonist
   d. Inhibits monoamine oxidase

2. Serotonin reuptake inhibitor is mechanism of action of which of the following drugs?
   a. Amoxapine
   b. Amitriptyline
   c. Imipramine
   d. Fluoxetine

3. Dopamine antagonist is mechanism of action of which of the following drugs?
   a. Tranylcypromine
   b. Fluoxetine
   c. Amoxapine
   d. Maprotiline

4. To which of the following drugs does this statement refer to: Tertiary amines are more potent against serotonin reuptake, and secondary amines are more potent against norepinephrine reuptake.
   a. Desipramine
   b. Tranylcypromine
   c. Phenelzine
   d. Amoxapine

5. Monoamine reuptake inhibitor is mechanism of action of which of the following drugs?
   a. Phenelzine
   b. Tranylcypromine
   c. Fluoxetine
   d. Paroxetine

6. In which synthesis is Curtius rearrangement employed?
   a. Pargyline
   b. Tranylcypromine
   c. Fluoxetine
   d. Imipramine

QUESTIONS

1. What are the two primary neurotransmitters involved in antidepressant action?
2. What are the three categories of antidepressants and what are their primary neuronal actions?

3. Codeine is a well-known analgesic drug. Why is it not appropriate to prescribe codeine to patients who are being treated with antidepressant drugs such as fluoxetine?

4. Give the synthesis of phenalzine from phenylacetonitrile.

5. What are monoamine oxidase inhibitors?

SOLUTION TO MULTIPLE-CHOICE QUESTIONS

1. c;
2. d;
3. c;
4. a;
5. a;
6. b.

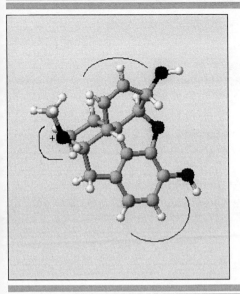

CHAPTER 11

Narcotic Analgesics

LEARNING OBJECTIVES

- Define analgesics and their utility
- List the subtypes of opioid receptor and their functions
- Describe the mode of action of narcotic analgesics
- Describe the pharmacophoric requirement for a molecule to act as opioid antagonist
- Describe the development of morphine and their derivatives
- Illustrate the structure-activity relationship of various narcotic analgesics
- Define narcotic antagonists and their utility

11.1 INTRODUCTION

Analgesics are agents that relieve pain by acting centrally to elevate pain threshold without disturbing consciousness or altering other sensory modalities. Pain is a very important protective phenomenon that accompanies many pathological conditions. However, in fulfilling its function of signalling, it can, upon excessive intensity, in turn aggravate the course of the primary disease, and in some cases such as severe trauma can facilitate the development of shock. Analgesics are divided into two groups: opioids (morphine-like substances), which predominantly influence the central nervous system (CNS), and nonopioids (nonsteroidal anti-inflammatory or fever-reducing drugs—NSAID), which act predominantly on the peripheral nervous system.

Narcotic analgesic literally means that the agents cause sleep in conjunction with their analgesic effect. If a narcotic is used for a long time, it may become habit-forming (causing mental or physical dependence). Physical dependence may lead to withdrawal side-effects when you stop taking the medicine.

The effect of narcotic analgesics is usually explained in terms of their interactions with specific opiate receptors (opioid receptors are a group of G-protein-coupled receptors with opioids as ligands), which are located in the medial thalamus and processes deep, chronic, burning pain. Among the several receptor subtypes identified so far, three are selectively involved in their action.

1. Mu (μ) receptors are responsible for supraspinal analgesia, physical dependence, respiratory depression, miosis, euphoria, reduced GI motility, and physical dependence.
2. Kappa (κ) receptors appear to mediate spinal analgesia, sedation, miosis, and inhibition of antidiuretic hormone release.
3. Delta (δ) receptors may be responsible for analgesia, euphoria, and physical dependence.

Sigma receptors (σ) were once considered to be opioid receptors, but are not usually currently classified as such.

11.2 MECHANISM OF ACTION

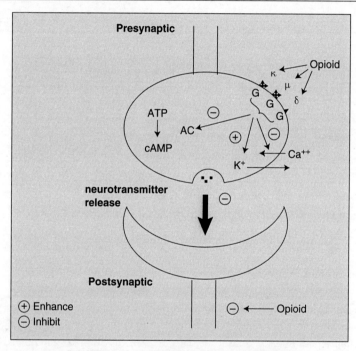

FIGURE 11.1 Mechanism of action of opiates.

Narcotic analgesics have been proposed to inhibit neurotransmitter release by inhibiting calcium entry, by enhancing outward movement of potassium ions, which makes neurons less excitable, or by inhibiting adenylate cyclase (AC), the enzyme that converts adenosine triphosphate (ATP) to cyclic adenosine monophosphate (cAMP), which is required for the release of neurotransmitter; this results in blockage of inter-neuronal transmission of pain impulses at different levels of CNS integration.

11.3 PHARMACOPHORE REQUIREMENT

A schematic for an analgesic receptor site may look as shown in the given figure on morphine. Three areas are needed:

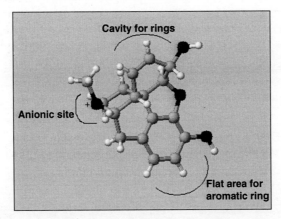

FIGURE 11.2 Pharmacophore for opiates.

- A flat area to accommodate a flat nonpolar aromatic ring,
- A cavity to accept another series of rings perpendicular
- An anionic site for polar interaction of the amine group

11.4 CLINICAL USES

1. Management of acute pain, chronic pain, severe pain of acute myocardial infarction, obstetric analgesia
2. Cough suppression (Codeine, Dextromethorphan)
3. Treatment of diarrhoea (Diphenoxylate, Loperamide)
4. Management of acute pulmonary oedema
5. Preoperative medication and intraoperative adjunctive agents in anaesthesia (Fentanyl, Alfentanyl, Sufentanyl)
6. Maintenance programmes for addicts (Methadone)

Side effects: Narcotic analgesics cause side-effects that limit their use. They include respiratory depression, nausea, vomiting, constipation, a heightened level of blood pressure, urine retention, perspiration, and itching; of course, the most dangerous of these is respiratory depression. Opioids cause dependency and addiction.

11.5 CLASSIFICATION

1. Morphine and its analogues

Chapter 11: Narcotic Analgesics

	R_1	R_2	R_3	Other Changes*
Morphine	—OH	—OH	—CH$_3$	—
Heroin	—OCOCH$_3$	—OCOCH$_3$	—CH$_3$	—
Hydromorphone	—OH	=O	—CH$_3$	(1)
Oxymorphone	—OH	=O	—CH$_3$	(1), (2)
Levorphanol	—OH	—H	—CH$_3$	(1), (3)
Levallorphan	—OH	—H	—CH$_2$CH=CH$_2$	(1), (3)
Codeine	—OCH$_3$	—OH	—CH$_3$	—
Hydrocodone	—OCH$_3$	=O	—CH$_3$	(1)
Oxycodone	—OCH$_3$	=O	—CH$_3$	(1), (2)
Buterphanol	—OH	—H	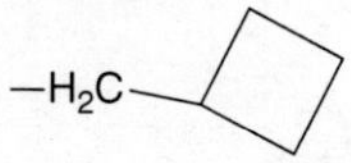 —H$_2$C—◻	(2), (3)
Nalbuphine	—OH	—OH	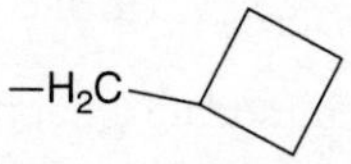 —H$_2$C—◻	(1), (2)

*1. Single instead of double bond between C$_7$ and C$_8$
2. OH added to C$_{14}$
3. No oxygen between C$_4$ and C$_5$
2. Phenyl (ethyl) piperidines

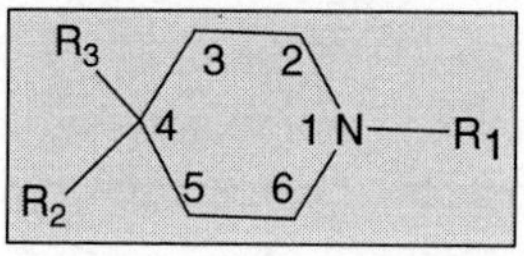

	R_1	R_2	R_3	Any Other	Analgesic Activity
Meperidine	—CH$_3$	—C$_6$H$_5$	—COOC$_2$H$_5$	—	1.0
Bemidone	—CH$_3$	—(m)C$_6$H$_4$OH	—COOC$_2$H$_5$	—	1.5
Properidone	—CH$_3$	—C$_6$H$_5$	—COOCH(CH$_3$)$_2$	—	15.0
Ketobemidone	—CH$_3$	—(m)C$_6$H$_4$OH	—COC$_2$H$_5$	—	6.2
Alphaprodine	—CH$_3$	—C$_6$H$_5$	—OCOC$_2$H$_5$	3-CH$_3$	5.0
Anileridine	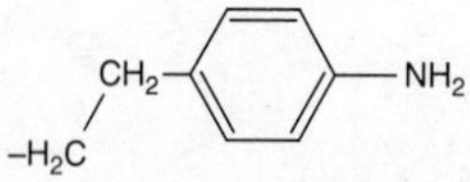 —H$_2$C—CH$_2$—C$_6$H$_4$—NH$_2$	—C$_6$H$_5$	—COO C$_2$H$_5$	—	3.5
Fentanyl	—H$_2$C—CH$_2$—C$_6$H$_5$	—H	N(C$_2$H$_5$)—CO—C$_6$H$_5$	—	940

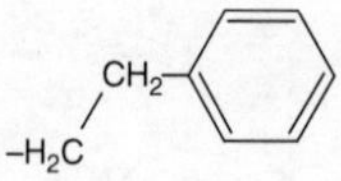

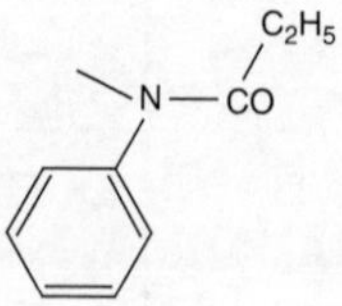

(Continued)

	R_1	R_2	R_3	Any Other	Analgesic Activity
Lofentanil	–H₂C–CH₂–C₆H₅	–COOCH₃	–N(C₂H₅)–CO–C₆H₅	3-CH₃	8400
Sufentanil	–H₂C–CH₂–(2-thienyl)	–CH₂OCH₃	–N(C₂H₅)–CO–C₆H₅	—	Gen. anaesthetic
Alfentanil	–H₂C–CH₂–N(tetrazolone-N-C₂H₅)	–CH₂OCH₃	–N(C₂H₅)–CO–C₆H₅	—	Gen. anaesthetic
Diphenoxylate	(C₆H₅)₂C(CN)–CH₂–CH₂–	–C₆H₅	–COO C₂H₅	—	Antidiarrhoeal
Loperamide	(C₆H₅)₂C[CON(CH₃)₂]–CH₂–CH₂–	–(p) C₆H₄Cl	–OH	—	Antidiarrhoeal

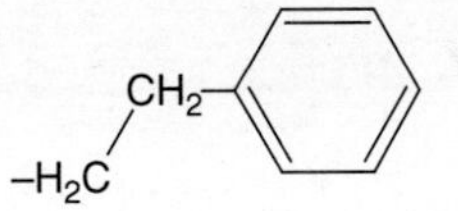

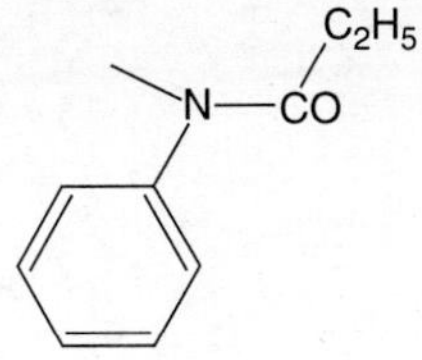

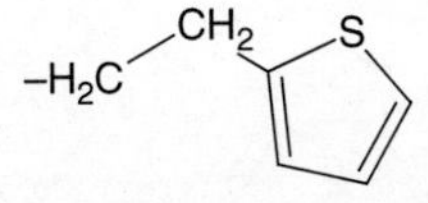

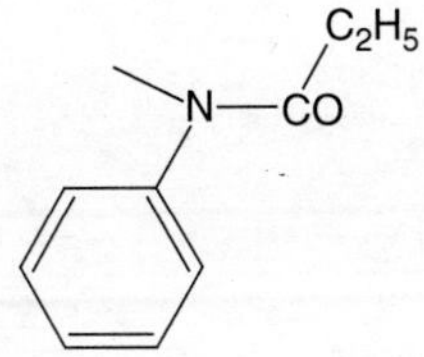

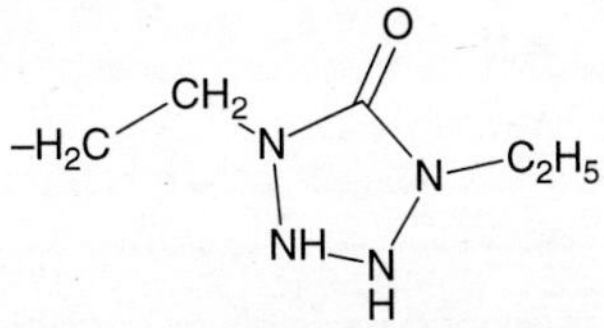

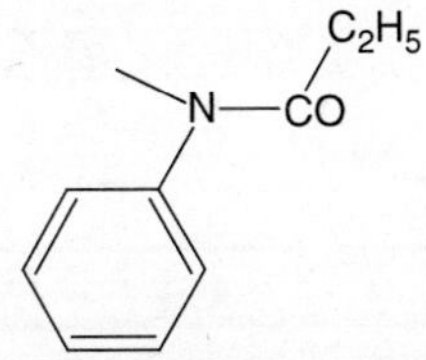

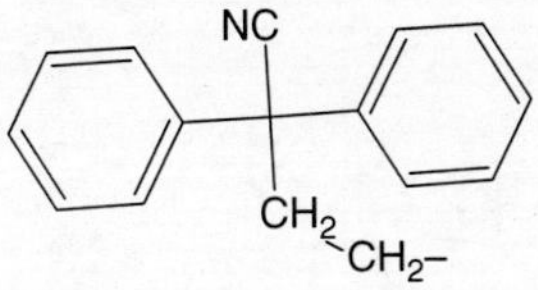

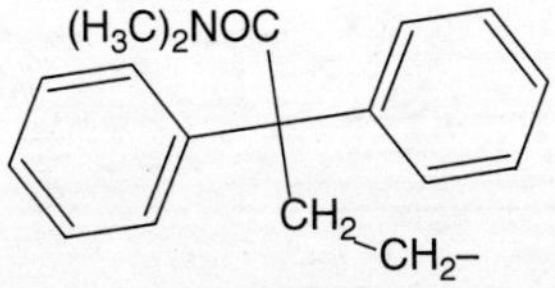

3. Diphenyl heptanones

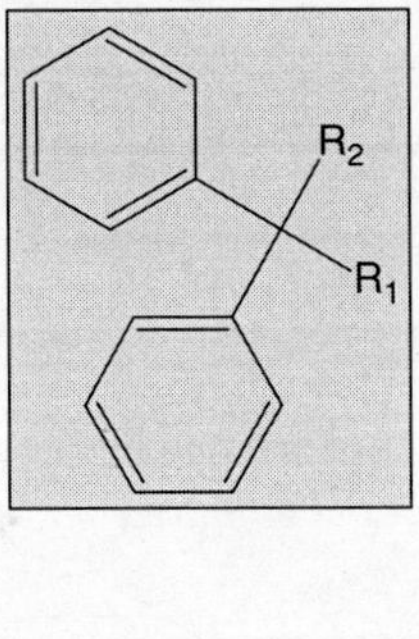

	R_1	R_2	Analgesic Activity
Methadone	—COC$_2$H$_5$	—CH$_2$CH(CH$_3$)N(CH$_3$)$_2$	1.0
Isomethadone	—COC$_2$H$_5$	—CH(CH$_3$)CH$_2$N(CH$_3$)$_2$	0.65
Normethadone	—COC$_2$H$_5$	—CH$_2$CH$_2$N(CH$_3$)$_2$	0.44
Dipanone	—COC$_2$H$_5$	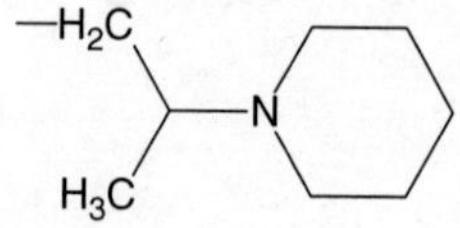 —H$_2$C—CH(CH$_3$)—N(piperidine)	0.80
Phenadoxone	—COC$_2$H$_5$	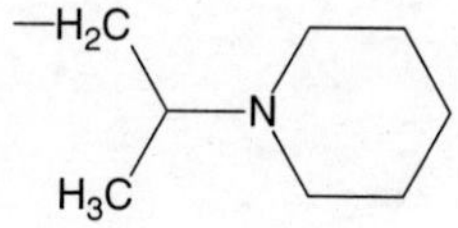 —H$_2$C—CH(CH$_3$)—N(morpholine)	1.4
Dextromoramide	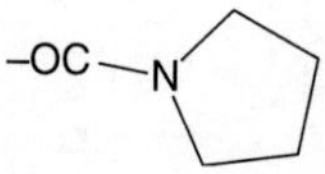 —OC—N(pyrrolidine)	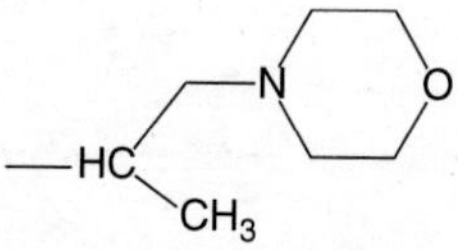 —HC(CH$_3$)—CH$_2$—N(morpholine)	13

4. Benzazocin derivatives

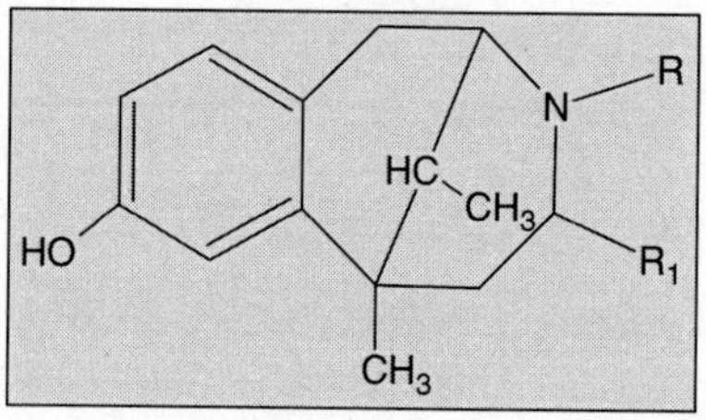

	R	R_1
Pentazocin	—CH$_2$CH=C(CH$_3$)$_2$	H
Phenazocin	—CH$_2$CH$_2$C$_6$H$_5$	H
Cyclazocin	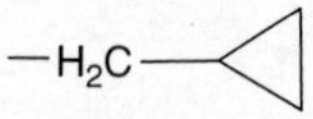 —H$_2$C—(cyclopropyl)	H
Ketazocin	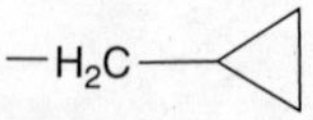 —H$_2$C—(cyclopropyl)	=O

11.6 MORPHINE AND ITS DERIVATIVES

Morphine is the prototype opioid selective for the mu opioid receptors. The naturally occurring isomer of morphine is levo [(1) or (−)] rotatory. (+)-Morphine has been synthesised, and it is devoid of analgesic and other opioid activities. (*See Fig 11.3 in the coloured set of pages.*)

The table here shows the selected SAR for (−)-morphine.

11.6.1 SAR of Morphine

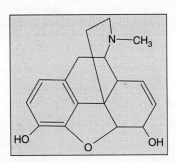

Position	Modification (On Morphine Unless Otherwise Indicated)	Effects
Phenolic hydroxyl	—OH to —OCH₃ (codeine)	Less analgesic, cough suppression
	—OH to —OC₂H₅ (ethyl morphine)	Less analgesic, cough suppression
	—OH to —O-CH₂CH₂-morpholine (Phocodeine)	Less analgesic, cough suppression
Alcoholic hydroxyl	—OH to —OCH₃ (heterocodeine)	5 times more active than morphine
	—OH to —OC₂H₅	2.4 times more active than morphine
	—OH to —OCOCH₃	4.2 times more active than morphine
	—OH to =O (morphinone)	Less active than morphine
Alicyclic unsaturated linkage	—CH=CH— to –CH₂CH₂— (dihydromorphine)	1.2 times more active than morphine
Tertiary nitrogen	N—CH₃ to NH (normorphine)	Less active than morphine
	N—CH₃ to NCH₂CH₂Ph	14 times more active than morphine
	N—CH₃ to N-allyl, methallyl, propyl	Morphine antagonist
	N—CH₃ to N⁺(CH₃)₂ Cl⁻	Strong curare action and no analgesic activity

Levallorphan (−)-3-Hydroxy-N-allylmorphinane

Synthesis

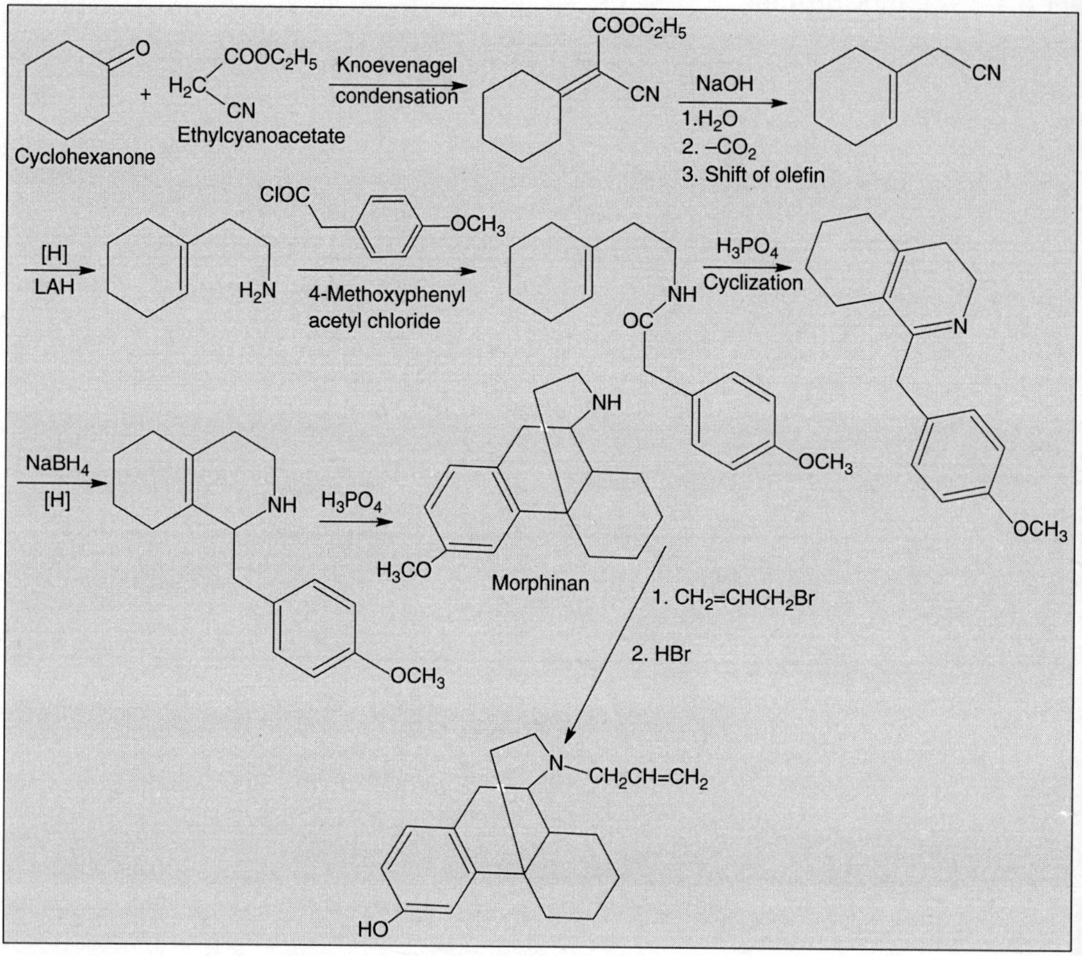

Levallorphan is synthesised from cyclohexanone by its condensation with ethyl cyanoacetate (Knoevenagel reaction), to form alkylidene derivative, which undergoes simultaneous hydrolysis, decarboxylation, olefin shift occurs, forming 1-cyclohexenylacetonitrile. Reduction of the nitrile group by hydrogen in the presence of Raney cobalt gives 2-(1-cyclohexenyl) ethylamine. The resulting amine is further acylated by 4-methoxyphenylacetyl chloride, which forms the amide 2-(1-cyclohexenyl) ethyl-4-methoxyphenylacetamide. Cyclization of the last using phosphorous oxychloride or phosphoric acid leads to the formation of 1-(4-methoxybenzyl)-3,4,5,6,7,8-hexahydroquinolin (Bischler-Napieralski reaction). The imine bond in the obtained compound is hydrogenated in the presence of sodium borohydride, forming 1-(4-methoxybenzyl)-1,2,3,4,5,6,7,8-octahydroquinolin, which undergoes cyclization, allylation and hydrolysis of ether to afford levallorphan.

Use: Opiate antagonist used for the management of acute opioid overdoses.

Levorphanol 3-Hydroxy-*N*-methylmorpinane

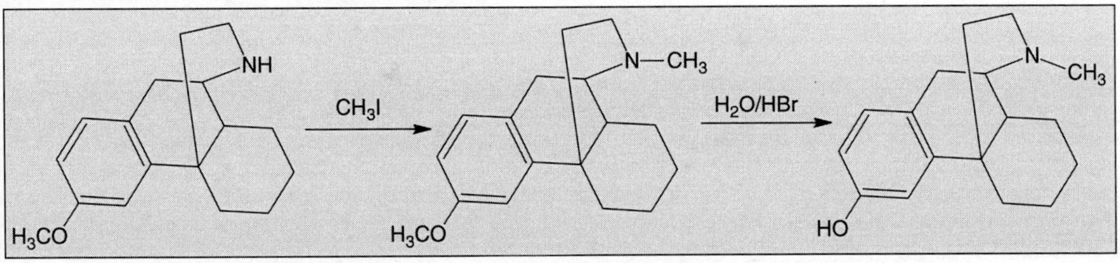

3-Methoxy morpinane is *N*-methylated followed by hydrolysis with hydro bromic acid affords levorphanol.

Use: Very potent analgesic agent. It is about 6 to 8 times as potent as morphine in human.

Buprenorphine   17-(Cyclopropylmethyl)-α-(1,1-dimethylethyl)-4,5-epoxy-18,19-dihydro-3-hydroxy-6-methoxy-α-methyl-6,14-ethenomorphinan-7-methanol

Buprenorphine is a more complex molecule than morphine which would interact with the opioid receptor (analgesic) and because of its complex structure, would not interact with other receptors that produce side-effects.

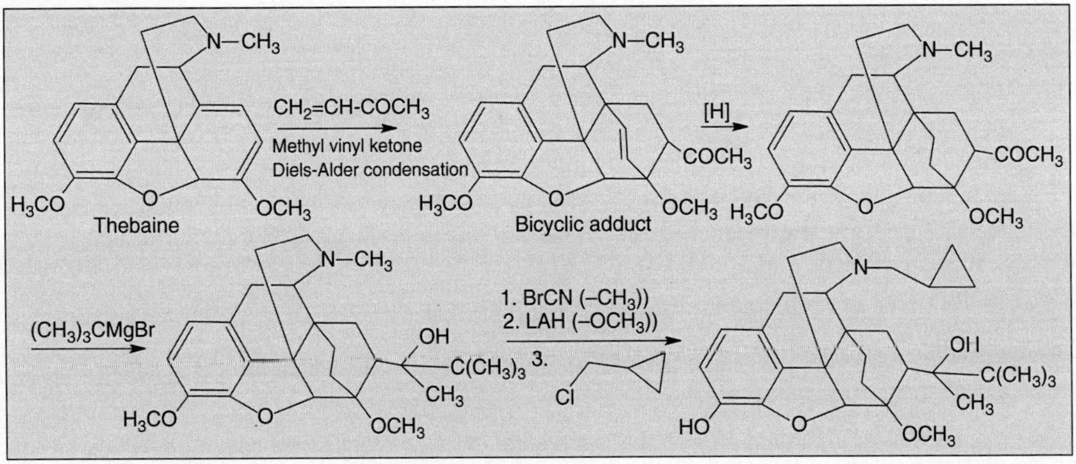

Buprenorphine is synthesised from thebaine. Synthesis of buprenorphine begins on the basis of the reaction product of 4+2 cycloaddition of thebaine and methyl vinyl ketone. The resulting product 7-acetyl-6,14-endoethanotetrahydrothebaine is further hydrogenated using a palladium on carbon catalyst into 7-acetyl-6,14-endoethanotetrahydrothebaine. This is reacted with *tert*-butyl-magnesium bromide to form 6,14-endoethano-7-(2-hydroxy-3,3-dimethyl-2-butyl)-tetrahydrothebaine. The product is demethylated using cyanogen bromide, giving 6,14-endoethano-7-(2-hydroxy-3,3-dimethyl-2-butyl)-tetrahydronorthebaine; this on demethoxylation followed by alkylation with cyclopropyl methyl chloride affords buprenorphine.

Etorphine 6,14-Endoetheno-7 a (1-(R)-hydroxy-1 methylbutyl)-tetrahydro-nororipavine

Etorphine is 1,000 times more potent than morphine, which could be interpreted as a better or tighter fit to the receptor. It is used primarily in veterinary medicine to immobilize large animals.

7-Acetyl-6,14-endoethanotetrahydrothebaine on reaction with propyl magnesium bromide and O-demethylation affords etorphine.

11.7 PHENYL (ETHYL) PIPERIDINES

Analgesic compounds in the 4-phenyl piperidine class may be viewed as A, D ring analogues of morphine. Meperidine proved to be a typical mu agonist with about one-fourth the potency of morphine on weight basis.

SAR

1. Replacement of the 4-phenyl group of meperidine by hydrogen, alkyl, other aryl, arylalkyl, and heterocyclic groups reduces analgesic activity.
2. Introduction of *m*-hydroxyl group on the phenyl ring increases activity. The effect is more pronounced on the keto compound than on meperidine (bemidone vs ketobemidone).
3. Replacement of the carbethoxy group in meperidine by acyloxy group gave better analgesic, as well as spasmolytic (alphaprodine).
4. Replacement of phenyl group by phenylethyl derivative is seen to be about 3 times as active as meperidine. The amino congener, anileridine, is about 4 times more active.
5. Enlargement of piperidine ring to a seven-membered hexahydroazepine is less active but has low incidence of side-effects.
6. Contraction of piperidine ring to the pyrrolidine gives more active compound but causes abuse liability.
7. In fentanyl, the phenyl and acyl groups are separated by nitrogen. It is 50 times stronger than morphine with minimal side-effects. Its short duration of action makes it well suited for use in anaesthesia.
8. The 3-methyl analogue with an ester group at the 4-position like lofentanil was 8,400 times more potent than meperidine as an analgesic.
9. Diphenoxylate, a structural hybrid of meperidine and methadone types, lacks analgesic activity. It is effective as an intestinal spasmolytic and is used for the treatment of diarrhoea.
10. The related *p*-chloro analogue (loperamide) has been shown to bind to opiate receptors in the brain, but not to penetrate the blood-brain barrier sufficiently to produce analgesia.

Meperidine (Pethidine) 1-Methyl-4-phenylpiperidine-4-carboxylic acid
Synthesis
Method I

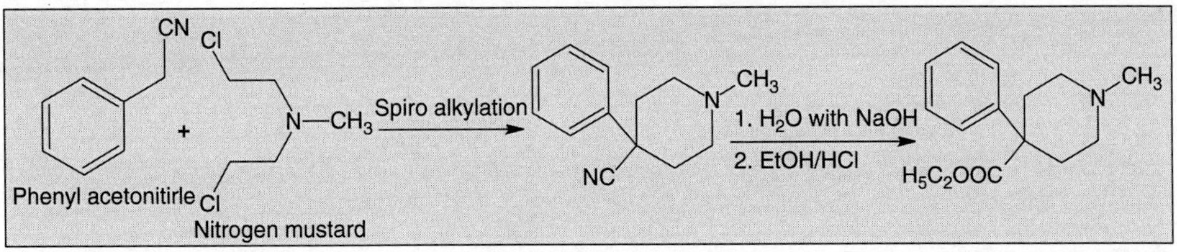

Meperidine synthesis is accomplished by the alkylation of benzyl cyanide using *N,N-bis*-(2-chlorethyl)-*N*-methylamine in the presence of sodium amide, which forms 1-methyl-4-phenyl-4-cyanopiperidine, and its subsequent acidic ethanolysis into meperidine.

Method II

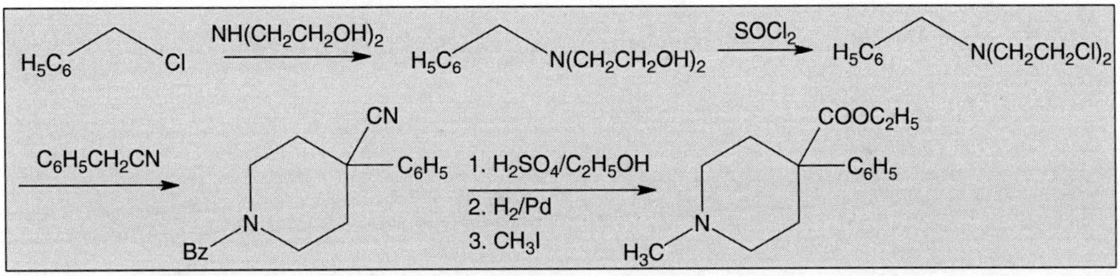

Benzyl chloride on reaction with *bis*(2-hydroxyethyl)amine forms benzylamino derivative, which is converted to chloro ethyl derivative by reaction with thionyl chloride. This in turn reacts with benzyl cyanide, which forms 1-benzyl-4-phenyl-4-cyanopiperidine, and its subsequent acidic ethanolysis, debenzylation, and *N*-methylation into meperidine.

Anileridine Ethyl 1-[2-(4-aminophenyl)ethyl]- 4-phenyl-piperidine-4-carboxylate

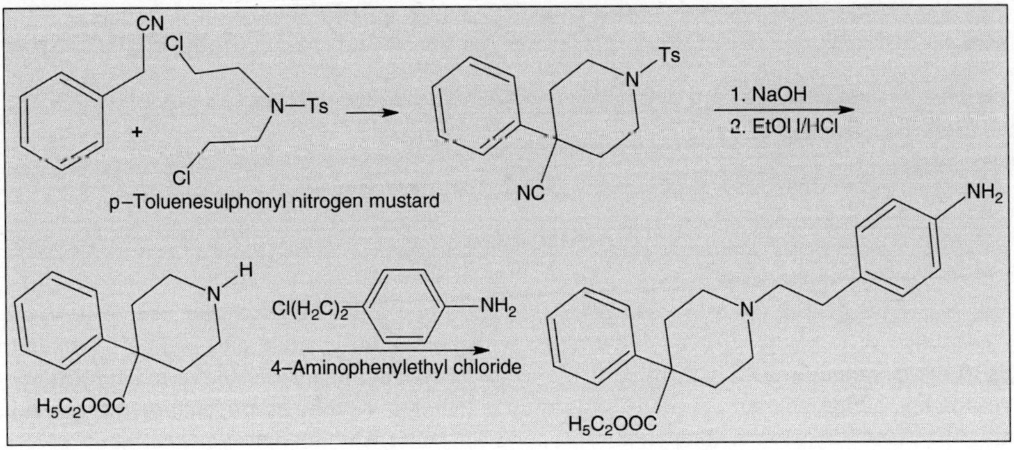

Alkylation of normeperidine nitrogen with 4-aminophenyl ethyl chloride affords anileridine.

Ketobemidone 1-[4-(3-Hydroxyphenyl)-1-methyl-4-piperidyl]propan-1-one

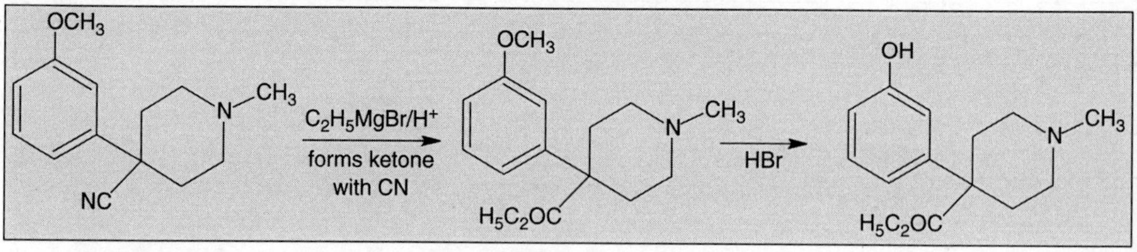

1-Methyl-4-(*m*-methoxy)phenyl-4-cyanopiperidine on reaction with ethyl magnesium bromide forms ketone intermediate, which on demethylation with hydro bromic acid affords ketobemidone.

Fentanyl and alfentanyl For synthesis, refer to General Anaesthetics (Chapter 5).

Lofentanil Methyl (3S,4R)-1-(2-cyclohexylethyl)-4 -(cyclohexyl-propanoylamino)-3-methylpiperidine-4-carboxylate

It is 8,400 times more potent than meperidine.

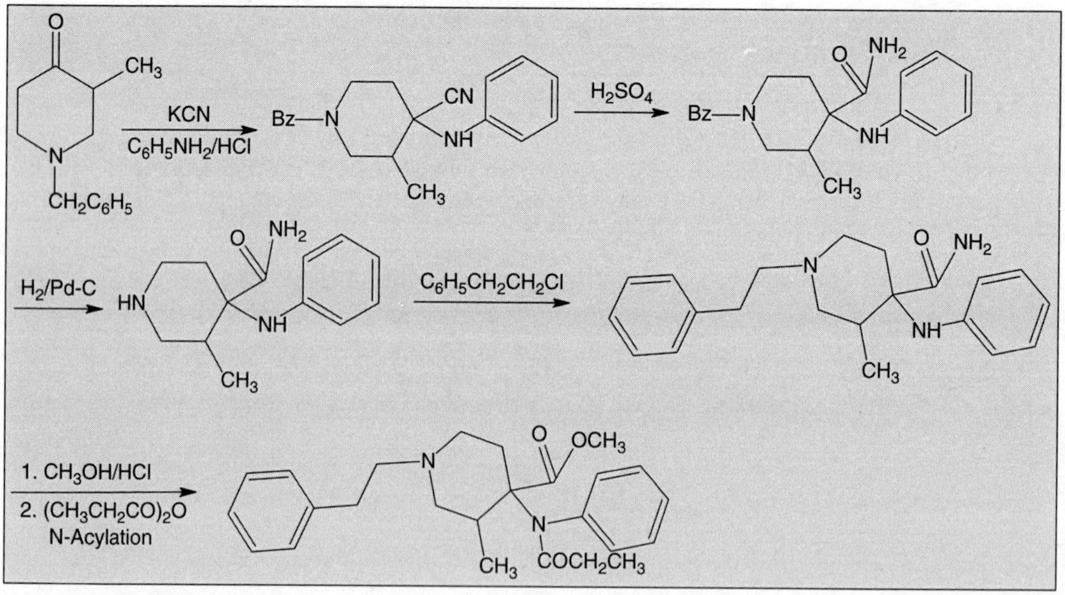

Reaction of 3-Methyl-*N*-benzylpiperidone with potassium cyanide and aniline hydrochloride leads to the α-aminonitrile. Hydrolysis of the cyano group by means of sulphuric acid affords the corresponding amide; the benzyl-protecting group is then removed by hydrogenation over palladium to give the secondary amine. Phenylethyl group is introduced by alkylation at the more basic ring nitrogen with 2-phenylethyl chloride. Interchange with ethanolic hydrogen chloride then converts the amide to an ethyl ester; acylation of the remaining secondary amine with propionic anhydride affords lofentanil.

Diphenoxylate Ethyl 1-(3-cyano-3,3-diphenylpropyl)-4-phenylpiperidine-4-carboxylic acid

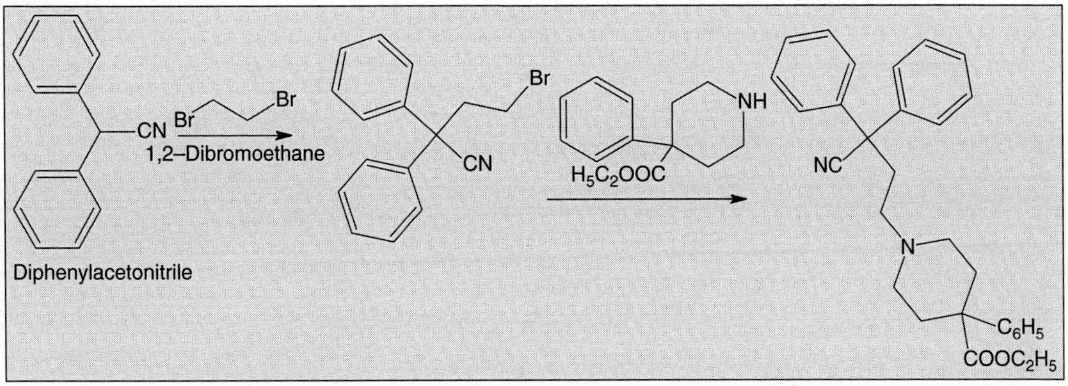

Alkylation of diphenyl acetonitrile 1,2-dibromoethane gives 2,2-diphenyl-4-bromobutyronitrile, which in turn reacts with normeperidine affords diphenoxylate.

Use: Diphenoxylate inhibits intestinal motility and acts as an antidiarrhoeal agent.

11.8 DIPHENYL HEPTANONES

11.8.1 SAR (Structure Activity Relationship)

1. The *levo* isomer of methadone and isomethadone were twice as effective as their racemic mixtures.
2. Removal of any one phenyl group sharply decreased the activity.
3. Replacement of terminal dimethylamino group by piperidine group decreased activity.
4. Replacement of propionyl group by hydrogen, hydroxyl, or acetoxyl led to decreased activity, whereas amide analogue, pyrrolidinoyl, and terminal morpholino moiety enhanced activity several times (e.g., dextromoramide).

Methadone 6-(Dimethylamino)-4,4-diphenylheptan-3-one
Synthesis

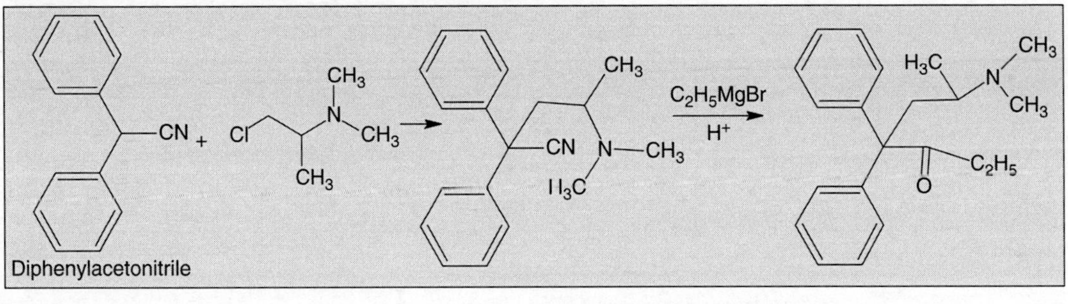

Methadone is synthesised by alkylation of diphenylacetonitrile using 2-dimethylaminopropylchloride in the presence of sodamide. The resulting 4-dimethylamino-2,2-dephenylvaleronitrile is reacted with the ethylmagnesiumbromide to afford methadone.

Methadone, which looks structurally different from other opioid agonists, has steric forces that produce a configuration that closely resembles that of other opiates. Methadone is more active and more toxic than morphine. It can be used for the relief of many types of pain. In addition, it is used as a narcotic substitute in addiction treatment because it prevents morphine abstinence syndrome.

Dextromoramide (3R)-3-Methyl-4-morpholin-4-yl-2,2-diphenyl-1-pyrrolidin-1-yl-butan-1-one

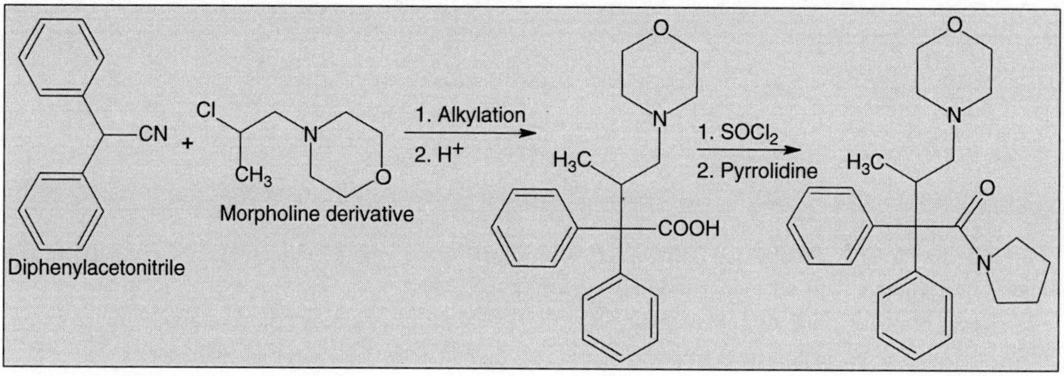

Dextromoramide is synthesised by alkylation of diphenylacetonitrile using 1-morpholinyl-2-chloropropane in the presence of sodamide. The nitrile group is hydrolysed to acid, which is converted to acid chloride by reaction with thionyl chloride. Acid chloride on reaction with pyrrolidine affords amide dextromoramide.

11.9 BENZAZOCIN (BENZOMORPHAN) DERIVATIVES

Pentazocine 1,2,3,4,5,6-Hexahydro-6,11-dimethyl-3-(3-methyl-2-butenyl)-2,6-methano-3-benzazocin-8-ol

Synthesis

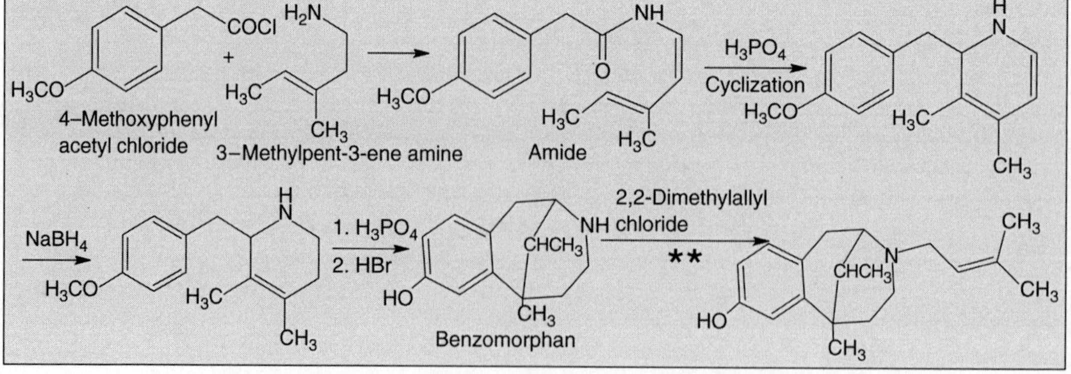

Acylation of aliphatic amine with 4-methoxyphenylacetyl chloride gives the corresponding amide. It cyclizes to the dihydropyridine when treated under the condition of Bischler-Napieralski reaction. Reaction with sodium borohydride results in reduction of the enamine double bond. Cyclization of this intermediate with strong acid proceeds to the benzomorphan, which on allylation with 2,2-dimethylallyl chloride affords pentazocine.

Cyclazocin — alkylation with cyclopropylmethyl chloride in the last step.

Phenazocin — alkylation with phenylethyl chloride in the last step.

11.10 NARCOTIC ANTAGONISTS

The euphoria accompanying the use of heroin and other narcotics reinforces repeated drug-seeking behaviour as physical dependence develops. Once tolerance develops, the opiate-dependent individual avoids painful withdrawal symptoms by continuously increasing the amounts of opiates consumed.

Narcotic antagonists prevent or abolish excessive respiratory depression caused by the administration of morphine or related compounds. They act by competing for the same analgesic receptor sites. They are structurally related to morphine with the exception of the group attached to nitrogen.

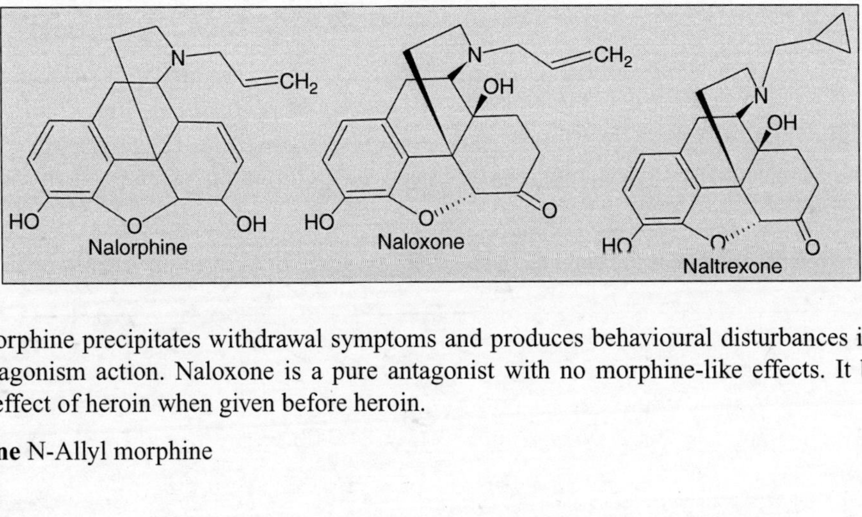

Nalorphine precipitates withdrawal symptoms and produces behavioural disturbances in addition to the antagonism action. Naloxone is a pure antagonist with no morphine-like effects. It blocks the euphoric effect of heroin when given before heroin.

Nalorphine N-Allyl morphine

Synthesis

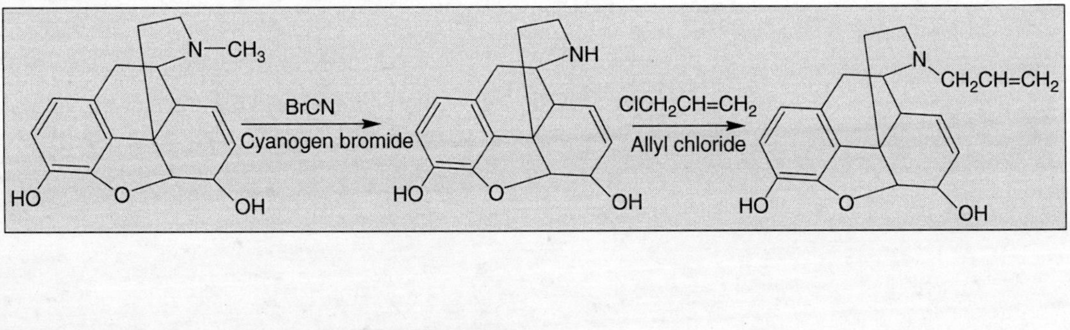

Morphine is demethylated by treatment with cyanogens bromide, followed by N-allylation by reaction with allyl chloride.

Use: Partial opioid receptor agonist and antagonist.

Naloxone (−)-17-(Allyl)-4,5-epoxy-3,14-dihydroxymorphinan-6-one

Synthesis

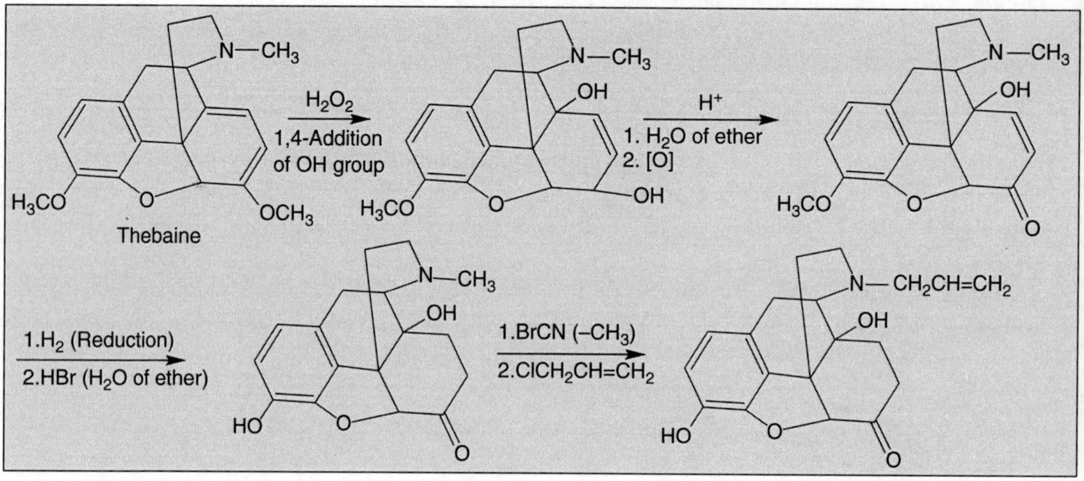

Naloxone is synthesised by the alkylation of 14-hydroxydihydronormorphinane by allylbromide.

Use: Opioid antagonist. At higher doses, naloxone may be useful in the treatment of shock and spinal cord injury

Naltrexone: (−)-17-(Cyclopropylmethyl)-4,5-epoxy-3,14-dihydroxymorphinan-6-one

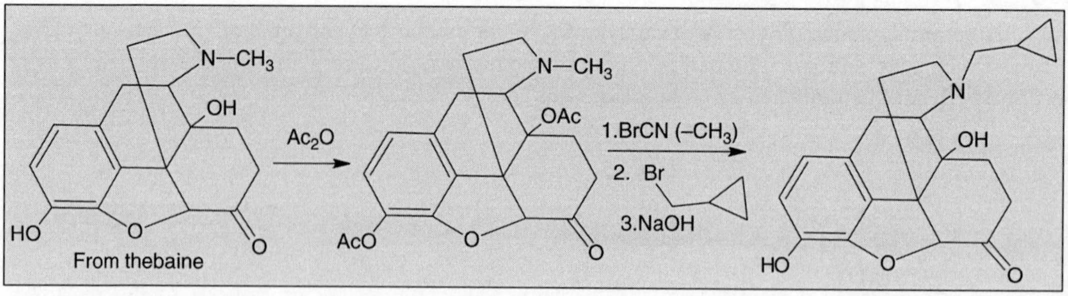

Naltrexone is an *N*-cyclopropylmethyl derivative of oxymorphone. One of the methods of synthesis is analogous to the synthesis of naloxone, which consists of using cyclopropylmethylbromide instead of allylbromide.

Naltrexone became clinically available in 1985 as a new narcotic antagonist. Its actions resemble those of naloxone, but naltrexone is well absorbed orally and is long-acting, necessitating only a dose of 50 mg to 100 mg. Therefore, it is useful in narcotic treatment programmes where it is desired to maintain an individual on chronic therapy with a narcotic antagonist. In individuals taking naltrexone, subsequent injection of an opiate will produce little or no effect. Naltrexone appears to be particularly effective for the treatment of narcotic dependence in addicts who have more to gain by being drug-free rather than drug-dependent. Naltrexone is at least 17 times more potent than nalorphine in morphine-dependent humans and twice as potent as naloxone in precipitating withdrawal symptoms.

11.11 NEWER DRUGS

Remifentanil   Methyl   1-(2-methoxycarbonylethyl)-4-(phenyl-propanoyl-amino) -piperidine-4-carboxylate

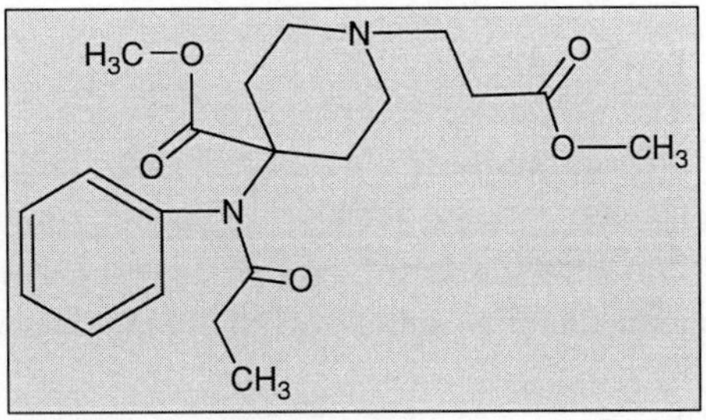

It is a potent ultrashort-acting synthetic opioid analgesic drug. It is given to patients during surgery to relieve pain and as an adjunct to an anaesthetic. Remifentanil is used for sedation as well as combined with other medications for use in general anaesthesia. Remifentanil has a similar potency to fentanyl.

Ohmefentanyl   N-[(3R,4S)-1-[(2S)-2-Hydroxy-2-phenyl-ethyl] -3-methyl-4-piperidyl]-N-phenyl-propanamide

It (β-hydroxy-3-methylfentanyl) is an extremely potent analgesic drug that selectively binds to the μ-opioid receptor. In mouse studies, the most active isomer 3R,4S,βS-ohmefentanyl was 28 times more powerful as a painkiller than fentanyl, the chemical from which it is derived, and 6,300 times more effective than morphine.

Meptazinol   3-(3-Ethyl-1-methylazepan-3-yl)phenol

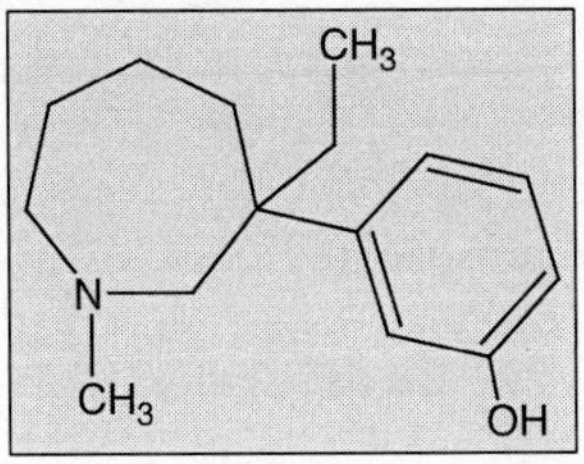

It is an opioid analgesic for use with moderate to severe pain, most commonly used to treat pain in obstetrics (childbirth). A partial μ-opioid receptor agonist, its mixed agonist/antagonist activity affords it a lower risk of dependence and abuse than full μ agonists like morphine. Meptazinol exhibits a short onset of action, but also a shorter duration of action relative to other opioids such as morphine, pentazocine, or buprenorphine.

Tramadol (±)cis-2-[(Dimethylamino)methyl]-1-(3-methoxyphenyl) cyclohexanol

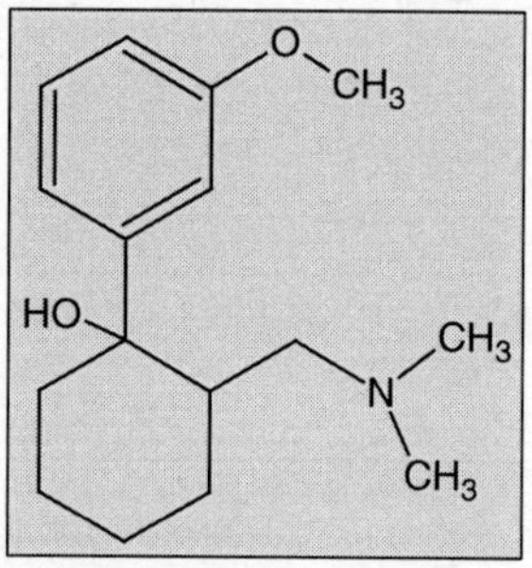

It is an atypical opioid that is a centrally acting analgesic, used for treating moderate to severe pain. It is a synthetic agent, as a 4-phenyl-piperidine analogue of codeine, and appears to have actions on the GABAergic, noradrenergic, and serotonergic systems.

Tapentadol 3-[(1R,2R)-3-(Dimethylamino)-1-ethyl-2-methylpropyl]phenol

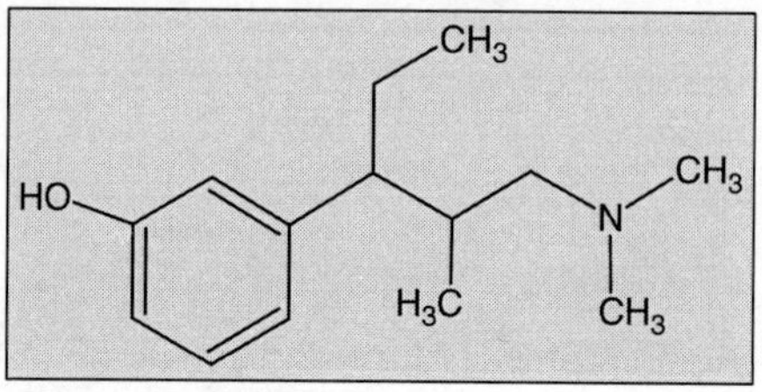

It is a centrally acting analgesic with a unique dual mode of action as an agonist at the μ-opioid receptor and as a norepinephrine reuptake inhibitor. It is considered to have a potency between morphine and tramadol. Its dual mode of action provides analgesia at similar levels of more potent narcotic analgesics such as hydrocodone, oxycodone, and morphine with a more tolerable side-effect profile.

FURTHER READINGS

1. Jane V. Aldrich and Sandra C. Vigil-Cruz, 2003. *Narcotic Analgesics in Burger's Medicinal Chemistry and Drug Discovery*, 6th Ed., D. Abraham (ed.), John Wiley, Inc.: New York, pp. 329–481.
2. Martin, T.J. and Eisenach, J.C., 'Pharmacology of opioid and nonopioid analgesics in chronic pain states', J. Pharmacol. Exp. Ther, 299, 811817(2001).

3. Aldrich, J.V, 'Narcotic analgesics', Am. J Pharm. Educ., 57, 153161(1993).
4. Casy, A.F. and Parfitt, R.T., 1986. *Opioid Analgesics*, Plenum Press, New York NY.

MULTIPLE-CHOICE QUESTIONS

1. One of the following is an ester:
   a. Morphine
   b. Phenazocin
   c. Heroin
   d. Ketobemidone

2. Dextromoramide belongs to
   a. Morphine
   b. Phenyl (ethyl) piperidines
   c. Diphenyl heptanones
   d. Benzazocin

3. When the alcoholic OH is converted to ether, it is
   a. Useful as cough suppressant
   b. More active than morphine
   c. Less active than morphine
   d. Strong curare action and no analgesic action

4. Substitution in the phenyl ring of phenyl (ethyl)piperidines shows activity in the order of
   a. H > OH > =O
   b. =O > H > OH
   c. OH > =O > H
   d. =O > OH > H

5. Pethidine is same as
   a. Morphine
   b. Heroin
   c. Meperidine
   d. Phenazocin

6. Synthesis of levallorphan involves
   a. Diels Alder condensation
   b. Mannich reaction
   c. Knoevenagel condensation
   d. Polonovski's reaction

7. Synthesis of buprenorphine involves
   a. Diels Alder condensation
   b. Mannich reaction
   c. Knoevenagel condensation
   d. Polonovski's reaction

8. The drug that is 1,000 times more potent than morphine is
   a. Buprenorphine
   b. Meperidine
   c. Etorphine
   d. Pentazocin

9. The drug used as a narcotic substitute in addiction treatment because it prevents morphine abstinence syndrome is
   a. Nalorphine
   b. Methadone
   c. Pethidine
   d. Codeine

10. A structural hybrid of meperidine and methadone is
    a. Diphenoxylate
    b. Lofentanil
    c. Loperamide
    d. Pentazocin

11. The most potent narcotic antagonist is
    a. Naloxone
    b. Naltrexone
    c. Nalorphine
    d. Pethidine

QUESTIONS

1. What are the general structural requirements of opioids? Explain with morphine as example.
2. With the help of a general structure, discuss the SAR of narcotics.
3. Discuss the general drug action of narcotic analgesics and narcotic antagonists.
4. Give the structures of narcotic analgesics such as morphine and codeine derived from opium as well as synthetics such as meperidine and propoxyphene. Rank the narcotic analgesics as to effectiveness.
5. Give the synthesis of meperidine, diphenoxylate, and nalorphine.

SOLUTION TO MULTIPLE-CHOICE QUESTIONS

1. c;
2. c;
3. b;
4. d;
5. c;
6. c;
7. a;
8. c;
9. b;
10. a;
11. b.

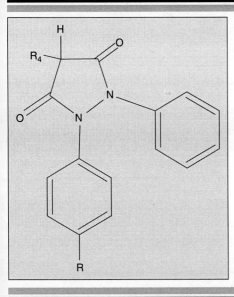

CHAPTER 12

Antipyretics and Non-Steroidal Anti-Inflammatory Drugs

LEARNING OBJECTIVES

- Define pyrexia, inflammation, their causes and modes of treatment options
- Classify analgesics and anti-inflammatory drugs
- Define eicosanoids and their role in inflammation
- Describe the mode of action and complications of anti-inflammatory agents
- Describe the importance of structure on activity of important classes of anti-inflammatory agents
- List the synthesis and utility of other miscellaneous classes of drugs
- Describe some drugs useful for the treatment of gout
- Define COX-2 inhibitors and their significance
- List some newer drugs in the class of analgesics and anti-inflammatory agents

12.1 INTRODUCTION

Fever usually results from microbes such as bacteria or viruses triggering the body's defence mechanisms. This activates certain types of cells, some of which release the substance interleukin. Prostaglandin is another chemical released by the body that plays a part in this process. Prostaglandin is induced by bacterial pyrogens and is produced in the CNS. Interleukin affects the hypothalamus, which is the part of the brain that regulates body temperature, signalling it to raise the temperature by a few degrees. The hypothalamus works like a thermostat while the interleukin that is released serves to raise its preset temperature. A huge quantity of drugs belonging to various classes of compounds exhibit analgesic, anti-fever, and anti-inflammatory action. In addition, they are devoid of many undesirable effects that accompany opioid analgesics (respiratory depression, addiction, etc.).

Antipyretics are the drugs that reduce the elevated body temperature. However, they will not affect the normal body temperature if one does not have fever. Antipyretics cause the hypothalamus to override an interleukin-induced increase in temperature. The body will then work to lower the temperature and the result is a reduction in fever.

Anti-inflammatory agents are used to cure or prevent inflammation caused by prostaglandin (PGE_2). These drugs are widely utilized for the alleviation of minor aches, pains, fever, and symptomatic treatment of rheumatic fever, rheumatoid arthritis, and osteoarthritis. Non-steroidal anti-inflammatory drugs, usually abbreviated to NSAIDs, are drugs with analgesic, antipyretic, and anti-inflammatory effects—they reduce pain, fever, and inflammation. The term 'non-steroidal' is used to distinguish these drugs from steroids.

12.2 CLASSIFICATION

1. Salicylic acid derivatives: Aspirin, Diflunisal, Salsalate, Sulfsalazine
2. p-Aminophenol derivatives: Paracetamol, Phenacetin
3. Pyrazolidinedione derivatives: Phenylbutazone, Oxyphenbutazone, Sulfinpyrazone
4. Anthranilic acid derivatives: Mefenemic acid, Flufenemic acid, Meclofenamate
5. Aryl alkanoic acid derivatives:
    (i) Indoleacetic acid: Indomethacin
    (ii) Indeneacetic acid: Sulindac
    (iii) Pyrroleacetic acid: Tolmetin, Zomipirac
    (iv) Phenylacetic (propionic) acid: Ibuprofen, Diclofenac, Naproxen, Caprofen, Fenoprofen, Ketoprofen, Flubiprofen, Ketoralac, Etodolac
6. Oxicams: Piroxicam, Meloxicam, Tenoxicam
7. Miscellaneous: Nambumetone, Nimesulide, Analgin
8. Anti-gout drugs: Colchicine, Allopurinol, Probenecid, Sulfipyrazone
9. Selective COX II inhibitors: Celecoxib, Rofecoxib, Valdecoxib

12.3 BIOSYNTHESIS OF EICOSANOIDS

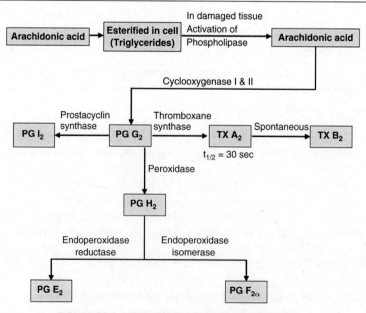

FIGURE 12.1 Biosynthetic pathway of eicosanoids.

Prostaglandin E_2 appear to play a major role in the inflammatory processes particularly it causes pain, vasodilatation and edema.

12.4 MECHANISM OF ANTI-INFLAMMATORY ACTION

Most NSAIDs act as non-selective inhibitors of the enzyme cyclo-oxygenase, inhibiting both the cyclo-oxygenase-1 (COX-1) and cyclo-oxygenase-2 (COX-2) isoenzymes. Cyclo-oxygenase catalyses the formation of prostaglandins, prostacyclin and thromboxane from arachidonic acid (itself derived from the cellular phospholipid bilayer by phospholipase A_2). Prostaglandins act (among other things) as messenger molecules in the process of inflammation.

12.5 SIDE-EFFECTS

1. **In stomach:** NSAIDs blocks biosynthesis of PG, especially PGE_2 and PGI_2, which serve as cytoprotective agents in the gastric mucosa. These PGs inhibit acid secretion by the stomach, enhance mucosal blood flow, and promote the secretion of cytoprotective mucus in the GIT. Inhibition of the PG synthesis may reduce the stomach more susceptible to damage and leads to **gastric ulcer**.
2. **In platelets:** Platelet function gets disturbed because NSAIDs prevent the formation of TXA_2 in platelets, as TXA_2 is a potent platelet-aggregating agent. This accounts for the tendency of these drugs to **increase the bleeding time**.
   This side effect has been exploited in the prophylactic treatment of thromboembolic disorders.
3. **In uterus:** NSAIDs **prolong gestation** because of the inhibition of $PGF_{2\alpha}$ in uterus. $PGF_{2\alpha}$ is a potent uterotropic agent and their biosynthesis by uterus increases dramatically in the hours before parturition. Accordingly, some anti-inflammatory drugs have been used as tocolytic agent to inhibit preterm labour.
4. **In kidney:** NSAIDs decrease renal blood flow and the rate of glomerular filtration in patients with congestive heart failure, hepatic cirrhosis, and chronic renal disease. In addition, they prolong the retention of salt and water, and this may cause oedema in some patients.

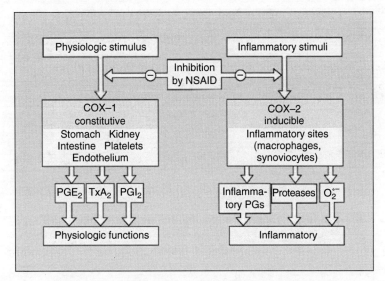

FIGURE 12.2 Differentiation of cox enzymes w.r.t function.

12.6 SALICYLIC ACID DERIVATIVES

Salicylates not only possess antipyretic, analgesic, and anti-inflammatory properties, but also other actions that have been proven to be therapeutically beneficial. Salicylates promote the excretion of uric acid and they are useful in the treatment of gouty arthritis. More recently, attention has been given to the ability of salicylates (aspirin) to inhibit platelet aggregation, which may contribute to heart attacks and strokes and, hence, aspirin reduces the risk of myocardial infarction. Also, a recent study suggested that aspirin and other NSAIDs might be protective against colon cancer.

12.6.1 Structure–Activity Relationship (SAR) of Salicylates

1. Substitution on either the carboxyl or phenolic hydroxy groups may affect the potency and toxicity.
2. Reducing the acidity of the COOH, e.g., the corresponding amide (salicylamide), retains the analgesic action of salicylic acid but is devoid of anti-inflammatory properties.
3. Placing the phenolic hydroxyl group *meta* or *para* to the carboxyl group abolishes the activity.
4. Substitution of halogen atoms on the aromatic ring enhances potency and toxicity.
5. Substitution of aromatic rings at the 5-position of salicylic acid increases anti-inflammatory activity (diflunisil).

Aspirin Acetylsalicylic acid
Synthesis

Aspirin is synthesised by the acetylation of salicylic acid using acetic anhydride or acetyl chloride. It is indicated for the relief of minor aches and mild to moderate pain (325 mg–650 mg every 4 hours), for arthritis (3.2 g–6 g/day), for reducing the risk of transient ischemic attacks in men (1.3 g/day), and for myocardial infarction prophylaxis (40 mg–325 mg/day). The anti-aggregatory effect of aspirin is explained by the irreversible inability to synthesise thromboxane A_2 in the thrombocytes.

Salsalate 2-(2-Hydroxybenzoyl)oxybenzoic acid

Salsalate, salicylsalicylic acid, is a dimer of salicylic acid. It is insoluble in gastric juice but is soluble in the small intestine, where it is partially hydrolysed to two molecules of salicylic acid and absorbed. It does not cause GI blood loss.

It is prepared by esterifying benzyl salicylate with benzyloxy benzoyl chloride, followed by debenzylation with hydrogenolysis.

Sulphasalazine 2-Hydroxy-5-[4-(pyridin-2-ylsulfamoyl)phenyl]azo-benzoicacid

Sulphasalazine is a mutual prodrug. In large intestines it gets activated to liberate 5-aminosalicylic acid, which in turn inhibits PG synthesis, and the sulphapyridine is useful for the treatment of infection. Hence, sulphasalazine is used in the treatment of ulcerative colitis.

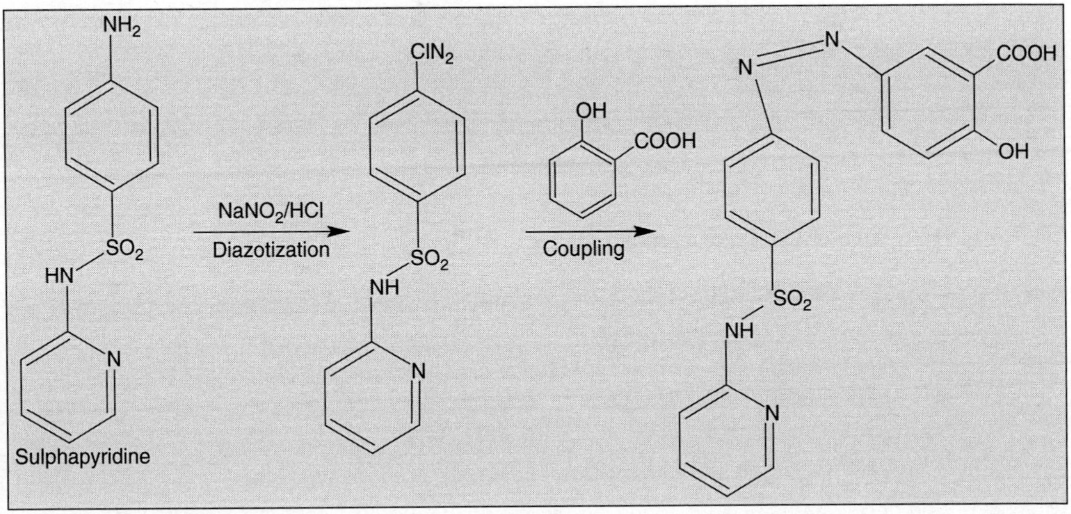

On treatment of sulphapyridine with sodium nitrite and hydrochloric acid, the primary amino group is diazotized to form diazonium salt intermediate. This, on coupling with salicylic acid, affords sulphasalazine.

Diflunisal 5-(2, 4-Diflurophenyl)salicylic acid

It is more potent than aspirin but produces fewer side-effects, and has a biological half-life 3 to 4 times greater than that of aspirin.

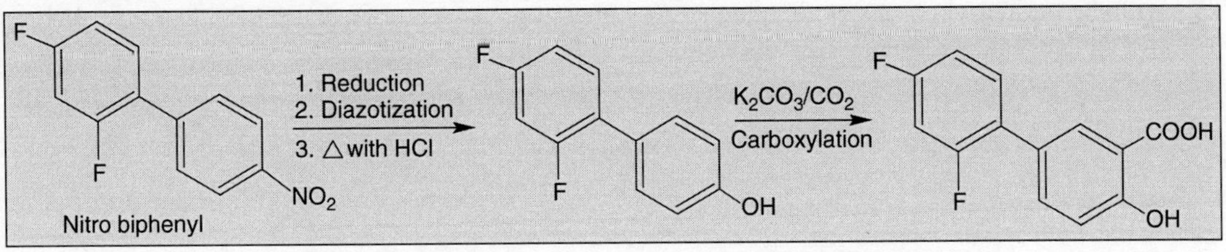

Nitro biphenyl derivative undergoes reduction, and diazotization followed by heating in acidic media yield 4-(2,4-difluorophenyl)-phenol. This product is reacted with carbon dioxide in the presence of a base according to the Kolbe–Schmitt phenol carboxylation method, giving diflunisal.

12.7 P-AMINOPHENOL DERIVATIVES

These derivatives possess analgesic and antipyretic actions but lack anti-inflammatory effects. Acetanilide was introduced into the therapy 1886 as an antipyretic–analgesic agent, but was subsequently found to be too toxic, having been associated with methemoglobinemia and jaundice. Phenacetin was introduced in the following year and was widely used, but was withdrawn recently because of its nephrotoxicity. Acetaminophen (paracetamol) was introduced in 1893 and it remains the only useful agent of this group used as an antipyretic–analgesic agent.

12.7.1 Structure–Activity Relationship

1. Etherification of the phenolic function with methyl or propyl groups produces derivatives with greater side-effects than ethyl derivative.
2. Substituents on the nitrogen atom, which reduce the basicity, also reduce activity unless the substituent is metabolically labile—e.g., acetyl.
3. Amides derived from aromatic acid—e.g., N-phenyl benzamides—are less active or inactive.

Paracetamol N-(4-Hydroxyphenyl)acetamide
Paracetamol produces antipyresis by acting on the hypothalamic heat-regulating centre, and analgesia by elevating the pain threshold. Hepatic necrosis and death have been observed following over-dosage; hepatic damage is likely if an adult takes more than 10 grams in a single dose or if a two-year-old child takes more than 3 grams. Paracetamol is metabolised primarily in the liver, where its major metabolites include inactive sulphate and glucuronide conjugates, which are excreted by the kidneys. Only a small, yet significant, amount is metabolized via the hepatic cytochrome P_{450} enzyme system, which is responsible for the toxic effects of paracetamol due to a minor alkylating metabolite N-acetyl-p-benzo-quinone imine (NAPQI). At usual doses, the toxic metabolite NAPQI is quickly detoxified by combining irreversibly with the sulfhydryl groups of glutathione. At higher dose, because of exhaust of glutathione, NAPQI reacts with liver cell and causes necrosis.

Phenol is nitrated using sulphuric acid and sodium nitrate (as phenol is highly activated, its nitration requires very mild conditions compared to the oleum-fuming nitric acid mixture required to nitrate benzene). The *para* isomer is separated from the ortho isomer by fractional distillation (there will be little of *meta*, as OH is *o, p* directing). The 4-nitrophenol is reduced to 4-aminophenol using a reducing agent such as sodium borohydride in basic medium. 4-aminophenol is reacted with acetic anhydride to give paracetamol.

> Paracetamol consists of a benzene ring core, substituted by one hydroxyl group and the nitrogen atom of an amide group in the para (1,4) pattern. The amide group is acetamide (ethanamide). It is an extensively conjugated system, as the lone pair on the hydroxyl oxygen, the benzene pi cloud, the nitrogen lone pair, the *p* orbital on the carbonyl carbon, and the lone pair on the carbonyl oxygen is all conjugated.

12.8 PYRAZOLIDINEDIONE DERIVATIVES

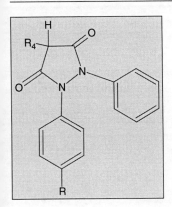

	R	R_4
Phenylbutazone	H	—C_4H_9
Oxyphenbutazone	OH	—C_4H_9
Sulfinpyrazone	H	-$(CH_2)_2SOC_6H_5$

12.8.1 Structure–Activity Relationship

1. In 3,5-pyrazolidinedione derivatives, pharmacological activities are closely related to their acidity (the acidic H at 4-position). The dicarbonyl functions at the 3- and 5-positions enhance the acidity of hydrogen atom at the 4-position.
2. Decreasing or eliminating acidity by removing the acidic proton at the 4-position (e.g., 4,4-dialkyl derivatives) abolishes anti-inflammatory activity. Thus, if the hydrogen atom at the 4-position of phenylbutazone is replaced by a substituent such as a methyl group, anti-inflammatory activity is abolished.
3. If acidity is enhanced too much, anti-inflammatory and sodium-retaining activities decrease, while other properties such as the uricosuric effect increase.
4. A single alkyl group at the 4-position enhances anti-inflammatory activity. Although *n*-butyl group enhances activity most, propyl and allyl analogues also possess anti-inflammatory activity.
5. Introduction of polar functions in these alkyl groups gives mixed results. The γ-hydroxy-*n*-butyl derivative possesses pronounced uricosuric activity but gives fewer anti-inflammatory effects.
6. Substitution of 2-phenylthioethyl group at the 4-position produces anti-gout drug (sulfinpyrazone).

7. The presence of both phenyl groups is essential for both anti-inflammatory analgesic activity.
8. Various substituents in the *para* position of one or both aromatic ring do not drastically affect activity. A *p*-hydroxy group present in oxyphenbutazone, the major metabolite of phenylbutazone, contributes therapeutically useful anti-inflammatory activity. Other derivatives such as methyl, chloro, or nitro groups also possess activity.

Phenylbutazone 4-*n*-Butyl-1,2-diphenyl-3,5-pyrazolidinedione

Synthesis

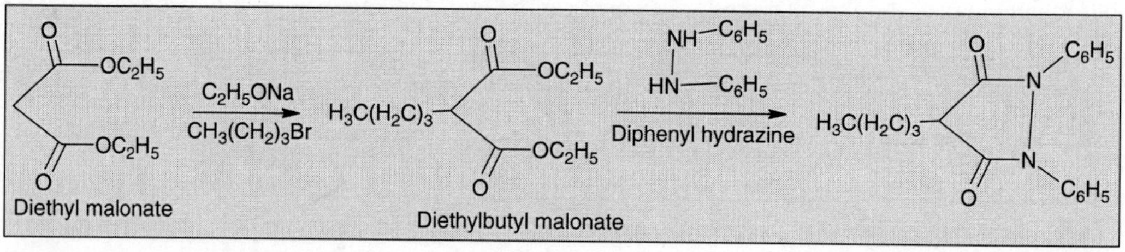

Diethyl malonate undergoes mono alkylation by treatment with *n*-butyl bromide in presence of sodium ethoxide. Phenylbutazone is obtained by condensation of diethyl-*n*-butylmalonate with hydrazobenzene in the presence of base.

Oxyphenbutazone 4-Butyl-1-(4-hydroxyphenyl)-2-phenylpyrazolidine-3,5-dione

Oxyphenbutazone, the major metabolite of phenylbutazone, differs only in the *para* location of one of its phenyl groups, where a hydrogen atom is replaced by a hydroxyl group.

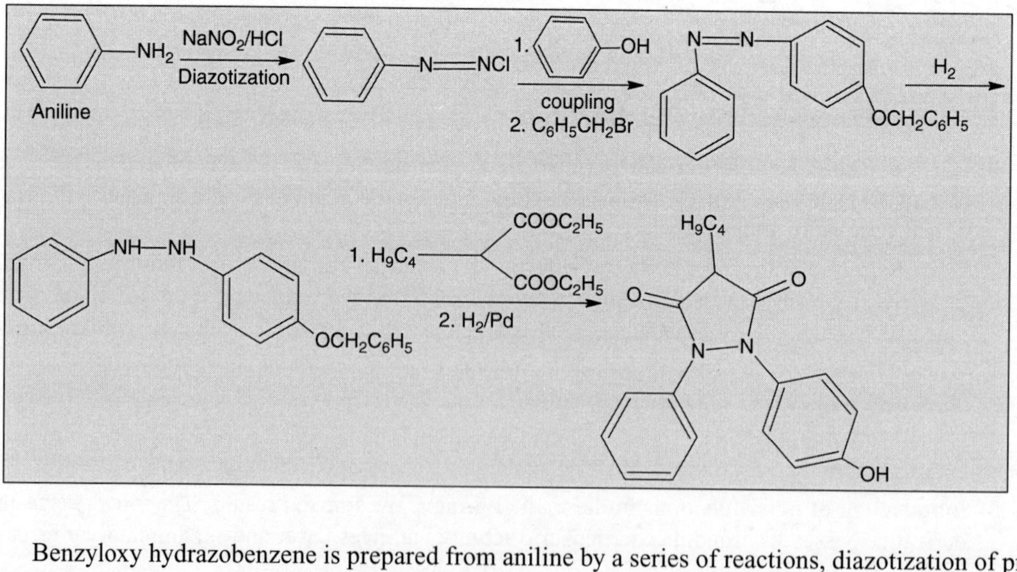

Benzyloxy hydrazobenzene is prepared from aniline by a series of reactions, diazotization of primary amino group, coupling with phenol, and protection of phenolic hydroxyl with benzyl group and reduction of azo group. Oxyphenylbutazone is obtained by condensation of diethyl-*n*-butylmalonate with benzyloxy hydrazobenzene in the presence of base, followed by debenzylation.

Sulfinpyrazone 1,2-Diphenyl-4-(2-phenylsulfinylethyl) pyrazolidine-3,5-dione
Sulfinpyrazone is a uricosuric medication used to treat gout.

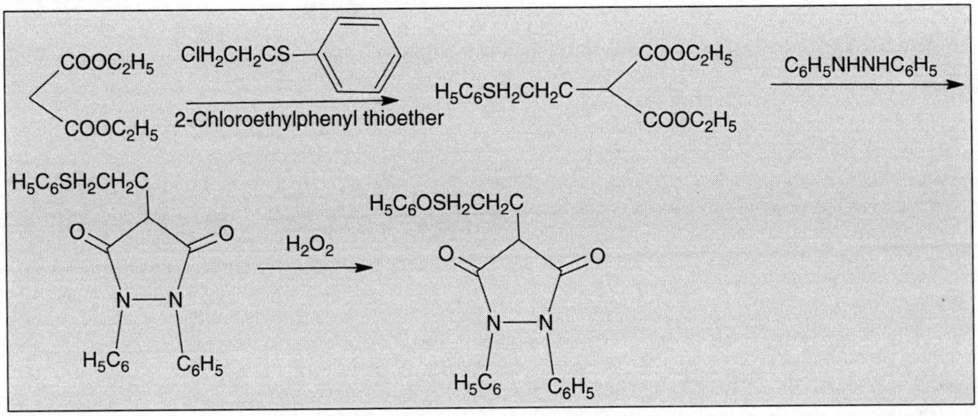

Sulfinpyrazone is an analogue of phenylbutazone that is synthesised in the analogous manner of condensing hydrazobenzene with 2-(2-phenylthioethyl)malonic ester into pyrazolidinedione, and the subsequent oxidation of thiol ether by hydrogen peroxide in acetic acid into the sulfoxide, sulfinpyrazone.

12.9 ANTHRANILIC ACID DERIVATIVES (FENAMATES)

The anthranilic acid class NSAIDs result from the application of classic medicinal chemistry bioisosteric drug design concept, as these derivatives are nitrogen isosteres of salicylic acid.

12.9.1 Structure–Activity Relationship

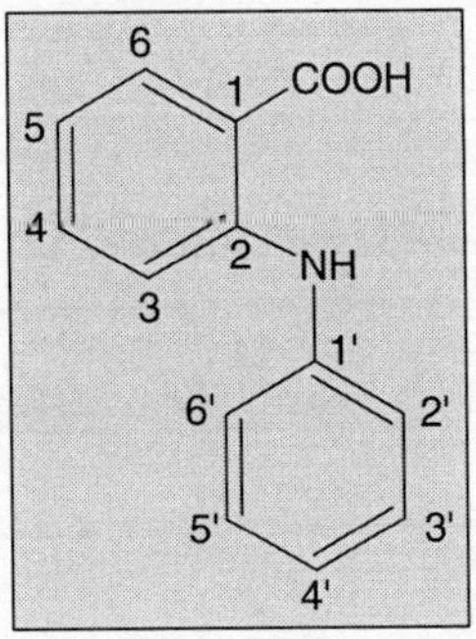

1. Substitution on the anthranilic acid ring generally reduces the activity.
2. Substitution on the N-aryl ring can lead to conflicting results. In the UV erythema assay for anti-inflammatory activity, the order of activity was generally 3'>2'>4' for mono substitution, with CF3 group (flufenamic acid) being particularly potent. The opposite order of activity was observed in rat paw oedema assay, the 2'-Cl derivative being more potent than 3'-Cl analogue.
3. In di-substituted derivatives, where the nature of the two substituents is the same, 2',3'-di-substitution appears to be the most effective (mefenemic acid).
4. The NH moiety of anthranilic acid appears to be essential for activity since replacement of NH function with O, CH$_2$, S, SO$_2$, N-CH$_3$, or NCOCH$_3$ functionalities significantly reduced the activity.

5. Finally, the position of the acidic function is critical for activity. Anthranilic acid derivatives are active, whereas the *m*- and *p*-aminobenzoic acid analogues are not. Replacement of carboxylic acid functions with the isosteric tetrazole has little effect on activity.

Flufenamic acid 2-[[3-(Trifluoromethyl)phenyl]amino]benzoic acid
Synthesis

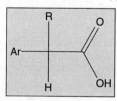

Flufenamic acid is synthesised by the reaction of 2-iodobenzoic acid with 3-trifluoromethylaniline in the presence of potassium carbonate and copper filings.

Mefenemic acid It is obtained through analogous approach by reaction of *o*-chlorobenzoic acid with 2,3-dimethylaniline.

Meclofenamate It is obtained by Ullman condensation employing 2-iodobenzoic acid and 2-chloro-3-methylaniline.

12.10 ARYL ALKANOIC ACID DERIVATIVES

The largest group of NSAIDs is represented by this class, as typified by the given chemical structure.

12.10.1 Structure–Activity Relationship

1. All agents possess a centre of acidity, which can be represented by a carboxylic acid, an enol, a hydroxamic acid, a sulphonamide, or a tetrazole.
2. The centre of acidity is generally located one carbon atom adjacent to a flat surface represented by an aromatic or heteroaromatic ring.
3. The distance between these centres is critical because increasing this distance to two or three carbons generally diminishes activity.
4. Substitution of a methyl group on the carbon atom separating the aromatic ring tends to increase anti-inflammatory activity.

Indole acetic acid derivatives

Structure–activity relationship

1. Replacement of the carboxyl group with other acidic functionalities decreases activity; also, amide analogues are inactive.

2. Acylation of the indole nitrogen with aliphatic carboxylic acid or arylalkyl carboxylic acids results in the decrease of activity.
3. The N-benzoyl derivatives substituted in the p-position with F, Cl, CF$_3$, and S-CH$_3$ groups are the most active.
4. The 5-position of the indole when substituted with OCH$_3$, F, N(CH$_3$)$_2$, CH$_3$, and COCH$_3$ groups was more active than the unsubstituted indole analogue.
5. The presence of indole ring nitrogen is not essential for activity because the corresponding 1-benzylidenylindene analogue (Sulindac) was active.
6. Alkyl groups, especially methyl group, at 2-position are much more active than aryl-substituted analogues.
7. Substitution of a methyl group at the α-position of the acetic acid side chain leads to equiactive analogues.
8. Anti-inflammatory activity is displayed only by the (S) (+) enantiomer.

Indomethacin 1-(p-Chlorobenzoyl)-5-methoxy-2-methylindole-3-acetic acid

Indomethacin is used in rheumatoid arthritis, non-specific infectious polyarthritis, gouty arthritis, osteoarthritis, ankylosing spondylitis, arthrosis, back pain, neuralgia, myalgia, and other diseases accompanied by inflammation.

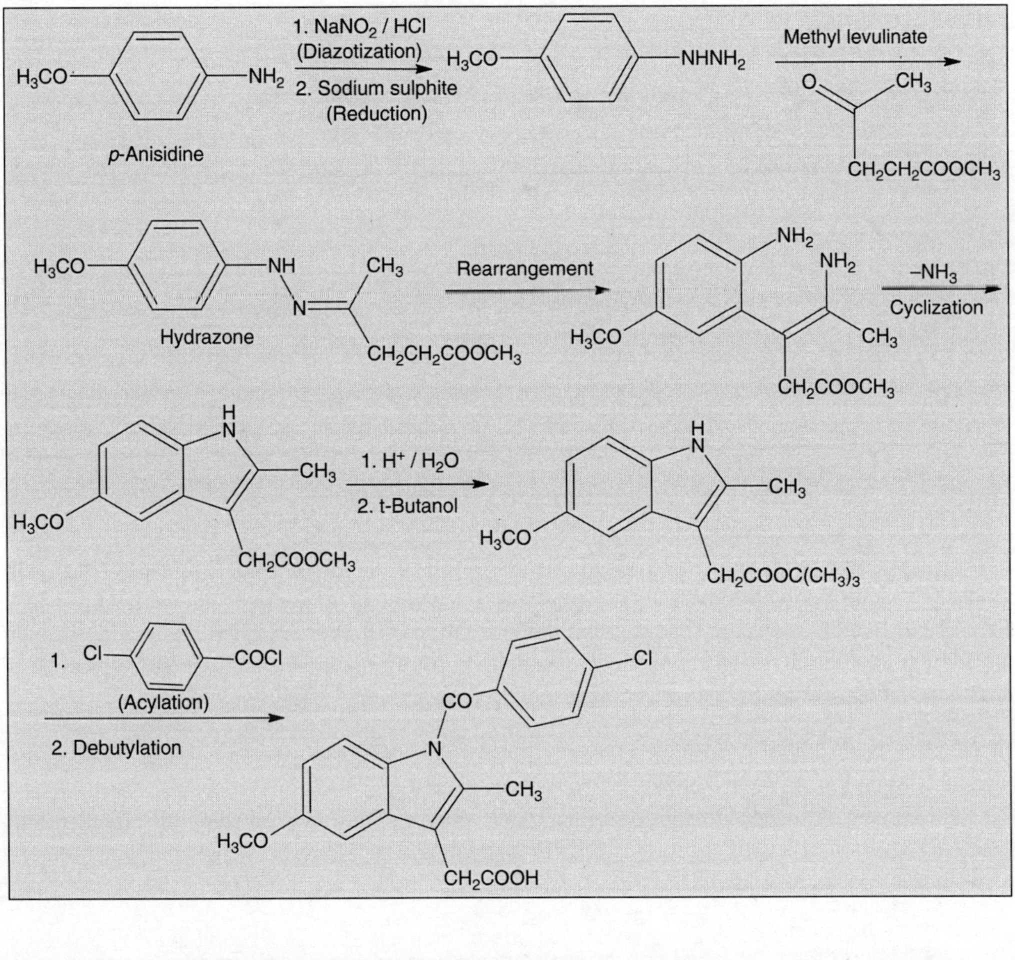

Indomethacin synthesis starts with 4-methoxyphenylhydrazine. According to this method, a reaction is done to make indole from phenylhydrazone by Fischer's method, using levulinic acid methyl ester as a carbonyl component, hydrogen chloride as a catalyst, and ethanol as a solvent, to give the methyl ester of 5-methoxy-2-methyl-3-indolylacetic acid. This product is hydrolysed by an alkali into 5-methoxy-2-methyl-3-indolylacetic acid, from which *tert*-butyl ester of 5-methoxy-2-methyl-3-indolylacetic acid is formed by using *tert*-butyl alcohol and zinc chloride in the presence of dicyclohexylcarbodiimide. The resulting product undergoes acylation at the indole nitrogen atom by *p*-chorobenzoyl chloride in dimethylformamide, using sodium hydride as a base. The resulting *tert*-butyl ester of 1-(*p*-chlorobenzoyl)-5-methoxy-2-methyl-3-indolylacetic acid further undergoes thermal decomposition to the desired acid, indomethacin.

Indene acetic acid derivative

Sulindac 5-Fluoro-2-methyl-1-[(4-methylsulfinyl)phenylmethylene]indene-3-acetic acid

Sulindac is a prodrug derived from sulfinylindene that is converted in the body to an active NSAID. More specifically, the agent is converted by liver enzymes to a sulphide that is excreted in the bile and then reabsorbed from the intestine. This is thought to help maintain constant blood levels with reduced gastrointestinal side-effects.

The (Z)-isomer of sulindac showed much more potent anti-inflammatory activity than the corresponding (E)-isomer.

Synthesis

The indanone ring is formed by internal Friedel—Crafts acylation of 4'fluoro-2-methyl phenylpropionic acid by means of polyphosphoric acid. Reformatskii reaction on this ketone with methyl bromoacetate and zinc leads to addition to the ketone group and formation of β-hydroxy ester; this readily dehydrates followed by condensation with 4-thiomethylbenzaldehyde gives benzylidene compound. Hydrolysis of the ester and oxidation of sulphur by reaction with sodium metaperiodate affords sulindac.

Pyrrole acetic acid derivatives

Tolmetin 1-Methyl-5-(4-methylbenzoyl)pyrrole-2-acetic acid

SAR: Replacement of the *p*-toluoyl group with a *p*-chlorobenzoyl moiety produced little effect on activity, whereas introduction of a methyl group in the 4th position and *N*-*p*-chlorobenzoyl analogue (zomepirac) was ~4 times as potent as tolmetin.

Synthesis

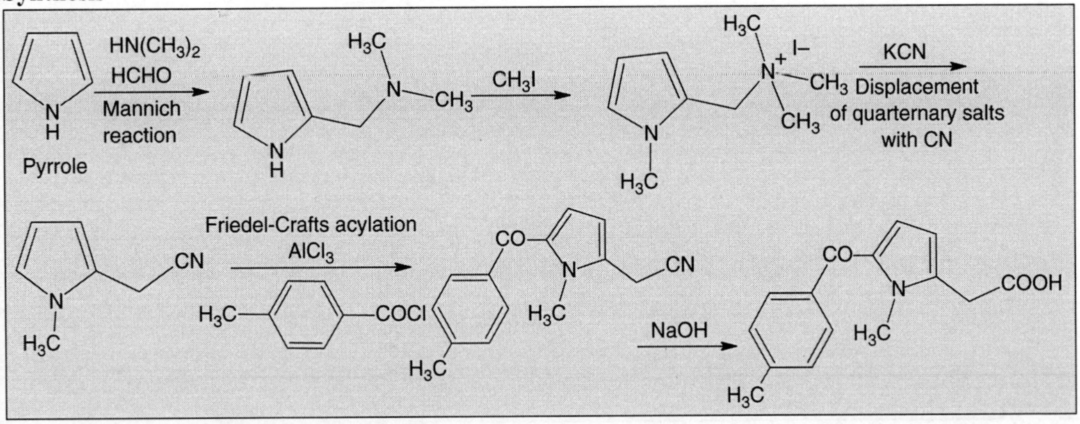

Tolmetin is synthesised from pyrrole, which is aminomethylated using formaldehyde and dimethylamine, forming 2-dimethylaminomethyl pyrrole. The product is methylated by methyl iodide, giving the corresponding quaternary salt. Reaction of the product with sodium cyanide gives 1-methylpyrrole-2-acetonitrile, which is acylated at the free α-position of the pyrrole ring by 4-methylbenzoylchloride in the presence of aluminium chloride. The resulting 1-methyl-5-*p*-toluylpyrrol-2-acetonitrile undergoes further alkaline hydrolysis, giving the corresponding acid, tolmetin.

Zomepirac 2-[5-(4-Chlorobenzoyl)-1,4-dimethyl-pyrrol-2-yl]acetic acid

Synthesis

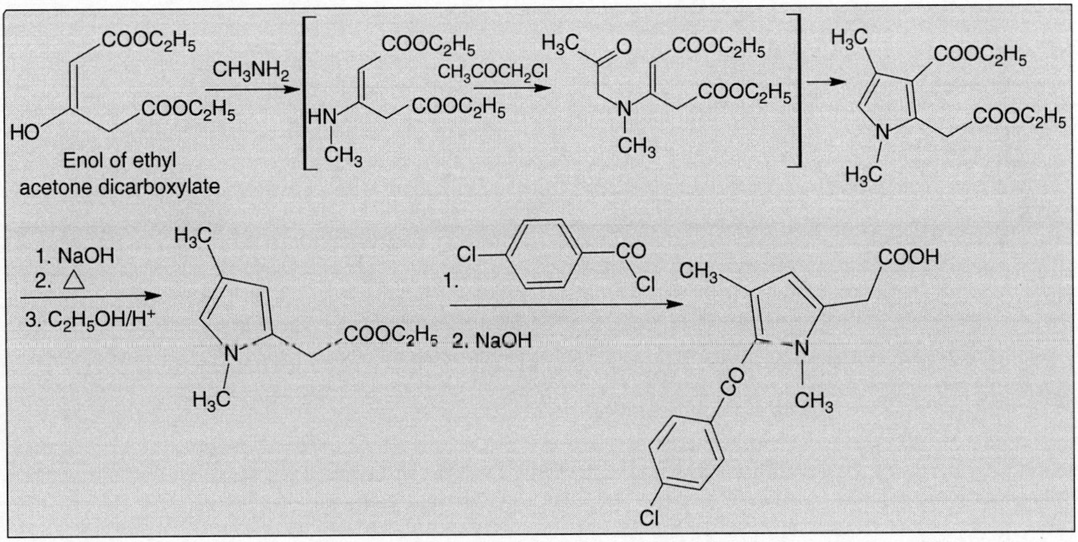

Enol form of ethylacetonedicarboxylate reacts with methylamine and chloroacetone to give pyrrole. The reaction can be rationalized by assuming formation of eneamine as the first step; alkylation with chloroacetone will then give ketoester. Internal aldol condensation leads to pyrrole derivative. This compound is then saponified to a dicarboxylic acid; heating of this product leads to loss of the ring carboxyl group. The remaining acid is then re-esterified. Friedel-Crafts acylation with the 4-chlorobenzoyl chloride followed by hydrolysis of ester affords zomepirac.

Aryl- and heteroaryl acetic/propionic acid derivatives
Ibuprofen 2-(*p*-Isobutylphenyl)propionic acid
Synthesis

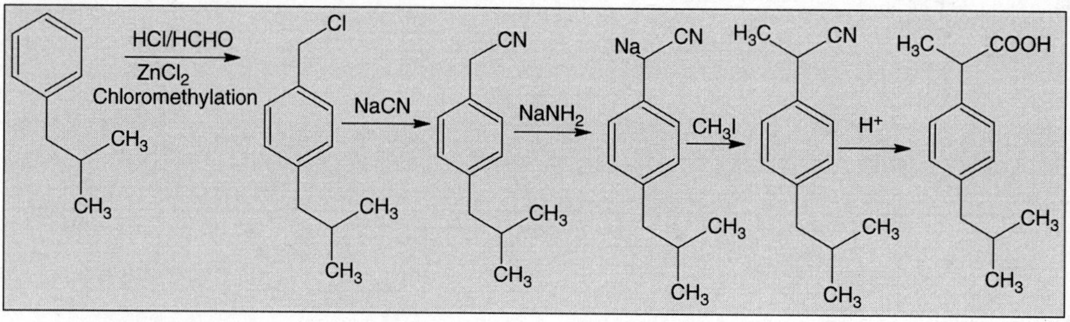

Chloromethylation of *iso*-butylbenzene gives 4-*iso*-butylbenzylchloride. This product is reacted with sodium cyanide, making 4-*iso*-butylbenzyl cyanide, which is alkylated in the presence of sodamide by methyl iodide into 2-(4-*iso*-butylbenzyl)propionitrile. Hydrolysis of the resulting product in the presence of a base produces ibuprofen.

Diclofenac 2-[(2,6-Dichlorophenyl)amino] benzene acetic acid
Synthesis

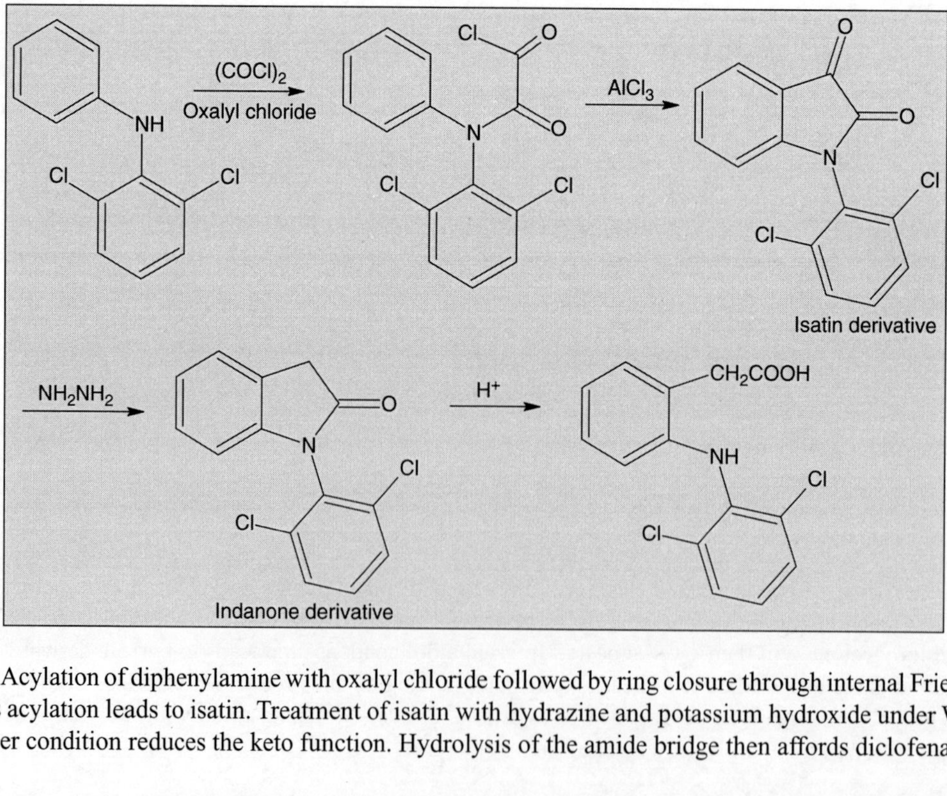

Acylation of diphenylamine with oxalyl chloride followed by ring closure through internal Friedel–Crafts acylation leads to isatin. Treatment of isatin with hydrazine and potassium hydroxide under Wolff-Kishner condition reduces the keto function. Hydrolysis of the amide bridge then affords diclofenac.

Naproxen (+)-(S)-2-(6-Methoxynaphthalen-2-yl)propanoic acid
Synthesis

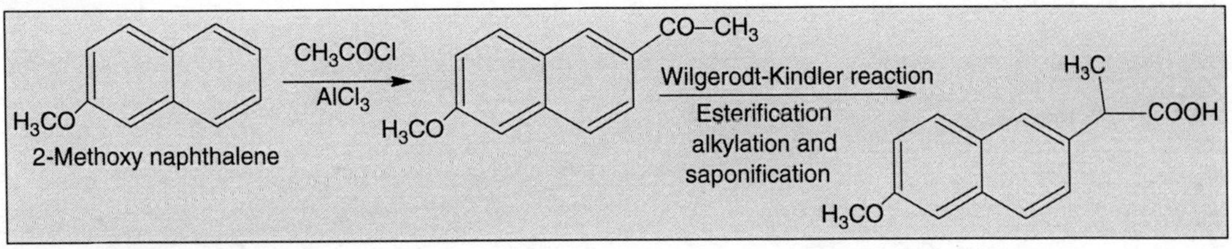

Acetylation of 2-methoxynaphthalene gives ketone, which is then converted to the acetic acid by the Wilgerodt reaction. Esterification and alkylation of the carbanion with methyl iodide and finally saponification afford naproxen.

Fenoprofen 2-(3-Phenoxyphenyl)propionic acid
Synthesis

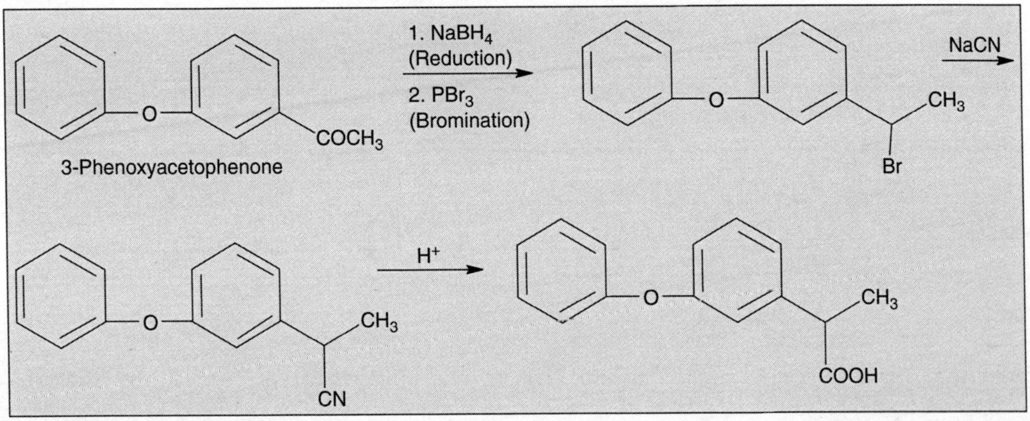

Fenoprofen is synthesised from 3-hydroxyacetophenone, which is esterified by bromobenzene in the presence of potassium carbonate and copper filings, forming 3-phenoxyacetophenone. The carbonyl group of the resulting product is reduced by sodium borohydride, and the resulting alcohol is brominated by phosphorous tribromide. The reaction of the resulting bromo derivative with sodium cyanide gives 2-(3-phenoxyphenyl)propionitrile, which is hydrolysed into the desired fenoprofen.

Ketoprofen 2-(3-Benzoylphenyl)propionic acid
Synthesis

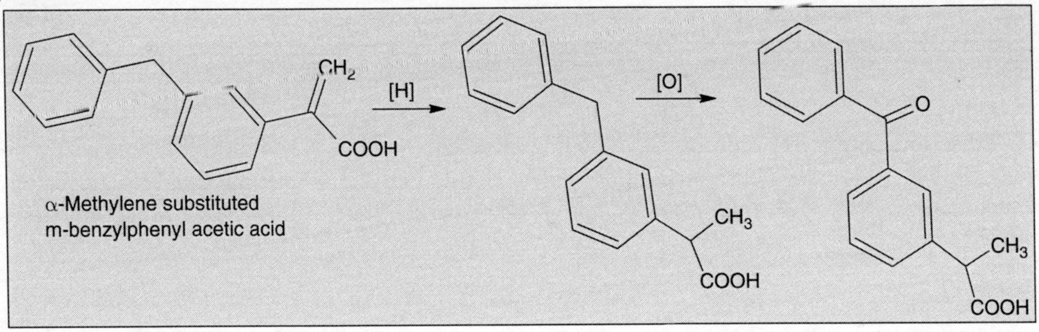

Reduction of 2-methylene substituted 3-benzylphenyl acetic acid followed by oxidation affords ketoprofen.

Flubiprofe 2-(2-Fluoro-4-biphenyl)propionic acid

Synthesis

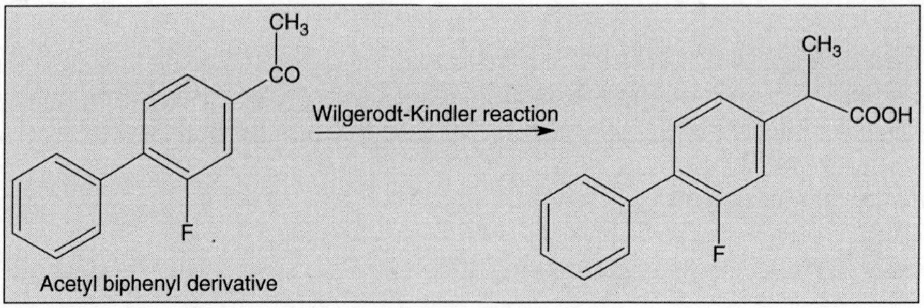

Acetyl biphenyl derivative → (Wilgerodt-Kindler reaction) → flurbiprofen

Synthesis is similar to that of naproxen.

Caprofen 6-Chloro-α-methylcarbazole-2-acetic acid

Synthesis

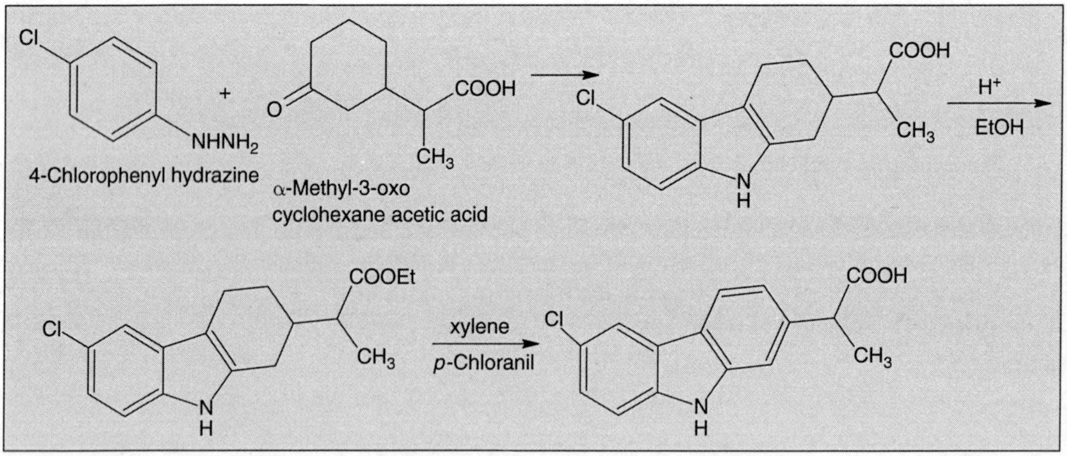

4-Chlorophenyl hydrazine + α-Methyl-3-oxo cyclohexane acetic acid → tetrahydrocarbazole intermediate → (H⁺, EtOH) → ethyl ester → (xylene, p-Chloranil) → caprofen

Reaction between 4-chlorophenyl hydrazine and α-methyl-3-oxo cyclohexane acetic acid under Skraup indole synthesis condition gives 2-(6-chloro-2,3,4,9-tetrahydro-1H-2-carbazolyl) propanoic acid, which further esterified to protect carboxylate followed by reduction affords caprofen.

Ketorolac 5-Benzoyl-2,3-dihydropyrrolizine-1-carboxylic acid

Synthesis

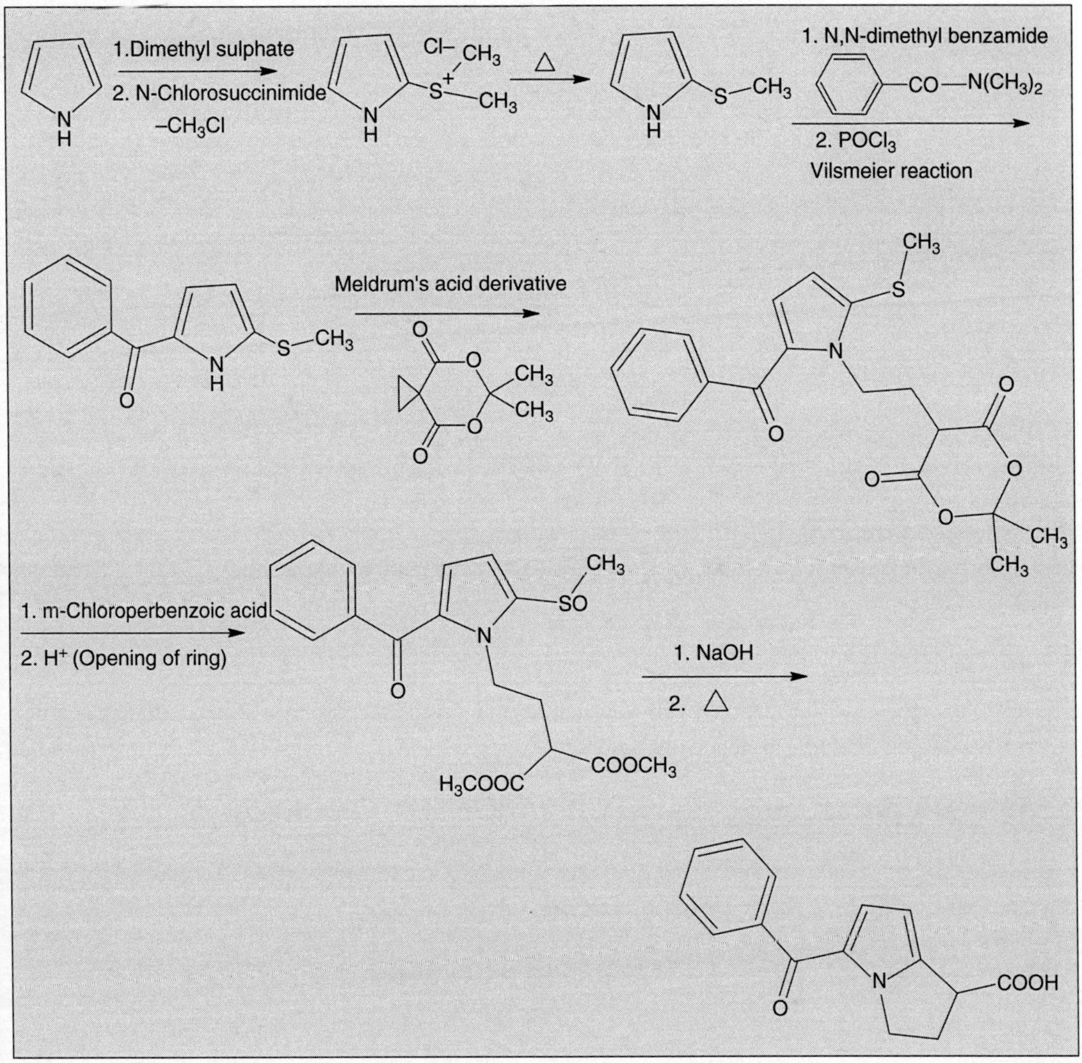

Pyrrole is electrophillically substituted by reaction with adduct of *N*-chlorosuccinimide and dimethyl sulphide to give the intermediate, which on thermolysis dealkylates to 2-thiomethylpyrrole. Under Vilsmeier–Haack reaction condition, acylation to 5-acyl pyrrole derivative takes place using *N,N*-dimethylbenzamide. *N*-Alkylation with Meldrum's acid derivative followed by oxidation and treatment with acid gives dimethylmalonate derivative. Heating this with base leads to saponification, decarboxylation, and cyclization to afford ketorolac.

Etodolac 2-(1,8-Diethyl-4,9-dihydro-3*H*-pyrano[3,4-b]indol-1-yl)acetic acid

Synthesis

Hydroxy group of 7-ethyl-3-(hydroxyethyl) indole is added to the keto function of 3-ketopentanoic acid to form hemiketal. This, in turn, is deoxygenated and cyclization in presence of *p*-toluenesulphonic acid affords etodolac.

12.11 OXICAMS

The term 'oxicam' describes the relatively new enolic acid class of 4-hydroxy-1,2-benzothiazine carboxamides with anti-inflammatory and analgesic properties.

12.12 SAR (STRUCTURE ACTIVITY RELATIONSHIP OF OXICAMS)

Oxicams are represented by the structure given here:

1. Optimum activity was observed when R' was a methyl substituent.
2. The carboxamide substituent R is generally an aryl or heteroaryl substituent because alkyl substituents are less active.
3. N-Heterocyclic carboxamides (piroxicam) are generally more acidic than the corresponding N-aryl carboxamides.
4. Interchanging of benzene ring with thiophene (tenoxicam) gives biologically active compounds.

Piroxicam: 4-Hydroxy-2-methyl-N-2-pyridinyl-1,2-benzothiazine-3-carboxamide-1,1-dioxide.

Synthesis

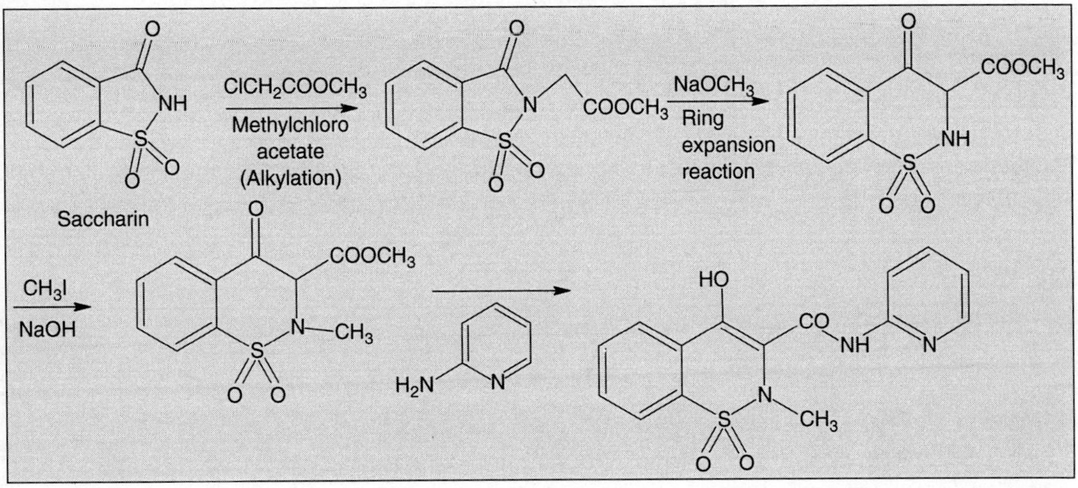

The reaction of saccharin with sodium hydroxide results in substitution of the imide hydrogen atom of saccharin with sodium, giving a sodium salt. The resulting product is reacted with methyl chloroacetate, giving the saccharin-substituted acetic acid methyl ester. Upon reaction with sodium methoxide in dimethylsulfoxide, the product undergoes rearrangement into 1,1-dioxide 3-methoxycarbonyl-3,4-dihydro-2-*H*-1,2-benzothiazin-4-enol. This product is methylated at the nitrogen atom using methyl iodide, and finally, reaction of the resulting product with 2-aminopyridine gives piroxicam.

Two closely related analogues are obtained by varying the heterocyclic amine used in the last step; 2-amino thiazole thus leads to **sudoxicam**, while 3-amino-5-methylisoxazole affords **isoxicam**.

Tenoxicam (3E)-3-[Hydroxy-(pyridin-2-ylamino)methylidene]-2-methyl-1,1-dioxothieno[2,3-e]thiazin-4-one

Synthesis

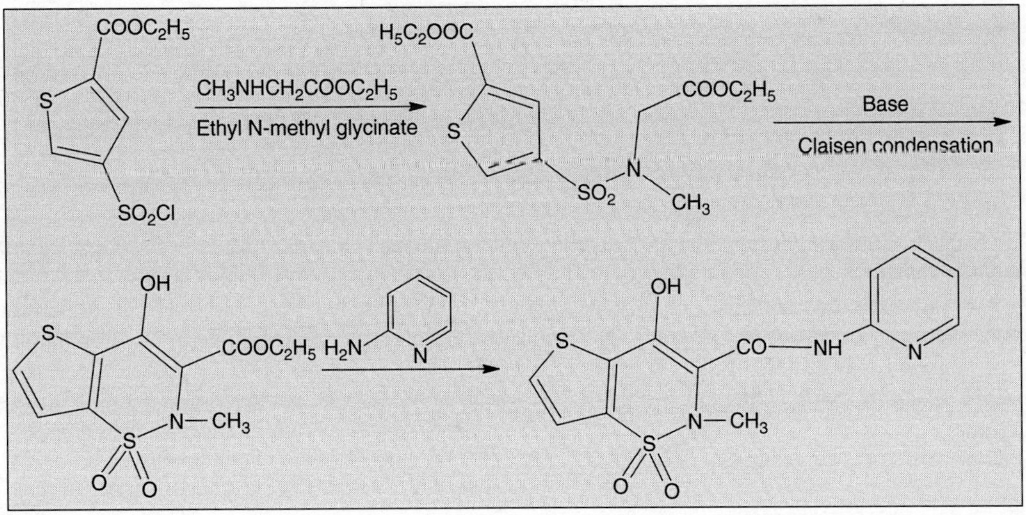

Reaction of thiophen sulphonyl chloride with ethyl-N-methyl glycinate gives sulphonamide. Treatment of this with base leads to intramolecular Claisen condensation and thus formation of the β-ketoester. Amide-ester interchange with 2-aminopyridine affords tenoxicam.

12.13 MISCELLANEOUS DRUGS

Nambutone 4-(6-Methoxy-2-naphthyl)-2-butanone
Nambutone is a non-acidic compound, and because of this nature it produces minimum gastrointestinal side-effects.

Synthesis

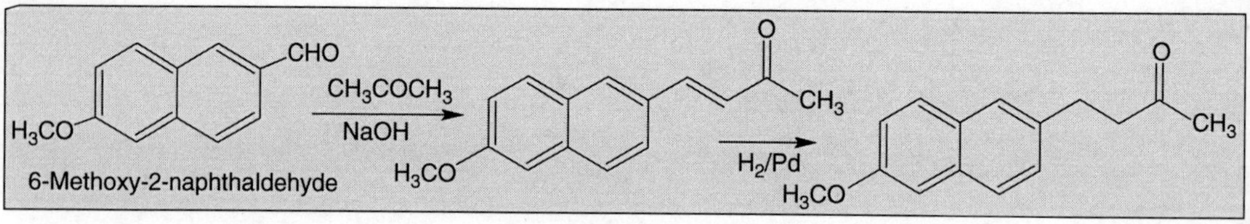

Condensation of 6-methoxy-2-naphthaldehyde with acetone in presence of base gives chalcone; this in turn reduced with hydrogen-palladium affords nambutone.

Nimesulide 4-Nitro-2-phenoxymethane sulphonamide
Nimesulide contains a sulphonamide moiety as the acidic group rather than carboxylic acid. It shows moderate incidence of gastric side-effects because it exhibits significant selectivity towards COX-2.

Synthesis

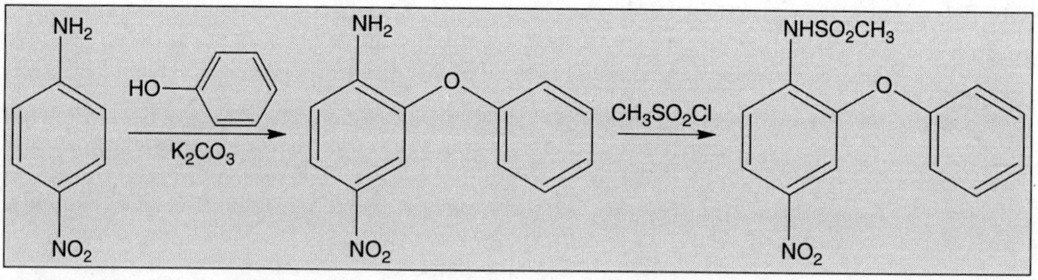

4-Nitro-2-phenoxyaniline on reaction with methane sulfonyl chloride affords sulphonamide derivative nimesulide.

Due to concerns about the risk of hepatotoxicity, nimesulide has been withdrawn from market in many countries. The patient information leaflet informs that the use of nimesulide in children under the age of 12 is contraindicated.

Analgin Sodium N-(2,3-dihydr-1,5-dimethyl-3-oxo-2-phenyl-pyrazol-4-yl)-N-methylamino methane sulfonate

Synthesis

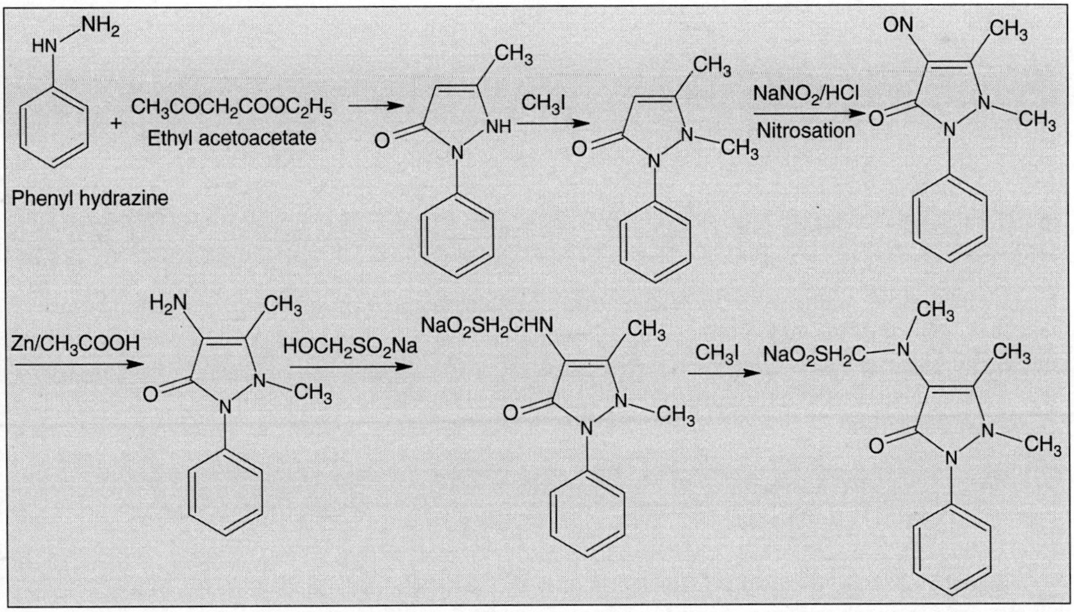

Condensation of ethyl acetoacetate with phenyl hydrazine gives the pyrazolone. Methylation by means of methyl iodide affords the antipyrine. Reaction of this compound with nitrous acid gives nitroso derivative, which on reduction gives aminopyrine. Treatment with hydroxymethyl sodium sulfonate and methyl iodide affords analgin.

12.14 ANTI-GOUT DRUGS

An acute attack of gout occurs as a result of an inflammatory reaction to crystals of sodium urate (the end product of purine metabolism in human beings) that is deposited on the articular cartilage of joints, tendons, and surrounding tissues due to elevated concentrations of uric acid in the blood stream. This provokes an inflammatory reaction of these tissues. Drugs used to treat gout may act in the following ways:

1. By inhibiting uric acid synthesis: Allopurinol
2. By increasing uric acid excretion: Probenecid, Sulfinpyrazone
3. Miscellaneous: Colchicine (alkaloid obtained from *Colchicum autumnale*)

Gout is characterised by excruciating, sudden, unexpected, burning pain, as well as swelling, redness, warmth, and stiffness in the affected joint. This occurs most commonly in men's toes but can appear in other parts of the body and affect women as well.

Allopurinol Pyrazolopyrimidine-4-ol
Synthesis

Malononitrine condenses with triethyl ortho formate and gives the starting material ethoxymethylene malononitrile. This undergoes addition–elimination reaction by treatment with hydrazine and gives imidazole derivative. Controlled hydrolysis of nitrile group to carboxamide, followed by reaction with 1C donor formamide, affords pyrazolopyrimidine derivative allopurinol.

Mechanism of action:

Adenine → (Adenine deaminase) → Hypoxanthine → (Xanthine oxidase) → Xanthine → (Xanthine oxidase) → Uric acid

In human beings, uric acid is formed primarily by the metabolism of adenine. Adenine is converted to hypoxanthine which in turn to kanthine and uric acid by the enzyme xanthine oxidase. At low concentrations, allopurinol is a for xanthine oxidase competitive inhibitor of the enzyme; at high concentration, it is a non-competitive inhibitor.

Probenecid 4-(Dipropylsulfamoyl)benzoic acid

Probenecid is a uricosuric agent that increases the rate of excretion of uric acid and used for the treatment of chronic gout. The oral administration of probenecid in conjugation with penicillin G results in higher and prolonged concentrations of the antibiotic in plasma than when penicillin is given alone.

Synthesis

Reaction of 4-cyano benzene sulphonyl chloride with dipropylamine, followed by hydrolysis of the nitrile group, affords probenecid.

12.15 SELECTIVE COX-2 INHIBITORS

The prostaglandins that mediate inflammation, fever, and pain are produced solely via COX-2 (highly inducible by inflammatory response), and the prostaglandins that are important in gastrointestinal, platelet, uterus, and renal function are produced solely via COX-1 (constitutively expressed). Selective COX-2 inhibitors (celecoxib, rofecoxib, and valdecoxib) are devoid of side-effects like gastric ulcer and do not affect the normal functioning of platelet, uterus, and renal system.

On September 27, 2004, vioxx (rofecoxib) was withdrawn voluntarily from the market due to an increased risk of myocardial infarction and stroke. At present, it is unclear whether this adverse effect pertains also to other drugs of this group or is specific for Vioxx.

The chief mechanism proposed to explain rofecoxib's cardiotoxicity is the suppression of prostacyclin, an anti-clotting agent in the blood. Prostacyclin production can lead to decrease in endothelial cells and cause inefficiency in declumping and vasodilatation.

Celecoxib 4-[5-(4-Methylphenyl)-3-(trifluoromethyl) pyrazol-1-yl]benzenesulphonamide (diarylpyrazole derivative)

Celecoxib is used to treat arthritis, pain, menstrual cramps, and colonic polyps. Celecoxib is used for the relief of pain, fever, swelling, and tenderness caused by osteoarthritis, rheumatoid arthritis, and ankylosing spondylitis. In familial adenomatous polyposis, patients develop large numbers of polyps in their colons, and the polyps invariably become malignant. The cramping and pain during menstrual periods is due to prostaglandins, and blocking the production of prostaglandins with celecoxib reduces the cramps and pain.

Synthesis

Claisen condensation of 4-methyl acetophenone and ethyltrifluoromethyl acetate in presence of base gives β-diketone. The enol form of diketone condenses with 4-sulphamoyl phenyl hydrazine to afford diarylpyrazole derivative celecoxib.

Rofecoxib 4-[4-(Methylsulphonyl)phenyl]-3-phenyl-2(5H)-furanone (diarylfuran derivative)
Rofecoxib is used to relieve the pain, tenderness, inflammation (swelling), and stiffness caused by arthritis, and to treat painful menstrual periods and pain from other causes.

Synthesis

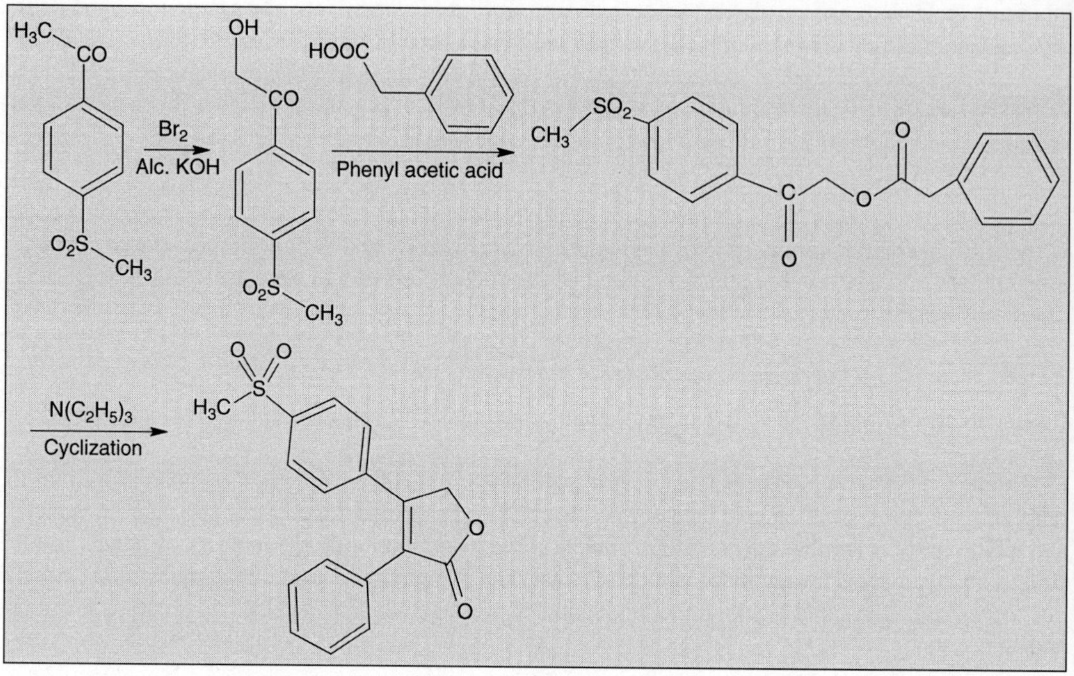

1-[4-(Methylsulphonyl)phenyl]-1-ethanone undergoes bromination to form bromo acetophenone derivative, which is converted to alcohol by treatment with alcoholic potassium hydroxide. This on reaction with phenyl acetic acid forms ester derivative, which undergoes cyclization to afford diarylfuran derivative rofecoxib.

Valdecoxib 4-(5-Methyl-3-phenylisoxazol-4-yl)benzenesulphonamide (Diaryl isoxazole derivative)

Synthesis

Deoxybenzoin reacts with hydroxyl amine to form oxime, which is condensed with ethylacetate to form isoxazoline derivative. Treatment with chlorosulphonic acid and ammonium hydroxide introduces sulphamoyl moiety in one of the phenyl rings. Dehydration affords diaryl isoxazole derivative valdecoxib.

12.16 NEWER DRUGS

Aceclofenac 2-[2-[2-[(2,6-Dichlorophenyl)amino]phenyl]acetyl]oxyacetic acid

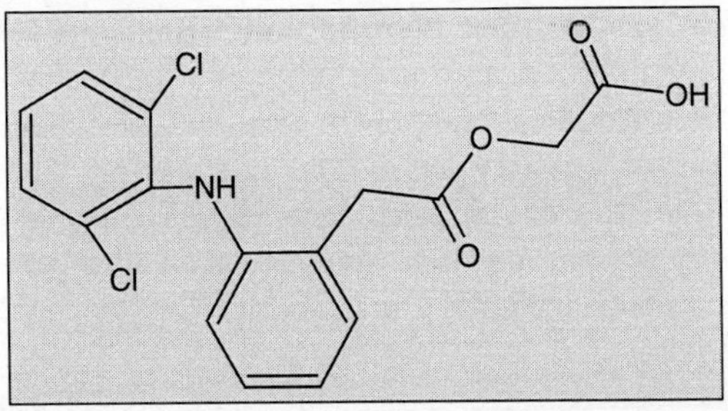

It directly blocks PGE$_2$ secretion at the site of inflammation by inhibiting IL-β & TNF in the inflammatory cells (intracellular action). Aceclofenac has been demonstrated to inhibit cyclo-oxygenase (COX) activity and to suppress the PGE$_2$ production by inflammatory cells, which are likely to be a primary source of PGE$_2$.

Etoricoxib 5-Chloro-6'-methyl-3-[4-(methylsulphonyl)phenyl]-2,3'-bipyridine

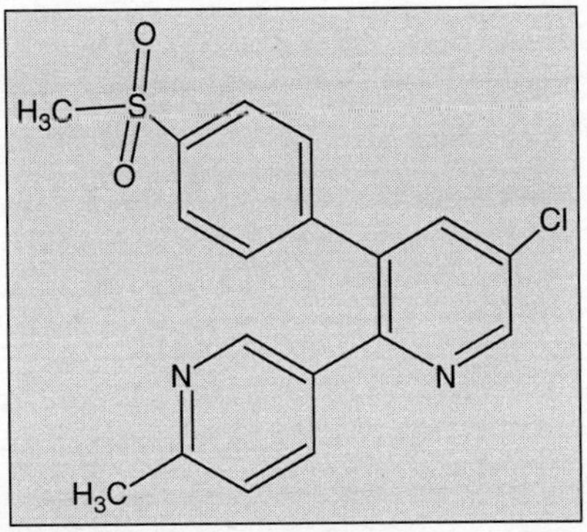

It is a new COX-2 selective inhibitor (approx. 106.0 times more selective for COX-2 inhibition over COX-1). Currently, it is approved in more than 60 countries worldwide but not in the United States, where the Food and Drug Administration (FDA) requires additional safety and efficacy data for etoricoxib before it will issue approval. Current therapeutic indications are: treatment of rheumatoid arthritis, osteoarthritis, ankylosing spondylitis, chronic low back pain, acute pain, and gout.

Lumiracoxib {2-[(2-Chloro-6-fluorophenyl)amino]-5-methylphenyl}acetic acid

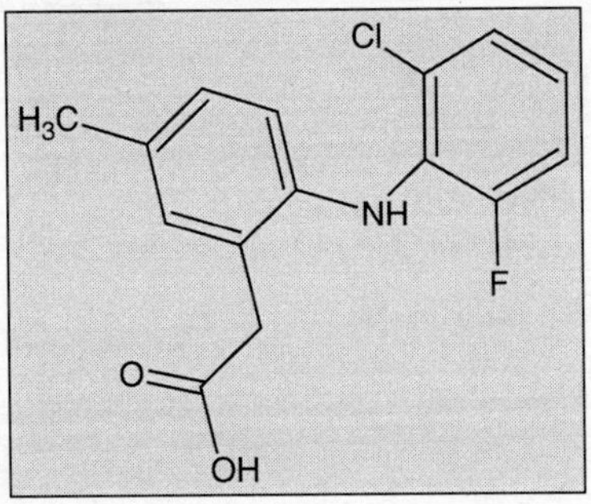

It is a COX-2 selective inhibitor non-steroidal anti-inflammatory drug. In November 2006, it received marketing approval for all European Union countries through a common procedure called MRP. As of 2007, the Food and Drug Administration (FDA) has not yet granted approval for its sale in the United States. Lumiracoxib has a different structure from the standard COX-2 inhibitors (e.g., celecoxib). It more closely resembles the structure of diclofenac (one chlorine substituted by fluorine, the phenylacetic acid has another methyl group in meta position), making it a member of the arylalkanoic acid family of NSAIDs. It binds to a different site on the COX-2 receptor than the standard COX-2 inhibitors. It displays extremely high COX-2 selectivity.

Parecoxib *N*-{[4-(5-Methyl-3-phenylisoxazol-4-yl)phenyl]sulphonyl}propanamide
It is a water-soluble and injectable prodrug of valdecoxib.

FURTHER READINGS

1. Williams, David A. and Lemke, Thomas L., 2001. *Foye's Principles of Medicinal Chemistry*, 15th Ed., Lippincott Williams & Wilkins: New York.
2. Wolff, Manfred E. (ed.), May 1997. Burger's Medicinal Chemistry and Drug Discovery.
3. Alex Gringauz, January 1997. *Introduction to Medicinal Chemistry: How Drugs Act and Why*.
4. FOYE - 'Principles of Medicinal Chemistry,' 4th Edition, 1995.

MULTIPLE-CHOICE QUESTIONS

1. Acetaminophen is used for
   a. Fever
   b. Inflammation
   c. Anti-platelet
   d. Cancer chemoprevention

2. NSAID is class of which of the following drugs:
   a. Imipramine
   b. Acetaminophen
   c. Alfentanyl
   d. Morphine

3. Which of the following statements is true about the NSAIDs?
   a. Are strong inhibitors of phospholipase A2
   b. Are water-soluble carboxylic acid derivatives
   c. Are salicylic acid derivatives
   d. All of the above

4. Indomethacin is
   a. Less acidic
   b. Optically active
   c. Selective to COX-II
   d. Producer of gastric ulcer

5. Which of the following statements about therapy with non-steroidal anti-inflammatory drugs (NSAIDs) is false?
   a. Aspirin is associated with a dose-dependent increased risk of upper gastrointestinal (GI) bleeding.
   b. Cyclo-oxygenase-2 (COX-2) inhibitors are associated with minimal or no risk of cardiovascular events.

c. Aspirin reduces the GI benefits of COX-2 inhibitors over non-selective NSAIDs.
d. Compared with non-selective NSAIDs, COX-2 inhibitors may have a reduced risk of adverse events in the lower GI tract.

6. PGG_2 is converted to PGI_2 by
   a. Phopholipase
   b. Prostacyclin synthase
   c. Cyclo-oxygenase
   d. Peroxidase

7. When acidic nature of salicylic acid is reduced,
   a. Analgesic activity reduced
   b. Analgesic activity retained
   c. Anti-inflammatory activity increases
   d. Anti-inflammatory and analgesic activity improves

8. One of the following is a prodrug:
   a. Aspirin
   b. Sulphasalazine
   c. Sulindac
   d. Naproxen

9. Nitrogen isosteres of salicylates are
   a. *p*-Aminophenol derivatives
   b. Anthranilic acid derivatives
   c. Aryl alkanoic acid derivatives
   d. Pyrazolidinedione derivatives

10. One of the following is synthesised using Ullman's condensation reaction:
    a. Anthranilic acid derivatives
    b. *p*-Aminophenol derivatives
    c. Aryl alkanoic acid derivatives
    d. Pyrazolidinedione derivatives

11. One of the following is not a sulphonamide drug:
    a. Nimesulide
    b. Valdecoxib
    c. Rofecoxib
    d. Celecoxib

12. One among the following is not a uricosuric agent:
    a. Probenecid
    b. Sulfinpyrazone
    c. Allopurinol
    d. Tolmetin

13. Celecoxib is a
    a. Diarylisoxazole derivative
    b. Diarylfuran derivative
    c. Diaryloxazole derivative
    d. Diarylpyrazole derivative

14. Analgin is a
    a. Oxazole derivative
    b. Isoxazole derivative
    c. Imidazole derivative
    d. Pyrazole derivative

15. 'Interchanging of benzene ring with thiophene gives biologically active compounds.' – This statement is true with respect to:
    a. Aryl alkanoic acid derivatives
    b. Oxicams
    c. *p*-Aminophenol derivatives
    d. Anthranilic acid derivatives

16. Side-effects of non-steroidal anti-inflammatory drugs (NSAIDs) include except
    a. Peptic ulcer
    b. Reduced kidney function
    c. GIT bleeding
    d. Seizures

QUESTIONS

1. NSAIDs such as ibuprofen can produce gastrointestinal damage. Why?
2. Differentiate between non-narcotic and narcotic analgesics.
3. What is gout and how is it treated?

4. What are the advantages of selective COX-2 inhibitors over non-selective NSAIDS? Give examples of COX-2 selective NSAIDS.

5. Give the names of the heterocyclic ring system present in the following:

   a. Analgin
   b. Celecoxib
   c. Rofecoxib
   d. Valdecoxib
   e. Tenoxicam

6. Discuss the synthesis of the following with suitable scheme.

   a. Aspirin
   b. Celecoxib
   c. Allopurinol
   d. Mefenemic acid
   e. Ketorolac

SOLUTION TO MULTIPLE-CHOICE QUESTIONS

1. a;
2. b;
3. d;
4. d;
5. b;
6. b;
7. b;
8. b;
9. b;
10. a;
11. c;
12. d;
13. d;
14. d;
15. b;
16. d.

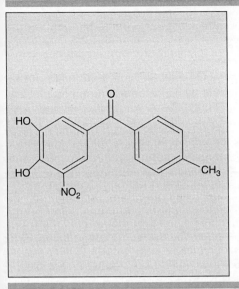

CHAPTER 13

Miscellaneous CNS Agents

LEARNING OBJECTIVES

- Define Parkinsonism and its possible causes
- Categorize various drugs effective in the treatment of Parkinsonism disorder
- Describe the mode of action of various anti-Parkinsonism drugs of various classes
- Describe the utility of centrally acting muscle relaxant
- Classify various muscle relaxants based on the chemical structure
- Define Alzheimer's disorder and its possible causes
- Categorize anti-Alzheimer drugs based on their mode of action
- Define CNS stimulants and their uses

13.1 ANTI-PARKINSONISM AGENTS

13.1.1 Introduction

Parkinsonism is a clinical syndrome comprising four cardinal features: bradykinesia (slowness and poverty of movement), muscular rigidity, resting tremor (which usually abates during voluntary movement), and an impairment of postural balance leading to disturbances of gait and falling.

The most common cause of Parkinsonism is a loss of the dopaminergic neurons (and hence dopamine) of the substantia nigra pars compacta which provide dopaminergic innervation to the striatum.

For more than a century, treatment of Parkinsonism was based on use of central anti-cholinergic substances. Up until recent times, various alkaloid drugs of belladonna, which have a characteristic cholinergic action (i.e., the ability to reduce sensitivity to acetylcholine, a neurotransmitter of cholinergic synapses), have been used for Parkinsonism. Currently, a sufficient quantity of facts have been established that support the idea that Parkinsonism is a consequence of an imbalance between dopaminergic and cholinergic systems, and that treatment of Parkinsonism should consist of either blocking excessive stimulation of the cholinergic system or normalizing functional activity of the dopaminergic system. Consequently, one of the approaches of Parkinsonism pharmacotherapy may include eliminating the deficit of dopamine. Because dopamine itself does not penetrate through the blood-brain barrier (BBB), either a dopamine precursor (levodopa), drugs that release dopamine, dopamine receptor agonists, or inactivation inhibitors of dopamine are used. On the other hand, during treatment of Parkinsonism, anticholinergic drugs should be used. From the information cited above, treatment of Parkinsonism should be based on using two groups of substances: drugs that stimulate the dopaminergic system of the brain and substances that inhibit the cholinergic system of the brain.

> List of potential causes: AIDS; corticobasal degeneration; Creutzfeldt-Jakob disease; diffuse Lewy body disease; drug-induced Parkinsonism (due to drugs such as antipsychotics and metoclopramide); encephalitis lethargica; multiple-system atrophy; pantothenate kinase-associated neurodegeneration; progressive supranuclear palsy; toxicity due to substances such as carbon monoxide, carbon disulphide, manganese, paraquat, hexane, rotenone, and toluene; Wilson's disease; paraneoplastic syndrome; neurodegeneration with brain iron accumulation

13.1.2 Classification

1. Dopamine agonist: Levodopa, Pergolide
2. DOPA decarboxylase inhibitor: Carbidopa
3. Dopamine-releasing agent: Amantadine
4. Synthetic anti-cholinergics: Benztropine, Trihexylphenidyl, Procyclidine, Biperidone
5. Miscellaneous: Phenothiazines: Ethopropazine

Biosynthesis of dopamine

L-Tyrosine →(Hydroxylase)→ L-DOPA →(DOPA decarboxylase)→ Dopamine

Levodopa is formed from L-tyrosine as an intermediate metabolite in the biosynthesis of dopamine, which is formed from the levodopa by the cytoplasmic enzyme DOPA decarboxylase.

13.1.3 Dopamine Agonist

Levodopa 3-(3,4-Dihydroxyphenyl) alanine

Synthesis

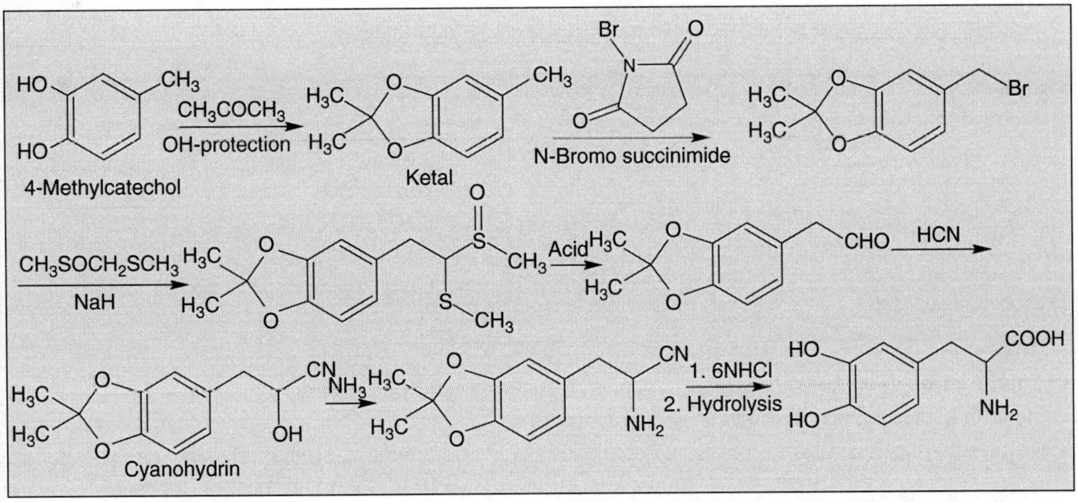

Two hydroxyl groups of 4-methyl catechol are protected as ketal by reaction with acetone and treatment with N-bromosuccinimide gives bromo methyl derivative. This on reaction with (methylsulphinyl)(methylthio)methane in presence of base followed by acid treatment gives phenylacetaldehyde derivative. Treatment with hydrogen cyanide gives cyanohydrin, which is converted to amino nitrile by treatment with ammonia. Hydrolysis of nitrile group and deprotection affords levodopa.

Mechanism of action: Dopamine does not cross the blood-brain barrier, and hence has no therapeutic utility. However, its immediate metabolic precursor, levodopa, is transported into the brain by neutral amino acid transporters and permeates into striatal tissues, where it is decarboxylated to dopamine. The dopamine produced is responsible for the therapeutic effectiveness of the drug in Parkinson's disease.

Pergolide 7-Methyl-9-[(methylsulphanyl)methyl]-4,6,6a,7,8,9,10,10a-octahydroindolo[4,3-*fg*]quinoline

Synthesis

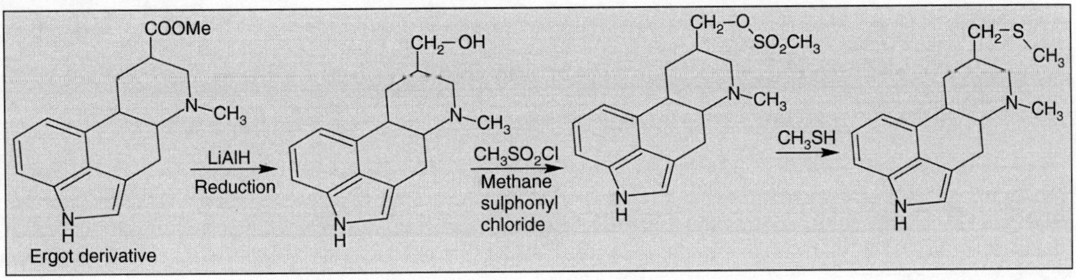

Reduction of ester of ergoline derivative with lithium aluminium hydride gives the corresponding alcohol; this is then converted to its mesylate. Displacement with methylmercapto group affords pergolide.

Mechanism of action: Pergolide is an ergot derivative that acts as agonist at both D_1 and D_2 receptors in the striatal region. It is effective in relieving the clinical symptoms of Parkinsonism.

13.1.4 DOPA Decarboxylase Inhibitor

Carbidopa α-Hydrazino-3,4-dihydroxy- α-methyl hydrocinnamic acid

Synthesis

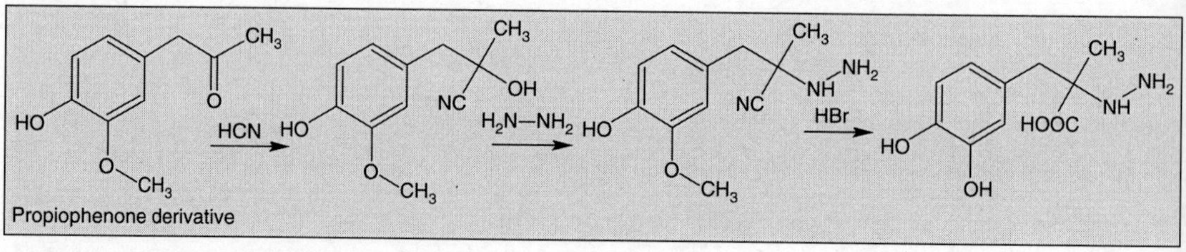

4-Hydroxy-3-methoxy propiophenone reacts with hydrogen cyanide to form cyanohydrin, which in turn reacts with hydrazine and hydrolysis affords carbidopa.

Mechanism of action: Carbidopa and levodopa are given concomitantly. In humans, dopa decarboxylase activity is greater in the liver, heart, lungs, and kidneys than in the brain. Therefore, ingested levodopa is converted into dopamine in the peripheral parts in preference to the brain. It is thought that in humans levodopa enters the brain only when administered in dosage high enough to overcome losses caused by peripheral metabolism. Inhibition of peripheral dopa decarboxylase by carbidopa (which does not cross BBB) increases the proportion of levodopa that crosses BBB.

13.1.5 Dopamine-Releasing Agent

Amantadine Tricyclodecan-1-amine

Synthesis

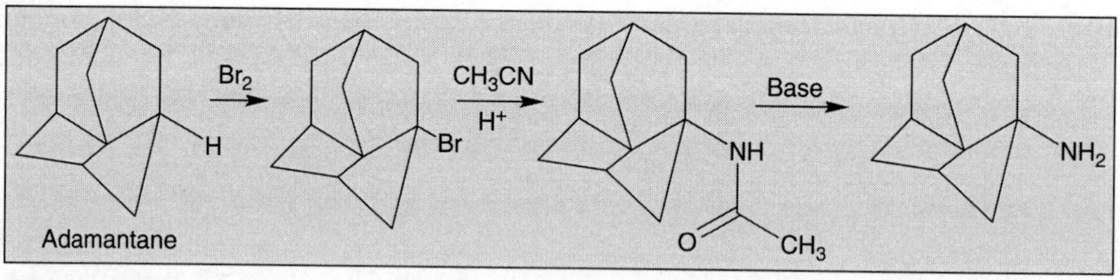

Amantadine is synthesised from adamantane. It is directly brominated to 1-bromadamantane, which in Ritter reaction conditions when heated with a mixture of acetonitrile and concentrated sulphuric acid transforms into 1-acetylaminoadamantane. Hydrolysis of this product using alkali leads to the formation of amantadine.

Mechanism of action: In the brain, it appears to block the reuptake of dopamine at dopaminergic nerve terminals, thus increasing the concentration of dopamine at the synapses. This facilitates the function of the remaining nigrostriatal neuronal pathways in patients with Parkinsonism.

It also prevents the uncoating of certain viruses (parainfluenzae A) and, hence, also acts as an antiviral agent.

13.1.6 Synthetic Anti-Cholinergics

Muscarinic receptor antagonists are administered with levodopa, which in addition to diminishing the cholinergic striatal effects (suppress tremor and rigidity) may inhibit the reuptake and storage of dopamine at the striatum.

Benztropine 3-(Diphenylmethoxy)-8-azabicyclooctane

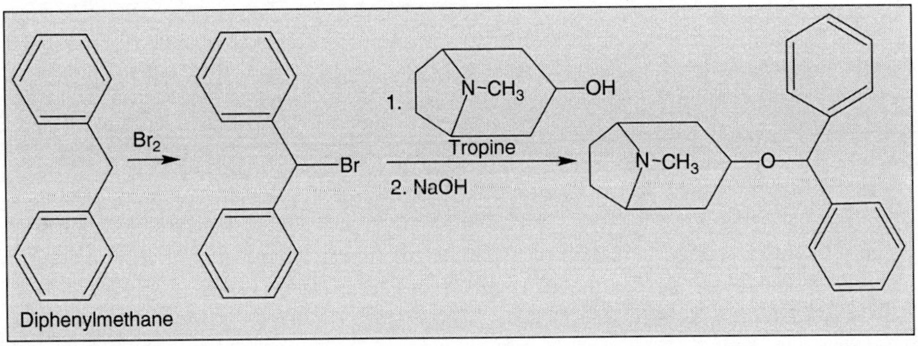

Synthesis
Bromination of diphenylmethane in presence of light followed by reaction of tropine in presence of alkali affords benztropine. The structure of benztropine resembles both atropine and anti-histaminics of the diphenhydramine.

Trihexyphenidyl α-Cyclohexyl- α-phenyl-1-piperidine propanol

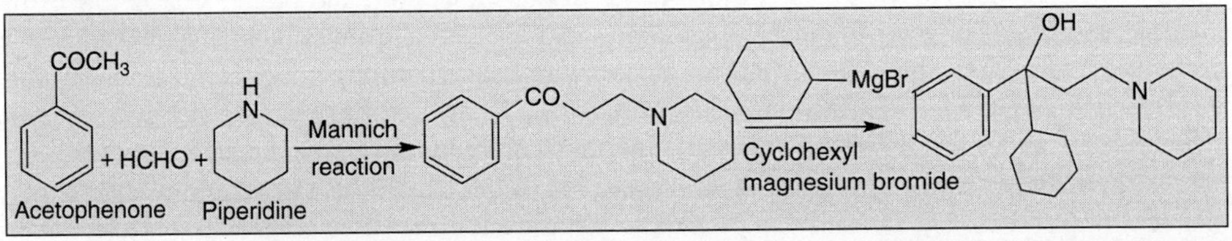

Trihexyphenidyl is synthesised by the reaction of 2-(1-piperidino)propiophenone with cyclohexylmagnesiumbromide. The initial 2-(1-piperidino)propiophenone is synthesised in turn by the aminomethylation of benzophenone using paraformaldehyde and piperidine.

Procyclidine 1-Cyclohexyl-1-phenyl-3-pirrolidinopropan-1-ol
Synthesis is similar to that of trihexyphenidyl, except for the use of pyrrolidine instead of piperidine.

Biperidine 1-(5-Norbornen-2-yl)-1-phenyl-3-piperidinopropan-1-ol

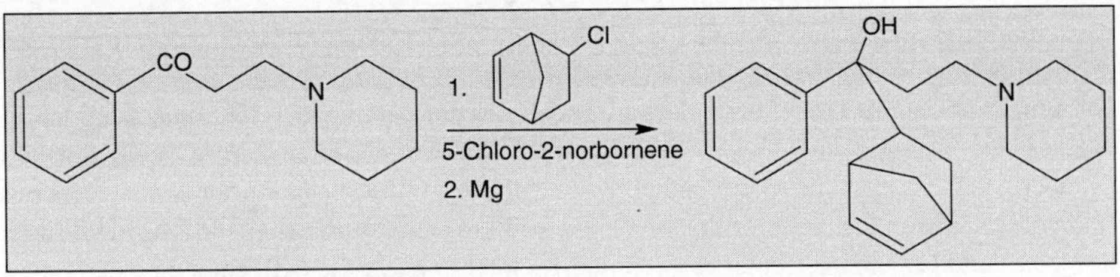

Biperiden is also synthesised according to the method of making trihexyphenidyl, except by reacting 2-(1-piperidino)propiophenone with 5-norbornen-2-ylmagnesiumbromide.

Phenothiazine derivative:
Ethopropazine 10-[2-(Diethylamino)propyl] phenothiazine

Synthesis

Ethopropazine is synthesised by alkylation of phenothiazine using 1-diethylamino-2-propylbromide in the presence of sodium amide. It is used in the management of Parkinsonism, especially for control of rigidity.

13.2 MUSCLE RELAXANTS

Centrally acting skeletal muscle relaxants selectively depress the CNS that controls muscle tone. It is used to treat muscle spasm of local origin such as strains, sprains, and lumbago. It also relieves pain and spasm associated with fibrositis, bruisitis, spondylytis, and muscle injury.

A quite large, diverse group of substances can affect skeletal muscle by acting both at the level of neuromuscular junctions as well as at various levels of the spinal cord and brain stem. A few of them influence transmission of nerve impulses at neuromuscular contacts and are capable of paralysing skeletal muscle. They are used mainly as adjuvant substances in anaesthesia during minor surgical interventions.

Depending on the localization and mechanism of action, myorelaxants can be classified as peripherally acting myorelaxants, direct-acting muscle relaxants, and centrally acting muscle relaxants.

The activity of peripherally acting myorelaxants is exhibited in the area of neuromuscular contacts, which results in a weakened transmission from motor neuron endings to the membranes of skeletal muscle cells. These drugs, in turn, can be subdivided into neuromuscular transmission blockers, which include anti-depolarizing drugs (tubocurarine, atracurium, gallamine) and depolarizing drugs (succinylcholine).

Direct-acting muscle relaxants directly block the process of contraction of the muscle fibres themselves. Of all the direct-acting myotropic drugs, only dantrolene is used in medical practice.

Centrally acting muscle relaxants are also widely used (baclofen, cyclobenzaprin, carisoprodol, methocarbamol, chlorphenesin, chlorzoxazone, orphenadrine, and diazepam), which suppress motor impulse transmission in inter-neuronal synapses of the CNS.

Most neuromuscular blockers function by blocking transmission at the end plate of the neuromuscular junction. Normally, a nerve impulse arrives at the motor nerve terminal, initiating an influx of calcium ions which causes the exocytosis of synaptic vesicles containing acetylcholine. Acetylcholine then diffuses across the synaptic cleft. It may be hydrolysed by acetylcholine esterase (AchE) or bind to the nicotinic receptors located on the motor end plate. The binding of two acetylcholine molecules results in a conformational change in the receptor that opens the sodium-potassium channel of the nicotinic receptor. This allows Na^+ and Ca^{2+} ions to enter the cell and K^+ ions to leave the cell, causing a depolarization of the end plate and resulting in muscle contraction.

Normal end-plate function can be blocked by two mechanisms. Non-depolarizing agents like tubocurarine block the agonist, acetylcholine, from binding nicotinic receptors and activating them, thereby preventing depolarization. Alternatively, depolarizing agents such as succinylcholine are nicotinic receptor agonists who mimic Ach, and block muscle contraction by depolarizing to such an extent that it desensitizes the receptor and it can no longer initiate an action potential and cause muscle contraction. These neuromuscular blocking drugs are structurally similar to acetylcholine, the endogenous ligand, in many cases containing two acetylcholine molecules linked end-to-end by a rigid carbon ring system, as in pancuronium.

> The earliest known use of muscle relaxant drugs dates back to the 16th century, when European explorers encountered natives of the Amazon Basin in South America using poison-tipped arrows that produced death by skeletal muscle paralysis. This poison, known today as curare, led to some of the earliest scientific studies in pharmacology. Its active ingredient is tubocurarine.

13.2.1 Classification

1. Glycerol monoethers and its derivatives: Mephenesin, Chlorphenesin carbamate, Methocarbamol
2. Substituted alkanediols and its derivatives: Meprobamate, Carisoprodol, Metaxolone, Lorbamate
3. Benzodiazepines: Diazepam, Fletazepam
4. Benzazole: Chlorzoxazone
5. Miscellaneous: Orphenadrine, Baclofen, Dandrolene, Cyclobenzaprine

13.2.2 Glycerol Monoether and its Derivatives

Mephenesin 3-(2-Methylphenoxy)-1,2-propanediol

Synthesis

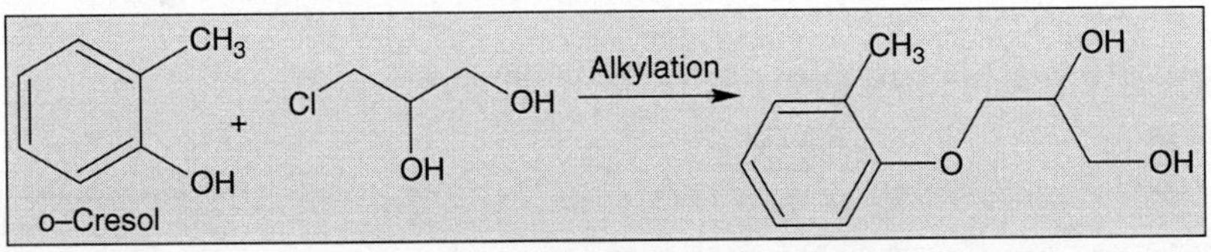

Alkylation of *o*-cresol with 1-chlorpropan-2,3-diol affords mephenesin.

Chlorphenesin carbamate 3-(4-Chlorophenoxy)-1,2-propanediol-1-carbamate

Synthesis

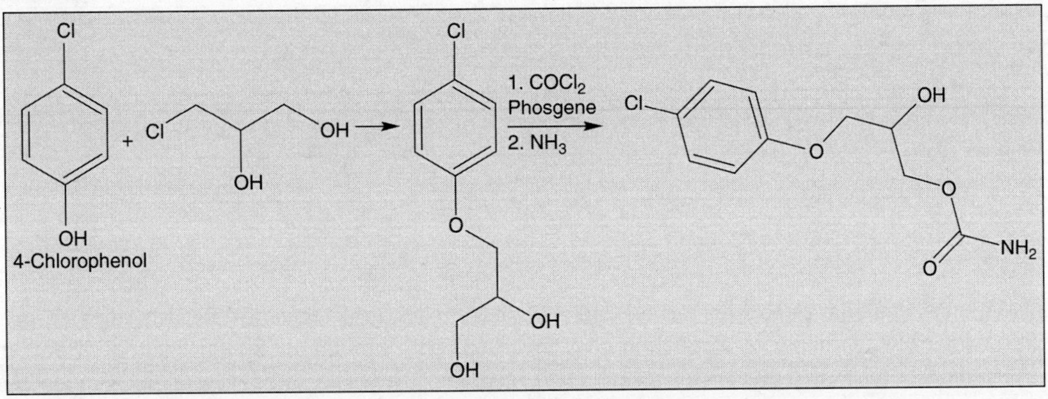

Alkylation of *o*-cresol with 1-chlorpropan-2,3-diol, followed by reaction with phosgene and ammonia, affords chlorphenesin carbamate.

Methocarbamol 3-(2-Methoxyphenoxy)-1,2-propanediol-1-carbamate

Synthesis

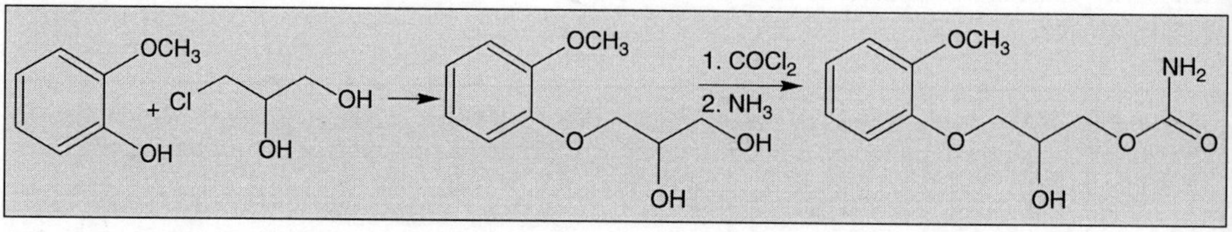

Synthesis is similar to chlorphenesin carbamate, with 2-methoxyphenol as starting material.

Methocarbamol is a milder muscle relaxant for minor musculoskeletal spasm or in patients who are intolerant to stronger muscle relaxants.

13.2.3 Structure Activity Relationship of Glycerol Derivatives

1. The *ortho*-substituted benzene derivatives are generally more potent than the corresponding *para* or *meta* isomers.
2. Substitution of a small alkyl or alkoxy group in the benzene ring increases activity.
3. Substitution at the benzene ring by OH, NH$_2$, COCH$_3$, etc., decreases activity.
4. Mephenesin is weakly active and short-lived largely due to facile metabolism of the primary hydroxyl group. Carbamylation of this group increases activity. *p*-Chlorination increases the lipid/water partition coefficient and seals off the *p*-position from metabolic hydroxylation.

13.2.4 Substituted Alkane Diols and their Derivatives

Carisoprodol [2-Methyl-2-(1-methylethylcarbamoyloxymethyl) pentyl]aminomethanoate

Synthesis

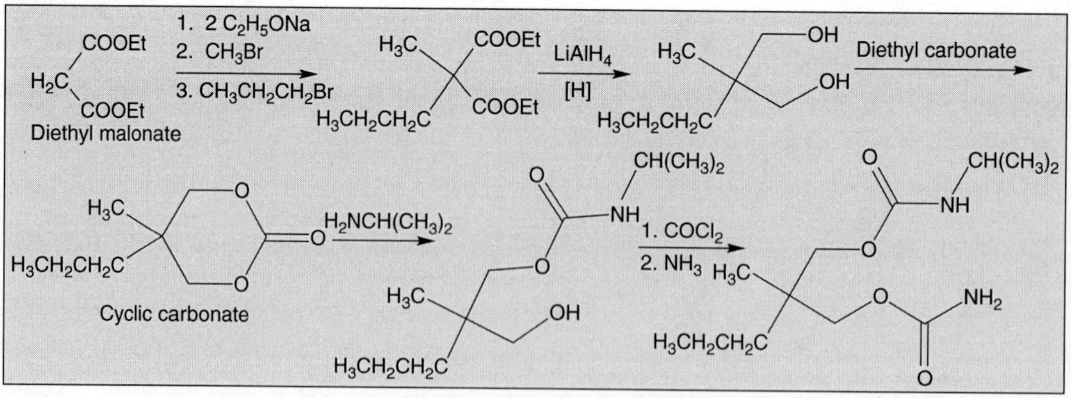

Alkylation of malonic ester with methyl and propyl bromides in presence of sodium ethoxide gives methyl ethyl diethyl malonate. Reduction of ester groups with lithium aluminium hydride affords the corresponding diol, which on reaction with diethyl carbonate affords the cyclic carbonate. Ring opening of this compound by means of isopropylamine gives the monocarbamate. The remaining hydroxyl group on reaction with phosgene and ammonia affords carisoprodol.

Carisoprodol is a moderately acting strong muscle relaxant.

Metaxalone 5-[(3,5-Dimethylphenoxy)methyl]oxazolidin-2-one

Synthesis

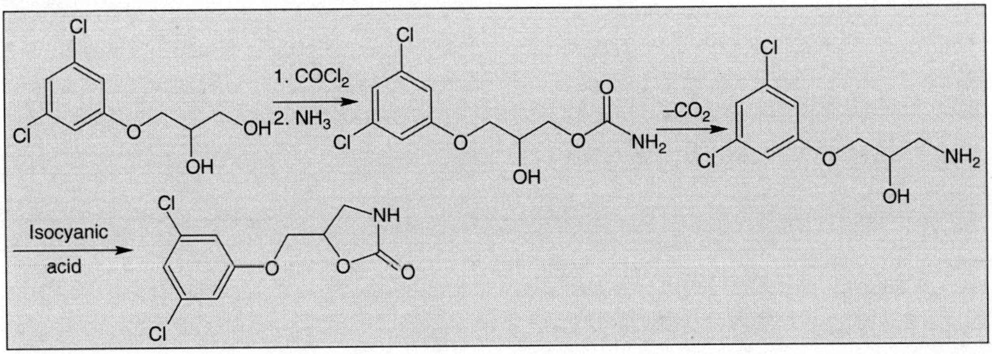

3-(3,5-Dichlorophenoxy)-1,2-propanediol reacts with phosgene and ammonia to form mono carbamate, which undergoes decarboxylation to form amino alcohol. Treatment with isocyanic acid affords metaxalone.

Lorbamate [2-Methyl-2-(1-cyclopropylcarbamoyloxymethyl) pentyl]aminomethanoate

Synthesis

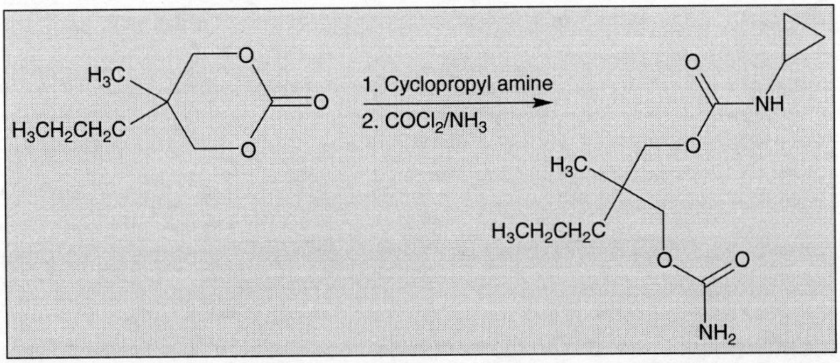

The starting material is prepared by reducing diethyl methyl propyl malonate with lithium aluminium hydride, followed by treatment with diethylcarbonate to give cyclic carbonate derivative. This in turn reacts with two moles of phosgene and one mole each of cyclopropyl amine and ammonia to afford lorbamate.

Benzazoles

Chlorzoxazone 5-Chloro-2-benzoxazolione

Synthesis

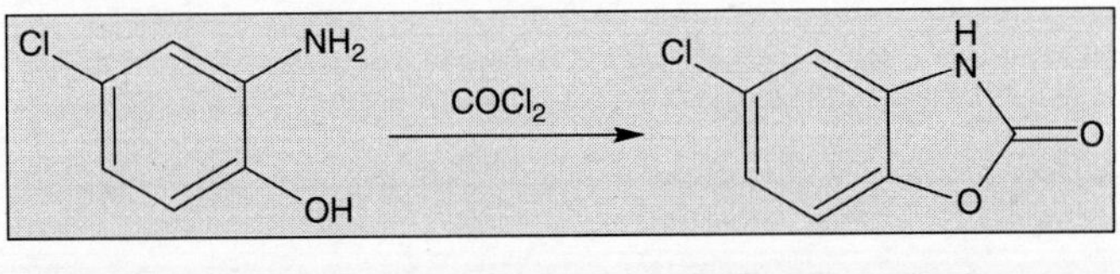

Chlorzoxazone is synthesised by a heterocyclization reaction of 2-amino-4-chlorphenol with phosgene.

Chlorzoxazone is another milder muscle relaxant that is not as strong as carisoprodol or cyclobenzaprine, and is also less sedating.

Benzodiazepines

Fletazepam 7-Chloro-5-(2-fluorophenyl)-1-(2,2,2-trifluoroethyl)-2,3-dihydro-1,4-benzodiazepine

Synthesis

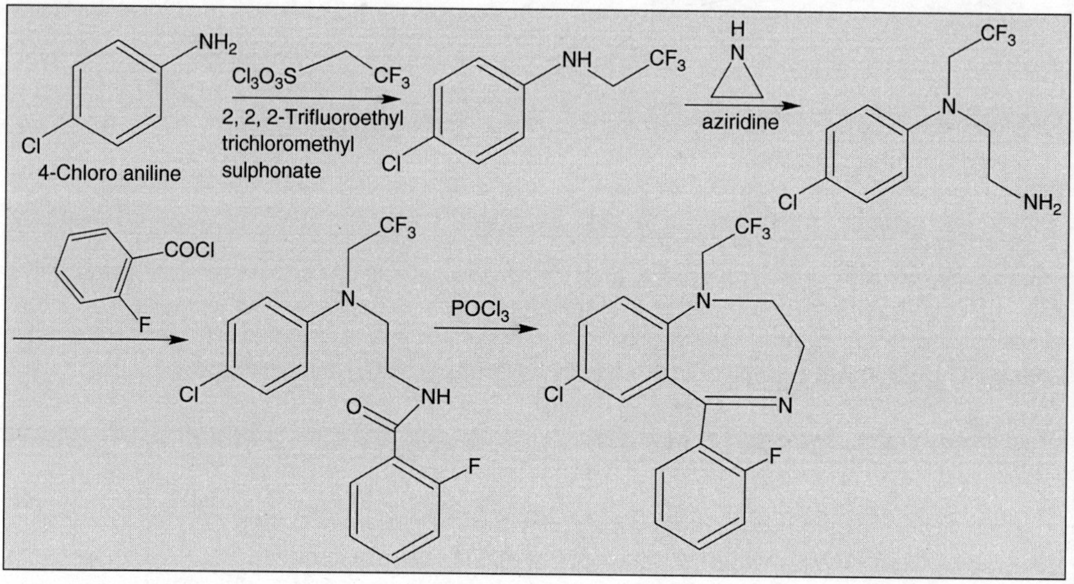

Alkylation of 4-chloroaniline by the trichloromesyl ester of trifluoroethanol gives secondary amine. This undergoes alkylation by aziridine to produce diamine. Acylation with 2-fluorobenzoyl chloride produces the secondary amide, which undergoes Bischler–Napieralski cyclodehydration with phosphorous oxy chloride to afford fletazepam.

13.2.5 Miscellaneous

Orphenadrine N,N-Dimethyl-2-[(2-methylphenyl)- phenyl-methoxy]-ethanamine

Synthesis

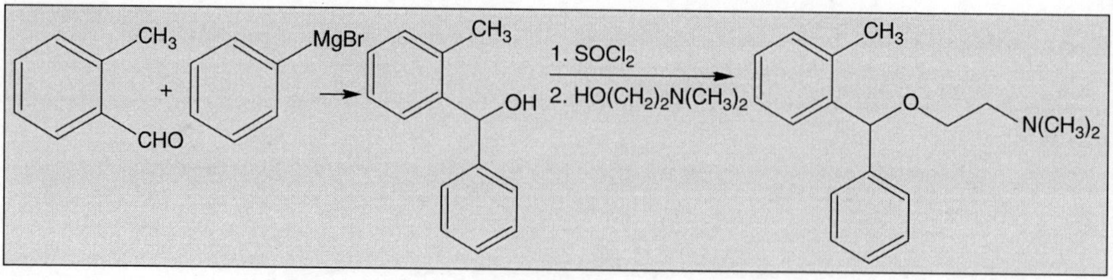

It is produced by reacting dimethylaminoethanol with 2-methylbenzhydryl chloride. The 2-methylbenzhydryl chloride can be formed *via* a Grignard reaction.

Orphenadrine is a methylated derivative of diphenhydramine, and thus belongs to the ethanolamine family of antihistamines.

Baclofen 4-Amino-3-(4'-chlorophenyl) butyric acid

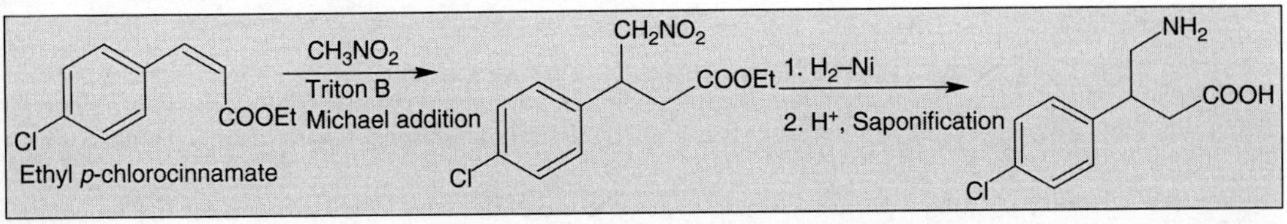

Ethyl 4-chlorocinnamate undergoes Triton B-catalysed Michael addition with niromethane, followed by Raney nickel reduction and saponification to afford baclofen.

Dantrolene 1-[[[5-(4-Nitrophenyl)-2-furanyl]methylene]amino]-2,4-imidazolidinedione

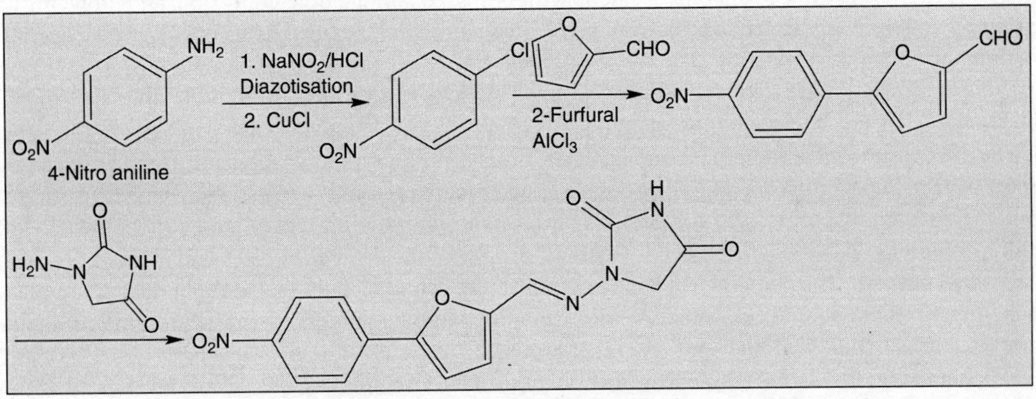

Dantrolene is synthesised by reacting 4-nitrophenyldiazonium chloride with furfural, forming 5-(4-nitrophenyl)-2-furancarboxaldehyde, which is reacted further with 1-aminohydantoin to give the corresponding hydrazone, dantrolene.

Cyclobenzaprine *N,N*-Dimethyl-3-(dibenzo[a,d]cyclohepten-5-ylidene) propylamine

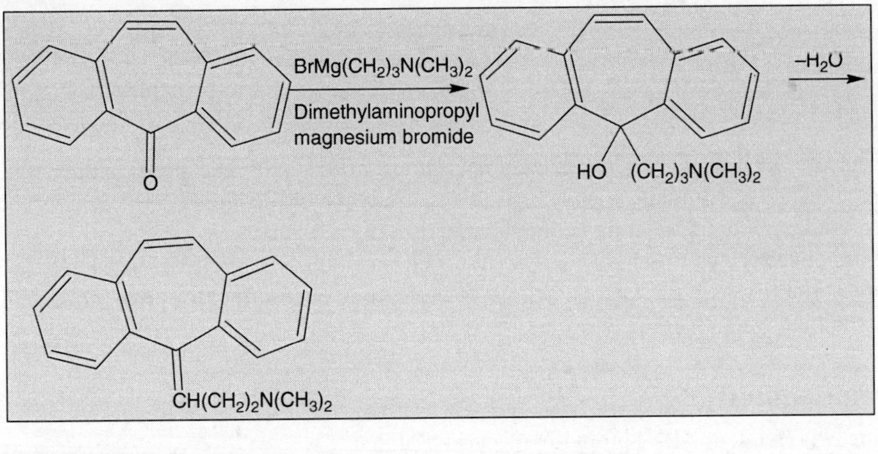

Cyclobenzaprine is synthesised by reacting 5*H*-dibenzo[a,d]cyclohepten-5-one with 3-dimethylaminopropylmagnesium bromide and subsequent dehydration of the resulting carbinol in acidic conditions into cyclobenzaprine.

Cyclobenzaprine is one of the strongest muscle relaxants. It has been shown to be beneficial in fibromyalgia.

13.3 DRUGS FOR ALZHEIMER'S DISEASE TREATMENT

Dementia, a progressive brain dysfunction, leads to a gradually increasing restriction of daily activities. The most well-known type of dementia is Alzheimer's disease.

Alzheimer's disease (AD) is a progressive neurodegenerative disease of the brain, and is the most common form of dementia among the elderly population. AD is characterised by more than just memory loss; it also results in other cognitive and behavioural symptoms that progressively impair function in activities of daily living.

Symptoms: The cognitive symptoms include memory loss, disorientation, confusion, and problems with reasoning and thinking. Behavioural symptoms include agitation, anxiety, delusions, depressions, hallucination, insomnia, and watering. The earliest evidence of AD is the onset of chronic, insidious memory loss that is slowly progressive over several years. This loss can be associated with slowly progressive behavioural changes. Although other neurologic systems (e.g., extra-pyramidal, cerebellar systems) can also be affected, the most prominent finding as the disease progresses to its moderate and severe stages is progressive memory impairment. Other common neurologic presentations include changes in language ability (e.g., anomia, progressive aphasia), impaired visuospatial skills, and impaired executive function.

Pathophysiology: AD is characterised by marked atrophy of the cereberal cortex and loss of cortical and subcortical neurons. The pathological hallmarks of AD are senile plaques that are spherical accumulation of the protein β-amyloid accompanied by degenerating neuronal processes, and neurofibrillary-tangles, composed of paired helical filaments and other proteins.

Neurochemistry: In AD, the severe loss of cholinergic neurons in the nucleus basalis and associated area that form the cholinergic forebrain area, and their projections to the cerebral are marked with decreased levels of acetyl choline and its rate-limiting synthetic enzyme, choline acetyl transferase, in the cortex. (*See Fig. 13.1 in the coloured set of pages.*)

13.3.1 Treatment Strategies

The dramatic cholinergic cell losses in the AD brain, agents that augment the cholinergic system have received the greatest attention with regard to drug development and intervention. In this regard, three strategies have been investigated to increase cholinergic transmission. These have involved the use of

1. Precursors, such as choline, to augment neurotransmitter synthesis at pre-synaptic terminals
2. Direct agonist to stimulate post-synaptic muscarinic and/or nicotinic cholinergic receptors – e.g., besiperidine, linopirdine, milameline, fuzomeline, quilostigmine
3. Anti-cholinesterases to inhibit the enzymes that metabolise naturally released acetylcholine and thereby amplify its post-synaptic action – e.g., tacrine, phenserine, tolserine, donepezil, icopezil, rivastigmine, galanthamine.

Cholinergic agonist
Besipirdine *N*-(1*H*-1-Indolyl)-*N*-propyl-*N*-(4-pyridyl)amine

Synthesis

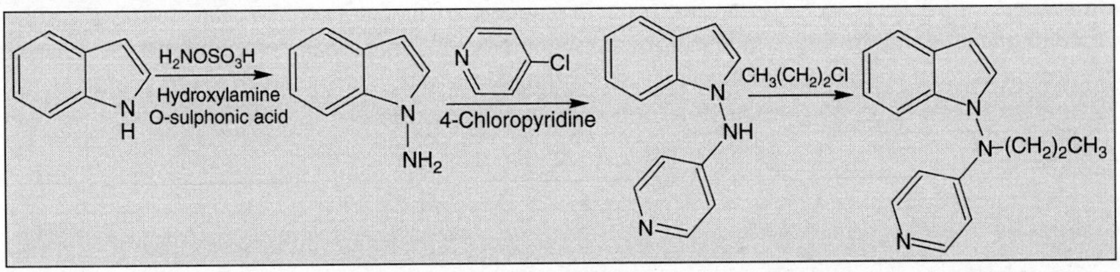

Treatment of indole with hydroxylamine O-sulphonic acid leads to displacement of sulphonate and formation of a new N-N bond to give the hydrazine. Reaction of this compound with 4-chloropyridine leads to nucleophilic displacement of chloro group to give the pyridylated hydrazine. Alkylation of the secondary nitrogen with propylchloride leads to besipirdine.

Linopirdine 1-Cyclohexyl-3,3-bis(pyridin-4-ylmethyl)indol-2-one

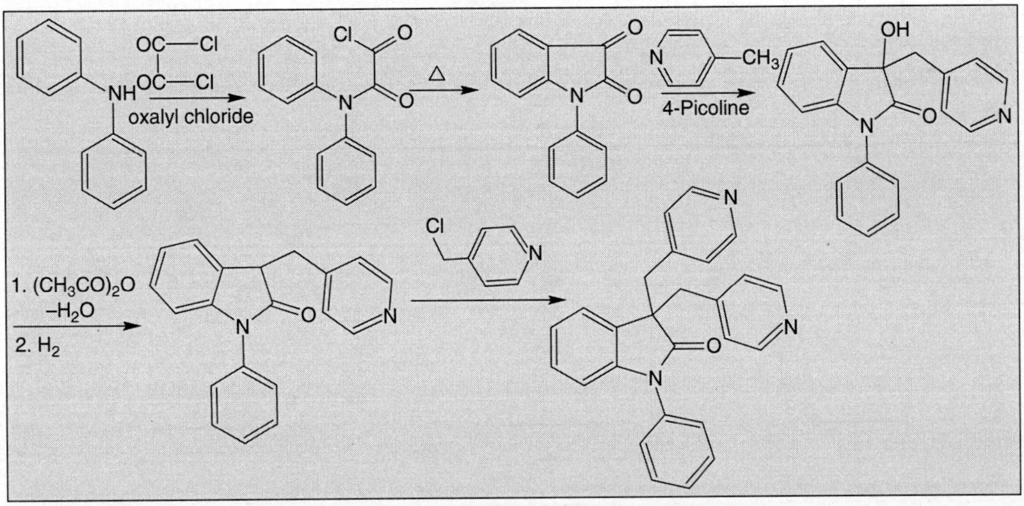

Acylation of diphenylamine with oxalyl chloride leads to the acyl acid chloride. This cyclizes on heating to give 1-phenylisatin. Condensation of this with 4-picoline under phase-transfer condition leads to addition of the anion to the carbonyl group to give the alcohol. Heating in acetic anhydride causes dehydration; hydrogenation of the resulting olefin gives 2-indolone derivative. Alkylation with 4-chloromethyl pyridine in presence of base affords linopirdine.

Linopirdine is a compound based on the indolone nucleus, which enhances cognition by enhancing the release of the neurotransmitter in the brain.

Milameline 1-Methyl-1,2,5,6-tetrahydro-3-pyridinecarbaldehyde O3-methyloxime

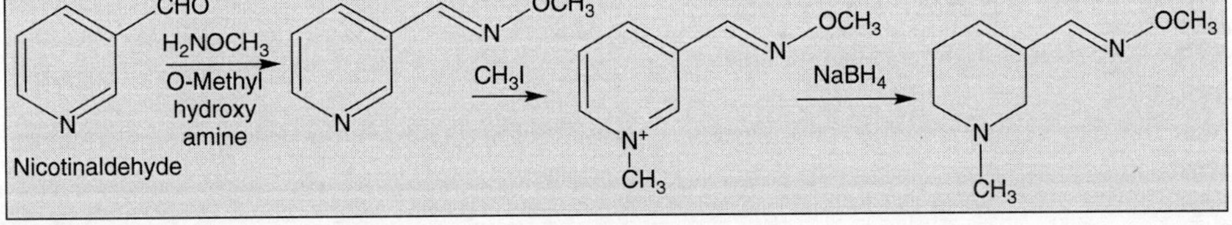

Reaction between pyridine-3-carbaldehyde and O-methyl hydroxylamine gives oxime; alkylation of this with methyl iodide leads to the quaternary salt. Treatment of this salt with sodium borohydride leads initially to the dihydropyridine, which results from reduction of the charged iminium function; the resulting enamine is itself susceptible to borohydride, leading to the tetrahydro derivative milameline.

Tazomeline 3-(Hexyloxy)-4-(1-methyl-1,2,5,6-tetrahydro-3-pyridinyl)-1,2,5-thiadiazole

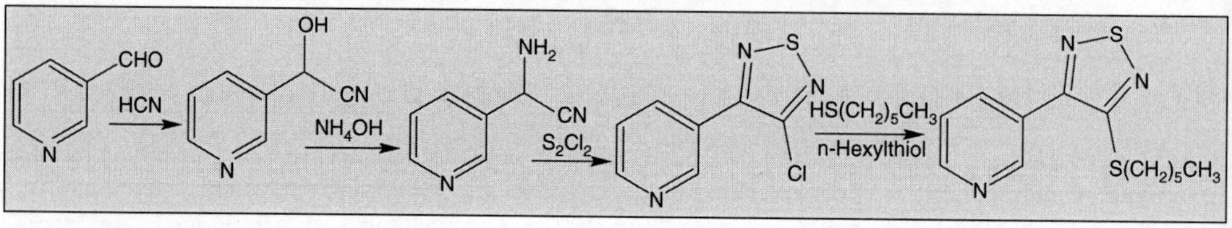

Nicotinaldehyde reacts with hydrogen cyanide to form cyanohydrin, which on reaction with ammonium hydroxide leads to the formation of the α-aminonitrile. This intermediate cyclizes to the thiadiazole on treatment with sulphur chloride. Displacement of chloro group with n-hexylthiol leads to the thioether tazomoline.

Quilostigmine 1,3a,8-Trimethyl-1,2,3,3a,8,8a-hexahydropyrrolo[2,3-b]indol-5-yl1-oxo-1,2,3,4-tetrahydro-2-isoquinolinecarboxylate

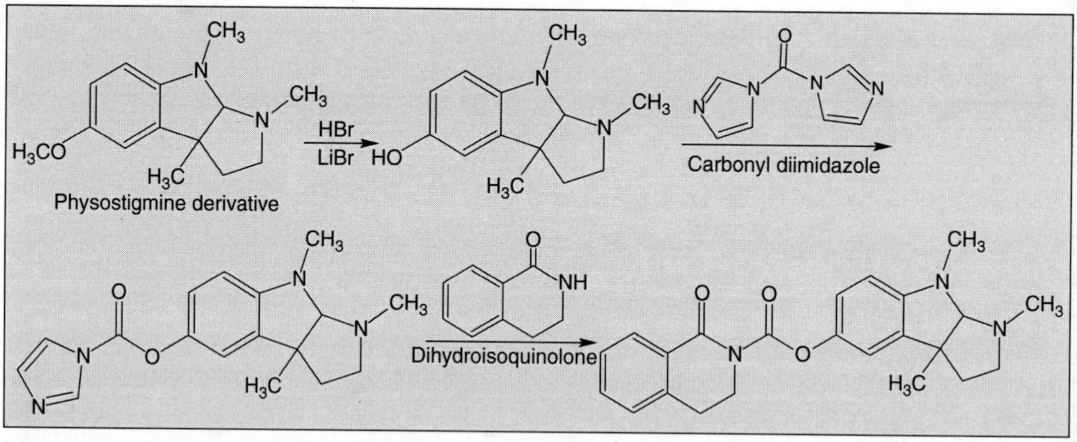

Physostigmine derivative 1,3a,8-trimethyl-1,2,3,3a,8,8a-hexahydropyrrolo[2,3-b]indol-5-ylmethyl ether undergoes demethylation by reaction with lithium bromide and hydrogen bromide. Treatment of this compound with carbonyl diimidazole leads to corresponding imidazolide; reaction with dihydroisoquinolone leads to the displacement of imidazole by amide nitrogen to afford quilostigmine.

Acetyl choline esterase inhibitors

Tacrine 1,2,3,4-Tetrahydroacridin-9-amine
Synthesis
It was the first centrally-acting cholinesterase inhibitor approved for the treatment of Alzheimer's disease.

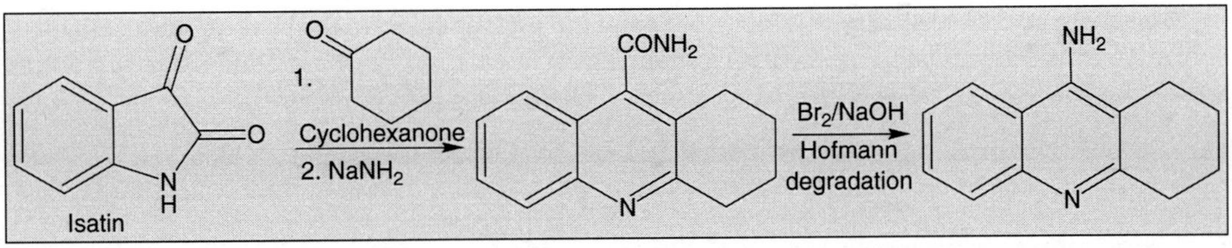

Sodamide-catalysed reaction of isatin and cyclohexanone gives tetrahydroacridin-9-carboxamide. Reaction of this with bromine in the presence of sodium hydroxide leads to the degradation of this side chain to an amine.

Phenserine 1,3a,8-Trimethyl-1,2,3,3a,8,8a-hexahydropyrrolo[2,3-b]indol-6-yl N-phenylcarbamate

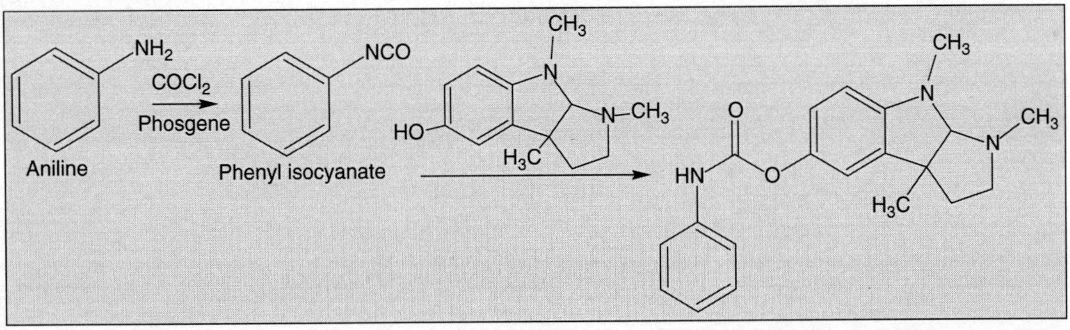

Aniline is converted to phenyl isocyanate by reaction with phosgene, which on reaction with 1,3a,8-trimethyl-1,2,3,3a,8,8a-hexahydropyrrolo[2,3-b]indol-6-ol gives the carbamate phenserine.

Tolserine Synthesis is similar to phenserine, but the starting compound is 3-methylaniline (3-toluidine).

Donepezil 2-[(1-Benzyl-4-piperidyl)methyl]- 5,6-dimethoxy-2,3-dihydroinden-1-one

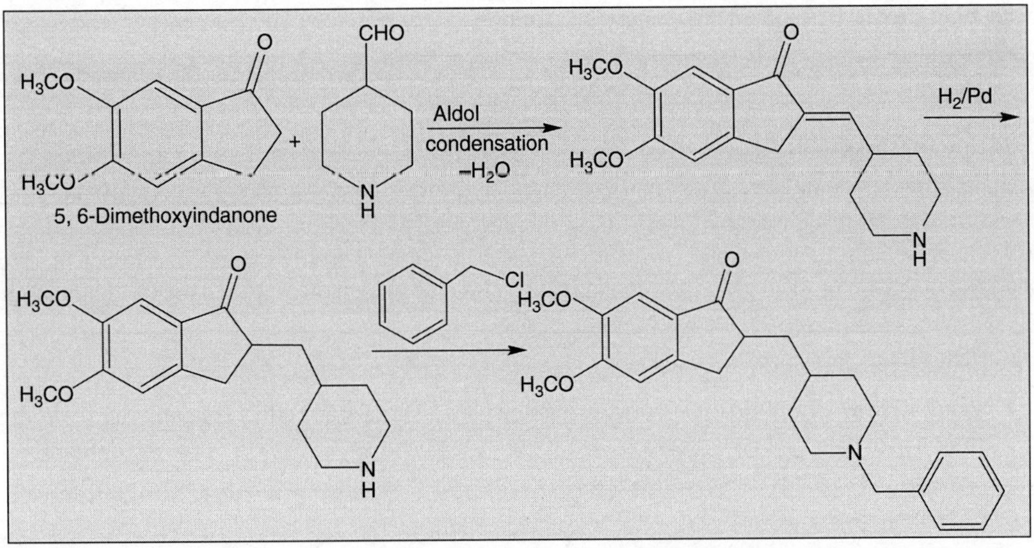

5,6-Dimethoxyindanone undergoes aldol condensation 4-piperidinecarbaldehyde to form alkylidene derivative, which is reduced with hydrogen and palladium to afford 2-(4-piperidyl)methyl- 5,6-dimethoxy-2,3-dihydroinden-1-one. Benzylation at piperidine nitrogen affords donepezil.

Icopezil 3-[2-(1-Benzyl-4-piperidyl)ethyl]-6,7-dihydro-5H-isoxazolo[4,5-f]indol-6-one

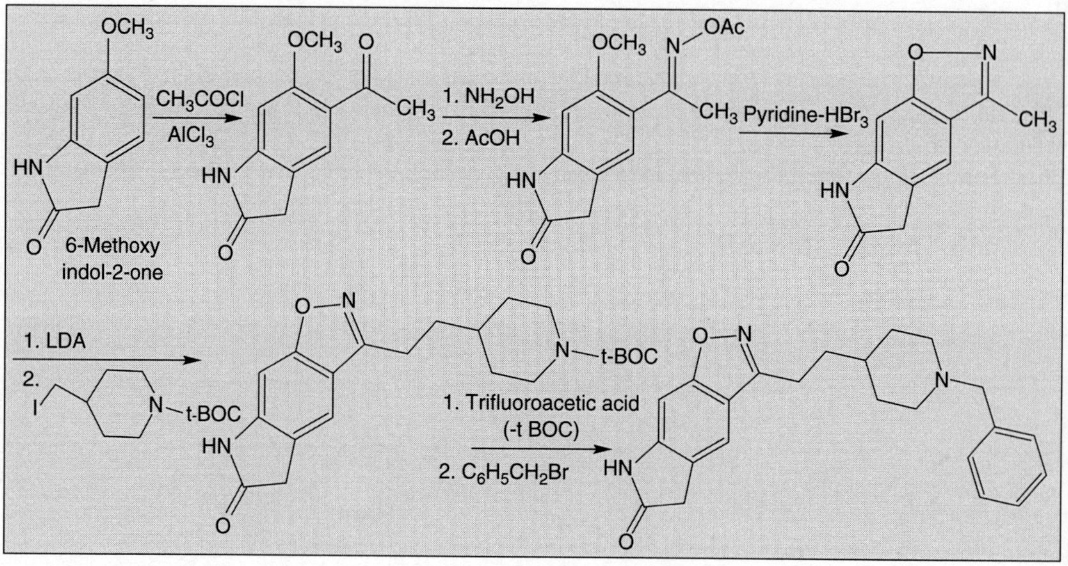

Friedel—Crafts acylation of 6-methoxyindol-2-one with acetyl chloride leads to the methyl ketone. The ketone is converted to oxime and this is acylated with acetic anhydride to give the intermediate, which on treatment with pyridine-hydrogen bromide leads to cleavage of methyl ether and displacement of acetyl group, resulting in benzisoxazole moiety. The methyl group on that newly formed ring reacts with t-BOC-protected 4-iodomethyl piperidine, followed by treatment with trifluoroacetic acid to afford icopezil.

13.4 CNS STIMULANTS

A large number of biologically active substances exhibit stimulatory action on the central nervous system (CNS). Substances that increase vigilance and reduce the need for sleep are considered as CNS stimulants or psychostimulants. In other words, there are drugs capable of temporarily keeping one awake, elevating mood and maintaining adequate perception of reality, reducing outer irritability and the feeling of fatigue, and elevating the physical and mental capacity of work.

CNS stimulants can be classified as:

1. **Psychomotor stimulants:** Compounds that display a stimulatory effect primarily on brain functions and activate mental and physical activity of the organism. They are made up of: methylxan-

thines (caffeine, theophylline, pentoxifyllin), amphetamines (dextroamphetamine, methamphetamine), and also methylphenidate and pemoline

2. **Respiratory stimulants or analeptics:** Compounds that cause certain activations of mental and physical activity of the organism, and primarily excite the vasomotor and respiratory centres of the medulla (doxapram, almitrine)
3. **Drugs that suppress appetite or anorectics:** Drugs that activate mental and physical activity of the organism, but primarily accentuate the excitatory centre of satiation in the hypothalamus (phentermine, diethylpropion)

Caffeine 1,3,7-Trimethylxanthine
In small doses, caffeine is a relatively weak psychostimulant and is used for increasing awareness as well as for relieving headaches associated with blood-flow disorders of the brain. Caffeine has a stimulatory effect on the respiratory and vasomotor centres, and it stimulates centres of the vagus nerve. It has a direct stimulatory effect on the myocardium, and in large doses can cause tachycardia and arrhythmia.

Caffeine is found in varying quantities in the beans, leaves, and fruits of some plants, where it acts as a natural pesticide that paralyses and kills certain insects feeding on the plants.

Amphetamines 2-Amino-1-phenylpropane
It is a powerful synthetic psychostimulant with a high potential of addiction. It increases vigilance and the ability to concentrate, temporarily elevates mood, and stimulates motor activity. It is believed that the mechanism of action of amphetamines lies in their ability to release epinephrine (adrenaline) and dopamine from pre-synaptic nerve endings, which stimulate the corresponding receptors in the CNS. It is also possible that they reduce neuronal uptake of amines as well as inhibit their degradation by monoaminoxidase (MAO).

Methylphenidate Methyl ester α-phenyl-2-piperidilacetic acid

Arylation of benzylcyanide by 2-chloropyridine in the presence of a base gives α-phenyl-α-(2-pyridil) acetonitrile. Sulphuric acid hydrolysis of the nitrile group and subsequent esterification with methanol gives the methyl ester of α-phenyl-α-(2-pyridylacetic acid). The pyridine moiety is reduced into a piperidine by hydrogen over platinum, giving methylphenidate.

Pemoline 2-Amino-5-phenyl-2-oxazolin-4-one

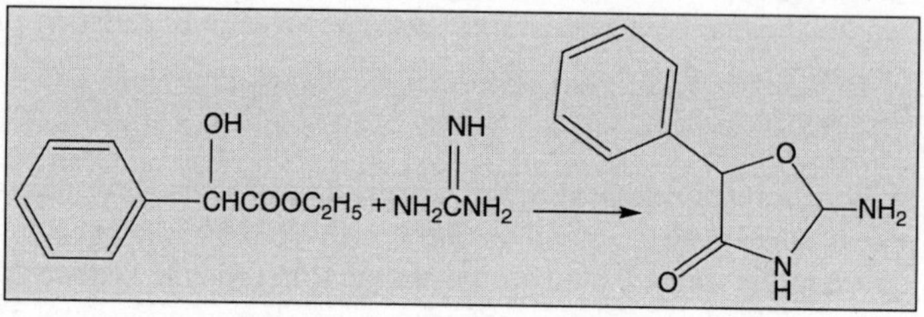

It is synthesised by the condensation of the ethyl ester of mandelic acid with guanidine.

Doxapram 1-Ethyl-4-(2-morpholinoethyl)-3,3-diphenyl-2-pyrrolidinone

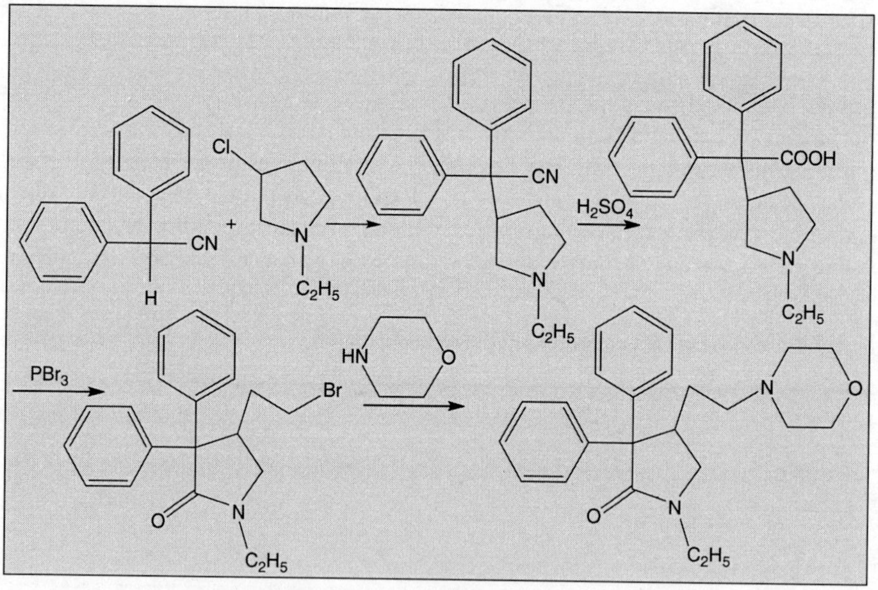

Diphenylacetonitrile in the presence of sodium amide is alkylated with 1-ethyl-3-chlorpyrrolidine, giving (1-ethyl-3-pyrrolidinyl) diphenylacetonitrile. Acidic hydrolysis of the nitrile group gives (1-ethyl-3pyrrolidinyl)diphenylacetic acid. Reacting this with phosphorous tribromide leads to rearrangement with an opening of the pyrrolidine ring and the subsequent closing of the pyrrolidinone ring, forming 1-ethyl-4-(2-bromoethyl)-3,3-diphenyl-2-pyrrolidinone. Substitution of the bromine atom with a morpholine group gives doxapram.

Almitrine 2,4-*bis* (Allylamino)-6-[4-[*bis*-(*p*-fluorophenyl)methyl]-1-piperazinyl]-*s*-triazine

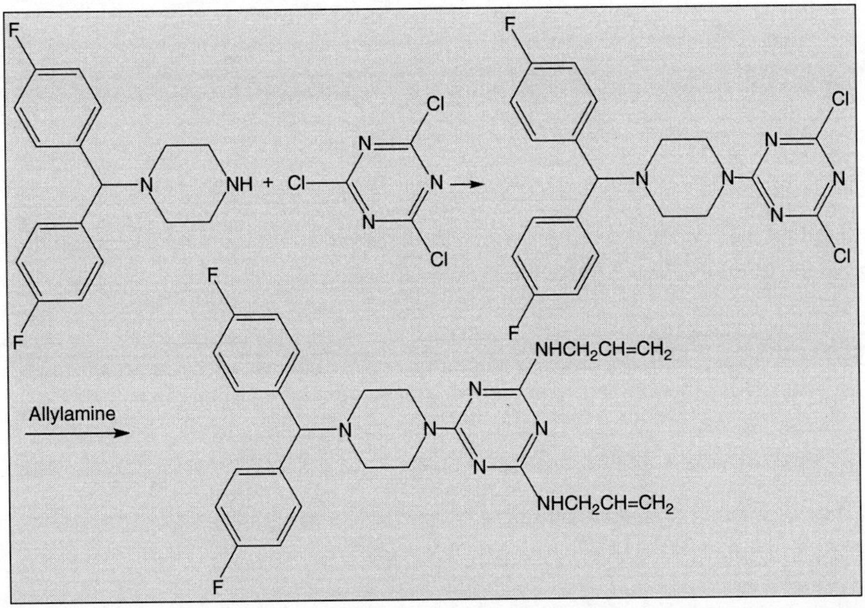

It is synthesised by reacting 1-[*bis*-(p-fluorophenyl)methyl]piperazine with cyanuric chloride, which gives 2,4-dichloro-6-[4-[*bis*-(p-fluorophenyl) methyl]-1-piperazinyl]-*s*-trazine. Reacting this with allylamine gives almitrine.

Phentermine α, α-Dimethylphenylethylamine
It differs from amphetamine in the presence of an additional methyl group in the α-position of the amino group.

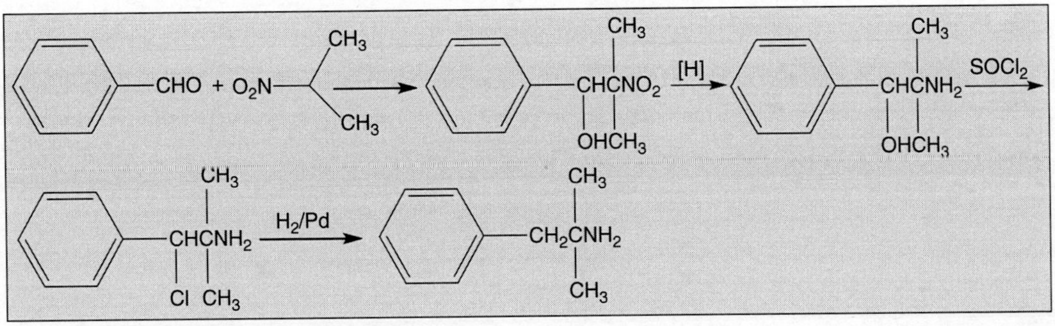

It is synthesised from benzaldehyde, the condensation of which with 2-nitropropane gives carbinol. Reduction of the nitro group of this product gives 2-amino-2-methyl-1-phenylpropanol. The hydroxyl group is replaced with a chlorine atom upon reaction with thionyl chloride, giving 2-amino-2-methyl-1-phenylpropylchloride. Reducing this with hydrogen using a palladium on calcium carbonate catalyst gives phentermine.

Diethylpropion 1-Phenyl-2-diethylaminopropanon

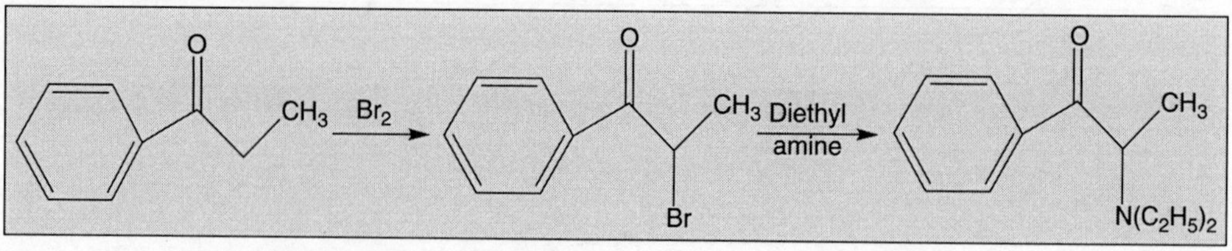

It is synthesised by the bromination of propiophenone into α-bromopropiophenone and the subsequent substitution of the bromine atom with a diethylamino group.

13.5 NEWER DRUGS

Pramipexole: (6S)-N^6-Propyl-4,5,6,7-tetrahydro-1,3-benzothiazole-2,6-diamine

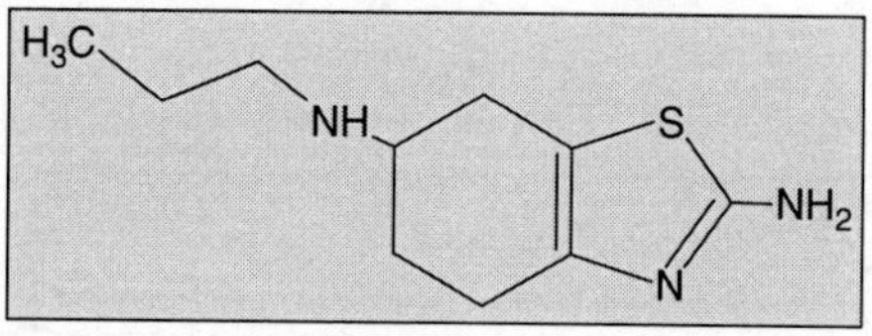

It is a medication indicated for treating Parkinson's disease. By acting as an agonist to the dopamine receptors, pramipexole may directly stimulate dopamine receptors in the striatum, thereby restoring the dopamine signals needed for proper functioning of the basal ganglia.

Selegiline: N-Methyl-N-(1-methyl-2-phenyl-ethyl)-prop-2-yn-1-amine

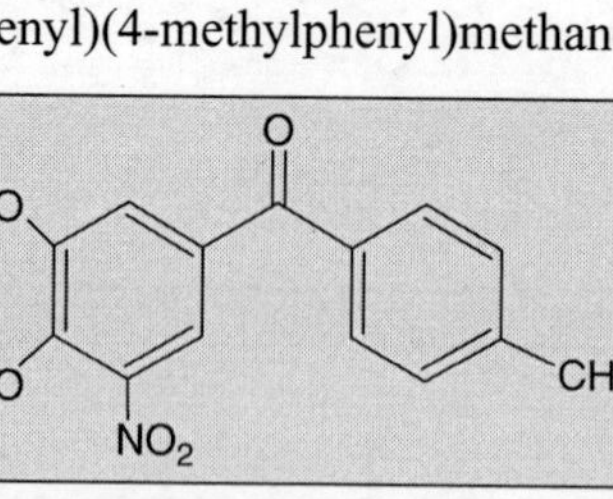

It is a selective inhibitor of MAO-B; MAO-B metabolises dopamine. Selegiline exhibits little therapeutic benefit when used independently, but enhances and prolongs the anti-Parkinson effects of levodopa.

Tolcapone (3,4-Dihydroxy-5-nitrophenyl)(4-methylphenyl)methanone

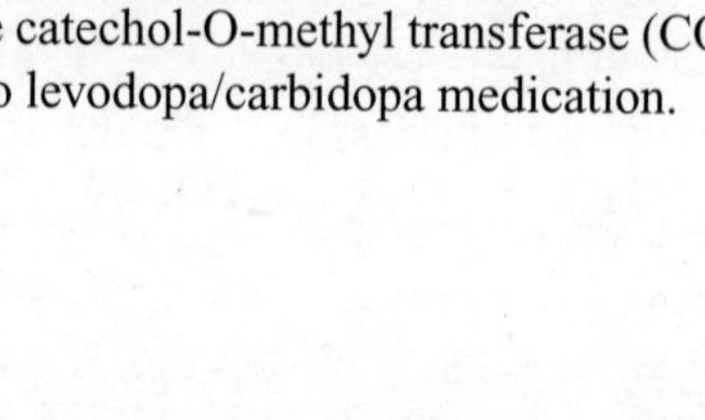

It is a drug that inhibits the enzyme catechol-O-methyl transferase (COMT). It is used in the treatment of Parkinson's disease as an adjunct to levodopa/carbidopa medication.

FURTHER READINGS

1. *Foye's Principles of Medicinal Chemistry*, Williams, D.A. and Lemke, T.L. (ed.), 5th Ed., 2002.
2. *Therapy of Parkinson's Disease*, Rajesh Pahwa, Kelly Lyons, William C. Koller, and Marcel Dekker (ed.), Vol. 63, 3rd Ed., 2004.
3. Serge Gauthier, 1998. Pharmacotherapy of Alzheimer's disease, Taylor & Francis.

MULTIPLE-CHOICE QUESTIONS

1. Which area is involved in Parkinson's disease?
   a. Neostriatum
   b. Substantia nigra
   c. Pallidum
   d. Amygdala

2. Dopamine is biosynthesised from
   a. L-Alanine
   b. L-Tyrosine
   c. L-Phenylalanine
   d. L-DOPA

3. Which one of the following is an ergot derivative?
   a. Pergolide
   b. Levodopa
   c. Benztropine
   d. Biperidine

4. Anti-Parkinson's drug that is also useful as antiviral:
   a. Levodopa
   b. Amantadine
   c. Benztropine
   d. Biperidine

5. The phenothiazine derivative used for Parkinson's disease is
   a. Prochlorperazine
   b. Promethazine
   c. Ethopropazine
   d. Carphenazine

6. The muscle relaxant with less sedation effect is
   a. Diazepam
   b. Carisoprodol
   c. Chlorzoxazone
   d. Cyclobenzaprine

7. Chlorphenesin
   a. Has increased partition coefficient than mephensin
   b. Has decreased partition coefficient than mephensin
   c. Gets easily metabolised by *p*-hydroxylation
   d. None of the above

8. Memory loss is associated with
   a. Parkinson's disease
   b. Alzheimer's disease
   c. Psychosis
   d. Arthritis

9. Which of the following is not a cholinergic agonist?
   a. Besperidine
   b. Galanthamine
   c. Fuzomeline
   d. Quilostigmine

10. An indole derivative useful for the treatment of Alzheimer's disease is
    a. Donepezil
    b. Tacrine
    c. Besiperidine
    d. Milameline

11. The anti-Alzheimer's drug synthesised starting from isatin is
    a. Donepezil
    b. Tacrine
    c. Besiperidine
    d. Milameline
12. The drug used for fibromyalgia is
    a. Benztropine
    b. Cyclobenzaprine
    c. Tacrine
    d. Donepezil

QUESTIONS

1. Why is carbidopa concomitantly administered with levodopa?
2. Discuss the SAR of glycerol monoethers.
3. Derive the synthetic protocol for the following:
    a. Tacrine
    b. Donepezil
    c. Bacolfen
    d. Besperidine
    e. Carisoprodol
4. Discuss briefly Alzheimer's disease neuropathology, symptoms, and therapeutic strategies.

SOLUTION TO MULTIPLE-CHOICE QUESTIONS

1. b;
2. b;
3. a;
4. b;
5. c;
6. c;
7. a;
8. b;
9. b;
10. c;
11. b;
12. b.

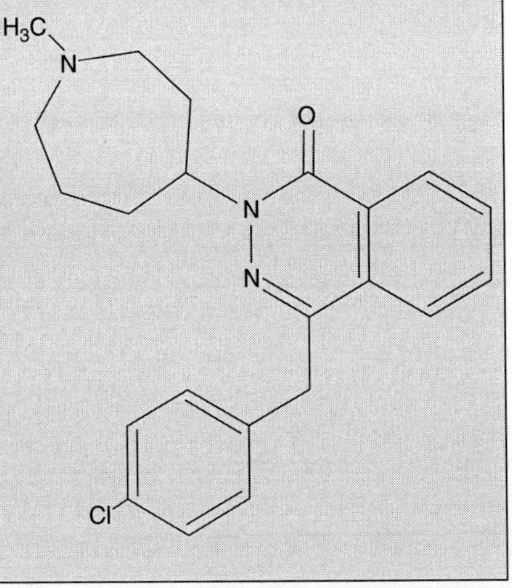

CHAPTER 14

Antihistamines and Anti-Ulcer Agents

LEARNING OBJECTIVES

- Define histamines and their receptor functionality
- Define H_1 receptor antagonists and their utility
- Categorize antihistamines on the basis of the chemical nucleus
- Describe the basic structural requirement for antihistamines
- Define non-sedative antihistamines and their utility
- Define H_2 receptor antagonists and their utility
- Describe the structural requirements of a H_2 receptor antagonists
- Define H_3 and H_4 receptors and their functional utility
- Define proton pump inhibitors
- List some new drugs against histamine receptors

14.1 INTRODUCTION

Histamine [2-(imidazol-4-yl)ethylamine], which is biosynthesised by decarboxylation of the basic amino acid histidine, is found in all organs and tissues of the human body.

Histamine is an important chemical messenger communicating information from one cell to another, and is involved in a variety of complex biologic actions. It is mainly stored in an inactive bound form, from which it is released as a result of an antigen–antibody reaction initiated by different stimuli such as venoms, toxins, proteolytic enzymes, detergents, food materials, and numerous chemicals. Histamine exerts its biologic function by interacting with at least three distinctly specific receptors H_1, H_2, and H_3. It was the mediator associated with allergic manifestations. It binds to and activates specific receptors in the nose, eyes, respiratory tract, and skin, causing characteristic allergic signs and symptoms. It also causes contraction of smooth muscles, relaxation of capillaries, and gastric acid secretion.

277

Historically, the term 'antihistamine' has been used to describe drugs that act as H_1 and H_2 receptor antagonists.

14.2 H_1 RECEPTOR ANTAGONIST

Allergic illnesses are a complex collection of disturbances with chronic and severe effects ranging from slight reddening, rashes, and runny nose to severe and even fatal anaphylaxis. It has been shown that around 10 per cent of the population may be prone to some form of allergy. Therapy directed towards removing the source of allergen is not always successful. In a number of cases, the allergen itself is never found. Therefore, symptomatic treatment using H_1 antihistamines is carried out.

Histamine H_1 receptors are metabotropic G-protein-coupled receptors expressed throughout the body, specifically in smooth muscles, on vascular endothelial cells, in the heart, and in the central nervous system. The H_1 receptor is linked to an intracellular G-protein (G_q) that activates phospholipase C and the phosphatidylinositol (PIP2) signalling pathway.

H_1 antihistamines are clinically used in the treatment of histamine-mediated allergic conditions. Specifically, these indications may include allergic rhinitis, allergic conjunctivitis, allergic dermatological conditions (contact dermatitis), pruritus (atopic dermatitis, insect bites), anaphylactic or anaphylactoid reactions.

Antihistamines are reversible competitive antagonists of histamine at H_1 receptor sites. They do not prevent histamine release or bind to the histamine that has already been released. The H_1 receptor blockade results in decreased vascular permeability, reduction of pruritus, and relaxation of smooth muscle in the respiratory and gastrointestinal tracts. H_1 receptor antagonists have been used clinically to treat various allergic disorders such as seasonal or perennial allergic rhinitis and chronic urticaria.

> The most common adverse effect of the first-generation H_1 anti-histamines is sedation; this is due to their relative lack of selectivity for the peripheral H_1 receptor.

14.2.1   Classification

1. Amino alkyl ethers

$$Ar-\underset{Ar_1}{\overset{R}{C}}-O-CH_2CH_2-N(CH_3)_2$$

	Ar	Ar₁	R
Diphenhydramine	phenyl	phenyl	H
Doxylamine	phenyl	2-pyridyl	CH₃
Carbinoxamine	4-chlorophenyl	2-pyridyl	H

(Continued)

	Ar	Ar₁	R
Medrylamine	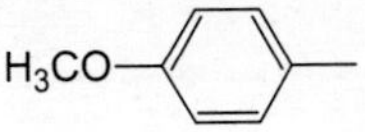 H₃CO—C₆H₄—	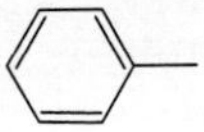 C₆H₅—	H
Clemastine	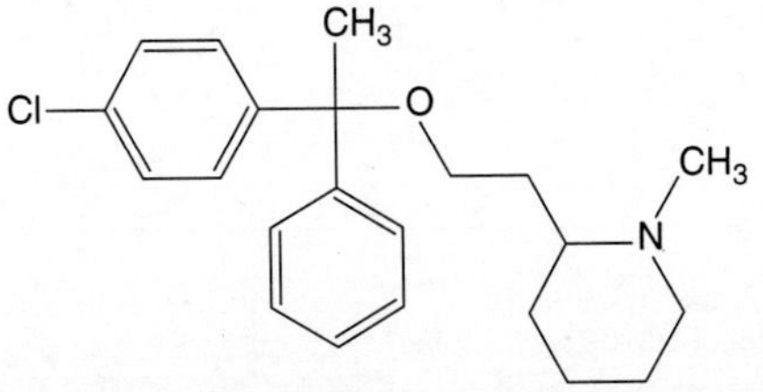 (see structure below)		

Clemastine structure: 4-Cl-C₆H₄–C(CH₃)(C₆H₅)–O–CH₂CH₂–(2-position of N-methylpiperidine)

2. **Ethylenediamine derivatives**

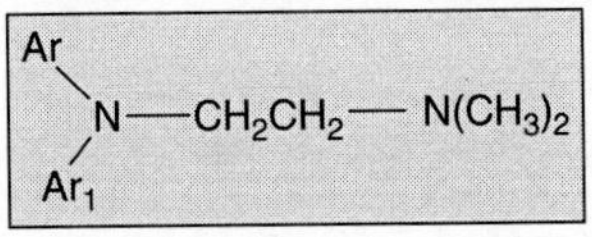

$$\text{Ar}\diagdown\!\!\!\underset{\text{Ar}_1}{\text{N}}\!\!-CH_2CH_2-N(CH_3)_2$$

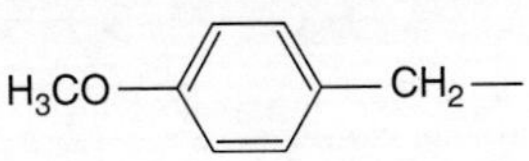

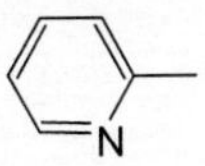

	Ar	Ar₁
Pyrilamine	H₃CO—C₆H₄—CH₂—	2-pyridyl
Tripelennamine	C₆H₅—CH₂—	2-pyridyl
Methapyrilene	2-thienyl—CH₂—	2-pyridyl
Thonzylamine	H₃CO—C₆H₄—CH₂—	2-pyrimidinyl

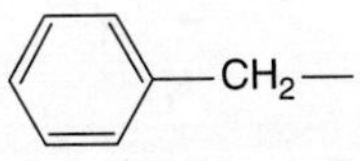

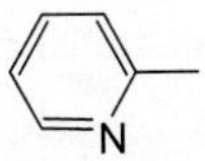

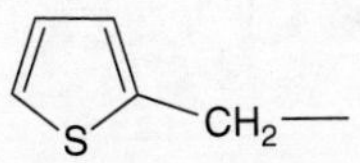

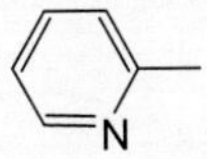

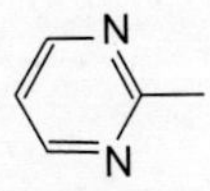

3. **Propylamine derivatives**
   (i) Saturated analogues

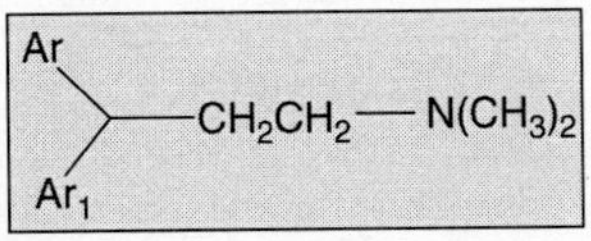

$$\text{Ar}\diagdown\!\!\!\underset{\text{Ar}_1}{\text{CH}}\!\!-CH_2CH_2-N(CH_3)_2$$

Chapter 14: Antihistamines and Anti-Ulcer Agents

	Ar	Ar₁
Pheniramine	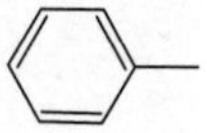 phenyl	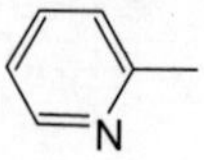 2-pyridyl
Chlorpheniramine	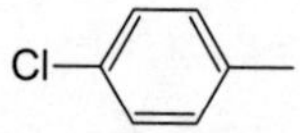 4-Cl-phenyl	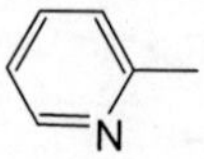 2-pyridyl

(ii) Unsaturated analogues

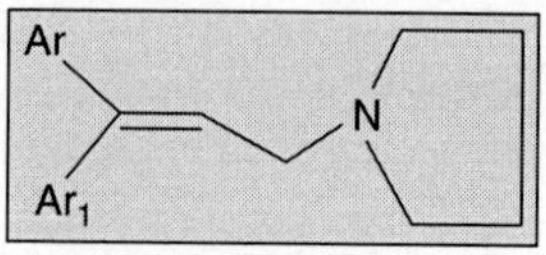

[Structure: Ar and Ar₁ on C=C–CH₂–N(pyrrolidine)]

	Ar	Ar₁
Pyrrobutamine	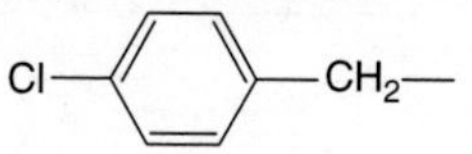 4-Cl-C₆H₄–CH₂–	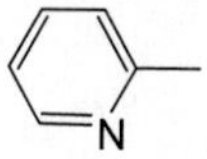 2-pyridyl
Triprolidine	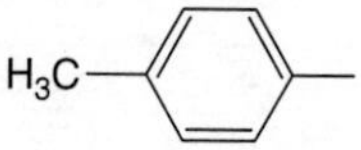 4-H₃C-C₆H₄–	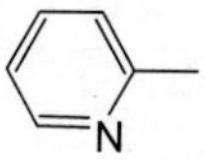 2-pyridyl

4. Phenothiazine derivatives

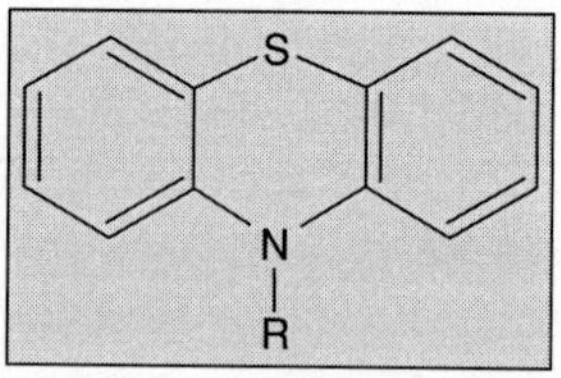

[Phenothiazine core with N–R]

	R
Promethazine	–H₂C–CH(CH₃)–N(CH₃)₂
Trimeprazine	–H₂C–CH(CH₃)–CH₂N(CH₃)₂
Methadilazine	–H₂C–(pyrrolidine with N–CH₃)

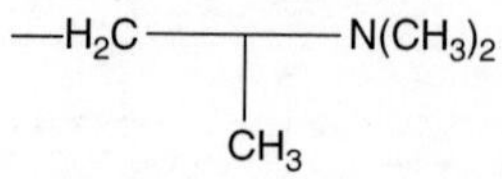

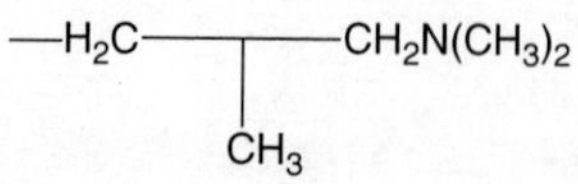

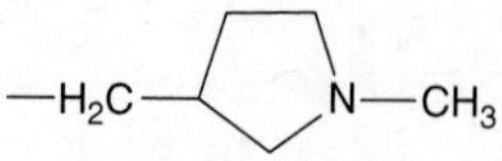

5. Piperazine derivatives

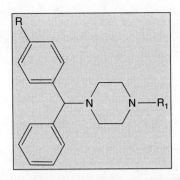

	R	R_1
Cyclizine	H	CH_3
Chlorcyclizine	Cl	CH_3
Hydroxyzine	Cl	—$CH_2CH_2OCH_2CH_2OH$
Meclizine	Cl	—CH_2—C₆H₄(m-CH₃)
Buclizine	Cl	—H_2C—C₆H₄(p-C(CH₃)₃)
Cinnarizine	H	—H_2C—CH=CH—C₆H₅

6. Miscellaneous: Cyproheptadine, Azatadine, Dimethindene, Antazoline
7. Non-sedative antihistamines (second generation): Terfenadine, Cetrizine, Acrivastine, Astemizole, Loratidine

14.2.2 Mechanism of Action

H_1 receptor antagonists act by competitively antagonising the effects of histamine at receptor sites; they do not block the release of histamine and, hence, offer only palliative relief of allergic symptoms.

14.2.3 Uses

1. Effective in perennial and seasonal allergic rhinitis, vasomotor rhinitis, allergic conjunctivitis, urticaria and angioedema, allergic reaction to blood and plasma, and as adjuncts to conventional therapy in anaphylactic reactions.
2. A few antihistamines are effective in mild, local allergic reaction to insect bites.
3. Selected antihistamines (e.g., diphenhydramine) reduce rigidity and tremors in Parkinson's disease, and in drug-induced extra-pyramidal symptoms.

Chapter 14: Antihistamines and Anti-Ulcer Agents

4. Some antihistamines (e.g., buclizine, cyclizine, and diphenhydramine) are also effective in the active and prophylactic management of motion sickness.
5. The phenothiazine antihistaminics (e.g., promethazine) are useful for pre-operative and postoperative vomiting and obstetric sedation.

14.2.4 Side-Effects

The most common side-effect of antihistamines is sedation, evidenced principally by drowsiness, plus a diminished alertness and inability to concentrate. Less common effects include dryness of mouth, blurred vision, vertigo, and gastrointestinal distress.

14.2.5 Structural Requirements

Most of the H_1 receptor antagonists may be described by the general structure:

$$\begin{array}{c} Ar \\ \diagdown \\ X\!-\!(CH_2)_n\!-\!N \\ \diagup \diagdown \\ Ar_1 CH_3 \end{array} \begin{array}{c} CH_3 \\ \\ \\ \end{array}$$

where Ar = a phenyl group, Ar_1= a phenyl, benzyl, pyridyl, or thenyl group, X=O, C, or N, and n the number (usually 2) so that the distance between the centres of the aromatic rings and the aliphatic N may be 5–6 Å.

Substitutions on the aryl groups, replacement of aliphatic dimethylamino group with small basic heterocyclic ring (piperidine and piperazine), increased branching on (—CH_2—)$_n$, and substitution between X and N, all serve to modify the potency, metabolism, ability to reach the site of action, and toxicity *in vivo*.
Structure–activity relationship: In general, the terminal nitrogen atom should be a tertiary amine for maximal activity. However, the terminal nitrogen atom may be part of a heterocyclic structure, as in cyclizine, and may still result in a compound of high antihistaminic potency.

Extension or branching of the 2-aminoethyl side chain results in a less active compound (exemption: promethazine).

14.2.6 Aminoalkyl Ether Derivatives

Diphenhydramine 2-(Diphenyl methoxy)-N, N-dimethyl ethanamine
Synthesis

Diphenyl methane $\xrightarrow[hv]{Br_2}$ CHBr $\xrightarrow[\text{2. }Na_2CO_3]{\text{1. }HOCH_2CH_2N(CH_3)_2 \text{ Dimethylaminoethanol}}$ CHOCH$_2$CH$_2$N(CH$_3$)$_2$

Diphenhydramine is synthesised by a simple reaction of benzhydrylbromide and 2-dimethyl-aminoethanol.

Besides antihistamine activity, diphenhydramine exhibits a local anaesthetic effect, relaxes smooth muscle, and has sedative and soporific action.

Doxylamine 2-[α-[2-(Dimethyl amino)ethoxy]- α-methylbenzyl]pyridine.
Synthesis

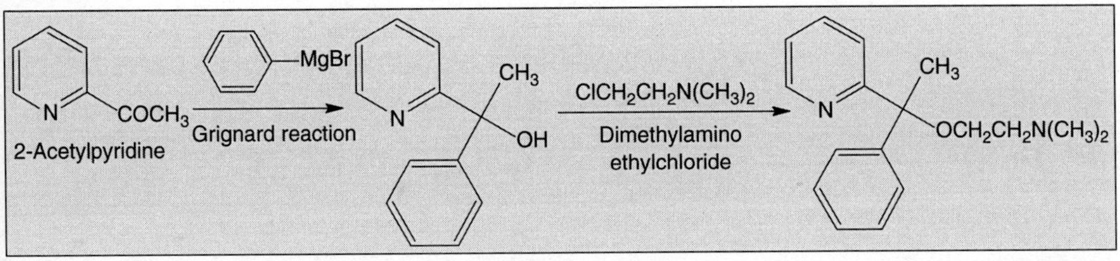

Reaction of 2-acetylpyridine with phenylmagnesium bromide gives the tertiary alcohol. Alkylation of these as their sodium salt with dimethylamino ethyl chloride affords doxylamine.

Carbinoxamine 2-[p-Chloro- α-[2-(dimethylamino)ethoxy]benzyl]pyridine
Synthesis

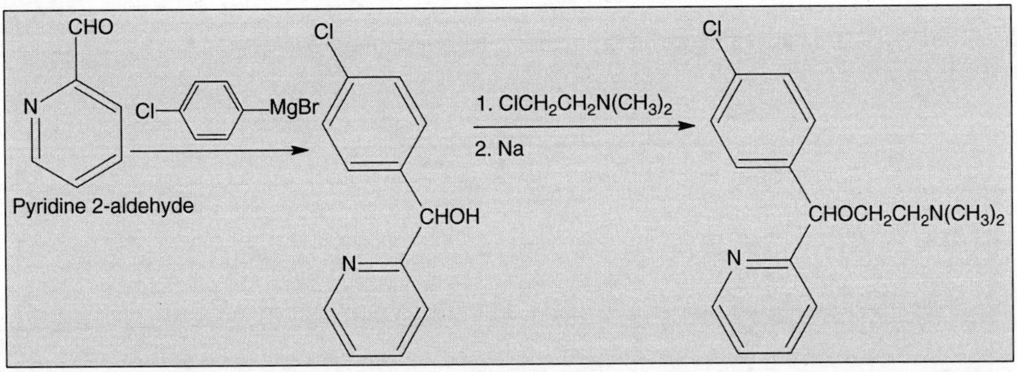

Synthesis is similar to doxylamine, with pyridine-2-aldehyde and 4-chlorophenylmagnesium bromide.

Medrylamine 2-[(4-Methoxyphenyl)-phenyl-methoxy]-N,N-dimethyl-ethanamine
Synthesis

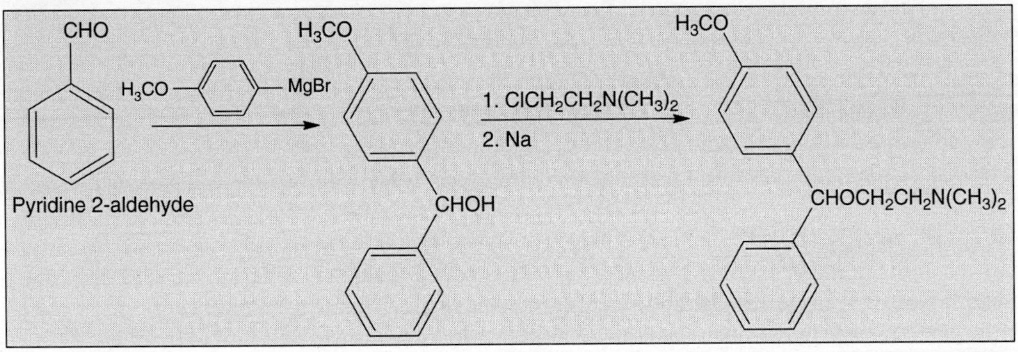

Synthesis is similar to doxylamine, with benzaldehyde and 4-methoxyphenylmagnesium bromide.
Clemastine 2-[2-[1-(4-Chloro phenyl)-1 phenyl ethoxy]ethyl]-1-methyl pyrrolidine.
Synthesis

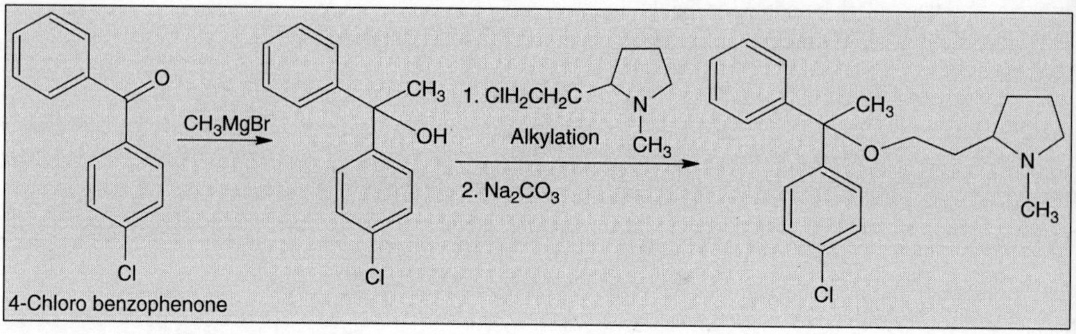

Clemastine is synthesised by reacting 1-(4-chlorophenyl)-1-phenylethanol with 2-(2-chlorethyl)-1-methylpyrrolidine using sodium amide as a base. The starting 1-(4-chlorophenyl)-1-phenylethanol is synthesised by reacting 4-chlorobenzophenone with methylmagnesium bromide.

14.2.7 Ethylenediamine Derivatives

Pyrilamine 2-[[2-(Dimethyl amino)ethyl](p-methoxybenzyl)amino]pyridine
Synthesis

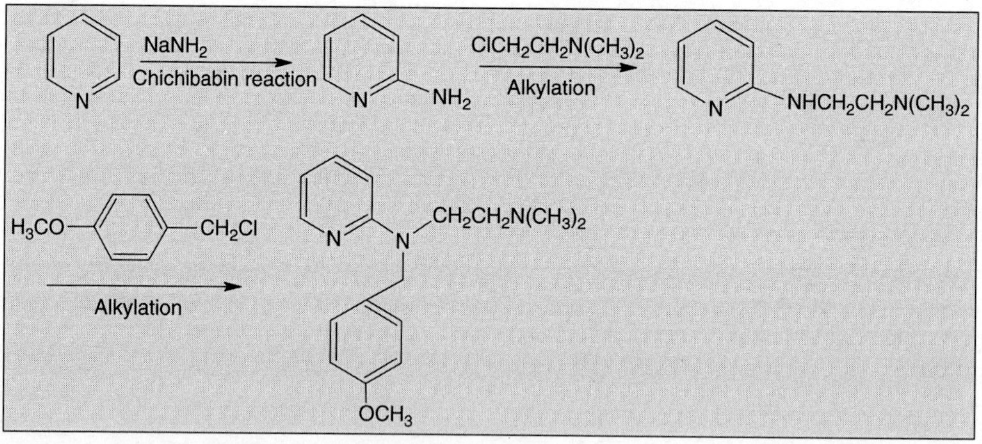

Pyrilamine is synthesised by bis-alkylating 2-aminopyridine with 2-dimethylaminoethylchloride in the presence of sodium amide, followed by 4-methoxybenzyl chloride.
Tripelennamine 2-[Benzyl[2-(dimethylaminoethyl]amino]pyridine
Synthesis: Similar to pyrilamine, involving alkylation with benzyl chloride instead of 4-methoxy benzyl chloride
Methapyrilene 2-[[2-(Dimethylamino) ethyl]-2-thenylamino]pyridine
Synthesis: Similar to pyrilamine, involving alkylation with 2-thenyl chloride
Thonzylamine 2-{[2-(Dimethylamino)ethyl] (p-methoxybenzyl amino}pyrimidine
Synthesis: Similar to pyrilamine, involving 2-aminopyrimidine instead of 2-amino pyridine

14.2.8 Propylamine Derivatives

Pheniramine: (2-[α-[2-Dimethylaminoethyl]benzyl]pyridine) and **Chlorpheniramine**
Synthesis

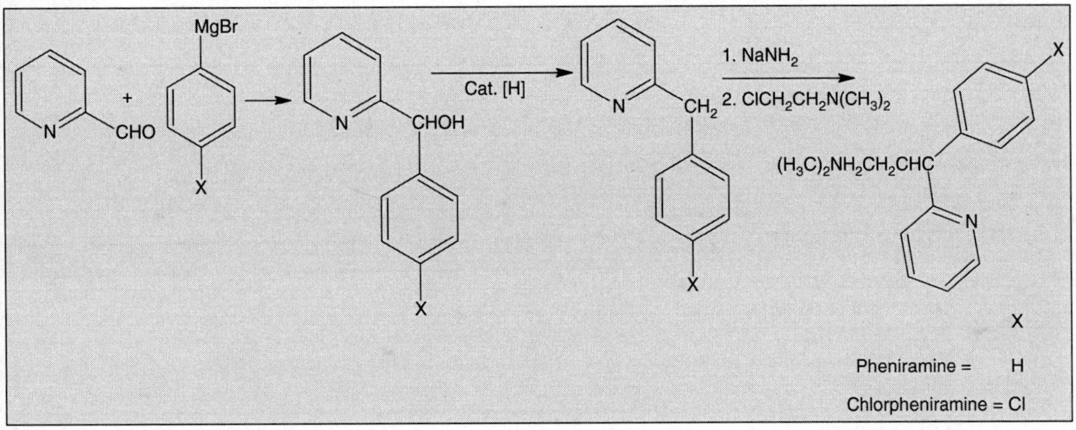

Pheniramine = X: H
Chlorpheniramine = X: Cl

Pheniramine is synthesised by reacting pyridine-2-aldehyde and phenylmagnesium bromide to give the tertiary alcohol. Reduction of this alcohol followed by alkylation with dimethylamino ethyl chloride in presence of sodamide affords pheniramine.

Chlorination of pheniramine in the *para* position of the phenyl ring gives a ten-fold increase in potency. It has a half-life of 12 to 15 hours. Introduction of bromo group as in brompheniramine increases the half-life to 25 hours.

Dexchlorpheniramine is D-isomer synthesised by separating the racemate obtained from the synthesis of chlorpheniramine using D-phenylsuccinic acid. Activity of this drug is approximately twice that of chlorpheniramine.

Triprolidine 2-[3-(1-Pyrrolidinyl)-1-*p*-tolylpropenyl]pyridine
Synthesis

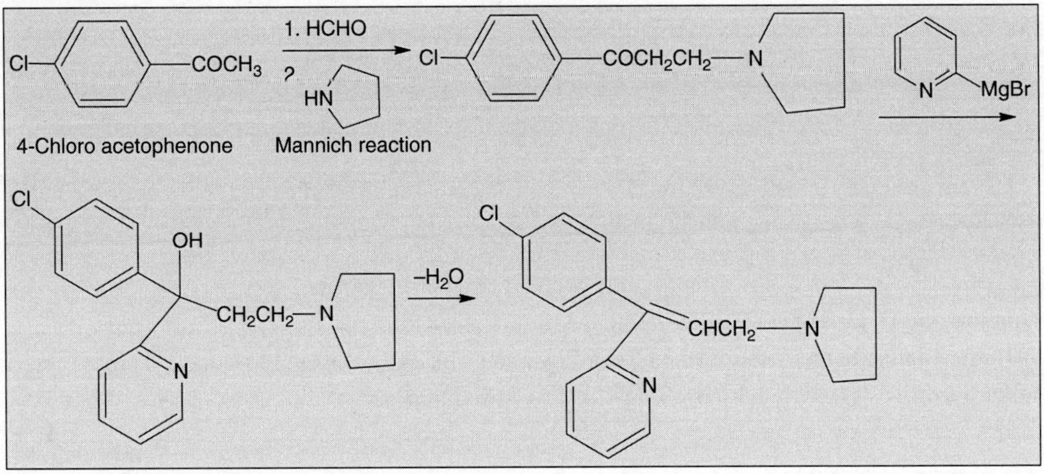

Mannich reaction of 4-chloroacetophenone, formaldehyde, and pyrrolidine affords the amino ketone. Reaction with an organometallic reagent from 2-bromopyridine gives tertiary alcohol. Dehydration leads to triprolidine.

Pyrrobutamine 1-[4-(4-Chlorophenyl)-3-phenyl-2-butenyl]pyrrolidine

Synthesis

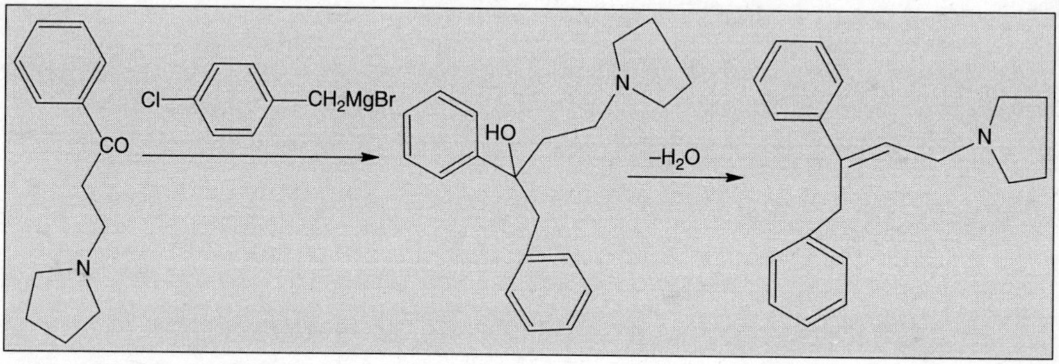

Reaction of the Mannich product from acetophenone with 4-chlorophenylmagnesium bromide gives tertiary alcohol. Dehydration leads to pyrrobutamine.

14.2.9 Phenothiazine Derivatives

Promethazine 10-[2-(Dimethylamino)propyl]phenothiazine

Synthesis

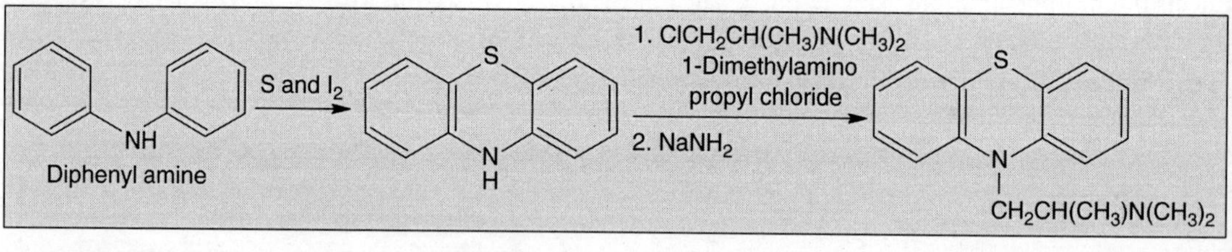

Promethazine is synthesised by alkylating phenothiazine with 1-dimethylamino-2-propylchloride in the presence of sodium amide.

Promethazine is a first-generation H_1 receptor antagonist antihistamine and anti-emetic medication. Promethazine also has strong anticholinergic and sedative/hypnotic effects. Previously it was used as an antipsychotic, although it is generally not administered for this purpose now; promethazine has only approximately one-tenth of the antipsychotic strength of chlorpromazine.

Trimeprazine 10-[3-(Dimethylamino)-2methyl propyl]phenothiazine

Synthesis: Similar to promethazine involving alkylation with 3-dimethylamino-2-methyl propyl chloride

Methadilazin 10-[(1-Methyl-3-pyrrolidinyl)methyl]phenothiazine

Synthesis

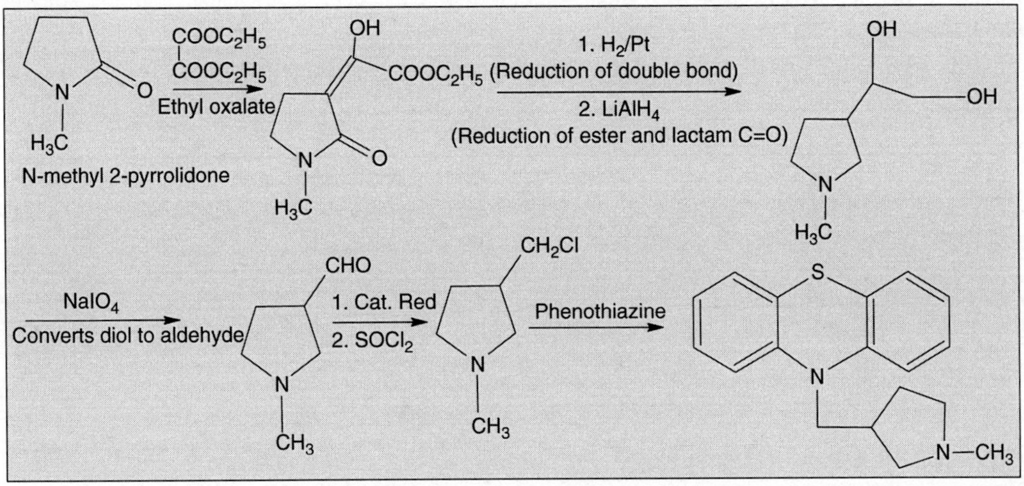

N-Methyl-2-pyrrolidinone condenses with ethyl oxalate to give an intermediate, which on reduction with hydrogen/palladium followed by lithium aluminium hydride gives diol. Oxidation of this compound gives 3-aldehyde derivative; this is further reduced and halogenated to afford 3-chloromethyl derivative. Alkylation of phenothiazine with this intermediate affords methadilazine.

14.2.10 Piperazine Derivatives

Cyclizine 1-(Diphenylmethyl)-4-methylpiperazine and **Chlorcyclizine**
Synthesis

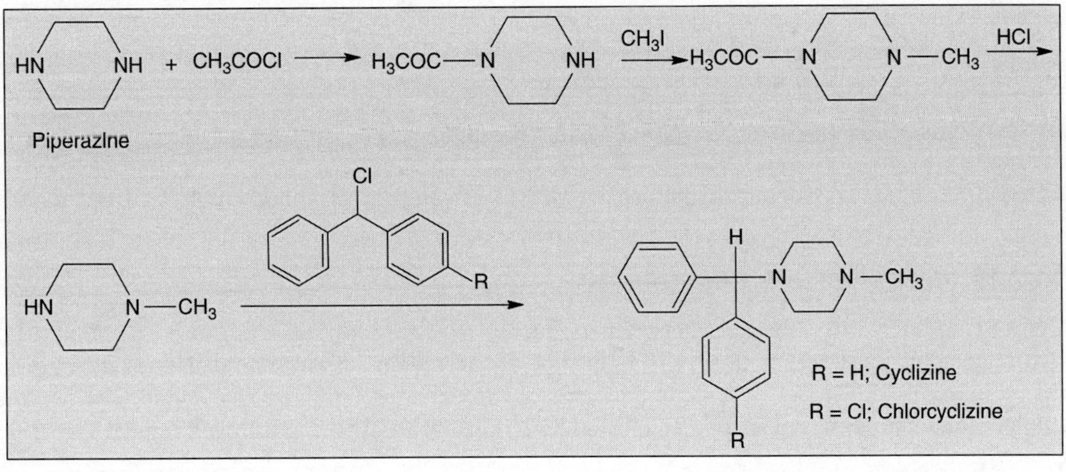

1-Methylpiperazine is prepared from piperazine by protecting one of the nitrogen with acetyl chloride, methylating the second nitrogen with methyl iodide, and then deprotecting acetyl group. Alkylating 1-methylpiperazine with benzhydrylbromide affords cyclizine.

Meclizine 1-[(4-Chlorphenyl)methyl]-4-[(3-methylphenyl)phenyl]piperazine

Synthesis

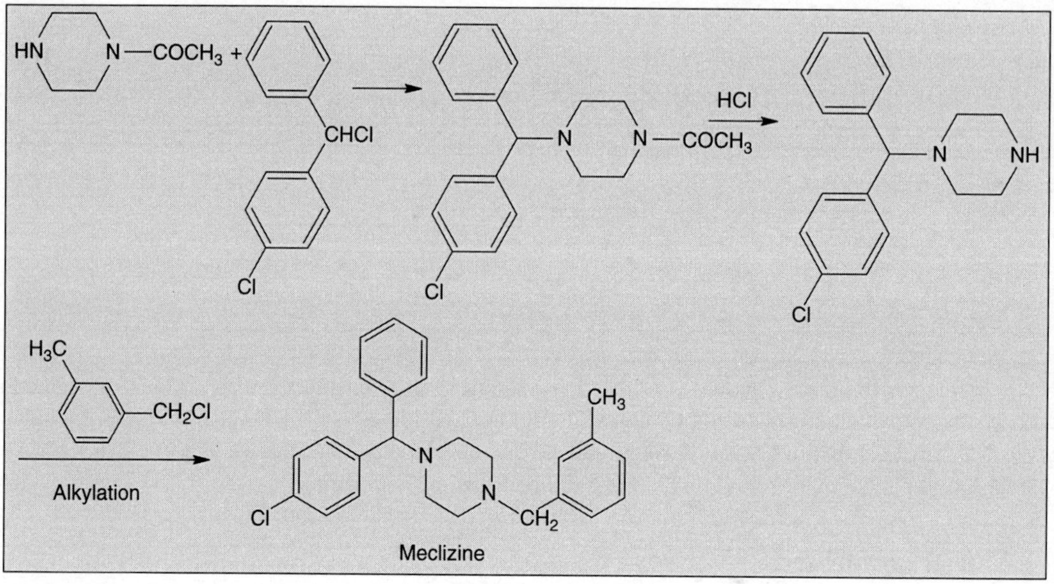

Alkylation of N-acetylpiperazine with 4-chlorobenzhydrylchloride followed by deacetylation gives mono substituted piperazine. The second nitrogen is alkylated with 3-methylbenzyl chloride to afford meclizine.

Buclizine 1-[(4-Chlorophenyl)- phenyl-methyl]-4- [(4-*tert*-butylphenyl) methyl] piperazine

Synthesis

It is synthesised by alkylation with *p-tert*-butylbenzylchloride

Hydroxyzine 2-(2-{4-[(4-Chlorophenyl)(phenyl)methyl]piperazin-1-yl}ethoxy)ethanol

Synthesis

It is synthesised by alkylation with chloro derivative of diethylene glycol ($ClCH_2CH_2OCH_2CH_2OH$)

14.2.11 Miscellaneous Compounds

Cyproheptadine 4-(5H-Dibenzocyclohepten-5-ylidene)-1-methyl piperidine

Cyproheptadine possesses both antihistamine and antiserotonin activity, and is used as an antipruritic agent. This dibenzocycloheptene may be regarded as a phenothiazine analogue in which the sulphur atom has been replaced by an isosteric vinyl group and the ring nitrogen replaced by a sp^2 carbon atom.

Synthesis

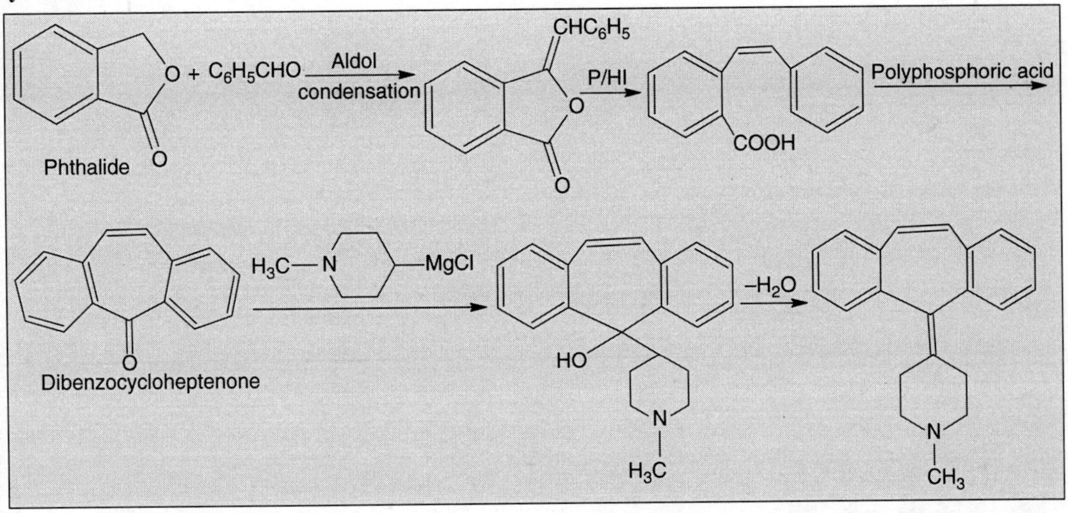

Cyproheptadine is synthesised by reacting 1-methyl-4-magnesiumchloropiperidine with 5H-dibeno[a,d]cycloheptene-5-one, which forms carbinol, the dehydration of which in an acidic medium leads to the formation of cyproheptadine. The ketone is prepared from phthalide by aldol condensation with benzaldehyde followed by reduction, and cyclization of acid.

Antazoline 2-[(N-Benzylanilino)-methyl]-2-imidazoline

Antazoline, similarly to the ethylene diamines, contains an N-benzylanilino group linked to basic nitrogen through a two-carbon chain.

Synthesis

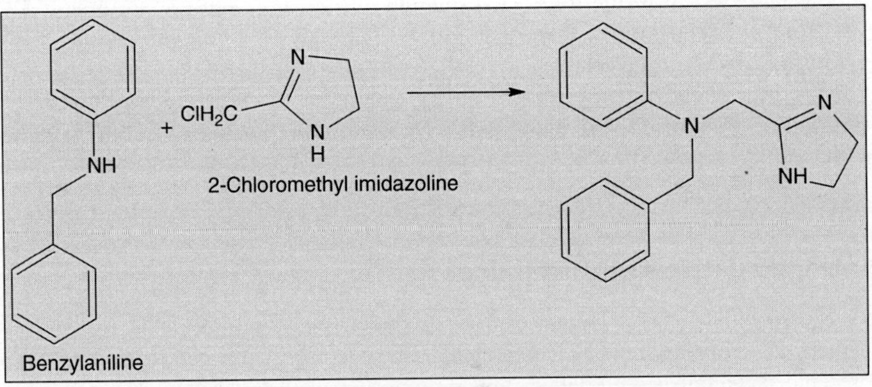

Alkylation of benzylaniline with 2-chloromethylimidazoline affords antazoline.

Antazoline is an antihistamine used to relieve nasal congestion and in eye drops, usually in combination with naphazoline, to relieve the symptoms of allergic conjunctivitis.

Azatadine 6,11-Dihydro-11-(1-methyl-4-piperidylidene)-5H-benzocycloheptapyridine

Azatadine is an aza isostere of cyproheptadine in which the 10, 11-double bonds is reduced.

Synthesis

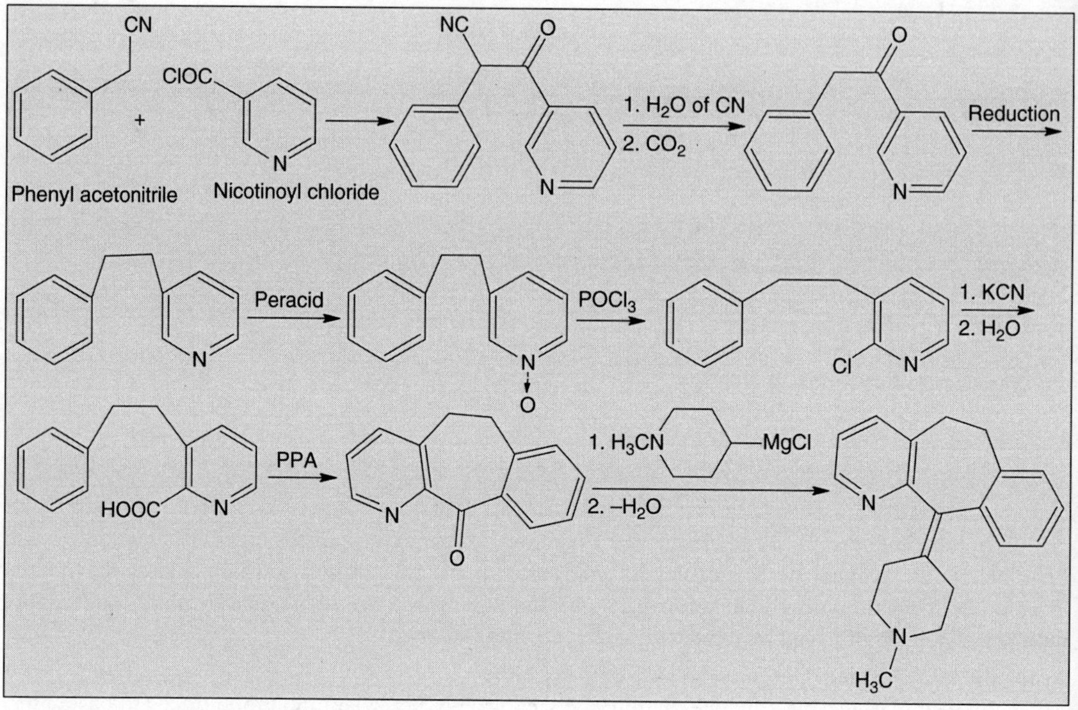

Azatadine is synthesised by reacting 1-methyl-4-magnesiumchloropiperidine with 6,11-dihydro-5H-benzo[5,6]cyclohepta[b]pyridin-11-one, which forms carbinol, the dehydration of which in an acidic medium leads to the formation of cyproheptadine. Cyclic ketone is prepared by acylation of the anion from phenylacetonitrile with nicotinoyl chloride to give cyanoketone. Hydrolysis of the nitrile followed by decarboxylation gives ketone, and reduction then leads to diaryl ethane. Functionality is introduced into pyridine ring by treatment with peracid followed by treatment with phosphorous oxy chloride, and leads to the corresponding 2-chloropyridine. Displacement of halogen with cyanide followed by hydrolysis gives carboxylic acid. Cyclization by means of poly phosphoric acid yields the ketone.

14.2.12 Non-Sedative Anti-Histamines

The major side-effect of the classic H_1 antagonists is sedation, possibly attributable to occupation of cerebral H_1 receptors. CNS effects are to be anticipated because they cross the blood-brain barrier. During the 1980s, H_1 receptor antagonists with better clinical properties were discovered. These substances have a relative low affinity for central H_1 receptors, a limited ability to penetrate the CNS, or both. Consequently, they are largely free of side-effects, especially sedation.

Second-generation H_1-antihistamines are newer drugs that are much more selective for peripheral H_1 receptors in preference to the central nervous system histaminergic and cholinergic receptors. This selectivity significantly reduces the occurrence of adverse drug reactions compared with first-generation agents, while still providing effective relief of allergic conditions.

Third-generation H_1-antihistamines are the active enantiomer (levocetirizine) or metabolite (desloratadine and fexofenadine) derivatives of second-generation drugs intended to have increased efficacy with fewer adverse drug reactions. Indeed, fexofenadine is associated with a decreased risk of cardiac arrhythmia compared to terfenadine. However, there is little evidence for any advantage of levocetirizine or desloratadine, compared to cetirizine or loratadine, respectively.

Terfenadine α-[4-(1,1-Dimethylethyl)-phenyl]-4-(hydroxydiphenylmethyl)-1-piperidine butanol

Synthesis

Terfenadine is synthesised by reacting piperidine-4-magnesiumchloride with benzophenone, giving (4-piperidyl)diphenylcarbinol. This product is alkylated by either 1-(4-*tert*-butylphenyl)-4-chlorobutanol, which forms terfenadine, or by alkylating with (4-*tert*-butylphenyl)-3-chloropropiophenone, which forms the product, the carbonyl group of which is reduced to an alcohol group, thus giving the desired terfenadine.

Fexofenadine 2-[4-[1-Hydroxy-4-[4-(hydroxy-diphenyl-methyl)-1-piperidyl]butyl]phenyl]-2- methylpropanoic acid

It is an antihistamine drug used in the treatment of hayfever and similar allergic symptoms. It was developed as a successor of and alternative to terfenadine in which one of the terminal methyl groups is converted to carboxylic acid. Fexofenadine is associated with a decreased risk of cardiac arrhythmia compared to terfenadine.

Cetrizine 2-[4-[(4-Chlorophenyl)phenylmethyl]-1-piperazinyl]ethoxy]-acetic acid

This is the principal metabolic product of hydroxyzine. The polar acid group prevents its penetration into the CNS.

Synthesis

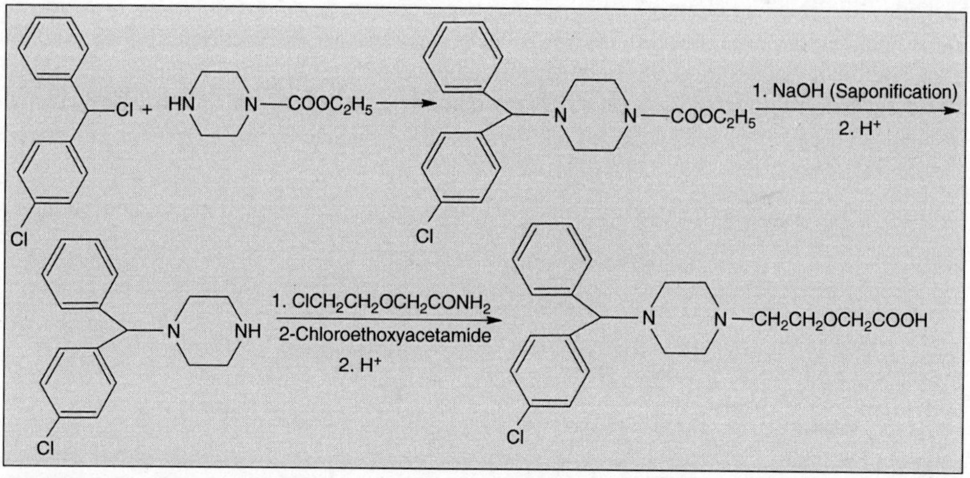

Alkylation of N-protected piperazine with 4-chlorobenzhydrylchloride followed by deprotection gives mono substituted piperazine. The second nitrogen is alkylated with 2-chloroethoxyacetamide, followed by hydrolysis of amide with acid to give acid cetrizine.

Levocetirizine 2-[2-[4-[(R)-(4-Chlorophenyl)-phenyl-methyl] piperazin-1-yl]ethoxy]acetic acid
It is a third-generation non-sedative antihistamine, developed from the second-generation antihistamine cetirizine. Chemically, levocetirizine is the active enantiomer of cetirizine. It is claimed to be more effective and with fewer side-effects than the second-generation drugs.

Acrivastine 3-[6-(4-Methylphenyl)-3-(1-pyrrolidinyl)-1-propenyl-2-pyridinyl]-2-propenoic acid

Synthesis

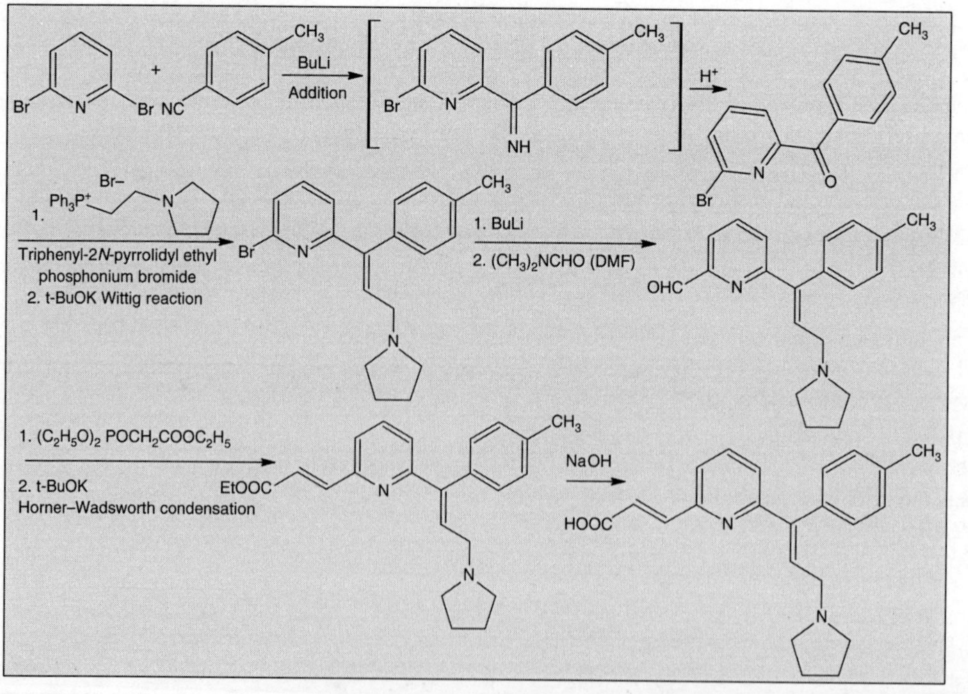

The synthesis of acrivastine starts by reaction of the mono lithio derivative from pyridine with 4-toluonitrile. Hydrolysis of the intermediate imine affords the benzophenone. Condensation of carbonyl group with the ylide from triphenyl(2-N-pyrrolidinoethyl)phosphonium chloride gives the alkene intermediate. The remaining halogen on the pyridine is converted to lithio reagent and condenses with dimethylformamide to give aldehyde. This is subjected to Wittig-type reaction by reacting with triphenylphosphono acetate to give unsaturated ester. Saponification of ester group affords acrivastine.

Astemizole 1-(4-Fluorobenzyl)-2-((1-(4-methoxyphenyl)-4-piperidyl)amino) benzimidazole

Synthesis

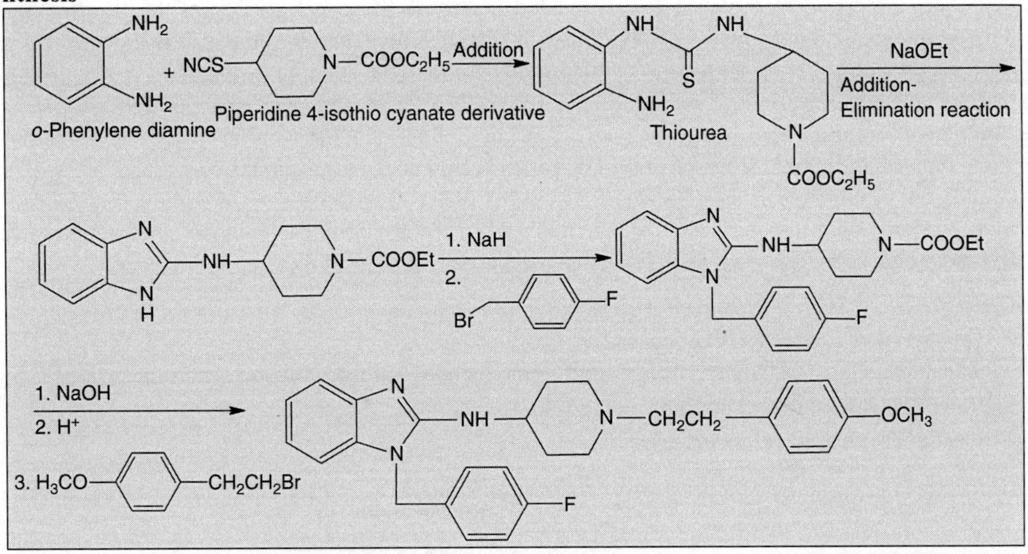

o-Phenylenediamine reacts with piperidine-4-isothiocyanate derivative to form thiourea, which in presence of base undergoes addition-elimination reaction to form N-[1-[2-(4-carethoxy)]-4-piperidinyl]benzimidazol-2-amine. The obtained aminobenzimidazole derivative is alkylated with 4-fluorobenzylbromide into 1-[(fluorophenyl)methyl]-N-[1-[2-(4-carbethoxy)]-4-piperidinyl] benzimidazol-2-amine. The carbethoxyl group of the resulting compound is hydrolysed, and decarboxylates to give piperidine, the alkylation of which with 2-(4-methoxyphenyl)ethylbromide leads to the formation of astemizole.

Loratidine 4-(8-Chloro-5,6-dihydro-11H-benzocycloheptapyridine-11-ylidene)-1-piperidine carboxylic acid ethyl ester

Synthesis

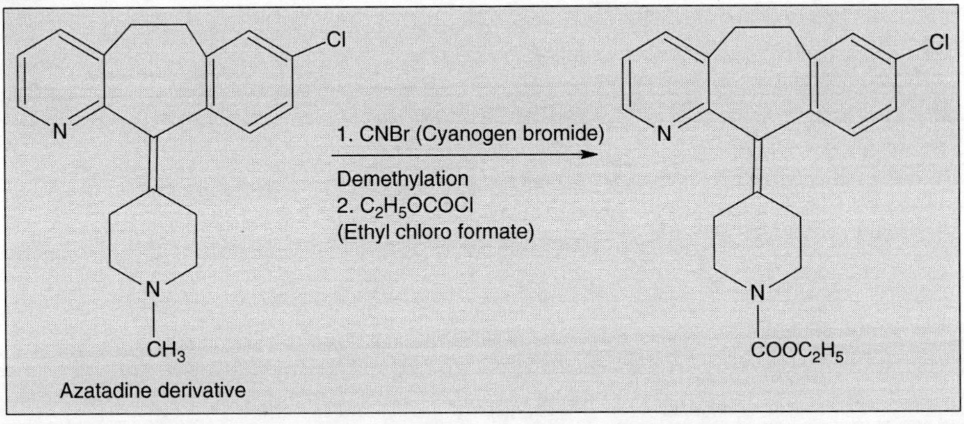

Azatadine derivative is demethylated at N-position by reaction with cyanogens bromide; this, followed by reaction with ethylchloro formate, affords loratidine.

Desloratidine 8-Chloro-6,11-dihydro-11-(4-piperdinylidene)- 5H-benzo[5,6]cyclohepta[1,2-b]pyridine
It is the major metabolite of loratadine. It is desethylester derivative of loratidine.

14.3 H_2 RECEPTOR ANTAGONIST

Gastric acid is secreted from parietal cells located mainly in the upper portion of the stomach and is stimulated by three endogenous substances: gastrin, acetylcholine, and histamine. It is thought that gastrin and acetylcholine act on mast cells to release histamine; the histamine then acts on the H_2 receptors on parietal cells to stimulate acid secretion.

Mechanism of action: The H_2 antagonists are competitive inhibitors of histamine at the parietal cell H_2 receptor. They suppress the normal secretion of acid by parietal cells and the meal-stimulated secretion of acid. They accomplish this by two mechanisms: histamine released by ECL cells in the stomach is blocked from binding on parietal cell H_2 receptors that stimulate acid secretion, and other substances that promote acid secretion (such as gastrin and acetylcholine) have a reduced effect on parietal cells when the H_2 receptors are blocked.

Uses
1. Treatment of gastric and duodenal ulcer
2. The management of hypersecretory conditions, such as Zollinger-Ellison syndrome, systemic mastocytosis, and multiple endocrine adenomas
3. Gastrooesophageal reflux diseases

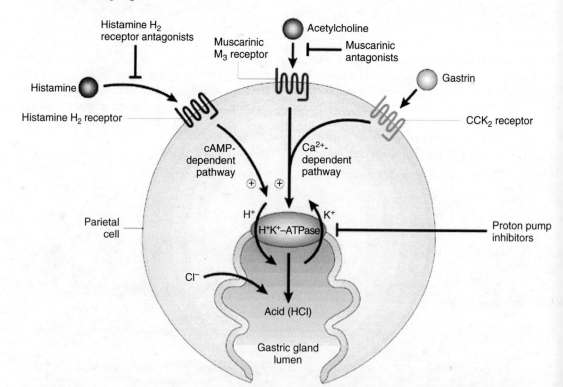

FIGURE 14.1 Mechanism of gastric and secretion.

Structural requirements and SAR: The H_2 receptor antagonists were the result of the intentional modification of the histamine structure and the deliberate search for a chemically related substance that would act as competitive inhibitor of the H_2 receptors. Most of the clinically useful H_2 antagonists can be represented by the following general formula:

Basic heterocycle group	Flexible chain or aromatic ring	Polar group system

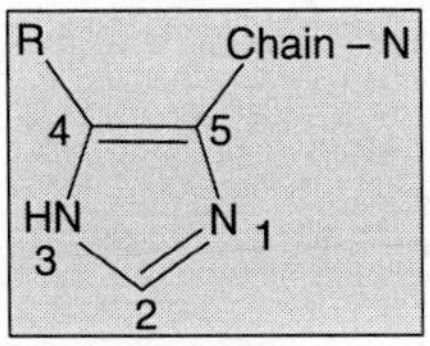

- The imidazole ring of histamine is not required for competitive antagonism of histamine at H_2 receptors. Other heterocyclic rings (furan, thiophene, thiazole, etc.) may be used.
- Separation of the ring and the nitrogen group with the equivalent of a four-carbon chain appears to be necessary for optimum antagonist activity. The isosteric thioether link is acceptable.
- The terminal nitrogen group should be polar, non-basic substituents for maximal antagonist activity. In general, antagonist activity varies inversely with the hydrophilic character of the nitrogen group (exception: ranitidine and nizatidine).

Cimetidine N"-Cyano-N-methyl-N'-[2-[[(5-methylimidazol-4-yl)methyl]thio]ethyl] guanidine

Synthesis

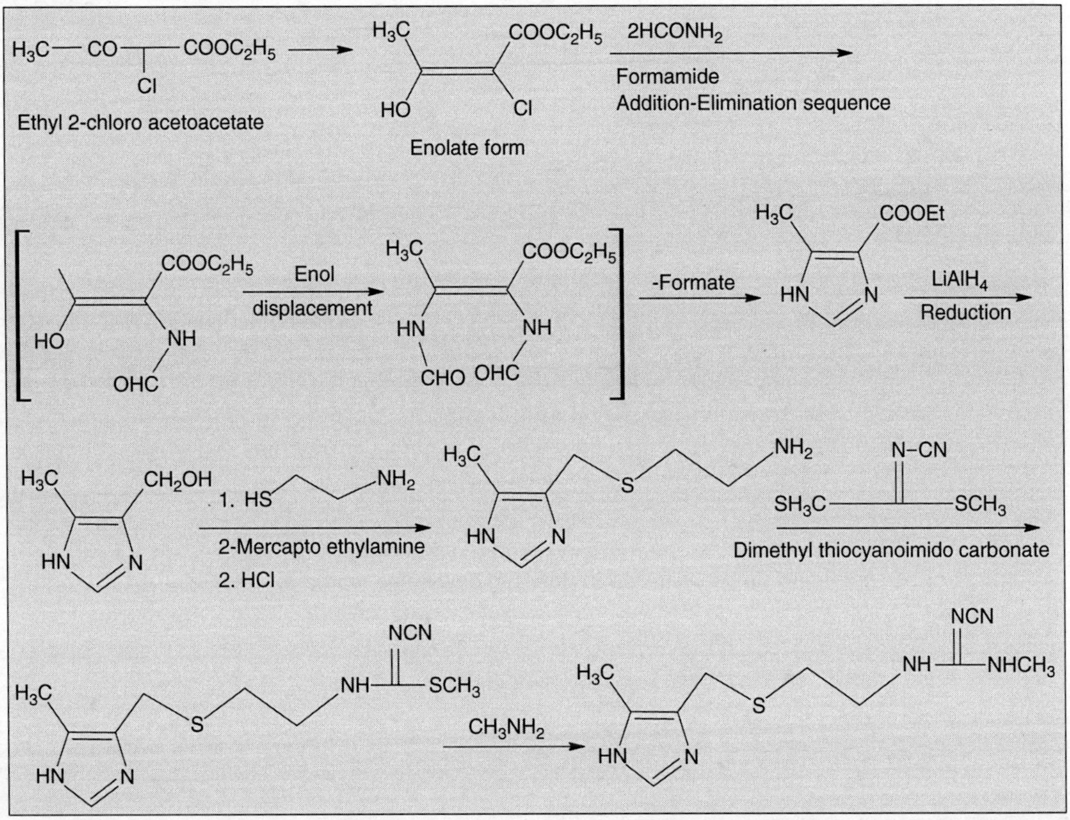

Cimetidine is synthesised by reacting 2-chloroacetoacetate with two moles of formamide to give 4-carbethoxy-5-methylimidazol. Reduction of the carbethoxy group of this with lithium aluminium hydride gives 4-hydroxymethyl-5-methylimidazol. The hydrochloride of the resulting alcohol is reacted with 2-mercaptoethylamine hydrochloride to produce 4-(2-aminomethyl)-thiomethyl-5-methylimidazol dihydrochloride. This is reacted with N-cyanimido-S,S dimethyldithiocarbonate to give a thiourea derivative, which upon reaction with methylamine turns into cimetidine.

Ranitidine   N-[2-{[[5-(Dimethylaminomethyl)-2-furanyl]methyl]thio}ethyl]-N'-methyl-2-nitro-1,1-ethenediamine

In ranitidine, the imidazole ring of cimetidine was replaced by furan in conjunction with some rearrangement of the terminal functionality.

Synthesis

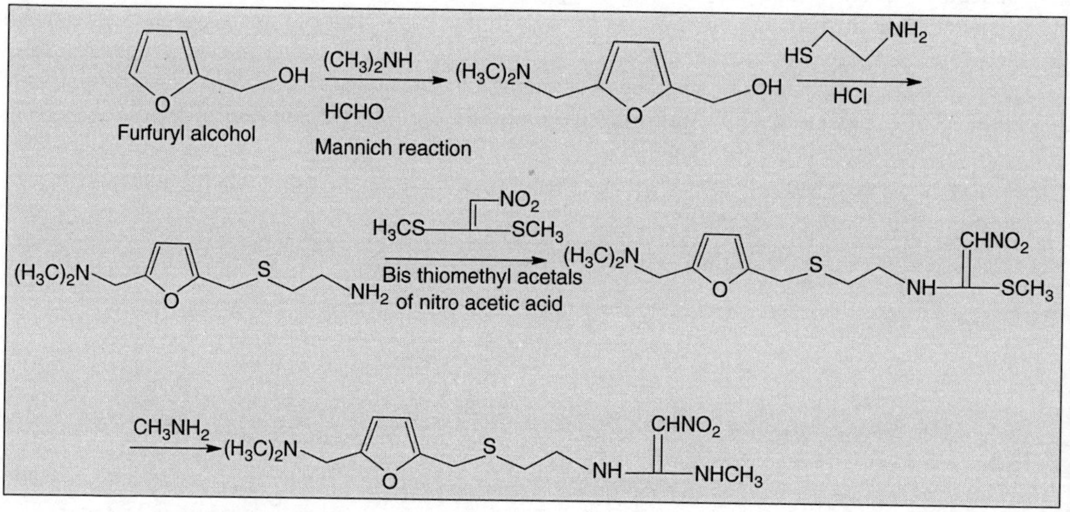

Ranitidine is synthesised from furfuryl alcohol, which undergoes aminomethylation reaction using dimethylamine and paraform aldehyde, which form 5-(dimethylaminomethyl)furfuryl alcohol. Further reaction with 2-mercaptoethylamine hydrochloride gives a product of substitution of the hydroxyl group. This treated with N-methyl-1-methylthio-2-nitroethenamine gives 5-dimethylaminomethyl-2-(2'-aminoethyl)thiomethylfurane. Reacting this with N-methyl-1-methylthio-2-nitroethenamine and methylamine gives ranitidine.

> Ranitidine was developed by Glaxo by replacing the imidazole ring of cimetidine with a furan ring with a nitrogen-containing substituent, and in doing so developed ranitidine. Ranitidine was found to have a far-improved tolerability profile (i.e., fewer adverse drug reactions), longer-lasting action, and ten times the activity of cimetidine.

Famotidine N¹–(Aminosulphonyl)-3-[{[2-[(diaminomethylene)amino]-4-thiazolyl]methyl} thio]-propanimidamide

Synthesis

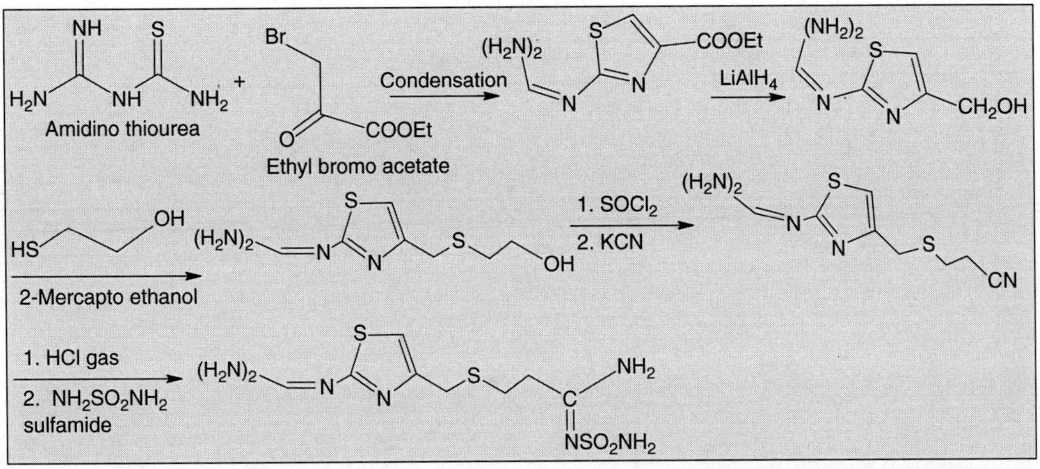

Condensation of amidinothiourea with ethyl bromo acetate gives thiazole derivative. Ethyl ester group is reduced to alcohol with lithium aluminium hydride. Further reaction with 2-mercaptoethylamine gives thioether. This is reacted with thionyl chloride and potassium cyanide to give cyanoethyl derivative, which upon reaction with sulphamide in presence of acid turns into famotidine.

Nizatidine N-[2-[{[2-[(Dimethylamino)methyl]-4-thiazolyl]-methyl}thio]-ethyl]-N-methyl-2-nitro-1,1-ethene diamine

Synthesis

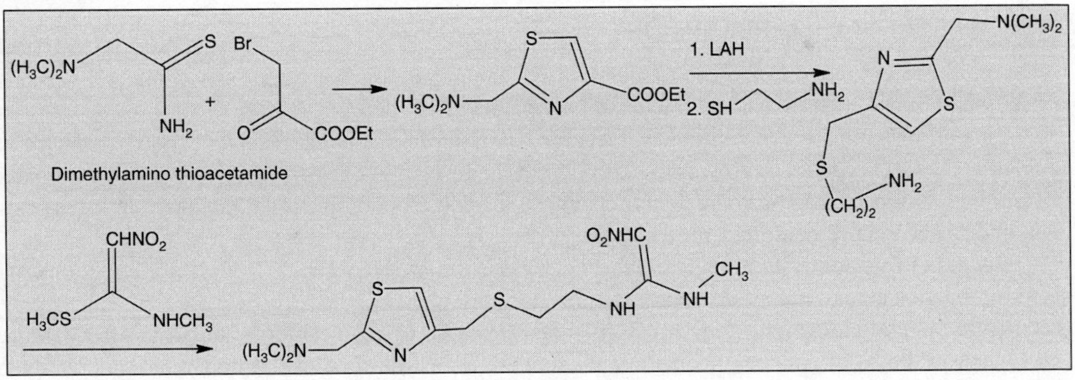

Condensation of thioamide with ethyl bromo acetate gives thiazole derivative. Reduction of the ester group by means of lithium aluminium hydride leads to corresponding alcohol, and this is then converted to thioether by reaction with 2-mercaptoethylamine. This in turn reacts with *N*-methyl-1-methythio-2-nitroethenamine to afford nizatidine.

Roxatidine 3-[3-(1-Piperidylmethyl)phenoxy]propylcarbamoylmethyl acetate
It is a simple phenolic ether histamine H_2 antagonist, representing one of the farthest departures from the prototype.

Synthesis

Reductive alkylation of 3-hydroxybenzaldehyde with piperidine affords the piperidyl derivative. Alkylation of the phenoxide from that intermediate with 1-(*N*-phthalimido)-3-bromopropane leads to the phthalimide; the free primary amine is obtained by treatment with hydrazine. Acylation of the amine with glycolyl chloride followed by acetylation with acetic anhydride affords roxatidine.

14.4 HISTAMINE H_3 RECEPTORS

H_3 receptors are expressed in the central nervous system and, to a lesser extent, the peripheral nervous system, heart, lungs, gastrointestinal tract, and endothelial cells.

Like all histamine receptors, the H_3 receptor is a G-protein coupled receptor. The H_3 receptor is coupled to the G_i G-protein, so it leads to inhibition of the formation of cAMP. Also, the β and γ subunits interact with N-type voltage-gated calcium channels, to reduce action potential mediated influx of calcium and, hence, reduce neurotransmitter release.

Because of its ability to modulate other neurotransmitters, H_3 receptor ligands are being investigated for the treatment of numerous neurological conditions, including obesity (because of the histamine/orexinergic system interaction), movement disorders (because of H_3 receptor-modulation of dopamine and GABA in the basal ganglia), schizophrenia, and ADHD (again because of dopamine modulation), and research is even underway as to whether H_3 receptor ligands could be useful in modulating wakefulness (because of effects on noradrenaline, glutamate, and histamine).

H_3-receptor agonists: Currently, no therapeutic products are selective for H_3 receptors. Some, though not totally selective, are: R-α-methylhistamine and immepip, and addition of methyl groups to the α and side chain of histamine can result in potent H_3 receptor agonists.

H_3-receptor antagonists: These include ABT-239, GT-2331, thioperamide, and clobenpropit.

14.5 HISTAMINE H₄ RECEPTORS

The histamine H_4 receptor is, like the other three histamine receptors, a member of the G protein-coupled receptor superfamily. H_4 is highly expressed in bone marrow and white blood cells, and regulates zymosan-induced neutrophil release from bone marrow and subsequent infiltration in the pleurisy model along with L-selectin. It is also expressed in the colon, liver, lung, small intestine, spleen, testes, thymus, tonsils, and trachea.

It has been shown to mediate mast cell chemotaxis. It seems to do this by the mechanism of G_1-coupled decrease in cAMP.

By inhibiting the H_4 receptor, asthma and allergy may be treated.

H_4 receptor antagonists: The following substances have been found to bind with high affinity to this receptor: amitriptyline, chlorpromazine, doxepin, promethazine, and cinnarizine.

14.6 PROTON PUMP INHIBITORS

Proton pump inhibitors represent a newer class of drugs that is likely to be useful for control of gastric acidity and the treatment of peptic ulcer. Proton pump inhibitors are used to heal stomach and duodenal ulcers. This includes stomach ulcers caused by taking non-steroidal anti-inflammatory drugs. They are also used to relieve symptoms of oesophagitis (inflammation of the oesophagus) and severe gastro-oesophageal reflux, a condition where acid leaks up from the stomach.

Mechanism of action: The ultimate mediator of acid secretion is the H^+- K^+ ATPase ('proton pump') of the apical membrane of the parietal cell. Since the pump is unique to parietal cells, newer substituted benzimidazoles as specific inhibitors have been developed. These agents are useful in patients with peptic ulcer.

All the proton pump inhibitors contain a sulphinyl group in a bridge between substituted benzimidazole and pyridine rings. At neutral pH these agents are chemically stable, lipid-soluble, weak bases that are devoid of inhibitory activity. These neutral weak bases reach parietal cells from the blood and diffuse into the secretory canaliculi, where the drugs become protonated and thereby trapped. The protonated agent rearranges to form a sulphenic acid and a sulphenamide. The sulphenamide interacts covalently with sulphhydryl groups of cysteine at critical sites in the extra-cellular domain of the H^+- K^+ ATPase, and inhibits irreversibly and thereby blocks gastric acid secretion.

The activation of omeprazole is illustrated thus:

Chapter 14: Antihistamines and Anti-Ulcer Agents

Omeprazole 5-Methoxy-2-(((4-methoxy-3,5-dimethyl-2-pyridinyl)methyl)sulphinyl)-1H-benzimidazole
Synthesis

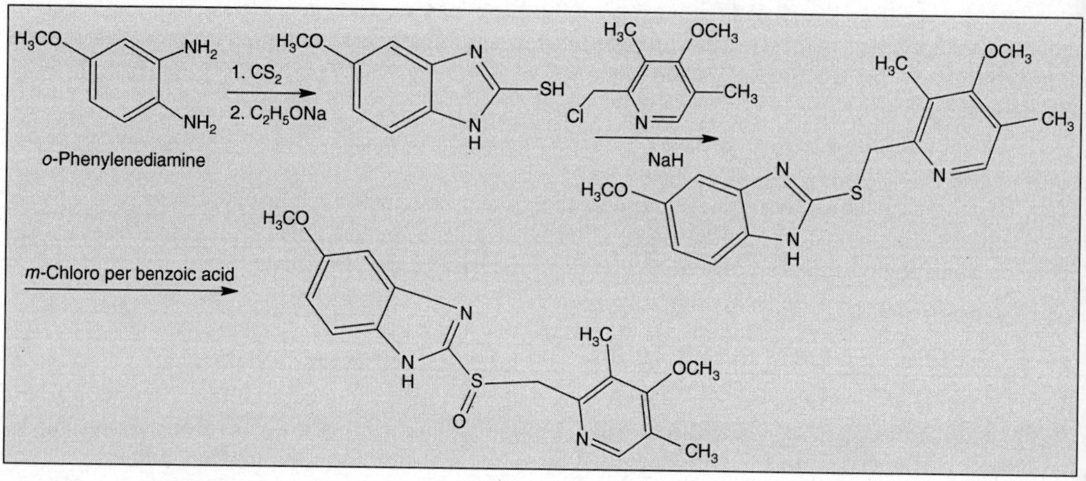

Reaction of 4-methoxy o-phenylenediamine with carbon disulphide and sodium ethoxide gives the 2-mercapto benzimidazole derivative. Reaction of 2-chloromethyl-3,5-dimethyl-4-methoxy pyridine with thiol leads to the formation of thioether. Controlled oxidation of sulphur with peracid leads to the sulphinyl derivative omeprazole.

> Omeprazole is a racemate. It contains a tricoordinated sulphur atom in a pyramidal structure and, therefore, can exist in equal amounts of both the S and R enantiomers. In the acidic conditions of the stomach, both are converted to achiral products.

Esomeprazole is the *S*-enantiomer of omeprazole.

Pantoprazole 5-Difluoromethoxy-2-[[(3,4-dimethoxy-2-pyridinyl)methyl]sulphinyl]-1H- benzimidazole
Synthesis

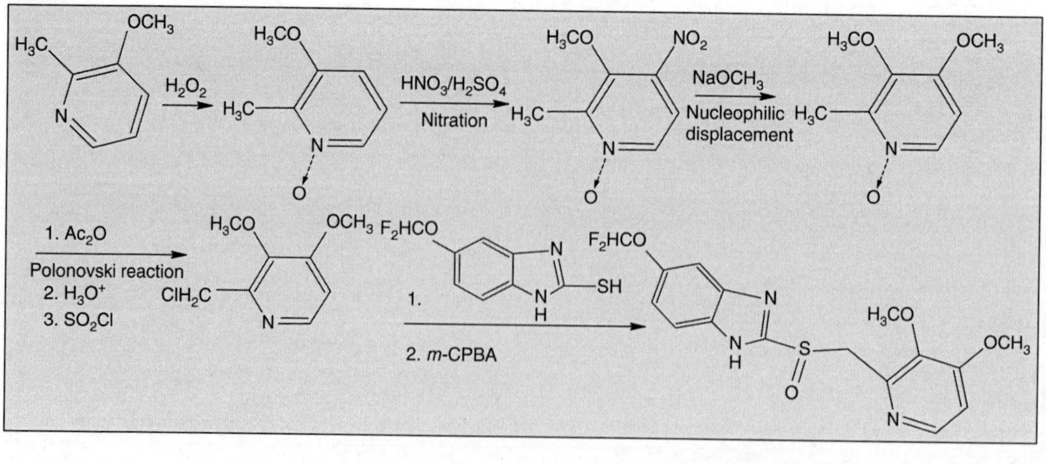

Synthesis is similar to omeprazole. The pyridine derivative is prepared by aromatic nitration of pyridine N-oxide and leads to 4-nitro derivative. The presence of N-oxide at the 4th position activates the nitro group toward nucleophilic displacement. A transformation is achieved by reaction with sodium methoxide. Polonovski rearrangement and the shift of oxygen from the ring nitrogen to the methyl group take place, and is converted to the chloride.

Lanzoprazole 2-[{[3-Methyl-4-(2,2,2-trifluoromethoxy)-2-pyridinyl]methyl} sulphinyl]-1H-benzimidazole

Synthesis

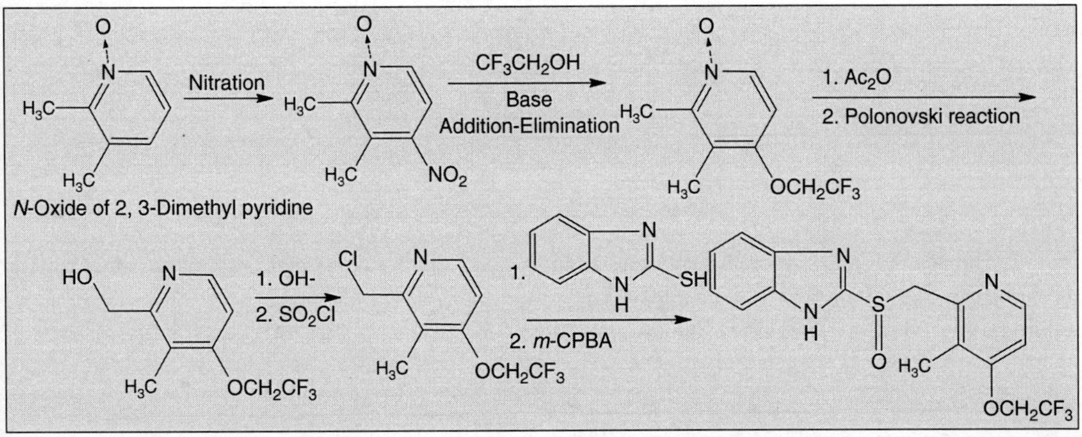

Synthesis is similar to pantoprazole.

Rabeprazole 2-[(4-(3-Methoxypropoxy)- 3-methyl-pyridin-2-yl) methylsulphinyl]- 1H-benzoimidazole

Synthesis

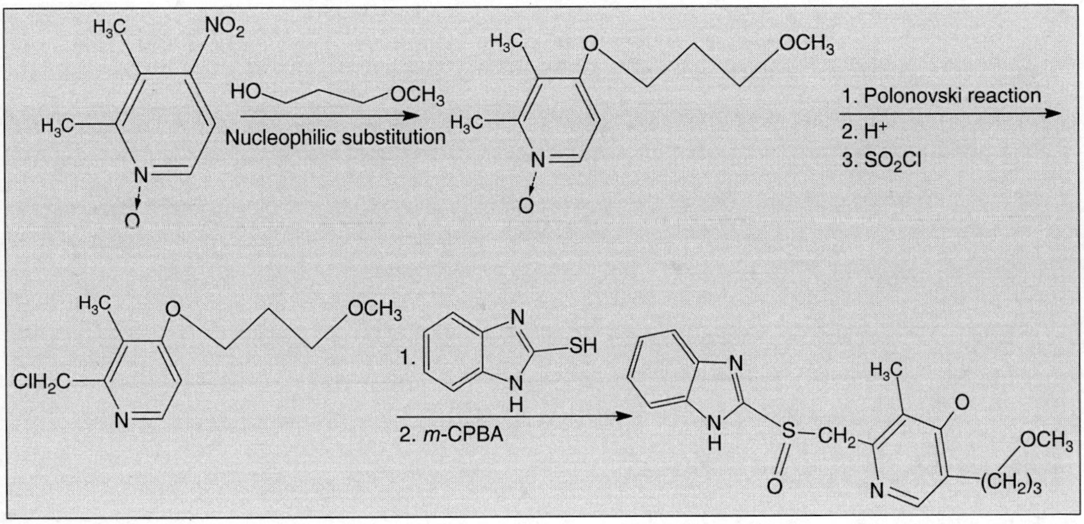

Synthesis is similar to pantoprazole.

14.7 NEWER DRUGS

Azelastine 4-[(4-Chlorophenyl)methyl]-2- (1-methylazepan-4-yl)-phthalazin-1-one

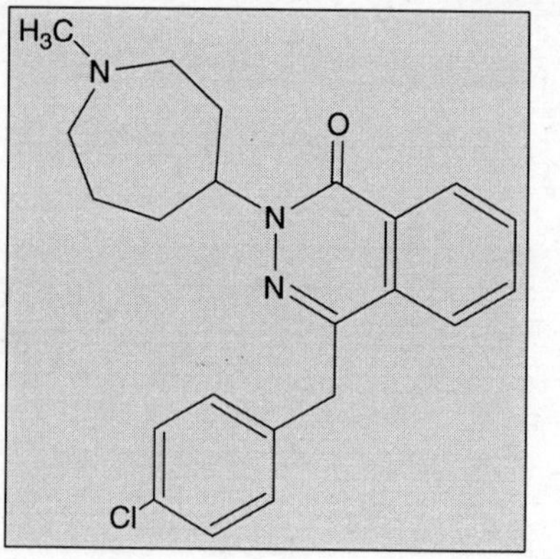

It is topical second-generation antihistamine and mast cell stabilizer available as a nasal spray for hay fever and as eye drops for allergic conjunctivitis.

Olopatadine {(11Z)-11-[3-(Dimethylamino)propylidene]-6,11-dihydrodibenzo[b,e]oxepin-2-yl}acetic acid

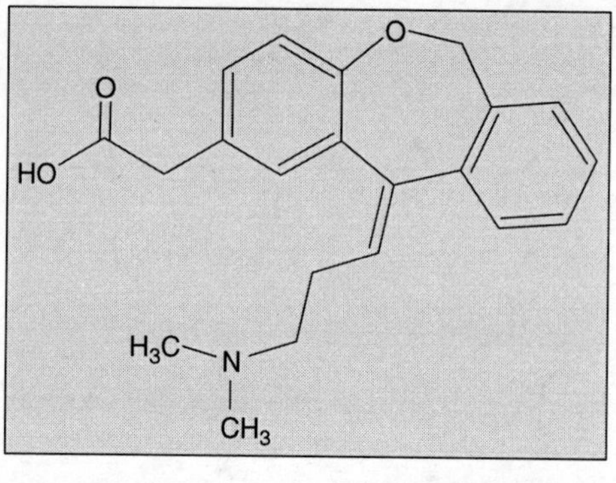

It is an antihistamine and mast cell stabilizer. It is used to treat itching associated with allergic conjunctivitis (eye allergies).

FURTHER READINGS

1. Estelle, F., R. Simons, and Marcel Dekker, 2002. *Histamine and H1-Antihistamines in Allergic Disease*.
2. Silverman, Richard B., 2004. *The Organic Chemistry of Drug Design and Drug Action*, Elsevier.

MULTIPLE-CHOICE QUESTIONS

1. Famotidine
   a. Inactivates pepsin
   b. Competitively antagonises H_2 receptor
   c. Adsorbs bile salts
   d. Neutralizes gastric acid

2. Clemastine belongs to
   a. Ethylenediamine derivatives
   b. Aminoalkyl ether derivatives
   c. Propylamine derivatives
   d. Piperazine derivatives
3. The antihistaminic drug with no heteroaryl ring system in its structure is
   a. Doxylamine
   b. Triprolidine
   c. Medrylamine
   d. Meclizine
4. The antihistamine that is also an anti-emetic is
   a. Cetrizine
   b. Pyrilamine
   c. Promethazine
   d. Loratidine
5. Non-sedative antihistamines include all except
   a. Cyclizine
   b. Cetrizine
   c. Loratidine
   d. Astemizole
6. For a molecule to exhibit antihistaminic activity, the distance between the aryl and aliphatic N should be
   a. 5-6 Å
   b. 4-5 Å
   c. 3-4 Å
   d. 6-7 Å
7. Antihistamine synthesised starting from phthalide is
   a. Promethazine
   b. Cyproheptadine
   c. Cyclizine
   d. Doxylamine
8. The drug useful for the treatment of Zollinger-Ellison syndrome is
   a. Cetrizine
   b. Promethazine
   c. Famotidine
   d. Omeprazole
9. The separation of the ring and the side-chain nitrogen should be ... carbon for optimal H2 antagonist activity.
   a. 2
   b. 3
   c. 4
   d. 5
10. Famotidine contains
    a. Furan ring
    b. Imidazole ring
    c. Thiazole ring
    d. Pyrrole ring

QUESTIONS

1. The antihistamines, such as promethazine, make you very tired when you take them, whereas newer compounds such as loratidine do not. Why not?
2. What are proton pump inhibitors? Give examples.
3. Give the structures of the following:
   a. Ranitidine
   b. Omeprazole
   c. Cetrizine
   d. Cyclizine
   e. Doxylamine
4. Write the synthetic scheme for the following:
   a. Cimetidine
   b. Loratidine
   c. Pantaprazole
   d. Pyrilamine
5. Discuss the mechanism of omeprazole with suitable structures.
6. Name the heterocyclic ring system present in cimetidine, ranitidine, famotidine, rabeprazole, roxatidine, and cyproheptadine.

SOLUTION TO MULTIPLE-CHOICE QUESTIONS

1. b;
2. b;
3. c;
4. c;
5. a;
6. a;
7. b;
8. c;
9. c;
10. c.

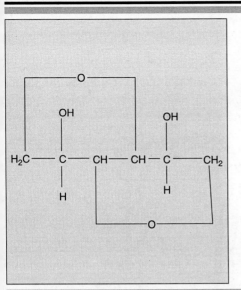

CHAPTER 15

Diuretics

LEARNING OBJECTIVES

- Define diuretics and their action in various regions of the urinary system
- Classify diuretics and explain their mode of action
- Describe the importance of structural modification with respect to diuretic potential
- Describe various synthetic modes for the preparation of thiazide diuretics
- Describe the synthesis, utility, and duration of action of various diuretics

15.1 INTRODUCTION

Diuretics are chemicals that increase secretion of excess water and salt that accumulate in tissues and urine. An excess quantity of intercellular fluid is formed in the organism as a result of an inability of the kidneys to release sodium ions fast enough to ensure that a sufficient quantity of water is excreted along with them. Therefore, efficacy of a diuretic depends first and foremost on its ability to release sodium ions from the body, since they are accompanied by an osmotically equivalent amount of water that is released from interstitial fluids.

15.2 USES

1. Diuretics are used primarily to prevent and alleviate edema and ascitis. These conditions occur in diseases of the heart, kidney and liver. Consequently, diuretics are used in the treatment of oedema associated with chronic congestive heart failure, acute pulmonary oedema, oedema of pregnancy, brain oedema, and cirrhosis associated with ascitis.

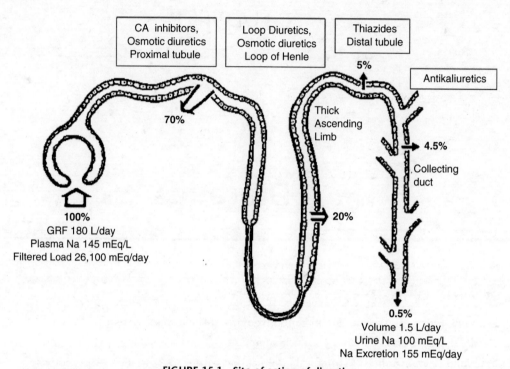

FIGURE 15.1 Site of action of diuretics.

2. They are also used in hypertension, diabetes insipidus, renal calculi, hypocalcaemia, acute renal failure, and the nephritic syndrome.
3. Some diuretics have highly specialized uses in glaucoma, hyperpotassemia, bromide intoxication, anginal syndrome, epilepsy, and migraine, and in premenstrual depression conditions in which oedema is not present.

15.3 CLASSIFICATION

1. Osmotic diuretics: Isosorbide, Mannitol, Glycerol, Urea
2. Carbonic anhydrase inhibitors: Acetazolamide, Methazolamide, Dichlorphenamide
3. Thiazide derivatives:
   (i) Chlorthiazide and analogues

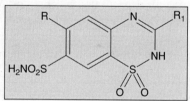

	R	R_1
Chlorthiazide	—Cl	H
Benzthiazide	—Cl	—$CH_2SCH_2C_6H_5$

`(ii) Hydrochlorothiazide and analogues

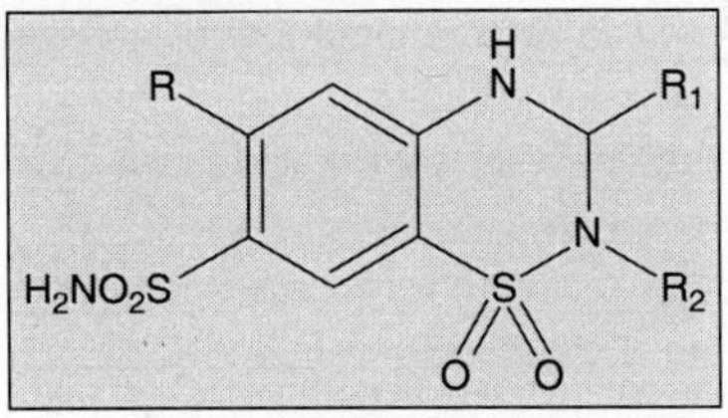

	R	R_1	R_2
Hydrochlorothiazide	—Cl	—H	—H
Hydroflumethiazide	—CF_3	—H	—H
Bendroflumethiazide	—CF_3	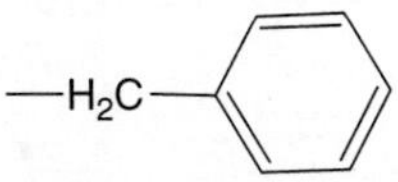 —H_2C—(phenyl)	—H
Trichlormethiazide	—Cl	—$CHCl_2$	—H
Methylclothiazide	—Cl	—CH_2Cl	—CH_3
Polythiazide	—Cl	—$CH_2SCH_2CF_3$	—CH_3
Cyclothiazide	—Cl	—H_2C—(norbornenyl)	—H
Cyclopenthiazide	—Cl	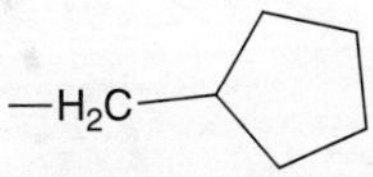 —H_2C—(cyclopentyl)	—H

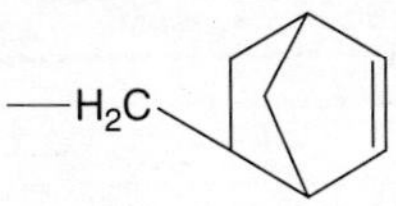

4. Potassium-sparing diuretics: Amiloride, Triamterene, Spironolactone
5. Loop diuretics: Furosemide, Bumetanide, Ethacrynic acid
6. Miscellaneous: Indapamide, Xipamide, Clopamide, Quinethazone, Metolazone, Chlorthalidone, Clorexolone

15.4 OSMOTIC DIURETICS

Mechanism of action: Osmotic diuretics are filtered at the glomerulus and are not reabsorbed by the renal tubules. Because of its osmotic action in the proximal tubules, these prevent the absorption of water and impair sodium reabsorption by lowering the concentration of sodium in the tubular fluid. In the loop of Henle, these reduce medullary hypertonicity by increasing medullary blood flow. In the collecting duct, these reduce sodium and water reabsorption because of papillary washout and high flow rate. The primary sites of action for osmotic diuretics are the Loop of Henle and the proximal tubule where the membrane is most permeable to water.

Uses

1. Osmotic diuretics are often used in the management of cerebral oedema
2. Diagnostic agent for kidney function (mannitol).
3. To increase urine volume in some patients with acute renal failure caused by ischemia, nephrotoxins, haemoglobinuria, and myoglobinuria
4. For the reduction of intra-ocular pressure (glycerine, isosorbide)
5. Reduction of cerebrospinal fluid pressure and volume. Also used in the reduction of intracranial pressure before and after neurosurgery and in neurological conditions

Preparation

Glycerin

Glycerin is an orally active diuretic and obtained from the production of soaps and fatty acids through hydrolysis or by hydration of propylene.

Mannitol

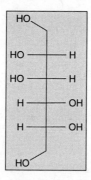

Mannitol is also orally active and is produced by catalytic or electrolytic reduction of certain monosaccharides such as mannose and glucose.

Isosorbide

Isosorbide is prepared by acid dehydration of sorbitol.

15.5 CARBONIC ANHYDRASE INHIBITORS

Carbonic anhydrase (CA) inhibitors are mild diuretics that decrease the acidity of urine, and the action is limited by metabolic acidosis.

Mechanism of action: This class of diuretics inhibits carbonic anhydrase in the membrane and cytoplasm of the epithelial cells. The primary site of action is in proximal tubules. Carbonic anhydrase is an enzyme present in the renal tubular cells, and CO_2 produced metabolically in the cells of the renal tubules is converted immediately to carbonic acid by the enzyme.

$$CO_2 + H_2O \underset{}{\overset{CA}{\rightleftharpoons}} H^+ + HCO_3^-$$

The hydrogen ion derived from carbonic acid formed in the proximal tubular cells is exchanged for sodium ions.

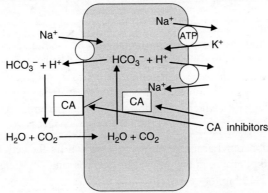

FIGURE 15.2 Mechanism of action of CA inhibitors.

In the proximal tubule, Na^+—H^+ antiport in the apical membrane of epithelial cells transports H^+ into tubular lumen in exchange for Na^+ movement into the cytoplasm. Na^+ in the cytoplasm is pumped out to the interstitium by sodium pump. H^+ in the lumen reacts with HCO_3^- to form H_2CO_3. H_2CO_3 is dehydrated to CO_2 and H_2O. This reaction is catalysed by carbonic anhydrase in the luminal membrane. Both CO_2 and H_2O can permeate into cells, and rehydrate to form H_2CO_3. The rehydration is catalysed by the cytoplasmic carbonic anhydrase. H_2CO_3 dissociates to form H^+, which is secreted into lumen, and HCO_3^-, which is transported into interstitium. Inhibition of anhydrase thus inhibits HCO_3^- reabsorption. Accumulation of HCO_3 in the tubular lumen subsequently inhibits Na^+—H^+ exchange and Na^+ reabsorption.

When CA is inhibited via adenyl cylase stimulation, the amount of hydrogen ions available for exchange with sodium is decreased; the excess of sodium ions retained in the tubule combine with bicarbonate and is excreted by the kidney with an increased volume of water and a loss of potassium. The increase in sodium concentration in the tubular fluid may be compensated partially by increased NaCl reabsorption in later segments of the tubule. Thus, the diuretic effect of the carbonic anhydrase inhibitors is mild.

Uses

1. Treatment of cystinuria, and enhanced excretion of uric acid and other organic acids
2. Useful in metabolic alkalosis condition
3. Adjuvant treatment of oedema due to congestive heart failure and drug-induced oedema
4. Used for the treatment of glaucoma
5. Also used for epilepsy and acute altitude sickness (acute mountain sickness)

Acetazolamide: N-[5-(Aminosulphonyl)-1,3,4-thiadiazol-2-yl] acetamide
Synthesis

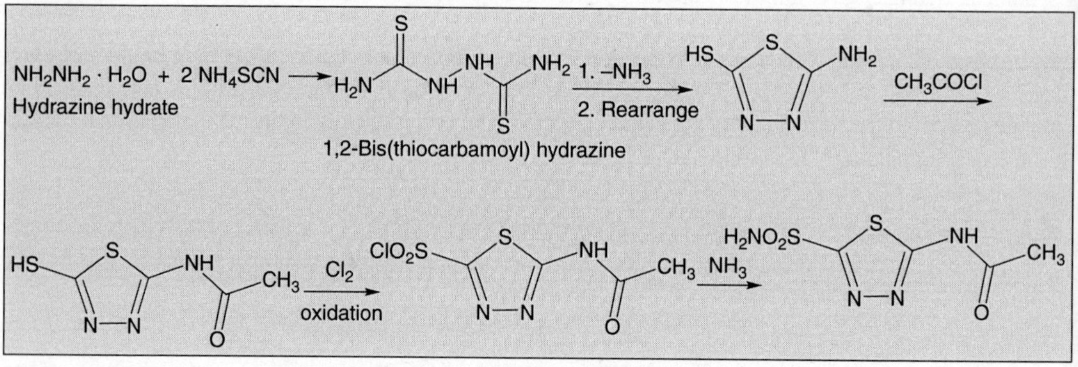

The synthesis of acetazolamide is based on the production of 2-amino-5-mercapto-1,3, 4-thiadiazole, which is synthesised by the reaction of ammonium thiocyanate and hydrazine, forming hydrazino-N,N-*bis*-(thiourea), which cyclize into thiazole. Acylation of thiazole derivative with acetic anhydride gives 2-acetyl-amino-5-mercapto-1,3,4-thiadiazol. The obtained product is chlorinated to give 2-acetylamino-5-mercapto-1,3,4-thiadiazol-5-sulphonylchloride, which is transformed into acetazolamide upon reaction with ammonia.

SAR

1. The unsubstituted sulphamoyl group is absolutely essential for the activity.
2. Substitution of a methyl group on one of acetazolamide's ring nitrogens yields methazolamide, a product that retains CA inhibitory activity.
3. The moiety to which the sulphamoyl group is attached must possess aromatic character.

Methazolamide: N-[5-(Aminosulphonyl)-3-methyl-1,3,4-thiadiazol-2-yliden] acetamide

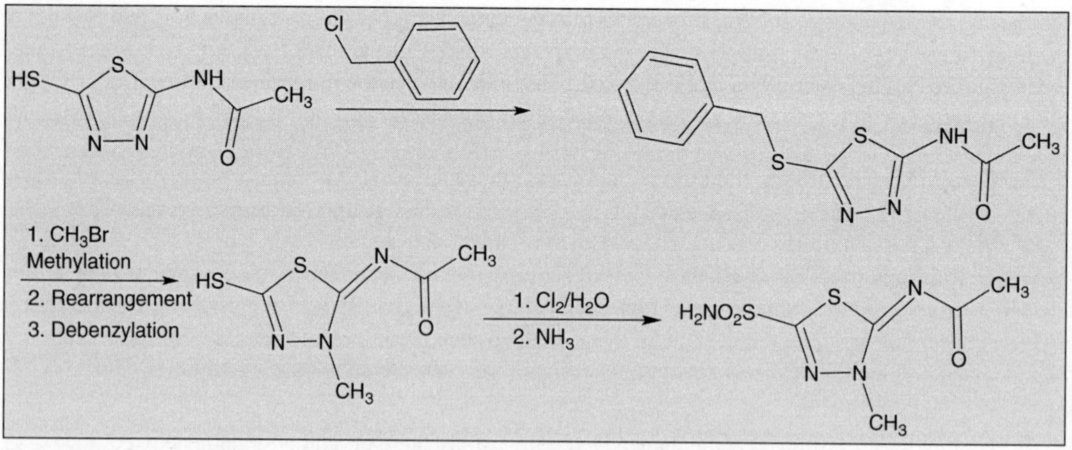

Methazolamide is made by an intermediate product of acetazolamide synthesis, 2-acetylamino-5-mercapto-1,3,4-thadiazol. This is benzylated with benzylchloride at the mercapto group, forming 2-acetyl-amino-5-benzylthio-1,3,4-thiadiazole. Further methylation of the product with methyl bromide leads to the formation of N-(4-methyl-2-benzylthio-1,3,4-thiadiazol-5-yliden)acetamide. Oxidation and simultaneous chlorination of the resulting product with chlorine in an aqueous solution of acetic acid, and reacting the resulting chlorosulphonic derivative with ammonia gives methazolamide.

Dichlorphenamide: 4,5-Dichloro-1,3-benzenedisulphonamide

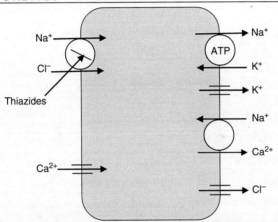

Dichlorphenamide is made from 2-chlorophenol. 2-Chlorophenol undergoes sulphochlorination by chlorosulphonic acid, forming 4-hydroxy-5-chlorobenzol-1,3-disulphonylchloride. The hydroxyl group is replaced by a chlorine atom using phosphorous pentachloride, giving 4,5-dichlorobenzol-1,3-disulphonylchoride, the reaction of which with ammonia gives the desired dichlorphenamide.

15.6 THIAZIDE DIURETICS

FIGURE 15.3 Mechanism of action of thiazides.

Mechanism of action: Thiazide diuretics act mainly to block sodium and chloride reabsorption at the first portion of the distal tubules. Thiazides inhibit a Na⁺—Cl⁻ symport in the luminal membrane of the epithelial cells in the distal convoluted tubule. Thus, thiazides inhibit NaCl reabsorption in the distal convoluted tubule, and may have a small effect on the NaCl reabsorption in the proximal tubule. Thiazides enhance Ca^{++} reabsorption in the distal convoluted tubule by inhibiting Na^+ entry and thus enhancing the activity of Na^+—Ca^{++} exchanger in the basolateral membrane of epithelial cells. They also have a mild anti-carbonic anhydrase effect. The resulting diuretics are accompanied by increased excretion of potassium, bicarbonates, chloride, and water.

The anti-hypertensive action of the thiazide is attributable to two factors:

a. depletion of sodium and subsequent reduction in plasma volume, and
b. a decrease in peripheral resistance.

Uses

1. As adjunctive therapy in oedema associated with congestive heart failure, hepatic cirrhosis, and renal dysfunction, and corticosteroid and oestrogen therapy
2. Treatment of hypertension

3. Prevents the formation and recurrence of calcium stones and nephrolithiasis in hypercalciuric patients
4. Nephrogenic diabetes insipidus

SAR

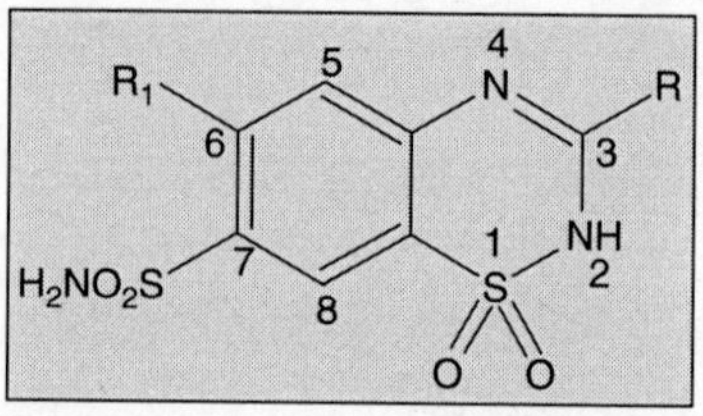

1. The position 2 can tolerate the presence of relatively small alkyl groups such as CH_3.
2. Substituents in the 3-position play a dominant role in determining the potency and duration of action of thiazide diuretics.
3. Loss of the carbon-carbon double bond between the 3- and 4-position of nucleus increases the diuretic potency approximately 3–10 fold.
4. Direct substitution of the 4-, 5-, or 8-position with an alkyl group usually results in diminished diuretic activity.
5. Substitution of the 6-position with an activating group is essential for diuretic activity. The best substituents include Cl, Br, CF_3, and NO_2 groups.
6. The sulphamoyl group in the 7-position is a prerequisite for diuretic activity.

Synthesis
Method A

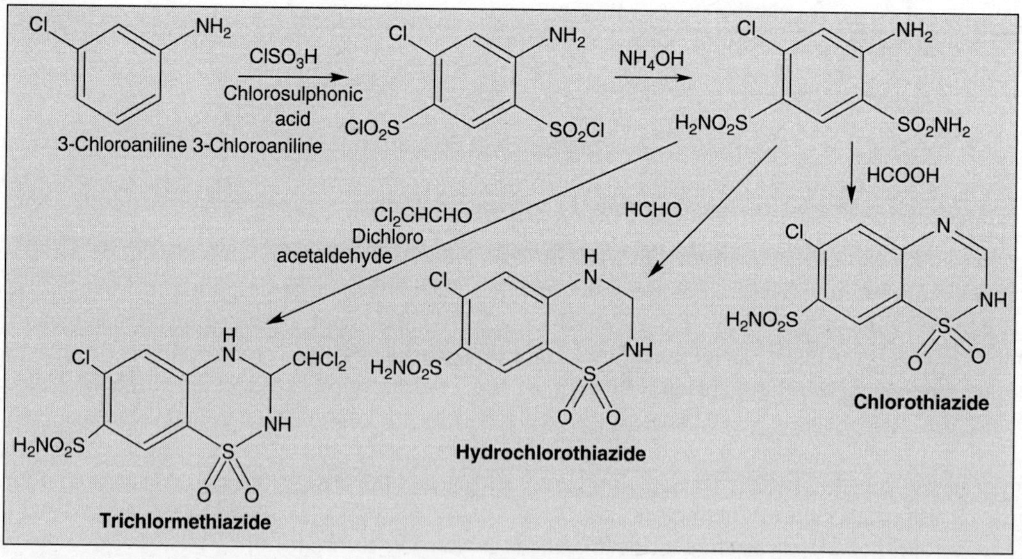

3-Chloroaniline undergoes sulphoylchlorination by chlorosulphonic acid, forming 4,6-sulphonochloride-3-chloroaniline, the reaction of which with ammonia gives 4,6-sulphonylamido-3-chloroaniline. Heating this with formic acid, formaldehyde, and dichloroacetaldehyde leads to formation of chlorothiazide (1,1-dioxide 6-chloro-2H-1,2,4-benzothiadiazin-7-sulphonamide), hydrochlorothiazide (1,1-dioxide 6-chloro-3,4-dihydro-2H-1,2,4-benzothiadiazin-7-sulphonamide), and trichlormethiazide (6-chloro-3-(dichloromethyl)-1,1-dioxo-3,4-dihydro-2H-benzo[e][1,2,4]thiadiazine-7-sulphonamide), respectively.

Method B

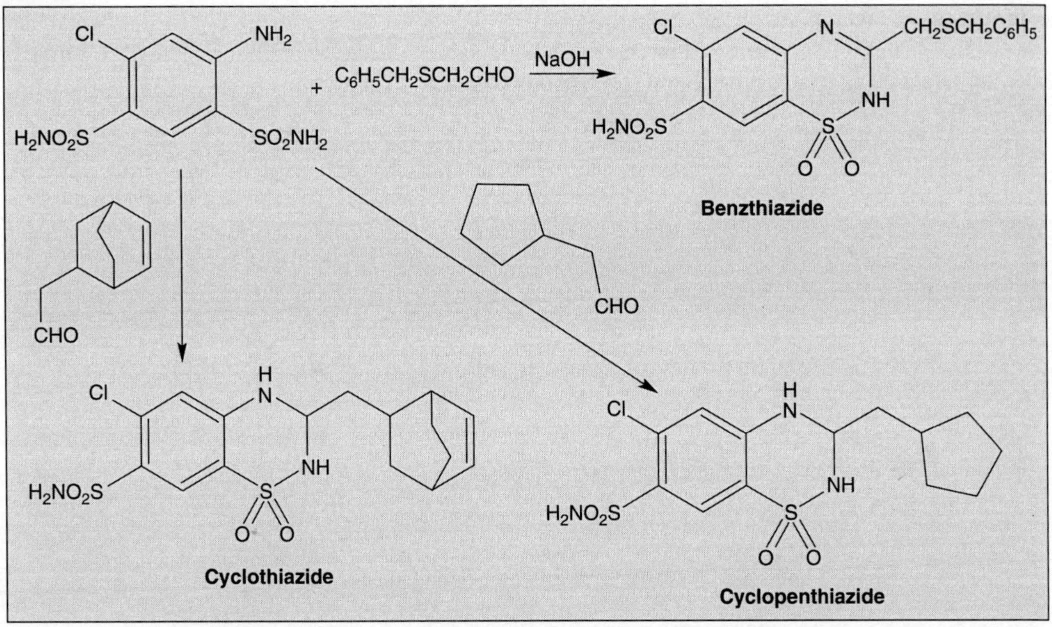

Benzthiazide

Cyclothiazide

Cyclopenthiazide

Heating 4,6-sulphonylamido-3-chloroaniline with benzyl-2-mercaptoacetaldehyde, cyclopentylacetaldehyde, and 3,4-dihydro-3-(5-norbornen)-2-acetaldehyde leads to formation of benzthiazide (1,1-dioxide 3,4-dihydro-3-benzylthiomethyl-6-chloro-2H-1,2,4-benzothiadiazin-7-sulphonamide), cyclopenthiazide, and cyclothiazide(1,1-dioxide 3,4-dihydro-3-(5-norbornen-2-yl)-6-chloro-2H-1,2,4-benzothiadiazin-7-sulphonamide), respectively.

Method C

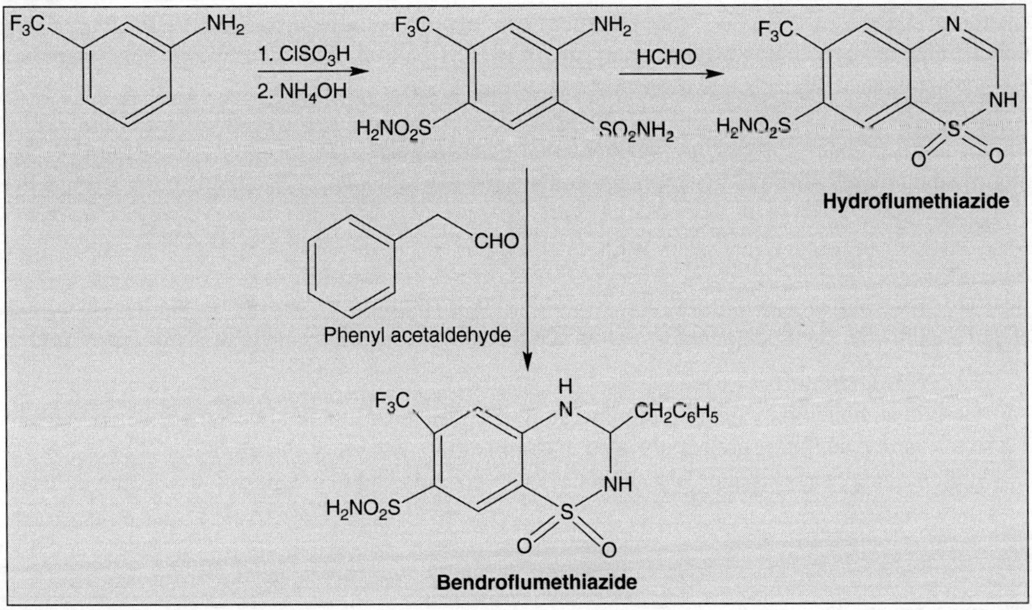

Hydroflumethiazide

Phenyl acetaldehyde

Bendroflumethiazide

Heating 4,6-sulphonylamido-3-trifluoromethylaniline with formaldehyde and phenyl acetaldehyde leads to formation of hydroflumethiazide (1,1-dioxide 3,4-dihydro-6-trifluoromethyl-2H-1,2,4-benzothiadiazin-7-sulphonamide) and bendroflumethiazide (1,1-dioxide 3-benzyl-6-(trifluoromethyl)-3,4-dihydro-2H-1,2,4-benzothiadiazin-7-sulphonamide), respectively.

Method D:

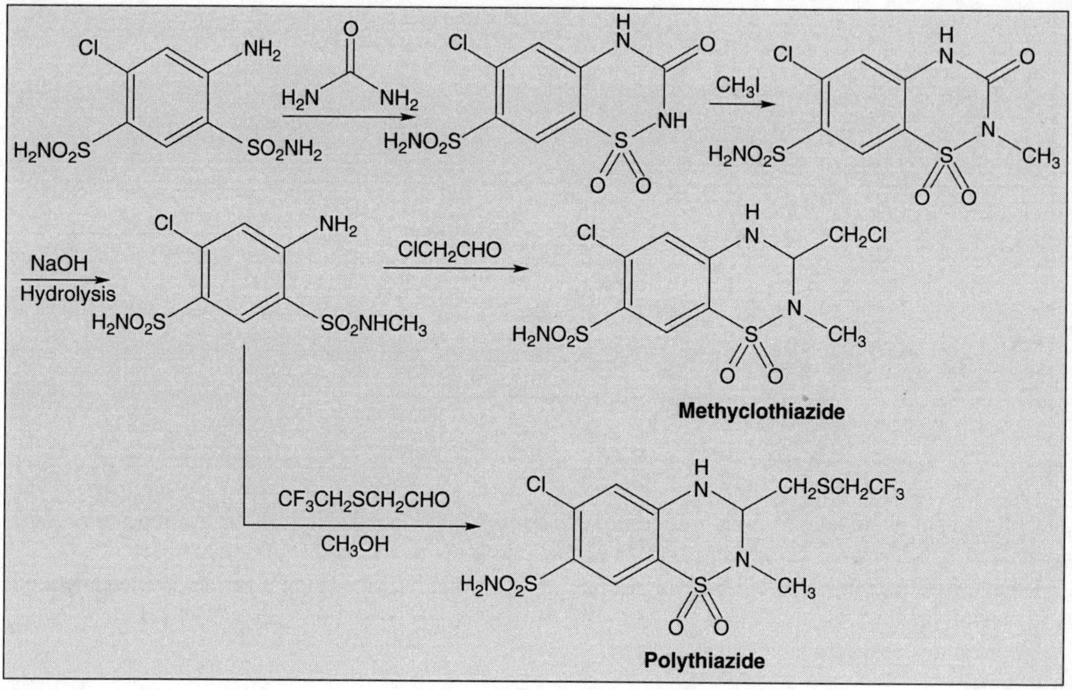

Reaction of 4,6-sulphonylamido-3-chloroaniline with urea leads to the benzothiazide nucleus. The ring sulphonamide nitrogen is methylated by means of methyl iodide; base hydrolysis of this intermediate leads to ring opened intermediate. Reaction of this intermediate with chloroacetaldehyde and 2,2,2-trifluoroethylmercapto acetaldehyde affords methyclothiazide (1,1-dioxide 3,4-dihydro-3-(chloromethyl)-6-chloro-2H-1,2,4-benzothiadiazin-7-sulphonamide) and polythiazide (1,1-dioxide 2-methyl-3-(2,2,2-trifluoroethylthiomethyl)-6-chloro-3,4-dihydro-2H-1,2,4-benzothiadiazin-7-sulphonamide), respectively.

15.7 POTASSIUM-SPARING DIURETICS

Potassium-sparing diuretics are mild diuretics that inhibit sodium reabsorption in the late distal tubule and, thus, indirectly spare potassium excretion. They tend to cause bicarbonate loss, but not chloride.

Uses

1. Used in combination with loop diuretics and thiazides in treatment of oedema and hypertension
2. Na^+ channel inhibitors are mainly used in combination with other classes of diuretics such as loop diuretics and thiazides in order to enhance Na^+ excretion and to counteract K^+ wasting induced by these diuretics

3. Pseudo-hyperaldosteronism (Liddle's syndrome)
4. Treatment of oedema associated with secondary hyperaldosteronism (such as cardiac failure, hepatic cirrhosis, and nephrotic syndrome). Spironolactone is the diuretic of choice in patients with hepatic cirrhosis
5. Amiloride is used to treat Lithium-induced nephrogenic diabetes insipidus by blocking Li^+ transport into tubular epithelial cells
6. Amiloride also inhibits Na^+ channel in airway epithelial cells, and is used to improve mucociliary clearance in patients with cystic fibrosis

15.7.1 Sodium Channel Inhibitors

Amiloride and triamterene are the only two drugs in this class.
Mechanism of action: They inhibit the sodium channel in the luminal membrane of the collecting tubule and collecting duct. This sodium channel is critical for Na^+ entry into cells down the electrochemical gradient created by sodium pump in the basolateral membrane, which pumps Na^+ into interstitium. This selective transepithelial transport of Na^+ establishes a luminal negative transepithelial potential, which in turn drives secretion of K^+ into the tubule fluid. The luminal negative potential also facilitates H^+ secretion via the proton pump in the intercalated epithelial cells in collecting tubule and collecting duct. These interfere with the sodium absorption in the late distal tubules and cortical collecting ducts, thereby promoting sodium excretion while conserving potassium.

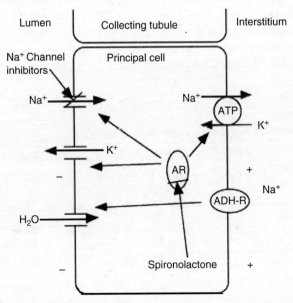

FIGURE 15.4 Mechanism of action of Sodium channel inhibitors.

Amiloride N-Amidino-3,5-diamino-6-chlorpyrazine carboxamide
Synthesis

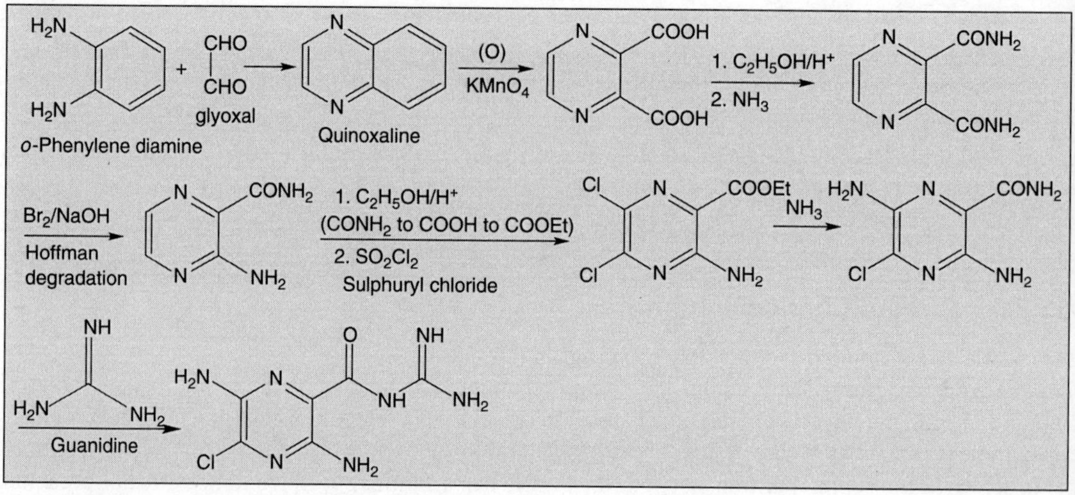

Condensation of o-phenylenediamine and glyoxal leads to quinoxaline. Treatment of quinoxaline with oxidizing agent potassium permanganate leads to selective cleavage of the benzene ring and formation of dicarboxylic acid. Heating of this with ethanol and ammonia affords bis-amide. Exposure of bis-amide to one mole of bromine in base leads to selective Hoffmann rearrangement of one of the two amide groups and forms amino-amide. Alcoholysis of this leads to conversion of the carboxyl group to its ester; reaction with sulphuryl chloride results in chlorination of the two open ring positions to the dichloro compound. Reaction of this with ammonia leads to displacement of the halogen para to the carboxyl group and forms diamine; ammonia concurrently converts the ester group to amide. Treatment of this with guanidine leads to formation of an acyl-guanidine function by an exchange reaction to afford amiloride.

Triamterene 2,4,7-Triamino-6-phenylpteridine
Synthesis

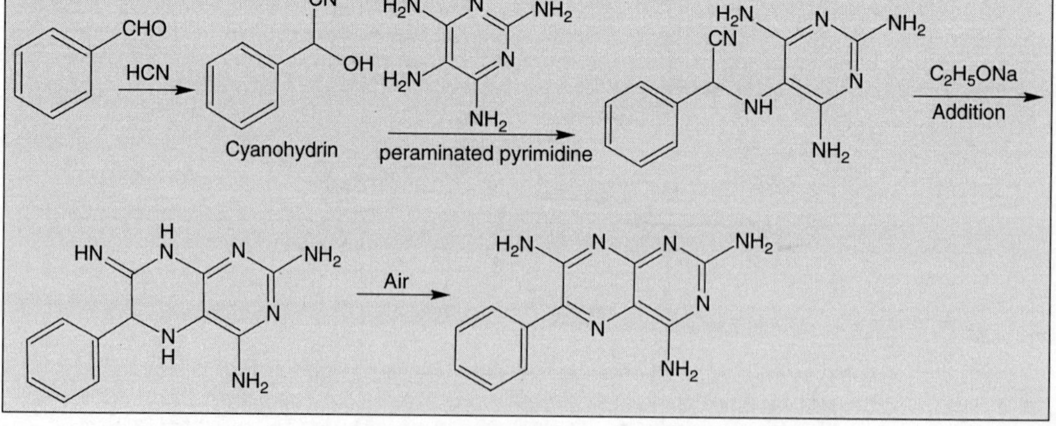

The reaction of benzaldehyde with hydrogen cyanide gives cyanohydrin, which reacts with peraminated pyrimidine to form an intermediate. Treatment of this with base leads to addition of the amine to the nitrile, to give the dihydropteridine. Exposure to air leads to dehydrogenation and formation of triamterene.

15.7.2 Aldosterone Antagonist

Spironolactone is the only available aldosterone antagonist. A metabolite of spironolactone, canrenone, is also active and has a half-life of about 16 hours.
Mechanism of action: Aldosterone, by binding to its receptor in the cytoplasm of epithelial cells in collecting tubule and duct, increases expression and function of Na^+ channel and sodium pump, and thus enhances sodium reabsorption. Spironolactone competitively inhibits the binding of aldosterone to its receptor and abolishes its biological effects. Hence, it is effective only when aldosterone is present.
Spironolactone 17-Hydroxy-7 α-mercapto-3-oxo-17 α-preg-4-ene-21-carboxylic acid-γ-lactone acetate

Synthesis

Spironolactone is synthesised from dehydroepiandrosterone, which undergoes ethynylation by reaction with lithiumacetylide to form 17α-ethynyl-3β-,17β-dihydroxy-5-androstene. Subsequent reaction of this with methylmagnesiumbromide and then with carbon dioxide gives the corresponding propenal acid. Reduction of the triple bond in this product with hydrogen using a palladium on calcium carbonate catalyst forms the corresponding acrylic acid derivative, which is treated with acid without being isolated, and leads to cyclization into an unsaturated lactone derivative. The double bond is reduced by hydrogen, in this case using a palladium on carbon catalyst. The resulting lactone undergoes oxidation in an Oppenauer reaction, giving an unsaturated keto-derivative-4-androsten-3,17-dione. Further oxidation of the product using chloranil gives dienone, which when reacted with thioacetic acid gives the desired spironolactone.

15.8 LOOP DIURETICS

Two major classes of loop diuretics are: 1) sulphonamide derivatives such as furosemide and bumetanide, and 2) non-sulphonamide loop diuretics such as ethacrynic acid.

Uses

1. Acute pulmonary oedema
2. Chronic congestive heart failure when diminution of extracellular fluid volume is desirable to reduce venous and pulmonary oedema
3. Treatment of hypertension when patients do not response satisfactorily to thiazide diuretics and anti-hypertensive drugs
4. Hypercalcaemia
5. Treatment of hyperkalemia in combination with isotonic NaCl administration
6. Used in acute renal failure to increase the urine flow and K^+ secretion
7. Treatment of toxic ingestions of bromide, fluoride, and iodide (with simultaneous saline administration)

Mechanism of action: Loop diuretics inhibit reabsorption of NaCl and KCl by inhibiting the Na^+—K^+-$2Cl^-$ symport in the luminal membrane of the thick ascending limb (TAL) of loop of Henle. As TAL is responsible for the reabsorption of 35 per cent of filtered sodium, and there are no significant downstream compensatory reabsorption mechanisms, loop diuretics are highly efficacious and are thus called high-ceiling diuretics.

As the Na^+—K^+—$2Cl^-$ symport and sodium pump together generate a positive lumen potential that drives the reabsorption of Ca^{++} and Mg^{++}, inhibitors of the Na^+—K^+—$2Cl^-$ symport also inhibit reabsorption of Ca^{++} and Mg^{++}.

By unknown mechanisms (possibly prostaglandin-mediated), loop diuretics also have direct effects on vasculature including increase in renal blood flow and increase in systemic venous capacitance.

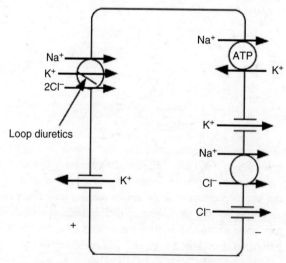

FIGURE 15.5 Mechanism of action of loop diuretics.

Bumetanide 3-(Butylamino)-4-phenoxy-5-sulphamoylbenzoic acid

Synthesis

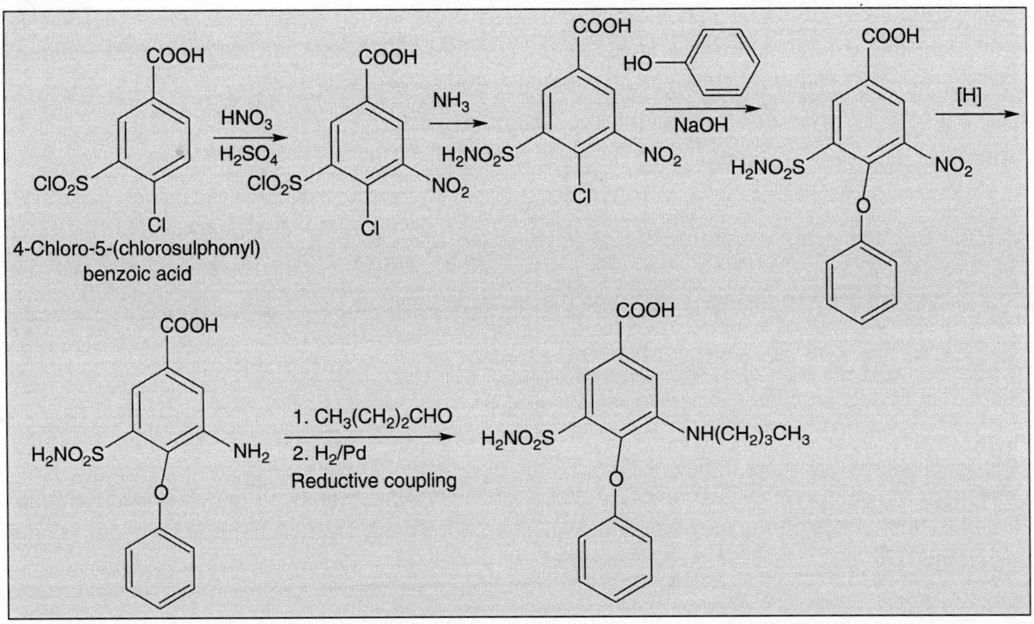

In the first stage of synthesis, 4-chloro benzoic acid undergoes sulphonylchlorination by chlorosulphonic acid, forming 4-chloro-3-chlorosulphonylbenzoic acid, which is further nitrated with nitric acid to 4-chloro-3-chlorosulphonyl-5-nitrobenzoic acid. Reacting this with ammonia gives 5-aminosulphonyl-4-chloro-3-nitrobenzoic acid, which when reacted with sodium phenolate is transformed into 5-amino-sulphonyl-3-nitro-5-phenoxybenzoic acid. Reduction of the nitro group in this product by hydrogen using a palladium on carbon catalyst gives 3-amino-5-aminosulphonyl-5-phenoxybenzoic acid. Finally, reacting this with butyrylaldehyde in the presence of hydrogen/palladium gives the desired bumetanide.

Ethacrynic acid [2,3-Dichloro-4-(2-methylenebutyryl)phenoxy] acetic acid

Synthesis

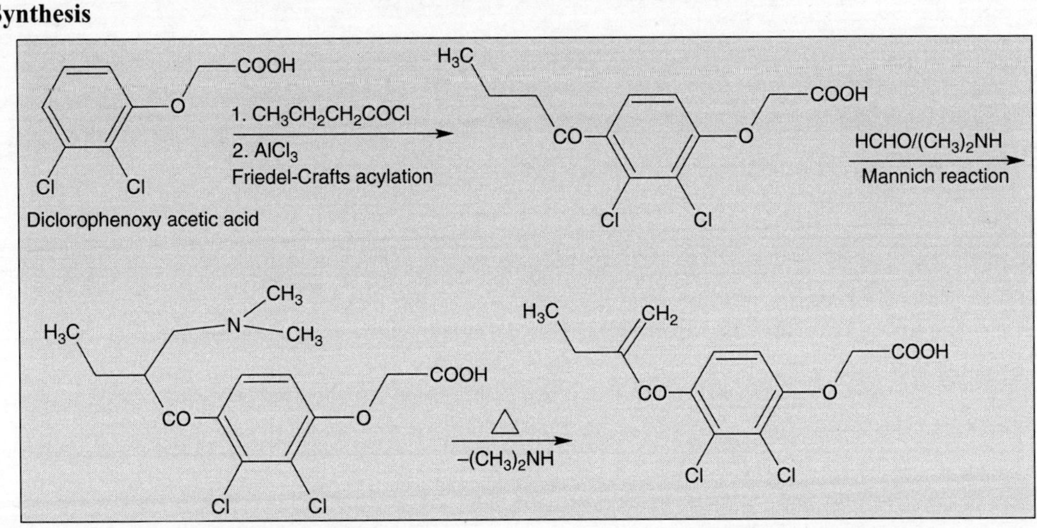

Ethacrynic acid is synthesised from 2,3-dichlorophenoxyacetic acid. This is acylated with butyroyl chloride, forming 4-butyroyl-2,3-dichlorophenoxyacetic acid, which is further aminomethylated under Mannich reaction conditions using dimethylamine and formaldehyde. The resulting product undergoes further thermal degradation, forming an unsaturated ketone, ethacrynic acid.

Furosemide 4-Chloro-N-furfuryl-5-sulphamoylanthranilic acid
Synthesis

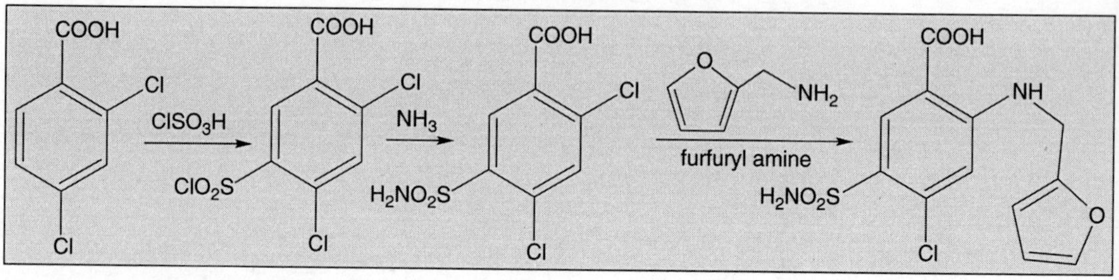

Furosemide (Lasix) is synthesised from 2,4-dichlorobenzoic acid, which is converted into 5-aminosulphonyl-4,6-dichlorobenzoic acid during subsequent reaction with chlorosulphonic acid and ammonia. Reacting this with furfurylamine gives furosemide.

> The name Lasix is derived from 'lasts six (hours)', referring to its duration of action.

15.9 MISCELLANEOUS DRUGS

Indapamide 4-Chloro-N-(2-methyl-1-indolinyl)-3-sulphamoyl benzamides
Synthesis

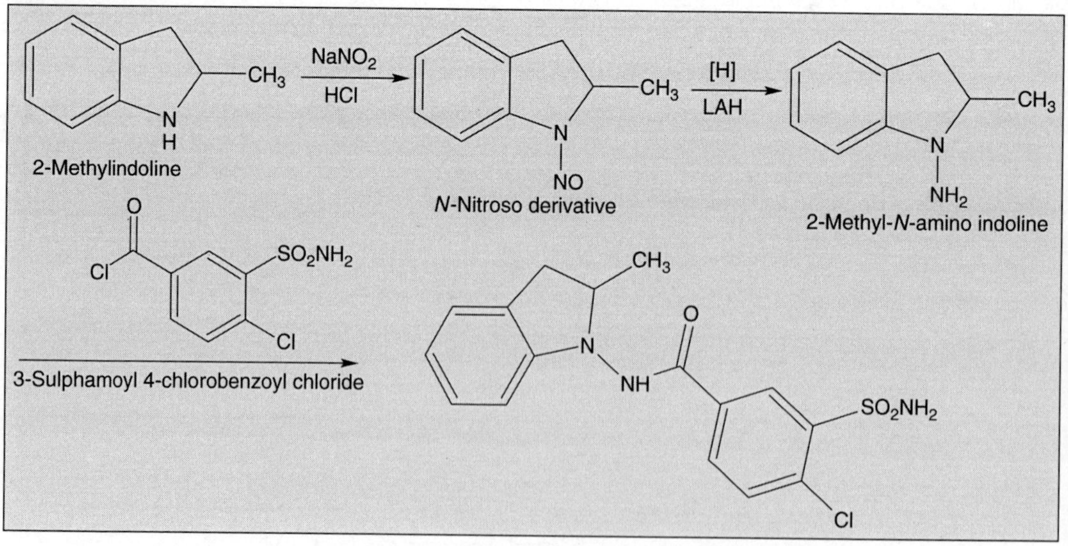

Indapamide is synthesised from 2-methylendoline, the nitrosation of which gives 2-methyl-1-nitrosoindoline. Reducing this with lithium aluminium hydride leads to formation of 1-amino-2-methylindoline. Acylating this with 3-sulphonylamino-4-chlorbenzoic acid chloride leads to indapamide.
Use: Oedema associated with congestive heart failure and hypertension.
Xipamide 4-Chloro-5-sulphamoyl-2',6'-salicyloxylidide
Diuretic activity can be retained by replacement of one of the sulphonamide groups by a carboxamide group, as in the case of xipamide.

Synthesis

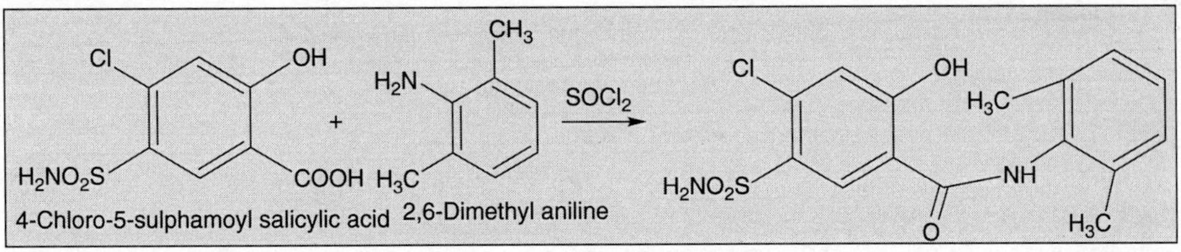

Xipamide is prepared by coupling of 4-chloro-5-sulphamoyl salicylic acid with 2,6-dimethylaniline in presence of thionyl chloride.
Use: Treatment of hypertension and oedema.
Clopamide 3-(Aminosulphonyl)-4-chloro-N-(2,6-dimethyl-1-piperidinyl) benzamide
It is a diuretic with properties similar to those of the thiazide diuretics even though it does not contain a thiazide ring system.

Synthesis

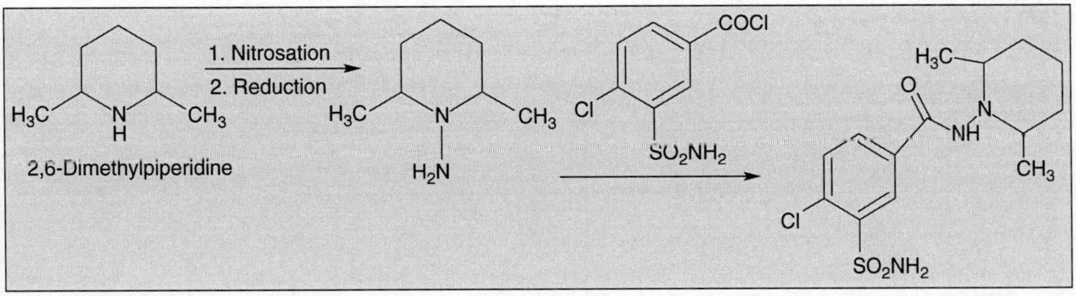

2,6-Dimethylpiperdine on nitrosation followed by reduction gives N-amino derivative, which on coupling with 4-chloro-3-sulphamoylbenzoyl chloride affords clopamide.
Quinethazone 7-Chloro-2-ethyl-1,2,3,4-tetrahydro-4-oxo-6-quinazoline sulphonamide
Replacement of SO_2 group of thiazide by carbonyl group yields quinazolones, and the compound has the same diuretic response as the thiazides. Quinethazone lowers blood pressure and gets rid of extra salt and water in the body by acting on the kidneys.

Synthesis

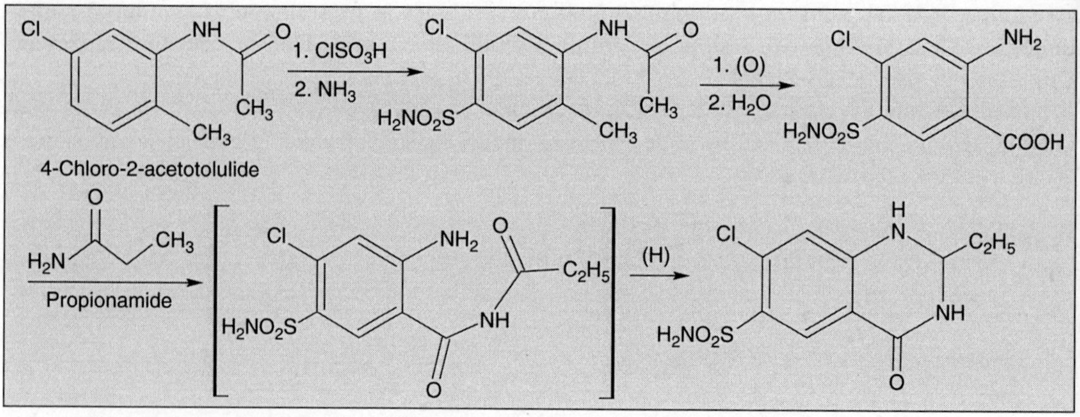

N1-(5-chloro-2-methylphenyl)acetamide on reaction with chlorosulphonic acid and ammonia affords N1-[4-(aminosulphonyl)-5-chloro-2-methylphenyl]acetamide. This intermediate on oxidation (methyl group to acid) followed by hydrolysis (amide to amine) gives anthranilic acid derivative. Reaction with propionamide leads to the quinazolone. Reduction of this gives quinethazone.

Metolazone 7-Chloro-1,2,3,4-tetrahydro-2-methyl-4-oxo-3-o-toluyl-6-quinazoline sulphonamide
Although it is not a true thiazide, metolazone is chemically related to the thiazide class of diuretics, and works in a similar manner.

Synthesis

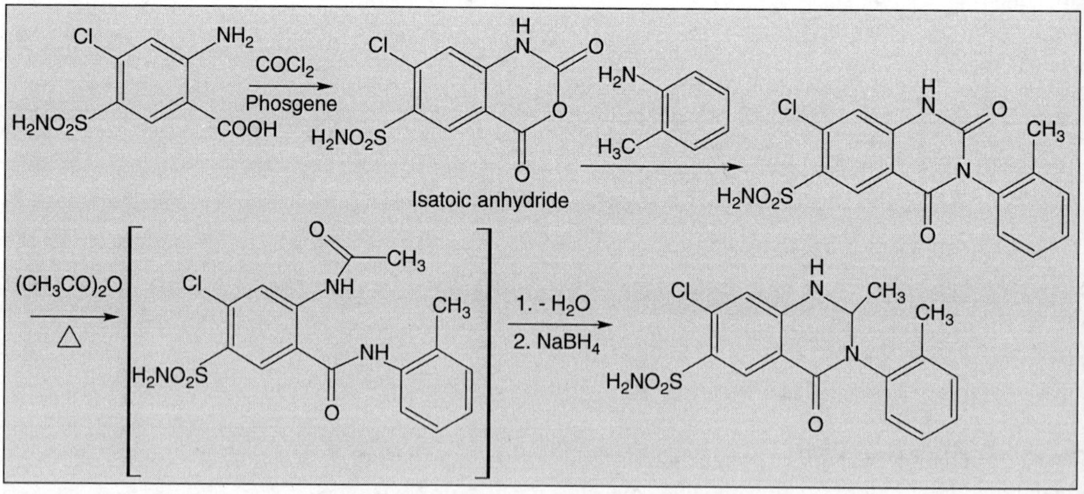

Reaction of anthranilic acid derivative with phosgene gives isatoic anhydride. Condensation of this compound with 2-methylaniline leads to intermediate compound, which on treatment with acetic anhydride opens the ring with the loss of carbon dioxide. The ring is further formed by keto-enol tautomerization and further reduction gives metolazone.

Chlorthalidone 2-Chloro-5-(1-hydroxy-3-oxo-1-isoindolinyl) benzene sulphonamide
Synthesis

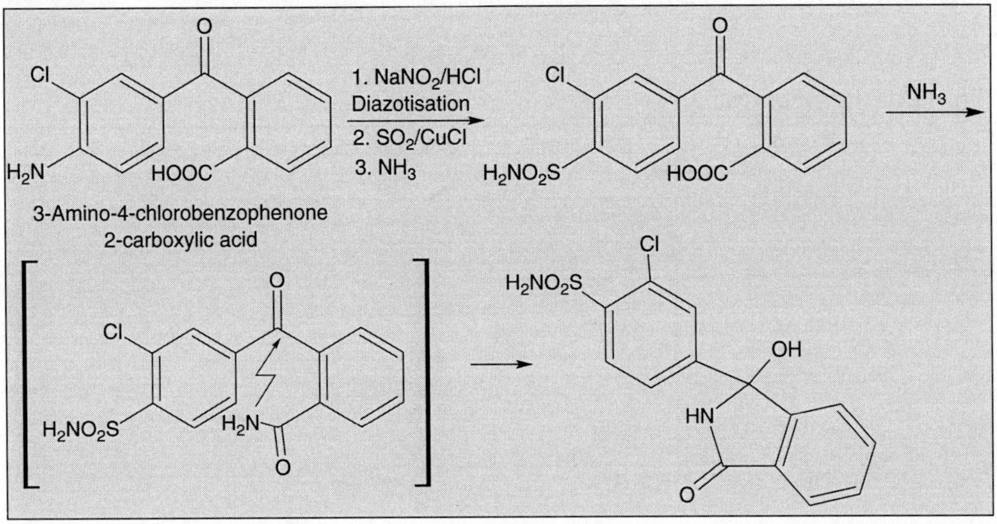

The amino group of benzoylbenzoic acid is converted to sulphonamide by a sequence of reactions: a. diazo compound (diazotisation); b. sulphonyl chloride (Sandmeyer reaction); and c. sulphonamide. The carboxylic acid group is then converted to amide, and amide nitrogen adds to the benzoyl group to give the isoindolone, chlorthalidone.

Uses Chlorthalidone is used to treat high blood pressure and fluid retention caused by various conditions, including heart disease. Chlorthalidone may also be used to treat patients with diabetes insipidus and certain electrolyte disturbances, and to prevent kidney stones in patients with high levels of calcium in their blood.

Clorexolone 6-Chloro-3-oxo-2-phenyl-5-isoindolinesulphonamide
Synthesis

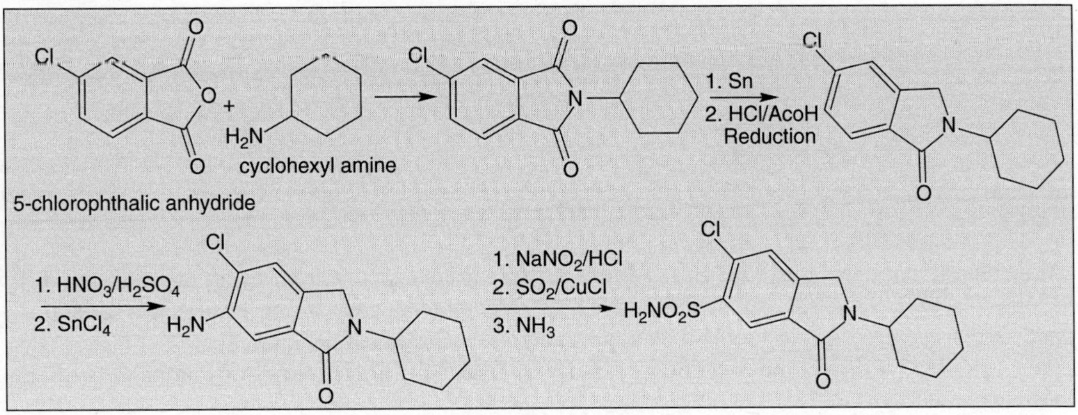

Reaction of 5-chlorophthalimide with cyclohexylamine provides phthalimido derivative. One of the carbonyl is reduced with tin and hydrochloric acid; and then amino group is converted to sulphonamide by reaction with nitrous acid, sulphurdioxide/cuprous chloride, and ammonia.

15.10 NEWER DRUGS

Torasemide 1-[4-(3-Methylphenyl) aminopyridin-3-yl] sulphonyl-3-isopropan-2-yl-urea

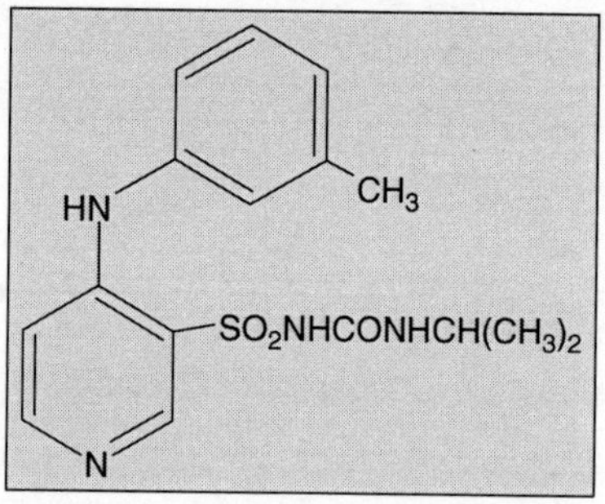

It is a pyridine-sulphonylurea-type loop diuretic mainly used in the management of oedema associated with congestive heart failure. It is also used at low doses for the management of hypertension. Compared to other loop diuretics, torasemide has a more prolonged diuretic effect than equipotent doses of furosemide and relatively decreased potassium loss.

Eplerenone Pregn-4-ene-7,21-dicarboxylic acid, 9,11-epoxy-17-hydroxy-3-oxo, γ-lactone, methyl ester (7α, 11α, 17α)

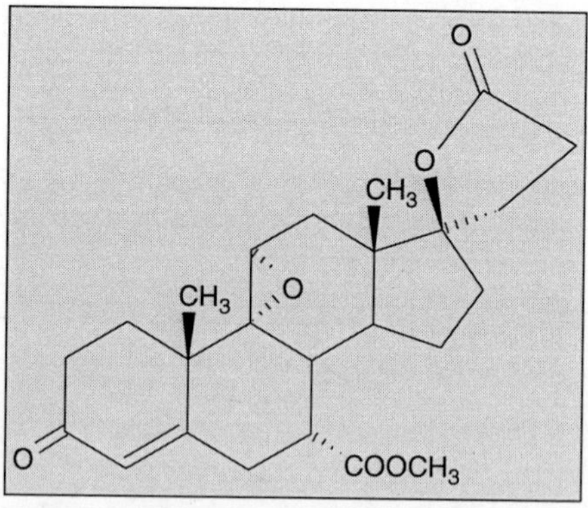

It is an aldosterone antagonist used as an adjunct in the management of chronic heart failure. It is similar to spironolactone, though it may be more specific for the mineralocorticoid receptor and is specifically marketed for reducing cardiovascular risk in patients following myocardial infarction.

Dorzolamide 2-Ethylamino-4-methyl-5, 5-dioxo-5(6),7-dithiabicyclo[4.3.0] nona-8,10-diene-8-sulphonamide

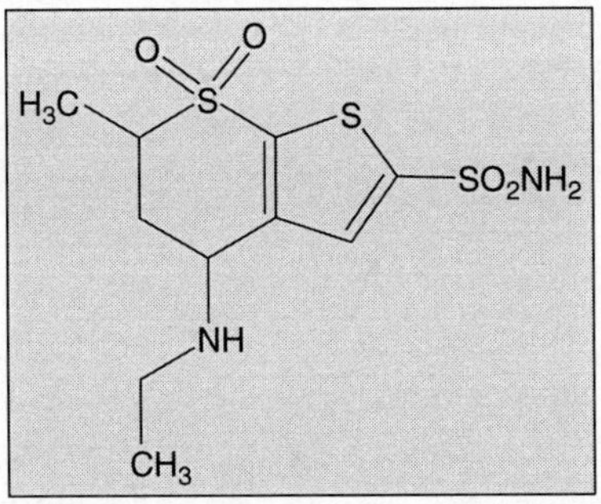

It is a carbonic anhydrase inhibitor. It is an anti-glaucoma agent and topically applied in the form of eye drops. It is used to lower increased intraocular pressure in open-angle glaucoma and ocular hypertension.

FURTHER READINGS

1. Nishimori I., Vullo D., Innocenti A., Scozzafava A., Mastrolorenzo A., Supuran C.T., *Carbonic anhydrase inhibitors: Inhibition of the transmembrane isozyme XIV with sulfonamides*, Bioorg Med Chem Lett. 2005 September 1, 15(17):3828-33.
2. 'Molecule of the month', *Drug News Perspect.*, 2005 March, 18(2):141.
3. Vasudevan, A., Mantan, M., Bagga, A., 'Management of Edema in Nephrotic Syndrome', *Indian Pediatrics*, 2004, 41:787-795.
4. Rapkin A., 'A review of treatment of premenstrual syndrome and premenstrual dysphoric disorder', *Psychoneuroendocrinology*, 2003 August 28, Suppl. 3:39-53.

MULTIPLE-CHOICE QUESTIONS

1. One among the following is not an osmotic diuretic:
    a. Urea nitrate
    b. Glycerol
    c. Mannitol
    d. Isosorbide
2. One of the following diuretics acts on the loop of Henle:
    a. Spironolactone
    b. Ethacrynic acid
    c. Clorexolone
    d. Dichlorphenamide
3. Thiazides are the most frequently used diuretics for treating hypertension. They lower blood pressure by:
    a. Decreasing the intravascular volume
    b. Decreasing the peripheral vascular resistance by direct action on vascular smooth muscle
    c. Decreasing the responsiveness of smooth muscles to nor-adrenaline
    d. All of the above
4. Which is the most appropriate diuretic for treating acute pulmonary oedema?
    a. Loop diuretics
    b. Thiazide diuretics
    c. Potassium-sparing diuretics
    d. Osmotic diuretics

5. Thiazide diuretics will have a beneficial effect in all these conditions, except
   a. Osteoporosis
   b. Gout
   c. Diabetes insipidus
   d. Calcium nephrolithiasis
6. Which diuretic competes with aldosterone and is used for treatment of hyperaldosteronism?
   a. Hydrochlorthiazide
   b. Isosorbide
   c. Furosemide
   d. Spironolactone
7. Which of the following actions is related to thiazide diuretics?
   a. Hyperuricemia
   b. Hyperkalemia
   c. Hypoglycaemia in Diabetics
   d. Hypercalcaemia
8. All the following diuretics will be effective in a person suffering from Addison's disease, except
   a. Clorthiazide
   b. Urea
   c. Spironolactone
   d. Furosemide
9. Which of these is not true with regard to loop diuretics?
   a. Can cause hyperuricemia (rarely leading to gout)
   b. Can cause hyperglycaemia (may precipitate diabetes mellitus)
   c. Increase plasma level of LDL cholesterol
   d. Increase plasma level of HDL cholesterol
10. Primary site of action of thiazide diuretics in the nephron is:
    a. Proximal tubule
    b. Loop of Henle
    c. Distal tubule
    d. Convoluted tubules
11. Which class of diuretics is useful in reducing intra-ocular pressure during acute attacks of glaucoma?
    a. Loop diuretics
    b. Thiazide diuretics
    c. Potassium-sparing diuretics
    d. Osmotic diuretics
12. When resistance develops to loop diuretics in congestive heart failure, the following class of diuretics can be used:
    a. Potassium-sparing diuretics
    b. Osmotic diuretics
    c. Thiazide diuretics
    d. Carbonic anhydrase inhibitors
13. Thiazide diuretics can have beneficial effect in osteoporosis/calcium nephrolithiasis due to which effect?
    a. Hypercalciuria
    b. Hypocalciuria
    c. Hyperuricemia
    d. Hyperkalemia
14. Which of the following statements refers to methazolamide?
    a. Reduces calcium excretion – hence use in nephrolithiasis (and possibly in prevention of osteoporosis)
    b. Acts chiefly on distal convoluted tubules
    c. Increases GFR (hence use in renal failure)
    d. Seldom used as diuretic – except in metabolic alkalosis
15. Mechanism of action of methazolamide is
    a. Inhibits Na-Cl transporter
    b. Inhibits Na-K-2Cl transporter
    c. Inhibits carbonic anhydrase
    d. Osmotic diuretic
16. Which of the following statements refers to bendroflumethiazide?
    a. Most effective when aldosterone levels are high (cirrhotics)
    b. Inhibits (weakly) carbonic anhydrase
    c. Weak carbonic anhydrase inhibitors
    d. Possesses anti-hypertensive effect independent from diuretic effect
17. Mechanism of action of furosemide is:
    a. Inhibits carbonic anhydrase
    b. Inhibits Na-K-2Cl transporter
    c. Osmotic diuretic
    d. Aldosterone antagonist

18. Class of bumetanide is
    a. Carbonic anhydrase inhibitors
    b. Osmotic diuretics
    c. Loop diuretics
    d. Potassium-sparing diuretics

QUESTIONS

1. Explain, with the aid of structures, why sulphonamide derivatives inhibit carbonic anhydrase to exert its diuretic action.
2. Give the general structure of thiazides and discuss the SAR.
3. Write a short note on furosemide.
4. Thiazide diuretics are used, among other things, for the treatment of hypertension. How do they work at the cellular level?
5. Write the synthesis of ethacrynic acid and chlorthiazide with structures.
6. Write a short note on Hoffman's degradation.

SOLUTION TO MULTIPLE-CHOICE QUESTIONS

1. a;
2. b;
3. d;
4. a;
5. b;
6. d;
7. a;
8. c;
9. a;
10. c;
11. d;
12. c;
13. b;
14. d;
15. c;
16. c;
17. b;
18. c.

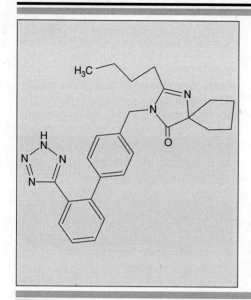

CHAPTER 16

Antihypertensive Agents

LEARNING OBJECTIVES

- Define antihypertensive agents and their utility in various conditions
- Describe the mechanism of maintenance of blood pressure and their disorder
- Categorize antihypertensive drugs based on the target on which they act
- Describe the modes of action and their importance over other classes of antihypertensive agents
- Compare α and β-adrenergic blockers
- Describe the importance of angiotensin system in the treatment of hypertension
- List some newer drugs for the treatment of hypertension

16.1 INTRODUCTION

The body controls blood pressure by a complex feedback mechanism between baroreceptors and effector nerves, primarily adrenergic in nature. This system is modulated by peptide systems (angiotensin/renin). **Antihypertensive drugs** are used in the treatment of high blood pressure (hypertension). Hypertension is defined conventionally as blood pressure ≥140/90. Hypertension is called 'the silent killer' because it usually has no symptoms. It is linked with the hardening of the arteries, a condition called atherosclerosis, and is a factor in 75 per cent of all strokes and heart attacks. When hypertension is not found and treated, it can cause:

- The heart to get larger, which may lead to heart failure
- Small bulges to form in blood vessels. Common locations are the main artery from the heart, arteries in the brain, legs, intestines, and spleen
- Blood vessels in the kidney to narrow, which may cause kidney failure

- Arteries throughout the body to 'harden faster, especially those in the heart, brain, kidney, and legs. This can cause heart attack, stroke, kidney failure, or amputation of part of the leg
- Blood vessels in the eyes to burst or bleed, which may cause vision changes

Evidence suggests that reduction of the blood pressure by 5–6 mmHg can decrease the risk of stroke by 40 per cent and of coronary heart disease by 15–20 per cent, and reduces the likelihood of dementia, heart failure, and mortality from cardiovascular disease.

16.1.1 Categories of B.P. Levels in Adults (in mmHg)

Category	Systolic	Diastolic
Normal	Less than 120	Less than 80
Pre-hypertension	120–139	80–89
Hypertension		
Stage I	140–159	90–99
Stage II	160 or higher	100 or higher

> The risk of hypertension is five times higher in the obese as compared to those of normal weight, and up to two-thirds of cases can be attributed to excess weight. More than 85 per cent of cases occur in those with a body mass index greater than 25.

Hypertension originating from these latter conditions is called secondary hypertension. The main systems controlling the body's arterial blood pressure are the central nervous system (CNS), sympathetic ganglia, adrenergic nerve endings, vascular smooth musculature, kidneys and arterioles, and lastly, the renin-angiotensin system. Lowering arterial blood pressure can be accomplished by affecting vascular smooth musculature using hydralazine, diazoxide, minoxidil, sodium nitroprusside, diuretics, and calcium channel blockers, which relax vascular smooth musculature, thus lowering both systolic and diastolic blood pressure.

Cholinoblockers (ganglioblockers) such as mecamylamine and trimethaphan act on autonomic ganglia to reduce blood pressure.

Lowering arterial blood pressure by acting on the adrenergic system can be accomplished by stimulating α-adrenoreceptors (clonidine, guanabenz, guanacin, and methyldopa), which leads to a reduction of sympathetic impulses to vessels and the heart, thus reducing cardiac output and heart rate, which consequently lowers arterial blood pressure; by blocking α_1-adrenoreceptors (prazosin, terazosin), the main importance of which is dilating peripheral vessels, which leads to reduced blood pressure; and by blocking β-adrenoreceptors (propranolol, atenolol, nadolol, and others), which reduce cardiac output and peripheral resistance of vessels, resulting in lower blood pressure.

Lowering blood pressure can also be done by acting on the renin-angiotensin system by using angiotensin-converting enzyme (capopril, enalapril). These drugs block action of the angiotensin-converting enzyme, which results in less production of angiotensin II and inhibits its vasoconstricting action on arterial and venous blood vessels. Diuretics can act on the kidneys and arterioles for the purpose of lowering blood pressure. Finally, calcium channel blockers can act on smooth musculature in order to lower blood pressure (verapamil, diltiazem, and nifedipine).

16.2 CLASSIFICATION

1. Peripheral antiadrenergic drugs: Prazosin, Terazosin, Trimazosin, Doxazosin
2. Centrally acting agents: Clonidine, Methyldopa, Guanabenz
3. Direct vasodilators: Diazoxide, Hydralazine, Minoxidil
4. Ganglionic blocking agents: Mecamylamine, Trimethaphan
5. β-Adrenergic blockers: Atenolol, Propranolol, Timolol, Labetalol, Esmolol
6. Calcium channel blockers: Nifedipine, Taludipine, Felodipine, Nimodipine, Verapamil, Diltiazem
7. Angiotensin-converting enzyme (ACE) inhibitors: Captopril, Enalapril, Lisinopril, Spirapril, Fosinopril
8. Angiotensin II antagonists: Losartan, Eprosartan, Valsartan, Candesartan
9. Miscellaneous: Thiazide diuretics, Loop diuretics

16.3 PERIPHERAL ANTI-ADRENERGIC DRUGS

Mechanism: These drugs lower blood pressure by blocking α_1-adrenoceptors that subserve vasoconstrictor functions. α_1-Receptors act by phospholipase C activation, which, in turn, forms inositol trisphosphate, which together with diacylglycerol are second messenger molecules used in signal transduction in biological cells. In blood vessels, these cause vasoconstriction.

Advantage over other α-adrenergic blockers: These drugs differ from the classic α-blocking agents, in that they do not block α_2-adrenoceptors on the adrenergic nerve terminals, which serve a negative feedback function to limit the release of noradrenaline. Conventional α-blockers block these receptors and cause an excessive continuing release of noradrenaline, which in the heart gives rise to tachycardia and palpitation.

Uses

1. Used alone in the treatment of mild to moderate essential hypertension and with other drugs in severe hypertension
2. These drugs are useful in the preoperative management of pheochromocytoma
3. These drugs relax spastic digital arterioles in patients with Raynaud's phenomenon

Prazosin 1-(4-Amino-6,7-dimethoxy-2-quinazolinyl)-4-(2-furoyl) piperazine

Synthesis

Prazosin is synthesised from 2-amino-4,5-dimethoxybenzamide, which upon reaction with urea undergoes heterocyclation into quinazolodione derivative. Substituting keto (enol) groups of this compound with chlorine atoms by reaction with thionyl chloride, or a mixture of phosphorous oxychloride with phosphorous pentachloride, gives 2,4-dichloro-6,7-dimethoxyquinazoline. Upon subsequent reaction with ammonia, the chlorine atom at C4 of the pyrimidine ring is replaced with an amino group, which leads to the formation of 4-amino-2-chloro-6,7-dimethoxyquinazoline. Introducing this into a reaction with 1-(2-furoyl)piperazine gives prazosin.

Terazosin 1-(4-Amino-6,7-dimethoxy-2-quinazolinyl)-4-(2-tetrahydrofuroyl)-Piperazine

Synthesis

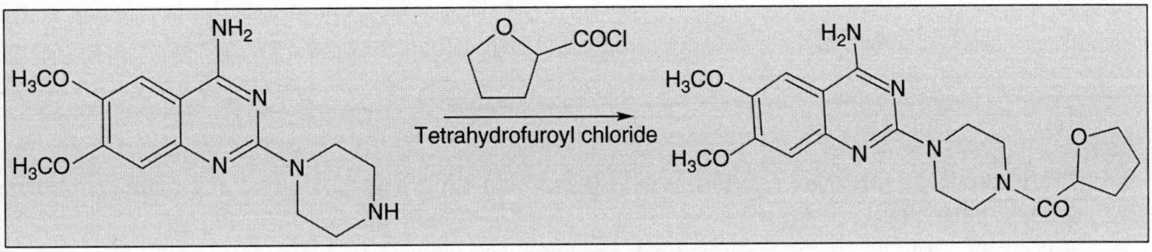

Terazosin only differs from prazosin in that the furyl moiety is replaced with a tetrahydrofuryl moiety. It is synthesised in exactly the same manner of prazosin except in the aspect of using 1-(2-tetrahydrofuroyl) piperazine instead of 1-(2-furoyl)piperazine.

It has longer half-life (12 hrs) and longer duration of action (24 hrs) than does Prazosin.

Trimazosin 4-(4-Amino-6,7,8-trimethoxy-2-quinazolinyl)-1-piperazine-carboxylic acid, 2-hydroxy-2-methyl propyl ester

Synthesis

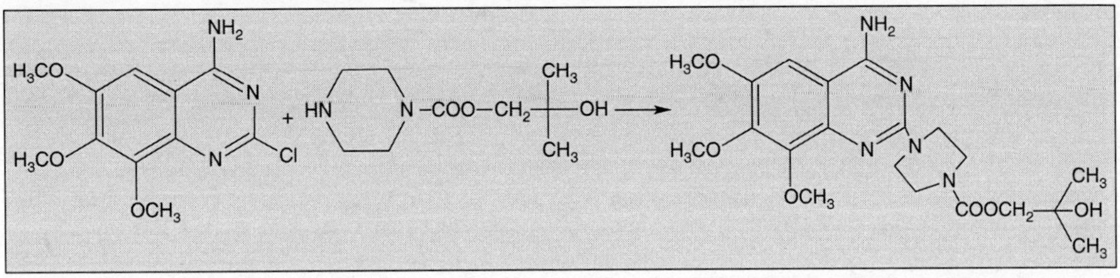

Synthesis is similar to prazosin, except in starting material (2-amino-3,4,5-trimethoxybenzamide) and with amine (1-piperazine-carboxylic acid, 2-hydroxy-2-methyl propyl ester) in the last step.

Doxazosin 1-(4-Amino-6,7-dimethoxy-2-quinazolinyl-4-[(2,3-dihydro-1,4-benzodioxinyl-2-yl)carbonyl] piperazine

Synthesis

Synthesis is similar to prazosin, except in the last-step amination with (2,3-dihydro-1,4-benzodioxinyl-2-yl)carbonyl piperazine.

It is a water-soluble analogue of prazosin and terazosin. Doxazosin differs from prazosin in that its long half-life (24 hours) enables once-a-day oral administration.

16.4 CENTRALLY ACTING ANTIHYPERTENSIVE DRUGS

Mechanism: The mechanism of action of these drugs is caused by stimulation of α_2-adrenoreceptors in the inhibitory structure of the brain. It is believed that interaction of these drugs with α_2-adrenergic receptors is expressed in the suppression of vasomotor centre neurons of the medulla, and reduction of hypothalamus activity, which leads to a decline in sympathetic impulses to the vessels and the heart. In summary, cardiac output and heart rate are moderately reduced, and consequently arterial pressure is reduced.

Clonidine 2-(2,6-Dichloroanilino)-2-imidazoline

Synthesis

Clonidine is synthesised from 2,6-dichloroaniline, the reaction of which with ammonium thiocyante gives N-(2,6-dichlorophenyl)thiourea. Methylation of this product followed by subsequent reaction with ethylendiamine gives clonidine.

Uses

1. Primary hypertension
2. Post-menopausal vasomotor instability, dysmenorrhoea, and in the prophylaxis of migraine and cluster headaches

Methyldopa 1–3-(3,4-Dihydroxyphenyl)-2-methylalanine

Synthesis.

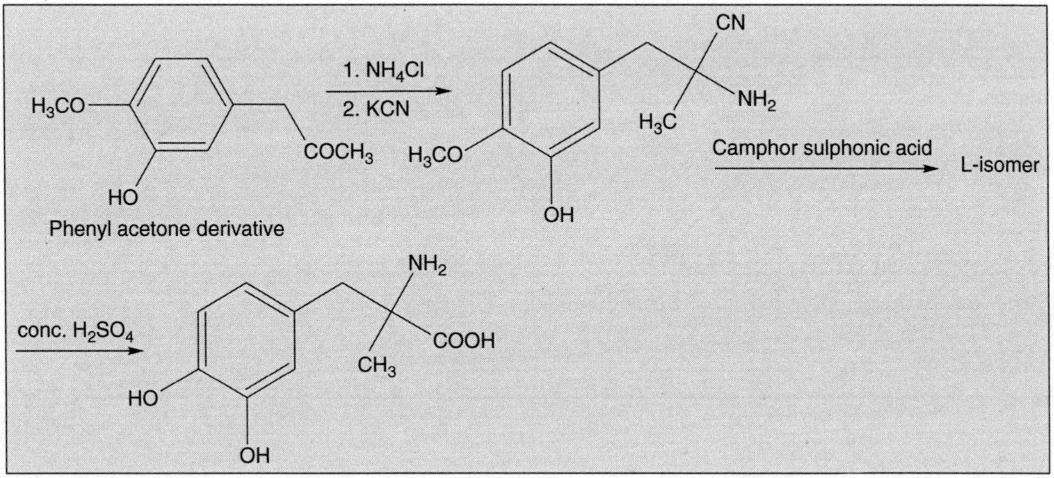

[Synthesis scheme showing phenyl acetone derivative reacting with 1. NH₄Cl, 2. KCN to form aminonitrile, then with camphor sulphonic acid to give L-isomer, then conc. H₂SO₄ to give methyldopa]

Reaction of 3,4-dihydroxyphenylacetone with ammonium chloride and potassium cyanide affords corresponding α-aminonitrile. The *L* isomer is then separated from the racemate and treated with concentrated sulphuric acid to hydrolyse the nitrile to the acid methyldopa.

Mechanism:

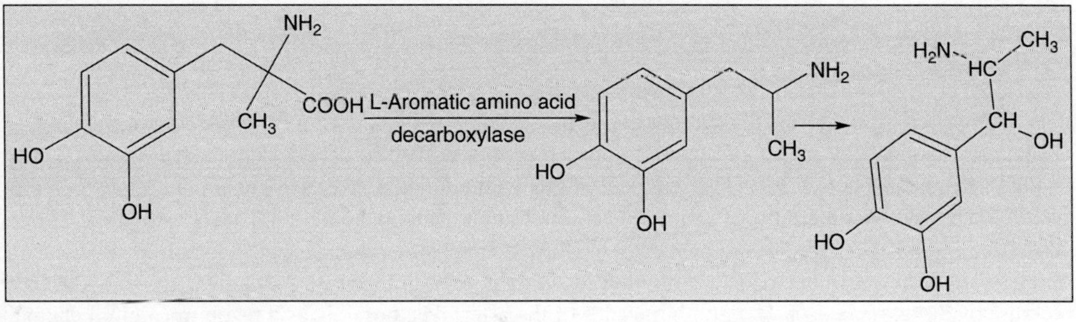

[Mechanism scheme showing methyldopa converted by L-Aromatic amino acid decarboxylase to α-methyldopamine, then to α-methylnorepinephrine]

Methyldopa is metabolised to α-methylnorepinephrine and the metabolite acts in the brain to inhibit adrenergic neuronal outflow from the brainstem, and this central effect is principally for its antihypertensive action.

Uses: Antihypertensive, and also has some usefulness in the treatment of pheochromocytoma and carcinoid tumour

Guanabenz (2,6-Dichlorobenzylidene) aminoguanidine

Synthesis

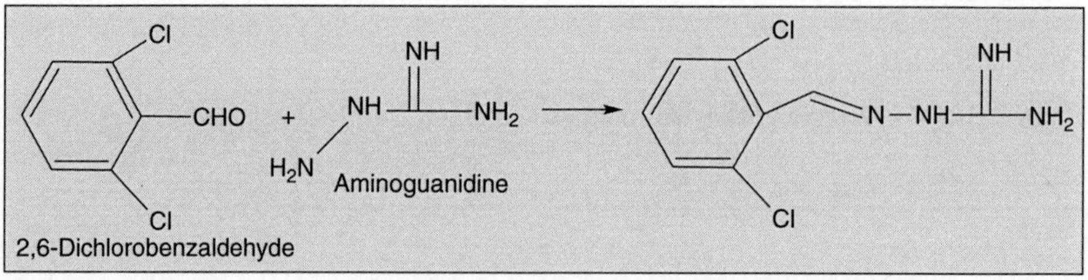

Guanabenz is synthesised in one step by reacting 2,6-dichlorobenzaldehyde with aminoguanidine. It appears to act as an α_2-adrenergic agonist at pre-synaptic nerve terminals in the CNS, thus decreasing the release of neurotransmitter.

16.5 DIRECT VASODILATORS

Diazoxide 7-Chloro-3-methyl-1,2,4-benzothiadiazine-1, 1-dioxide

Synthesis

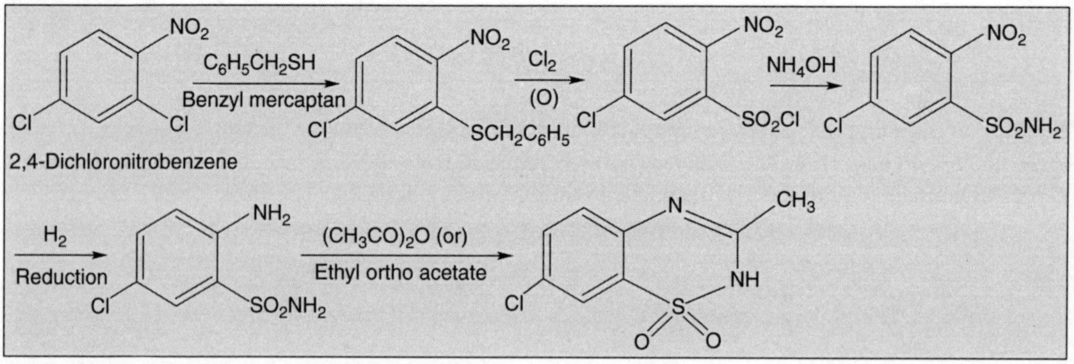

It is synthesised fron 2,4-dichloronitrobenzene by reacting with benzyl mercaptide to form thioether. Oxidation of sulphur by means of aqueous chlorine results as well in debenzylation; the overall reaction finally yields the sulphonyl chloride. Treatment of this with ammonia gives sulphonamide; catalytic hydrogenation then converts the nitro to amine. Reaction of amino sulphonamide with acetic anhydride affords diazoxide.

Mechanism of action: Diazoxide is a nondiuretic derivative of thiazides that dramatically reduces blood pressure by direct relaxation of smooth muscles of the arterioles, possibly as a result of calcium channel activation of smooth musculature in arterioles. It has a weak effect on the venous system and on the heart. In addition to hypotensive action, diazoxide causes a sharp increase in the level of glucose in the blood as a result of the inhibition of insulin release from adrenal glands.

Uses: Used intravenously in acute hypertensive cases. Injection of an i.v. bolus lowers blood pressure within 30 seconds, and a maximum effect is achieved within 3 to 5 minutes.

Hydralazine 1-Hydrazinophthalazine

Synthesis

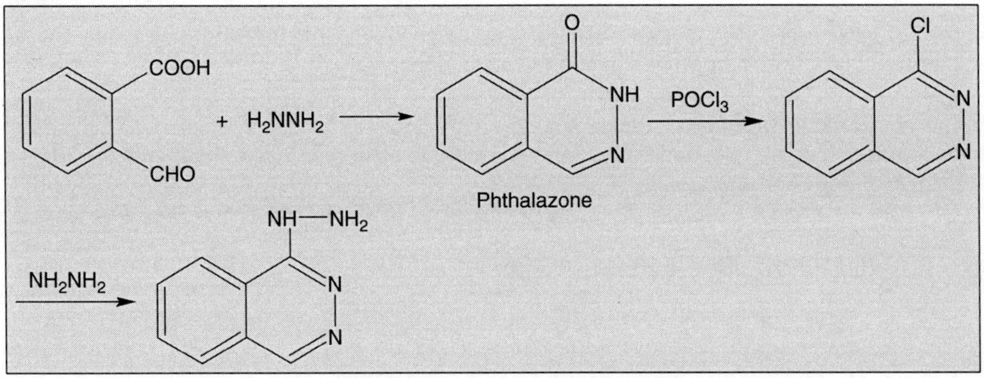

Reaction of acid aldehyde with hydrazine gives phthalazone; this undergoes a reaction with phosphorous oxychloride, forming 1-chlorophthalazine, in which substitution of the chlorine atom with hydrazine gives the desired hydralazine.

Mechanism of action: It causes vasodilatation by stimulating guanylate cyclase in arteriolar smooth muscle; the stimulant appears to be nitric oxide (NO) from the local oxidation of the hydrazine moiety. NO is a natural, endothelin-derived relaxing factor.

Uses: Treatment of hypertensive emergencies, toxemia of pregnancy, acute congestive heart failure, or after myocardial infarction

Minoxidil 2, 6- Diamino-6-piperidinopyrimidine 3-oxide

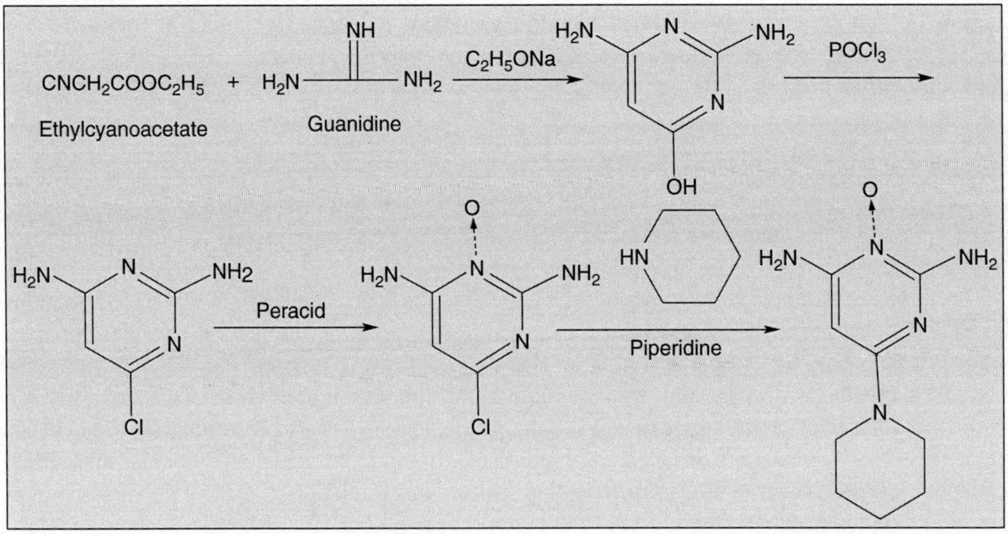

Ethylcyanoacetate condensed with guanidine in presence of base yields 2,4-diamino-6-hydroxypyrimidine; this in turn reacts with phosphorous oxychloride to give 6-chloro derivative. Oxidation of this

product with 3-chloroperbenzoic acid gives 2,4-diamino-6-chloropyrimidine-3-oxide. Reaction with piperidine replaces chloro group and affords minoxidil.

Mechanism: It dilates arterioles by opening potassium channels, which causes hyperpolarization and relaxation of smooth muscle. This lowers the total peripheral vascular resistance and, hence, the blood pressure.

Uses: All types of hypertension. Further studies on minoxidil showed that the hair growth stimulation could be obtained in appropriate cases when the drug was administered topically. Consequently, the drug is used topically to restore hair growth in androgenetic alopecia

> Upjohn Corporation produced a topical solution that contained 2 per cent minoxidil to be used to treat baldness and hair loss, under the brand name Rogaine in the United States and Canada, and Regaine in Europe and the Asia-Pacific.

16.6 GANGLIONIC BLOCKING AGENTS

These drugs block sympathetic ganglia, interrupting adrenergic control of arterioles, and results in vasodilatation, improved peripheral blood flow in some vascular beds, and a fall in blood pressure.

Uses: Hypertension, vasospastic disorders, and peripheral vascular diseases. Most ganglionic blockers are now more often used as a research tool

16.7 β-ADRENERGIC BLOCKERS

These drugs reversibly bind with β-adrenergic receptive regions and competitively prevent activation of these receptors by catecholamines released by the sympathetic nervous system, or externally introduced sympathomimetics.

These drugs block the β-receptors of the heart, slow the heart, reduce the force of contraction, and reduce the cardiac output. They also inhibit the secretion of renin by the juxtaglomerular apparatus of the kidney and, hence, decrease the plasma levels of angiotensin II, a potent vasoconstrictor and sensitizer to the sympathetic nervous activity, and stimulant of the secretion of aldosterone, an antisaluretic.

β-Adrenoreceptors are subdivided into $β_1$-adrenoreceptors, which are predominantly found in cardiac muscle, and β2-adrenoreceptors, which are predominantly found in bronchial and vascular muscles. Thus, β-adrenoblocking substances are classified by their selectivity in relation to these receptors. Compounds that exhibit roughly the same affinity to $β_1$- and $β_2$-receptors independent of dosage such as nadolol, propranolol, pindolol, timolol, and labetalol (combined α- and β-adrenoblocker) are classified as non-selective blockers. Drugs that in therapeutic doses have higher affinity to $β_1$-receptors than to $β_2$-receptors - such as acebutol, atenolol, metoprolol, and esmolol - are called selective or cardioselective β-adrenoblockers.

Propranolol 1- [(1-Methylethyl)amino] -3-(1-naphthalenyloxy) -2-propanol
Synthesis

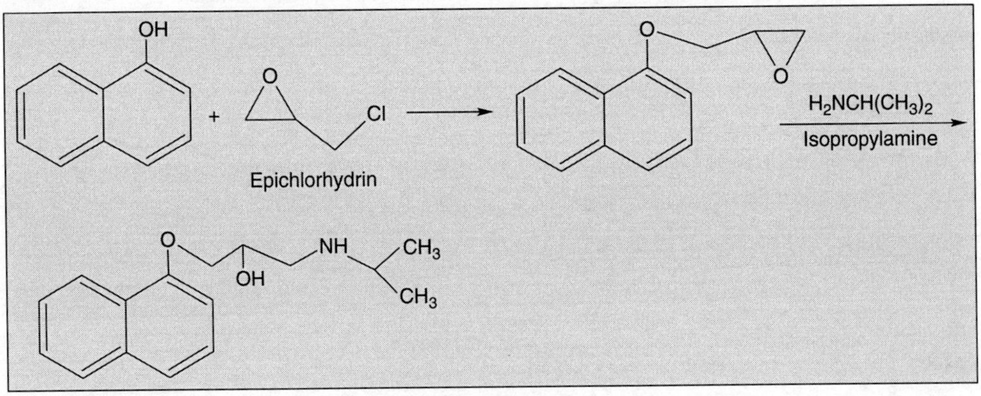

Propranolol is synthesised by reacting 1-naphthol with epichlorhydrin. Opening of the epoxide ring gives 1-chloro-3-(1-naphthyloxy)-2-propanol, which is reacted further with *iso*-propylamine, giving propranolol.

It is a non-selective β-antagonist and used for the treatment of hypertension, arrhythmias, myocardial infarction, pheochromocytoma, angina pectoris, tremors, and thyrotoxicosis.

Atenolol 2-[*p*-[2-Hydroxy-3-(isopropylamino)propoxy]phenyl]acetamide
Synthesis

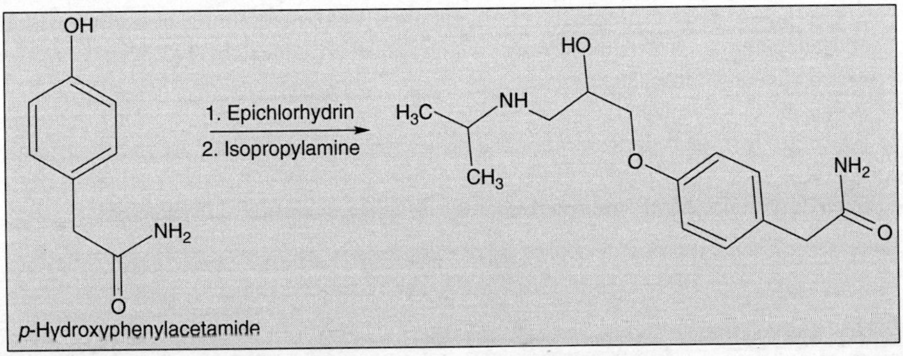

Synthesis is similar to that of propranolol, with 4-hydroxyphenylacetamide as the starting material.

It is a selective $β_1$-antagonist with very weak $β_2$-antagonist activity. It is used for the treatment of angina pectoris and hypertension.

Esmolol Methyl 4-[2-hydroxy-3-[(1-methylethyl)-amino]propoxy]benzene propionic acid
Synthesis

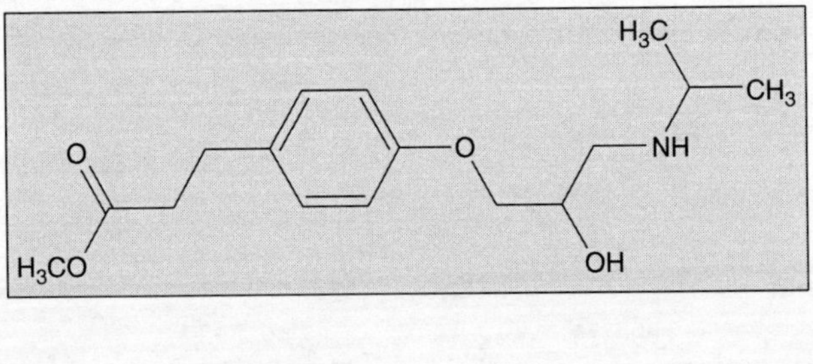

It is a selective β_1-receptor antagonist prepared from methyl 4-hydroxy phenyl propionic acid similar to propranolol.

Labetalol 5-[1-Hydroxy-2-[(1-methyl-3-phenylpropyl)amino]ethyl]salicylamide

Synthesis

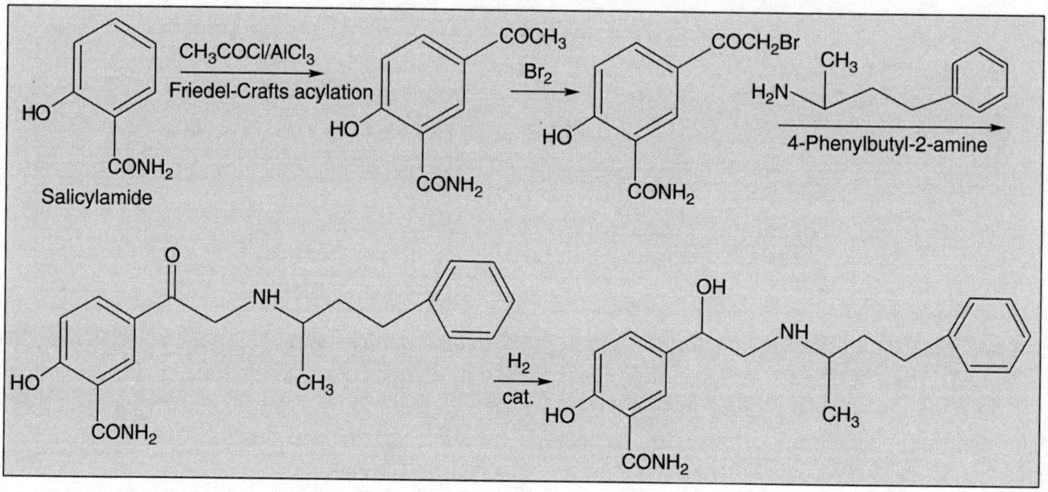

Labetalol is synthesised by the acylation of salicylamide with acetylchloride to give acetophenone derivative; this on bromination gives 5-bromoacetylsalicylamide. N-Alkylation of 4-phenylbutyl-2-amine with 5-bromoacetylsalicylamide forms aminoketone, which is further reduced by hydrogen using a palladium-platinum on carbon catalyst into labetalol.

It combines both non-selective β- and α-antagonist activity, in a ratio of 3:1 by the oral route and 7:1 by the I.V. route. It is used in the treatment of hypertension, toxemia of pregnancy, angina pectoris, and pheochromocytoma.

Timolol 1-(*tert*-Butylamino)-3-[(4-morpholino-1,2,5-thiadiazol-3-yl)oxy]-2-propanol

Synthesis

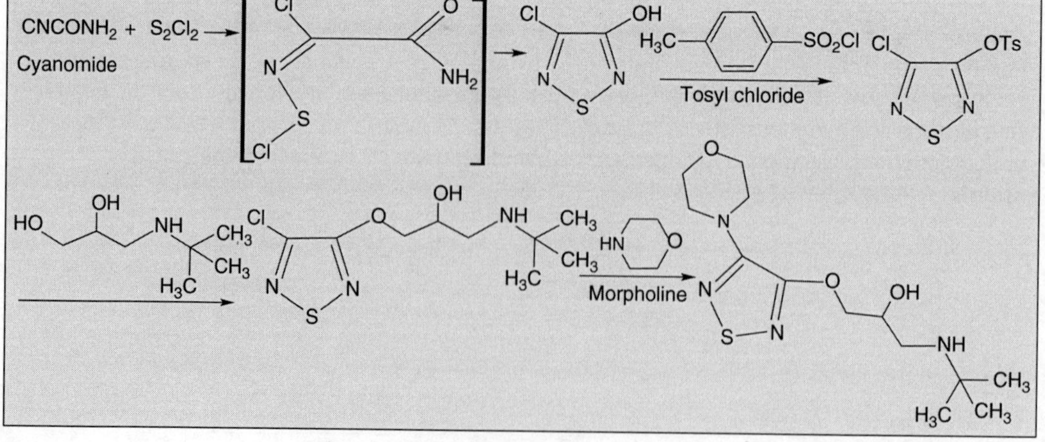

Reaction of cyanomide with sulphur monochloride leads directly to the (sub)thiadiazole. Hydroxyl group is activated by converting it into tosyl derivative. Displacement of good leaving tosyl group with terminal alkoxide from diol gives intermediate; this is followed by reaction with morpholine, which displaces remaining leaving chloro group on the heterocycle and affords timolol.

It is a non-selective β-adrenoreceptor antagonist used for the treatment of hypertension, angina pectoris, and glaucoma.

16.8 CALCIUM CHANNEL BLOCKERS

Calcium ions (Ca^{2+}) play a pivotal role in vascular smooth muscle contraction. Calcium enters cells through specialized pores in the membrane wall called calcium channels, activated by membrane depolarization (voltage-operated). Calcium channel-blocking drugs block voltage-dependent calcium channels, especially the calcium channels in cardiac and smooth muscles. By decreasing calcium influx during action potential in a frequency- and voltage-dependent manner, they reduce systolic intracellular calcium concentration and muscle contractility. Because of this arteriole vasodilatation, they decrease blood pressure.

Uses: Hypertension, angina pectoris, and arrhythmias

Nifedipine Dimethyl-1,4-dihydro-2,6-dimethyl-4-(2-nitrophenyl)-3,5-pyridine dicarboxylate

Synthesis

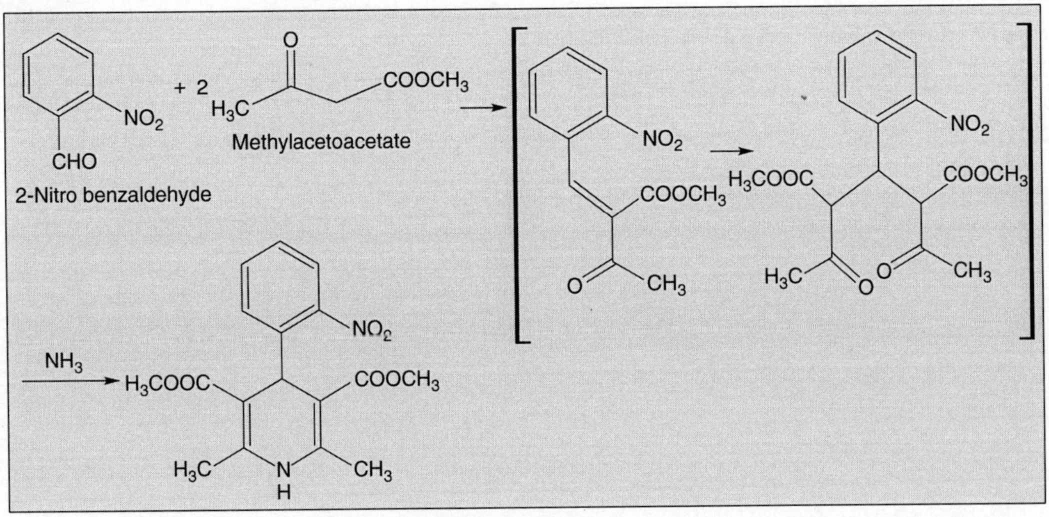

Nifedipine is synthesised by a Hantsch synthesis from two molecules of a β-dicarbonyl compound methyl acetoacetate, using as the aldehyde component 2-nitrobenzaldehyde and ammonia. The sequence of the intermediate stages of synthesis has not been completely established.

Taludipine Ethyl 4–2-[3-(*tert*-butoxy)-3-oxo-1-propenyl]phenyl-2-[(dimethylamino)methyl]-6-methyl-5-propionyl-1,4-dihydro-3-pyridinecarboxylate

Synthesis

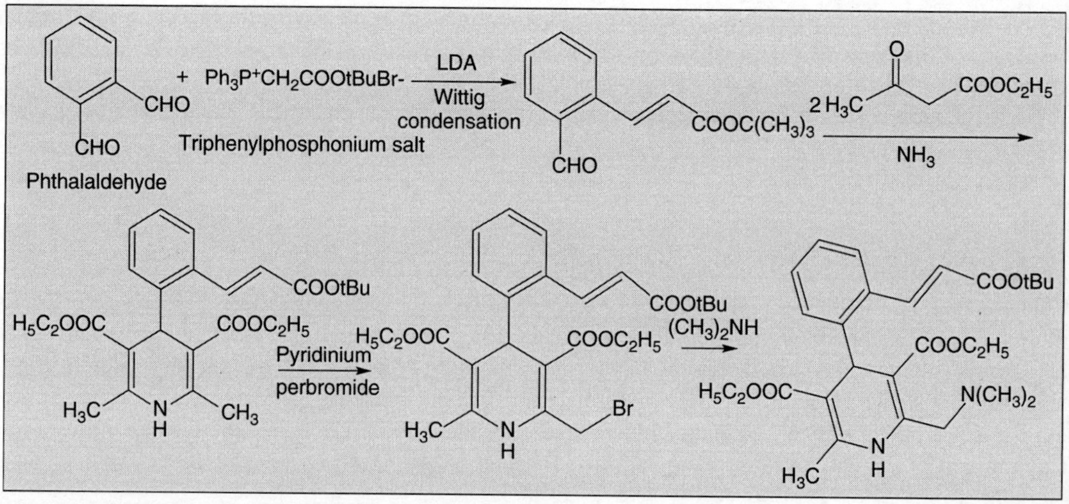

Wittig condensation of phthalaldehyde with *tert*-butyl-2-triphenylphosphonium acetate gives alkene. This compound is then condensed with ethylaceto acetate and ammonia leads to dihydropyridine derivative. Reaction of this product with pyridinium bromide hydrobromide results in halogenation of the allylic methyl group, and reaction with dimethylamine affords taludipine.

Side-effects of these synthetic drugs may includes dizziness, headache, redness in the face, fluid buildup in the legs, rapid heart rate, constipation, and gingival overgrowth.

16.8.1 Other Calcium Channel Blockers

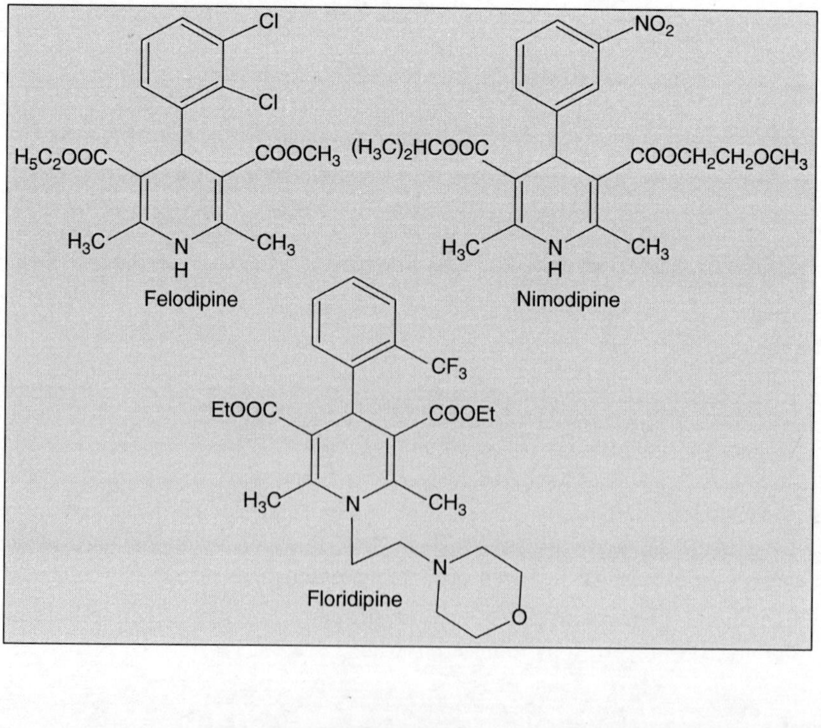

Felodipine

Nimodipine

Floridipine

Verapamil: 5-[3,4-Dimethoxyphenylethyl)methylamino]-2-(3,4-dimethoxyphenyl)-2-isopropylvaleronitrile

Synthesis

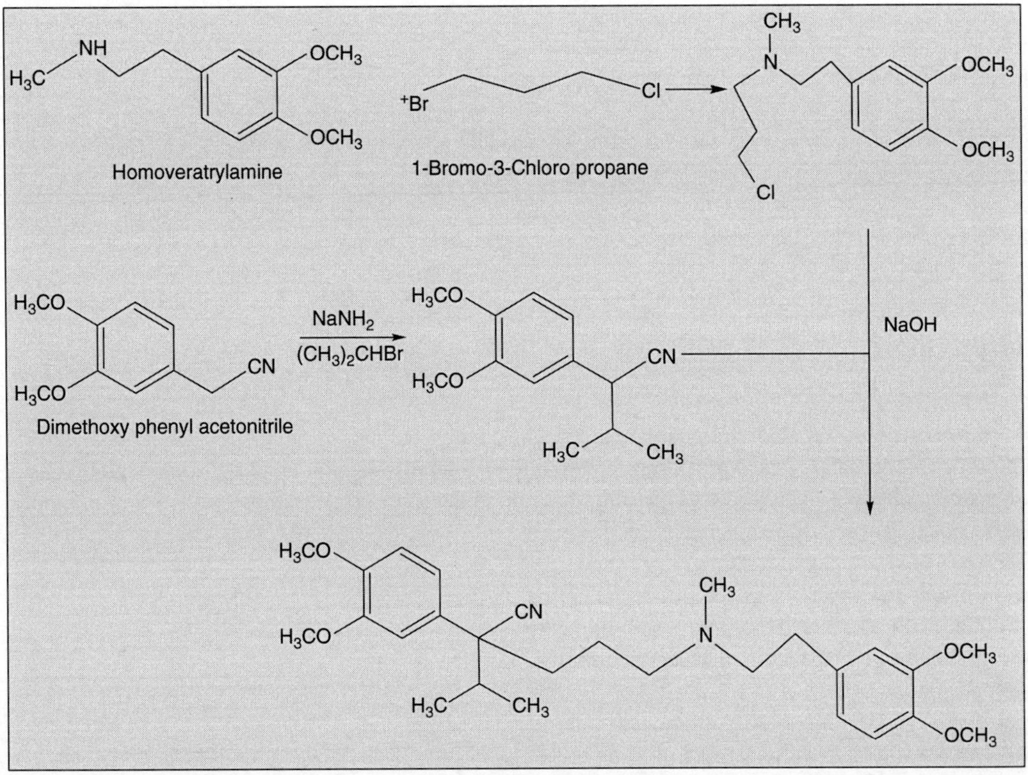

Verapamil is synthesised by using 3,4-dimethoxyphenylacetonitrile as the initial substance. The synthesis of the final product is accomplished by alkylating 2-(3,4-dimethoxyphenyl)-3-methylbutyronitrile with N-[2-(3,4-dimethoxyphenyl)-ethyl]-N-3-(chloropropyl)-N-methylamine. The initial 2-(3,4-dimethoxyphenyl)-3-methylbutyronitrile is synthesised by alkylating 3,4-dimethoxyphenylacetonitrile with isopropyl chloride in the presence of sodium amide. The alkylating agent, N-[2-(3,4-dimethoxyphenyl)-ethyl]-N-3-(chloropropyl)-N-methylamine, is also synthesised from 3,4-dimethoxyphenylacetonitrile followed by reduction into 3,4-dimethoxyphenylethylamine, with subsequent methylation into N-methyl-N-3,4-dimethoxyphenylethylamine. Next, the resulting N-[2-(3,4-dimethoxyphenyl)-ethyl] -N-methylamine is alkylated by 1-chloro-3-bromopropane into the desired N-[2-(3,4-dimethoxyphenyl)-ethyl]-N-3-(chloropropyl)-N-methylamine, which is alkylated by 2-(3,4-dimethoxyphenyl)-3-methylbutyronitrile to give the final product, verapamil.

Diltiazem: (+)-*cis*-3-(Acetoxy)-5-[2-(dimethylamino)ethyl]-2,3-dihydro-2-(4-methoxyphenyl)-1,5-benzothiazepin-4-one

Chapter 16: Antihypertensive Agents

Synthesis

[Synthesis scheme for diltiazem showing: Nitrothiophenol + Glycidic ester (H₃COOC) → SnCl₄ → intermediate → NaOH → intermediate → 1. H₂ 2. Δ → intermediate → 1. Cl-CH₂CH₂-N(CH₃)₂ 2. (CH₃CO)₂O → Diltiazem]

Diltiazem is synthesised in the following manner. The condensation of 4-methoxybenzaldehyde with methylchloroacetate in the presence of sodium methoxide in Darzens reaction conditions gives methyl ester of 3-(4-methoxyphenyl)-glycidylic acid (starting material in the scheme). Reacting it with 2-aminothiophenol with the opening of epoxide ring gives methyl ester of 2-hydroxy-3-(2'-aminophenylthio)-3-(4"-methoxyphenyl)propionic acid. Hydrolysis of the resulting compound with alkali leads to the formation of the corresponding acid in the form of a racemic mixture, which on interaction with (+)-α-phenylethylamine gives threo-(+)-2-hydroxy-3-(2'-aminophenylthio)-3-(4"-methoxyphenylpropionic) acid. Boiling this in a mixture of acetic anhydride/dimethylformamide/pyridine system brings to cyclization to the thiazepine ring and simultaneously acylates the hydroxyl group, forming (+)-cis-2-(4-methoxyphenyl)-3-acetoxy-2,3-dihydro-1,5-benzothiazepin-4-(5H)-one. Alkylation of the resulting product with 2,2-dimethylaminoethylchloride forms diltiazem.

16.9 ANGIOTENSIN-CONVERTING ENZYME (ACE) INHIBITORS

16.9.1 Renin-Angiotensin System

Renin converts angiotensinogen (a peptide that contains 14 amino acids) to the decapeptide angiotensin I (which is biologically inactive); ACE acts

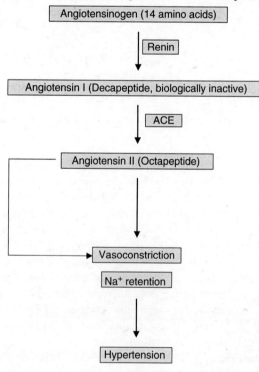

FIGURE 16.1 Mechanism of action of ace inhibitors.

upon angiotensin I to give the octapeptide angiotensin II, which is a potent vasoconstrictor, and also causes Na⁺ retention. These effects lead to an elevation of blood pressure. ACE inhibitors competitively inhibit ACE and block the conversion of angiotensin I to II, and act as antihypertensive agents.

Adverse effects of ACE inhibitors: ACE inhibitors not only prevent the conversion of angiotensin I to II, but also prevent ACE-mediated degradation of bradykinin and substance P. Because of the elevation of bradykinin and substance P, they cause angiooedema (also known by its eponym Quincke's oedema, angiooedema is the rapid swelling [edema] of the skin, mucosa, and submucosal tissues) and cough (a persistent dry cough is a relatively common adverse effect believed to be associated with the increases in bradykinin levels produced by ACE inhibitors).

Captopril 1-[(2S)-3-Mercapto-2-methylpropionyl]-L-proline

Synthesis

Hydrochloric acid is added across the double bond of methacrylic acid; acid group is activated as acid chloride by reaction with thionyl chloride. Acid chloride reacts with amino group of proline to form amide. Reaction with ammonium hydrogen sulphide affords captopril.

Captopril is also commonly associated with side effects like rash and taste disturbances (metallic or loss of taste), which are attributed to the unique sulphhydryl moiety.

Enalapril (S)-1-[N-[1-(Ethoxycarbonyl)-3-phenylpropyl]-L-alanyl]-L-proline

Enalapril is a prodrug that is hydrolysed in the liver by a serum esterase to the active parent dicarboxylic acid enalaprilate.

Synthesis

Amino group of alanine is protected with benzyloxycarbonyl (BOC) group, followed by reaction with proline in presence of dicyclohexyl carbodiimide (DCC) as carboxylic acid activator to give amide.

BOC group is removed by hydrogenation/palladium and reaction with ethyl-2-oxophenylbutyrate in presence of sodium cyanoborohydride undergoes reductive amination to afford enalapril.

Lisinopril 1-[6-Amino-2- (1-carboxy-3-phenyl-propyl) amino-hexanoyl]pyrrolidine- 2-carboxylic acid dehydrate

Synthesis

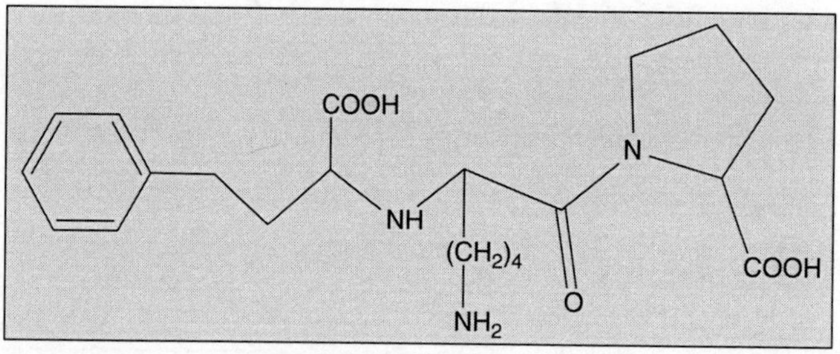

The replacement of alanine residue of enalapril by a lysine moiety leads to lisinopril, which is administered in this case as a free dicarboxylic acid. The synthesis is quite analogous to the one shown for enalapril.

Fosinopril 4-Cyclohexyl-1-[2-[[(2-methyl-1-propanoyloxy-propoxy)-(4-phenylbutyl)phosphoryl]acetyl] -pyrrolidine-2-carboxylic acid

It is a phosphate-containing prodrug cleaved at the ester moiety by hepatic esterases, and transformed into fosinoprilate.

Synthesis

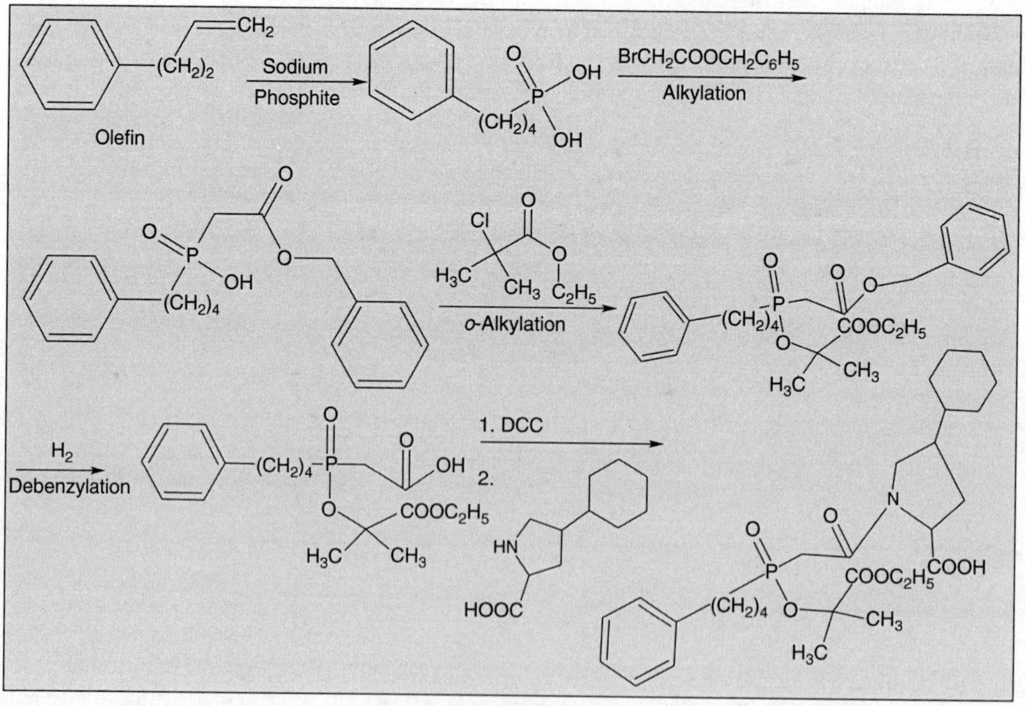

Alkylation of phosphinic acid with benzyl bromoacetate in presence of base gives *P*-alkylation product. Reaction of the alkoxide from this intermediate with chloroester derivative leads to alkylation on oxygen and forms ether. This product is then debenzylated by treatment with hydrogen/palladium. Coupling of this acid with proline in presence of DCC affords fosinopril.

16.10 ANGIOTENSIN II ANTAGONISTS (AT$_2$ ANTAGONISTS)

These compounds compete with angiotensin II at the receptor site and block the contractile effect of angiotensin II in all vascular smooth muscles. Unlike ACE inhibitors, these drugs do not cause cough, and have not been associated with angioneurotic oedema. They do not inhibit the breakdown of bradykinin or other kinins, and are thus only rarely associated with the persistent dry cough and/or angiooedema that limit ACE inhibitor therapy.

> There are three functional groups that are the most important parts for the bioactivity.
> The first one is the imidazole ring that binds to amino acids in helix 7 (Asn^{295}). The second group is the biphenyl-methyl group that binds to amino acids in both helices 6 and 7 (Phe^{301}, Phe^{300}, Trp^{253} and His^{256}). The third one is the tetrazole group that interacts with amino acids in helices 4 and 5 (Arg^{167} and Lys^{199}).

Losartan (1-((2'-(2*H*-Tetrazol-5-yl)biphenyl-4-yl)methyl)-2-butyl-4-chloro-1*H*-imidazol-5-yl)methanol

Synthesis

Alkylation of 2-butyl-6-chloro-5-hydroxymethyl imidazole with bromomethylbiphenyl derivative gives the *N*-alkylated imidazole. Treatment of this compound with sodium azide in presence of acid converts the nitrile group to a tetrazole, losartan.

Candesartan 3-((2'-(2*H*-Tetrazol-5-yl)biphenyl-4-yl)methyl)-2-ethoxy-3*H*-benzo[d]imidazole-4-carboxylic acid

Synthesis

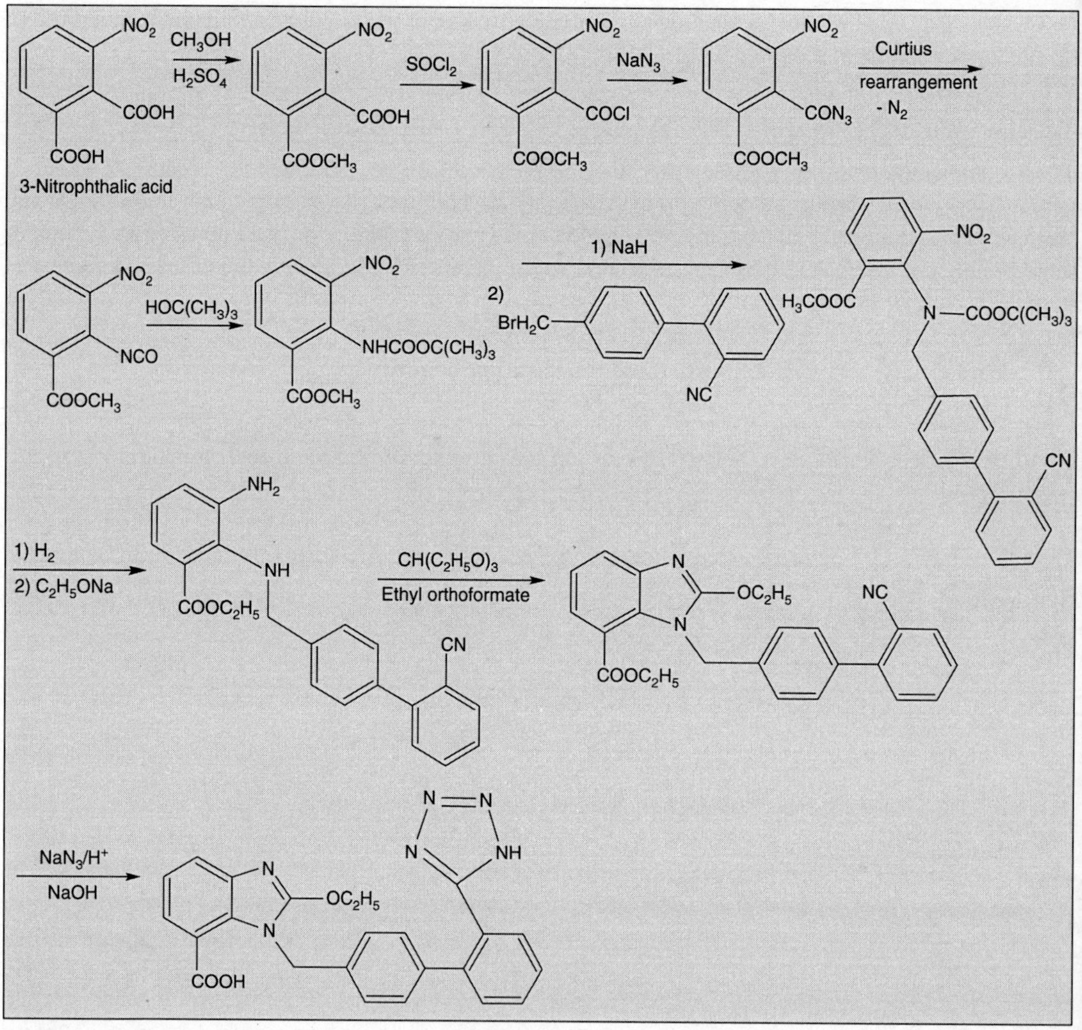

Esterification of 3-nitrophthalic acid takes place at the more open carboxyl group to give the methyl ester. The free carboxyl group is then converted to its acid chloride and then treated with sodium azide which undergo Curtius rearrangement to an isocyanate, which in presence of isopropanol gives carbamate derivative. This is treated with base and then allowed to react with the bromomethylbiphenyl derivative to give N-alkylated intermediate. Nitro group is reduced; sodium ethoxide treatment cleaves carbamate to amine. Reaction of this diamine with ethyl orthoformate cyclizes to the benzimidazole ring. Conversion of the nitrile group to tetrazole with sodium azide followed by saponification affords candesartan.

Valsartan 3-Methyl-2-[pentanoyl-[[4-[2-(2H-tetrazol-5-yl)phenyl]phenyl] methyl]amino]-butanoic acid

Synthesis

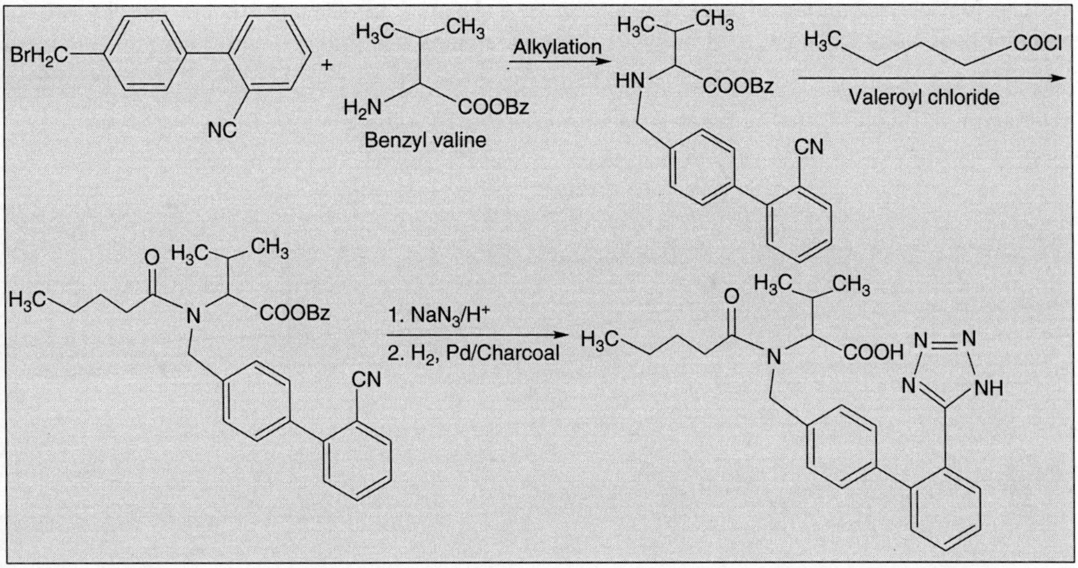

Alkylation of the benzyl valine with bromomethylbiphenyl derivative gives *N*-alkyl derivative, which in turn condenses with valeroyl chloride to afford the amide. The nitrile group is then converted to the corresponding tetrazole, and debenzylation affords valsartan.

16.11 MISCELLANEOUS DRUGS

Thiazide and high-ceiling diuretics lower blood pressure of persons with essential hypertension (see Chapter 15 on Diuretics).

16.12 NEWER DRUGS

Eprosartan 4-[[2-Butyl-5-(2-carboxy-3-thiophen-2-yl-prop-1-enyl)-imidazol-1-yl]methyl]benzoic acid

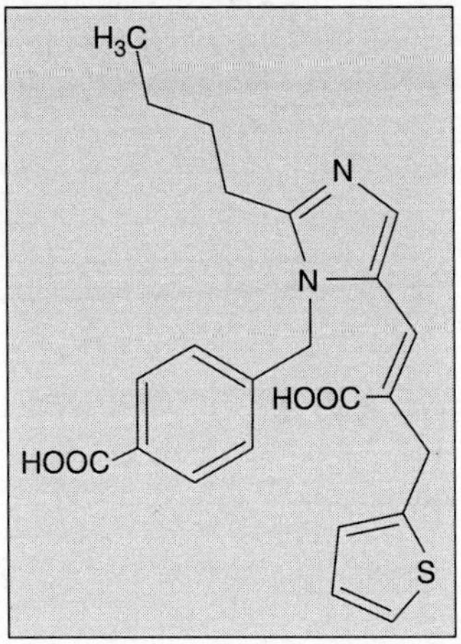

This drug acts on the renin-angiotensin system in two ways to decrease total peripheral resistance. First, it blocks the binding of angiotensin II to AT_2 receptors in vascular smooth muscle, causing vascular dilatation. Second, it inhibits sympathetic norepinephrine production, further reducing blood pressure.

Irbesartan 2-Butyl-3-[*p*-(*o*-1*H*-tetrazol-5-ylphenyl)benzyl]-1,3-diazaspiro[4.4]non-1-en-4-one

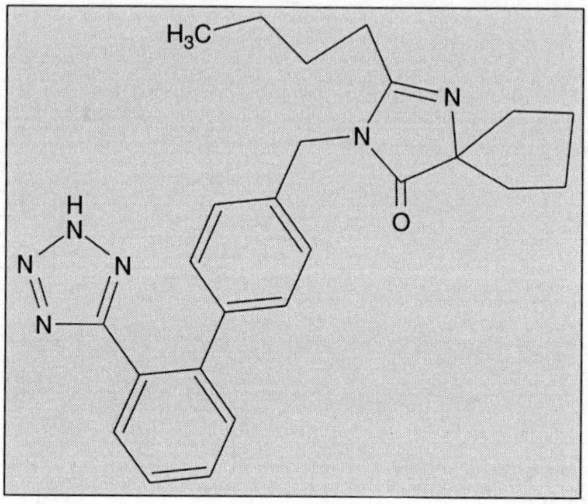

As with all angiotensin II receptor antagonists, irbesartan is indicated for the treatment of hypertension. Irbesartan may also delay progression of diabetic nephropathy and is also indicated for the reduction of renal disease progression in patients with type 2 diabetes.

> **Blood pressure vaccine:** Blood pressure vaccinations are being trialed and may become a treatment option for high blood pressure in the future. Research published in *The Lancet* on 8 March, 2008, titled 'Vaccination against high blood pressure: A new strategy', showed patients experienced a drop in systolic and diastolic blood pressure after taking the vaccine. Effective blood pressure vaccines would assist those people who forget to take their medication. It would also help those who stop taking their medication due to side-effects or falsely believing they do not need them anymore once their blood pressure is lowered.

FURTHER READINGS

1. Van Zwieten, 1997. *Antihypertensive Drugs*, Taylor & Francis.
2. Wilson and Giswold's Textbook of Organic Medicinal & Pharmaceutical Chemistry, Block, J.H. and Beale, J.M., Jr (ed.), 11th Ed., 2004.

MULTIPLE-CHOICE QUESTIONS

1. One of the following antihypertensives is a prodrug:
   a. Verapamil
   b. Sodium nitroprusside
   c. Losartan
   d. Minoxidil

2. From the given structures, identify which one belongs to calcium channel blockers.

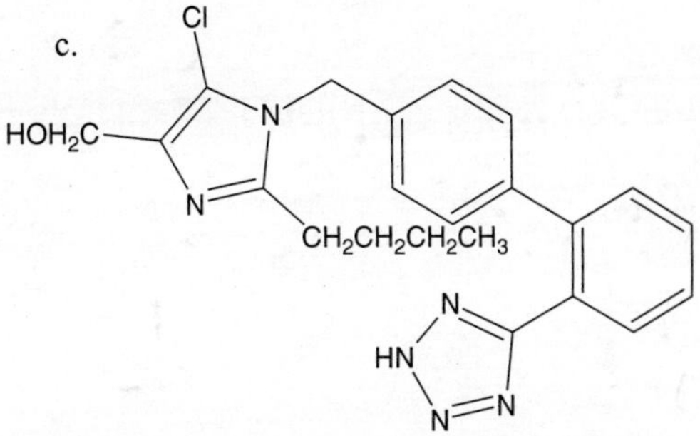

3. The normal blood pressure is
   a. 120/90
   b. 120/80
   c. 139/80
   d. 140/90

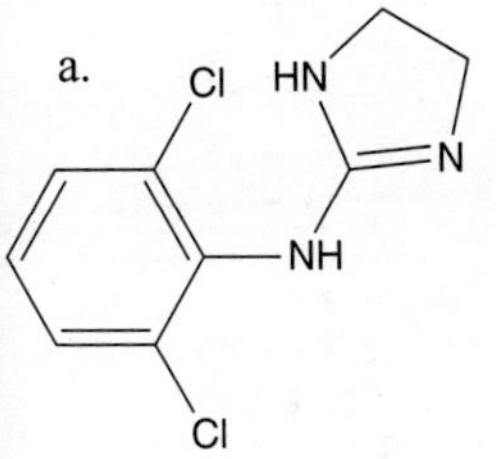

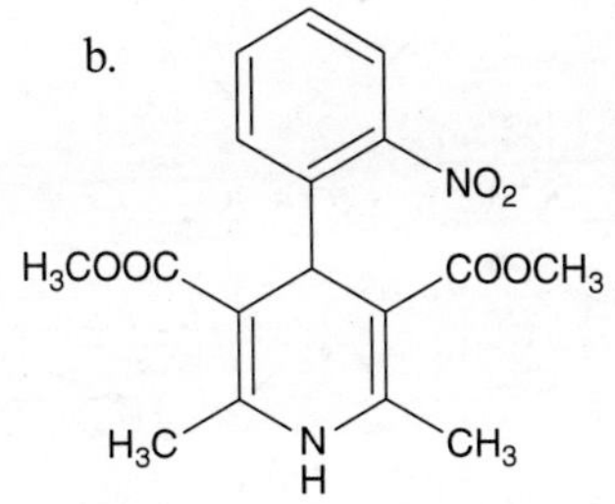

4. The water-soluble analogue of prazosin is
   a. Terazosin
   b. Trimazosin
   c. Doxazosin
   d. Hydralazine

5. The antihypertensive agent used in the prophylaxis of migraine and cluster headaches is
   a. Verapamil
   b. Clonidine
   c. Methyldopa
   d. Diltiazem

6. The drug that acts through mediation of nitric oxide is
   a. Atenolol
   b. Propanolol
   c. Hydralazine
   d. Clonidine

7. The antihypertensive with both non-selective β- and α-antagonist activity is
   a. Propanolol
   b. Timolol
   c. Esmolol
   d. Labetolol

8. The antihypertensive drug that is a potassium channel blocker is
   a. Verapamil
   b. Minoxidil
   c. Propanolol
   d. Diltiazem

9. The antihypertensive drug with a tetrazole nucleus is
   a. Diazoxide
   b. Valsartan
   c. Taludipine
   d. Fosinopril

10. The antihypertensive also useful topically to stimulate hair growth is
    a. Losartan
    b. Minoxidil
    c. Nifedepine
    d. Esmolol

QUESTIONS

1. Classify antihypertensives and give two structural examples for each class.
2. Discuss the mechanism of action of hydralazine and propranolol.
3. What are the current targets for blood pressure treatment?
4. Explain the mechanism of action of ACE inhibitors.
5. Give the synthetic protocol for nifedepine, propranolol, and esmolol.

SOLUTION TO MULTIPLE-CHOICE QUESTIONS

1. d;
2. b;
3. b;
4. c;
5. b;
6. c;
7. d;
8. b;
9. b;
10. b.

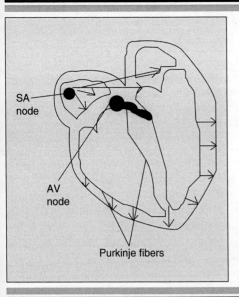

CHAPTER 17

Antiarrhythmic Drugs

LEARNING OBJECTIVES

- Define antiarrhythmic drugs and their utility
- Categorize antiarrhythmics based on the mode of action
- Describe the synthesis, mode of action, and dose administered from each class of antiarrhythmics

17.1 INTRODUCTION

Antiarrhythmic drugs are medicines that correct irregular heartbeats and slow down hearts that beat too fast. The heart is a muscle that works like a pump. It is divided into four chambers, two atria at the top of the heart and two ventricles at the bottom. The heart beats (contracts) when an electrical impulse from the heart's 'natural pacemaker"—the sinoatrial (SA) node—moves through it. The normal sequence begins in the right atrium, spreads throughout the atria, and to the atrioventricular (AV) node. From the AV node, the impulse travels down a group of specialized fibres (His—Purkinje system) to all parts of the ventricles. This exact route must be followed for the heart to pump properly. As long as electric impulse is transmitted normally, the heart pumps and beats at a regular pace. A normal heart beats 60–100 times a minute.

The term arrhythmia refers to any change from the normal sequence of electrical impulses, causing abnormal heart rhythm. This can cause the heart to pump less effectively.

The rhythm of heart contractions depends on many parameters: condition of pacemaker cells and the conduction system, myocardial blood flow, and other factors; consequently, arrhythmia can originate for different reasons that are caused by disruptions in electrical impulse generation or conduction. They can be caused by heart disease, myocardial ischemia, electrolytic and acid–base changes, heart innervations problems, intoxication of the organism, and so on.

351

Some arrhythmias are life-threatening medical emergencies that can result in cardiac arrest and sudden death. Others cause aggravating symptoms such as an abnormal awareness of heart beat (palpitations), and may be merely annoying. Others may not be associated with any symptoms at all, but predispose towards potentially life-threatening stroke or embolus.

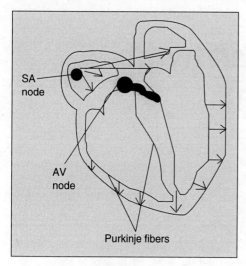

FIGURE 17.1

Drugs used for treating arrhythmia can have an effect on the electrical conduction system of the heart, its excitability, automatism, the size of the effective refractory period, and adrenergic and cholinergic heart innervations. Accordingly, compounds of various chemical classes can restore heart-rate disturbances. As already noted, arrhythmia originates from problems forming electric impulses and propagating them to the heart, or when both of these happen simultaneously, which can be accomplished by transferring Na^+, K^+, and Ca^{2+} ions through cell membranes. Therefore, the mechanism of action of many antiarrhythmic drugs consists of blocking Na^+ and Ca^+ ion channels of the myocardium, which prolongs the time necessary for restoration after being activated by these channels, and which in turn acts on the electrical conduction system of the heart, its excitability, automatism, and so on.

17.2 CLASSIFICATION

Antiarrhythmic drugs are classified according to their electrophysiological properties.

1. **Class I agents**:
   A. Class IA: Quinidine, Procainamide, Disopyramide, Cifenline
   B. Class IB: Lidocaine, Phenytoin, Tocainide, Mexiletine
   C. Class IC: Encainide, Flecainide, Propafenone
2. **Class II agents**: β-adrenoreceptor antagonists
3. **Class III agents**: Amiodarone, Bretylium, Acecainide

17.2.1 Class I Agents

Mechanism of action: Drugs belonging to this group form complexes with lipoproteins of cell membranes of the myocardia, thus blocking Na^+ channel conductivity and the flow of Na^+ into the cell, and facilitate the release of K^+ from myocardial cells, which as a result leads to a weak suppression of depolarization of myocardial cells, reduction of repolarization time, and a slowing of the propagation of excitation.

- Drugs such as encainide, flecainide, and propafenone in class IC slowly dissociate from the Na^+ channel, causing slowing down of the conduction time of the impulse through the heart.

Medicinal Chemistry

- Class IB drugs, which include lidocaine, phenytoin, tocainide, and mexiletine, have the property of rapidly dissociating from the Na⁺ channels. These drugs shorten the action potential duration.
- Class IA drugs, which include quinidine, procainamide, and disopyramide, have an intermediate rate of dissociation from the Na⁺ channels. These drugs lengthen the refractory period of cardiac tissue to cause cessation of arrhythmias.

Procainamide *p*-Amino-N-[2-(diethylamino)ethyl]benzamide

Synthesis

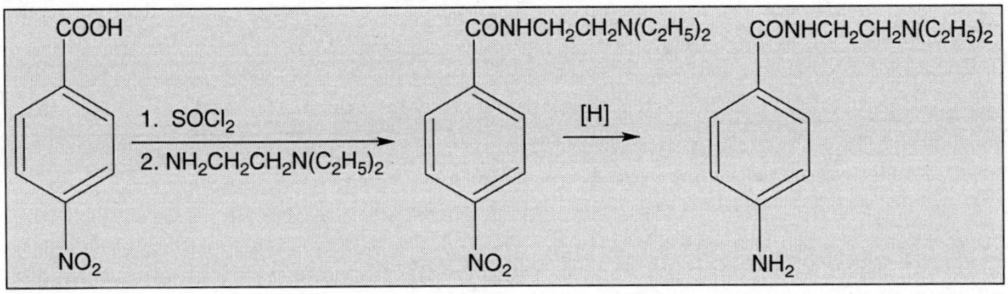

Procainamide is synthesised by reacting 4-nitrobenzoic acid chloride with N,N-(diethyl)ethylendiamine and subsequent reduction of the nitro group of the resulting 4-nitro-N-[2-(diethylamino)ethyl]benzamide into an amino group.

Procainamide is used to treat a variety of arrhythmias including atrial fibrillation and ventricular tachycardia.

Disopyramide α-[2-(Diisopropylamino)ethyl]-α-phenyl-2-pyridineacetamide

Synthesis

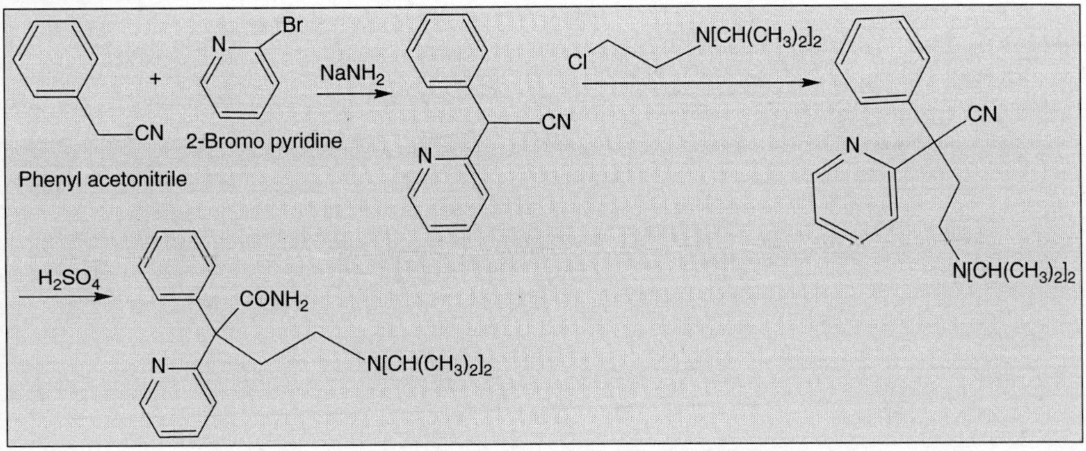

Disopyramide is synthesised by arylating benzylcyanide with 2-chloropyridine in the presence of sodium amide and subsequent alkylation of the resulting α-phenyl-α-(2-pyridyl) acetonitrile with 2-diisopropylaminoethylchloride using sodium amide. Sulphuric acid hydrolysis of the resulting nitrile leads to the formation of α-(2-diisopropylaminoethyl)-α-phenyl-2-pyridineacetamide, disopyramide.

Disopyramide is used to treat a variety of arrhythmias including atrial fibrillation and ventricular arrhythmias.
Tocainide 2-Amino-N-(2,6-dimethyl phenyl)propanamide

Synthesis

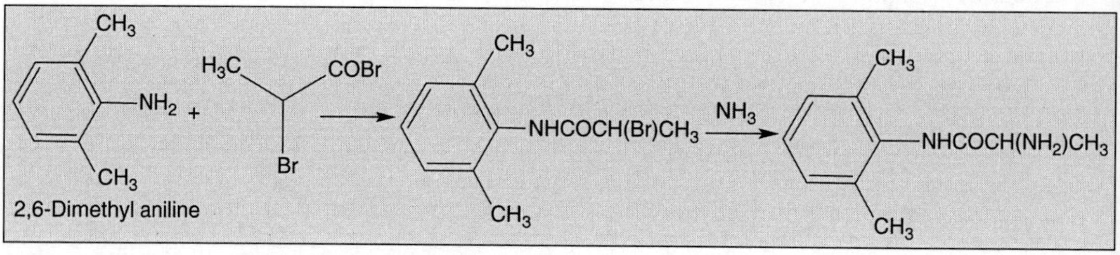

Tocainide is synthesised by reacting 2,6-dimethylaniline with 2-bromopropionic acid bromide and subsequent substitution of the bromine atom in the resulting amide with an amino group.
Mexiletine 1-Methyl-2-(2,6-xylyloxy)ethylamine

Synthesis

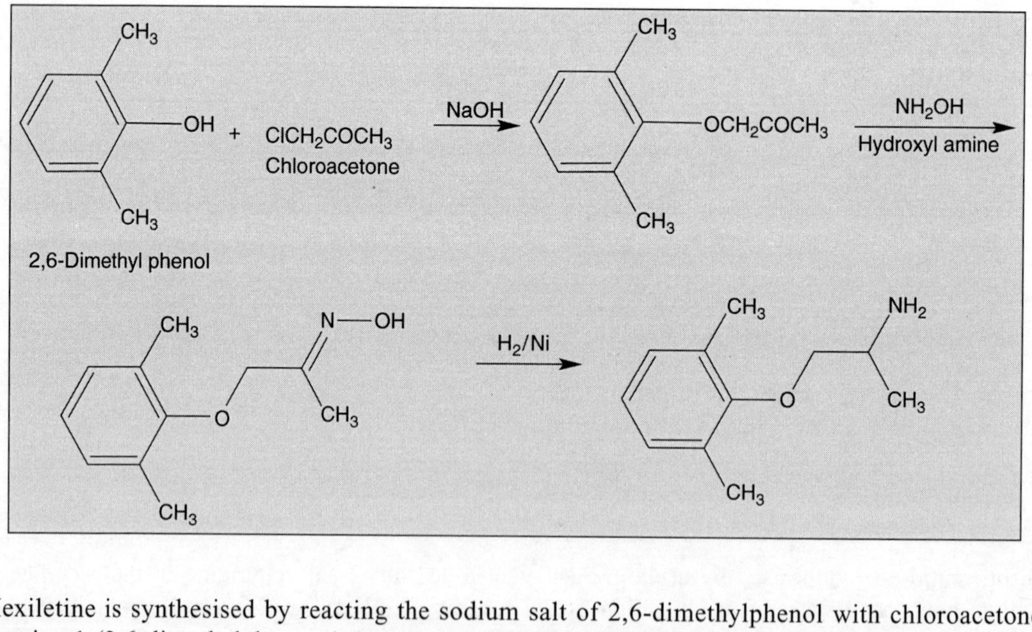

Mexiletine is synthesised by reacting the sodium salt of 2,6-dimethylphenol with chloroacetone, forming 1-(2,6-dimethylphenoxy)-2-propanone. Reacting this with hydroxylamine gives the corresponding oxime. Reduction of the oxime group using hydrogen over Raney nickel gives mexiletine.
Encainide 4-Methoxy-N-[2-(1-methyl-2-piperidinyl)ethyl]phenylbenzamide

Synthesis

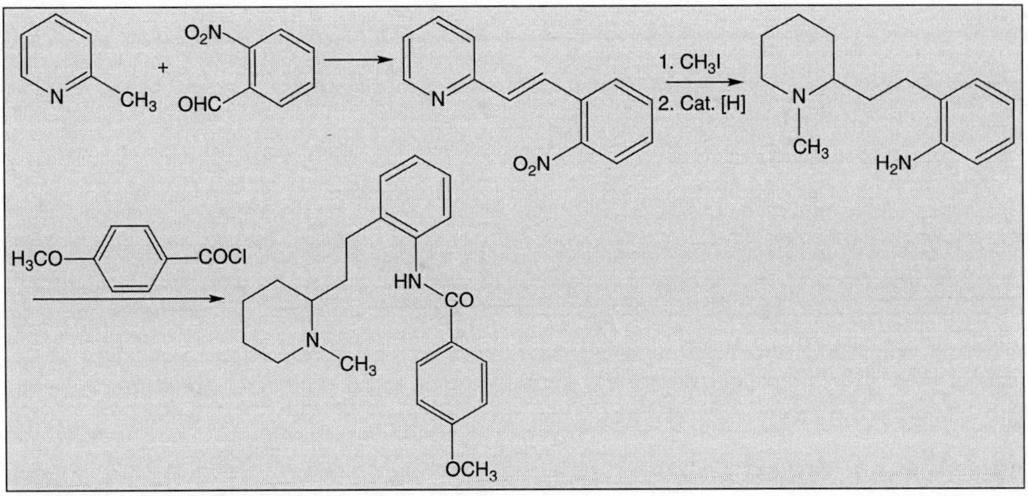

Condensation of 2-picoline and 2-nitrobenzaldehyde gives benzylidene derivative. N-Methylation followed by catalytic reduction gives piperidine derivative. Acylation with 4-methoxybenzoyl chloride affords encainide.

Flecainide N-(2-Piperidinylmethyl)-2,5-bis(2,2,2-trifluoromethoxy) benzamide

Synthesis

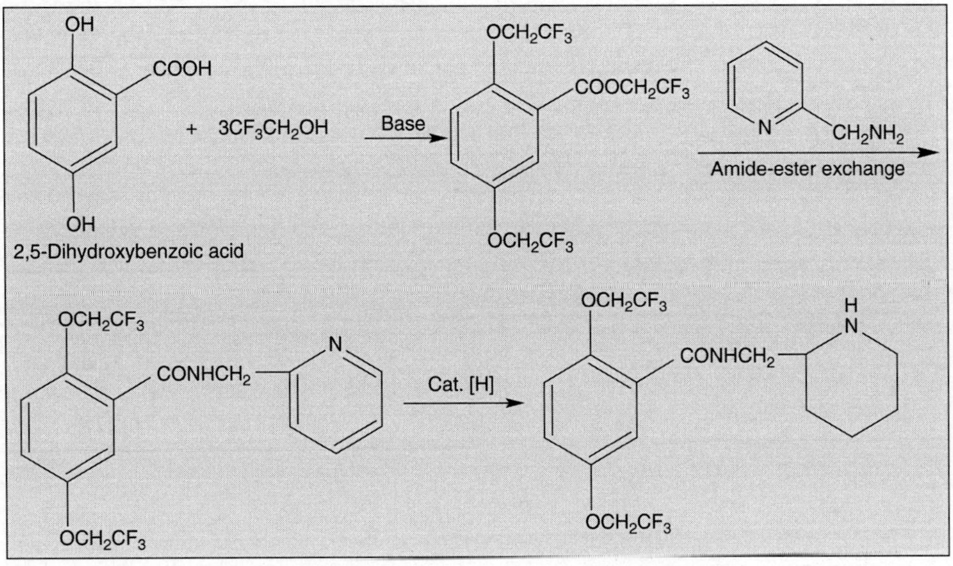

Flecainide is synthesised from 2,5-dihydroxybenzoic acid. Reacting this with 2,2,2-trifluoroethanol gives 2.2.2-trifluoroethoxylation of all three groups, to produce 2,2,2-trifluoroethyl ester of 2,5-bis-(2,2,2-trifluoroethoxy)benzoic acid. Reacting this with 2-aminomethylpiridine gives the corresponding amide, which upon reduction of the pyridine ring with hydrogen gives flecainide.

Propafenone 1-[2-(2-Hydroxy-3-isopropylamino-propoxy)phenyl]-3-phenyl-propan-1-one
Synthesis

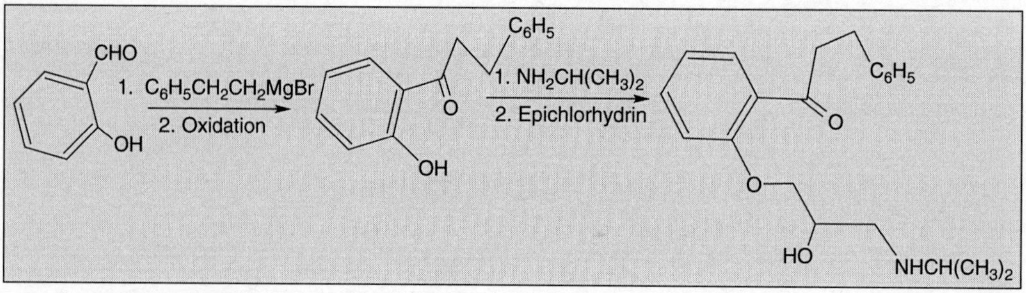

Reaction of salicylaldehyde and phenylethyl magnesium bromide gives secondary alcohol, which on oxidation gives phenylpropanone derivative. Hydroxyl group reacts with epichlorhydrin and isopropyl amine to afford propafenone.

17.3 CLASS II AGENTS

β-Adrenergic blocking agents (propranolol) have the property of causing membrane-stabilizing or depressant effects on myocardial tissues. However, their antiarrhythmic properties are considered to be principally due to inhibition of adrenergic stimulation of the heart. (Refer to Chapter 16 on Antihypertensive Agents)

17.4 CLASS III AGENTS

Drugs in this class cause several different electrophysiological changes on the myocardial tissue, but share one common effect—that of prolonging the duration of the membrane action potential without altering the phase of depolarization or the resting membrane potential. Drugs in this class have been shown to:

- Increase refractoriness without reducing conduction velocity in tissues
- Have little or no negative inotropic activity
- Have high efficacy in the treatment of refractory cardiac arrhythmias

Amiodarone 2-Butyl-3-benzofuranyl-4-[2-(diethylamino)ethoxy]-3,5-diiodophenyl ketone
Synthesis

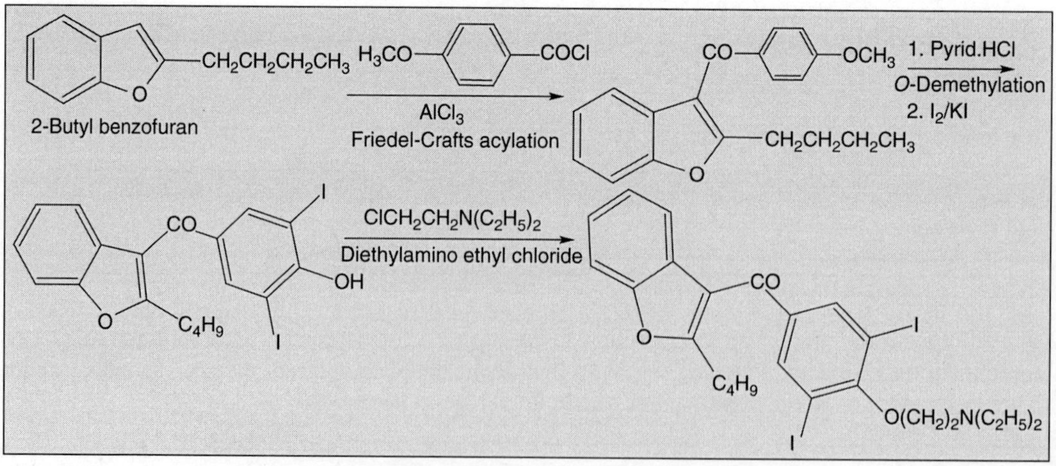

Amiodarone is synthesised from 2-butylbenzofurane. This is acylated with 4-methoxybenzoic acid chloride, giving 2-butyl-3-(4-methoxybenzoyl)benzofuran, which undergoes demethylation by pyridine hydrochloride, forming 2-butyl-3-(4-hydroxy-benzoyl)-benzofuran. The resulting product is iodized in the presence of potassium iodide, forming 2-butyl-3-benzofuranyl-4-(2-hydroxy-3,5-diiodophenyl) ketone, which is reacted further with 2-diethylaminoethylchloride, giving the desired amiodarone.

Bretylium (*o*-Bromobenzyl)ethyl dimethyl ammonium bromide

Synthesis

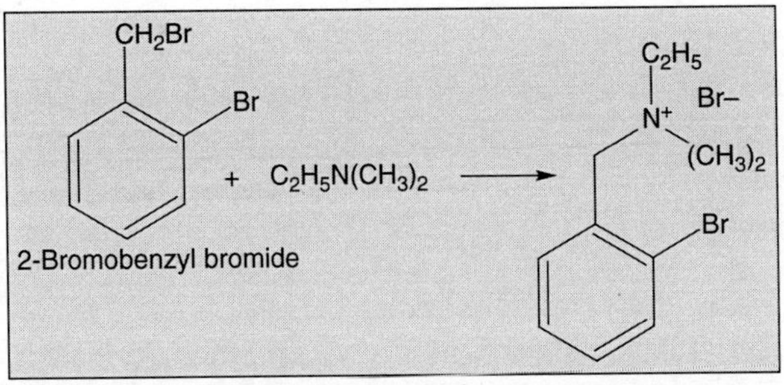

Bretylium is synthesised by reacting o-bromobenzyl bromide with ethyldimethylamine.

Acecainide 4-Acetylamino-N-[2-(2-diethylamino)ethyl]benzamide

Synthesis

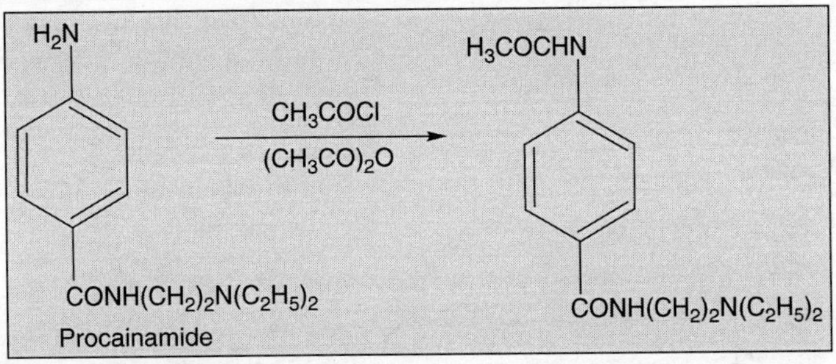

Acecainide is synthesised by acylating procainamide with acetyl chloride.

17.4.1 Calcium Channel Blockers

Calcium channel blockers like verapamil and diltiazem block the slow inward current of Ca^{2+} ions during phase-2 of the membrane action potential in cardiac cells. (Refer to Chapter 16, Antihypertensive agents) Calcium channel blockers inhibit slow transmembrane calcium ion flow in the cell of the conductive system of the heart during depolarization, which causes a slowing of atrioventricular conductivity and increased effective refractive period of atrioventricular ganglia, which eventually leads to the relaxation of smooth muscle of heart musculature and restores normal sinus rhythm during supraventricular tachycardias.

FURTHER READINGS

1. Richard N. Fogoros, 1997. *Antiarrhythmic drugs*, Blackwell Publishing.
2. Foye's Principles of Medicinal Chemistry, by Williams, David A. and Thomas L. Lemke (ed.), 5th Ed., Lippincott Williams & Wilkins, 2002.

MULTIPLE-CHOICE QUESTIONS

1. Mexiletine belongs to
   a. Class IA
   b. Class IB
   c. Class II
   d. Class III

2. Mechanism of action of propafenone is
   a. β-Blocker
   b. Blocks K^+ channel
   c. Blocks Na^+ channel
   d. Blocks Ca^{2+} channel

3. Flecainide belongs to
   a. Class IA
   b. Class IB
   c. Class IC
   d. Class III

4. Mechanism of action of bretylium is
   a. β-blocker
   b. Blocks K^+ channel
   c. Blocks Na^+ channel
   d. Blocks Ca^{2+} channel

5. Mechanism of action of verapamil is
   a. β-blocker
   b. Blocks K^+ channel
   c. Blocks Na^+ channel
   d. Blocks Ca^{2+} channel

6. The antiarrhythmic drug with local anaesthetic action is
   a. Verapamil
   b. Bretylium
   c. Procainamide
   d. Disopyramide

7. Acecainide is
   a. 4-Acetylamino-N-[2-(2-diethylamino)ethyl] benzamide
   b. 4-Amino-N-[2-(2-diethylamino)ethyl]benzamide
   c. 4-Acetylamino-N-[2-(2-amino)ethyl]benzamide
   d. 4-Acetylamino-N-[2-(2-diethylamino)] benzamide

8. A normal heart beats ...... times a minute
   a. 75
   b. 120
   c. 20
   d. 150

9. Tocainide is synthesised starting from
   a. 2,6-Dimethyl aniline
   b. 2,6-Dimethylphenol
   c. 2,6-Dimethylcyclohexylamine
   d. 2,6-dimethylacetophenone

10. Epichlorhydrine is used in the synthesis of
    a. Procainamide
    b. Propafenone
    c. Acecainide
    d. Amiodarone

QUESTIONS

1. How are antiarrhythmic drugs classified? Give at least two structural examples for each class.
2. Discuss the mechanism of action of Class I agents and how these are classified further according to the mode of action.
3. Write the synthetic protocol for acecainide, amiodarone, and tocainide.

SOLUTION TO MULTIPLE-CHOICE QUESTIONS

1. b;
2. c;
3. c;
4. b;
5. d;
6. c;
7. a;
8. a;
9. a;
10. b.

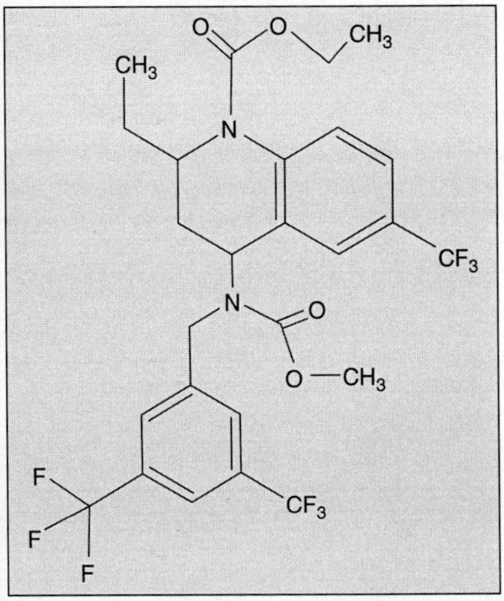

CHAPTER 18

Antihyperlipidemic Agents

LEARNING OBJECTIVES

- Define hyperlipidemia and atherosclerosis, and their causes
- Classify antihyperlipidemics based on the structure and mode of action
- Define HMG-CoA and its role in hyperlipidemia
- Define the pharmacophoric requirements of the statin class of antihyperlipidemics
- Describe some antihyperlipidemics obtained from fermentation process
- Describe the cholesterol biosynthetic pathway and its significance
- Define PPAR?-inhibitors and their utility
- Describe some new drug classes useful in the treatment of hyperlipidemia and atherosclerosis

18.1 INTRODUCTION

As a regulator of homeostasis, a precursor to the corticosteroids and sex hormones, and a critical factor in the maintenance of cell-wall integrity, cholesterol is essential to life. However, high levels of this lipophilic substance lead to atherosclerosis, a predisposing factor to the development of coronary artery disease (CAD). Hyperlipidemia is the most prevalent indicator for susceptibility to atherosclerotic heart disease; it is a term used to describe elevated plasma levels of lipids that are usually in the form of lipoproteins. It was demonstrated that there exists a link between serum cholesterol levels and risk to coronary heart disease. Hyperlipidemia may be caused by an underlying disease involving the liver, kidney, pancreas, or thyroid.

Atherosclerosis comes from the Greek words *athero* (meaning 'gruel' or 'paste') and *sclerosis* ('hardness'). It causes degenerative changes in the intima of medium and large arteries. The degeneration includes accumulation of fatty substances, cholesterol, cellular waste products, calcium, blood, and other substances, and is accompanied by the formation of fibrous tissues on the intima of the blood vessels. These deposits or plaques decrease the lumen of the artery, reduce its elasticity, and may create foci for thrombi and subsequent occlusion of the blood vessels.

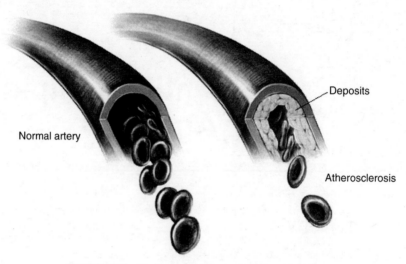

FIGURE 18.1 Normal and atherosclerotic artery.

The primary developmental mechanism of the atherosclerotic process is not completely understood. It seems likely that the development of atherosclerosis is preceded by metabolic abnormalities of the synthesis, transport, and utilization of lipids. Lipids such as triglycerides and cholesterol esters are circulated in the blood in the form of particles (lipoproteins) wrapped in hydrophilic membranes that are synthesised from phospholipids and free cholesterol. Cholesterol is transported by particles of various sizes synthesised from triglycerides, cholesterol esters, and phospholipids, each of which plays a very specific role.

The likely progression of the process of atherosclerotic plaque formation can be briefly described in the following manner. Over time, usually years, endothelium at a certain spot of a vessel is somehow damaged by turbulent blood flow. Thrombocytes are attracted to the damaged region. The combined action of endothelial growth, thrombocytes moving in, and growth factor attracting macrophages to the region causes an inflammatory reaction that leads to hypertrophy of middle artery muscles, which constricts the whole length of the vessel and forms plaque. Endothelia are never completely restored and can be a region of thrombosis formation in a constricted vessel.

Reducing the level of cholesterol in the organism mainly consists of either removing the excess amount from the plasma, or inhibiting low-density and very low-density lipoprotein (LDL and VLDL) synthesis. Hypolipidemic agents are accordingly subdivided into drugs that enhance catabolism and removal of atherogenic lipoproteins and lipids from the organism (colestipol and cholestyramine), and drugs that inhibit the formation of atherogenic lipoproteins fibrates (clofibrate and henfibrozil); natural compounds—statines (lovastatin, mevastatin, and their analogues), as well as probucol and nicotinic acid. In addition, drugs are sometimes used that lower cholesterol levels in the organism by mechanisms that are not completely explainable—dextrotiroxin and neomycin, for instance.

Medicinal Chemistry 361

18.2 CLASSIFICATION

1. HMG-CoA reductase inhibitors (statins): Lovastatin, Simvastatin, Dalvastatin, Pravastatin, and Fluvastatin
2. Aryloxyisobutyric acid derivatives (fibrates): Clofibrate, Fenofibrate, Ciprofibrate, Benzafibrate
3. Miscellaneous: Bile acid sequestrants, Probucol, Gemfibrosil, Niacin

18.3 HMG-CoA REDUCTASE INHIBITORS (STATINS)

Statins are most effective cholesterol-lowering drugs. Statins lower total cholesterol and LDL particles. These are competitive inhibitors. The HMG-CoA has a conformation similar to the lactone moiety of statins, resulting in binding at the same site without any productive effect.

The statin pharmacophore All statins consist of two specific structural components, a dihydroxyheptanoic acid unit and a ring system with lipophilic substituents. The dihydroxyheptanoic acid component is essential to HMGR-inhibiting activity. Since they bind to the same active site, the structure of the statin HMGR inhibitors resembles the endogenous substrate, HMG CoA.

Conformational similarity is depicted in the accompanying figure.

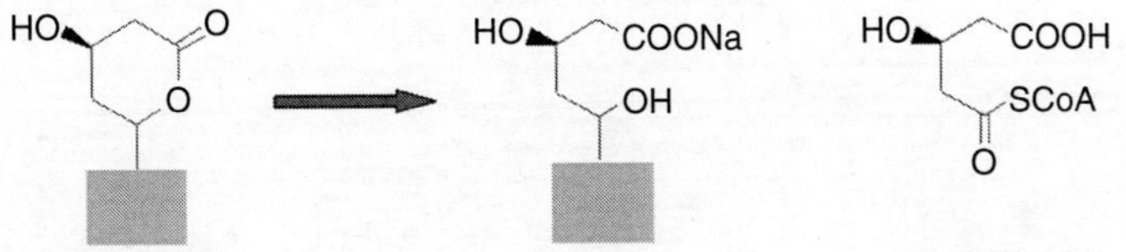

Mevastatin
Lovastatin

Pravastatin

HMG-CoA substrate

All statins contain a modified hydroxyglutaric acid component that mimics the 3-hydroxyglutaryl unit of both the substrate (HMG CoA) and the mevaloyl CoA transition state intermediate. Statins that are active and intact contain a free acidic functional group, which mimics the free COOH of HMG CoA. These compounds exist as anions at pH 7.4, and this is critical to their ability to compete for the HMGR active site by anchoring via an electrostatic bond to the cationic Lys735 of the enzyme. Compounds that have their carboxylic acid group 'tied up' in a lactone (cyclic ester) are prodrugs. The essential anionic carboxylate group must be liberated by hydrolysis before activity is realized. Of course, the enzymes that catalyse the activating hydrolysis reaction are esterases.

Fermentation-derived inhibitors:

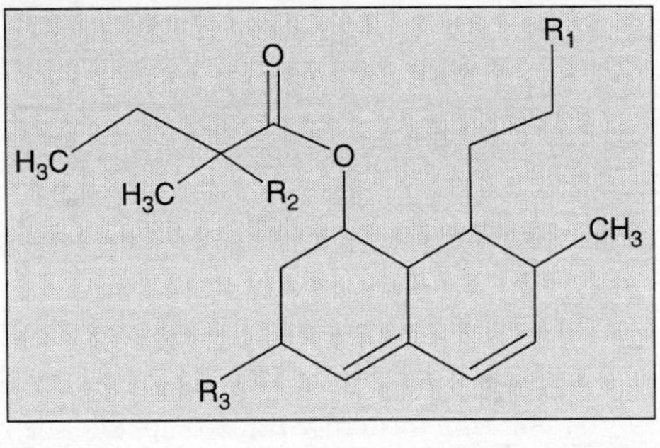

	R₁	R₂	R₃	Source
Mevastatin	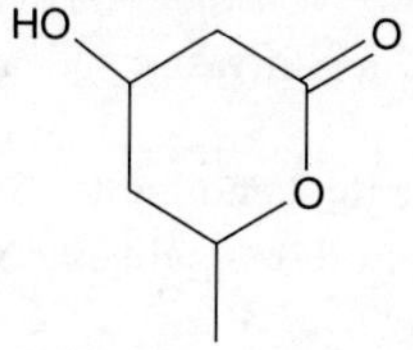 (hydroxy-lactone side chain)	H	H	Penicillium sps.
Lovastatin	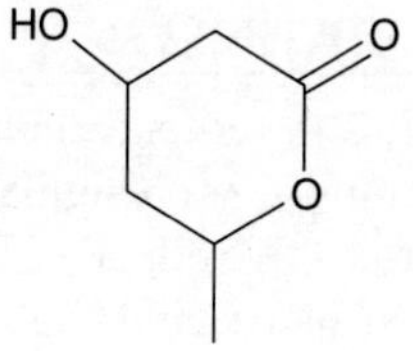 (hydroxy-lactone side chain)	H	CH₃	Aspergillus sps.
Simvastatin	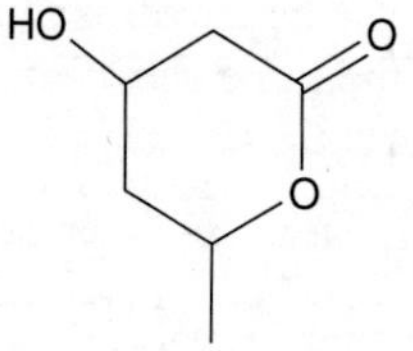 (hydroxy-lactone side chain)	CH₃	CH₃	Semi-synthetic
Pravastatin	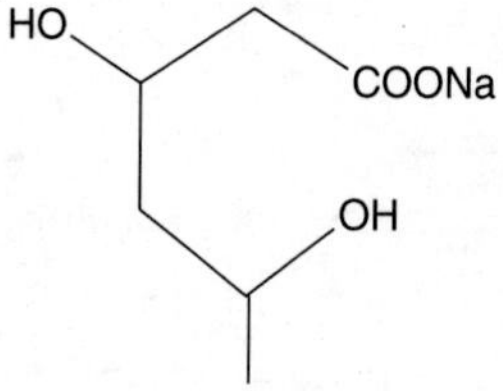 (open hydroxy acid, COONa, OH)	H	OH	Absidia corulea.

Synthetic inhibitors:

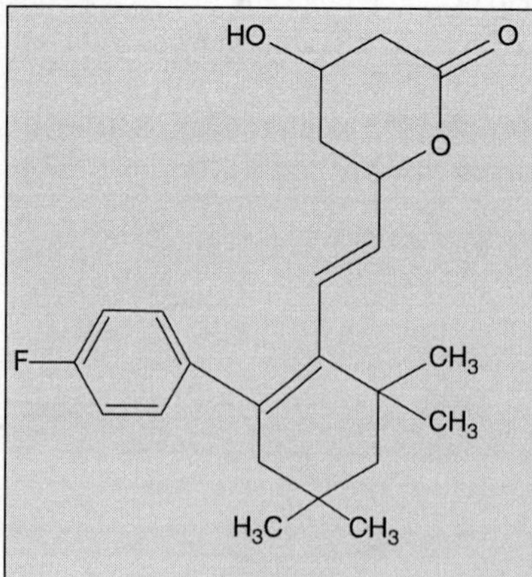

Dalvastatin

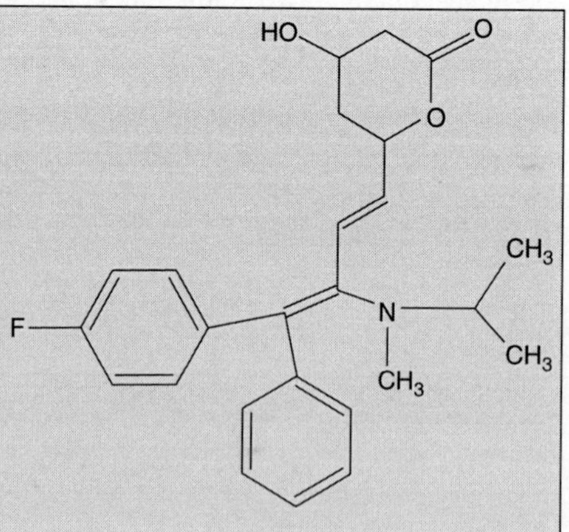

Fluvastatin

Fermentation-derived HMG-CoA reductase inhibitors (statins) are all closely related in structure. Each is composed of a hexahydronaphthalene ring system with two appendages: a methyl butyrate ester and a hydroxyl acid that can form a six-member lactone ring.

Mechanism of action:
Biosynthetic pathway of cholesterol

$$2\ CH_3COOH + 2\ CoA\text{-}SH \longrightarrow 2\ CH_3CO\text{-}SCoA \xrightarrow{\text{Thiolase}} CH_3COCH_2CO\text{-}SCoA$$
Acetate   Coenzyme A   Acetyl CoA   Acetoacetyl CoA

1. Acetyl CoA
2. HMG-CoA synthase

$$\xrightarrow{} HOOC\text{-}CH_2\text{-}\underset{OH}{\overset{CH_3}{C}}\text{-}CH_2CO\text{-}SCoA \xrightarrow{\text{HMG-CoA reductase}} HOOCCH_2\text{-}\underset{CH_3}{\overset{CH_3}{C}}\text{-}CH_2CH_2OH$$

3-Hydroxy-3-methylglutaryl CoA
(HMG-CoA)

Mevalonic acid

All the statins hydrolyse *in vivo* to hydroxyl carboxylic acid and inhibit HMG-CoA reductase found in the liver. This enzyme is the rate-limiting catalyst for the irreversible conversion of HMG CoA to mevalonic acid in the synthesis of cholesterol. The binding of the dihydroxyheptanoic acid portion of the statins to the HMG reductase precisely mimics that of the endogenous substrate HMG-CoA.

The statins take advantage of the flexibility of the HMGR receptor, enticing it to accommodate their large, lipophilic ring systems and substituents. These bulkier groups distort the active site cavity to form a shallow, hydrophobic pocket that binds tightly with these groups.

The first agent isolated was mevastatin, a molecule produced by Penicillium citrinum. The pharmaceutical company Merck & Co. showed an interest in the Japanese research in 1976, and isolated lovastatin, the first commercially marketed statin, from the mould Aspergillus terreus.

Lovastatin

8-[-[2R,4R]-[tetrahydro-4-hydroxy-6-oxo-2H-pyran-2-yl]ethyl]-1-naphthyl(S)-2-methylbutyrate] and simvastatin, which have their 3,5-dihydroxyheptanoic acid segment cyclized into a lactone ring, must be activated by hydrolysis to the active, anionic carboxylate form. Only when hydrolysed can these agents anchor to the cationic Lys735 of the HMGR enzyme.

Lovastatin R = H
Simvastatin R = CH$_3$

Active carboxylate metabolite

Fluvastatin 7-[3-(4-Fluorophenyl)-1-(1-methylethyl)-1H-indol-2-yl]-3, 5-dihydroxy-hept-6-enoic acid

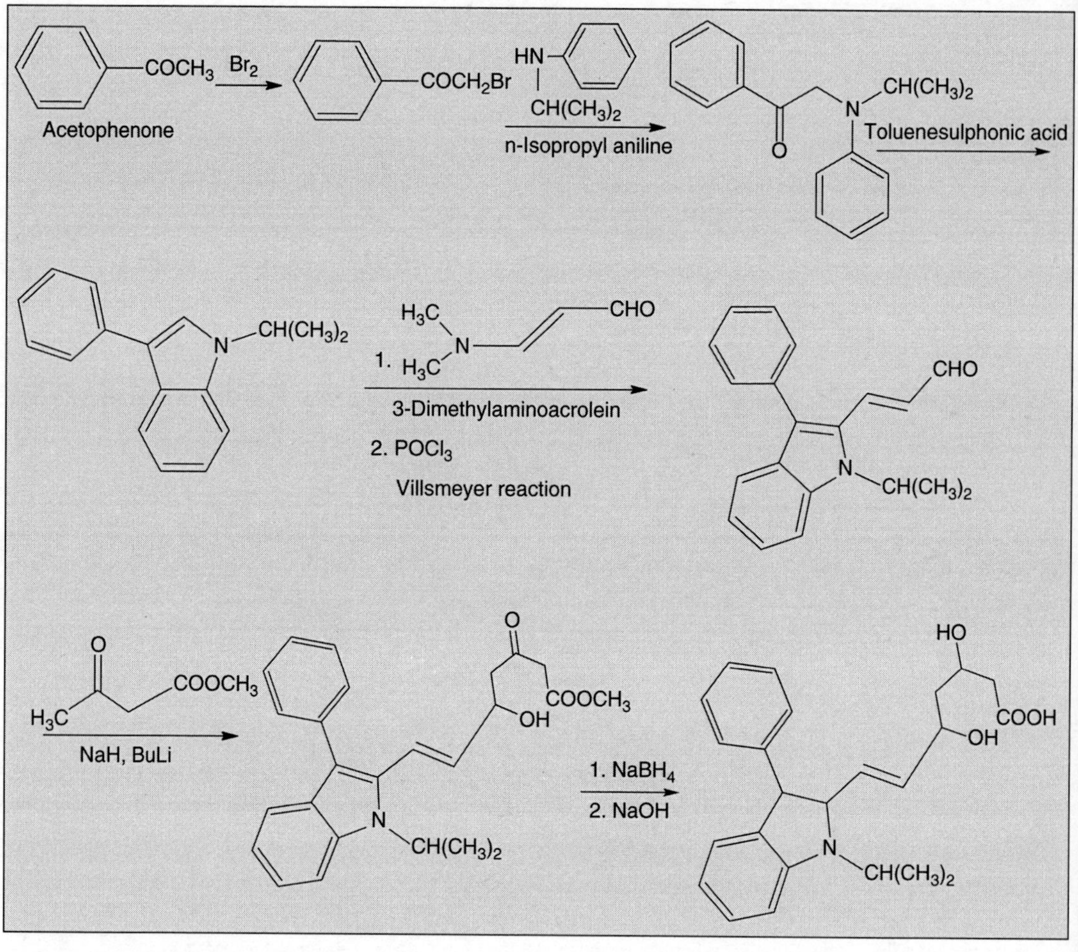

Acetophenone on reaction with bromine affords phenacyl bromide, which in turn reacts with N-isopropyl aniline to give N-alkylated compound. Cyclodehydration of the alkylation product in presence of p-toluenesulphonic acid affords the indole derivative. The side chain is added to the reactive 2-position by vinylogous Villsmeyer reaction by treatment with 3-dimethylaminoacrolein and phosphorous oxy chloride, to give the substituted acrolein derivative. The side chain is further extended by addition of the more nucleophilic terminal enolate of the acetoacetate dianion in presence of base. Carbonyl group is reduced to alcohol and saponification of ester affords fluvastatin.

Dalvastatin 6-(2-(2-(4-Fluoro-3-methylphenyl)-4,4,6,6-tetramethyl-1-cyclohexen-1-yl)ethenyl)tetrahydro-4-hydroxy-2H-pyran-2-one

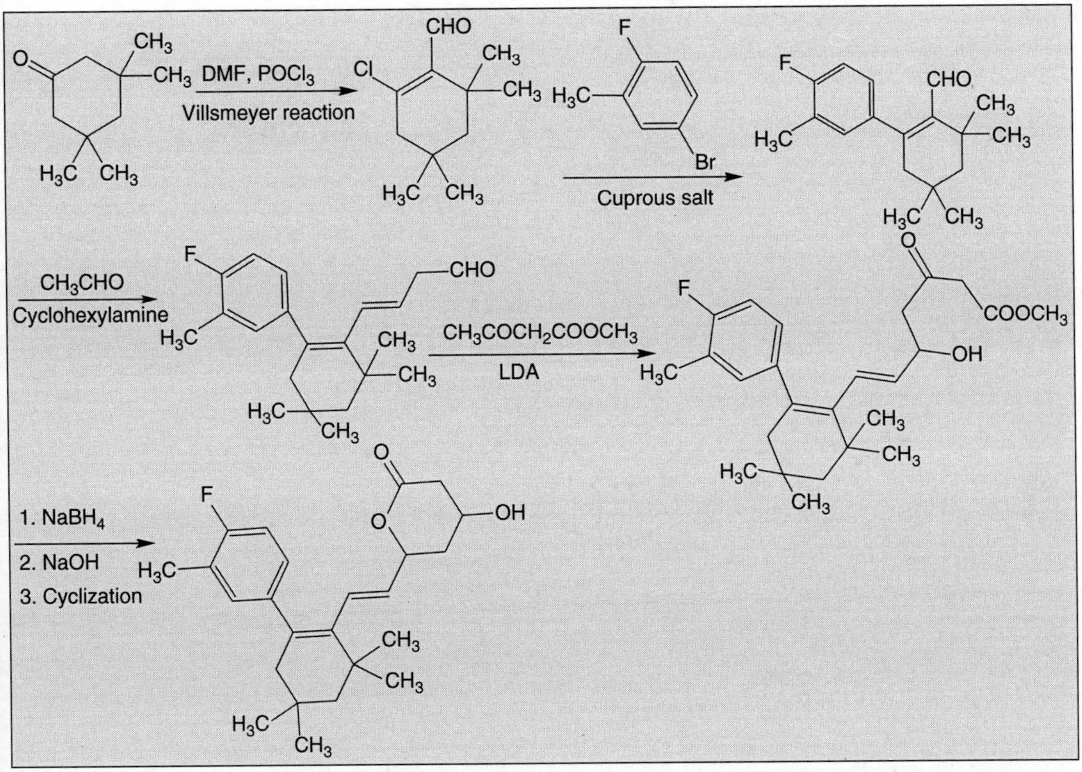

Reaction of cyclohexanone with dimethyl formamide and phosphorous oxychloride leads to Villsmeyer aldehyde; this is reacted with 2-fluoro-3-bromotoluene in the presence of cuprous salt to give arylated compound. The aldehyde is then extended by condensation with the carbanion from the cyclohexylamine Schiff base of acetaldehyde; treatment of the first formed product serves to hydrolyse the imine and to dehydrate the β-hydroxyl group, and the bishomologated product is obtained. Treatment of methyl acetoacetate gives a further chain-extended product, which on reduction, saponification, and cyclization affords dalvastatin.

18.4 ARYLOXYISOBUTYRIC ACID DERIVATIVES (FIBRATES)

Fibrates are agonists of the peroxisome proliferator-activated receptors (PPARα) which is present in muscle, liver, and other tissues (PPARs play essential roles in the regulation of cellular differentiation, development, and metabolism [carbohydrate, lipid, and protein] of higher organisms). Activation of PPARα signalling results in: (a) increased fatty oxidation in the liver, (b) decreased hepatic triglyceride secretion, (c) increased lipoprotein lipase activity, and thus increased VLDL clearance, (d) increased HDL, and (e) increased clearance of remnant particles.

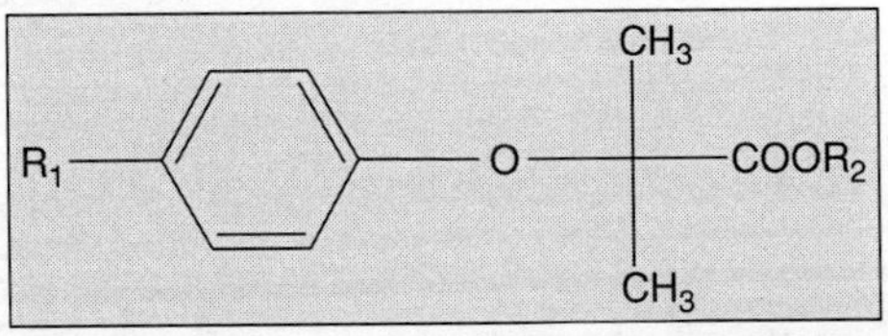

Chapter 18: Antihyperlipidemic Agents

	R_1	R_2
Clofibrate	Cl	—C_2H_5
Fenofibrate	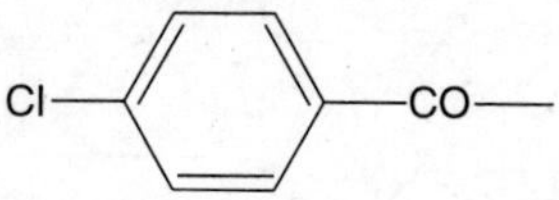 Cl—C₆H₄—CO—	—$CH(CH_3)_2$
Ciprofibrate	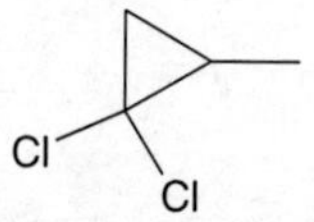 (2,2-dichlorocyclopropyl)	H
Bezafibrate	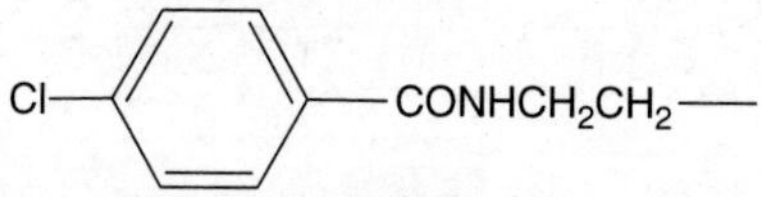 Cl—C₆H₄—CONHCH₂CH₂—	H

Clofibrate Ethyl-2-(*p*-chloro phenoxy)-2-methyl propionate
Clofibrate is metabolised to chlorophenoxyisobutyric acid (CPIB), which is the active form of the drug.

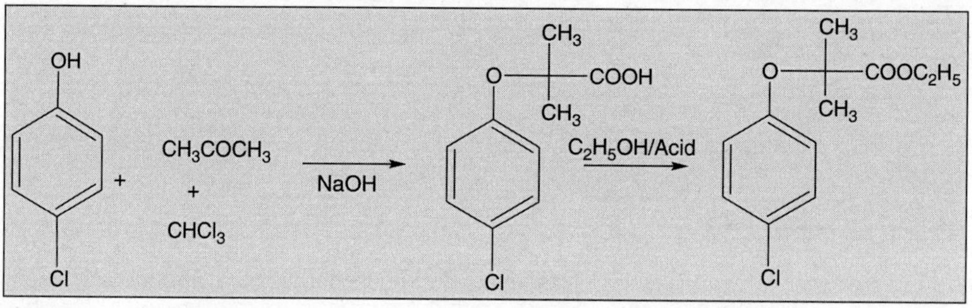

p-chlorophenol + CH_3COCH_3 + $CHCl_3$ →(NaOH)→ 2-(4-chlorophenoxy)-2-methylpropanoic acid →(C_2H_5OH/Acid)→ Clofibrate

Clofibrate is synthesised by esterifying 2-(4-chlorophenoxy)-*iso*-butyric acid with ethyl alcohol. This is synthesised in a single-stage reaction from 4-chlorophenol, acetone, and chloroform in the presence of an alkali, evidently by initial formation of chlorethone-trichloro-*tert*-butyl alcohol, which under the reaction conditions is converted into (4-chlorophenoxy)trichloro*tert*-butyl ether, and further hydrolysed to the desired acid, which is esterified with ethanol in the presence of inorganic acid.

Bezafibrate 2-[4-[2-[(4-Chlorobenzoyl)amino]ethyl]phenoxy]-2-methyl-propanoic acid

4-Chloro benzoyl chloride (COCl, Cl) + Tyramine ((CH₂)₂NH₂, OH) →(Pyridine, –HCl)→ intermediate amide (Cl—C₆H₄—CO—NH—CH₂CH₂—C₆H₄—OH) →(1. KOH; Br—C(CH₃)₂—COOC₂H₅; –HBr, Saponification)→ Bezafibrate

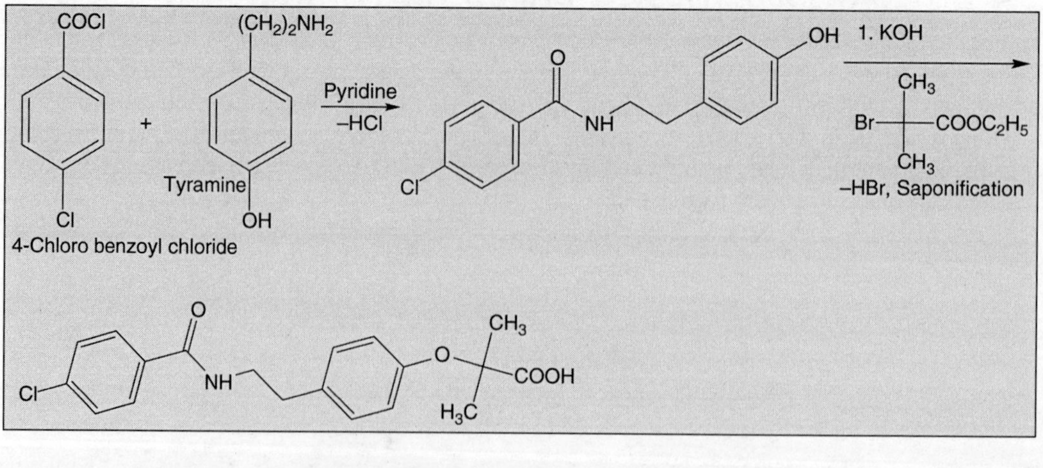

Acylation of tyramine with 4-chlorobenzoyl chloride affords amide derivative. The phenol on alkylation with ethyl 2-bromo-2-methylpropanoate in presence of base affords bezafibrate.

Ciprofibrate 2-[4-(2,2-Dichlorocyclopropyl)phenoxy]-2-methylpropanoic acid

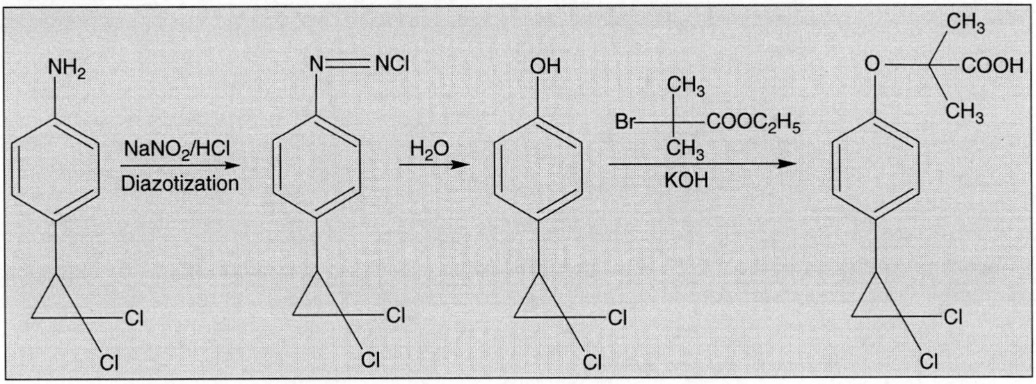

Sandmeyer replacement of amino group of 4-(2,2-dichlorocyclopropyl)aniline by a hydroxyl is obtained via the diazonium salt. The phenolic group on alkylation with ethyl 2-bromo-2-methylpropanoate in presence of base affords ciprofibrate.

Fenofibrate 1-Methylethyl2-[4-(4-chlorobenzoyl) phenoxy]- 2-methyl-propanoate

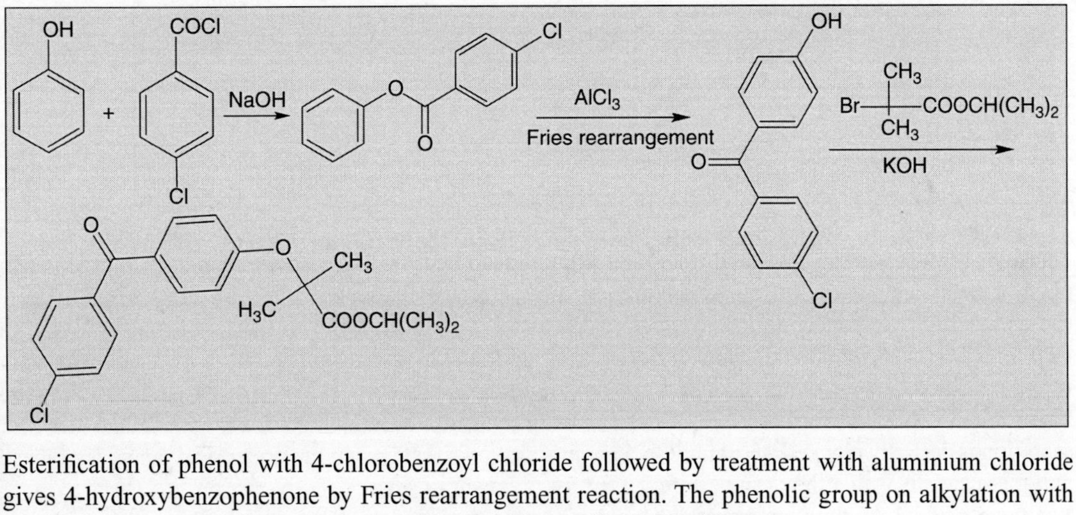

Esterification of phenol with 4-chlorobenzoyl chloride followed by treatment with aluminium chloride gives 4-hydroxybenzophenone by Fries rearrangement reaction. The phenolic group on alkylation with ethyl 2-bromo-2-methylpropanoate in presence of base affords fenofibrate.

18.5 MISCELLANEOUS DRUGS

Bile acid sequestrants Bile acids are secreted by the liver into the intestine, where they aid in the dissolution and absorption of lipids. Approximately 95 per cent of the bile acids that are secreted are reabsorbed and returned to the liver, while approximately 5 per cent is replaced by *de novo* synthesis from cholesterol. If this circulation is interrupted in some manner, there is a greater loss of bile acids from the body. To make up for this loss, the biosynthesis of bile acids from the cholesterol increases leading to partial depletion of hepatic cholesterol pool.

Cholestyramine resin

$$\left[\begin{array}{c} -CH_2-CH-CH_2- \\ | \\ \text{(phenyl)} \\ \\ -CH_2-CH-N^+(CH_3)_3\ Cl^- \\ | \\ \text{(phenyl)} \end{array} \right]_n$$

It is a copolymer of styrene and divinyl benzene with quaternary ammonium functional group. After oral ingestion it remains in the GIT, where it readily exchanges chloride ions for bile acids in the small intestine.

$$\left[\begin{array}{c} -N-CH_2CH_2-N- \\ | \quad\quad\quad\quad\quad | \\ CH_2 \quad\quad\quad CH_2 \\ | \quad\quad\quad\quad\quad | \\ CHOH \quad\quad CHOH \\ | \quad\quad\quad\quad\quad | \\ H_2C \quad\quad\quad CH_2 \\ | \quad\quad\quad\quad\quad | \\ -N-CH_2-CH_2-N- \end{array} \right]_n \quad\quad \left[\begin{array}{c} -HC-CONH- \\ | \\ CH_2 \\ | \\ CO \\ | \\ NH(CH_2)_3N^+(CH_3)_3 \end{array} \right]_n$$

Colestipol | Berlax

Colestipol and **cholestyramine** are anion exchange resins. The resins are water-insoluble, inert to digestive enzymes in the intestinal tract, and are not absorbed. Both resins are quaternized at stomach pH and exchange anions for bile acids, dramatically reducing the reabsorption of bile acids. The liver senses that bile acid concentrations have gone down and, hence, turns on cholesterol metabolism. Serum HDL and TG levels remain unchanged. LDL levels are found to decrease. The fall in LDL concentration is apparent in 4 to 7 days. The decline in serum cholesterol is usually evident by one month. When the resins are discontinued, the serum cholesterol usually returns to baseline within a month. When bile acid secretion is partially blocked, serum bile-acid concentration rises. One of the greatest advantages of these polymeric agents is that they can be safely used for pregnant women.

Probucol 4,4'-[(1-Methylidene)bis(thio)]bis[2,6-bis (1,1-dimethylethyl)] phenol

$$HO-C_6H_4-SH \xrightarrow[AlCl_3]{2\ (CH_3)_3CCl} \text{2,6-di-tert-butyl-4-mercaptophenol} \xrightarrow{CH_3COCH_3,\ \text{Acetone}} \text{Probucol}$$

Friedel-Crafts alkylation of 4-mercapto phenol with two moles of *tert*-butyl chloride in presence of Lewis acid gives 2,6-ditertiarybutyl-4-mercapto phenol. Reaction of two moles of this in acetone leads to the dithioketal probucol.

Mechanism of action: Oxidised LDL is taken by macrophages to convert it into foam cell. Groups of these foam cells constitute the earliest hallmark of atherosclerosis. The antioxidant effect of probucol inhibits the oxidation of LDL, thereby preventing its uptake by macrophages. This may help prevent the development of atherosclerosis.

Gemfibrosil 5-(2,5-Dimethylphenoxy)-2,2-dimethyl pentanoic acid

It is structurally related to clofibrate. It decreases the incorporation of long-chain fatty acids into triglycerides and thus decreases the hepatic synthesis of VLDL, and it also decreases the synthesis of VLDL carrier apo-lipoprotein.

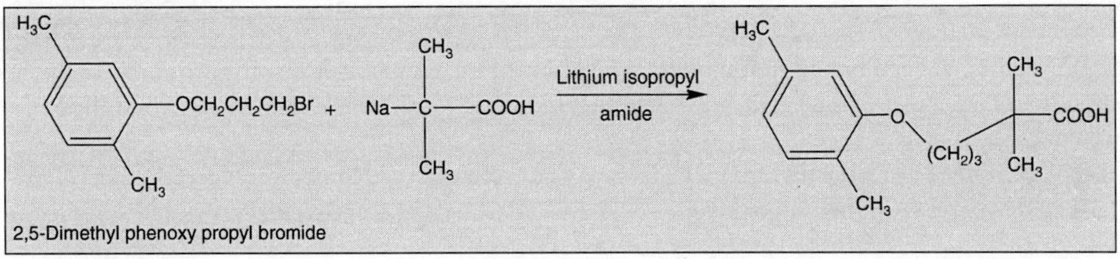

It is made readily by lithium diisopropylamide-promoted alkylation of sodium isopropionate with aryloxyalkyl bromide.

Niacin 3-Pyridine carboxylic acid

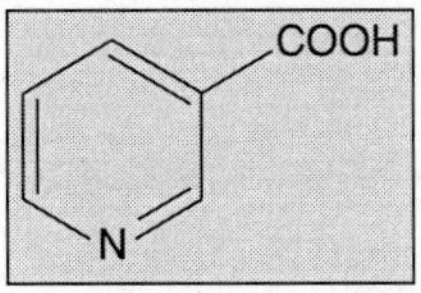

Niacin is rapidly absorbed. It reduces VLDL synthesis and subsequently its plasma products IDL and LDL. Plasma triglyceride levels are reduced because of the decreased VLDL production. Cholesterol levels are lowered in turn because of the decreased rate of LDL formation from VLDL.

18.6 NEWER DRUGS

Atorvastatin [R-(R*,R*)]-2-(4-fluorophenyl)-β,δ-dihydroxy-5-(1-methylethyl)-3-phenyl-4-[(phenylamino)carbonyl]-1H-pyrrole-1-heptanoic acid

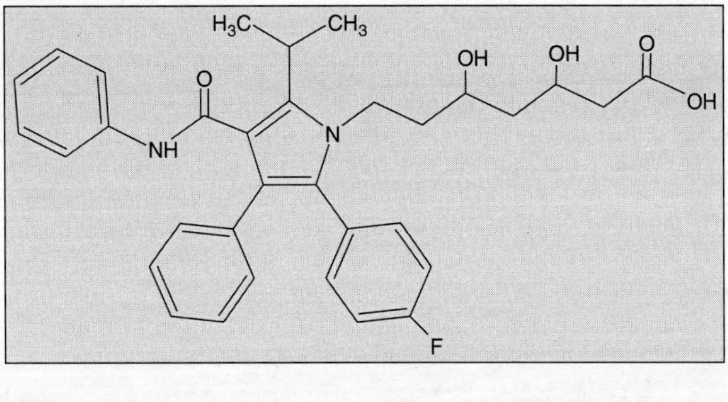

It is indicated as an adjunct to diet for the treatment of dyslipidemia, specifically hypercholesterolemia. With 2006 sales of US$12.9 billion, it is the largest-selling drug in the world.

Etofibrate 2-{[2-(4-chlorophenoxy)-2-methylpropanoyl]oxy}ethyl nicotinate

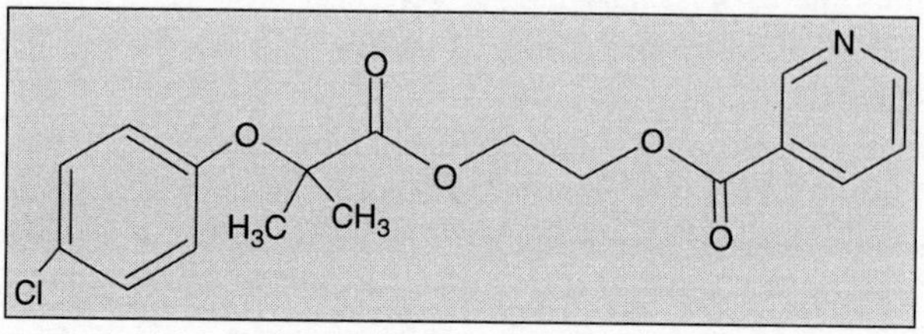

It is a fibrate. It is a combination of clofibrate and niacin, linked together by an ester bond. In the body, clofibrate and niacin separate and are released gradually, in a manner similar to controlled-release formulations.

Anacetrapib Ethyl 4-[[3,5-di(trifluoromethyl)benzyl](methoxycarbonyl)amino]-2-ethyl-6-(trifluoromethyl)-1,2,3,4-tetrahydro-1-quinolinecarboxylate

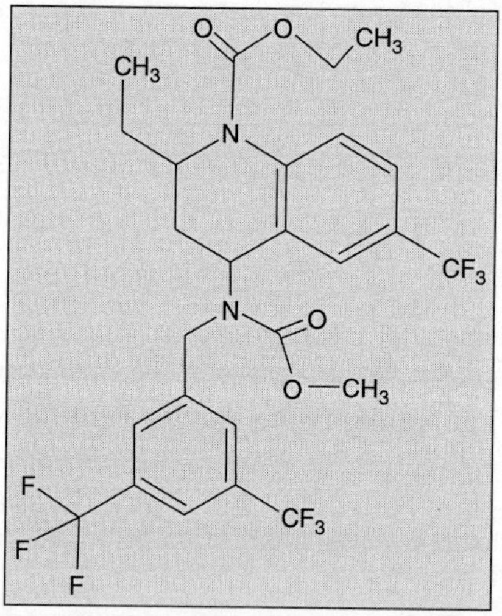

It is a CETP inhibitor being developed to treat hypercholesterolemia and prevent cardiovascular disease. It acts by inhibiting cholesterylester transfer protein (CETP), which normally transfers cholesterol from HDL cholesterol to very low-density or low-density lipoproteins (VLDL or LDL). Inhibition of this process results in higher HDL levels (the 'good' cholesterol-containing particle) and reduces LDL levels (the 'bad' cholesterol).

Ezetimibe (3R,4S)-1-(4-fluorophenyl)-3-((3S)-3-(4-fluorophenyl)-3-hydroxypropyl)-4-(4-hydroxphenyl)-2-azetidinone

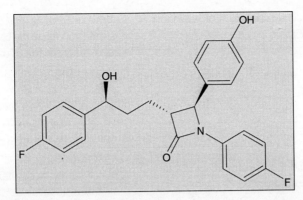

Ezetimibe localizes at the brush border of the small intestine, where it inhibits the absorption of cholesterol from the diet. Specifically, it appears to bind to a critical mediator of cholesterol absorption, the Niemann-Pick C1-Like 1 (NPC1L1) protein on the gastrointestinal tract epithelial cells as well as in hepatocytes. In addition to this direct effect, decreased cholesterol absorption leads to an increase in LDL-cholesterol uptake into cells, thus decreasing levels in the blood plasma.

FURTHER READINGS

1. *Practical Lipid Management Newsletter*, published by AstraZeneca. June 2003.
2. Harrold M., 'Antihyperlipoproteinemics and inhibitors of cholesterol biosynthesis'. In: Williams, D.A. and Lemke, T.L. (ed.), *Foye's Principles of Medicinal Chemistry*, 5th Ed., Baltimore, MD: Lippincott Williams & Wilkins; 2002:580-603.
3. Istvan, E.S. and J. Deisenhofer. 'Structural mechanism for statin inhibition of HMG-CoA reductase', *Science*, 2001; 292:1160-4.

MULTIPLE-CHOICE QUESTIONS

1. One of the following aminoacids is involved in the binding of HMG-CoA reductase:
   a. Lys559
   b. Asp735
   c. Lys735
   d. Asp684

2. Synthetic statin derivative is
   a. Lovastatin
   b. Simvastatin
   c. Fluvastatin
   d. Pravastatin

3. One of the following is not a prodrug:
   a. Clofibrate
   b. Lovastatin
   c. Simvastatin
   d. Ciprofibrate

4. One of the following drugs acts by sequestering the bile acid in the GIT:
   a. Gemfibrosil
   b. Colestipol
   c. Probucol
   d. Niacin

5. The drug that reduces the absorption of cholesterol from the GIT is
   a. Colestipol
   b. Niacin
   c. Ezetimibe
   d. Gemfibrosil

6. The anti-hyperlipidemic drug that is a benzophenone derivative is
   a. Ezetimibe
   b. Probucol
   c. Fenofibrate
   d. Dalvastatin

QUESTIONS

1. Discuss the sites and mechanisms of actions of the antihyperlipidemic agents.
2. What is meant by atherosclerosis and how would you relate the cholesterol level in the body to the health of the cardiac system?
3. Is lovastatin a prodrug? What is the structural basis for your answer?
4. Why is HMG-CoA reductase an attractive target in treating atherosclerosis?
5. What are fibrates? Give the structures of drugs in this class.
6. Write the synthesis of probucol, ciprofibrate, and fluvastatin.

SOLUTION TO MULTIPLE-CHOICE QUESTIONS

1. c;
2. c;
3. d;
4. b;
5. c;
6. c.

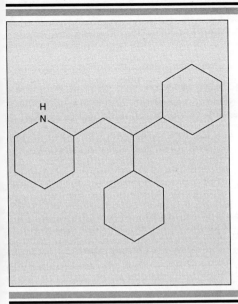

CHAPTER 19

Antianginal Drugs

LEARNING OBJECTIVES

- Define angina pectoris, their implications, and treatment options
- Categorize antianginal drugs based on the target on which they act
- Describe the mode of action of various antianginal drugs
- Define nitric oxide and its importance in angina
- Describe the importance of aspirin in the treatment of angina
- List some newer agents in the treatment of angina

19.1 INTRODUCTION

'Angina pectoris' is the medical term for chest pain or discomfort due to coronary heart disease. Angina is a symptom of a condition called myocardial ischaemia. It occurs when the heart muscle does not get as much blood (hence, as much oxygen) as it needs. This usually happens because one or more of the heart's arteries are narrowed or blocked. Angina also can occur in people with valvular heart disease, hypertropic cardiomyopathy (enlarged heart), or uncontrolled high blood pressure. Angina is a sign that someone is at increased risk of heart attack, cardiac arrest, and sudden cardiac death. **Antianginal drugs** are medicines that relieve the symptoms of angina pectoris.

Angina pectoris results from an imbalance between oxygen required and oxygen supplied to the ischemic region of the myocardium. Therefore, drugs that either reduce the need for oxygen in the myocardium or enhance oxygen supply are theoretically necessary for treating these states.

This can be accomplished by reducing the load on the heart, or by lowering systemic venous and arterial pressure (nitrates and nitrites), or by partial suppression of adrenergic innervation of the heart (β-adrenoreceptors), or by suppressing calcium ion transport in myocardial cells since the contraction of smooth muscle vessels is controlled by the concentration of calcium ions in the cytoplasm (Ca^{2+} channel blockers). The resulting effect of the aforementioned drugs is that they reduce the need for oxygen in the heart.

The main drugs used for myocardial ischema therapy and for relieving pain in angina pectoris are: nitrates and nitrites (nitroglycerin, isosorbide dinitrate, and pentaerythritol tetranitrate); substances that suppress adrenergic systems of the heart, β-adrenoblockers (atenolol, methoprolol, propranolol, and nadolol); and Ca^{2+} channel blockers (verapamil, diltiazem, and nifedipine). A few older drugs are used as well, in particular papaverine and dipyridamole.

19.2 CLASSIFICATION

1. Vasodilators (nitrites and nitrates): Glyceryl trinitrate, Amylnitrite, Isosorbide dinitrate, Erythrityl tetranitrate, Pentaerythritol tetranitrate, and Nicorandil
2. β-adrenergic antagonists: Propranolol, Atenolol, and Nadolol
3. Calcium channel blockers: Nifedipine, Verapamil, Diltiazem
4. Miscellaneous: Dipyridamole, Cyclandelate, and Aspirin

19.3 NITRATES/NITRITES

Organic nitrates act directly upon arterial and vascular smooth muscle to produce arterial and venous vasodilation. By reducing preload and afterload, both nitroglycerin and isosorbide dinitrate reduce oxygen consumption and restore the balance between oxygen supply and oxygen demand. Coronary blood flow is unaltered. Concomitant with a decrease in mean blood pressure is the activation of the sympathetic nervous system. Increases in heart rate and contractility partially reverse the decrease in oxygen consumption produced by arterial and venous vasodilation. In patients with variant angina, the organic nitrates can prevent or reverse coronary artery spasm.

The organic nitrates are rapidly and extensively metabolised by the liver. Only a small fraction of an orally administered dose appears unchanged in the peripheral blood. The drugs are lipid soluble and are well absorbed through the skin and mucous membranes.

Mechanism of action: These compounds have the common property of generating or releasing the unstable and lipophilic free radical nitric oxide (NO) *in situ*. The nitrogen oxide containing vasodilators first react with thiols in the cell to form unstable S-nitrosothiols, which break down to yield NO. The liberated NO activates guanylate cyclase and increases the cellular level of guanosine 3', 5'-monophosphate (cGMP). Cyclic GMP activates a cGMP-dependent protein kinase, which alters the phosphorylation state of several proteins. Included in this change is the dephosphorylation of the light chain myosin, which now cannot play a normal role in the contractile process of smooth muscle and the result is relaxation.

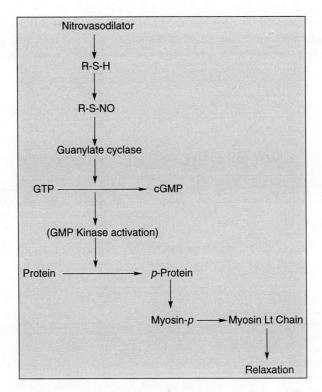

Amylnitrite 3-Methyl-1-nitrosooxybutane

It is a clear yellowish liquid prepared by esterification of amyl alcohol with nitrous acid.

It is quite volatile and is inhaled to obtain rapid effect, (Time of onset: 0.5 minutes). Amylnitrite is chemically related to nitroglycerin and has been used for many years to treat angina pectoris. It is also effective in the emergency management of cyanide poisoning by causing the oxidation of haemoglobin to the compound methemoglobin. Methemoglobin reacts with the cyanide ion to form cyanomethemoglobin, which has less affinity for oxygen, thus freeing haemoglobin to react with oxygen.

Glyceryl trinitrate 1,3-Dinitrooxypropan-2-yl nitrate

It is a colourless liquid prepared by nitrating glycerin with a mixture of nitric acid and sulphuric acid. By the sublingual route, the vasodilator effect of the drug appears in 2–3 minutes and lasts about 20 minutes.

Isosorbide dinitrate 1,4:3,6-Dianhydrosorbate-2,5-dinitrate

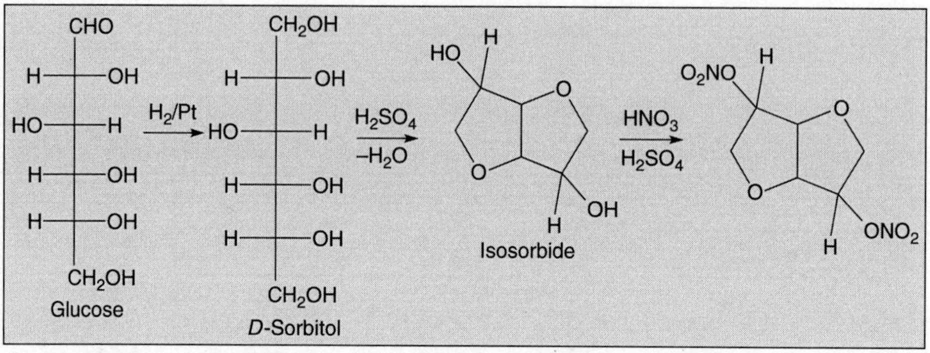

Isosorbide dinitrate is synthesised by intermolecular dehydration of D-sorbitol into isosorbide using p-toluenesulphonic acid/sulphuric acid and subsequent nitration of the two hydroxyl groups by nitric acid.

It is the long-acting organonitrate of choice with sublingual tablet. The onset effect is 2–5 minutes with offset of 4–6 hours.

Erythrityl tetranitrate

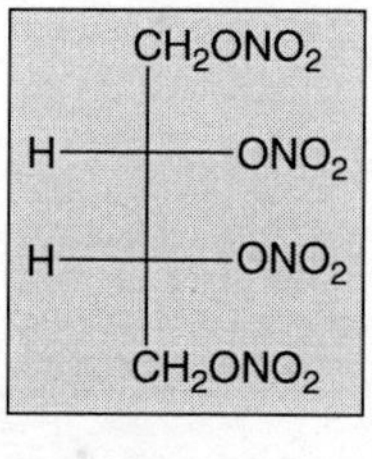

It is prepared by reacting erythritol with nitric acid in the presence of sulphuric acid.

By the sublingual route, it has a relatively long duration of action (about two hours).

Nicorandil N-[(2-Nitroxy)ethyl]-3-pyridine carboxamide

It acts differently from the other organonitrates. The proposed mechanisms of action include vascular smooth muscle hyperpolarization, inhibition of agonist-induced calcium release from intracellular stores, and stimulation of guanylate cyclase.

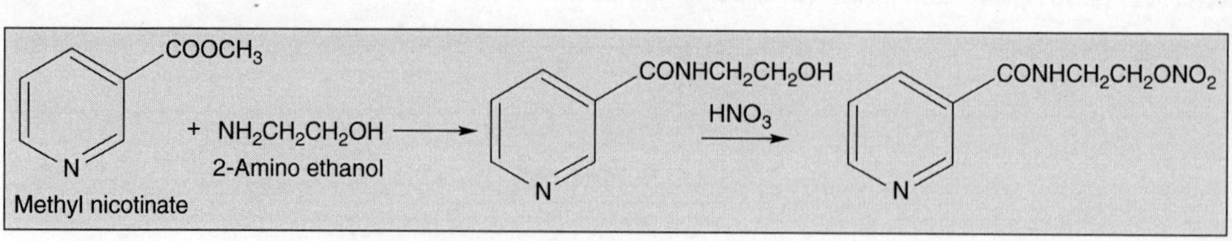

Reaction of methyl nicotinate with 2-aminoethanol gives amide and subsequent nitration of the hydroxyl group by nitric acid affords nicorandil.

19.4 β-ADRENERGIC BLOCKERS

Propranolol and other β-adrenoceptor antagonists increase exercise tolerance in angina because they improve blood flow to the vulnerable subendocardium, mostly by slowing the heart rate and increasing diastolic time, during which subendocardial profusion mainly occurs. Also, a decrease in heart rate decreases myocardial oxygen demand. Beta-blockers are first-line therapy for patients with effort-induced chronic stable angina; they improve exercise tolerance, relieve symptoms, reduce the severity and frequency of angina attacks, and increase the angina threshold.

β-Adrenergic receptor antagonists are used in combination with organic nitrates. The drugs antagonize the increased reflex sympathetic nervous system activity observed with the organic nitrates.

19.5 CALCIUM CHANNEL BLOCKERS

Calcium channel blockers, such as verapamil, diltiazem, and nifedipine, are used as an alternative to a beta-blocker to treat stable angina. These drugs directly slow the heart; this effect decreases myocardial oxygen demand as well as blunts reflex responses to arteriolar dilatation. Prevention of calcium influx into ischemic myocardial cells also may have a direct effect to decrease myocardial oxygen demand by preserving myocardial ATP. Calcium channel blockers are used to improve exercise tolerance in patients with chronic stable angina due to coronary atherosclerosis or with abnormally small coronary arteries and limited vasodilator reserve.

19.6 MISCELLANEOUS DRUGS

Dipyridamole 2,6-bis(Diethanolamino)-4,8-dipiperidinopyrimido pyrimidine
It inhibits phosphodiesterase, the synthesis of thromboxane-A_2, and the reuptake of adenosine, and promotes the synthesis of prostacyclin in vascular smooth muscles. In coronary blood vessels, all of these actions favour coronary vasodilatation.

Dipyridamole is easily synthesised from 2,4,6,8-tetrahydroxypyrimido[5,4-d]pyrimidine (5-nitroorotic acid on reduction gives 5-aminoorotic acid, which is reacted with urea or with potassium cyanide to give 2,4,6,8-tetrahydroxypyrimido[5,4-d]pyrimidine). This undergoes a reaction with a mixture of phosphorous oxychloride and phosphorous pentachloride, which forms 2,4,6,8- tetrachloropyrimido[5,4-d]pyrimidine. Reacting the resulting tetrachloride with piperidine replaces the chlorine atoms at C_4 and C_8 of the heterocyclic system with piperidine, giving 2,6-dichloropyrimido-4,8-dipiperidino[5,4-d]pyrimidine. Reacting the resulting product with diethanolamine gives dipyridamole.

Cyclandelate 3,3,5-Trimethylcyclohexyl mandelate

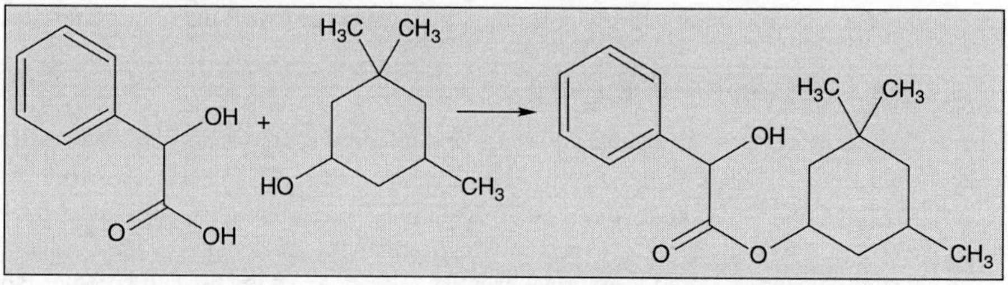

It is prepared by esterification of mandelic acid with 3,3,5-trimethylcyclohexanol.

Aspirin

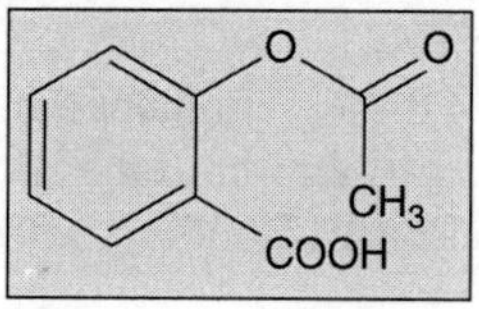

Aspirin may decrease the risk of heart attack and death in some people with stable angina. Aspirin at doses ranging from 75 to 325 mg daily irreversibly inhibits cyclo-oxygenase-1, blocking the formation of thromboxane A_2, thus reducing platelet aggregation. It has consistently reduced the risk of cardiac death.

19.7 NEWER DRUGS

Ivabradine (S)-3-(3-(((3,4-dimethoxybicyclo(4.2.0)octa-1,3,5-trien-7-l)methyl)methylamino)propyl)-1,3,4,5-tetrahydro-7,8-dimethoxy-2H-3-benzazepin-2-one

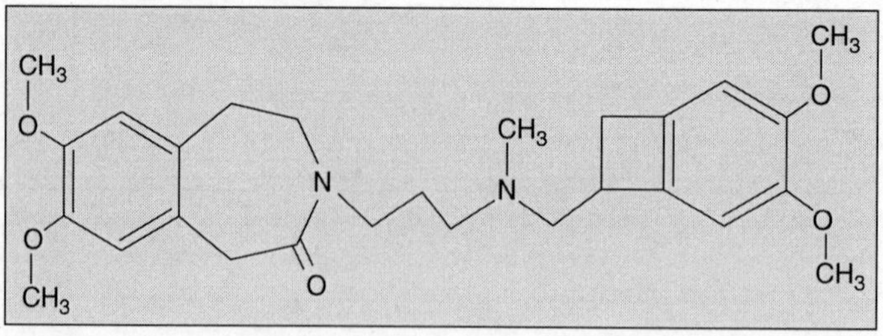

It is a novel medication used for the symptomatic management of stable angina pectoris.

Ivabradine acts on the I_f ion current, which is highly expressed in the sinoatrial node. I_f is a mixed Na^+–K^+ inward current activated by hyperpolarization and modulated by the autonomic nervous system. It is one of the most important ionic currents for regulating pacemaker activity in the sinoatrial (SA) node. Ivabradine selectively inhibits the pacemaker I_f current in a dose-dependent manner. Blocking this channel reduces cardiac pacemaker activity, slowing the heart rate and allowing more time for blood to flow to the myocardium.

Ranolazine N-(2,6-Dimethylphenyl)-2-[4-[2-hydroxy-3-(2-ethoxyphenoxy)propyl]piperazin-1-yl]acetamide

It is believed to have its effects via altering the transcellular late sodium current. It is acting by altering the intracellular sodium level that ranolazine affects the sodium-dependent calcium channels during myocardial ischemia. Thus, ranolazine indirectly prevents the calcium overload that causes cardiac ischemia.

Perhexiline 2-(2,2-Dicyclohexylethyl)piperidine

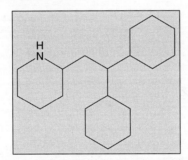

It is thought to act by inhibiting mitochondrial carnitine palmitoyltransferase-1 – this shifts myocardial metabolism from fatty acid to glucose utilization, which results in increased ATP production for the same O_2 consumption and consequently increases myocardial efficiency.

FURTHER READINGS

Wilson and Giswold's Textbook of Organic Medicinal & Pharmaceutical Chemistry, Block, J.H. and Beale, J.M., Jr. (ed.), 11th Ed., 2004.

MULTIPLE-CHOICE QUESTIONS

1. Mechanism of action of nitrates is
   a. Inhibits phosphodiesterase
   b. Stimulates guanylate cyclase
   c. B-blockers
   d. Calcium channel blockers

2. Mechanism of action of aspirin is
   a. Inhibits phosphodiesterase
   b. Stimulates guanylate cyclase
   c. β-blockers
   d. Antiplatelet agent

3. Which statement is false on organic nitrates?
   a. Organic nitrates can prevent or reverse coronary artery spasm.
   b. Organic nitrates are lipid-soluble and are well absorbed through the skin.
   c. These generate the unstable nitrate salts *in situ*.
   d. These can be administered by the sublingual route.
4. The antianginal drug useful for the emergency treatment of cyanide poisoning is
   a. Aspirin
   b. Glyceryl trinitrate
   c. Amylnitrate
   d. Dipyridamole
5. The antianginal drug that inhibits phosphodiesterase is
   a. Aspirin
   b. Dipyridamole
   c. Isosorbide dinitrate
   d. Nifedepine

QUESTIONS

1. Write a short note on organic nitrates.
2. Isosorbide dinitrate is a fully nitrated compound that can be metabolised to an active metabolite. Comment on the statement.
3. Classify antianginal drugs and discuss briefly their mode of action.
4. How useful are antiplatelet agents for angina treatment?
5. Compare isosorbide mononitrate with isosorbide dinitrate.
6. Give the synthetic protocol of isosorbide dintrate, dipyridamole, and nicorandil.

SOLUTION TO MULTIPLE-CHOICE QUESTIONS

1. b;
2. d;
3. c;
4. c;
5. b.

CHAPTER 20

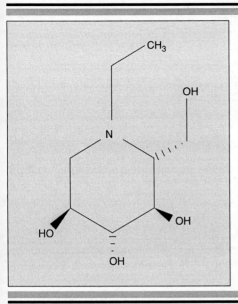

Insulin and Oral Hypoglycaemic Agents

LEARNING OBJECTIVES

- Define diabetes and its types and causes
- Describe the role of insulin in the body and the disease condition associated with it
- Define antidiabetics and categorize based on the chemical class
- Explain various strategies useful in the treatment of hyperglycaemia
- Describe the mechanism of action of various classes of oral hypoglycaemic agents
- Discuss the structure-activity relationship of sulphonyl ureas
- List the latest drugs in the class of antidiabetics

20.1 INTRODUCTION

The pancreas contains at least four different types of endocrine cells, including A (alpha, glucagon-producing), B (beta, insulin-producing), D (delta, somatostatin-producing), and F (PP, pancreatic polypeptide-producing). Of these, the B cells are predominant. The most common pancreatic disease requiring pharmacologic therapy is diabetes mellitus, a deficiency of insulin production or effect.

Type-I diabetes occurs when the pancreas cannot produce insulin, a hormone essential for moving glucose from the blood into cells. It is an autoimmune disorder, in which the body makes antibodies that attack the insulin-producing cells in the pancreas. It is usually diagnosed in children and young adults, and was called juvenile diabetes. People with type-I diabetes must supply insulin by injection.

Type-II diabetes is the most common disorder. In this, either the body does not produce enough insulin or the cells ignore the insulin. Sugar is the basic fuel for the cells in the body, and insulin takes the sugar from blood into the cells. When glucose builds up in the blood instead of going into cells, it can cause two problems:

1. Body cells may be starved for energy.
2. Over time, high blood glucose levels may hurt the eyes, kidneys, nerves, or heart.

Gestational diabetes is a form of diabetes that appears only during pregnancy and occurs in women with no previous history of diabetes.

Some diabetic symptoms are frequent urination, excessive thirst, extreme hunger, unusual weight loss, increased fatigue, irritability, and blurred vision.

Insulin therapy suffers from several problems. Insulin alleviates the symptoms associated with the cause and, hence, the disease persists endlessly. Insulin is a proteinaceous substance capable of inducing an immune response in the patient. Hypersensitivity to insulin is common. Because of its nature it has to be injected, causing some discomfort to the patient. Insulin resistance is another problem.

20.2 INSULIN

Insulin, a pancreatic hormone, is a specific antidiabetic agent, especially for type-I diabetes. Human insulin is a double-chain protein with molecular mass around 6000 that contains 51 amino acids (chain A-21 amino acids, and chain B-30 amino acids), which are bound together by disulphide bridges.

$$
\begin{array}{c}
\text{NH}_2\text{S}\text{—————————S} \quad\quad \text{NH}_2 \quad\quad \text{NH}_2 \\
\text{Gly-Ileu-Val-Glu-Glu-Cys-Cys-Thr-Ser-Ileu-Cys-Ser-Leu-Tyr-Glu-Leu-Glu-Asp-Tyr-Cys-Asp-NH}_2 \\
\text{NH}_2\,\text{NH}_2 \quad\quad \text{S————S} \quad\quad\quad\quad \text{S————————S} \\
\text{Phe-Val-Asp-Glu-His-Leu-Cys-Gly-Ser-His-Leu-Val-Glu-Ala-Leu-Tyr-Leu-Val-Cys-Glu-Gly-Arg-Gly-Phe-Phe-Tyr-Thr-Pro-Lys-Thr}
\end{array}
$$

20.2.1 Amino Acid Sequences (Structure) of Insulin

In the body, insulin is synthesised by β cells of Langerhans islets in the pancreas. In β cells, insulin is synthesised from the pro-insulin precursor molecule (pro-insulin consists of three domains: an amino-terminal B chain, a carboxy-terminal A chain, and a connecting peptide in the middle known as the C peptide) by the action of proteolytic enzymes, known as prohormone convertases (PC1 and PC2), as well as the exoprotease carboxypeptidase E. These modifications of pro-insulin remove the centre portion of the molecule (i.e., C peptide) from the C- and N-terminal ends of pro-insulin. The remaining polypeptides (51 amino acids in total), the B- and A-chains, are bound together by disulphide bonds.

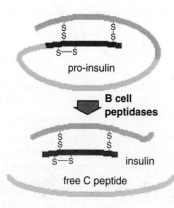

FIGURE 20.1 Biosynthesis of Insulin.

20.2.2 Mechanism of Action of Insulin

The mechanism of the hypoglycaemic action of insulin has proposed that insulin acts by binding with specific receptors on the surface of the insulin-sensitive tissues such as skeletal muscle, cardiac muscle, fatty tissue, and leukocytes. Insulin lowers the sugar content in the blood by turning glucose into glycogen. Using insulin in diabetes mellitus leads to lower levels of sugar in the blood, and a build-up of glycogen in tissues.

The initial sources of insulin for clinical use in humans were cow, horse, pig, or fish pancreases. Insulin from these sources is effective in humans as it is nearly identical to human insulin (three amino acid difference in bovine insulin, one amino acid difference in porcine). Synthetic 'human' insulin is now manufactured for widespread clinical use by means of genetic engineering techniques using recombinant DNA technology, which the manufacturers claim reduces the presence of many impurities. Eli Lilly marketed the first such insulin, Humulin, in 1982.

Unlike many medicines, insulin cannot be taken orally. Like nearly all other proteins introduced into the gastrointestinal tract, it is reduced to fragments (even single amino acid components), whereupon all 'insulin activity' is lost. Insulin is usually taken as subcutaneous injections by single-use syringes with needles. In 2006 the U.S. Food and Drug Administration approved the use of Exubera, the first inhalable insulin. It has been withdrawn from the market by its maker as of 3Q 2007, due to lack of acceptance.

A great deal of research towards the development of more effective ways of treating the disease has led to the development of orally active agents. **Oral hypoglycaemic agents** or **antidiabetic drugs** are drugs that lower the level of glucose (sugar) in the blood.

20.3 ORAL HYPOGLYCAEMIC AGENTS CLASSIFICATION

1. Sulphonyl ureas

R_1—⟨phenyl⟩—SO_2NH—$\overset{O}{\underset{\parallel}{C}}$—$NH$—$R$

	R	R_1
Carbutamide	—$CH_2CH_2CH_2CH_3$	—NH_2
Tolbutamide	—$CH_2CH_2CH_2CH_3$	—CH_3
Chlorpropamide	—$CH_2CH_2CH_3$	—Cl
Acetohexamide	—⟨cyclohexyl⟩	—$COCH_3$

(Continued)

Chapter 20: Insulin and Oral Hypoglycaemic Agents

	R	R_1
Glibenclamide	cyclohexyl	—(H$_2$C)$_2$HNOC—(2-OCH$_3$, 5-Cl-phenyl)
Glipizide	cyclohexyl	(2-CH$_3$-pyrimidin-5-yl)—CONH(CH$_2$)$_2$—

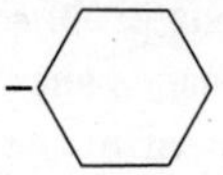

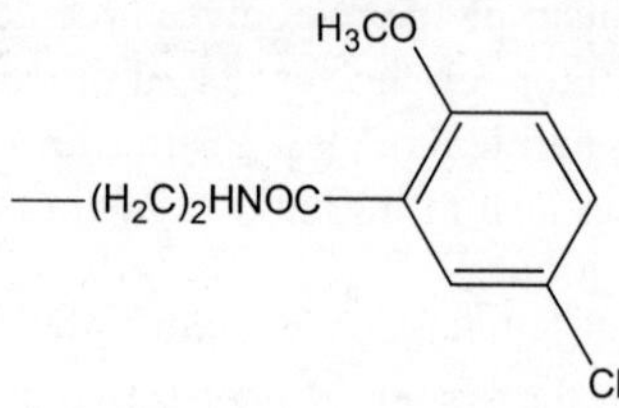

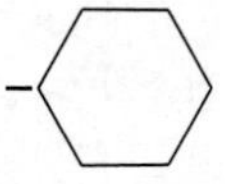

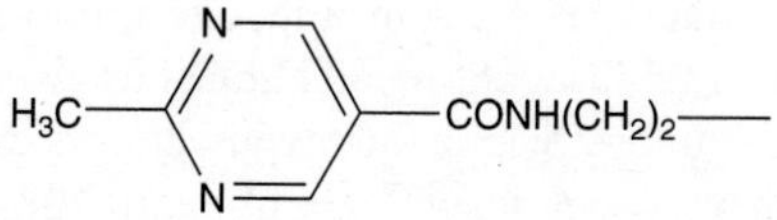

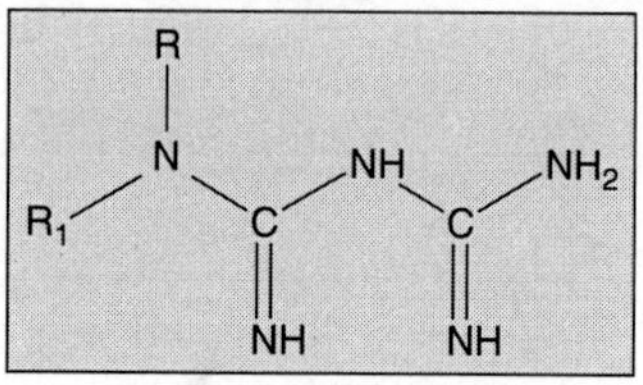

2. Biguanides

	R	R_1
Phenformin	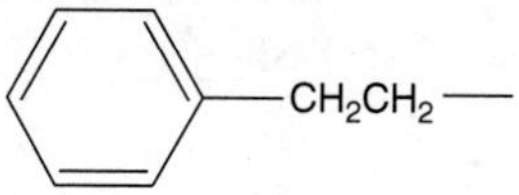 phenyl—CH$_2$CH$_2$—	H
Metformin	CH$_3$—	CH$_3$—
Buformin	CH$_3$CH$_2$CH$_2$CH$_2$—	H

3. Substituted benzoic acid derivatives (meglitinides)

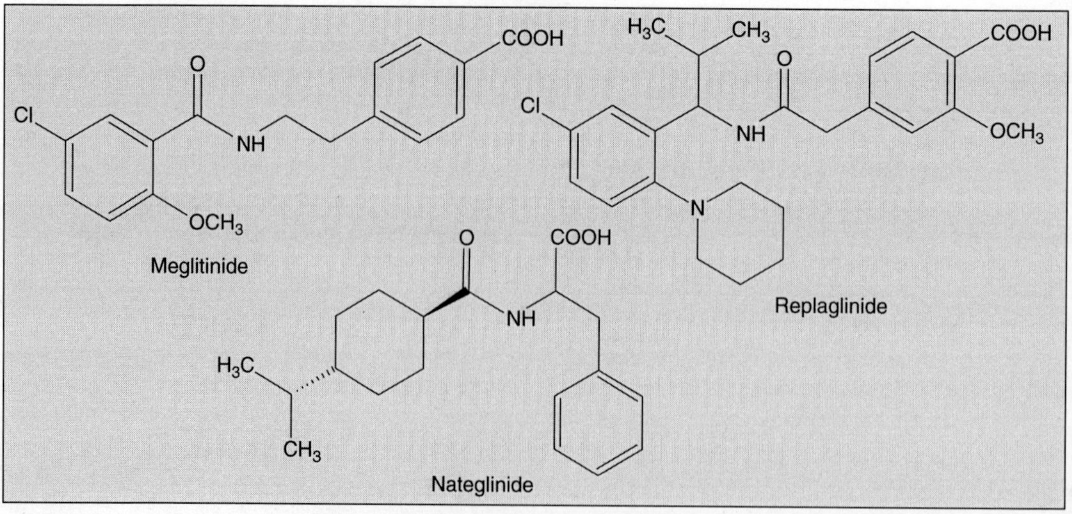

Meglitinide

Repaglinide

Nateglinide

4. Thiazolidinediones (glitazones)

	R
Pioglitazone	5-ethyl-2-pyridyl-CH$_2$– (H$_3$C on pyridine, CH$_2$– linker)
Ciglitazone	1-methylcyclohexyl-CH$_2$–
Rosiglitazone	2-(N-methyl-N-pyridyl-2-amino)ethyl–

4. Miscellaneous drugs: Linogliride, Palmoxirate sodium

20.4 STRATEGY FOR CONTROLLING HYPERGLYCAEMIA

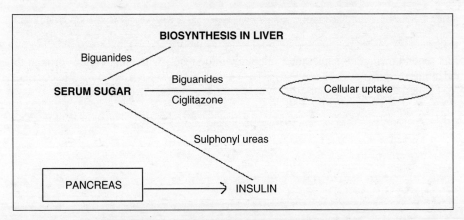

FIGURE 20.2 Mode of action of anti-diabetic drugs.

20.5 SULPHONYL UREAS

All members of this class differ by substitution at the *para* position on the benzene ring and at one nitrogen residue of the urea moiety. The discovery of this class was accidental. The compound 2-(*p*-aminobenzenesulphonamido)-5-isopropyl-thiadiazole (IPTD) was used in the treatment of typhoid fever in the early 1940s. However, many patients died from obscure causes while being treated with heavy doses of the drug. These deaths were eventually attributed to acute and prolonged hypoglycaemia. IPTD did not come to be used as hypoglycaemic agents because a second drug, carbutamide, was found to be an effective oral hypoglycaemic agent. Carbutamide was more active than IPTD and was the first sulphonylurea hypoglycaemic agent to be marketed.

Tolbutamide, chlorpropamide, and acetohexamide are first-generation sulphonyl ureas, while glibenclamide and glipizide are second generation.

Mechanism of action: Sulphonyl ureas bind to an ATP-dependent K$^+$ (K$_{ATP}$) channel on the cell membrane of pancreatic β cells. This inhibits a tonic, hyperpolarizing outflux of potassium, which causes the electric potential over the membrane to become more positive. This depolarization opens voltage-gated Ca^{2+} channels. The rise in intracellular calcium leads to increased fusion of insulin granulae with the cell membrane, and, therefore, increased secretion of (pro)insulin.

There is some evidence that sulphonyl ureas also sensitize β cells to glucose, that they limit glucose production in the liver, that they decrease lipolysis (breakdown and release of fatty acids by adipose tissue), and that they decrease clearance of insulin by the liver.

General methods for preparation Sulphonyl ureas are accessible by many methods that have been developed for the preparation of simpler ureas—for example, treatment of 4-substituted sulphanilamide with substituted isocyanate affords sulphonyl ureas.

In the second method, 4-substituted sulphanilamide reacts with ethyl chloro formate followed by substituted amine.

Structure–activity relationship

1. The benzene ring should contain a substituent, preferably in the *para* position. The substituents like methyl, acetyl, amino, chloro, bromo, trifluoromethyl, and thiomethyl were found to enhance the antihyperglycaemic activity.
2. When the *para* position of benzene is substituted with arylcarboxamidoalky group (second-generation sulphonyl ureas like glibenclamide), the activity was found to be more enhanced. It is believed that this is because of a specific distance between the nitrogen atom of the substituent and the sulphonamide nitrogen atom.
3. The size of the group attached to the terminal nitrogen is crucial for activity. The group should also impart lipophilicity to the compound. N-Methyl and ethyl substituents show no activity, whereas N-propyl and higher homologues were found to be active, and the activity is lost when the N-substituent contains 12 or more carbons.

20.6 BIGUANIDES

In 1918, guanidine was found to lower blood glucose levels in animals. However, it was found to be toxic precluding its use in routine medicine. In the 1950s phenformin was found to have antidiabetic properties. Two other biguanides, metformin and buformin, were also introduced though not in the market.

Mechanism of action: Biguanides are antihyperglycaemic, not hypoglycaemic. The exact mode of action of biguanides is not fully elucidated. However, in hyperinsulinemia, biguanides can lower fasting levels of insulin in plasma. Their therapeutic uses derive from their tendency to reduce gluconeogenesis in the liver, and as a result, reduce the level of glucose in the blood. Biguanides also tend to make the cells of the body more willing to absorb glucose already present in the blood stream, and thereby reduce the level of glucose in the plasma.

Phenformin 2-(*N*-Phenethylcarbamimidoyl)guanidine
It is no longer widely available because it is known to induce lactic acidosis.

Synthesis

Condensation of 2-phenylethylamine with dicyanamide affords phenformin.

20.7 SUBSTITUTED BENZOIC ACID DERIVATIVES (MEGLITINIDES)

The meglitinides are similar in structure to sulphonyl ureas. The sulphonyl urea and meglitinide classes of oral hypoglycaemic drugs are referred to as endogenous insulin secretagogues because they induce the pancreatic release of endogenous insulin. Repaglinide is a new non-sulphonyl urea insulin secretagogue agent, the first available from the meglitinide class. Nateglinide, the newest member of the class, has recently become available. Unlike the commonly used sulphonyl ureas, the meglitinides have a very short onset of action and a short half-life. Some potential advantages of this class of agents include a greater decrease in postprandial glucose and a decreased risk of hypoglycaemia.

Mechanism of action: The mechanism of action of the meglitinides closely resembles that of the sulphonyl ureas. The meglitinides stimulate the release of insulin from the pancreatic beta cells. However, this action is mediated through a different binding site on the 'sulphonyl urea receptor' of the beta cell, and the drug has somewhat different characteristics when compared with the sulphonyl ureas. However, meglitinides do exert effects on potassium conductance. Like the sulphonyl ureas, the meglitinides have no direct effects on the circulating levels of plasma lipids.

Meglitinide N_1-Phenethyl-5-chloro-2-methoxybenzamide

Synthesis

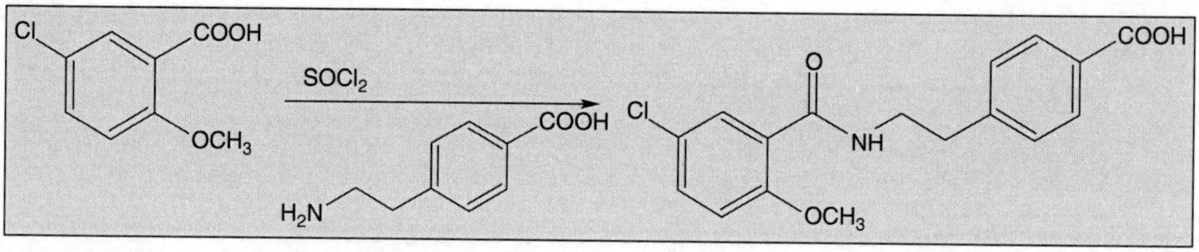

Reaction between 5-chloro-2-methoxybenzoic acid and phenylethylamine derivative affords the benzamide meglitinide.

Repaglinide S(+)2-Ethoxy-4(2((3-methyl-1-(2-(1-piperidinyl) phenyl)-butyl)amino)-2-oxoethyl)benzoic acid

Synthesis

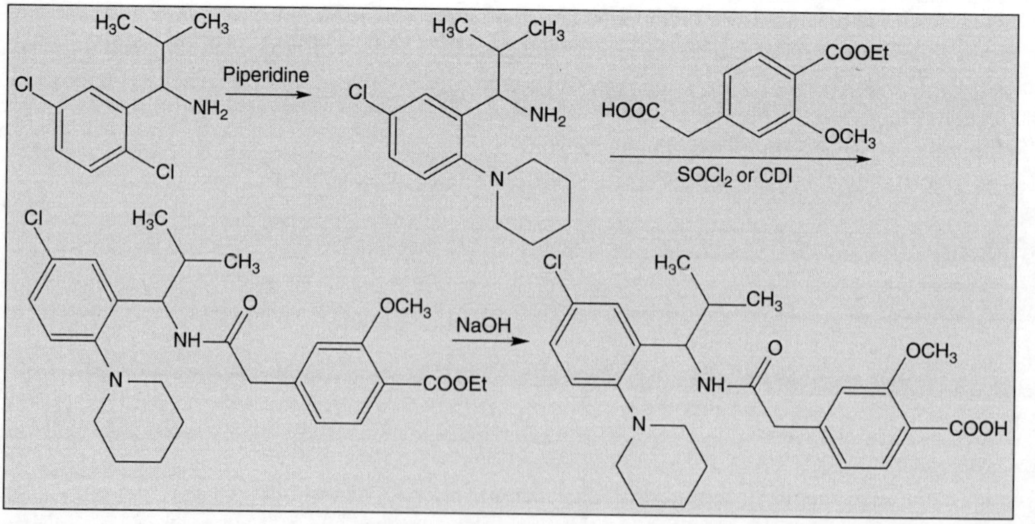

Reaction between equimolar ratio of 1,4-dichloro-2-(1,2-dimethylpropyl)benzene and piperidine gives 1-[4-chloro-2-(1,2-dimethylpropyl)phenyl]piperidine. This, in turn, reacts with phenylacetic acid derivative in presence of thionyl chloride/carbonyl diimidazole to afford amide. Saponification of ester gives free acid, repaglinide.

20.8 THIAZOLIDINEDIONES

Thiazolidinediones are a new class of oral antidiabetic agents (commercially known as glitazones) that enhance insulin sensitivity in peripheral tissues. Rosiglitazone and pioglitazone are now available for clinical use, and are extremely potent in reducing peripheral insulin resistance. Because these agents do not increase insulin secretion, hypoglycaemia does not pose a risk when thiazolidinediones are taken as monotherapy. Besides their effect in lowering the blood glucose levels, both drugs also have notable effects on lipids. The current data show that pioglitazone has a minimal effect on low-density lipoprotein (LDL) cholesterol levels and a favourable effect on high-density lipoprotein (HDL) cholesterol and triglyceride levels. Rosiglitazone has a favourable effect on HDL cholesterol levels but a negative effect on LDL cholesterol levels. The thiazolidinediones are relatively safe in patients with impaired renal function because they are highly metabolised by the liver and excreted in the faeces.

Mechanism of action: These compounds act by binding to peroxisome proliferator-activated receptors (PPARs), a group of receptor molecules inside the cell nucleus, specifically *PPARγ* (gamma). The normal ligands for these receptors are free fatty acids (FFAs) and eicosanoids. When activated, the receptor migrates to the DNA, activating transcription of a number of specific genes. By activating PPARγ: (a) insulin resistance is decreased; (b) adipocyte differentiation is modified; (c) VEGF-induced angiogenesis is inhibited; (d) leptin levels decrease (leading to an increased appetite); (e) levels of certain interleukins (e.g., IL-6) fall; and (f) adiponectin levels rise.

Ciglitazone 5-4-[(1-Methylcyclohexyl)methoxy]benzyl-1,3-thiazolidine-2,4-dione

Synthesis

Alkylation of 4-nitrophenol with 1-bromomethyl-1-methylcyclohexane gives 4-(1-methylcyclohexylmethoxy)nitrobenzene. Reduction of nitro group and diazotization of amino group followed by addition of methyl acrylate (Meerwein arylation) produce α-chlorinated ester. Reaction of this with thiourea probably proceeds through initial displacement of halogen by the nucleophilic sulphur; displacement of ethoxide by thiourea nitrogen leads, after bond reorganization, to the heterocycle ciglitazone.

Rosiglitazone 5-((4-(2-(Methyl-2-pyridinylamino) ethoxy)phenyl)methyl)- 2,4-thiazolidinedione

Synthesis

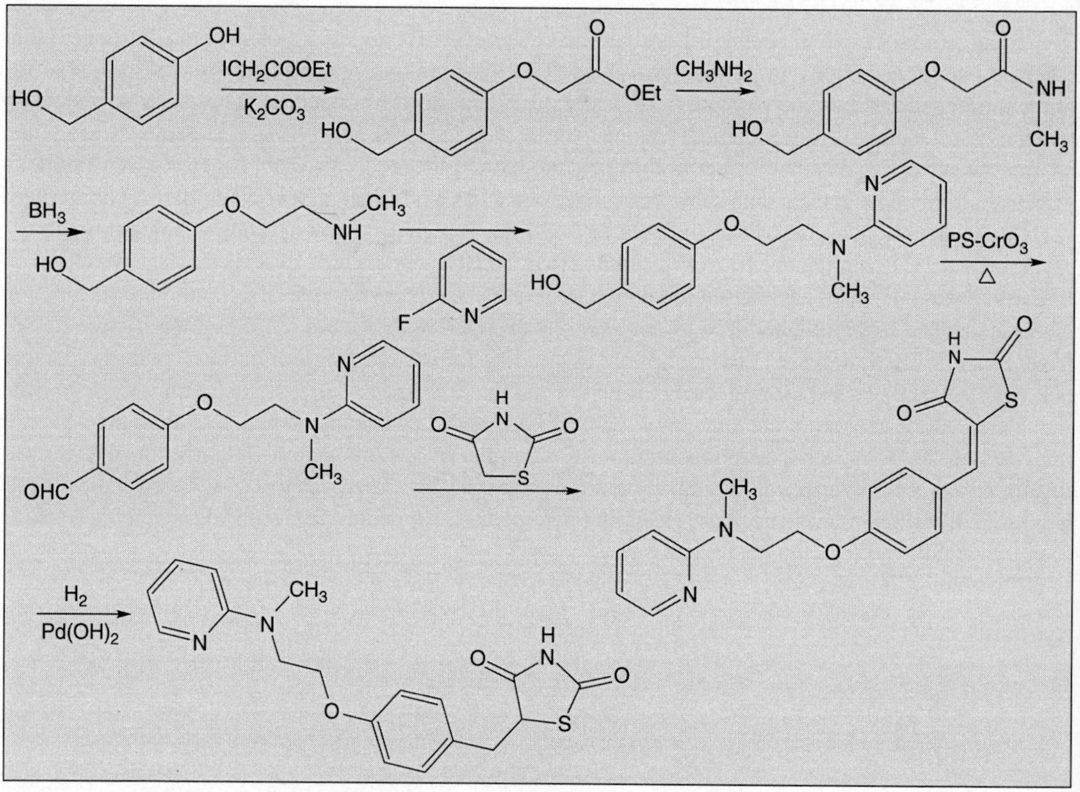

Alkylation of phenolic hydroxyl group with 2-iodoethyl acetate in presence of base gives phenoxy ester. Ester is interchanged to amide by reaction with methylamine, followed by reduction with boron trifluride to convert amide to amine. The secondary amine is further arylated by reaction with 2-fluoropyridine. Primary alcoholic group is oxidized to aldehyde, which in turn reacts with thiazolidine derivative to form benzylidene derivative; this on hydrogenation affords rosiglitazone.

A press release by GlaxoSmithKline in February 2007 noted that there is a greater incidence of fractures of the upper arms, hands, and feet in female diabetics given rosiglitazone compared with those given metformin.

20.9 MISCELLANEOUS DRUGS

Linogliride N_4-(1-Methyltetrahydro-1H-2-pyrrolyliden)-4-phenyltetrahydro-2H-1,4-oxazine-4-carboximidamide

Synthesis

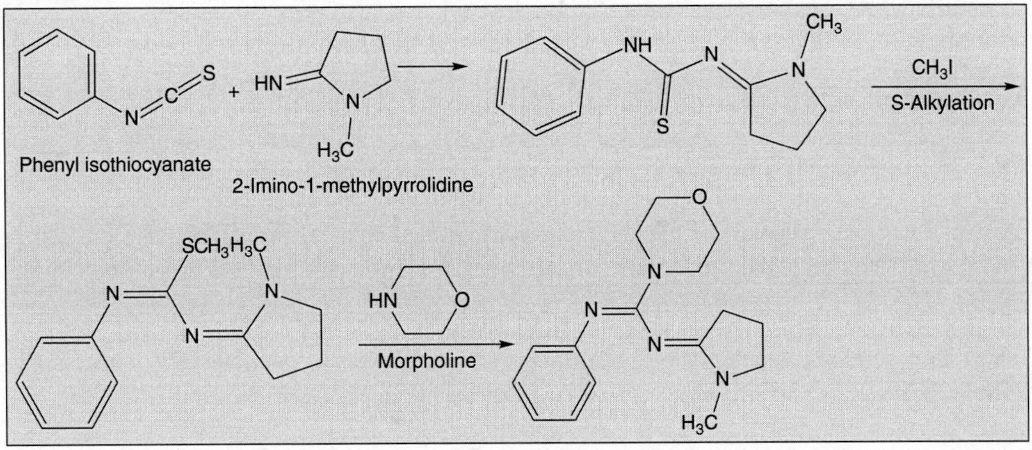

Condensation of phenyl isocyanate with 2-imino-1-methylpyrrolidine gives phenylthiourea derivative, which in turn undergoes *S*-alkylation with methyl iodide to give methyl mercapto derivative. An addition-elimination reaction with morpholine affords linogliride.

It works by an insulin secretagogue mechanism and stimulates the secretion of insulin in non-insulin-dependent patients.

20.10 NEWER DRUGS

Nateglinide 3-Phenyl-2-(4-isopropan-2-ylcyclohexyl) carbonylamino propanoic acid

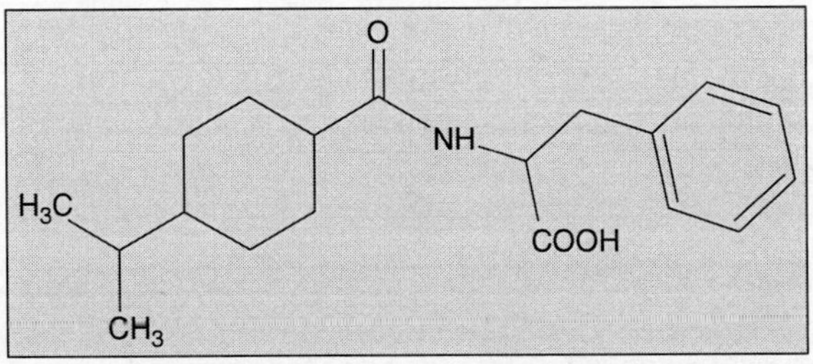

It is a drug for the treatment of type-II diabetes. Nateglinide belongs to the meglitinide class of blood glucose-lowering drugs.

Pioglitazone 5-4-[2-(5-Ethyl-2-pyridyl)ethoxy]benzyl-1,3-thiazolidine-2,4-dione

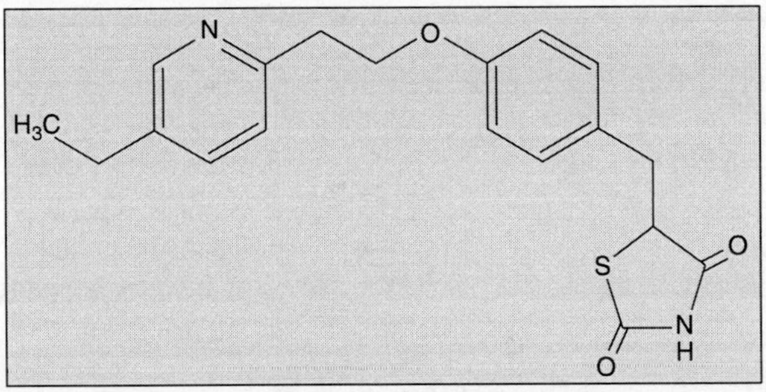

It is used for the treatment of diabetes mellitus type-II (non-insulin-dependent diabetes mellitus, NIDDM) in monotherapy, but usually in combination with sulphonyl urea, metformin, or insulin.

Acarbose (2R,3R,4S,5R,6R)-5-[(2R,3R,4S,5R,6R)-5-[(2R,3R,4S,5R,6R)-3,4-dihydroxy-6-methyl-5-[[(1S,4S,5S,6S)-4,5,6-trihydroxy-3-(hydroxymethyl)-1-cyclohex-2-enyl]amino]oxan-2-yl]oxy-3,4-dihydroxy-6-(hydroxymethyl)oxan-2-yl]oxy-6-(hydroxymethyl)oxane-2,3,4-triol

It is a carbohydrate drug and it inhibits enzymes (glycoside hydrolases) needed to digest carbohydrates—specifically, alpha-glucosidase enzymes in the brush border of the small intestines and pancreatic alpha-amylase. Pancreatic alpha-amylase hydrolyses complex starches to oligosaccharides in the lumen of the small intestine, whereas the membrane-bound intestinal alpha-glucosidases hydrolyse oligosaccharides, trisaccharides, and disaccharides to glucose and other monosaccharides in the small intestine. Inhibition of these enzyme systems reduces the rate of digestion of complex carbohydrates. Less glucose is absorbed because the carbohydrates are not broken down into glucose molecules. In diabetic patients, the short-term effect of these drug therapies is to decrease current blood glucose levels

Miglitol (2R,3R,4R,5S)-1-(2-hydroxyethyl)-2-(hydroxymethyl) piperidine-3,4,5-triol

It inhibits glycoside hydrolase enzymes called α-glucosidases. Since miglitol works by preventing digestion of carbohydrates, it lowers the degree of postprandial hyperglycaemia. It must be taken at the start of main meals to have maximal effect. Its effect will depend on the amount of non-monosaccharide carbohydrates in a person's diet.

FURTHER READINGS

1. Lebovitz, H.E., 2004. *Therapy for Diabetes Mellitus and Related Disorders*, 4th Ed., Alexandria:American Diabetes Association.
2. Williams, D.A. and Lemke, T.L., 2002, *Foye's Principles of Medicinal Chemistry*, 5th Ed.

MULTIPLE-CHOICE QUESTIONS

1. Glibenclamide belongs to the class
   a. Sulphonyl ureas
   b. Thiazolidinediones
   c. Benzoic acid derivatives
   d. Biguanides

2. Repaglinide belongs to the class
   a. Sulphonyl ureas
   b. Thiazolidinediones
   c. Benzoic acid derivatives
   d. Biguanides

3. The antidiabetic also effective in lowering the cholesterol level is
   a. Rosiglitazone
   b. Phenformin
   c. Chlorpropamide
   d. Repaglinide
4. The mechanism of action of rosiglitazone is
   a. Insulin-releasing agent
   b. PPARγ agonist
   c. Increases cellular uptake of glucose
   d. Reduces release of insulin
5. The terminal nitrogen in sulphonyl ureas should be substituted with
   a. At least 2 carbon chain
   b. At least 3 carbon chain
   c. At least 12 carbon chain
   d. At least 1 aryl unit

QUESTIONS

1. Classify oral hypoglycaemics with regard to the mode of action and give one structural example for each.
2. Derive the synthetic protocol for phenformin, ciglitazone, and glibenclamide.
3. Discuss briefly the SAR of sulphonyl urea derivatives as antihyperglycaemic agents.
4. Write a short note on thiazolidinediones.

SOLUTION TO MULTIPLE-CHOICE QUESTIONS

1. a;
2. c;
3. a;
4. b;
5. b.

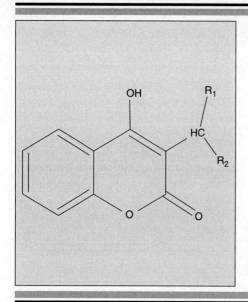

CHAPTER 21

Oral Anticoagulants

> **LEARNING OBJECTIVES**
>
> - Describe clotting of blood and its significance
> - Define anticoagulants and their utility
> - Classify anticoagulants based on the chemical structure
> - Discuss the mode of action of anticoagulants
> - Discuss the synthetic protocol of various anticoagulants
> - List some new anticoagulants

21.1 INTRODUCTION

The **coagulation** of blood is a complex process during which blood forms solid clots. It is an important part of haemostasis whereby a damaged blood vessel wall is covered by a fibrin clot to stop haemorrhage and aid repair of the damaged vessel. Disorders in coagulation can lead to increased haemorrhage and/or thrombosis and embolism.

Blood coagulation process requires **coagulation factors, calcium,** and **phospholipids**

- The coagulation factors (proteins) are manufactured by the liver.
- Ionized calcium (Ca^{++}) is available in the blood and from intracellular sources.
- Phospholipids are prominent components of cellular and platelet membranes. They provide a surface upon which the chemical reactions of coagulation can take place.

Coagulation can be initiated by either of two distinct pathways.

- The intrinsic pathway can be initiated by events that take place within the lumen of blood vessels. The intrinsic pathway requires only elements (clotting factors, Ca^{++}, platelet surface, etc.) found **within,** or **intrinsic to,** the vascular system.

- The extrinsic pathway is the other route to coagulation. It requires tissue factor (tissue thromboplastin), a substance that is 'extrinsic to', or not normally circulating in, the vessel. Tissue factor is released when the vessel wall is ruptured.

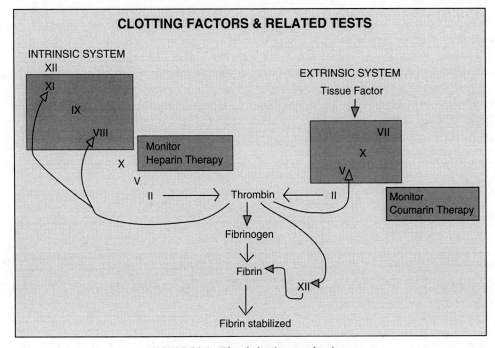

FIGURE 21.1 Blood clotting mechanism.

Regardless of whether the extrinsic or intrinsic pathway starts coagulation, completion of the process follows a common pathway. The common pathway involves the activation of factors: X, V, II, XIII, and I. Both pathways are required for normal haemostasis and there are positive feedback loops between the two pathways that amplify reactions to produce enough fibrin to form a life saving plug. Deficiencies or abnormalities in any one factor can slow the overall process, increasing the risk of haemorrhage.

The coagulation factors are numbered in the order of their discovery. There are 13 numerals but only 12 factors. Factor VI was subsequently found to be part of another factor. The following are coagulation factors and their common names:

- Factor I—fibrinogen
- Factor II—prothrombin
- Factor III—tissue thromboplastin (tissue factor)
- Factor IV—ionized calcium (Ca^{++})
- Factor V—labile factor or proaccelerin
- Factor VI—unassigned
- Factor VII—stable factor or proconvertin
- Factor VIII—antihaemophilic factor
- Factor IX—plasma thromboplastin component, Christmas factor

- Factor X—Stuart-Prower factor
- Factor XI—plasma thromboplastin antecedent
- Factor XII—Hageman factor
- Factor XIII—fibrin-stabilizing factor

The liver must be able to use Vitamin K to produce Factors II, VII, IX, and X. Dietary vitamin K is widely available from plant and animal sources.

Anticoagulants are agents that inhibit blood clot formation (blood clotting). Anticoagulants are used to treat blood clots, which appear especially frequenty in veins of the legs and pelvis in bedridden patients. Therapy helps to reduce the risk of clots reaching the lung, heart, or other organs. This is useful in primary and secondary prevention of deep vein thrombosis, pulmonary embolism, myocardial infarctions, and strokes in those who are predisposed.

21.2 CLASSIFICATION

1. Coumarin derivatives

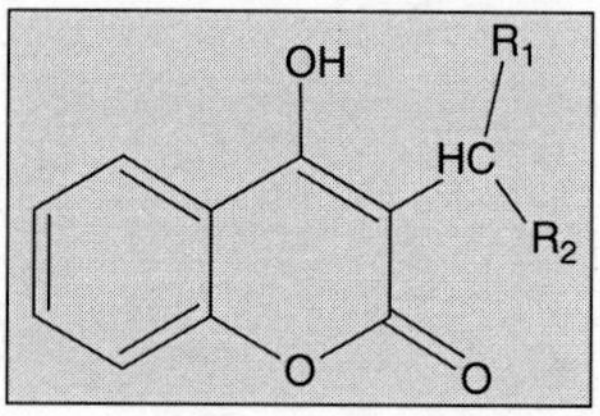

	R_1	R_2
Warfarin	—CH$_2$COCH$_3$	—C$_6$H$_5$
Dicoumarol	H	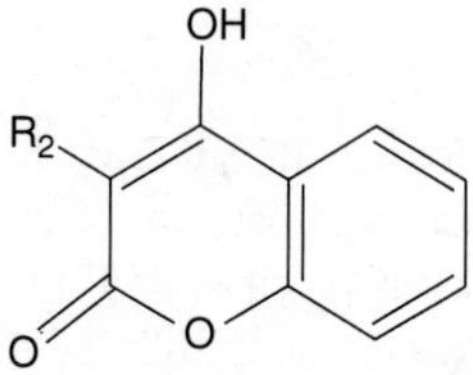 (4-hydroxycoumarin group)
Phenprocoumon	—C$_2$H$_5$	—C$_6$H$_5$
Acenocoumarin	—CH$_2$COCH$_3$	—C$_6$H$_4$—NO$_2$

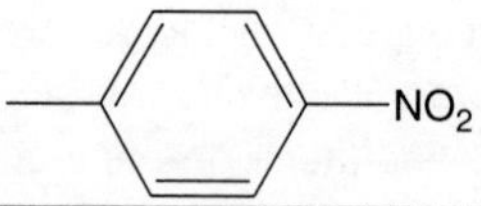

2. 1,3 – Indanedione derivatives

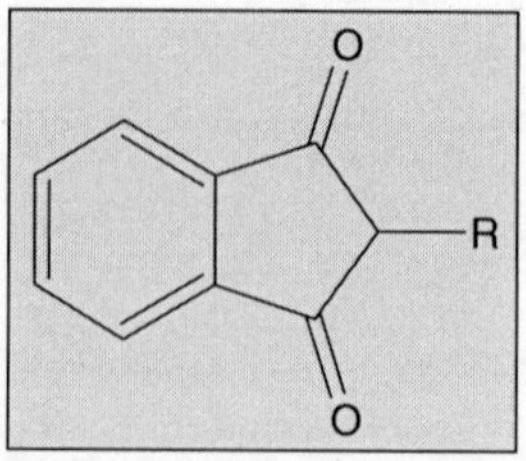

	R
Phenindione	—C_6H_5
Anisindione	H_3CO—⟨ ⟩—
Bromindione	Br—⟨ ⟩—
Diphenadione	—$COCH(C_6H_5)_2$

Mechanism of action: Oral anticoagulants produce their effect by interfering with the vitamin K cycle. Specifically, they interact with KO reductase enzyme so that vitamin KO cannot be recycled back to vitamin K. This leads to depletion of vitamin KH_2, thereby blocking the formation of prothrombin (factor II) from its precursor, and the plasma content of prothrombin is thus reduced and blood coagulation is impaired.

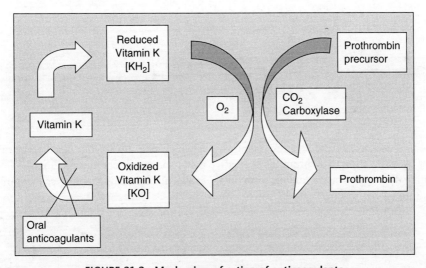

FIGURE 21.2 Mechanism of action of anticoagulants.

21.3 COUMARIN DERIVATIVES

The most widely used oral anticoagulants are coumarin derivatives. Their therapeutic action depends on the ability to suppress formation of a number of functional factors of blood coagulation in the liver. These factors are described as vitamin K-dependent factors since their biosynthesis by hepatocytes is partially linked with hepatic vitamin K metabolism. Oral anticoagulants are effective only *in vivo* because their principal effect is suppression of synthesis of prothrombin, proconvertin, and other blood coagulation factors in the liver. They are sometimes conventionally called vitamin K antagonists.

Generally, these anticoagulants are used to treat patients with deep-vein thrombosis (DVT), pulmonary embolism (PE), atrial fibrillation (AF), and mechanical prosthetic heart valves.

Warfarin 3-(α-Acetonyl benzyl)-4-hydroxy coumarin
Synthesis

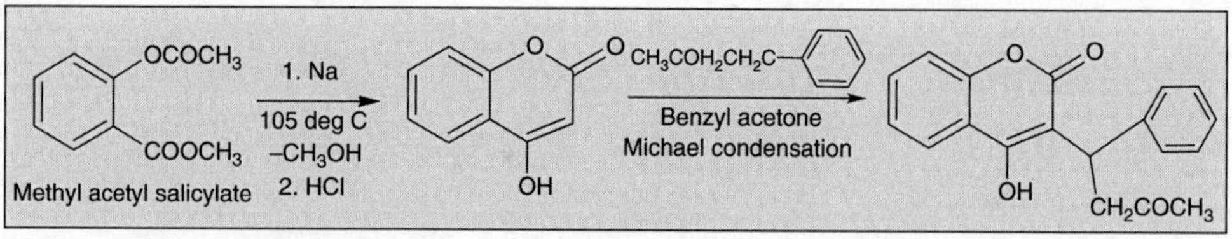

Warfarin is synthesised via Michael reaction by attaching 4-hydroxycoumarin to benzalacetone in the presence of pyridine. 4-Hydroxycoumarin is synthesised from salicylic acid methyl ester by cyclization to a chromone derivative using sodium or sodium methoxide.

Acenocoumarin: 3-(α-Acetonyl-p-nitrobenzyl)-4-hydroxycoumarin
Acenocoumarin is synthesised by a scheme completely analogous to making warfarin, but using p-nitrobenzalacetone.

Dicoumarol 3, 3'-Methylene bis-(4-hydroxy coumarin)
Synthesis

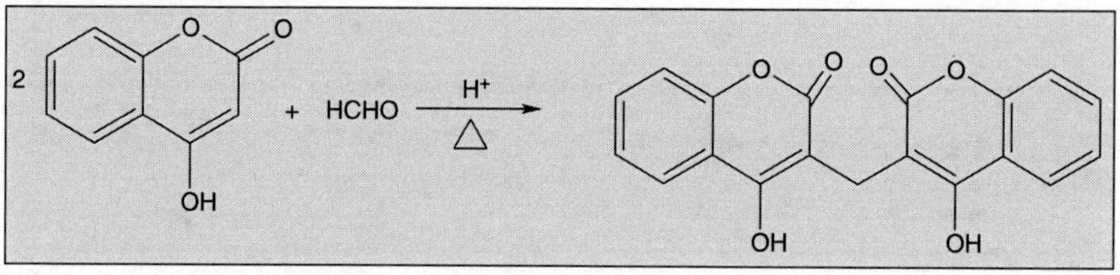

Dicoumarol is synthesised from 4-hydroxycoumarine, which is in turn synthesised from salicylic acid methyl ester by cyclization to a chromone derivative using sodium or sodium methoxide. Condensation of the resulting 4-hydroxycoumarin with formaldehyde as a phenol component gives dicoumarol.

Phenprocoumon 3-(α-Ethylbenzyl)-4-hydroxycoumarin
Synthesis

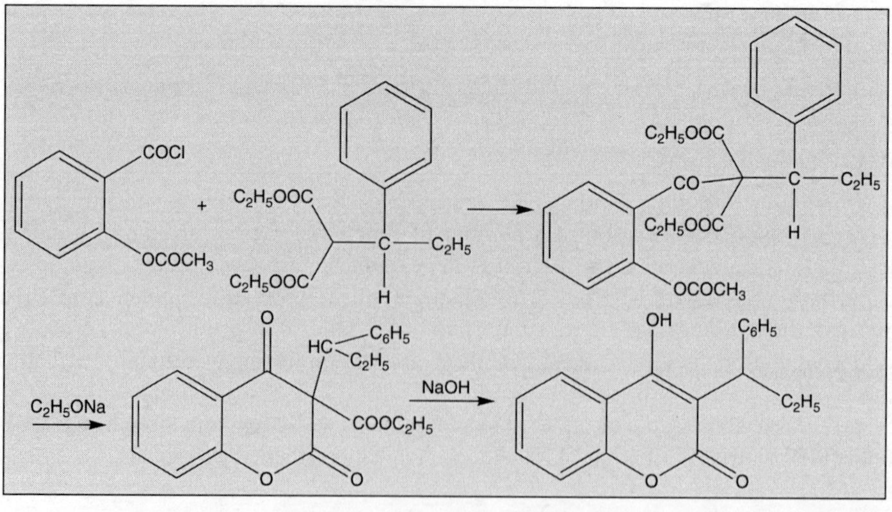

Phenprocoumon is synthesised by acylating sodium salts of diethyl ester (1-phenylpropyl)butyric acid with acetylsalicylic acid chloride, which forms the compound, which upon reaction with sodium ethoxide cyclizes to 3-(α-ethylbenzyl)-2-carboethoxy-4-hydroxycoumarin. Alkaline hydrolysis of this product and further decarboxylation gives phenprocoumon.

21.4 INDANEDIONE DERIVATIVES
General method of preparation

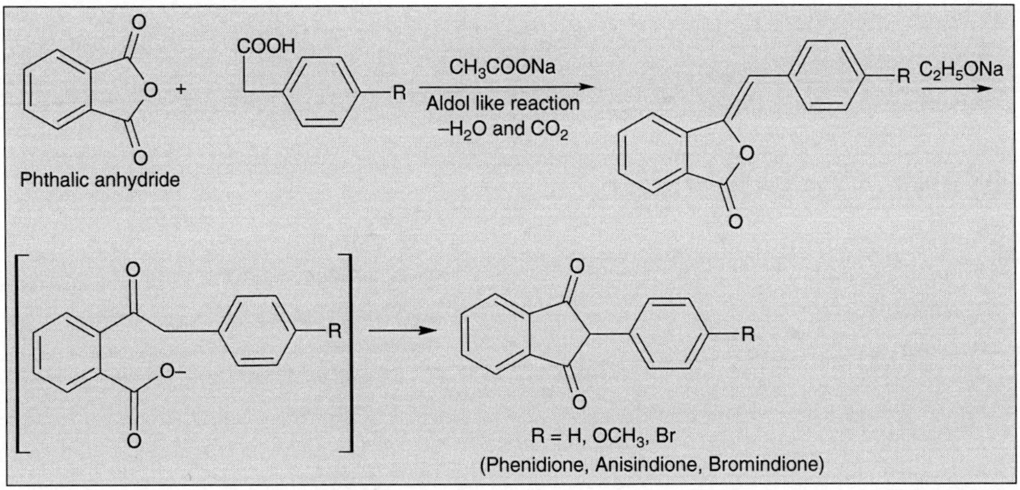

The method consists of condensation of (sub)phenylacetic acid with phthalic anhydride, forming phenyl-methylenphthalide, which rearranges further in the presence of sodium ethoxide to corresponding derivative.

21.5 NEWER DRUGS

Fondaparinux It is a synthetic pentasaccharide. Fondaparinux is identical to a sequence of five monomeric sugar units that can be isolated after either chemical or enzymatic cleavage of the polymeric glycosaminogly-cans heparin and heparan sulphate (HS). Within heparin and heparan sulphate, this monomeric sequence is thought to form the high-affinity binding site for the anti-coagulant factor antithrombin III (ATIII). Binding of heparin/HS to ATIII has been shown to increase the anticoagulant activity of antithrombin III 1,000-fold.

Ximelagatran Ethyl 2-[[(1R)-1-cyclohexyl-2-[(2S)-2-[[4-(N'-hydroxycarbamimidoyl) phenyl]methyl-carbamoyl]azetidin-1-yl]-2-oxo-ethyl]amino]acetate

It is an anticoagulant that has been investigated extensively as a replacement for warfarin that would overcome the problematic dietary, drug interaction, and monitoring issues associated with warfarin. Ximelagatran was the first member of the drug class of direct thrombin inhibitors that can be taken orally. It acts solely by inhibiting the actions of thrombin. Ximelagatran is a prodrug, being converted *in vivo* to the active agent melagatran. This conversion takes place in the liver and many other tissues through dealkylation and dehydroxylation (replacing the ethyl and hydroxylgroups with hydrogen).

FURTHER READINGS

1. David Green, 1994. *Anticoagulants,* CRC Press.
2. Indra K. Reddy, 2004. *Chirality in Drug Design and Development,* Marcel Dekker.

MULTIPLE-CHOICE QUESTIONS

1. Anticoagulants interfere with
   a. Prothrombin
   b. Vitamin K
   c. Factor V
   d. Factor XII
2. The most widely used anticoagulant is
   a. Warfarin
   b. Dicoumarol
   c. Phenidione
   d. Bromindione
3. The starting material for the synthesis of phenidione is
   a. 4-Methyl phthalic anhydride
   b. 4-Bromo phthalic anhydride
   c. Phthalic anhydride
   d. Benzoic acid
4. The anticoagulant synthesised using Michael condensation is
   a. Phenidione
   b. Warfarin
   c. Dicoumarol
   d. Bromindione

QUESTIONS

1. Sketch out the process of blood clotting schematically and discuss the intervention part of anticoagulants.
2. Discuss briefly the mechanism of action of coumarin derivatives.
3. How are the indanedione derivatives synthesised?
4. Sketch down the synthetic protocol of warfarin.

SOLUTION TO MULTIPLE-CHOICE QUESTIONS

1. b;   2. a;   3. c;   4. b.

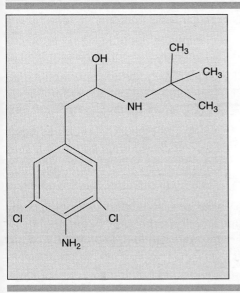

CHAPTER 22

Adrenergic Drugs

LEARNING OBJECTIVES

- Define adrenergic receptors and the drugs acting on them
- Describe the structural requirements of a sympathomimetics acting on α and β subtypes
- Classify sympathomimetics and describe their preparation method
- Describe the importance of specificity of the receptor environment based on optical isomers of ephedrine
- Define sympatholytics and their utility
- Classify sympatholytics, their uses and actions

22.1 INTRODUCTION

Adrenergic drugs are those chemical agents that exert their principal pharmacologic and therapeutic effects by acting at peripheral sites to either enhance or reduce the activity of components (adrenaline) of the sympathetic division of the autonomic nervous system. In general, those substances that produce effects similar to stimulation of sympathetic nervous activity are known as sympathomimetics, adrenomimetics, or adrenergic stimulants. Those that decrease sympathetic activity are referred to as sympatholytics, antiadrenergic, or adrenergic blocking agents.

Adrenergic receptors: In the adrenergic system there are two main types of receptors: α and β. There are two types of α-adrenoreceptors, α_1 and α_2. The α_1-adrenoreceptors subserve smooth muscle stimulant functions, adrenergic sweating, and salivation. The α_2-adrenoreceptors serve to inhibit the pre-synaptic release of noradrenaline and the post-synaptic activation of adenylate cyclase (and, hence, inhibit post-synaptic responses). The β-adrenoreceptors are subdivided into β_1 and β_2 adrenoreceptors, and perhaps more. β_1-adrenoreceptors effect cardiac stimulation and lipolysis; β_2-adrenoreceptors subserve adrenergic smooth muscle relaxation (vasodilatation, bronchodilatation, and intestinal and uterine relaxation) and glycolysis.

22.2 SYMPATHOMIMETIC DRUGS

Peripheral actions and uses of sympathomimetics are discussed below.

22.2.1 α-Adrenoreceptor Agonists

α-Agonists cause arteriolar and venous constriction and, hence, have an action to increase blood pressure. This vasopressor action is used to support blood pressure in hypotensive states, such as in orthostatic hypotension, carotid sinus syndrome, shock, and during spinal anaesthesia.

The systemic vasoconstrictor effects are also employed in the management of a variety of serious allergic conditions, such as giant urticaria, serum sickness, angioneurotic oedema, and anaphylaxis.

The α-agonists are applied topically to induce local vasoconstriction in nasopharyngeal, scleroconjunctival, and otic blood vessels in acute conditions of rhinitis, coryza, nasopharyngitis, sinusitis, conjunctivitis, and hay fever.

By inhalation, α-agonists may be used to suppress bronchial congestion in allergic and asthmatic conditions.

Structural requirements: The structural requirements for α-agonist activity are a phenylethylamine skeleton to which at least two hydroxyl groups are attached; the optimal positions are ring 3- and side chain L-2, but ring 4- and L-2 and ring 3, 4-dihydroxy compounds are active.

22.2.2 β$_1$-Adrenergic Agonists

The β$_1$-receptors are located in the heart. The β$_1$-agonists increase the heart rate, enhance atrioventricular conduction, and increase the strength of the heartbeat. These effects are achieved in part through the activation of the adenylyl cyclase system and intermediation of 3',5'-cyclic adenosine monophosphate (cAMP). They may be administered by intracardiac injection to restore the heartbeat in cardiac arrest and heart block with syncopal seizures. Sometimes, β$_1$-agonists also are used for their positive inotropic actions in the treatment of acute heart failure and in cardiogenic or other types of shock, in which contractility often is diminished.

22.2.3 β$_2$-Selective Adrenergic Agonists

The β$_2$-receptors are located in the lung and uterus. The β$_2$-agonists relax smooth muscle and induce hepatic and muscle glycogenolysis, also by activating the adenylyl cyclase system and increasing the intracellular levels of cAMP. Thus, they dilate the bronchioles, arterioles in vascular beds which are invested with β$_2$-receptors and veins, and they relax the uterus and intestines.

Some β$_2$-agonists are used as bronchodilators in the treatment of bronchial asthma, emphysema, and bronchitis. They are also used to relax the uterus and delay delivery in premature labour, and to treat dysmenorrhoeal problem.

Structural requirements: The structural requirements include an L-β-OH group, which is essential to both β_1 and β_2 activity. N-Alkyl substitution enhances both the activities, while isopropyl and *t*-butyl confer optimal activity. A ring hydroxyl group at the 3- or 4-position is required; the 3-OH appears to be more favourable for β_2- and the 4-OH for β_1-activity.

22.2.4 Classification of Sympathomimetics

(1) Phenylethanolamine derivatives

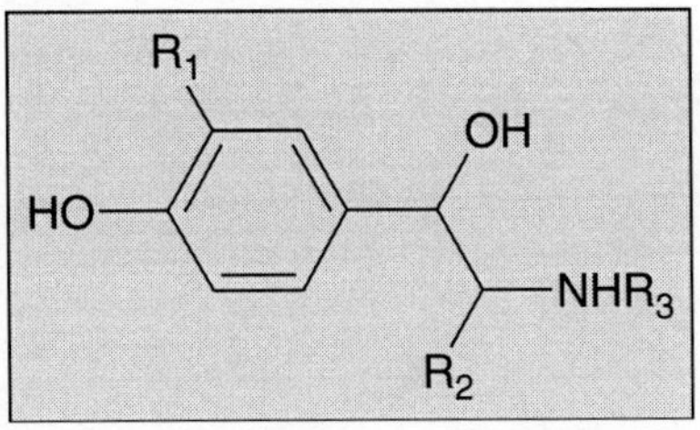

	R_1	R_2	R_3
Adrenaline	—OH	—H	—CH$_3$
Isoproterenol (Isoprenaline)	—OH	—H	—CH(CH$_3$)$_2$
Albuterol (Salbutamol)	—CH$_2$OH	—H	—C(CH$_3$)$_3$
Metarminol	—H	—CH$_3$	—H
Ephedrine	—H	—CH$_3$	—CH$_3$
Phenylephrine	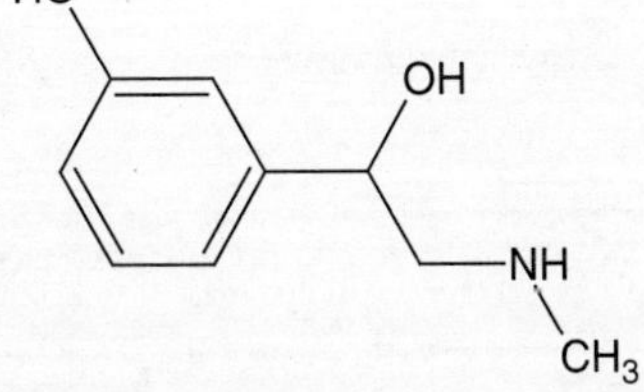 HO		

(2) Imidazoline derivatives

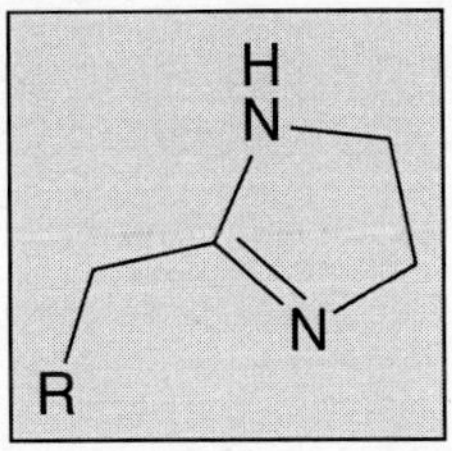

	R
Xylometazoline	(benzyl with (H₃C)₃C-, two CH₃ groups)
Oxymetazoline	(benzyl with HO, (H₃C)₃C-, two CH₃ groups)
Naphazoline	(naphthalene-methyl)

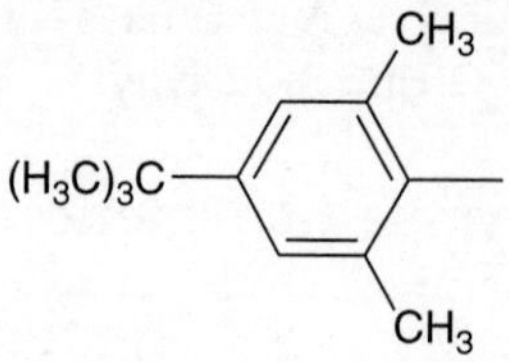

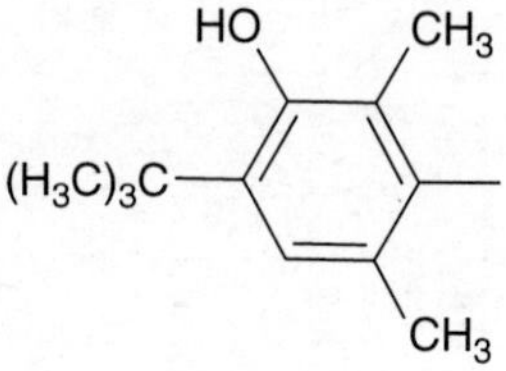

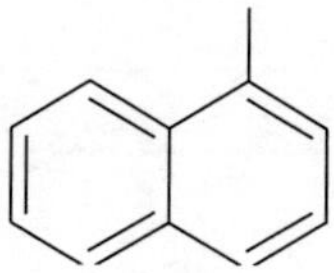

22.2.5 Phenylethanolamine Derivatives

Adrenaline (-)-3,4-Dihydroxy-α-[(methylamino)methyl] benzyl alcohol

It possesses all of strong α_1-, α_2-, β_1-, and β_2-agonist activities. It is the drug of choice in the management of allergic emergencies such as anaphylaxis, angioneurotic oedema, urticaria, and serum sickness.

Synthesis

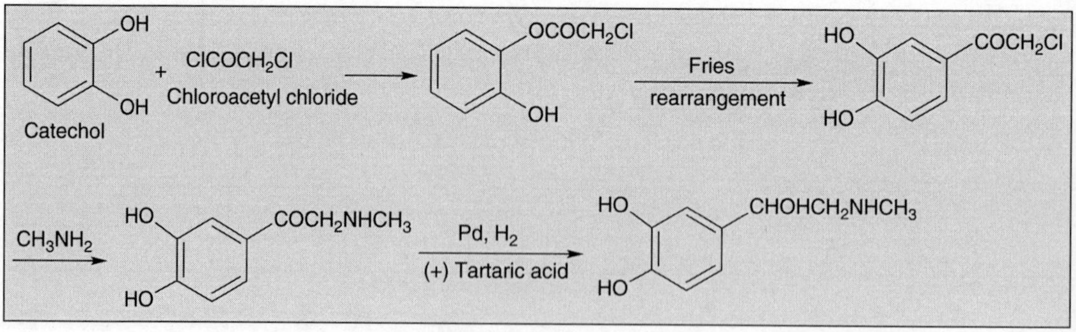

Catechol + ClCOCH₂Cl (Chloroacetyl chloride) → (aryl OCOCH₂Cl, OH) → Fries rearrangement → (HO, HO-aryl-COCH₂Cl) → CH₃NH₂ → (HO, HO-aryl-COCH₂NHCH₃) → Pd, H₂ / (+) Tartaric acid → (HO, HO-aryl-CHOHCH₂NHCH₃)

Adrenaline (epinephrine) is obtained from the adrenal glands tissue of livestock as well as in a synthetic manner. Epinephrine is synthesised by reaction between catechol and chloroacetyl chloride to give ester, which undergoes Fries rearrangement to form ω-chloro-3,4-dihydroxyacetophenone. Reaction of this with excess of methylamine gives ω-methylamino-3,4-dihydroxyacetophenone. Reduction of this gives D,L-epinephrine, which is separated into isomers using (-) tartaric acid.

> Epinephrine was isolated and identified in 1895 by Napoleon Cybulski, a Polish physiologist. In May 1896, William Bates reported the discovery of a substance produced by the adrenal gland in the *New York Medical Journal*.

Isoprenaline: 3, 4-Dihydroxy-α-[(isopropylamino)methyl]benzyl alcohol

Preparation: By the synthetic procedure given for adrenaline using isopropylamine in place of methylamine

It has strong $β_1$- and $β_2$-agonist activity but lacks α-activity. Its primary use is in the treatment of bronchial asthma.

Salbutamol: 4-Hydroxy-3-hydroxymethyl-α-[(*tert*-butylamino)methyl] benzyl alcohol.

Synthesis

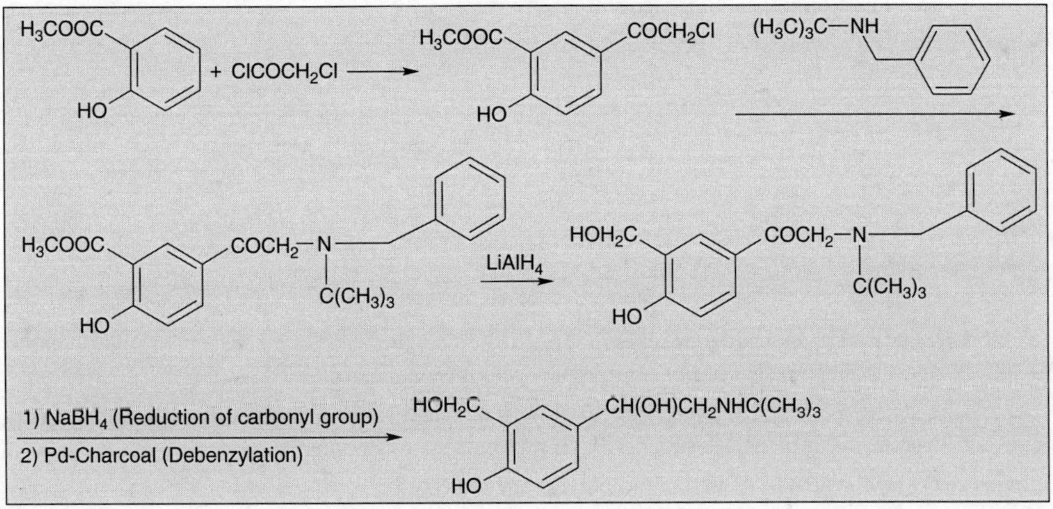

Acylation of methyl ester of salicylic acid using chloroacetyl chloride afford acetophenone derivative. This on reaction with N-benzyl-tert-butylamine, and the resulting product is completely reduced by lithium aluminium hydride into the N-benzyl-substituted salbutamol, the benzyl group of which is removed by hydrogen over a palladium catalyst to give the desired salbutamol.

The therapeutic action and uses of salbutamol are similar to those of isoprenaline.

Metarminol (-)-α-1-Aminoethyl-3-hydroxy benzyl alcohol

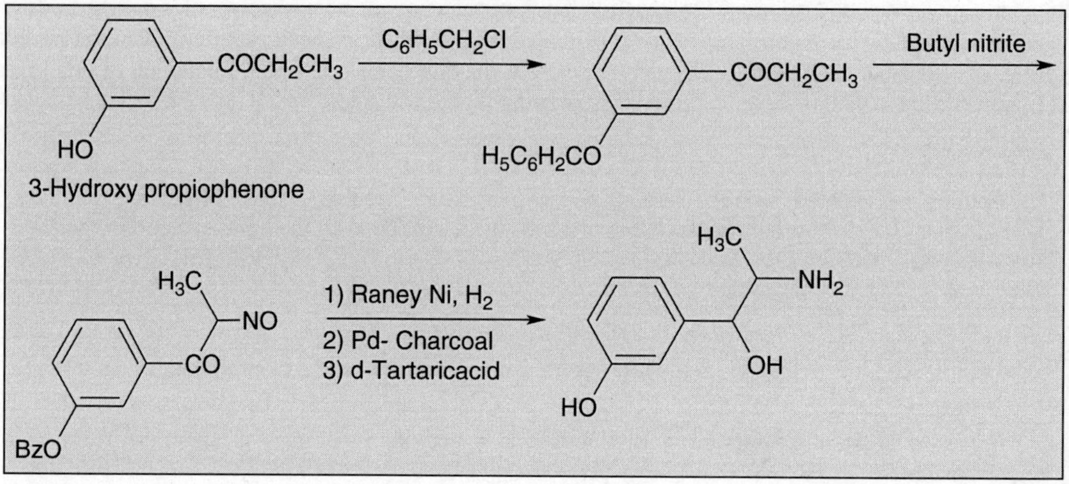

[Reaction scheme: 3-Hydroxy propiophenone + C₆H₅CH₂Cl → benzyl-protected propiophenone + Butyl nitrite → nitroso ketone intermediate → 1) Raney Ni, H₂; 2) Pd-Charcoal; 3) d-Tartaric acid → metarminol]

Reaction of 4-hydroxy propiophenone with benzyl chloride gives O-benzyl derivative, which on treatment with butyl nitrite undergoes nitrosation reaction at the α-position. Stepwise reduction of the nitrosoketone leads to amino alcohol metarminol.

Phenylephrine (-)-3-Hydroxy-α-[(methylamino)methyl] benzyl alcohol

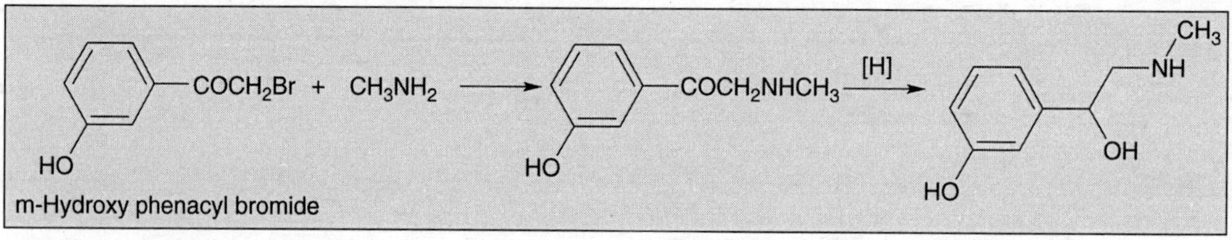

[Reaction scheme: m-Hydroxy phenacyl bromide + CH₃NH₂ → COCH₂NHCH₃ intermediate → [H] → phenylephrine]

It is synthesised by an analogous scheme of making epinephrine; however, instead of using ω-chloro-3,4-dihydroxyacetophenone, ω-chloro-3-dihydroxyacetophenone is used.

It has a strong α-agonist and negligible β-agonist activity. It is used in the treatment of paroxysmal supraventricular tachycardia and to support blood pressure.

Ephedrine (-)-Erythro-α-[1-(methylamino)ethyl] benzyl alcohol

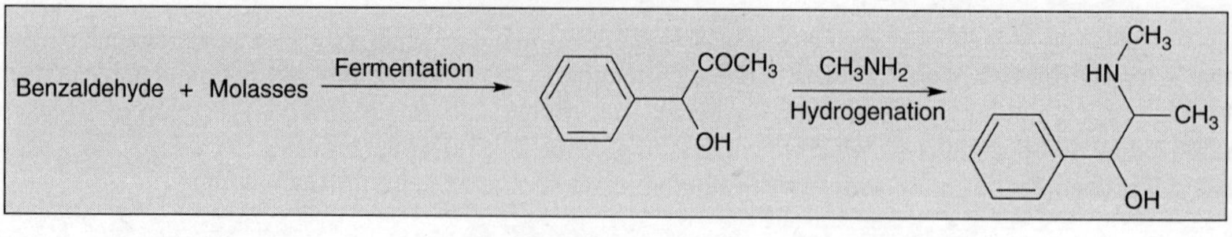

[Reaction scheme: Benzaldehyde + Molasses → Fermentation → phenyl-CH(OH)-COCH₃ → CH₃NH₂, Hydrogenation → ephedrine]

The method consists of the fermentation of glucose by yeast carboligase in the presence of benzaldehyde, which during the process turns into (-)-1-phenyl-2-ketopropanol. This is reduced by hydrogen in the presence of methylamine, to give the desired ephedrine.

Ephedrine, although 100 times less potent than epinephrine, has prolonged bronchodilatory effects after oral administration. Pseudoephedrine is the (-) optical isomer of (+) ephedrine. The activities of the optical isomers of ephedrine show the rather precise nature of the receptor–drug interactions. Pseudoephedrine is much less potent than ephedrine. The following figure depicts the importance of hydrogen bonding for activity.

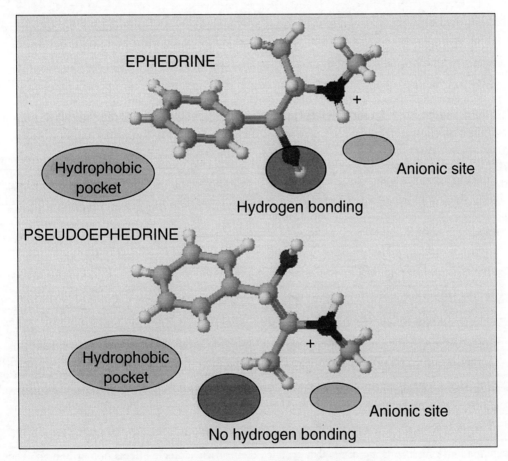

FIGURE 22.1 Binding site of ephedrine at the receptor.

Imidazoline derivatives

Xylometazoline 2-(4-*tert*-Butyl-2, 6-dimethylbenzyl)-2-imidazoline

Synthesis

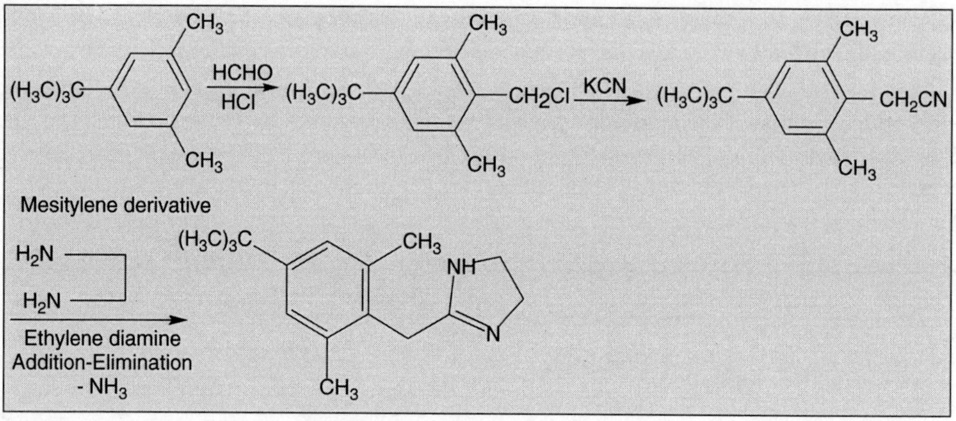

Xylometazoline is synthesised by chloromethylation of mesitylene derivative and the further transformation of the resulting chloromethyl derivative into a nitrile. The reaction of this with ethylenediamine gives xylometazoline.

Oxymetazoline 6-*tert*-butyl-3-(2-imidazolin-2-ilmethyl)-2,4-dimethylphenol

Synthesis

It is synthesised by chloromethylation of 6-*tert*-butyl-2,4-dimethylphenol and the further transformation of the resulting chloromethyl derivative into a nitrile. The reaction of this with ethylenediamine gives oxymetazoline.

Naphazoline 2-(1-Naphthylmethyl)-2-imidazoline

Synthesis

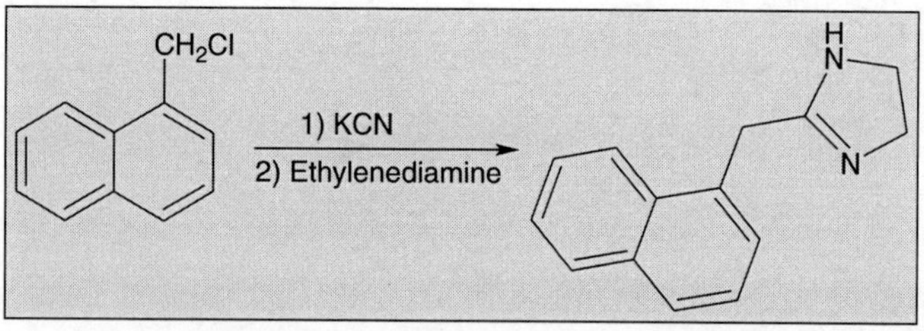

Naphazoline is synthesised by reaction with naphthyl methyl chloride and potassium cyanide to give (1-naphthyl)acetonitrile, which upon reaction with ethanol transforms into iminoester, and undergoes further heterocyclization into the desired imidozoline derivative upon reaction with ethylenediamine.

22.3 SYMPATHETIC (ADRENERGIC) BLOCKING AGENTS (SYMPATHOLYTICS)

Adrenergic blocking agents are drugs that produce their pharmacologic effects primarily by preventing the release of noradrenaline from sympathetic nerve terminals.

These drugs produce their effects by stabilization of the neuronal membrane or the membranes of the storage vesicles. This stabilization makes the membranes less responsive to nerve impulses, thereby inhibiting the release of noradrenaline into the synaptic cleft.

22.3.1 Classification, Action, and Uses

α-Adrenergic antagonists

1. **Non-selective α-antagonists:** Examples: Phenoxybenzamine, Phentolamine, Tolazoline

Medicinal Chemistry 409

This class produces α_1 and α_2 blockade. Antagonism of α_1-adrenergic impulses to the arterioles decreases vascular resistance, thus tending to lower the blood pressure, and causes a pink warm skin and nasal and scleroconjunctival congestion. α-Antagonism causes tachycardia, palpitations, and increased secretion of renin. They are used in the treatment of peripheral vascular disorders such as Raynaud's disease, acrocyanosis, frostbite, acute arteriolar occlusion, causalgia, and pheochromocytoma.

2. **Selective α-antagonists:** Theoretically, α_1-blockers (prazosin and terazosin) should be useful for the same disorders as are the non-selective α-blockers, but they are approved only for the treatment of hypertension.

Selective α_2-antagonists include yohimbine and rauwolscine, but there are presently no therapeutic application of α_2-blockade.

β-Adrenergic antagonists

1. **Non-selective β-antagonists:** Drugs such as Propranolol, Nadolol, Pindolol, and Timolol suppress both β_1- and β_2-adrenoreceptor-mediated responses almost equally. Blockade of myocardial β_1-receptors causes sinoatrial bradycardia, decreased force of myocardial contraction, slowing of atrioventricular conduction, and increased atrioventricular refractoriness.

Uses
- β-Blockers are of prophylactic value in the treatment of stable angina pectoris
- The effect to decrease sinoatrial rate is also used to suppress tachycardia in thyrotoxicosis and pheochromocytoma
- The effect to decrease atrioventricular nodal conduction is employed in the chronic management of paroxysmal supraventricular tachycardia
- All available β-antagonists are used in the treatment of hypertension
- β-Antagonists have usefulness in the prophylaxis of migraine headache

2. **Selective β_1-antagonists:** Examples: Acebutolol, Atenolol, Metoprolol, Practolol, and Tolamol
Selective β_1-antagonists can be used for all the purposes listed under the non-selective blockers. Advantages to selective β_1-blockade are lesser effect on bronchiolar airway resistance and diminished effect to increase insulin-induced hypoglycaemia.

3. **Partial agonist β-antagonist:** Examples: Oxprenolol, Acebutolol, and Pindolol
It also causes some stimulation of β-adrenoreceptors. This partial agonism acts as buffer to lessen the seriousness of the various adverse effects attributable to β-blockade.

22.3.2 Synthesis of α-Adrenergic Blockers

Phenoxybenzamine N-(2-Chloroethyl)-N-(1-methyl-2-phenoxyethyl)benzylamine

Phenoxybenzamine is synthesised by reacting phenol with propylene chlorohydrin, which forms 1-phenoxy-2-propanol, the chlorination of which with thionyl chloride gives 1-phenoxy-2-propylchloride. Reacting this with 2-aminoethanol leads to formation of 1-phenoxy-2-(2-hydroxyethyl)aminopropane. Alkylation of the secondary amino group gives *N*-(2-hydroxyethyl)-*N*-(1-methyl-2-phenoxyethyl)benzylamine, the hydroxyl group of which is chlorinated using thionyl chloride, giving phenoxybenzamine.

Phentolamine 3-[[(4,5-Dihydro-1H-imidazol-2-yl)methyl](4-methyl-phenyl)amino]phenol

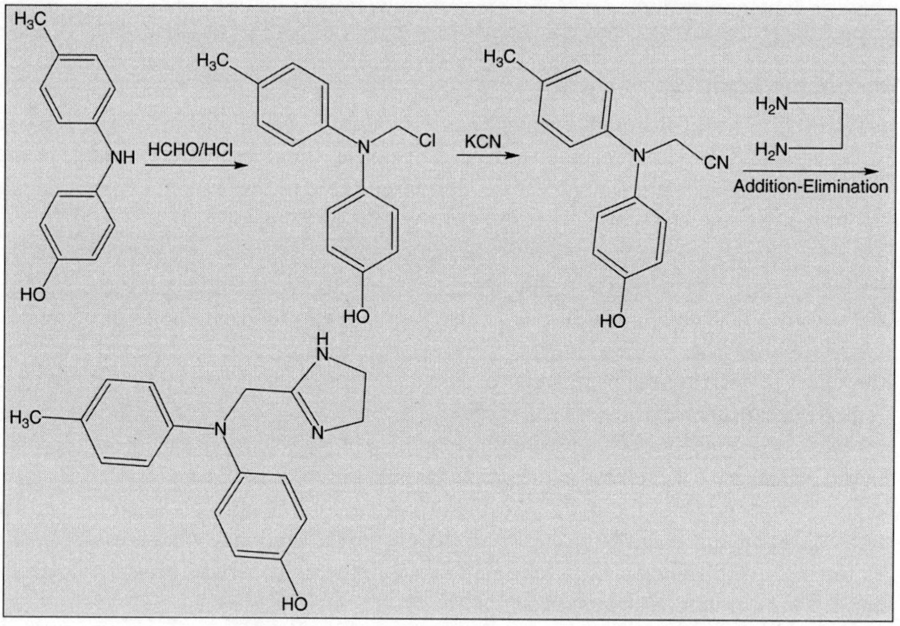

Chlormethylation of diphenylamine is followed by further transformation of the resulting chloromethyl derivative into a nitrile. The reaction of this with ethylenediamine gives phentolamine.

Tolazoline 2-Benzyl-2-imidazoline

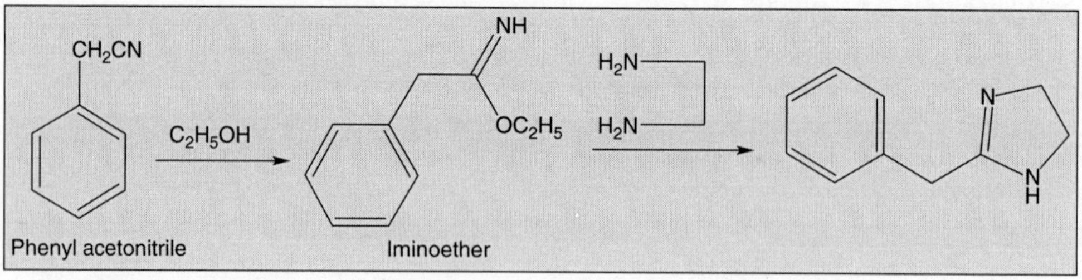

Tolazoline is synthesised by the heterocyclation of the ethyl ester of iminophenylacetic acid with ethylenediamine, which forms the desired tolazoline.

(Synthesis of **prazosin** and **terazosin** was discussed in Chapter 16 on 'Antihypertensive Agents'.)
(For synthesis of β-**Adrenergic blockers,** refer to Chapter 16 on 'Antihypertensive Agents'.)

22.4 NEWER DRUGS

Clenbuterol 1-(4-Amino-3,5-dichlorophenyl)-2-(tert-butylamino)ethanol

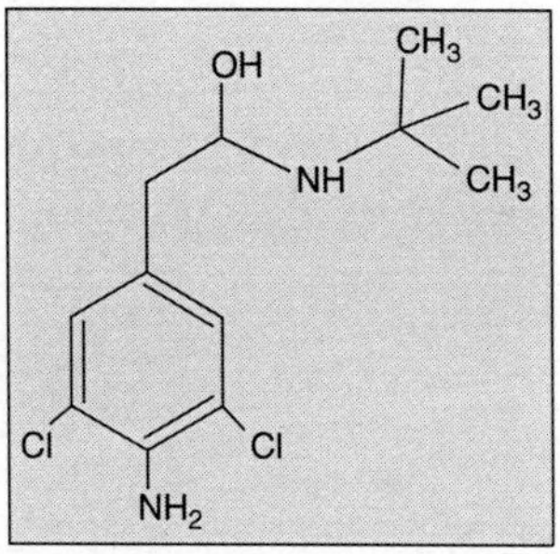

It is a drug prescribed to sufferers of breathing disorders as a decongestant and bronchodilator. Clenbuterol is a $β_2$-adrenergic agonist with some similarities to ephedrine, but its effects are more potent and longer lasting as a stimulant and thermogenic drug.

Salmeterol 2-(Hydroxymethyl)-4-{1-hydroxy-2-[6-(4-phenylbutoxy)hexylamino]ethyl}phenol

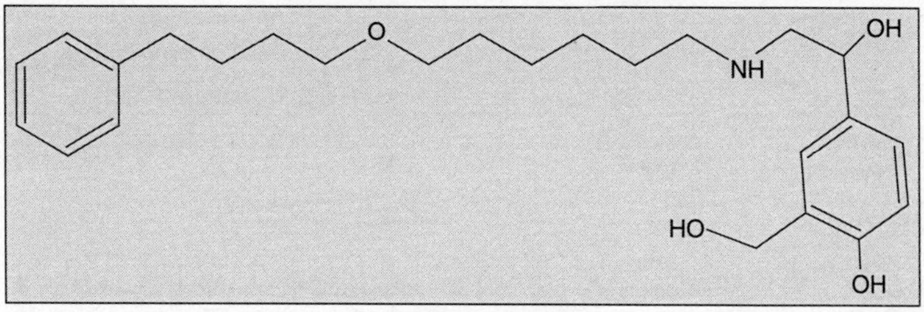

It is a long-acting $β_2$-adrenergic receptor agonist drug that is currently prescribed for the treatment of asthma and chronic obstructive pulmonary disease (COPD).

Formoterol N-[2-Hydroxy-5-[1-hydroxy-2-[1-(4-methoxyphenyl) propan-2-ylamino]ethyl] phenyl]formamide

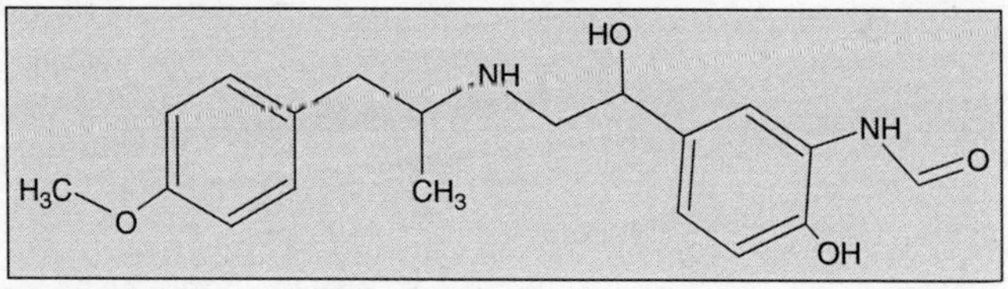

It is a long-acting $β_2$-agonist, which has an extended duration of action (up to 12 hours) compared to short-acting $β_2$-agonists such as salbutamol, which are effective for 4–6 hours. It is used in the management of asthma and/or COPD.

FURTHER READINGS

1. Choudhary, M. Iqbal, 1996. *Progress in Medicinal Chemistry*, Taylor & Francis (UK).
2. Silverman, Richard B., 2004. *The Organic Chemistry of Drug Design and Drug Action*, Elsevier.

MULTIPLE-CHOICE QUESTIONS

1. β-Adrenergic blockers are not used for
   a. Migraine headache
   b. Hypertension
   c. Angina pectoris
   d. Supra-ventricular bradycardia
2. Pindolol is a (an)
   a. α_1-Agonist
   b. β_1-Agonist
   c. β_1-Partial agonist antagonist
   d. α_1-Partial agonist antagonist
3. Propanolol is useful for
   a. Asthma
   b. Ventricular tachycardia
   c. Anaesthesia
   d. Atropine poison
4. The β_1-receptors are located in
   a. Heart
   b. Lungs
   c. Kidney
   d. Adrenal gland
5. Orthostatic hypotension is treated with
   a. β-Blockers
   b. β-Agonists
   c. α-Blockers
   d. α-Agonists
6. Drug of the following class is useful in the treatment of asthma:
   a. β_1-Blockers
   b. β_1-Agonists
   c. α_1-Blockers
   d. α_1-Agonists
7. The adrenergic agent obtained by fermentation procedure is
   a. Propanolol
   b. Adrenaline
   c. Ephedrine
   d. Phenylephrine
8. 3, 4-Dihydroxy-α-[(isopropylamino)methyl] benzyl alcohol is
   a. Adrenaline
   b. Propanolol
   c. Phenylephrine
   d. Isoprenaline
9. 3-[[(4,5-Dihydro-1H-imidazol-2-yl)methyl] (4-methyl-phenyl)amino]phenol is
   a. Phentolamine
   b. Tolazoline
   c. Naphazoline
   d. Phenoxybenzamine
10. Salbutamol is synthesised starting from
    a. Phenyl acetonitrile
    b. Methyl salicylate
    c. 4-Hydroxy propiophenone
    d. Mesitylene derivative

QUESTIONS

1. Classify sympathomimetics and discuss their therapeutic uses.
2. With the structure of epinephrine, discuss the design of agonists and antagonists based on the structural modification.
3. Write the synthetic scheme for salbutamol, isoprenaline, phentolamine, and ephedrine.
4. Classify antiadrenergic drugs with structural examples.
5. Write a short note on the therapeutic uses of β-blockers.

SOLUTION TO MULTIPLE-CHOICE QUESTIONS

1. d;
2. c;
3. b;
4. b;
5. d;
6. b;
7. c;
8. d;
9. a;
10. b.

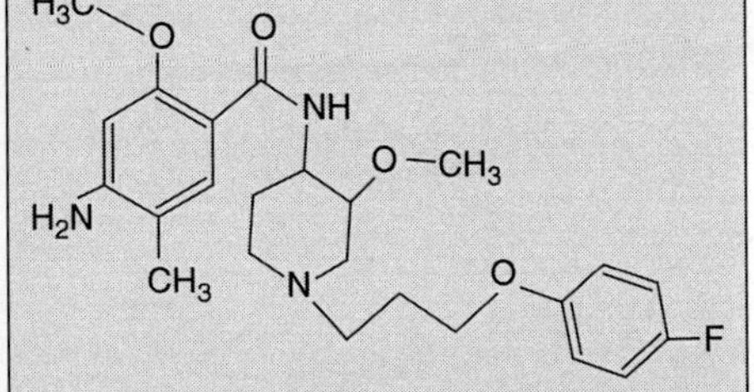

CHAPTER 23

Cholinergic Drugs

> **LEARNING OBJECTIVES**
>
> - Define acetyl choline and cholinergic receptors
> - Describe the actions and uses of cholinomimetics
> - Describe the structural requirements for cholinergic agonist
> - Define acetylcholinesterase and the utility of inhibitors
> - Define antimuscarinics and classify based on structure
> - Compare the tertiary and quarternary amines as antimuscarinics
> - List the newer drugs in the class of cholinergic agents

23.1 INTRODUCTION

Cholinergic nerves (parasympathetic nerves) are found in the peripheral and the central nervous system of humans. Cholinergic neurons synthesise, store, and release acetylcholine (ACh). Acetylcholine is prepared at the nerve ending by the transfer of an acetyl group from acetyl-CoA to choline. The reaction is catalysed by acetyltransferase. Acetylcholine is released from the synaptic vesicles by the nerve action potential. The major sites of activity of ACh in the periphery are as follows: the post-ganglionic parasympathetic receptors, the autonomic ganglia, and the skeletal neuromuscular junction.

> Acetylcholine (ACh) was first identified in 1914 by Henry Hallett Dale for its actions on heart tissue. It was confirmed as a neurotransmitter by Otto Loewi, who initially gave it the name 'vagusstoff' because it was released from the vagus nerve. Both received the 1936 Nobel Prize in Physiology or Medicine for their work. Acetylcholine was also the first neurotransmitter to be identified.

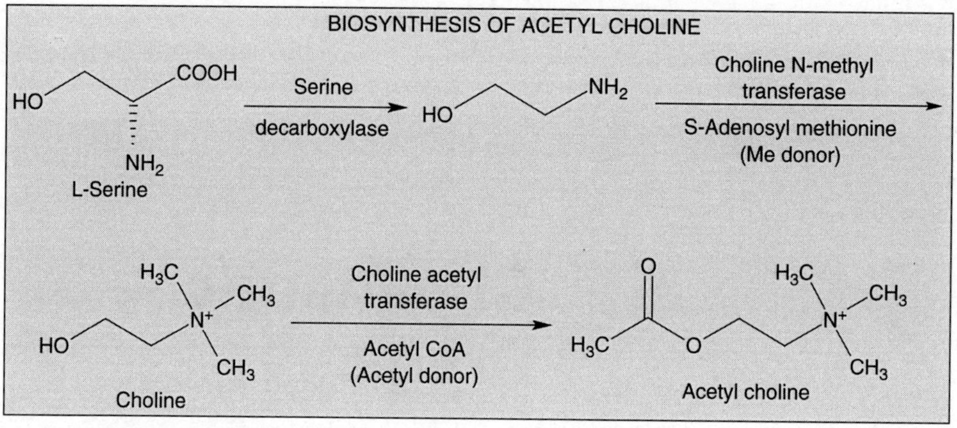

Chapter 23: Cholinergic Drugs

BIOSYNTHESIS OF ACETYL CHOLINE

[Reaction scheme: L-Serine → (Serine decarboxylase) → HO-CH2-CH2-NH2 → (Choline N-methyl transferase, S-Adenosyl methionine (Me donor)) → Choline → (Choline acetyl transferase, Acetyl CoA (Acetyl donor)) → Acetyl choline]

Cholinergic Receptors: There are two distinct receptors for ACh that differ in composition, location, and pharmacologic function, and have specific agonists and antagonists.

1. **Nicotinic receptors:** These are receptor sites acted upon by nicotine eliciting the same qualitative effects as acetylcholine. The receptor is a ligand-gated ion channel.

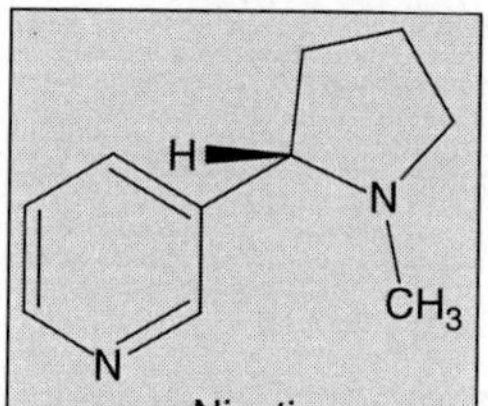

[Structure: Nicotine]

It is a glycoprotein embedded into the polysynaptic membrane. The receptor is believed to exist as a dimer of the two five-subunit polypeptide chain monomers linked together through a disulphide bond. When the neurotransmitter ACh binds to the nicotinic receptor, it causes a change in the permeability of the membrane to allow passage of small cations, Ca^{2+}, Na^+, and K^+. This leads to depolarization and results in muscular contraction at a neuromuscular junction. This is the target for autoimmune antibodies in myasthenia gravis and muscle relaxants used during the course of surgical procedures. Two subtypes of nicotinic receptors, N_1 and N_2, are located in somatic and neural, respectively.

2. **Muscarinic receptors:** These are receptor sites where muscarine binds, eliciting the same qualitative effects as acetylcholine. These are G-protein coupled receptors.

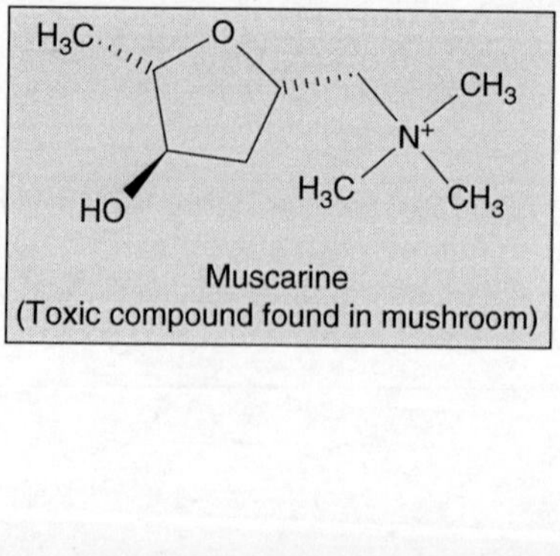

[Structure: Muscarine (Toxic compound found in mushroom)]

The action of ACh on muscarinic receptors can be either stimulatory or inhibitory. Acetylcholine stimulates secretion from salivary and sweat glands, and causes secretion and contraction of the gut, and constriction of the airways of the respiratory tract. It also inhibits the contraction of the heart and relaxes smooth muscles of blood vessels.

A number of cholinergic-induced responses depend on K^+ channels, although not always in the same way. In the heart, these channels are opened by ACh, and the result is hyperpolarization and inhibition. In other neurons, the channels are closed by ACh and the result is depolarization (e.g., contraction of smooth muscles).

There are two subtypes of muscarinic receptors, M_1 and M_2. M_1 receptors are located in autonomic ganglia and selected regions of the CNS. The M_2 receptors predominate in the periphery (e.g., heart, smooth muscles, and exocrine glands).

23.2 CHOLINOMIMETICS

Cholinomimetics are agents that bind to acetylcholine receptors and stimulate the parasympathetic system.

Actions of cholinomimetic drugs

1. Muscarinic drugs dilate nearly all blood vessels and cause smooth muscle relaxation.
2. In high doses, they may decrease heart rate and atrioventricular conduction velocity, and cause varying degrees of heart block; they also may decrease the strength of atrial contractions.
3. In therapeutic doses, usually only vasodilatation occurs and the heart rate and contractility actually may increase, because of sympathetically mediated reflexes to the hypotension caused by vasodilatation.
4. The muscarinic drugs stimulate gastrointestinal smooth muscles and increase peristalsis, thus decreasing bowel transit time.
5. These drugs contract the smooth muscle of detrusor muscle of the urinary bladder but relax that of the trigone sphincter, thus causing urination.
6. Stimulation of the sphincter of the iris causes miosis; stimulation of the ciliary body causes ciliary spasms and decrease in intraocular tension in glaucoma.
7. Most exocrine glands are stimulated and, thus, cause excessive salivation, rhinorrhoea, bronchorrhoea, increased gastric and pancreatic secretions, and copious sweating.

Uses

1. Used topically for the treatment of open-angle, acute congestive and narrow-angle glaucoma, and accommodative strabismus, and also used to antagonise mydriatics
2. Used in the treatment of atonic constipation, postoperative and postpartum adynamic intestinal ileus, and postvagotomy gastric atony
3. To treat functional urinary retention
4. Used for the treatment of paroxysmal atrial tachycardia

Structure–activity relationship The general structure of cholinergic agonist is represented thus:

$$H_3C-\underset{\underset{O}{\|}}{C}-O-CH_2-CH_2-N^+(CH_3)_3Cl^-$$

Acyloxy group • Ethylene bridge • Quaternary Ammonium group

1. **Acyloxy group**
   (i) The higher homologues of methyl groups like propionyl or butyryl groups are less active than ACh.
   (ii) Esters of aromatic or higher molecular weight acids possess cholinergic antagonist activity.
   (iii) The methyl ester is rapidly hydrolysed by cholinesterase to choline and acetic acid. To reduce susceptibility to hydrolysis, carbamate esters of choline (carbachol) were synthesised and were found to be more stable than carboxylate esters.
   (iv) Replacement of ester group with ether or ketone produces chemically stable and potent compounds.

2. **Ethylene bridge**
   (i) As the chain length is increased from two-carbon atoms to more than two-carbon atoms, the activity is reduced.
   (ii) Replacement of the hydrogen atoms of the ethylene bridge by methyl group leads to equal or greater activity. Groups larger than methyl lead to a decrease in the activity. The α- or β-methyl-substituted derivatives affect selectivity of receptors. Acetyl- β-methyl choline (methacholine) acts selectively on muscarinic receptors, while acetyl- α-methyl choline has greater nicotinic receptor affinity.

3. **Quaternary ammonium group**
   (i) It is essential for activity.
   (ii) Primary, secondary, or tertiary amines are less active.
   (iii) Replacement of methyl group by ethyl or large alkyl groups produces inactive compounds.

Acetylcholine chloride   2-(Acetyloxy)-N,N,N-trimethyl ethanaminium chloride

Synthesis

$$Cl\!-\!\!-\!\!-\!OH + (CH_3)_3N \longrightarrow \underset{H_3C}{\overset{H_3C}{\underset{|}{N^+}}}\!\!-\!\!-\!\!-\!OH \xrightarrow[\text{chloride}]{\text{Acetyl}} H_3C\!-\!\!\underset{\underset{O}{\|}}{C}\!-\!O\!-\!CH_2\!-\!CH_2\!-\!N^+(CH_3)_3Cl^-$$

2-Chloroethanol is reacted with trimethylamine, and the resulting N,N,N-trimethylethyl-2-ethanolamine hydrochloride, also called choline, is acetylated by acetic acid anhydride or acetylchloride, giving acetylcholine.

Acetyl choline is prototypical muscarinic (and nicotinic) agonist, but a poor therapeutic agent because of its low chemical and enzymatic stability with low availability due to poor absorption.

Use: In ocular surgery, it causes complete miosis in seconds. It is instilled directly in the anterior chamber.

Bethanechol   (2-Hydroxypropyl)trimethyl ammonium chloride

Synthesis

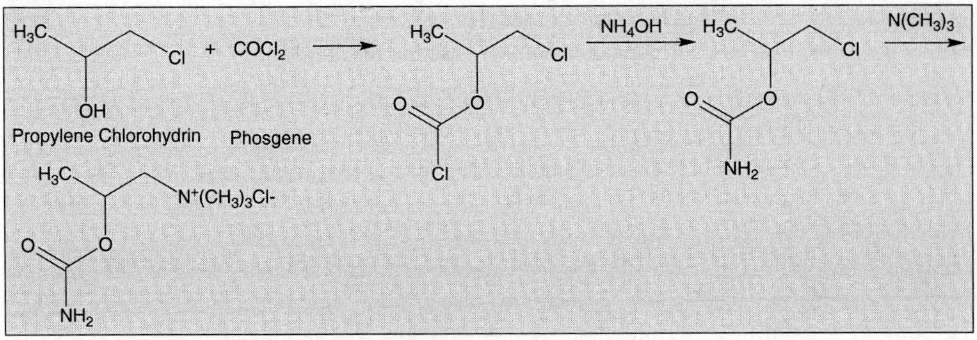

Propylene chlorohydrin on reaction with phosgene and ammonia gives carbamate derivative, which in turn reacts with triethylamine to afford bethanechol.

Bethanechol is a potent muscarinic agonist that is orally effective due to increased hydrolytic stability caused by carbamate linkage and steric bulk of methyl substituent.

Uses: It is used for the relief of post-surgical urinary retention and abdominal distention.

Carbachol   Carbamoyl choline
Synthesis

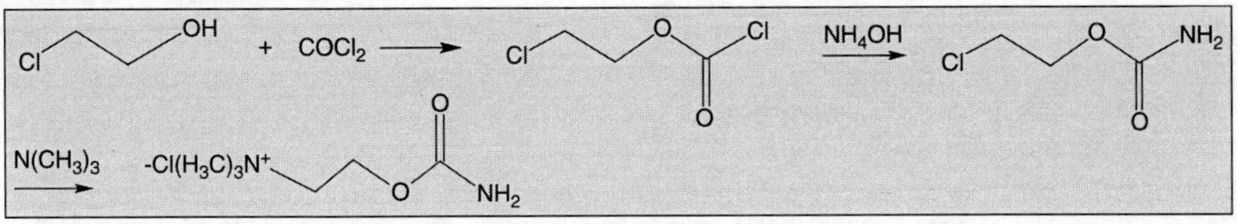

It is made by reacting 2-chloroethanol with phosgene, which forms 2-chloroethyl chloroformate. Upon reaction with ammonia, it turns into the corresponding amide, which is further reacted with an equimolar quantity of trimethylamine, giving carbachol.

Carbachol has increased hydrolytic stability compared to acetyl choline due to the fact that carbamates are more stable than esters.

Uses: Topically for glaucoma; intraocular for miosis in surgery

Pilocarpine

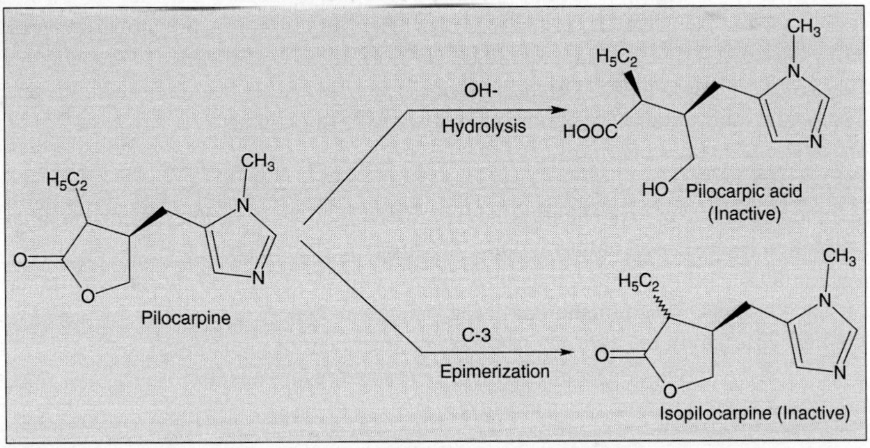

Pilocarpine is a natural product obtained from the leaves of *Pilocarpus jaborandi*. It is chemically unstable to alkali (hydrolysis) and basic conditions (epimerization).

Uses: It is used in the treatment of glaucoma and xerostomia (dry mouth).

23.3 ANTICHOLINESTERASES

Acetylcholinesterase (AChE) is a serine-dependent isoenzyme capable of hydrolysing ACh to choline and acetic acid. The active site of AChE comprises two distinct regions, an anionic site that possesses a glutamate residue and an esteratic site in which histidine imidazole ring and a serine -OH group are present. Catalytic hydrolysis occurs, whereby the acetyl group is transferred to the serine -OH group, leaving an acetylated enzyme molecule and a molecule of free choline. Spontaneous hydrolysis of the serine acetyl group occurs rapidly. AChE inhibitors are indirect acting cholinomimetics.

There are two main categories of AChE inhibitors:

1. The amine or ammonium AChE inhibitors react reversibly with the enzyme, and these compounds reversibly acylate the esteratic serine hydroxyl. Their duration of action are a few minutes to few hours.
2. The organophosphate type AChE inhibitors form an irreversible firm bond with the enzyme (esteratic site), and their duration of action are a few weeks to months.

Uses

1. Used for the treatment of myasthenia gravis (muscular fatigue or weakness) and intestinal distension
2. Edrophonium is used as an antiarrhythmic drug in paroxysmal atrial tachycardia
3. Physostigmine is employed to antagonise the toxic CNS effects of antimuscarinic agents, tricyclic antidepressants, H_1 antihistamines, and benzodiazepines. It is also used in the treatment of Alzheimer's disease. In the eye, it is used for the treatment of glaucoma
4. Also useful as agricultural insecticides and nerve-gas warfare agents

Ambenonium chloride Oxalylbis(iminoethylene)bis[(o-chlorobenzyl) diethylammonium] dichloride
Synthesis

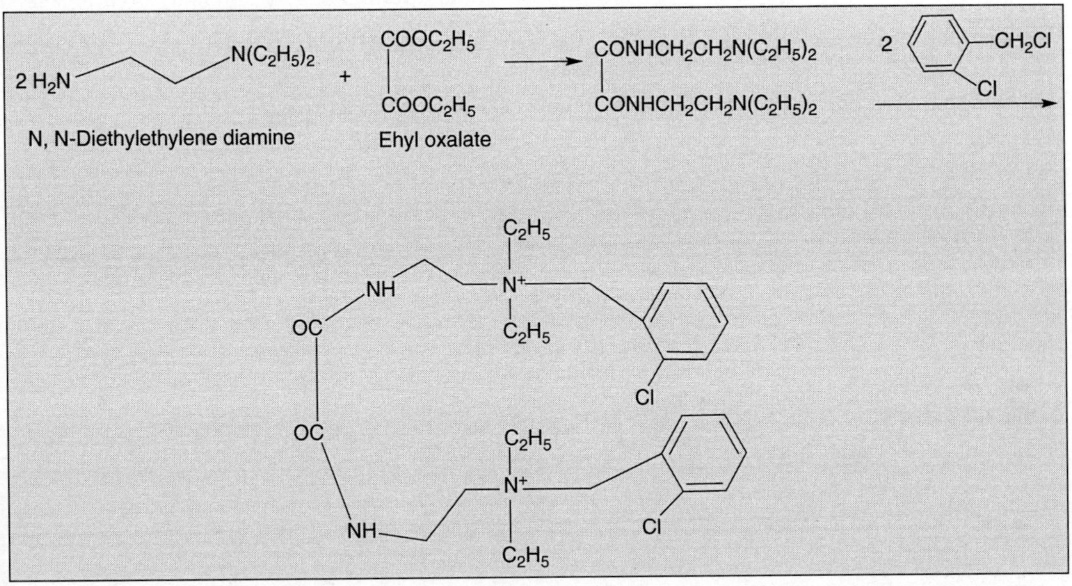

Ambenonium is made by reacting diethyloxalate with two moles of *N,N*-diethylethylendiamine, forming oxalyl-*bis*-(iminoethylen)-*bis*-*N,N*-diethylamine, which is alkylated by two moles of 2-chlorobenzylchloride, giving ambenonium.

Demecarium bromide (*m*-Hydroxyphenyl)trimethylammonium bromide decamethylene bis[methyl carbamate]
Synthesis

Demecarium is a disymmetrical compound that contains two ammonium and two carbamate groups. It is a reversible cholinesterase inhibitor that is longer lasting than the others.

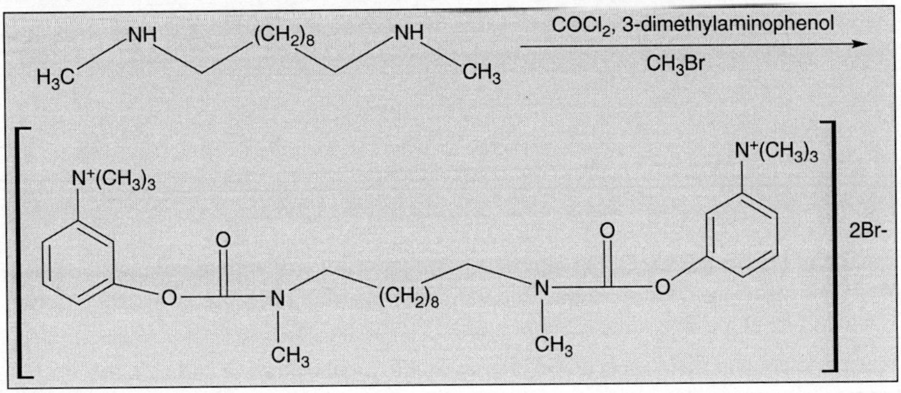

It is made by reacting two moles of phosgene with N,N'-dimethyldecamethylen-1,10-diamine, giving the *bis*-carbamoylchloride, which is transformed into *bis*-carbamoylester by reaction with two moles of the 3-dimethylaminophenol sodium salt. Reacting this with methylbromide gives demecarium.

Use: It is used topically for the treatment of glaucoma.

Echothiophate iodide   S-(2-Trimethylaminoethyl)-O,O-diethylthiophosphate

Synthesis

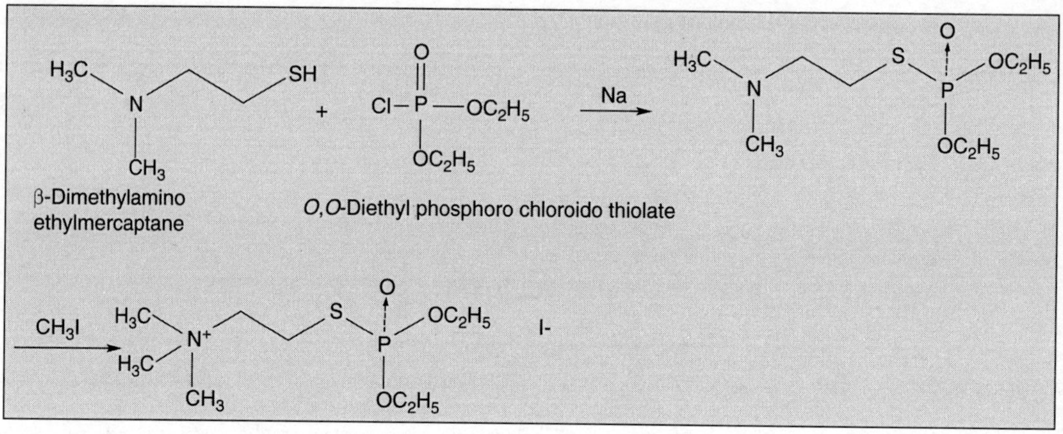

Echothiophate is made by reacting diethylchlorophosphoric acid with 2-dimethylaminoethylmercaptane, giving S-(2-dimethylaminoethyl)-O,O-diethylthiophosphate, which is alkylated by methyliodide, forming echothiophate.

Use: It is used for the treatment of glaucoma.

Edrophonium chloride Ethyl (m-hydroxyphenyl) dimethylammonium chloride

Synthesis

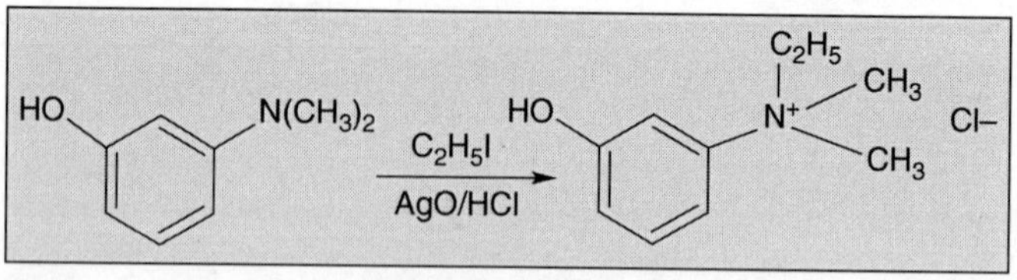

Edrophonium is made by reacting 3-dimethylaminophenol with ethyliodide, which forms ethyl(3-hydroxyphenyl)dimethylammonium iodide, the iodo atom of which is replaced with a chlorine atom by reacting it with silver chloride, giving edrophonium.

Uses: It is used to abolish neuromuscular paralysis due to *d*-tubocurarine and as a diagnostic agent for myasthenia gravis.

Isofluorophate Diisopropyl phosphorofluoridate
Synthesis

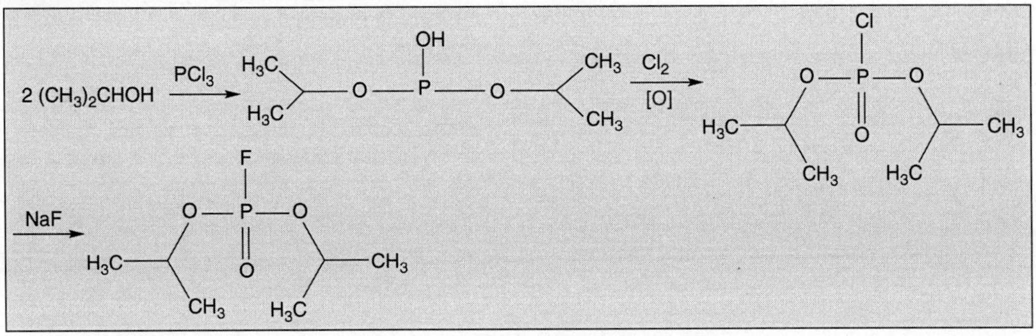

Isofluorophate is made by reacting *iso*-propyl alcohol with phosphorous trichloride, forming di-*iso*-propylphosphite, which is chlorinated and further reacted with sodium fluoride to replace the chlorine atom with fluorine, thus giving isofluorophate.

Mechanism of action: Compounds containing phosphoryl or phosphonic halides that can react with AChE form AChE-phosphate complexes stable to hydrolytic cleavage. These are irreversible inhibitors of AChE and, hence, toxic to human for internal use.

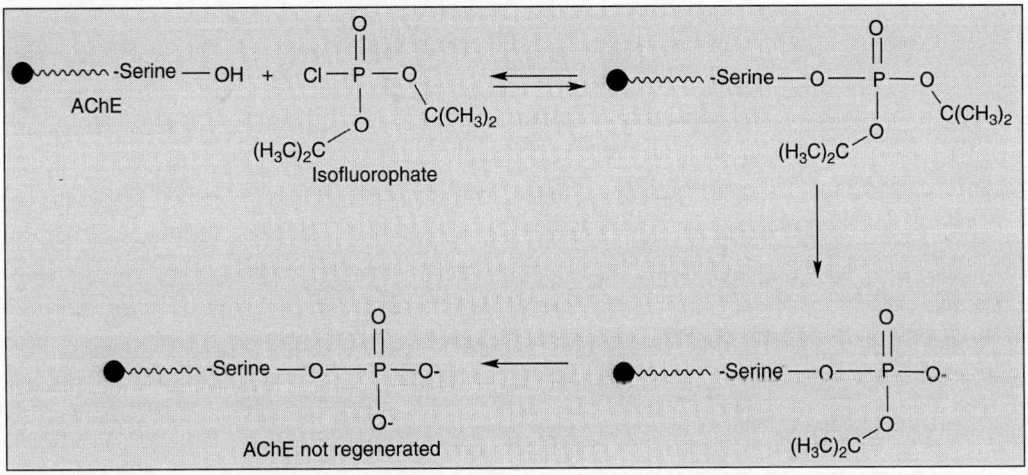

Use: It is used topically for the treatment of primary open-angle glaucoma.

Physostigmine

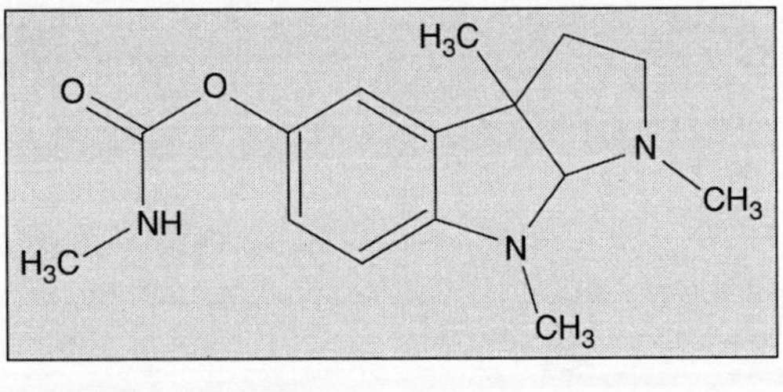

Physostigmine is a natural product obtained from the seeds of *Physostigma venenosum*.
Uses: It acts as an antidote for atropine poisoning and is used topically in the treatment of glaucoma.
Neostigmine (*m*-Hydroxyphenyl) trimethylammonium bromide dimethyl carbamate
Synthesis

It is made by reacting 3-dimethylaminophenol with *N*-dimethylcarbamoyl chloride, which forms the dimethylcarbamate, and its subsequent alkylation using dimethylsulphate/methyl bromide forms the neostigmine.

Uses: It is employed for the treatment of genitourinary, gastrointestinal, and neuromuscular disorders.

Pyridostigmine bromide 3-Hydroxy-1-methylpyridinium bromide dimethylcarbamate
Synthesis

It is synthesised from 3-hydroxypyridine, which is reacted with dimethylaminocarbamoyl chloride, which gives 3-(dimethylaminocarbamoyl)pyridine. This is reacted with methylbromide, giving pyridostigmine.

Both neostigmine and pyridostigmine possess more chemical stability than physostigmine.
Use: It is used in the treatment of myasthenia gravis.

23.4 CHOLINERGIC BLOCKING AGENTS (ANTIMUSCARINIC DRUGS)

Antimuscarinic drugs are competitive antagonists that act only on the cholinergic receptors at smooth muscles, secretory cells, and certain central synapses. Because of interruption of parasympathetic nerve stimulation, these agents cause decreased gastric secretion, dry mouth, drying of the mucous membranes in general, mydriasis, loss of accommodation, urinary retention, decreased sweating, bronchial and biliary dilatation, and tachycardia.

1. **Classification**
   (i) Tertiary antimuscarinics: Atropine sulphate, Cyclopentolate, Dicyclomine, Tropicamide
   (ii) Quaternary antimuscarinics: Anisotropine methyl bromide, Clidinium bromide, Glycopyrrolate, Ipratropium bromide

2. Uses
   (i) Used topically to dilate the pupil and paralyse accommodation
   (ii) Used as pre-anaesthetic medication, to inhibit excessive salivary and bronchial secretions, and to prevent bronchospasm and laryngospasm
   (iii) The antisecretory effects are also sought in the treatment of acute coryza, hay fever, and rhinitis
   (iv) Used in the treatment of bronchial asthma and peptic ulcers
   (v) Used in the treatment of Parkinson's disease

3. Differences between quaternary and tertiary antimuscarinics

Quaternary amines	Tertiary amines
1. These drugs do not pass through the blood-brain barrier and, hence, lack CNS actions	1. These drugs can penetrate cell membranes in the non-ionized form and, hence, pass through the blood-brain barrier. In the brain, they can exert both therapeutic and toxic actions
2. Poorly penetrate into the eye from the blood stream or cornea	2. These drugs penetrate through the cornea and cause mydriasis and cycloplegia
3. The quaternary compounds have greater affinity for nicotinic receptors, so that some degree of ganglionic blockade may result	3. No nicotinic receptor affinity
4. These are mostly excreted in the urine unchanged	4. These drugs are biotransformed in the liver

Anisotropine methyl bromide 8-Methyl tropinium bromide 2-propyl pentanoate
Synthesis

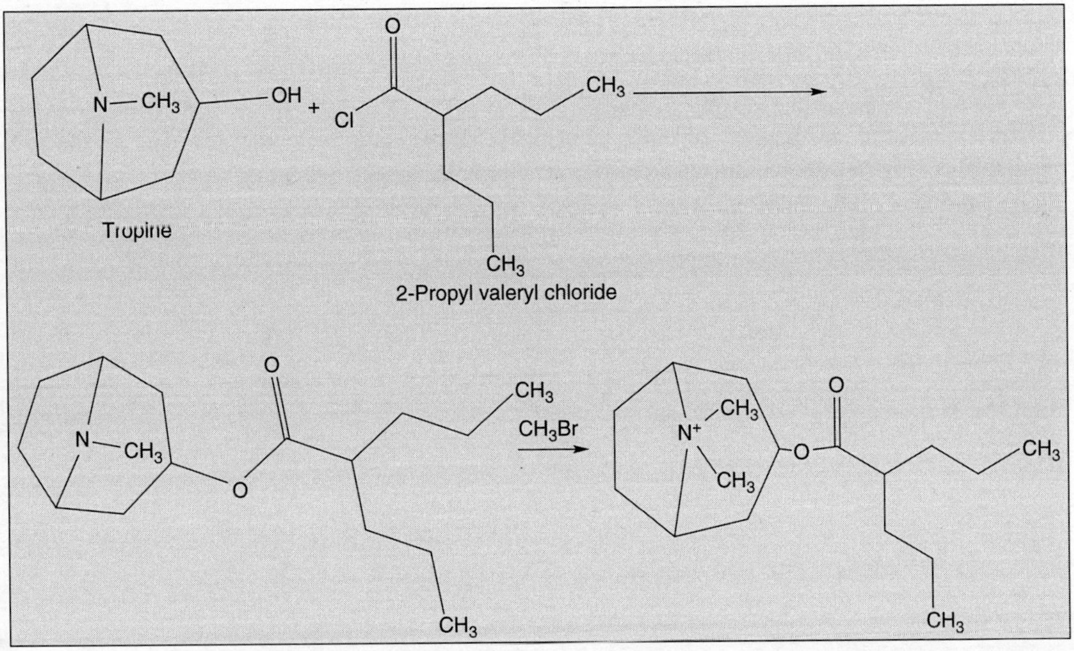

It has high selectivity for the GIT and is used for the treatment of spastic or hypermotile conditions of the GIT.

It is synthesised by the acidifying tropine with 2-propylvaleryl acid chloride, giving the ester, and subsequent reaction with methylbromide gives anisotropine.

Ipratropium bromide N-Isopropyl salt of atropine
Synthesis

The synthesis is similar to anisotropine, wherein the last step involves alkylation with isopropyl bromide instead of methyl bromide.

Use: It is used as bronchodilator in asthmatic conditions. It has longer-lasting effect compared to β-agonists.

Clidinium bromide 3-Hydroxy-1-methyl quinuclidinium bromide benzilate
Synthesis

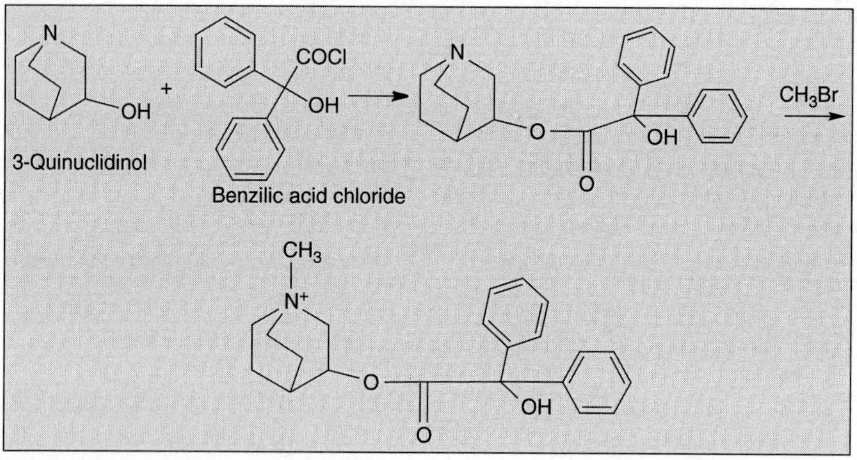

The use is same as that of Anisotropine.

It is synthesised by reacting 3-hydroxycynuclidine with benzilic acid chloride producing the ester, which further alkylated at the nitrogen atom by methylbromide gives clidinium.

Cyclopentolate 2-(Dimethylamino)ethyl-1-hydroxy-α-phenylcyclopentane acetate
Synthesis

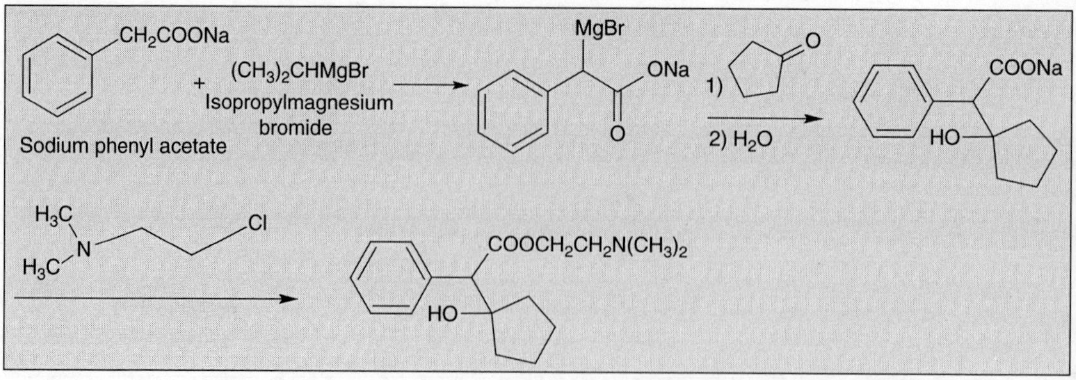

It is synthesised by the esterification of α-(1- hydroxycyclopentyl) phenylacetic acid using 2-dimethylaminoethylchloride, and α-(1- hydroxycyclopentyl) phenylacetic acid is synthesised by reacting the sodium salt of phenylacetic acid with cyclopentanone in the presence of isopropylmagnesium bromide. It is used for its ophthalmologic actions.

Dicyclomine 2-(Diethylamino)ethyl [bicyclohexyl]-1-carboxylate

Synthesis

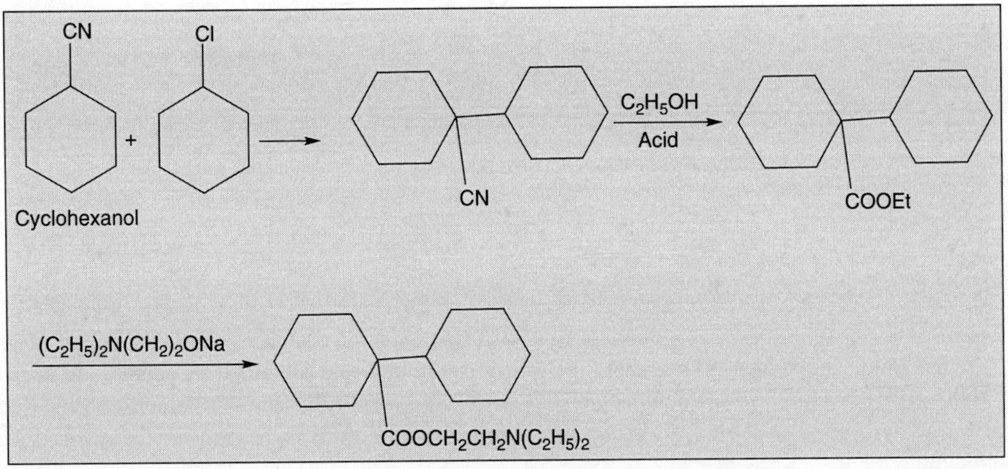

Synthesis of dicyclomine is started from cyanocyclohexane, which undergoes cycloalkylation by cyclohexylbromide, forming 1-cyanobicyclohexane. This undergoes alcoholysis, forming the ethyl ester of 1-bicyclohexanecarboxylic acid, which undergoes transesterification by 2-diethylaminoethanol in the presence of sodium.

It is used primarily for its antispasmodic properties.

Glycopyrrolate 3-Hydroxy-1,1-dimethylpyrrolidinium bromide α-cyclopentyl mandelate

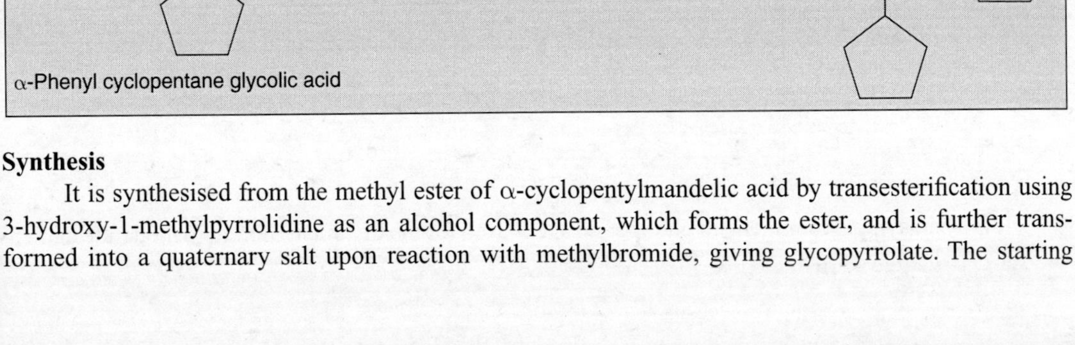

Synthesis

It is synthesised from the methyl ester of α-cyclopentylmandelic acid by transesterification using 3-hydroxy-1-methylpyrrolidine as an alcohol component, which forms the ester, and is further transformed into a quaternary salt upon reaction with methylbromide, giving glycopyrrolate. The starting methyl ester of α-cyclopentylmandelic acid is synthesised by reacting cyclopentylmagnesiumbromide with the methyl ester of phenylglyoxylic acid.

It is used for the treatment of peptic ulcer and as pre-anaesthetic medication.

Tropicamide N-Ethyl-2-phenyl-N-(4-pyridylmethyl) hydroxyl amide

Synthesis

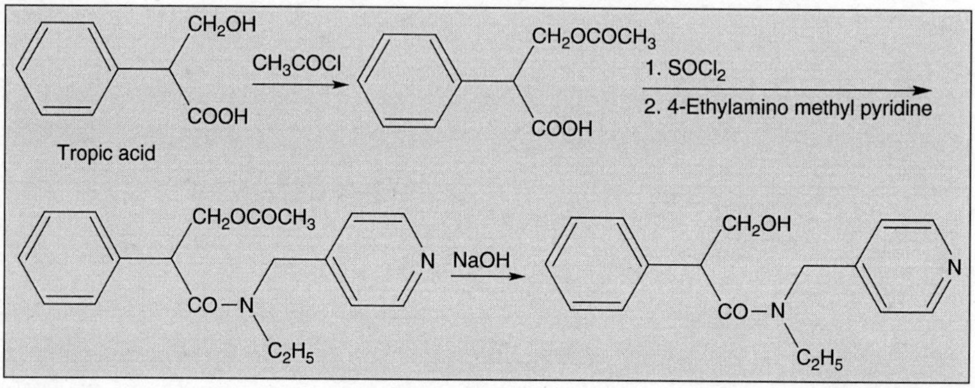

Tropicamide is synthesised by reacting O-acetyltropyl chloride (obtained by acetylation of tropic acid followed by reaction with thionyl chloride) with ethyl (4-pyridinylmethyl)amine and the subsequent acidic hydrolysis of the acetyl group in the resulting amide. It is used to induce mydriasis and cycloplegia.

23.5 NEWER DRUGS

Donepezil 2-[(1-Benzyl-4-piperidyl)methyl]- 5,6-dimethoxy-2,3-dihydroinden-1-one

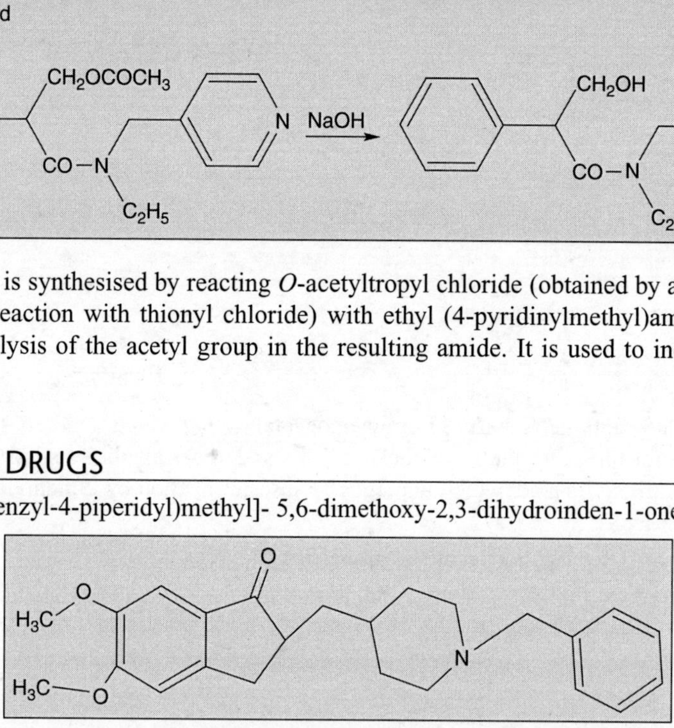

It is a centrally acting reversible acetylcholinesterase inhibitor. Its main therapeutic use is in the treatment of Alzheimer's disease, where it is used to increase cortical acetylcholine.

Rivastigmine (S)-N-Ethyl-N-methyl-3-[1-(dimethylamino)ethyl]-phenyl carbamate

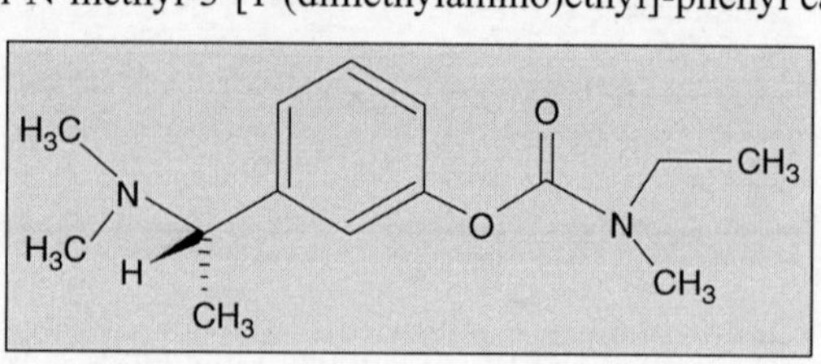

It is a parasympathomimetic or cholinergic agent used for the treatment of mild to moderate dementia of the Alzheimer's type.

Cisapride 4-Ano-5-chloro-N-[1-[3-(4-fluorophenoxy)propyl]-3-methoxy-4-piperidyl]-2-methoxybenzamide

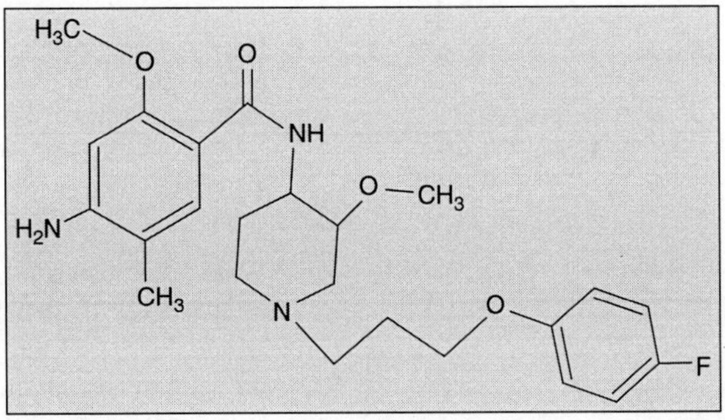

It is a parasympathomimetic that acts as a serotonin 5-HT$_4$ receptor agonist. Stimulation of the serotonin receptors increases acetylcholine release in the enteric nervous system. Cisapride increases muscle tone in the oesophageal sphincter in people with gastro-oesophageal reflux disease. It also increases gastric emptying in people with diabetic gastroparesis. It has been used to treat bowel constipation.

FURTHER READINGS

1. *Burger's Medicinal Chemistry and Drug Discovery*, Vol. 1–5, Wolff (ed.), 5th Ed.
2. *Principles of Medicinal Chemistry*, Foye, W.O., Lemke, T.L. and Williams, D.A. (ed.), 5th Ed.

MULTIPLE-CHOICE QUESTIONS

1. Acetyl choline is biosynthesised from
   a. L-Cysteine
   b. L Codeine
   c. L-Serine
   d. L-Cholic acid

2. Mechanism of action of neostigmine is
   a. Cholinomimetics
   b. Muscarinic antagonist
   c. Anticholinesterase
   d. Nicotinic antagonist

3. Identify the false statement.
   a. Nicotinic receptor is a ligand-gated ion channel.
   b. Cholinomimetics increase peristaltic movement.
   c. Acetyl choline is orally administered for therapy.
   d. Neostigmine is useful in myasthenia gravis

4. Toxic CNS effects of antimuscarinics could be treated with
   a. Carbochol
   b. Edrophonium chloride
   c. Physostigmine
   d. Glycopyrrolate

5. The cholinergic drug also useful in the treatment of cardiac arrhythmia is
   a. Physostigmine
   b. Edrophonium chloride
   c. Ipratropium bromide
   d. Neostigmine

6. The anticholinergic drug useful in the treatment of peptic ulcer is
   a. Carbochol
   b. Edrophonium chloride
   c. Physostigmine
   d. Glycopyrrolate
7. One of the following drugs is obtained from natural product:
   a. Edrophonium chloride
   b. Pilocarpine
   c. Neostigmine
   d. Echothiophate iodide
8. One of the following has a bicyclic nucleus in its structure:
   a. Clidinium bromide
   b. Pyridostigmine
   c. Echothiophate iodide
   d. Tropicamide
9. Anticholinergic useful as spasmolytic is
   a. Pyridostigmine
   b. Dicyclomine
   c. Tropicamide
   d. Glycopyrrolate
10. One of the following is a tertiary antimuscarinic agent:
    a. Clidinium
    b. Glycopyrrolate
    c. Atropine
    d. Ipratropium

QUESTIONS

1. What drugs are useful in the treatment of asthma? How do they act?
2. Discuss with structure of ephedrine how cholinergics interact in the binding site.
3. Discuss the SAR of cholinergic drugs.
4. Write the synthetic protocol for neostigmine, ipratropium bromide, tropicamide, and bethanechol.
5. What are the general uses of anticholinergic drugs?
6. Classify anticholinergics and give a few structural examples.
7. Differentiate between quaternary and tertiary antimuscarinics.
8. Write a short note on drugs useful for glaucoma treatment.

SOLUTION TO MULTIPLE-CHOICE QUESTIONS

1. c;
2. c;
3. c;
4. c;
5. b;
6. d;
7. b;
8. a;
9. b;
10. c.

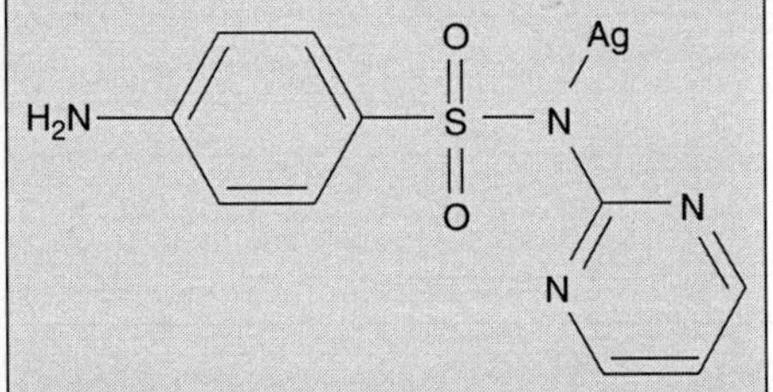

CHAPTER 24

Sulphonamides, Sulphones, and Dihydrofolate Reductase Inhibitors

LEARNING OBJECTIVES

- Define sulphonamides and their utility as chemotherapeutic agents
- Categorize and describe the mechanism of action of sulphonamides
- Describe the side-effects and structural requirement of sulphonamides
- Explain the synthetic protocol of sulpha drugs
- Define sulphones and their utility
- Describe the function of DHFR and its inhibitors
- List some new DHFR inhibitors

The sulphonamides are synthetic bacteriostatic antibiotics with a wide spectrum against most gram-positive and many gram-negative organisms. The sulphonamides and sulphone antibacterials as well as the 2,4-diaminopyrimidine antifolates continue to be successful chemotherapeutic agents.

24.1 SULPHONAMIDES

The sulphonamide drugs were the first effective chemotherapeutic agents to be employed systemically for the prevention of bacterial infections in human beings. These are totally synthetic substances that are produced by relatively simple chemical synthesis. Sulphonamide drugs are a group of synthetic antimicrobial drugs that have a broad spectrum of use with respect to gram-positive as well as gram-negative micro-organisms. They were introduced into medical practice even before the discovery of penicillins. Sulphonamide drugs are derivatives of sulphanilamide. The advent of penicillin and subsequently of other antibiotics has diminished the usefulness of sulphonamides.

Classification

1. Absorbed and excreted rapidly: Sulphisoxazole, Sulphamethoxazole, and Sulphadiazine
2. Poorly absorbed (active in bowel lumen): Sulphasalazine
3. Topically used: Sulphacetamide, Silver sulphadiazine, Mefenide
4. Long-acting: Sulphadoxine

Mechanism of action:

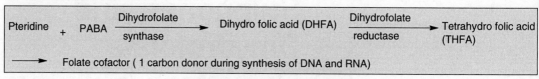

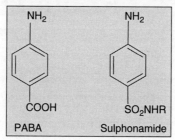

Sulphonamides are structurally similar to PABA. They compete with PABA for the enzyme dihydrofolate synthase and block the synthesis of DHFA (dihydrofolic acid), in turn tetrahydrofolic acid (THFA) and folate cofactor. Folate cofactor acts as 1-carbon donor for the synthesis of nucleic acid (DNA). The result of blocking the biosynthesis of folate coenzymes in bacteria, for example, is that growth and cell division are stopped. Sulphonamides are, thus, bacteriostatic drugs. Since mammalian cells use preformed folates from the diet, and since most bacteria cannot use preformed folates and must synthesise their own folic acid, the sulphonamides, therefore, demonstrate a selective toxicity to bacteria.

Uses

1. Because sulpha drugs concentrate in the urine before being excreted, treating urinary tract infections (UTI) is one of their most common uses. The sulphonamide antibacterials are primarily used for the treatment of uncomplicated urinary tract infections caused by *E. coli*, *Klebsiella*, *Enterobacter*, and *Proteus*, and only seldom for middle ear infections caused by *Haemophilus influenzae*. Sulphisoxazole and sulphamethoxazole are the major therapeutic drugs for UTI.
2. The sulphonamides are also used for the prophylaxis of recurrent rheumatic fever associated with streptococcal infection as well as for the prophylaxis of *Neisseria meningitis*.
3. Sulphonamides can be co-administered with antifolate antibacterials, usually with trimethoprim for antibacterial activity and with pyrimethamine for antimalarial activity.
4. Topically, sulphonamides (silver sulphadiazine) are useful in burn therapy.
5. Sulphasalazine is useful for the treatment of ulcerative colitis.

Side-effects

Sulphanilamide causes severe kidney damage from crystals of sulphanilamide forming in the kidney. Sulphanilamide and their metabolites are excreted entirely in the urine. Unfortunately, sulphanilamide is not very water-soluble and this leads to crystalluria. Crystalluria can be overcome by:

1. Greatly increasing urine flow
2. Raising the pH of the urine to 10.4 with sodium bicarbonate
3. Lowering pKa, which is achieved by attaching electron-withdrawing heterocycle at N^1 position of sulphanilamide

> Gerhard Domagk and Jacques and Therese Trefouel (1935) are generally credited with the discovery of sulphanilamide as a chemotherapeutic agent. Domagk was awarded the Nobel Prize for his work.

SAR of sulphonamide

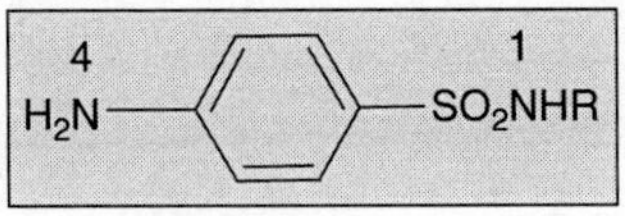

H_2N—(4)—benzene ring—(1)—SO_2NHR

1. The amino and sulphonyl groups on the benzene ring are essential and should be in 1,4-position.
2. The N^4 amino group could be modified to produce prodrugs, which are converted to free amino function *in vivo*.

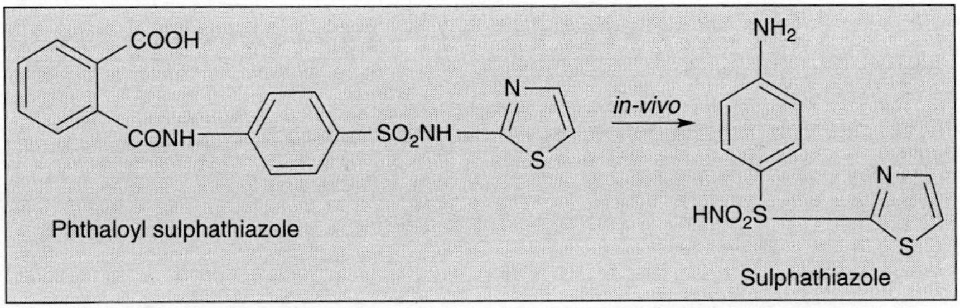

Phthaloyl sulphathiazole → in-vivo → Sulphathiazole

3. Replacement of benzene ring by other ring systems or the introduction of additional substituents on it decreases or abolishes activity.
4. Exchange of the -SO_2NH group by -$CONH$ reduces the activity.
5. Substitution of heterocyclic aromatic nuclei at N^1 yields highly potent compounds.
6. N^1-Disubstitution in general leads to inactive compounds.

Synthesis
General methods of preparation

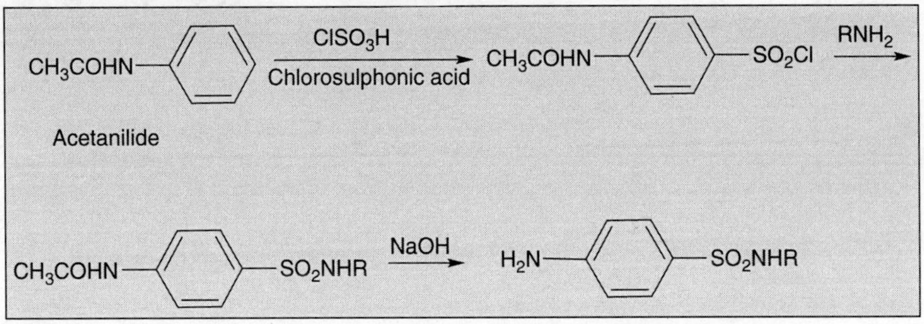

CH_3COHN—(Acetanilide) $\xrightarrow{ClSO_3H \text{ Chlorosulphonic acid}}$ CH_3COHN—SO_2Cl $\xrightarrow{RNH_2}$

CH_3COHN—SO_2NHR $\xrightarrow{NaOH}$ H_2N—SO_2NHR

Acetanilide on treatment with chlorosulphonic acid gives sulphonyl chloride derivative, which on reaction with appropriate primary amine gives sulphonamide derivative. Saponification gives the final target compound.

Sulphisoxazole N^1-(3,4-Dimethyl-5-isoxazolyl) sulphanilamide

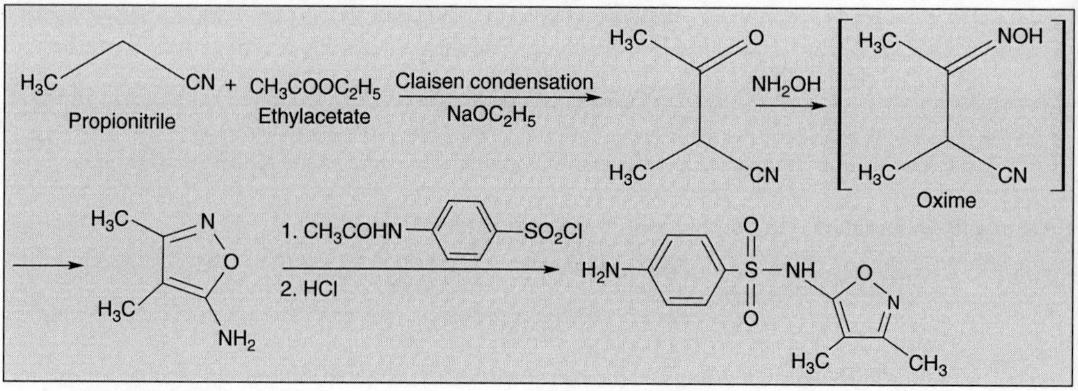

Sulphisoxazole is synthesised by reacting 4-acetylaminobenzenesulphonyl chloride with 5-amino-3,4-dimethylisoxazol, which is in turn synthesised by heterocyclization of 2-methylacetylacetonitrile (prepared by Claisen condensation of propionitrile and ethylacetate in presence of base) with hydroxylamine, and subsequent acidic hydrolysis (hydrochloric acid) of the protective acetyl group in the resulting product.

Sulphamethoxazole N1-(5-Methyl-3-isoxazolyl) sulphanilamide

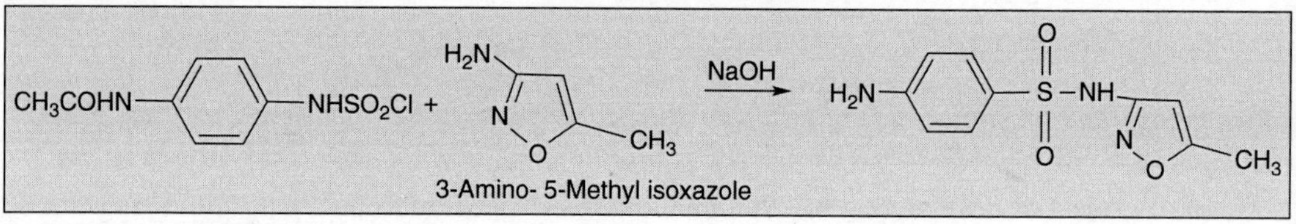

Sulphamethoxazole is synthesised by a completely analogous scheme, except by using 3-amino-5-methylisoxazol as the heterocyclic component.

It is used only in combination with trimethoprim (co-trimoxazole).

Sulphamoxole 4-Amino-N-(4,5-dimethyl-1,3-oxazol-2-yl)benzenesulphonamide

Sulphadiazine N^1-2-Pyrimidinyl sulphanilamide

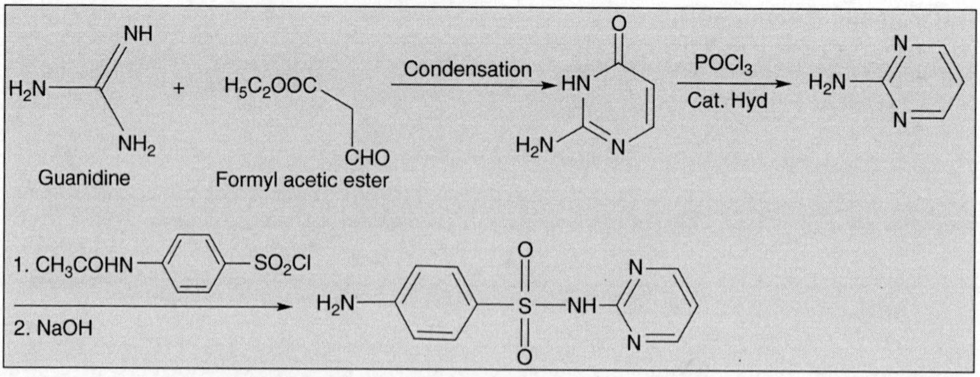

Sulphadiazine is synthesised by a completely analogous scheme, except by using 2-amino-pyrimidine as the heterocyclic component.

Sulphadimidin 4-Amino-*N*-(4,6-dimethylpyrimidin-2-yl)benzenesulphonamide
Sulphasalazine 2-Hydroxy-5-[[4-[(2-pyridinyl amino) sulphonyl] phenyl] azo] benzoic acid

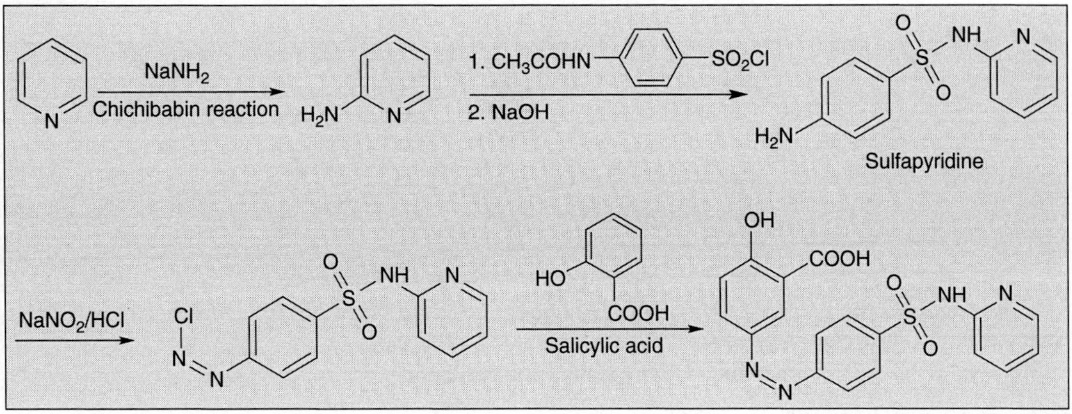

Sulphapyridine on treatment with sodium nitrite and hydrochloric acid gives diazonium salt intermediate. This, on coupling with salicylic acid, affords sulphasalazine.

Sulphasalazine is poorly absorbed from the small intestine, so that the drug passes into the colon where the bacterial enzymes release both 5-amino salicylic acid and sulphapyridine from the drug. It has a suppressive effect on ulcerative colitis. Sulphapyridine decreases anaerobic bacteria and 5-amino salicylate inhibits prostaglandin synthesis.

Sulphacetamide N_1-Acetylsulphanilamide

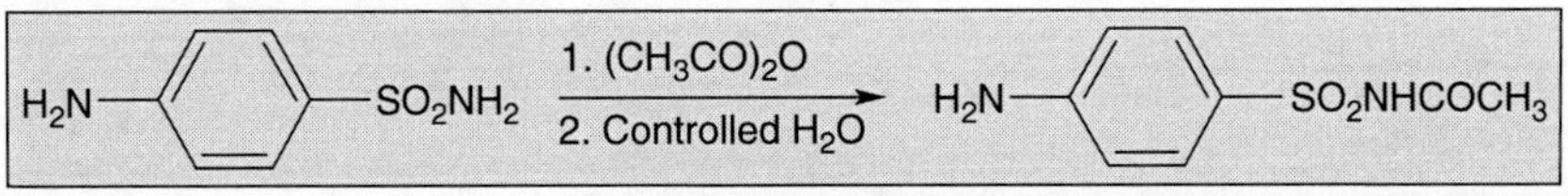

Sulphacetamide is synthesised by reacting 4-aminobenzenesulphonamide with acetic anhydride and subsequent selective, reductive deacylation of the resulting acetamide using a system of zinc-sodium hydroxide.

It is mainly used locally for ophthalmologic infections.

Sulphaguanidin 4-Amino-*N*-[amino(imino)methyl]benzenesulphonamide
Silver sulphadiazine 4-Amino-N-(2-pyrimidinyl)benzenesulphonamide silver salt

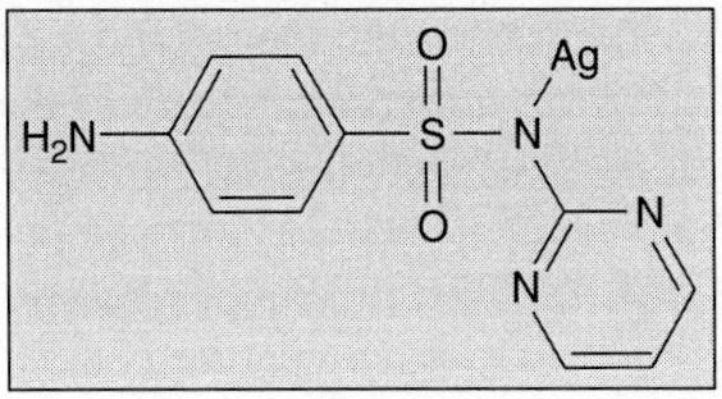

It combines in one compound the antibacterial properties of silver ion and sulphadiazine, and is especially effective in the treatment of burn infections. Silver sulphadiazine is typically delivered in a one-per-cent solution suspended in a water-soluble base.

Mefenide α-Amino-*p*-toluene sulphonamide

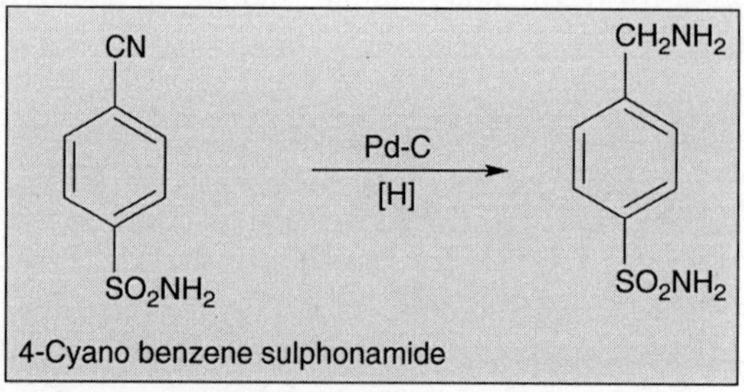

Mefenide is synthesised by reduction of 4-cyanobenzenesulphonamide.
It is applied locally for burn infections. Although it is a sulphonamide, the *para* substituent differs from the sulpha drug and its mechanism of action is much different.

Sulphadoxine N^1- (5,6- Dimethoxy-4-pyrimidinyl sulphanilamide)

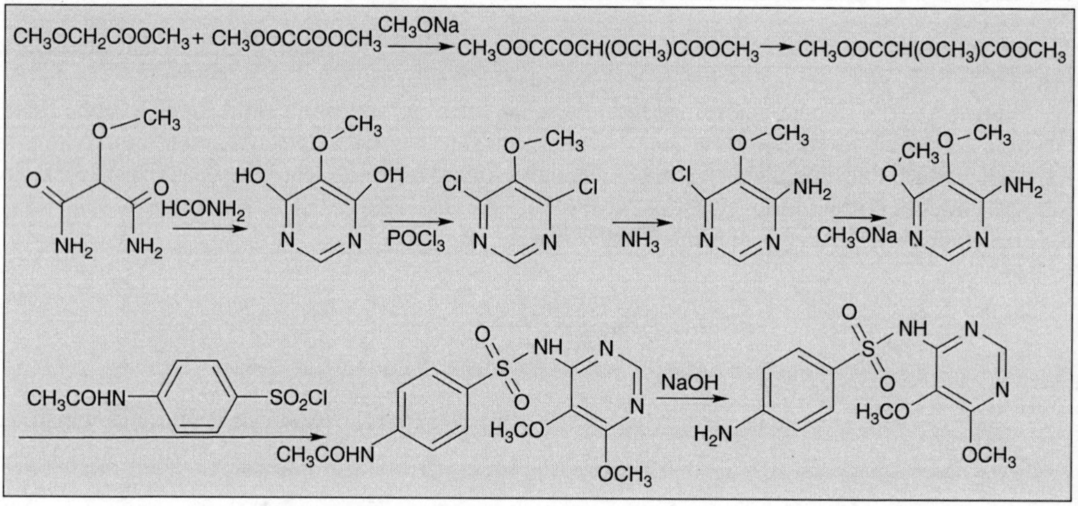

Sulphadoxine is synthesised by the standard scheme from 4-acetylaminobenzenesulphonyl chloride and 4-amino-5,6-dimethoxypyrimidine. However, 4-amino-5,6-dimethoxypyrimidine is synthesised from methyl ester of methoxyacetic acid. Interacting this with dimethyloxalate in the presence of sodium methoxide gives the methoxy derivative, and the pyrolysis of this compound gives the dimethyl ester of methoxymalonic acid. Reacting this with ammonia gives the diamide of methoxymalonic acid. Heterocyclization of the resulting product by a reaction with formamide in the presence of sodium ethoxide gives 4,6-dioxy-5-methoxypyrimidine, which is then transformed to 4,6-dichloro-5-methoxypyrimidine. The resulting 4,6-dichloro-5-methoxypyrimidine undergoes a reaction with ammonia to make 4-amino-6-chloro-5-methoxypyrimidine, and the resulting compound is then reacted with sodium methoxide to make the desired 5,6-dimethoxy-5-aminopyrimidine. Reacting this with 4-acetylaminobenzenesulphonyl chloride and subsequent hydrolysis of the acetyl group gives sulphadoxine.

Its principal use is in the prophylaxis or suppression of malaria caused by chloroquine-resistant *P. falciparum*. It is used only in combination with pyrimethamine.

24.2 SULPHONES

It has been estimated that there are about 11 million cases of leprosy in the world, of which about 60 per cent are in Asia (with 3.5 million in India alone). Lesions of the skin, loss of sensitivity to pain, and superficial nerves are the three main signs of leprosy, a disease caused by the mycobacteria *Mycobacterium leprae*. This disease is extremely infectious.

Diamino diphenyl sulphones (dapsone) is used for the treatment of infection caused by *Mycobacterium leprae*. Chemotherapy of leprosy consisted of taking dapsone, which gave good clinical results. However, because of the primary and secondary resistance that originated from prolonged use, it is now necessary to use a certain combination of drugs. Currently, dapsone is used along with rifampicin and clofazimine. Ethionamide is also prescribed.

Dapsone 4, 4'-Sulphonyl *bis* benzenamine

Reacting 4-chloronitrobenzene with sodium sulphide gives 4,4'-dinitrodiphenylthioester, and oxidation of the sulphur atom in this compound using potassium dichromate in sulphuric acid gives 4,4'-dinitrodiphenylsulphone. Reduction of the nitro group in the resulting compound using tin dichloride in hydrochloric acid makes the desired dapsone.

It has mechanism of action similar to that of sulphonamides. It is used in the treatment of both lepromatous and tuberculoid types of leprosy. Dapsone is used in combination with rifampicin and clofazimine. Dapsone is also the drug of choice for dermatitis herpetiformis, with pyrimethamine for the treatment of malaria, with trimethoprim for *Pneumocystis carinii* pneumonia (PCP), and has been used for rheumatoid arthritis.

> It is used prophylactically to prevent Pneumocystis pneumonia and toxoplasmosis in patients unable to tolerate trimethoprim with sulphamethoxazole. Dapsone is also used to treat Brown recluse spider bites. In December 2008, a 5 per cent dapsone gel called Aczone was intoduced to the prescription market as a treatment for moderate to severe acne.

24.3 DIHYDROFOLATE REDUCTASE (DHFR) INHIBITORS

2,4-Diamino pyrimidine derivatives like trimethoprim and pyrimethamine inhibit DHFR of bacteria and *Plasmodium*, respectively. The enzyme DHFR converts dihydrofolic acid to tetrahydrofolic acid, in turn to folate cofactors. These drugs block the DNA synthesis and cell division.

Sulphonamides and trimethoprim are used to treat and prevent infections. The sulphadiazine-and-trimethoprim combination is used to treat urinary tract infections. The sulphamethoxazole-and-trimethoprim (co-trimoxazole) combination is used to treat infections such as bronchitis, middle ear infection, urinary tract infection, and traveller's diarrhoea. It is also used for the prevention and treatment of *Pneumocystis carinii* pneumonia (PCP).

Trimethoprim   2,4-Diamino-5-(3',4',5'-trimethoxy benzyl)pyrimidine

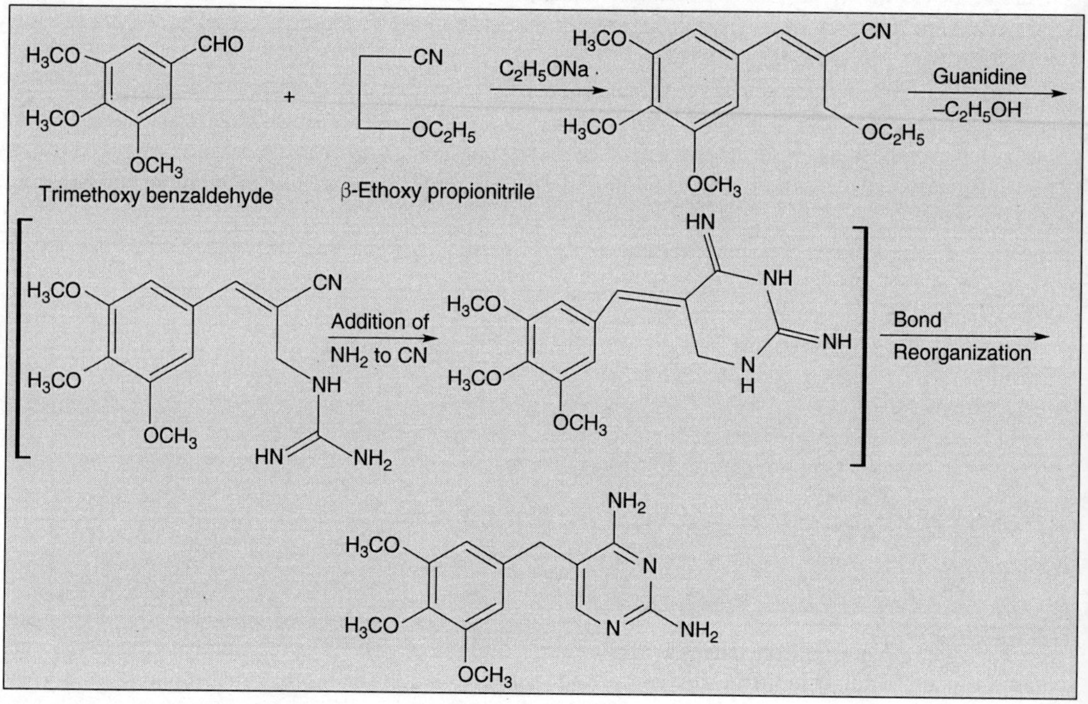

Condensation of 3,4,5-trimethoxybenzaldehyde with 3-ethoxypropionitrile gives the corresponding benzylidene derivative, which upon direct reaction with guanidine gives trimethoprim.

It is used along with sulphamethoxazole for bacterial infections.

Pyrimethamine   2,4-Diamino-5-(*p*-chlorophenyl)-6-ethylpyrimidine

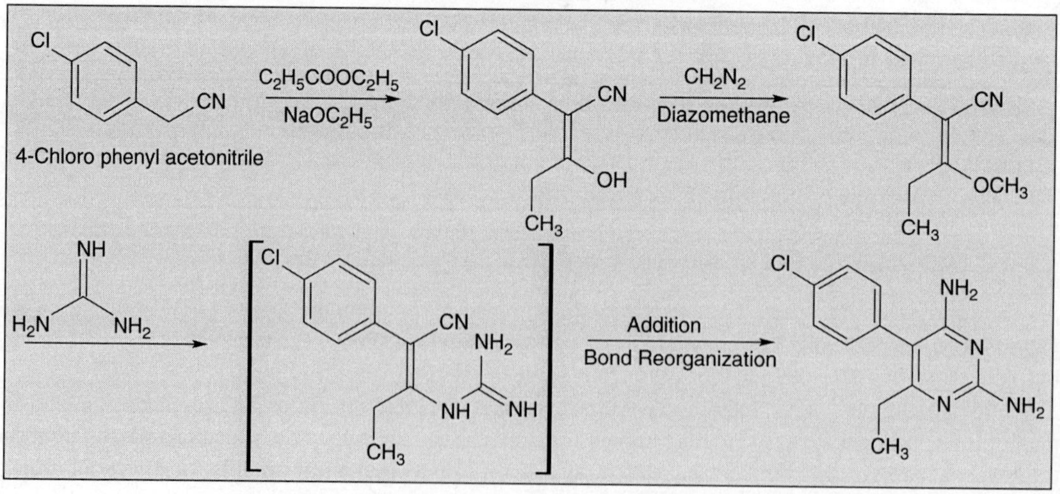

Pyrimethamine is synthesised from 4-chlorobenzycyanide, which upon condensation with ethyl ester of propionic acid in the presence of sodium methoxide gives the β-ketonitrile. Reacting this with diazomethane/ethylorthoformate gives a methoxymethylene derivative, which upon heterocyclization in pyrimidine using guanidine as the binucleophile forms the desired pyrimethamine.

It is used in combination with sulphadoxine for the treatment of malaria.

24.4 NEWER DRUG

Brodimoprim: 5-[(4-Bromo-3,5-dimethoxyphenyl)methyl] pyrimidine-2,4-diamine

It is a structural derivative of trimethoprim. In brodimoprim, the 4-methoxy group of trimethoprim is replaced with a bromine atom. As trimethoprim, brodimoprim is a selective inhibitor of bacterial dihydrofolate reductase.

FURTHER READINGS

1. *Annual Reports in Medicinal Chemistry*, September 30, 2003, Elsevier.
2. King, F.D., G. Lawton, and A.W. Oxford, May 2004. *Progress in Medicinal Chemistry*, Elsevier.
3. Martin, Yvonne Connolly, Eberhard Kutter, Volkhard Austel, and Marcel Dekker, 1989. *Modern Drug Research*.

MULTIPLE-CHOICE QUESTIONS

1. Sulphonamides block the synthesis of
    a. PABA
    b. DFA
    c. DHFA
    d. TFA
2. Topically used sulphonamide is
    a. Sulphadoxine
    b. Sulphamethoxazole
    c. Silver sulphadiazine
    d. Dapsone
3. Which of the following statements refers to dapsone?
    a. Not much used due to toxicity
    b. Inhibits folate synthesis in bacteria
    c. Is reduced by intestinal bacteria to a salicylate
    d. Is reduced by intestinal bacteria to a sulphonamide
4. Co-trimoxazole is a combination of
    a. Sulphadiazine and trimethoprim
    b. Sulphamethoxazole and trimethoprim
    c. Sulphamethoxazole and sulphadoxine
    d. Sulphamethoxazole and pyrimethamine
5. Dapsone is used primarily for the treatment of
    a. Tuberculosis
    b. Leprosy
    c. Malaria
    d. Urinary tract infection
6. One of the following does not contain pyrimidine in its structure:
    a. Sulphadiazine
    b. Sulphadoxine
    c. Pyrimethamine
    d. Dapsone
7. The long-acting sulphonamide is
    a. Sulphamethoxazole
    b. Sulphadiazine
    c. Sulphadoxine
    d. Sulphacetamide
8. Treatment of malaria is achieved with the combination of pyrimethamine with all of the following except:
    a. Sulphadoxine
    b. Dapsone

c. Trimethoprim
d. Sulphadiazine
9. Which of the following statements is true?
   a. Sulphanilamide is soluble in water.
   b. Sulphonamides are retained in urine before excretion.
   c. The terminal amino function is not required for activity.
   d. Sulphonamides are useful for antitubercular therapy.
10. Traveller's diarrhoea is treated with
    a. Sulphadiazine
    b. Co-trimoxazole
    c. Dapsone
    d. Pyrimethamine

QUESTIONS

1. What is crystalluria? How can it be prevented?
2. Discuss the mechanism of action of sulpha drugs.
3. Write short notes on combination of drugs and their uses.
4. Identify the heterocyclic ring system present in sulphadiazine, sulphasalazine, and sulphadoxine.
5. Explain how sulphasalazine exhibits its action.
6. Discuss briefly the SAR of sulphonamides.
7. Write the synthetic protocol for dapsone, pyrimethamine, sulphadiazine, and sulphamethoxazole.

SOLUTION TO MULTIPLE-CHOICE QUESTIONS

1. c;
2. c;
3. b;
4. b;
5. b;
6. d;
7. c;
8. c;
9. b;
10. b.

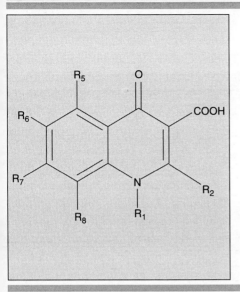

CHAPTER 25

Quinolone Antibacterials

LEARNING OBJECTIVES

- Define quinolones and the structural aspects
- Describe the mode of action and classify quinolones on the basis of treatment options
- Explain the structural modification of quinolones and their effect on biological functions
- Derive the synthetic routes for various fluoroquinolones
- List some newer quinolones

25.1 INTRODUCTION

Quinolones constitute a large class of synthetic antimicrobial agents that are highly effective in the treatment of many types of infectious diseases, particularly those caused by bacteria. Quinolones are potent, broad-spectrum antibacterial agents. The early congeners (non-fluorinated at C-6 like nalidixic acid) were limited to certain gram-negative infections such as urinary tract infections. However, the modern generation of fluoroquinolones containing C-6 fluoro substituent and a cyclic basic amine moiety at C-7 surpass their predecessors in terms of spectrum of activity and potency. This has allowed for their use against a variety of gram-negative as well as some gram-positive pathogens. Quinolones are relatively easily prepared and easily administered via parenteral and oral routes, and are well tolerated.

25.2 STRUCTURES OF FLUOROQUINOLONES

	R_1	R_7	R_5	R_8
Norfloxacin	—C_2H_5	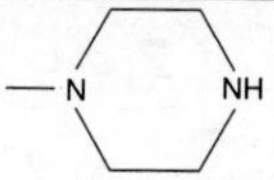 piperazinyl	H	H
Ciprofloxacin	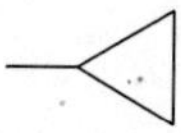 cyclopropyl	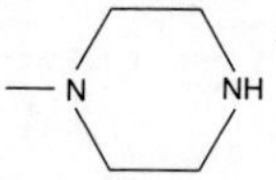 piperazinyl	H	H
Lomefloxacin	—C_2H_5	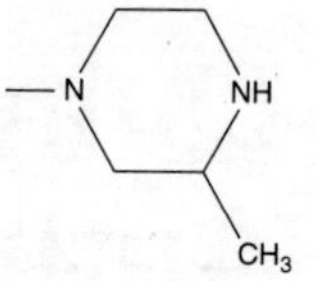 3-methylpiperazinyl	H	F
Pefloxacin	—C_2H_5	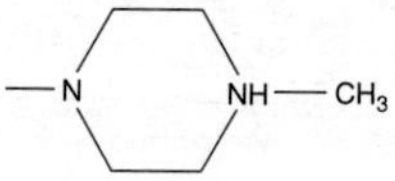 4-methylpiperazinyl	H	H
Gatifloxacin	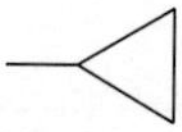 cyclopropyl	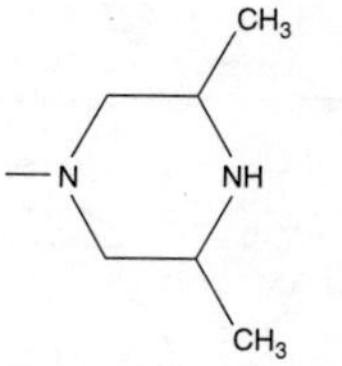 3-methylpiperazinyl	H	—OCH_3
Sparfloxacin	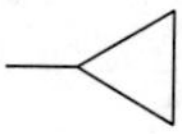 cyclopropyl	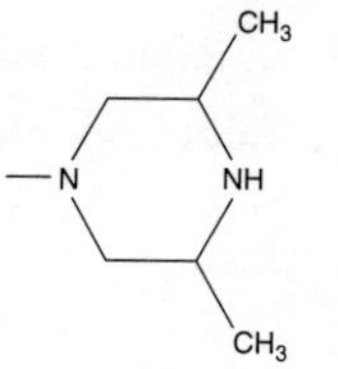 3,5-dimethylpiperazinyl	—NH_2	F
Moxifloxacin	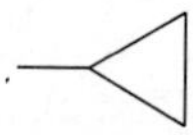 cyclopropyl	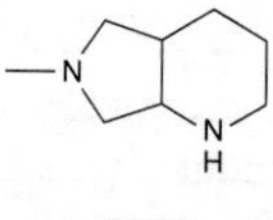 octahydropyrrolo[3,4-b]pyridinyl	H	OCH_3
Ofloxacin	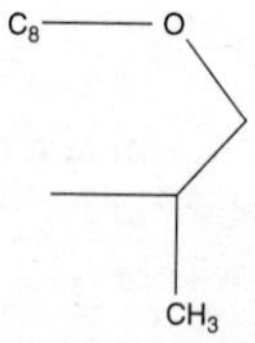 C_8—O—CH_2—CH(CH_3)—	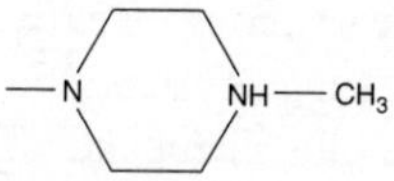 4-methylpiperazinyl	H	1, 8 bridge

25.3 MECHANISM OF ACTION

Quinolones inhibit the action of bacterial DNA gyrase enzyme. This enzyme is responsible to supercoil and compact bacterial DNA molecules into the bacterial cell during replication. This action is accomplished by modifying the topology of DNA via supercoiling and twisting of these macromolecules to permit DNA replication or transcription.

25.4 USES

Fluoroquinolones are used to treat upper and lower respiratory infections, gonorrhoea, bacterial gastroenteritis, skin and soft tissue infections, urinary tract infections, and bone and joint infections, and against tuberculosis.

25.5 CLASSIFICATION

1st generation: Cinoxacin, flumequine, nalidixic acid, oxolinic acid, piromidic acid, pipemidic acid
2nd generation: Ciprofloxacin, enoxacin, lomefloxacin, nadifloxacin, norfloxacin, ofloxacin, pefloxacin, rufloxacin
3rd generation: Balofloxacin, grepafloxacin, levofloxacin, pazufloxacin, sparfloxacin, temafloxacin, tosufloxacin
4th generation: Clinafloxacin, gemifloxacin, moxifloxacin, gatifloxacin, sitafloxacin, trovafloxacin
In development: Ecinofloxacin, prulifloxacin
Veterinary use: Danofloxacin, difloxacin, enrofloxacin, marbofloxacin, orbifloxacin, sarafloxacin

25.6 SAR OF QUINOLONES

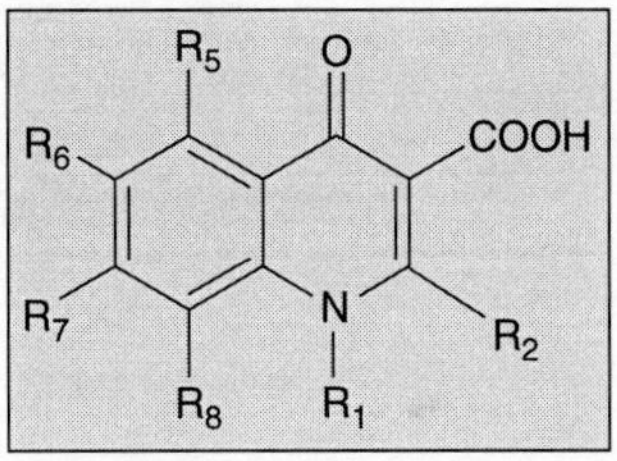

1. Substituent at N-1 position: A compilation of active N-1 quinolone substituents is shown below with an emphasis on overall *in vitro* potency.

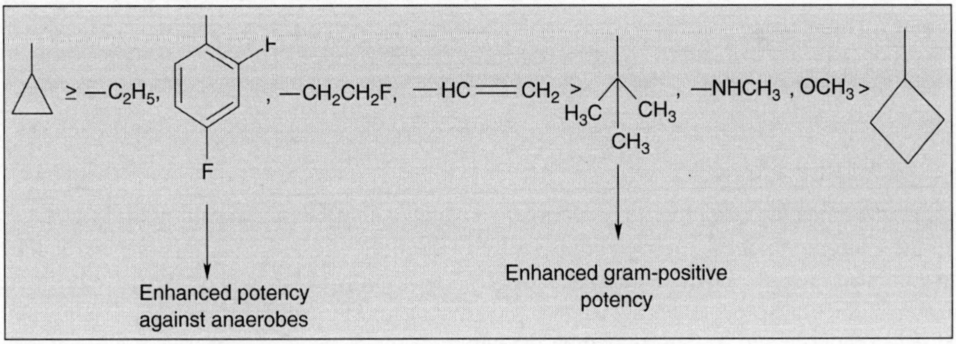

   Enhanced potency against anaerobes

   Enhanced gram-positive potency

2. The simple replacement of C-2 hydrogen has generally to be disadvantageous - e.g., C-2 methyl or hydroxyl groups. However, some derivatives containing a suitable C-1, C-2 ring have been shown to possess notable activity.

Prulifloxacin

3. Without doubt, the C-3 carboxylic acid moiety is most commonly encountered. Other acidic groups such as sulphonic acid, phosphonic acid, tetrazole, as well as derivatization as an ester results in a loss of antibacterial activity.
4. The C-4-oxo group of the quinolone nucleus appears to be essential for antibacterial activity. Replacement with 4-thioxo or sulphonyl group leads to a loss of activity.
5. The incorporation of an amino group at the C-5 position has proven beneficial in terms of antibacterial activity. The order of activity at R_5: NH_2, CH_3 > F, H > OH, OR, SH, SR.
6. The incorporation of a fluorine atom at the C-6 position of the quinolone is monumental. The order of activity at R_6: F > Cl, Br, CH_3 > CN.
7. The introduction of a piperazine moiety at C-7 was a landmark development. Other aminopyrrolidines also are compatible for activity.

8. A hydrogen atom at the C-8 or a nitrogen atom (a naphthyridone) is the most common. In general, a C-8 fluoro substituent offers good potency against gram-negative pathogens, while a C-8 methoxy moiety is active against gram-positive bacteria. The order of activity at R_8: F, Cl, OCH_3 > H, CF_3 > methyl, vinyl, propargyl.
9. A halogen (F or Cl) at the 8- position improves oral absorption.
10. The joining of N1-group to the C-8 position with oxazine ring leads to active ofloxacin.

25.7 ADVERSE EFFECTS

Some side-effects of the quinolones are class effects, and cannot be modulated by molecular variation. Most of the fluoroquinolones produce photosensitivity reactions and cause convulsions, particularly concurrent administration of NSAID fenbufen. This effect is strongly influenced by the C7 substituent, with simple pyrrolidines and piperazines the worst actors. Increasing steric bulk through alkylation ameliorates these effects. Phototoxicity is determined by the nature of the 8-position substituent with halogen causing the greatest photo reaction, while hydrogen and methoxy show little light induced toxicity. Arthralgia and joint swelling have developed in children receiving fluoroquinolones; therefore, these drugs are not generally recommended for use in prepubertal children or pregnant women. Genetic toxicity is controlled in additive fashion by the choice of groups at the 1, 7, and 8 positions.

25.8 SYNTHESIS OF FLUOROQUINOLONES

β-Keto ester is formed by the reaction between (sub) benzoyl chloride and magnesium salt of diethyl malonate. This keto ester on reaction with triethyl ortho formate gives methoxy methylene derivative, which in turn reacts with corresponding amine (ethylamine/cyclopropylamine) to give aminomethylene derivative. Base cyclization of this compound gives quinolone-3-carboxylic acid derivatives. Reaction with appropriate piperazine affords titled compound.

Norfloxacin, ciprofloxacin, lomefloxacin, gatifloxacin, and sparfloxacin are prepared by this method by using appropriate starting material, primary amine and piperazine derivatives.

Ofloxacin (+/-)-9-Fluoro-2, 3-dihydro-3-methyl-10-(4-methyl-1-piperazinyl)-7-oxo-7H-pyrido [1,2,3-de]-1, 4-benzoxazine-6-carboxylic acid

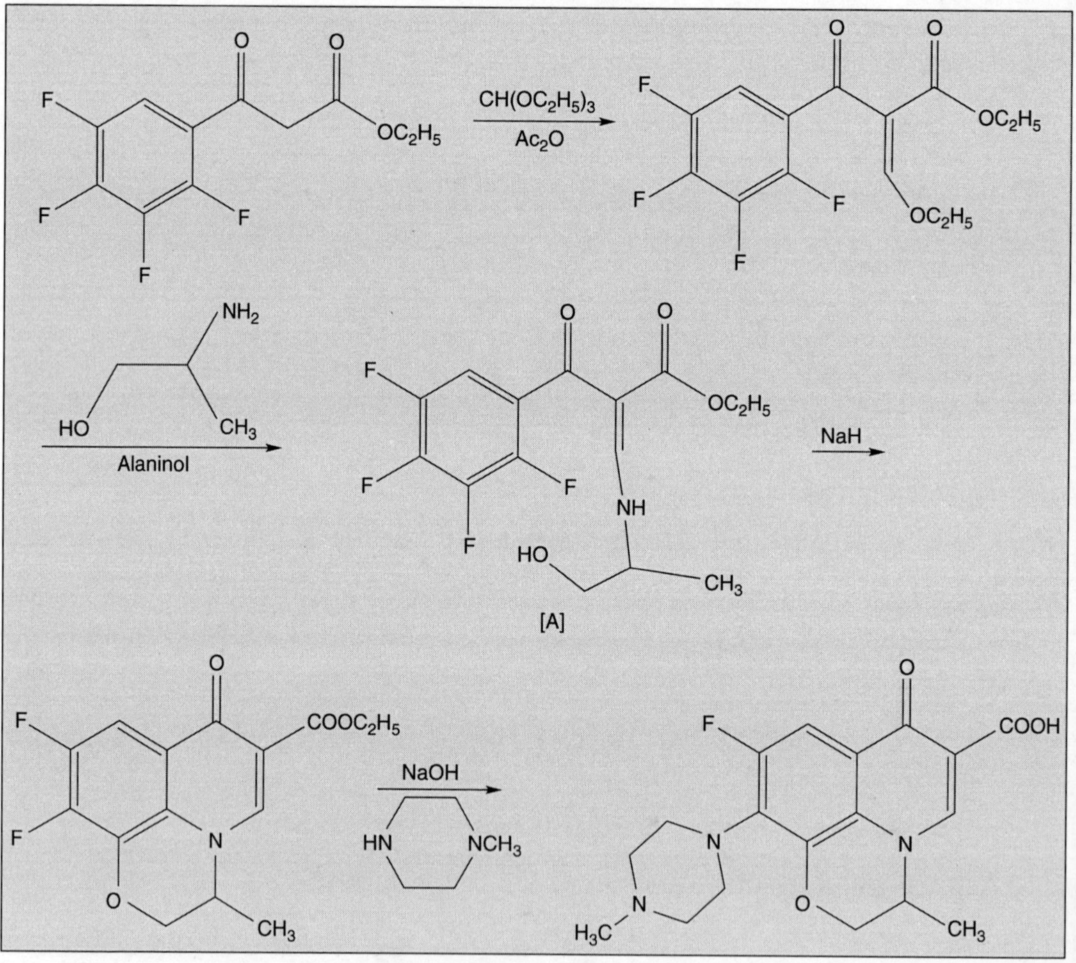

Tetrafluorobenzoic acid on reaction with 1,1'-carbonyldiimidazole in tetrahydrofuran afforded the corresponding imidazolide, which, in situ, was treated with neutral magnesium salt of ethyl potassium malonate in the presence of tri-ethyl amine to yield ethyl 3-(2,3,4,5-tetrafluorophenyl)-

3-oxopropanoate. Ethyl 2,3,4,5-tetrafluoro-α-[[(2-hydroxy-1-methyl-ethyl)amino]methylene]-β-oxo-benzenepropanoic acid (A) was prepared by a two-step one-pot reaction. First treatment of the keto ester with tri-ethyl orthoformate in acetic anhydride gave the one-carbon homologue enol ether intermediate ethyl α-(ethoxymethylene)-2,3,4,5-tetrafluoro-b-oxo-benzenepropanoic acid as an oil, which on reaction with (S)-(+)-2-amino-1-propanol at 0°C affords (A) as an oily residue. Compound (A) on cyclization with the base potassium carbonate/sodium hydride in dimethylsulphoxide yields ethyl 6,7,8-trifluoro-1,4-dihydro-1-(2-hydroxy-1-methylethyl)-4-oxo-3-quinolinecarboxylic acid. Further cyclization and hydrolysis of this compound is done by heating with 10 per cent aqueous sodium hydroxide in tetrahydrofuran, affording 9,10-difluoro-2,3-dihydro-3-methyl-7-oxo-7H-[1,4]oxazino[2,3,4-ij]quinoline-6-carboxylic acid. This on further reaction with N-methyl piperazine in dimethylsulphoxide affords ofloxacin.

Levofloxacin It is the *S* - enantiomer (L-isomer) of ofloxacin, and has approximately twice the potency of ofloxacin, because the *R* + enantiomer (D-isomer) of ofloxacin is essentially inactive. In addition, the *S* - enantiomer (L-isomer) of ofloxacin has substantially less toxicity.

Nalidixic acid 1-Ethyl-7-methyl-4-oxo-[1,8]naphthyridine-3-carboxylic acid

Condensation of ethoxymethylenemalonate with 2-amino-6-methyl pyridine proceeds directly to naphthyridine; the first step in this transformation probably involves an addition-elimination reaction to afford the β-hydroxy ester. N-Ethylation with ethyl iodide and base hydrolysis affords nalidixic acid.

It is especially used in treating urinary tract infections, for example, by *Escherichia coli*, *Proteus*, *Shigella*, *Enterobacter*, and *Klebsiella*. It selectively and reversibly blocks DNA replication in susceptible bacteria.

> Nalidixic acid is considered to be the predecessor of all members of the quinolone family, including the second, third, and fourth generations, commonly known as fluoroquinolones.

25.9 NEWER DRUGS

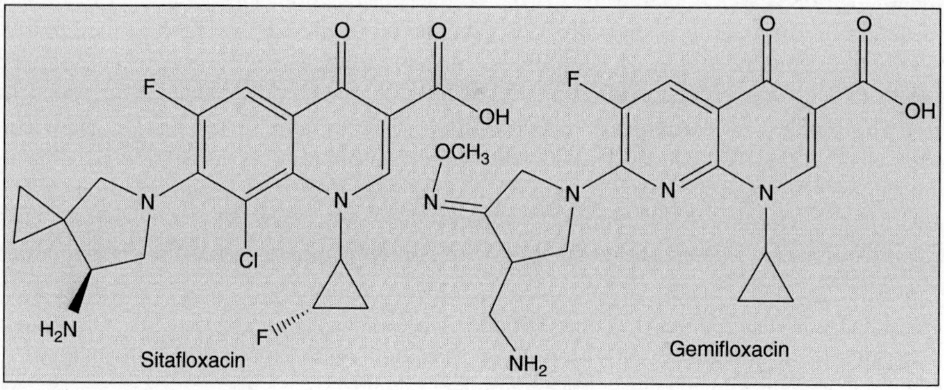

Sitafloxacin

Gemifloxacin

FURTHER READINGS

1. Andriole, V.T. *The Quinolones*, 3rd Ed., Academic Press.
2. Domagala, J.M., 1994. *Journal of Antimicrobial Chemotherapy*, Vol. 33, 685–706.

MULTIPLE-CHOICE QUESTIONS

1. The N-1 position of pefloxacin contains
   a. Cyclopropyl   b. Methyl
   c. Ethyl   d. Piperazine
2. The C-7 position of gatifloxacin contains
   a. Piperazine
   b. 4-Methylpiperazine
   c. 3,5-Dimethylpiperazine
   d. 3-Methylpiperazine
3. Fluoroquinolones are indicated for all of the following except:
   a. Urinary tract infection
   b. Tuberculosis
   c. Bone infection
   d. Bronchial asthma
4. The order of activity of I) cyclopropyl, II) methylamino, and III) cyclobutyl at N_1 position of quinolone is
   a. I > II > III   b. I > III > II
   c. II > III > I   d. III > I > II
5. One of the following is not a side-effect of fluoroquinolones:
   a. Phototoxcity   b. Ototoxicity
   c. Convulsion   d. Arthralgia

QUESTIONS

1. Write a short note on fluoroquinolone antibacterials with a few structural examples.
2. Discuss the SAR of fluoroquinolones and their effects on toxicity.
3. Give the synthetic protocol for ciprofloxacin, norfloxacin, and ofloxacin.
4. How do quinolones exhibit their antibacterial activity?

SOLUTION TO MULTIPLE-CHOICE QUESTIONS

1. c;
2. c;
3. d;
4. b;
5. b.

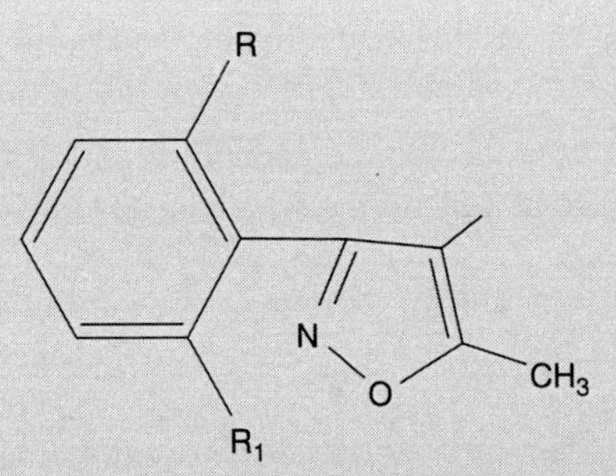

CHAPTER 26

Antibiotics

LEARNING OBJECTIVES

- Define antibiotics and developmental pathway of various antibiotics
- Define β-lactam antibiotics and their importance
- Categorize penicillins and describe how penicillins are degraded
- Discuss the structural importance of penicillins
- Define cephalosporins and classify according to their generation
- Describe the degradation pathway of cephalosporins
- Discuss the structural modifications and their effect on biological activity of cephalosporins
- Define tetracycline antibiotics and its utility
- Describe the structural importance of tetracycline and its mode of action
- Define macrolide antibiotics and its chemistry
- Describe the chemistry and utility of aminoglycoside antibiotics
- List some newer antibiotics

Antibiotics are substances produced by various species of microorganisms (bacteria, fungi, actinomycetes) that suppress the growth of other microorganisms and eventually may destroy them. Since their discovery in the 1930s, antibiotics have made it possible to cure diseases caused by bacteria such as pneumonia, tuberculosis, and meningitis – saving the lives of millions of people around the world. (*See Fig. 26.1 in the coloured set of pages.*)

Historically, the most common classification has been based on chemical structure and proposed mechanism of action, as follows:

1. Agents that inhibit the synthesis of bacterial cell walls; these include the penicillins and cephalosporins, which are structurally similar, and dissimilar agents such as cycloserine, vancomycin, bacitracin, and the imidazole antifungal agents
2. Agents that act directly on the cell membrane of the microorganisms affecting permeability and leading to leakage of intracellular compounds; these include polymyxin, polyene antifungal agents, nystatin and amphotericin B, which bind to cell wall sterols
3. Agents that affect the function of 30S and 50S ribosomal subunits to cause reversible inhibition of protein synthesis; these include tetracyclines, erythromycins, chloramphenicol, and clindamycin
4. Agents that bind to the 30S ribosomal subunit and alter protein synthesis, which eventually leads to cell death; these include aminoglycosides
5. Agents that affect nucleic acid metabolism such as rifamycins, which inhibit DNA-dependent RNA polymerase

26.1 β-LACTAM ANTIBIOTICS

Antibiotics that contain the β-lactam (a four-membered cyclic amide) ring structure constitute the dominant class of agents currently employed for the chemotherapy of bacterial infections. Penicillin, cephalosporin, and their semi-synthetic derivatives come under this class.

Mechanism of action: The cell wall of bacteria is essential for the normal growth and development. Peptidoglycan is a heteropolymeric component of the cell wall that provides rigid mechanical stability by virtue of its highly cross-linked lattice-wise structure.

The peptidoglycan is composed of glycan chains, which are linear strands of two alternating amino sugars (N-acetyl glucosamine and N-acetyl muramic acid) that are cross-linked by peptide chains.

The biosynthesis of peptidoglycan involves three stages.

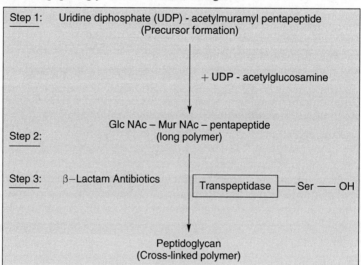

The last step in peptidoglycan synthesis is inhibited by β-lactam antibiotics. The transpeptidase enzyme contains serine, which is probably acylated by β-lactam antibiotics with cleavage of -CO-N- bond of the β-lactam ring. This renders the enzyme inoperative and inhibits peptidoglycan synthesis.

Penicillins: The penicillins are commonly named as penams, a designation in which the sulphur atom is given the top priority.

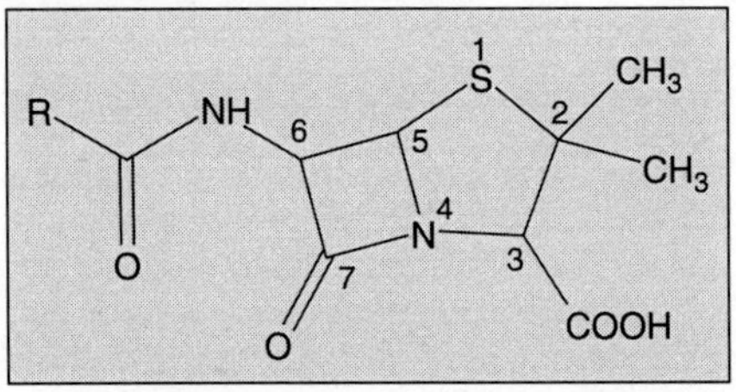

Using this nomenclature, the penicillins have a prerequisite carboxylic acid group placed at the C-3 position. The west-end substituents are joined to the C-6 centre and are usually substituted via acylation, thus constituting a variety of C-6 acylamido substituents. The β-lactam carbonyl centre is located at position 7, and the C-2 centre contains a geminal dimethyl substitution characteristic of penicillins.

Classification

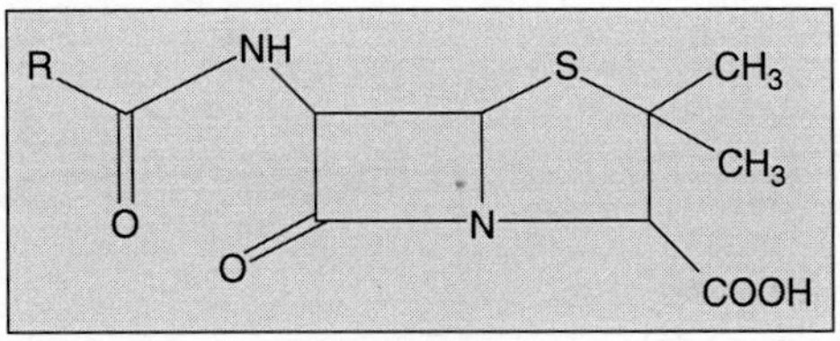

1. Early penicillins

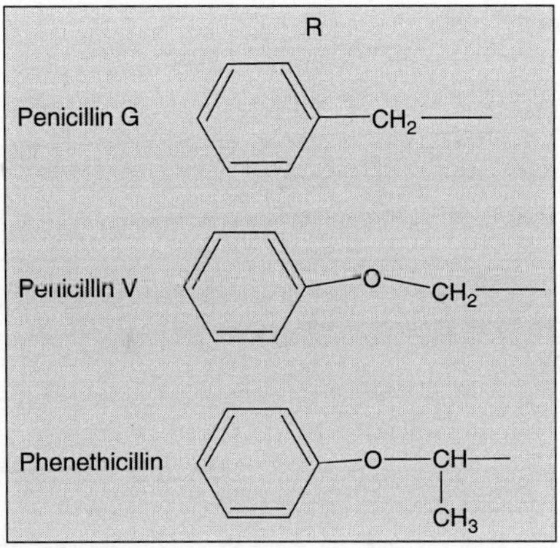

General impact on antibacterial activity:

- Excellent gram-positive potency against susceptible *Staphylococci, Streptococci*
- Useful against some gram-positive cocci
- Good oral absorption but relatively acid-labile
- Ineffective against gram-negative bacilli
- Susceptible by deactivation by penicillinase

2. Penicillinase-resistant penicillins

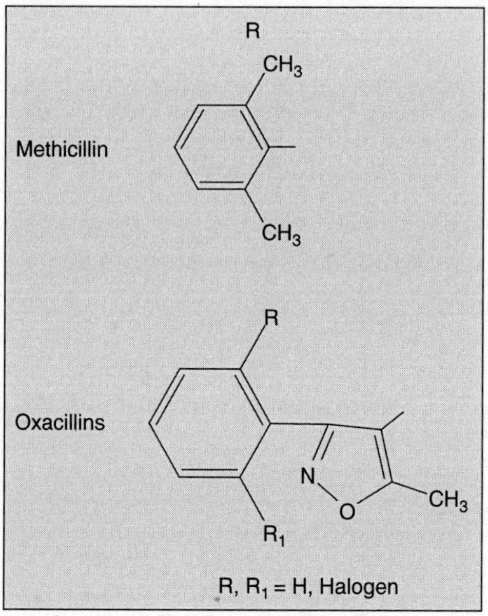

Methicillin: R group is a benzene ring with CH₃ groups at both ortho positions.

Oxacillins: R group is an isoxazole ring with a methyl substituent, connected to a phenyl ring bearing R and R₁ substituents at ortho positions.

R, R₁ = H, Halogen

General impact on antibacterial activity:

- Diminished susceptibility to many penicillinases; active against microorganisms resistant to early penicillins
- Oxacillins offer good oral activity
- Inadequate spectrum against many gram-negative species

3. Broad-spectrum penicillins

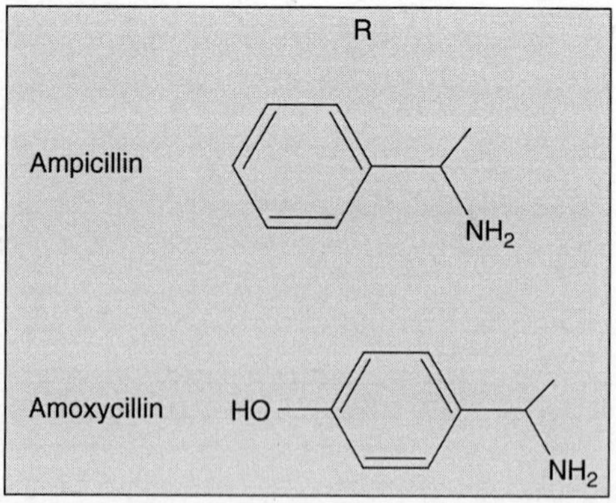

Ampicillin: R group is a phenyl ring with an α-NH₂ substituent.

Amoxycillin: R group is a 4-hydroxyphenyl ring with an α-NH₂ substituent.

- Extended spectrum of activity against some gram-negative bacteria; retention of gram-positive potency
- Generally well-absorbed orally; ampicillin can be dosed I.V. and I.M. as well; amoxycillin exceptional oral agent

- Prodrug esters (of ampicillin) enhance systemic drug levels
- Ineffective against *Pseudomonas aeruginosa*

4. Anti-pseudomonal penicillins

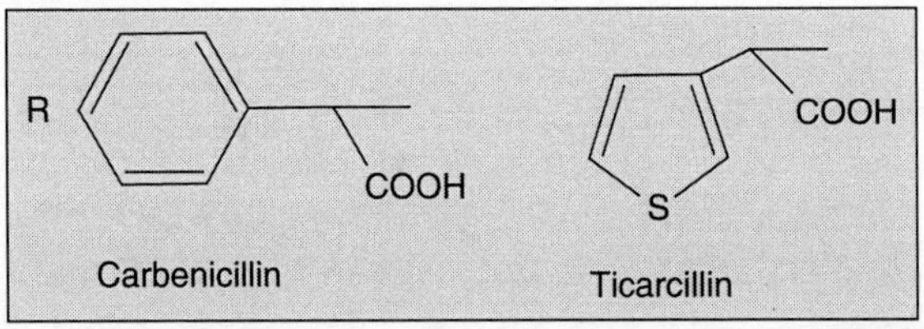

Carbenicillin, **Ticarcillin**

- Extended spectrum of activity against many pathogenic gram-negative bacteria; reduced gram-positive potency
- Good activity against *P. aeruginosa*
- Oral absorption, chemical stability problematic
- Prodrug esters (of carbenicillin) enhance systemic drug levels

5. Broad-spectrum ureido penicillins

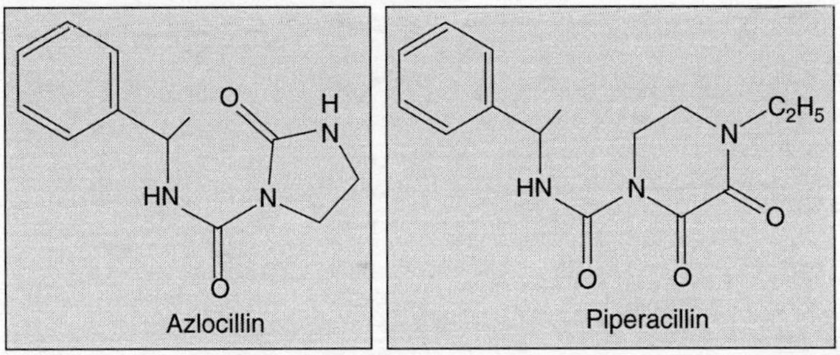

Azlocillin, **Piperacillin**

- Enhanced spectrum of activity against *P. aeruginosa*, expanded activity against *Klebsiella*, *Serratia*, *Proteus*
- Good potency against gram-positive bacteria, but generally not effective against penicillinase producers
- Good pharmacokinetic profile

6. Penicillin with a C-6 amidino west-end

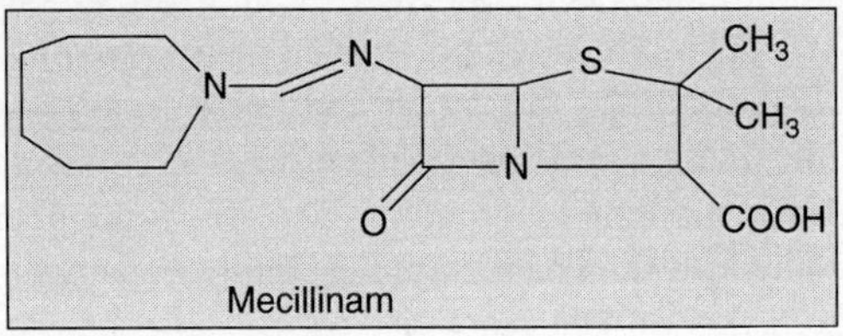

Mecillinam

- Good activity against *E. coli*, *Klebsiella*, *Shigella*, *Salmonella*, and many other resistant species
- Inactive against *P. aeruginosa*
- Prodrug esters enhance systemic drug levels

Mechanism: Refer to β-lactam antibiotics
Chemical degradation of penicillins

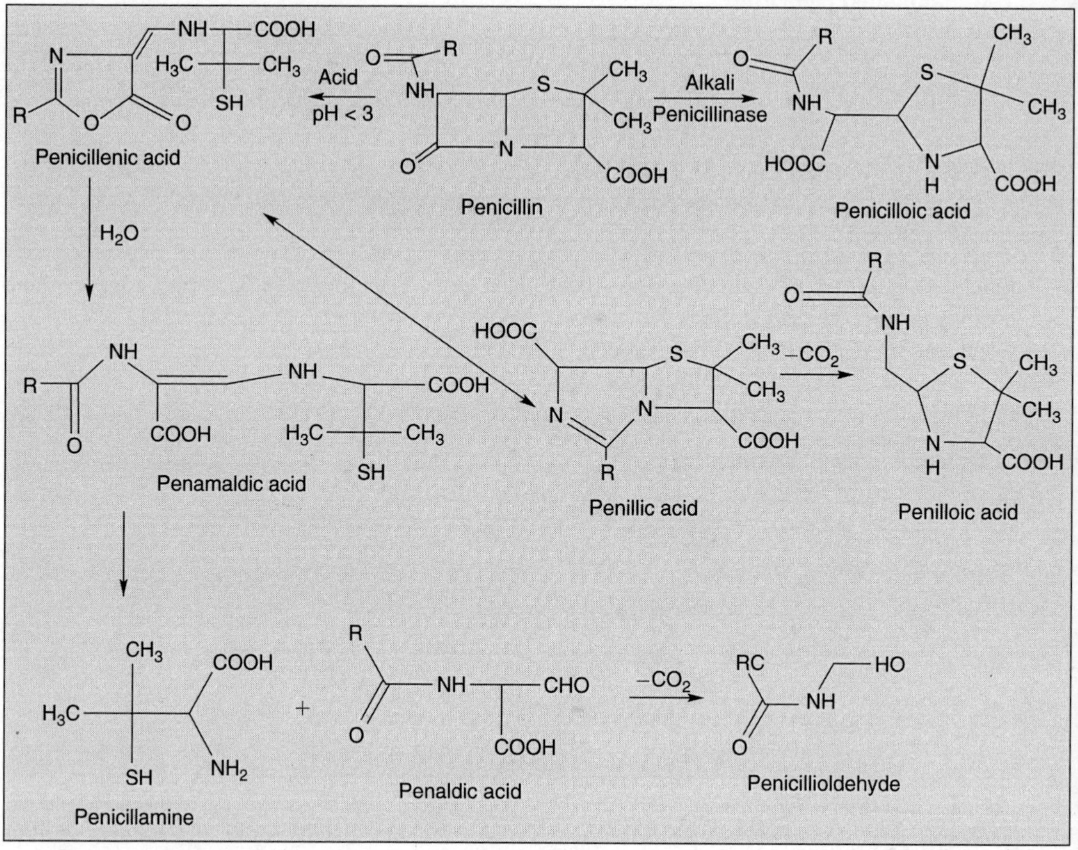

In strongly acidic solutions (pH<3), penicillin undergoes a complex series of reactions leading to a variety of inactive degradation products. Similarly, penicillinase enzyme hydrolyses the β-lactam ring to produce inactive penicilloic acid.

Acid-catalysed degradation in stomach contributes in a major way to the poor oral absorption of penicillin. Thus, efforts to obtain penicillins with improved pharmacokinetic and microbiologic properties have sought to find acyl functionalities that would minimize sensitivity of the β-lactam ring to acid hydrolysis and, at the same time, maintain antibacterial activity.

Substitution of an electron-withdrawing group in the α-position of benzyl penicillin has stabilized the penicillin to acid-catalysed hydrolysis. The increased stability imparted by such electron-withdrawing groups has been attributed to a decrease in reactivity of the side-chain amide carbonyl oxygen atom toward participation in β-lactam ring opening to form the penicillenic acid.

Adverse effects Penicillins cause hypersensitivity reactions (allergy) in several percentages of patients and this may be due to formation of antigenic penicilloyl proteins formed *in vivo* by the reaction of nucleophilic groups (e.g., ε-amino) on specific body proteins with the β-lactam carbonyl groups.

The first mode of β-lactam resistance is due to enzymatic hydrolysis of the β-lactam ring. If the bacterium produces the enzymes β-lactamase or penicillinase, these enzymes will break open the β-lactam ring of the antibiotic, rendering the antibiotic ineffective. The second mode of β-lactam resistance is due to possession of altered penicillin-binding proteins (PBP). β-lactams cannot bind as effectively to these altered PBPs, and, as a result, the β-lactams are less effective at disrupting cell-wall

26.1.1 SAR of Penicillins

1. C-6 Amino west-end substitution:
    (i) The design and development of the west-end substituents has been aimed at strengthening various weaknesses that have traditionally hampered penicillins in terms of activity, stability, resistance, and absorption/distribution.
    (ii) The C-6 amine moiety itself is necessary for appreciable antibacterial activity, but substitution of the amine via monoacylation can offer much more potent congeners.
    (iii) Only carboxamido-derived west-end moieties are tolerated; sulphonation- or phosphoramide-containing substituents are devoid of antibacterial activity. Similarly, imide- or carbamate-containing west-end are inferior.
    (iv) Agents that were stable to penicillinase enzymes were created by introducing a more crowded environment around β-lactam moiety. Methicillin contains 2, 6-dimethoxy benzamido west-end and the placement of methoxy groups on the aromatic ring is important; the bis *ortho* arrangement creates the most effective crowding around the β-lactam carbonyl centre, while retaining good activity.

    The oxacillins contain 5-methyl-3-phenyl-4-isoxazolyl west-end substituents that similarly impose crowding in proximity to the β-lactam ring. In these compounds, both the methyl and phenyl substituents are positioned closest to β-lactam system. Removal of either group increases susceptibility to penicillinases.
    (v) To expand the antibacterial spectrum of penicillins, more hydrophilic west-end was designed that can enhance potency against gram-negative pathogens. Ampicillin contains a D-α-aminophenylacetamido west-end and is most recognized as amino penicillins. In general, substituents on the phenyl ring are detrimental either due to decreased hydrophilicity or, conversely, due to adverse polar effects if an ionizable substituent is present. A notable balance of these opposing forces has been achieved with the placement of a *para*-hydroxyl group on to the phenyl ring. Amoxycillin is essentially comparable to ampicillin in terms of *in vitro* potency, but displaces better oral efficacy.
    (vi) The spectrum of activity was further expanded with introducing strongly acidic groups at the α-carbonyl centre of the side-chain. These imparted useful potency against *P. aeruginosa*. Carbenicillin possesses α-carboxy phenyl acetamido west-end.
    (vii) The acylation of the ampicillin west-end amine functionality with certain polar groups leads to cyclic urea derivatives, ureido penicillins (azlocillin), which contain a five-membered cyclic

urea system joined via N-acylation to the α-amino substituent of ampicillin. These are more active against *P. aeruginosa* than carbenicillin and possess potency against other gram-negative pathogenic species. The presence of urea group imparts improved penetration into these gram-negative species traditionally resistant to penicillins.

2. Substituents at sulphur:
Sulphur is the only atom at position 1 of the penicillin in order to retain appreciable antibacterial activity.

3. C-2 substituents:
The geminal dimethyl group at C-2 is characteristic of the penicillin.

4. C-3 substituents:
In general, derivatization of the C-3 carboxylic acid functionality is not tolerated unless the free penicillin carboxylic acid can be generated *Antibiotics*. Doubly activated penicillin esters such as alkanoyloxyalkyl congeners undergo rapid cleavage *Antibiotics* to generate active penicillin—e.g., pivampicillin and becampicillin.

5. Variation at N-4:
The nitrogen atom at the ring junction is vital for antibacterial activity; the nitrogen atom contributes to the reactivity of the β-lactam carbonyl centre.

Methicillin [2S-(2α,5α,6β)]-3,3-Dimethyl-7-oxo-6-(2,6-dimethoxybenzamido)-4-thia-1-azabicyclo[3.2.0]-heptan-2-carboxylic acid

Synthesis

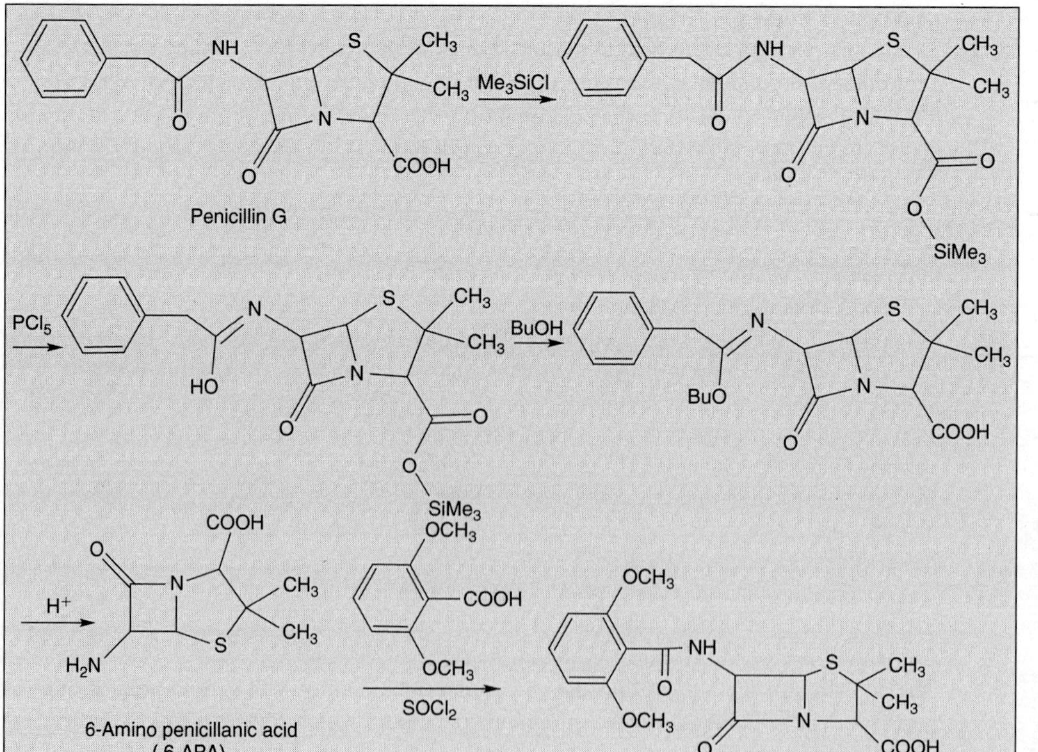

It is synthesised by acylating 6-APA with 2,6-dimethoxybenzoic acid in the presence of thionyl chloride.

Oxacillin [2S-(2α,5α,6β)]-3,3-Dimethyl-7-oxo-6-(5-methyl-3-phenyl-4-isoxazolcarboxamido)-4-thia-1-azabicyclo[3.2.0]-heptan-2-carboxylic acid

Synthesis

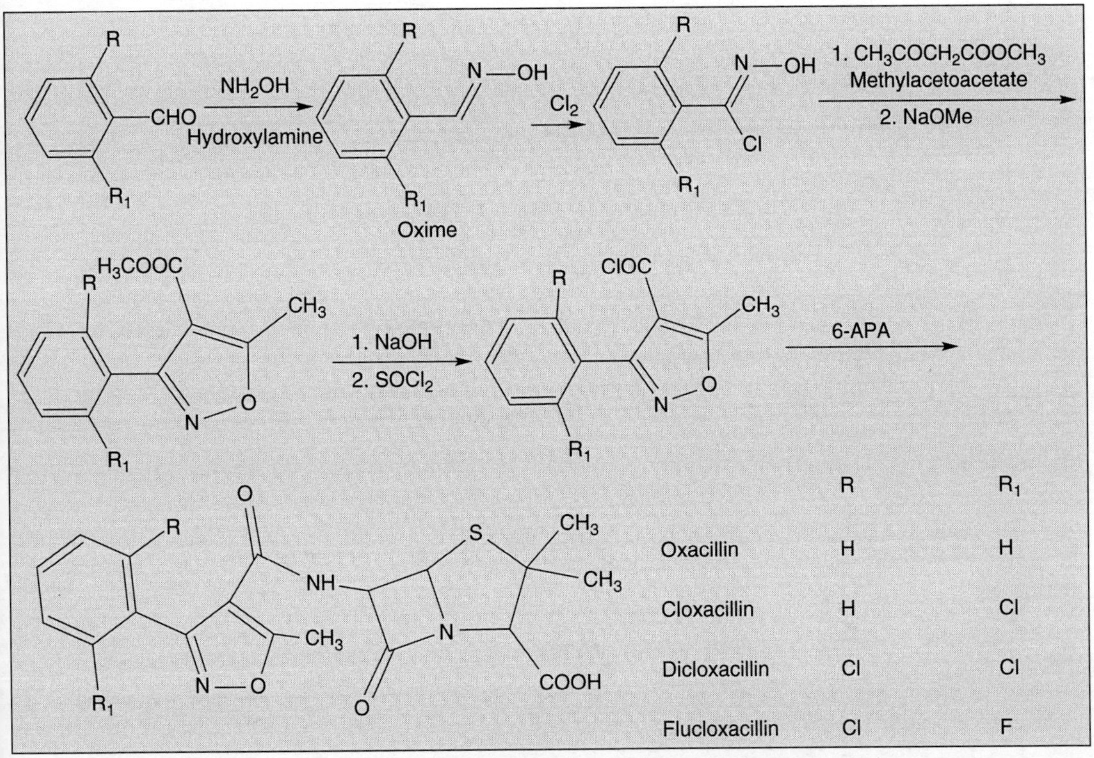

	R	R_1
Oxacillin	H	H
Cloxacillin	H	Cl
Dicloxacillin	Cl	Cl
Flucloxacillin	Cl	F

Oxacillin is synthesised by reacting 5-methyl-3-phenyl-4-isoxazolcarboxylic acid chloride with 6-APA in the presence of sodium bicarbonate. The 5-methyl-3-phenyl-4-isoxazolcarboxylic acid chloride is synthesised by the following method. Reacting benzaldehyde with hydroxylamine gives oxime, which when oxidized by chlorine gives benzhydroxamic acid chloride. This is reacted with methylacetoacetate in the presence of sodium methoxide, giving the methyl ester of 5-methyl-3-phenyl-4-isoxazolcarboxylic acid. Alkaline hydrolysis of the resulting ester gives the corresponding acid, which is reacted with thionyl chloride to give the acid chloride necessary for acylation.

Cloxacillin [2S-(2α,5α,6β)]-3,3-Dimethyl-7-oxo-6-[(5-methyl-3-(o-chlorphenyl)-4-isoxazol-carboxamido)]-4-thia-1-azabicyclo[3.2.0]-heptan-2-carboxylic acid

It is synthesised from o-chlorobenzaldehyde by the scheme described above.

Synthesis

Dicloxacillin [2S-(2α,5α,6β)]-3,3-Dimethyl-7-oxo-6-[(5-methyl-3-(2,6-dichlorophenyl)-4-isoxazol-carboxamido)]-4-thia-1-azabicyclo[3.2.0]-heptan-2-carboxylic acid.

It is also synthesised by the scheme described above, using 2,6-dichlorobenzaldehyde as the starting substance.

Ampicillin [2S-[2α,5α,6β(S)]]-3,3-Dimethyl-7-oxo-6-(2-amino-2-phenylacetamido)-4-thia-azabicyclo[3.2.0]-heptan-2-carboxylic acid
Synthesis

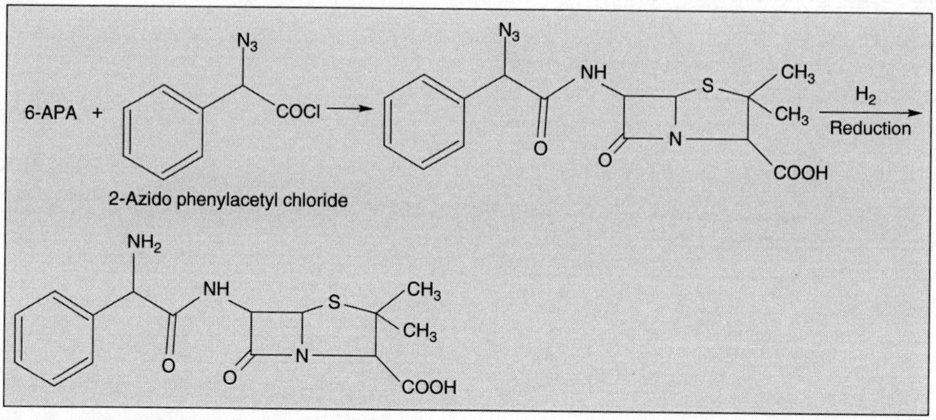

Acylation of 6-APA with the 2-azidophenylacetyl chloride gives the corresponding amide; catalytic reduction of azido group affords the amine ampicillin.

The corresponding product from acylation with 2-azido-4-hydroxyphenyl acetyl chloride is **amoxycillin**.

Pivampicillin 2,2-Dimethylpropanoyloxymethyl (2S,5R,6R)-6-{[(2R)-2-amino-2-phenylacetyl]amino}-3,3-dimethyl-7-oxo-4-thia-1-azabicyclo[3.2.0]heptane-2-carboxylate

Pivampicillin is a pivaloyloxymethylester of ampicillin. It is a prodrug that is thought to enhance the oral bioavailability of ampicillin because of its greater lipophilicity compared to that of ampicillin.
Synthesis

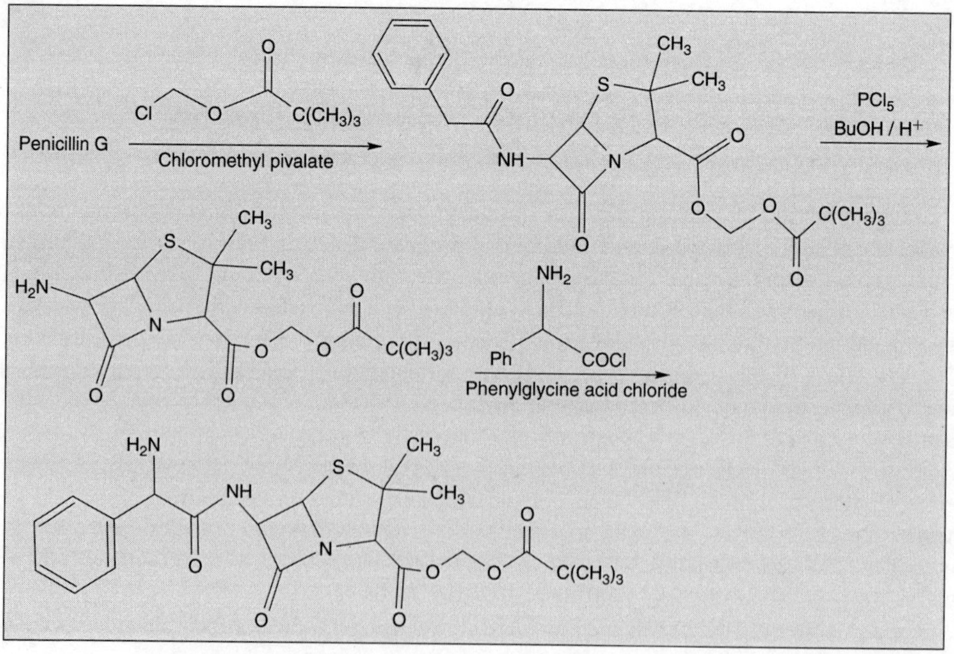

Benzyl penicillin is esterified with chloromethyl pivalate; acid hydrolysis of amide bond gives 6-APA ester derivative. Acylation of this with phenylglycine acid chloride affords pivampicillin.

Carbenicillin [2S-(2α,5α,6β)]-3,3-Dimethyl-7-oxo-6-(2-carboxy-2-phenylacetamido)-4-thia-1-azabicyclo[3.2.0]-heptan-2-carboxylic acid
Synthesis

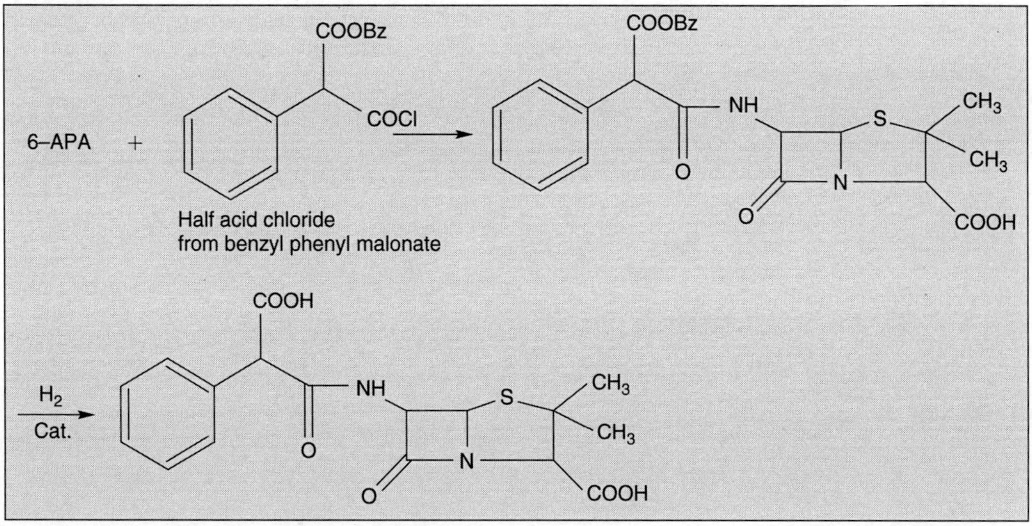

Carbenicillin is synthesised by direct acylation of 6-APA in the presence of sodium bicarbonate by phenylmalonic acid monobenzyl ester chloride, which forms the benzyl ester of carbenicillin, the hydrogenolysis of which using palladium on carbon or calcium carbonate as catalyst gives the desired product.

Ticarcillin [2S-(2α,5α,6β)]-3,3-Dimethyl-7-oxo-6-[2-carboxy-2-(3-thienyl)acetamido]-4-thia-1-azabicyclo[3.2.0]-heptan-2-carboxylic acid

It is also synthesised by direct acylation of 6-APA in the presence of sodium hydroxide, but with 3-thienylmalonic acid chloride, which gives ticarcillin.

Azlocillin (2S,5R,6R)-3,3-Dimethyl-7-oxo-6-[(R)-2-(2-oxoimidazolidin-1-carboxamido)-2-phenylacetamido]-4-thia-1-azabicyclo[3.2.0]-heptan-2-carboxylic acid
Synthesis

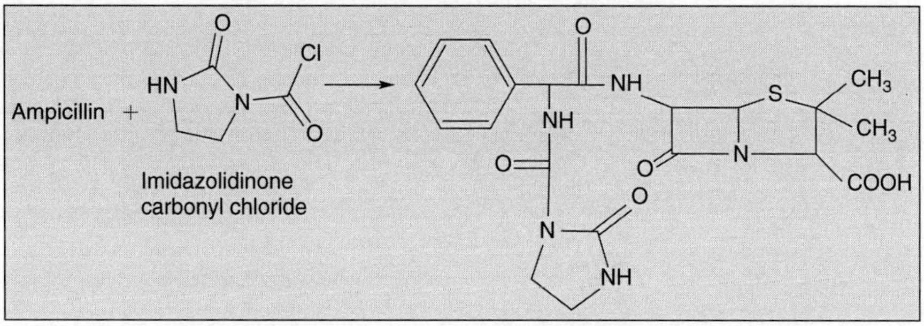

Azlocillin is synthesised by directly reacting 1-chlorocarbonyl-2-imidazolidinone (2-imidazolidinone is acylated with phosgene, forming 1-chlorocarbonyl-2-imidazolidinone) and ampicillin.

Synthesis
Piperacillin (2S,5R,6R)-3,3-Dimethyl-7-oxo-6-[(2R)-2-[(4-ethyl-2,3-dioxo-1-piperazinyl)formamido]-2-phenylacetamido]-4-thia-1-azabicyclo[3.2.0]-heptan-2-carboxylic acid
Synthesis

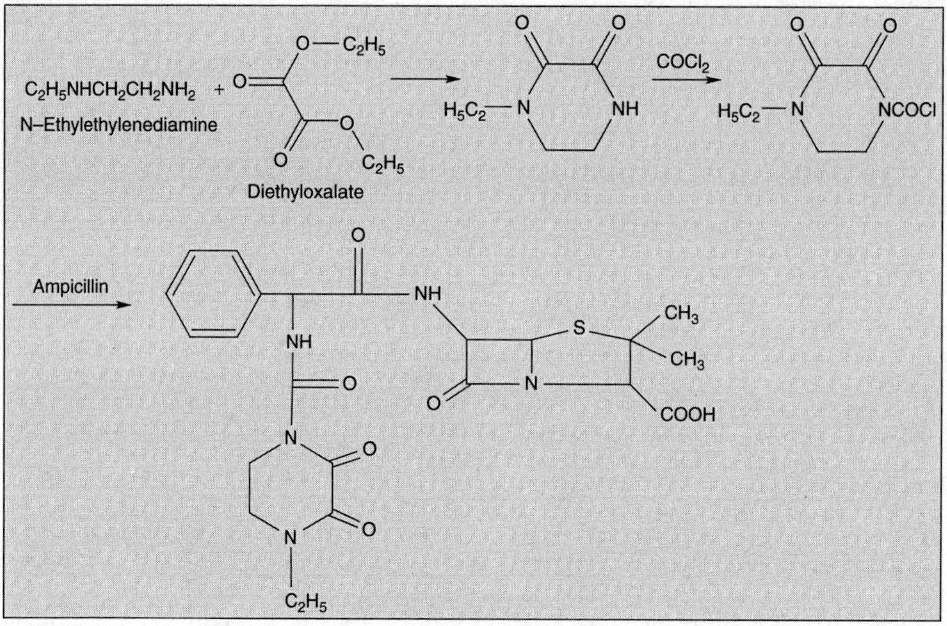

Piperacillin is also synthesised by acylating ampicillin, but with 1-chlorocarbonyl-4-ethylpiperazin-2,3-dione. The necessary 1-chlorocarbonyl-4-ethylpiperazin-2,3-dione is synthesised by reacting N-ethylethylenediamine with diethyloxalate, forming 4-ethylpiperazin-2,3-dione, and then acylating this with phosgene.

Mecillinam (Amdinocillin) (2S,5R,6R)-6-(azepan-1-ylmethylideneamino)-3,3-dimethyl-7-oxo-4-thia-1-azabicyclo[3.2.0]heptane-2-carboxylic acid

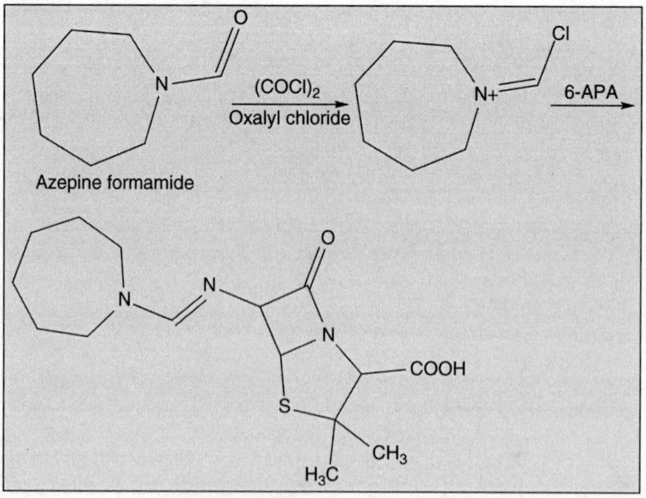

Reaction of azepine formamide with oxalyl chloride gives reactive derivative, which on condensation with 6-APA leads to the formation of the amidine mecillinam.

26.1.2 Cephalosporins

The cephalosporins are β-lactam antibiotics isolated from Cephalosporium species and/or prepared semisynthetically. This comes under the class of 7-amino cephalosporonic acid (7-ACA) derivatives and are much more acid-stable than the corresponding 6-APA compounds. The cephalosporins have a mechanism of action similar to that of penicillins - mainly, they inhibit the cross-linking of the peptidoglycan units in bacterial cell wall by inhibiting transpeptidase enzyme.

> The cephalosporin nucleus, 7-aminocephalosporanic acid (7-ACA), was derived from cephalosporin C and proved to be analogous to the penicillin nucleus 6-aminopenicillanic acid, but it was not sufficiently potent for clinical use. Modification of the 7-ACA side-chains resulted in the development of useful antibiotic agents, and the first agent cephalothin (cefalothin) was launched by Eli Lilly in 1964.

Classification

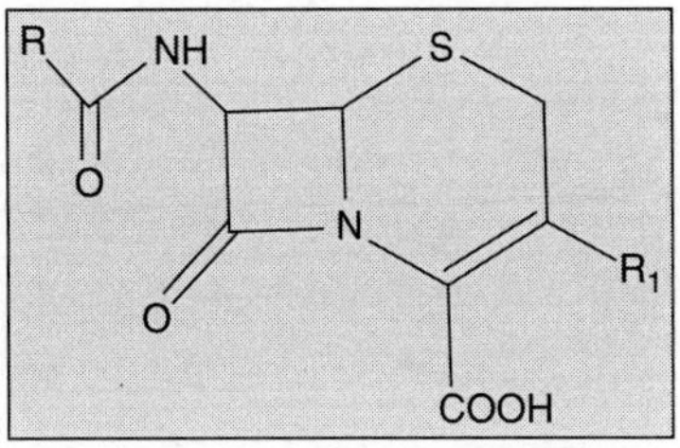

1. First-generation cephalosporins
These drugs have the highest activity against gram-positive bacteria and the lowest activity against gram-negative bacteria.

	R	R_1
Cephalexin	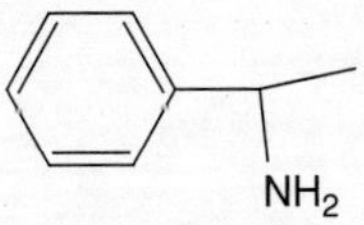 phenyl-CH(NH$_2$)-	CH$_3$
Cefadroxil	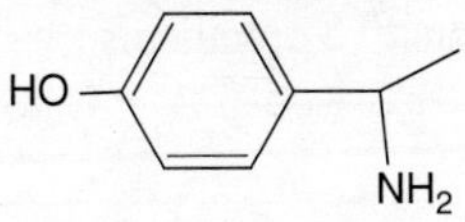 4-hydroxyphenyl-CH(NH$_2$)-	CH$_3$
Cephradine	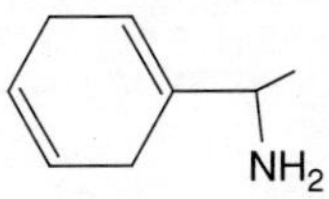 1,4-cyclohexadienyl-CH(NH$_2$)-	CH$_3$
Cephalothin	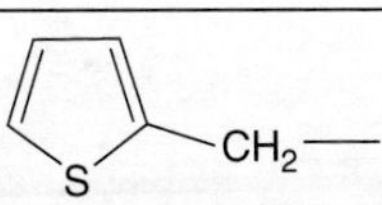 2-thienyl-CH$_2$-	—CH$_2$OCOCH$_3$

	R	R_1
Cephacetrile	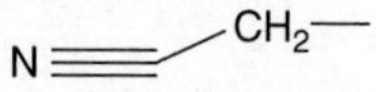 $N\equiv C-CH_2-$	$-CH_2OCOCH_3$
Cefazolin	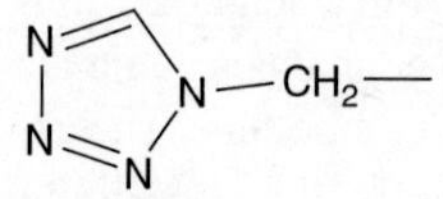 tetrazolyl-CH_2-	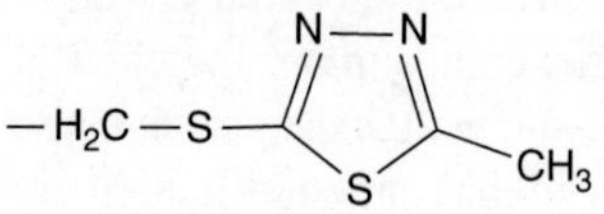 $-H_2C-S-$(thiadiazolyl-CH_3)

2. Second-generation cephalosporins

These drugs are more active against gram-negative bacteria and less active against gram-positive bacteria than first-generation members.

	R	R_1
Cefaclor	phenyl-$CH(NH_2)-$	Cl
Cefamandole	phenyl-$CH(OH)-$	$-H_2C-S-$(tetrazolyl-CH_3)
Cefuroxime	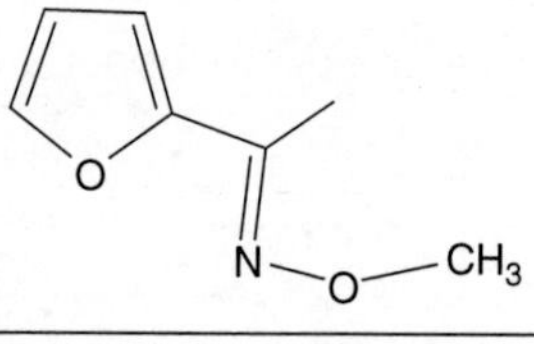 furyl-$C(=N-O-CH_3)-$	$-CH_2OCONH_2$

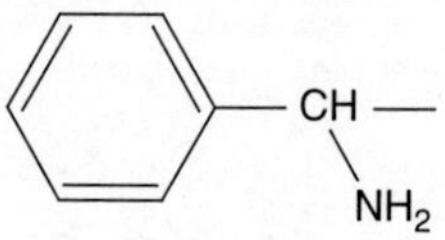

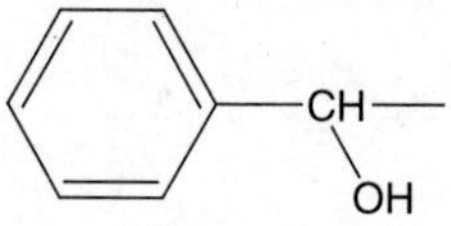

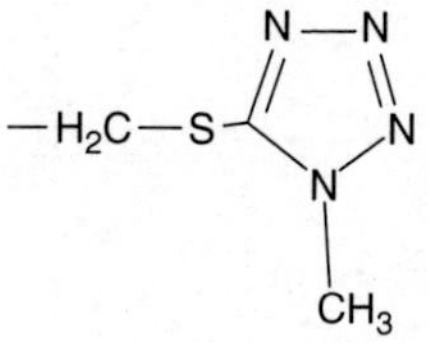

3. Third-generation cephalosporins

These drugs are less active than first-generation drugs against gram-positive bacteria, but have a much-expanded spectrum of activity against gram-negative organisms.

	R	R_1
Cefotaxime	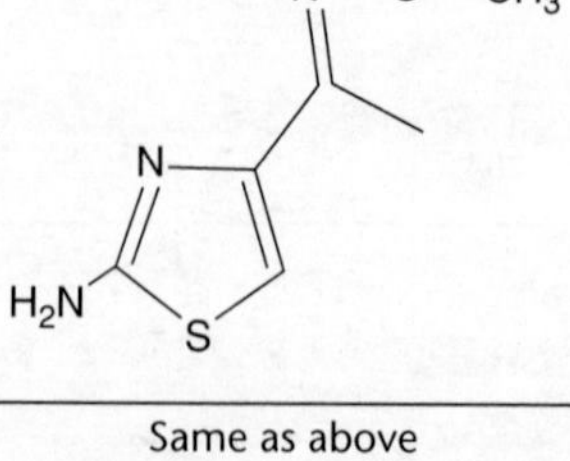 (2-aminothiazol-4-yl)-$C(=N-O-CH_3)-$	$-CH_2OCONH_2$
Ceftizoxime	Same as above	H

	R	R_1
Ceftriaxone	Same as above	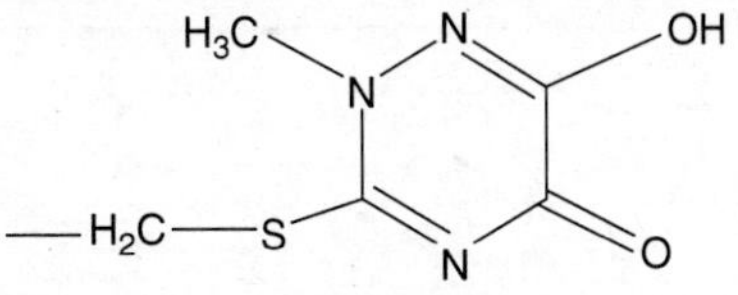 (methylene-thio linked to 2-methyl-5-hydroxy-6-oxo-1,2,4-triazin-3-yl: $-H_2C-S-$ connected to triazine bearing H_3C-N, N, OH, $=O$)
Ceftazidime	(2-aminothiazol-4-yl)=N—O—C(CH$_3$)$_2$—COOH	$-H_2C-N^+$ (pyridinium)
Cefoperazone	HO–C$_6$H$_4$–CH(NH–CO–N(piperazine with C$_2$H$_5$ and C=O))–	$-H_2C-S-$(1-methyl-1H-tetrazol-5-yl), N—N, N, N, CH$_3$

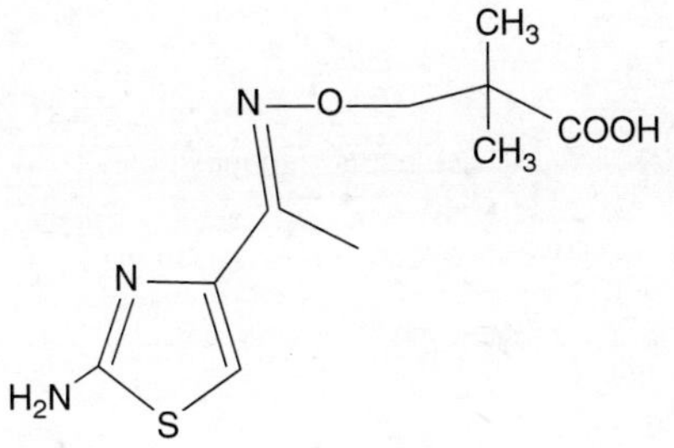

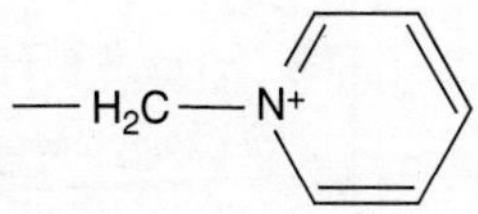

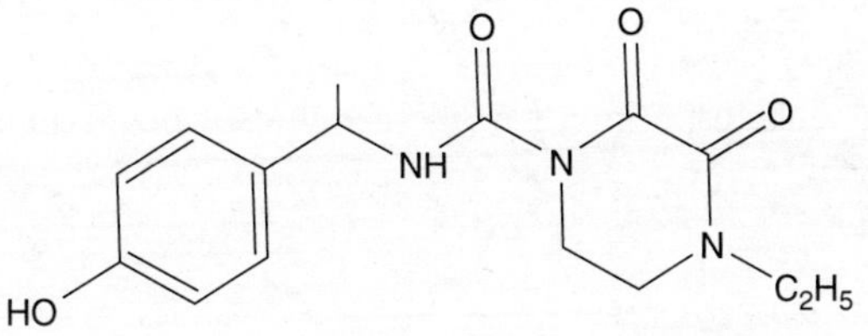

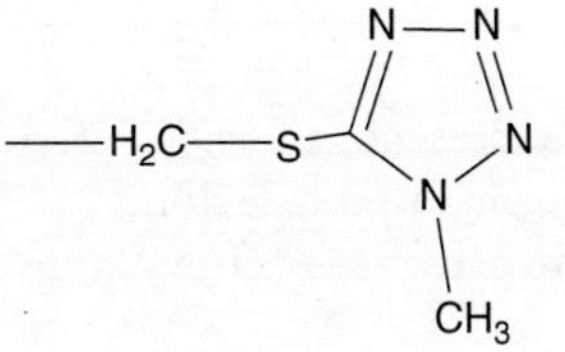

Degradation of cephalosporins Although cephalosporins are more stable to hydrolytic degradation reactions than penicillins, they experience a variety of chemical and enzymatic transformations, whose specific nature depends on the side-chain at C-7 and the substituent on C-3 atom.

The presence of a good leaving group at C-3 facilitates spontaneous expulsion of the 3'-substituent by concerted event due to hydrolysis of C-N bond of β-lactam nucleus by any general nucleophile or β-lactamase. Thus, desacetyl cefotoxime is more stable to hydrolysis in comparison to cefotoxime.

The absence of a leaving group at position 3 of cephalosporins makes them more acid-stable, thus rendering them suitable for oral consumption. Hence, cephalexin possessing methyl group at position 3 is much better absorbed than cephaloglycin having acetoxymethyl group at position 3, while both have identical phenylglycyl side-chain at the position 7. The nature of substituents at C-7 of cephalosporins plays an important role in determining the facility with which the reactive β-lactam bond is hydrolysed or broken either by chemical or enzymatic means of degradation of cephalosporin C.

1. In strong acid solution:

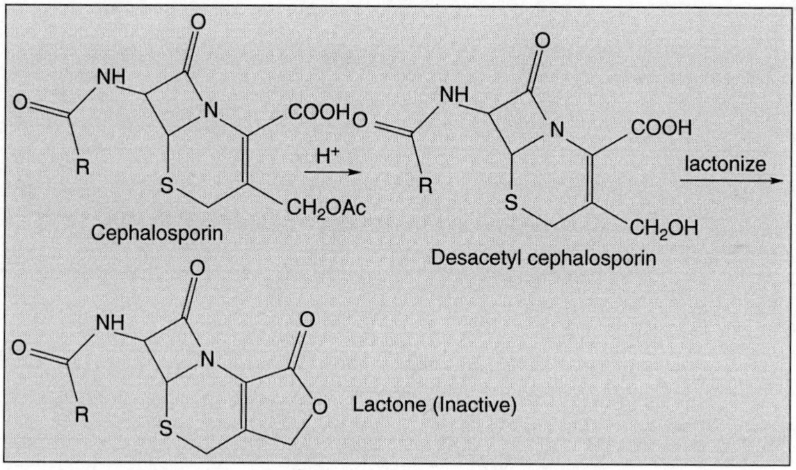

In the presence of esterase/acid, cephalosporin-C gave desacetyl cephalosporin and then inactive desacetyl cephalosporin lactone.

2. In the presence of β-lactamase:

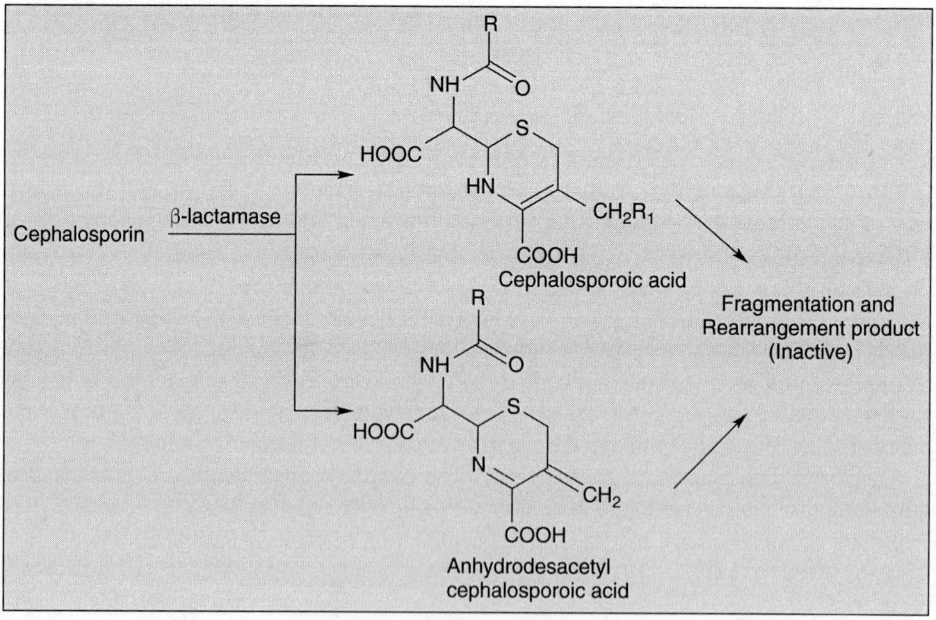

The enzyme β-lactamase or cephalosporanase degraded cephalosporin C into cephalosporoic acid, anhydrodesacetyl cephalosporoic acid, and desacetyl cephalosporoic acid. Further breakdown of these acidic products leads to many other fragmented and rearranged products.

3. In the presence of acylase

On enzymatic degradation in the presence of acylase, cephalosporin C gives 7-amino cephalosporanic acid, which in the presence of acid undergoes lactonization to give inactive des- acetyl-7-amino cephalosporanic acid lactone.

SAR of cephalosporins

1. **7-Acylamino substituents:**
   (i) Acylation of amino group generally increases the potency against gram-positive bacteria, but it is accompanied by a decrease in gram-negative potency.
   (ii) High antibacterial activity is observed only when the new acyl groups are derived from carboxylic acids for gram-positive bacteria.
   (iii) Substituents on the aromatic ring that increases lipophilicity provide higher gram-positive activity and generally lower gram-negative activity.
   (iv) The phenyl ring in the side-chain can be replaced with other heterocycles with improved spectrum of activity and pharmacokinetic properties, and these include thiophene, tetrazole, furan, pyridine, and aminothiazoles.

2. **C-3 substituents:**
The nature of C-3 substituents influences pharmacokinetic and pharmacological properties as well as antibacterial activity. Modification at C-3 position has been made to reduce the degradation (lactone of desacetyl cephalosporin) of cephalosporins.

   (i) The benzoyl ester displaces improved gram-positive activity but lower gram-negative activity.
   (ii) Pyridine and imidazole-replaced acetoxy groups show improved activity against *P. aeruginosa*. Displacement of acetoxy group by azide ion yields derivatives with relatively low gram-negative activity.
   (iii) Displacement with aromatic thiols of 3-acetoxy group results in an enhancement of activity against gram-negative bacteria with improved pharmacokinetic properties.
   (iv) Replacement of acetoxy group at C-3 position with —CH_3, Cl has resulted in orally active compounds.

Chapter 26: Antibiotics

3. **Cephamycins**: Introduction of C-7 α-methoxy group shows higher resistance to hydrolysis by β-lactamases.
4. Oxidation of ring sulphur to sulphoxide or sulphone greatly diminishes or destroys the antibacterial activity.
5. Replacement of sulphur with oxygen leads to oxacepam (latamoxef) with increased antibacterial activity, because of its enhanced acylating power. Similarly, replacement of sulphur with methylene group (loracarbef) has greater chemical stability and a longer half-life.
6. The carboxyl group of position-4 has been converted into ester prodrugs to increase bioavailability of cephalosporins, and these can be given orally as well. Examples include cefuroxime axetil and cefodoxime proxetil.
7. Olefinic linkage at C 3-4 is essential for antibacterial activity. Isomerization of the double bond to 2-3 position leads to great losses in antibacterial activity.

Synthesis of 7-aminocephalosporonic acid from cephalosporin C

Cephalosporin C is isolated on an industrial scale by fermentation using *Cephalosporium acremonium*.

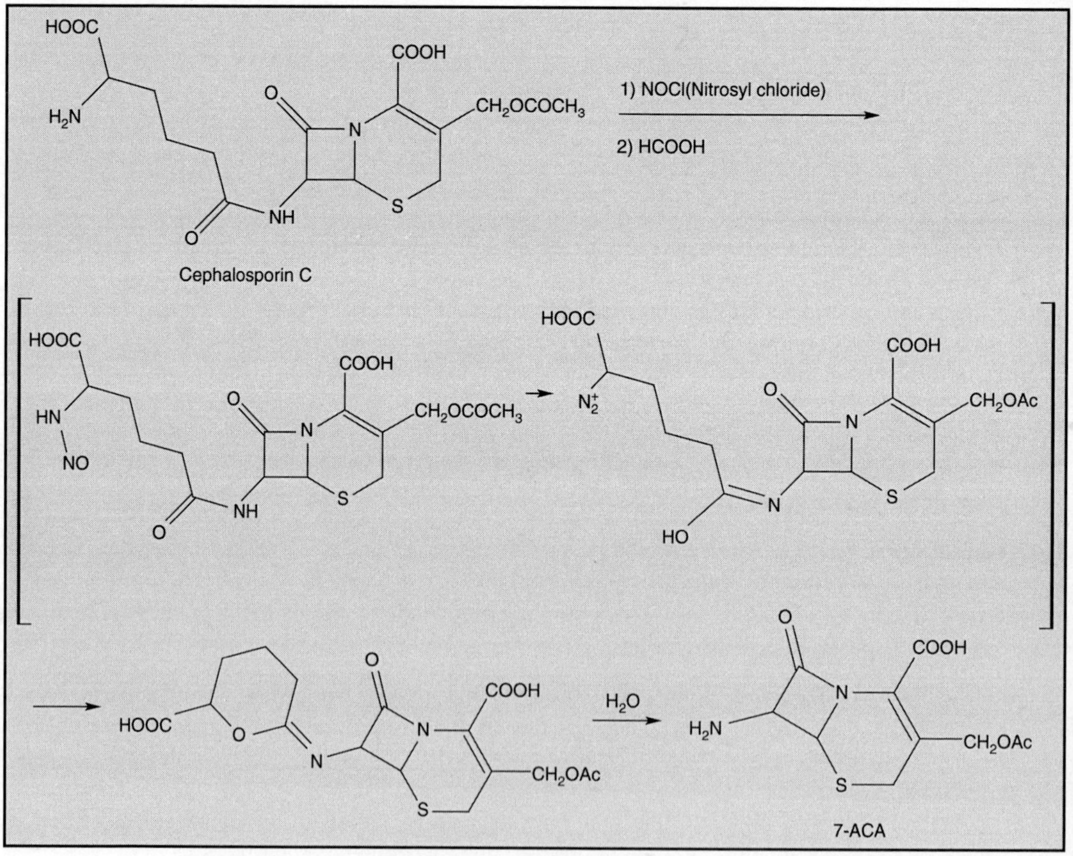

The key reaction, based on a method for removing glutamate residue in cephalosporin C, involves conversion of the primary amine in the molecule to a diazo function by reaction with nitrosyl chloride and formic acid. The diazo function can be displaced by oxygen from the enol form of amide at the 7-position to form iminolactone. Hydrolysis of the imine leads to 7-ACA.

Cephalexin [6R-[6α,7β(R)]]-3-Methyl-8-oxo-7-[(aminophenylacetyl)amino]-5-thia-1-azabicyclo[4.2.0]oct-2-en-2-carboxylic acid

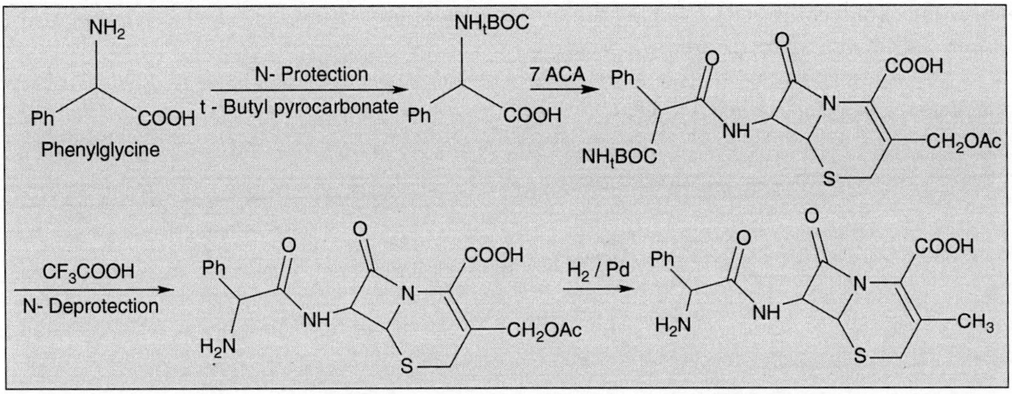

Amino group of phenylglycine is protected with t-BOC, and activated carboxylic acid reacts with 7-ACA to give amide derivative. t-BOC is deprotected by treatment with trifluoro acetic acid. Reducing this product with hydrogen using a palladium on barium sulphate catalyst results in the deacetoxylation at the third position of 7-aminocephalosporanic acid, making the desired cephalexin.

Cefadroxil [6R-[6α,7β(R)]]-3-Methyl-8-oxo-7-[[amino(4-hydroxyphenyl)acetyl]amino]-5-thia-1-azabicyclo[4.2.0]oct-2-en-2-carboxylic acid

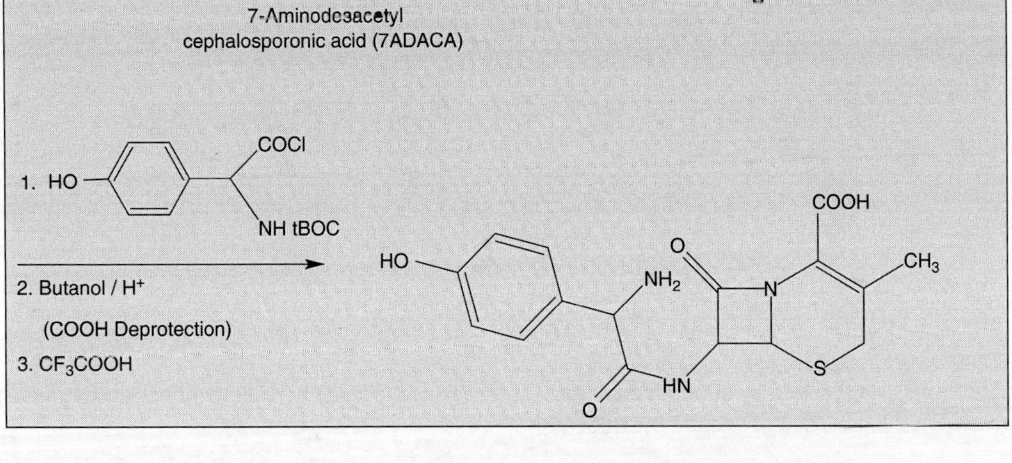

7-ACA is deacetylated by hydrogenolysis; carboxylate is protected with silyl derivative; and acylation with t-BOC-protected tyrosine and deprotection afford cefadroxil.

It is an analogue of cephalexin and differs only in the presence of a hydroxyl group in the fourth position of the phenyl ring of phenylglycine.

Cephradine [6R-[6α,7β(R)]]-3-Methyl-8-oxo-7-[(amino-1,4-cyclohexadien-1-ylacetyl)amino]-5-thia-1-azabicyclo[4.2.0]oct-2-en-2-carboxylic acid

It is a close analogue of cephalexin and differs in that the phenyl group in phenylglycine is partially hydrated to a 1,4-cyclohexadienyl moiety.

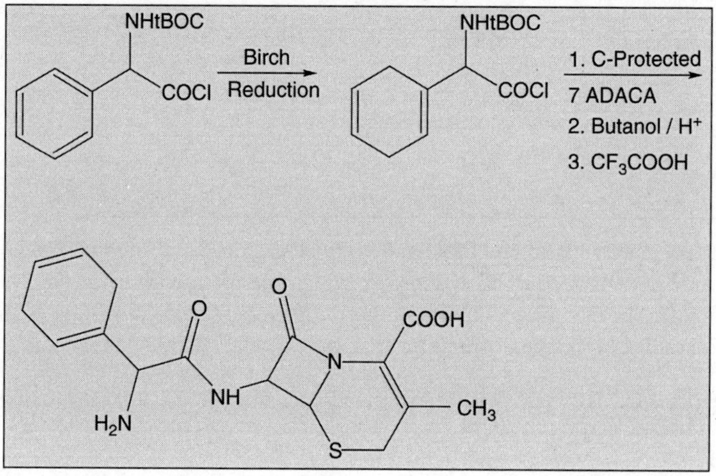

It is synthesised from phenylglycine, which is partially reduced by lithium in liquid ammonia, which forms 1,4-cyclohexadienylglycine, and the amino group in this compound is protected by reacting it with t-BOC. The resulting compound, on reaction with deacetoxylated 7-aminocephalosporanic acid, followed by deprotection gives cephradine.

Cephalothin (6R-trans)-3-[(Acetyloxy)methyl]-8-oxo-7-[(2-thienylacetyl)amino]-5-thia-1-azabicyclo[4.2.0]oct-2-en-2-carboxylic acid

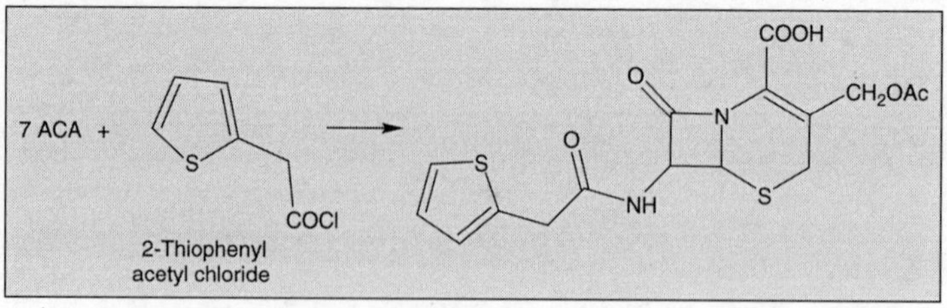

Cephalothin is synthesised by direct interaction of 2-thienylacetic acid chloride with 7-aminocephalosporanic acid in the presence of sodium bicarbonate.

Cefaclor (6R,7R)-7-[(R)-2-amino-2-phenylacetamido]-3-chloro-8-oxo-5-thia-1-azabicyclo[4.2.0]oct-2-en-2-carboxylic acid

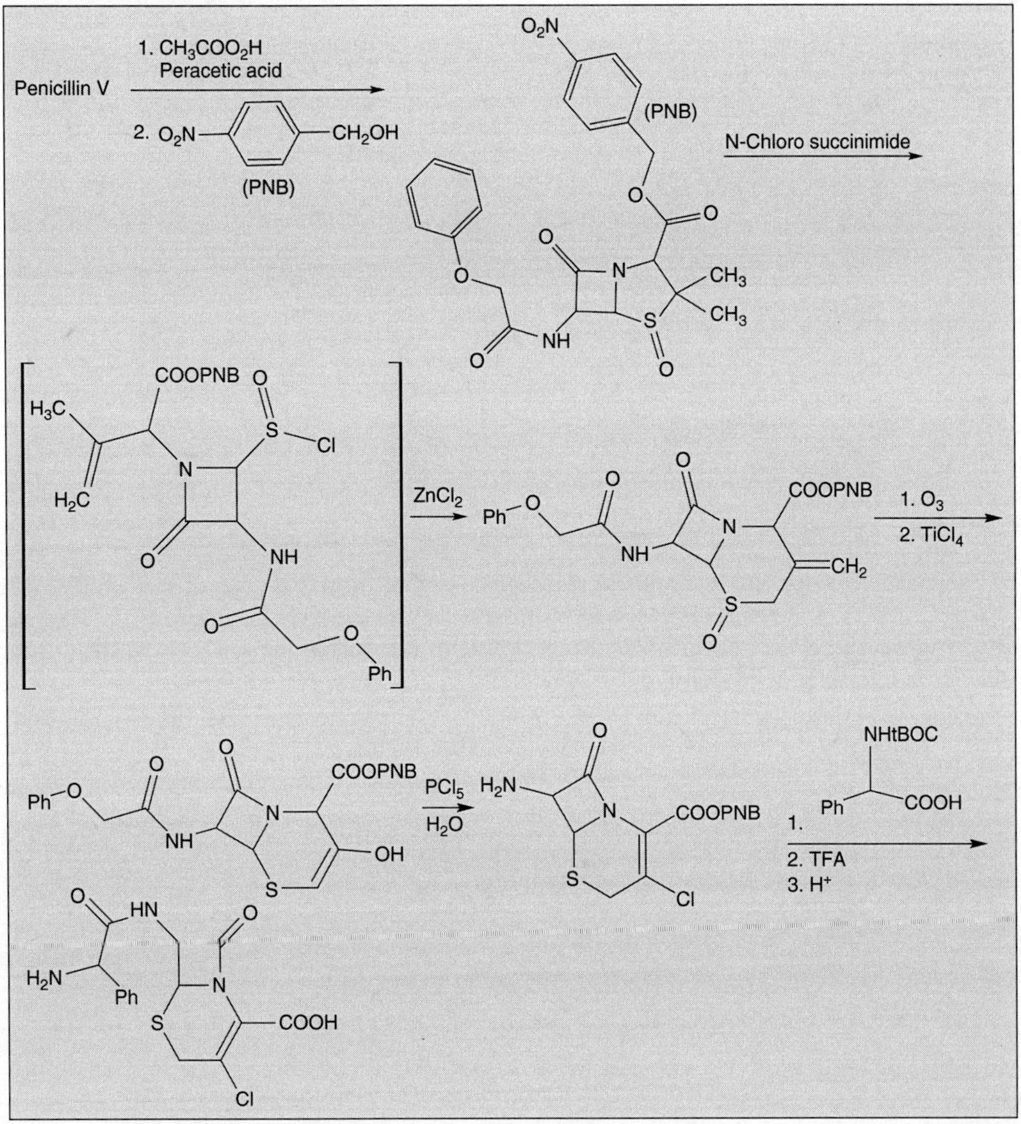

Penicillin V on treatment with peracetic acid gives sulphoxide, and carboxyl group is protected as 4-nitrobenzyl (PNB) ester. Reaction of this compound with a chlorinating agent such as N-chlorosuccinimide in the presence of acid results in chlorination on sulphur to form chlorosulphonium chloride; this compound then ring opens with the loss of hydrogen chloride to the unsaturated sulphonyl chloride. Treatment of this with Lewis acid leads to Friedel-Crafts-like attack on the olefin to afford a 6-membered cyclized product; the double bond concomitantly shifts to the exocyclic position. Ozonization followed by cleavage of the ozonide affords ring ketone; this exists as conjugated enol; sulphoxide is then reduced to cephem.

Reaction of this with phosphorous pentachloride converts the enol to chloride, which on hydrolysis gives 7-amino derivative. This is converted to cefaclor by reaction with phenylglycine by the standard protocol mentioned above.

Cefamandole 7-D-Mandelamido-3-[[(1-methyl-1*H*-tetrazol-5-yl)thio]methyl]-8-oxo-5-thia-1-azabicyclo[4.2.0]oct-2-en-2-carboxylic acid

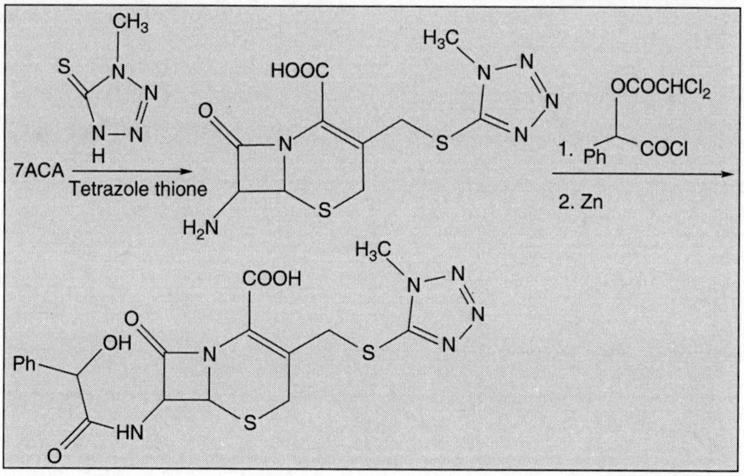

Cefamandole is synthesised from 7-aminocephalosporanic acid. Reaction of this with 1-methyl-1,2,3,4-tetrazol-5-thiol results in displacement of the acetoxy group to form the intermediate. This is then acylated with acid chloride from the dichloroacetyl ester of D-mandellic acid. The protecting group on the side-chain is then removed to afford cefamandole.

Cefuroxime (Z)-mono(O-methyloxim) (6R,7R)-7-[2-(2-furyl)glyoxylamido]-3-(hydroxymethyl)-8-oxo-5-thia-1-azabicyclo[4.2.0]oct-2-en-2-carboxylic acid carbamate

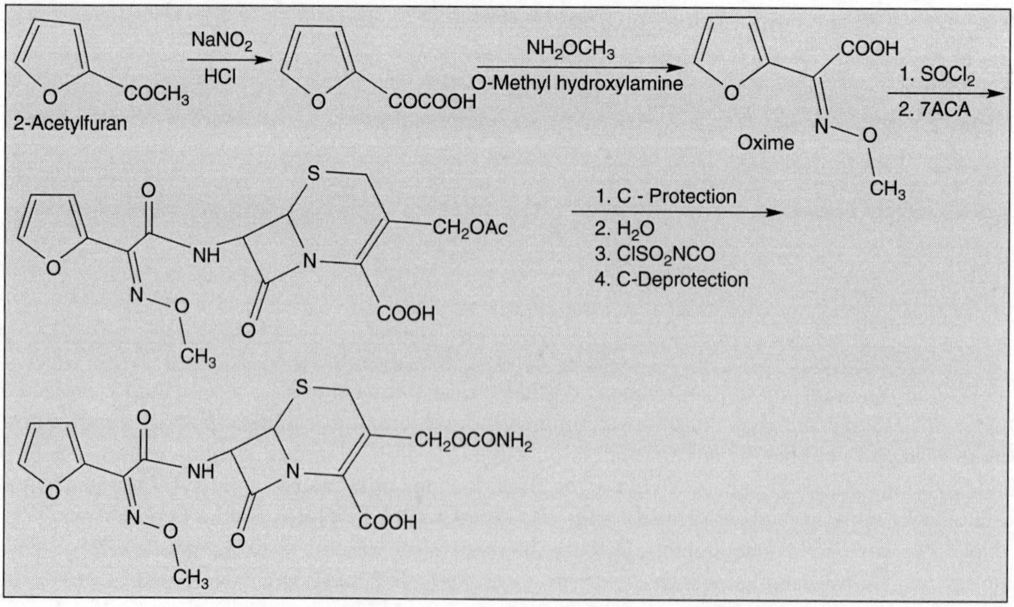

It is synthesised from 2-acetylfuran. Oxidizing this compound with nitrous acid gives 2-furylglyoxalic acid, which is reacted with methoxylamine to give the corresponding oxime, *syn-2*-methoxyamino-2-(2-furyl)acetic acid. Acid is converted to acid chloride by reaction with thionyl chloride, which on reaction with 7-ACA gives amide. Carboxyl group is protected (with benzhydryl) and enzymatic hydrolysis in an alkaline medium, in which the benzhydryl protection is not affected, and only the acetoxy group of the molecule at position C3 of the aminocephalosporanic acid is hydrolysed. The resulting product with a free hydroxymethyl group is reacted with chlorosulphonyl isocyanate, with intermediate formation of the corresponding N-chlorosulphonyl urethane, which is hydrolysed by water to the urethane. Finally, removal of the benzhydryl protection using trifluoroacetic acid gives the desired cefuroxime.

Cefotaxime α-*O*-Methyloxime acetate (6*R*, 7*R*)-7-[2-(2-amino-4-thiazolyl)-glyoxylamido]-3-(hydroxymethyl)-8-oxo-5-thia-1-azabicyclo[4.2.0]oct-2-en-2-carboxylic acid

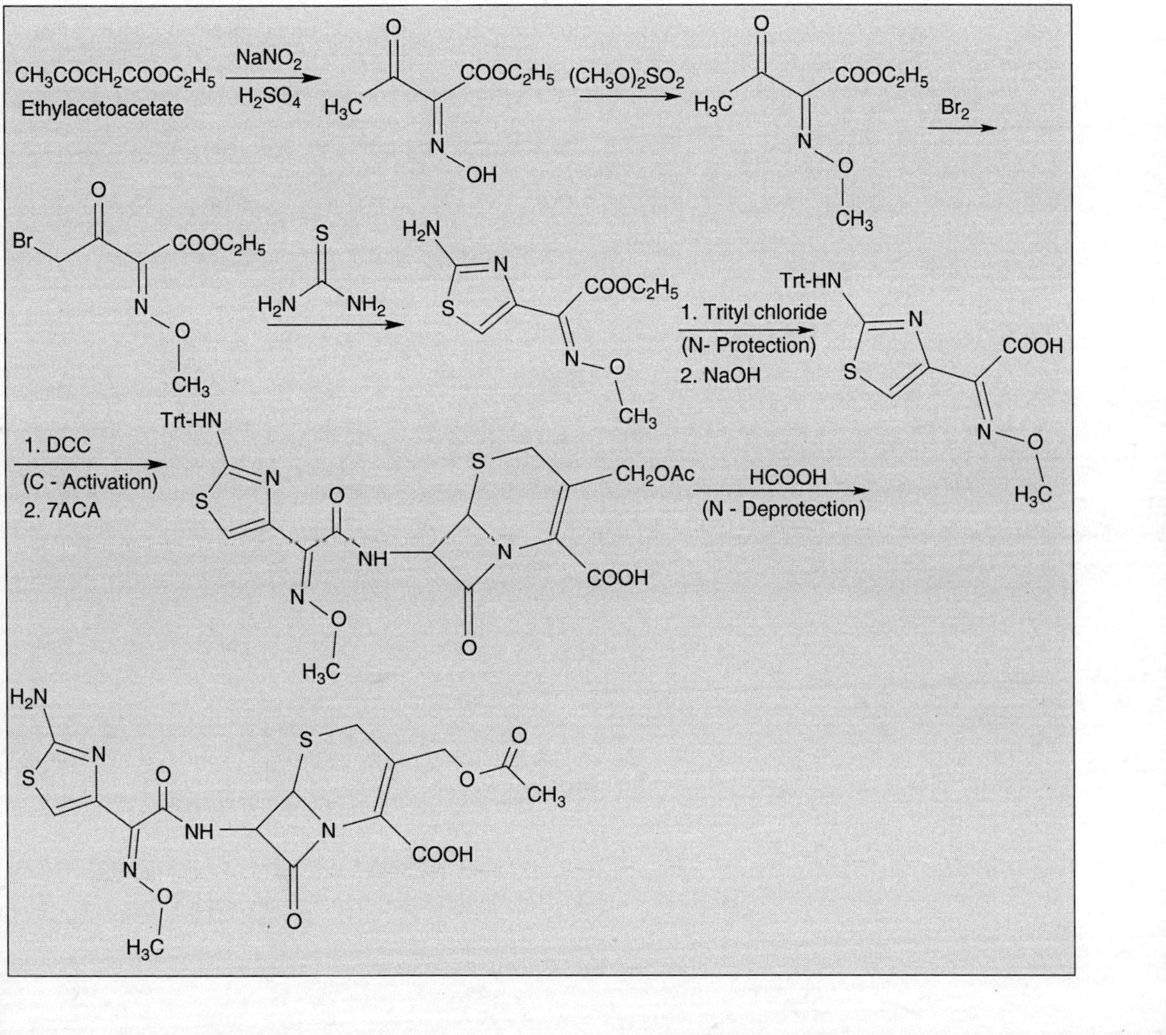

Cefotaxime is synthesised by acylating of 7-aminocephalosporanic acid with 2-(2-amino-4-thiazolyl)-2-methoxyiminoacetic acid, which is protected at the amino group by a trityl protection. After removing the trityl protection from the resulting product with dilute formic acid, the desired cefotaxime is formed. The ethyl ester of 2-(2-amino-4-thiazolyl)-2-methoxyiminoacetic acid necessary for this synthesis, as well as for the synthesis of a number of other antibiotics of the cephalosporin series, is synthesised from acetoacetic ester. Nitrosation of acetoacetic ester with nitrous acid gives isonitrosoacetoacetic ester. O-Methylation of the hydroxyl group of obtained product with dimethylsulphate in the presence of potassium carbonate gives ethyl 2-(methoxyimino) acetoacetate. Brominating the resulting product with bromine in methylene chloride in the presence of 4-toluenesulphonic acid gives 4-bromo-2-methoxyiminoacetoacetic ester. Reacting this with thiourea according to the classic scheme of preparing of thiazoles from α-bromocarbonyl compounds and thioamides gives the ethyl ester of 2-(2-amino-4-thiazolyl)-2-methoxyiminoacetic acid. Reacting this with triphenylchloromethane in the presence of triethylamine results in a trityl protection of the amino group, forming the ethyl ester of 2-(2-tritylamino-4-thiazolyl)-2-methoxyiminoacetic acid, which is hydrolysed to the acid using sodium hydroxide. The resulting acid, as was already stated, is used for acylating of 7-aminocephalosporanide acid in the presence of dicyclohexylcarbodiimide, giving tritylated cefotaxime, α-O-methyloxime acetate 7-[2-(2-tritylamino)-4-thiazolyl-glycoxylamido]-3-(hydroxymethyl)-8-oxo-5-thia-1-azabicyclo[4.2.0]oct-2-en-2-carboxylic acid. Finally, removing the trityl protection from the synthesised product using dilute formic acid gives cefotaxime.

Ceftizoxime α-O-Methyloxime of (6R,7R)-7-[2-(2-amino-4-thiazolyl)glyoxylamido]-8-oxo-5-thia-1-azabicyclo[4.2.0]oct-2-en-2-carboxylic acid

Acylation of 6-ACA derivative with thiazole derivative in presence of DCC, followed by deprotection of trityl group, affords ceftizoxime.

Ceftazidime 1-[[7-[[(2-Amino-4-thiazolyl)[(1-carboxy-1-methylethoxy)imino]acetyl]amino]-2-carboxy-8-oxo-5-thia-1-azabicyclo[4.2.0]oct-2-en-3-yl]methyl]pyridin-2-carboxylic acid

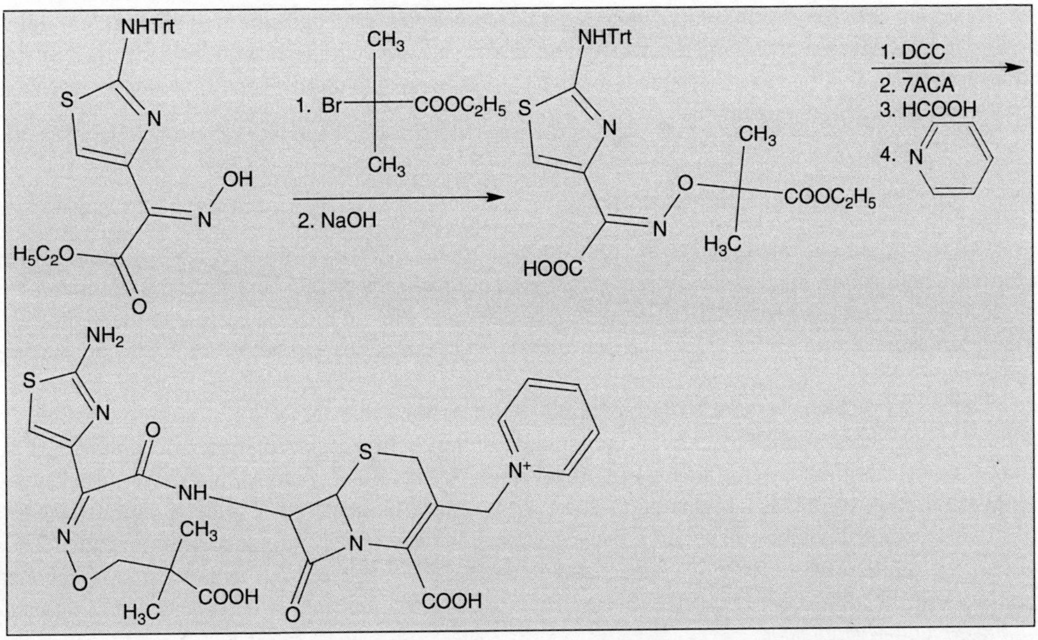

Reaction between ethyl 2-[2-(tritylamino)-1,3-thiazol-4-yl]-2-hydroxyiminoacetate and ethyl 2-bromo-2-methylpropanoate gives O-alkylated compound, and ethoxycarbonyl group in this molecule is hydrolysed by sodium hydroxide to acid. The acid is activated with DCC and amide formed by reaction with 7-ACA. Formic acid treatment removes the trityl group from amino group, and is followed by treatment with pyridine to replace the acetoxyl group with a pyridine group to afford ceftazidime.

Cefoperazone (6R,7R)-7-[(R)-2-(4-Ethyl-2,3-dioxo-1-piperazincarboxamido)-2-(p-hydroxyphenyl)acetamido]-3-[[(1-methyl-1H-tetrazol-5-yl)thio]methyl]-8-oxo-5-thia-1-azabicyclo[4.2.0]oct-2-en-2-carboxylic acid

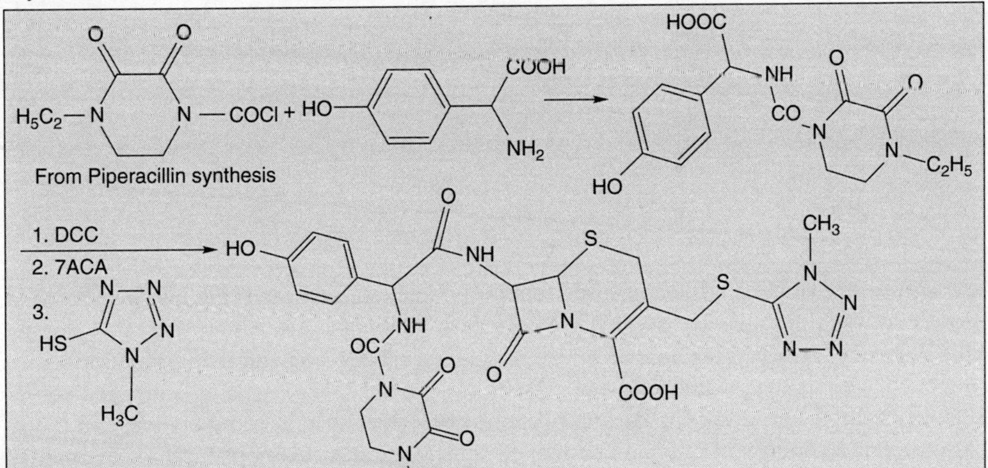

Reaction between 4-ethylpiperazin-2,3-dion-1-carboxylic acid chloride and the sodium salt of 4-hydroxyphenylglycine gives acid derivative. The acid is activated with DCC and amide formed by reaction with 7-ACA; this on reaction with 1-methyl-1,2,3,4-tetrazol-5-thiol results in displacement of the acetoxy group to form cefoperazone.

26.2 TETRACYCLINE ANTIBIOTICS

Tetracyclines are potent, broad-spectrum antibacterial agents effective against a host of gram-positive and gram-negative aerobic and anaerobic bacteria. As a result, the tetracyclines are drugs of choice or well-accepted alternatives for a variety of infectious diseases. Among these, their role in the treatment of sexually transmitted and gonococcal diseases, urinary tract infections, bronchitis, and sinusitis remains prominent.

The majority of the marketed tetracyclines (tetracycline, chlortetracycline, oxytetracycline, and demeclocycline) are naturally occurring compounds obtained from fermentation of *Streptomyces* spp broths. The semi-synthetic tetracyclines (methacycline, doxycycline, minocycline) have an advantage of longer duration of antibacterial action. However, all these tetracyclines exhibit a similar profile in terms of antibacterial potency. In general, the activity encompasses many strains of gram-negative *E.coli*, *Proteus*, *Klebsiella*, *Enterobacter*, *Niesseria*, and *Serratia spp.* as well as gram-negative *Streptococci* and *Staphylococci*. Of particular interest is the potency of tetracyclines against *Haemophilus*, *Legionella*, *Chlamydia*, and *Mycoplasma*.

Mechanism of action: The tetracyclines exhibit bacteriostatic effects on growing bacteria via the inhibition of protein synthesis. Their action occurs at the ribosomal level, where the drug binding to the 30S ribosomal subunit takes place on the ribosome-mRNA complex. This phenomenon stops the attachment of amino acylated t-RNA molecules and prevents peptide-chain growth.

26.2.1 Structural Features and SAR of Tetracyclines

The key structural feature is a linearly fused tetracyclic nucleus, and each ring needs to be six-membered and purely carbocyclic. The D-ring needs to be aromatic and the A-ring must be appropriately substituted at each of its carbon atoms for notable activity. The B-ring and the C-ring tolerate certain substituent changes as long as the keto-enol system (at C-11, 12, 12a) remains intact and conjugated to the phenolic D-ring. The D-, C-, B- ring phenol-, keto-, enol system is imperative, and the A-ring must also contain a conjugated keto-enol system. Specifically, the A-ring contains a tricarbonyl-derived keto-enol array at positions C-1, 2, and 3. Other structural requirements for good antibacterial activity include a basic amine function at the C-4 position of the A-ring.

C-1 substituents: The keto-enol system of the A-ring is indispensable for antibacterial activity. No variation at the C-1 position has been successful.

C-2 substituents: The carboxamide moiety is present in all naturally occurring tetracyclines and this group is crucial for antibacterial activity. The amide is best left unsubstituted, or mono-substitution is acceptable in the form of activated alkylaminomethyl amide (Mannich bases). An example is rolitetracycline. Large alkyl group on the carboxamide may alter the normal keto-enol equilibrium of the C-1, 2, and 3 conjugated system, and diminishes inherent antibacterial activity. The replacement of carboxamide group or dehydration of carboxamide to the corresponding nitrile results in the loss of activity.

C-3 substituents: In conjugation with the C-1 position, the keto-enol conjugated system is imperative for antibacterial activity.

C-4 substituents: The naturally occurring tetracyclines contain α-C-4 dimethylamino substituent that favourably contributes to the keto-enolic character of the A-ring. Replacement of dimethylamino group with a hydrazone, oxime, or hydroxyl group leads to a pronounced loss of activity, probably due to the increase in heteroatom basicity.

C-4a substituents: The α-hydrogen at C-4a position of tetracyclines is necessary for useful antibacterial activity.

C-5 substituents: Many naturally occurring antibacterial tetracyclines have an unsubstituted methylene moiety at the C-5 position. However, oxytetracycline contains C-5 α-hydroxyl group and was found to be a potent compound, and has been modified chemically to some semi-synthetic tetracyclines. Alkylation of the C-5 hydroxyl group results in a loss of activity.

Ester formation is only acceptable if the free oxytetracycline can be liberated *in vivo*; only small alkyl esters are useful.

C-5a substituents: The configuration of the naturally occurring tetracyclines places the C-5a hydrogen atom in an α-configuration. Epimerization is detrimental to antibacterial activity.

C-6 substituents: The C-6 position is tolerant of a variety of substituents. The majority of tetracyclines have an α-methyl group and a β-hydroxyl group at this position. Demeclocyclin is a naturally occurring C-6 demethylated chlortetracycline with an excellent activity. This C-6 methyl group contributes little to the activity of tetracycline. Similarly, the C-6 hydroxyl group also appears to offer little in terms of antibacterial activity; removal of this group affords doxycycline, which is a superb antibacterial.

C-7 and C-9 substituents: The nature of the aromatic D-ring predisposes the C-7 position to electrophilic substitution, and nitro and halogen groups have been introduced. Some C-7 nitro tetracyclines are among the most potent of all tetracyclines *in vitro*, but these compounds were potentially toxic/carcinogenic. Halogenated derivatives are less active. The C-7 acetoxy, azido, and hydroxyl tetracyclines are inferior in terms of antibacterial activity.

C-10 substituents: The C-10 phenolic moiety is absolutely necessary for antibacterial activity.

C-11 substituents: The C-11 carbonyl moiety is part of one of the conjugated keto-enol systems required for antibacterial activity.

C-11a substituents: In general, few modifications at the C-11a position of tetracycline have been tolerated. This is probably due to the detrimental effects exerted upon the keto-enol system, which is vital for magnesium cation binding and subsequent tetracycline uptake by the bacterial cell.

C-12 substituents: As with the C-11 position, the C-12 position is part of the keto-enol system vital for drug uptake, binding, and observed antibacterial activity.

C-12a substituents: The C-12a hydroxyl group is needed for antibacterial activity, although this moiety can be esterified to provide tetracycline with increased lipophilicity. Antibacterial properties are retained if the alkyl ester is small in size, and readily undergoes hydrolysis to liberate free tetracycline.

26.2.2 Structure of Tetracyclines

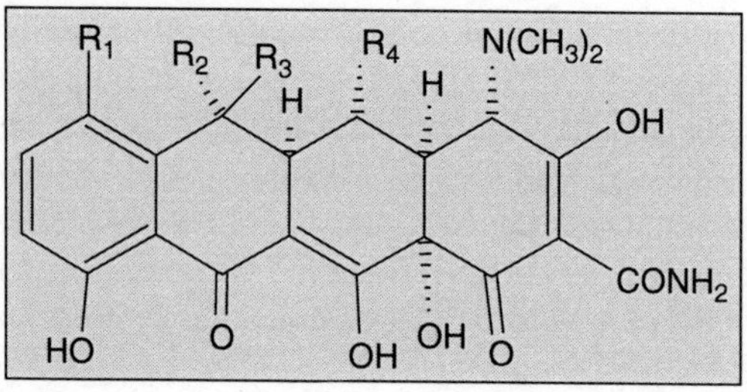

	R_1	R_2	R_3	R_4
Tetracycline	H	CH_3	OH	H
Chlortetracycline	Cl	CH_3	OH	H
Oxytetracycline	H	CH_3	OH	OH
Demeclocycline	Cl	H	OH	H
Methacycline	H	$=CH_2$	—	OH
Doxycycline	H	H	CH_3	OH
Minocycline	$-N(CH_3)_2$	H	H	H

Effect of pH on tetracyclines An interesting property of tetracyclines is their ability to undergo epimerization at C-4 in solutions of intermediate pH range. These isomers are called *epi*tetracyclines. Under the influence of the acidic conditions, an equilibrium is established in about one day consisting of approximately equal amount of isomers. Epitetracyclines exhibit much less activity than natural isomers.

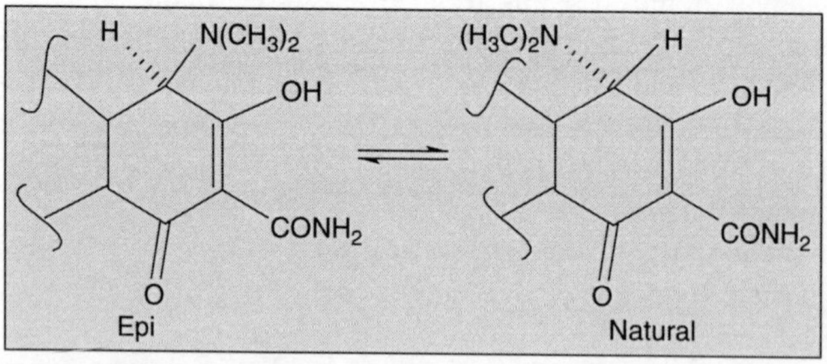

Strong acids and bases attack the tetracyclines having a hydroxyl group on C-6, causing a loss in activity through modification of C-ring. Strong acids produce dehydration through a reaction involving the C-6 hydroxyl group and C-5a hydrogen. The double bond formed between positions C-5a and C-6 induces a shift in the position of double bond between C-11a and C-12 to a position between C-11 and

C-11a, forming the more energetically favoured resonance of the naphthalene group found in the inactive anhydrotetracyclines.

[Chemical structures showing conversion of tetracycline to 5,6-Anhydrotetracycline (via H⁺) and to Isotetracycline (via OH⁻)]

Bases promote a reaction between the C-6 hydroxyl group and the ketone group at the C-11 position, causing the bond between the C-11 and C-11a atoms to cleave and to form the lactone ring found in the inactive isotetracycline.

> Tetracyclines can stain developing teeth (even when taken by the mother during pregnancy). Inactivated by Ca^{2+} ion, these are not to be taken with milk or yogurt. Inactivated by aluminium, iron, and zinc, these are not to be taken at the same time as indigestion remedies. These are inactivated by common antacids and over-the-counter heartburn medicines. These should be avoided during pregnancy as it may affect bone growth of foetus.

Effect of metals on tetracycline Stable chelate complexes are formed by tetracycline with many metals, including calcium, magnesium, and iron. Such chelates are insoluble in water, accounting for impairment in absorption of most tetracyclines in the presence of milk, calcium, magnesium, and aluminium-containing antacids and iron salts.

The affinity of tetracycline for calcium causes them to be laid down in newly formed bones and teeth as tetracycline-calcium orthophosphate complexes. Deposits of these antibiotics in teeth cause a yellow discolouration that darkens because of photochemical reaction. Tetracyclines are distributed into the milk of lactating mothers and also cross the placenta into the foetus. The possible effect of these agents on bones and teeth of the child should be taken into consideration before their use in pregnancy or in children under eight years of age.

Methacycline 4-Dimethylamino-1,4,4a,5,5a,6,11,12a-octahydro-3,6,10,12,12a-pentahydroxy-6-methylen-1,11-dioxo-2-naphthacencarboxamide

Synthesis

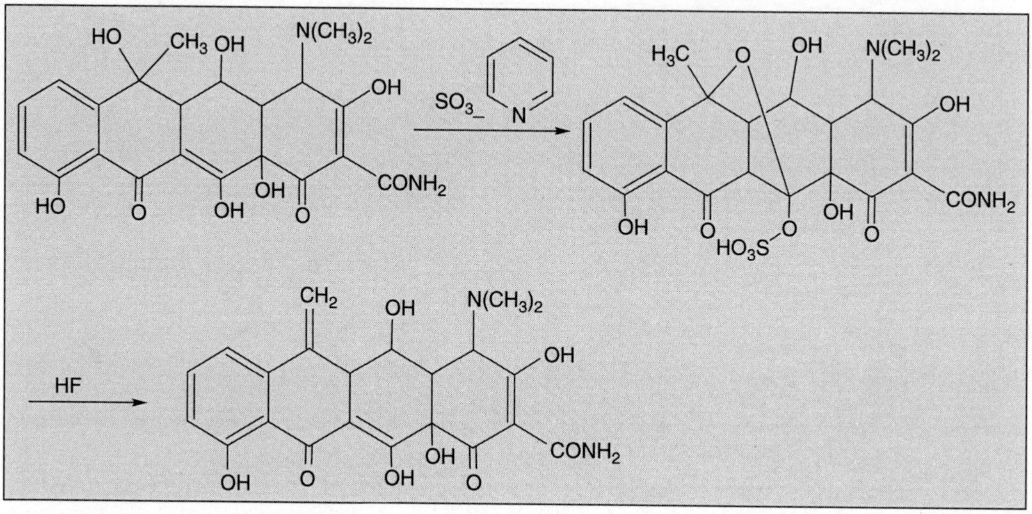

It is synthesised from oxytetracycline, which is reacted with a sulphur trioxide-pyridine complex, resulting in an oxidation reaction. Simultaneous sulphonation gives a naphthacen-sulphotetrahydrofuran derivative intermediate, which when reacted with hydrofluoric acid forms methacycline.

Doxycycline 4-Dimethylamino-1,4,4a,5,5a,6,11,12a-oxtahydro-3,5,10,12,12a-pentahydroxy-6-methyl-1,11-dioxo-2,naphthacencarboxamide

Doxycycline is an isomer of tetracycline that differs only in the placement of one hydroxyl group. Doxycycline can be formally viewed as the result of transferring the C6 hydroxyl group of tetracycline to C5.

Synthesis

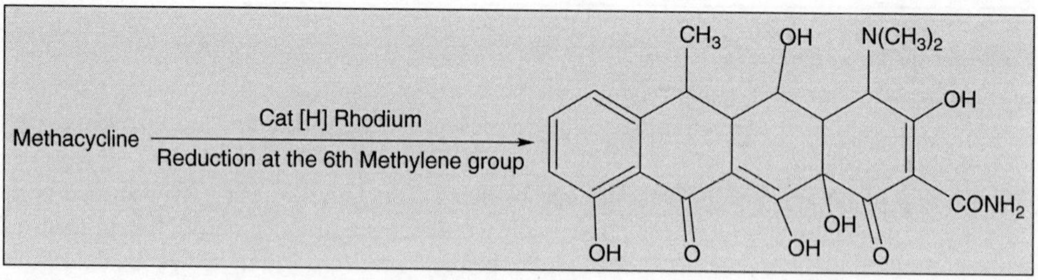

Doxycycline is synthesised in two different ways; one of the ways suggests dehydrating oxytetracycline at C6 by reducing the tertiary hydroxyl group with hydrogen using a rhodium on carbon catalyst. It is also synthesised by reducing methylene group of methacycline using rhodium on carbon catalyst.

Minocycline 4,7-*bis*(Dimethylamino)-1,4,4a,5,5a,6,11,12 α-octahydro-3,10,12,12a-tetrahydroxy-1,11-dioxo-2-naphthacencarboxamide

Synthesis

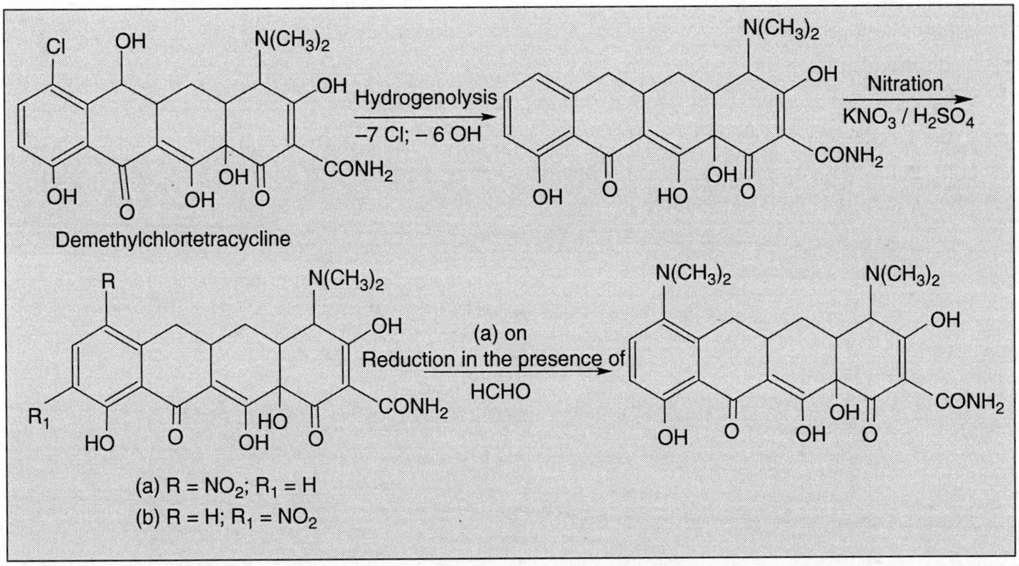

Hydrogenolysis of demethylchlortetracycline removes both the chlorine and the benzylic hydroxyl group. The product, 6-demethyl-6-deoxytetracycline, on nitration gives mixture of 7 (a)- and 9 (b)-nitro-6-demethyl-6-deoxytetracycline; (a) is reductively methylated by catalytic hydrogenation in the presence of formaldehyde to give minocycline.

Rolitetracycline (2Z,4S,4aS,5aS,6S,12aS)-4-Dimethylamino-6,10,11,12a-tetrahydroxy-2-[hydroxy-(pyrrolidin-1-ylmethylamino)methylidene]-6-methyl-4,4a,5,5a-tetrahydrotetracene-1,3,12-trione

It is a water-soluble prodrug of tetracycline.

Synthesis

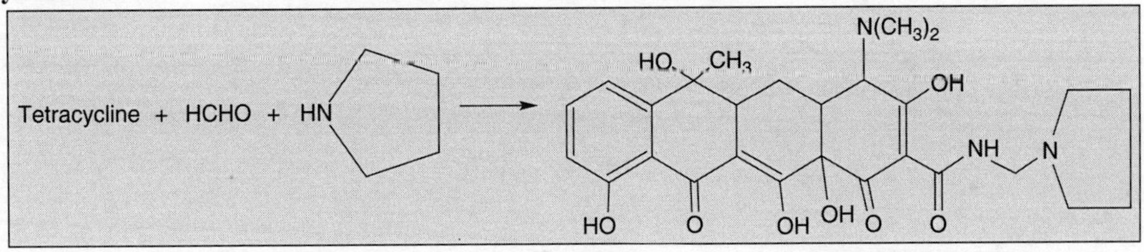

It is prepared by reacting tetracycline, formaldehyde, and pyrrolidine; amino methylation takes place in amide group (Mannich reaction).

26.3 MACROLIDE ANTIBIOTICS

The macrolide antibacterial agents are extremely useful chemotherapeutic agents for the treatment of a variety of infectious disorders and diseases caused by a host of gram-positive bacterial pathogens. These agents as exemplified by erythromycins are generally effective against *Streptococci*, *Staphylococci*,

Chlamydia, Legionella, and *Mycoplasma*. As a result, the macrolides are commonly administered for respiratory, skin and tissue, and genitourinary infections caused by these pathogens.

Chemistry The macrolide antibiotics have three common chemical characteristics:

1. A large lactone ring
2. A ketone group
3. A glycosidically linked amino sugar

Usually, the lactone ring has 12, 14, or 16 atoms in it and is often partially unsaturated, with an olefinic group conjugated with a ketone function. They may have, in addition to the amino sugar, a neutral sugar that is glycosidically linked to the lactone ring.

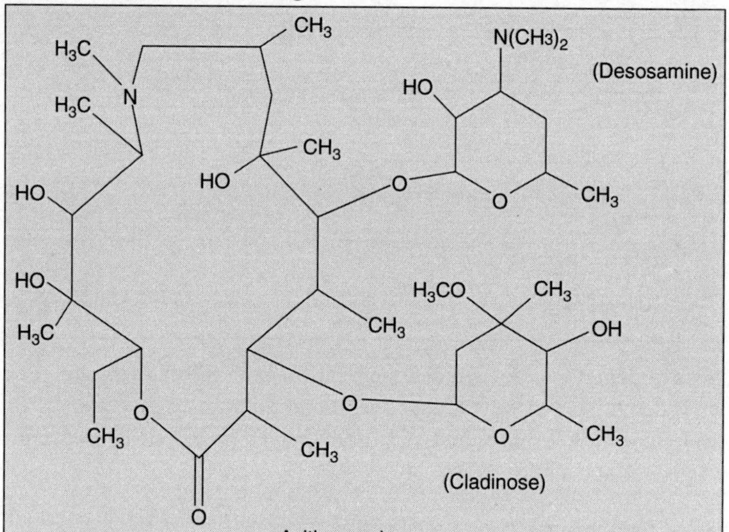

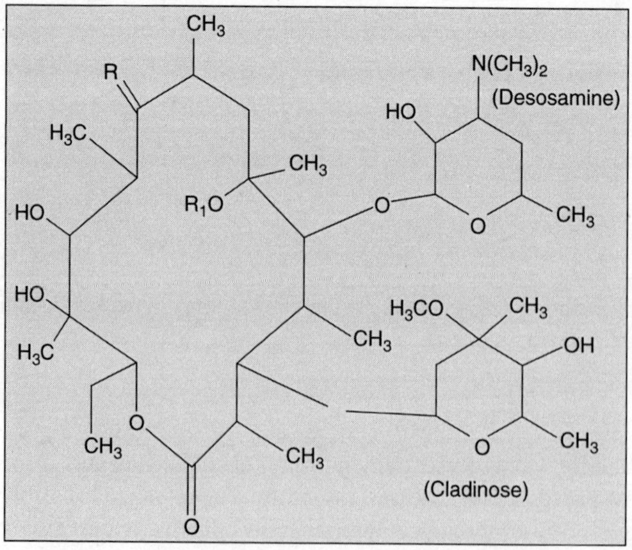

	R	*R*$_1$
Erythromycin	O	H
Roxithromycin	$CH_3OCH_2CH_2OCH_2ON=$	H
Clarithromycin	O	CH_3

Mecahnism of action The macrolides exert their antibacterial effects, which are usually bacteriostatic, via inhibition of bacterial protein biosynthesis. Specifically, the macrolides target the 50S ribosomal subunit; different members stop protein synthesis at varying stages of peptide-chain elongation. The macrolides inhibit ribosomal peptidyl transferase activity. Some macrolides also inhibit the translocation of the ribosome along the m-RNA template.

Acid degradation of erythromycin [(3*R*,4*S*,5*S*,6*R*,7*R*,9*R*,11*R*,12*R*,13*S*,14*R*)-4-[(2,6-dideoxy-3-*C*methyl-3-*O*-methyl-α-L-*ribo*-hexopyranosyl)-oxy]-14-ethyl-7,12,13-trihydroxy-3-,5,7,9,11,13-hexamethyl-6-[[3,4,6-trideoxy-3-(dimethylamino)-β-D-*xylo*-hexopyranosyl]oxy] oxacyclotetradecan-2,10-dione]:

Erythromycin
(Active)
(Desosamine)
(Cladinose)
H⁺
Acid
Desosamine
Cladinose
Hemiketal
(Inactive)
–H₂O
H⁺
Spiroketal
(Inactive)
Spiroketal + Desosamine + Cladinose

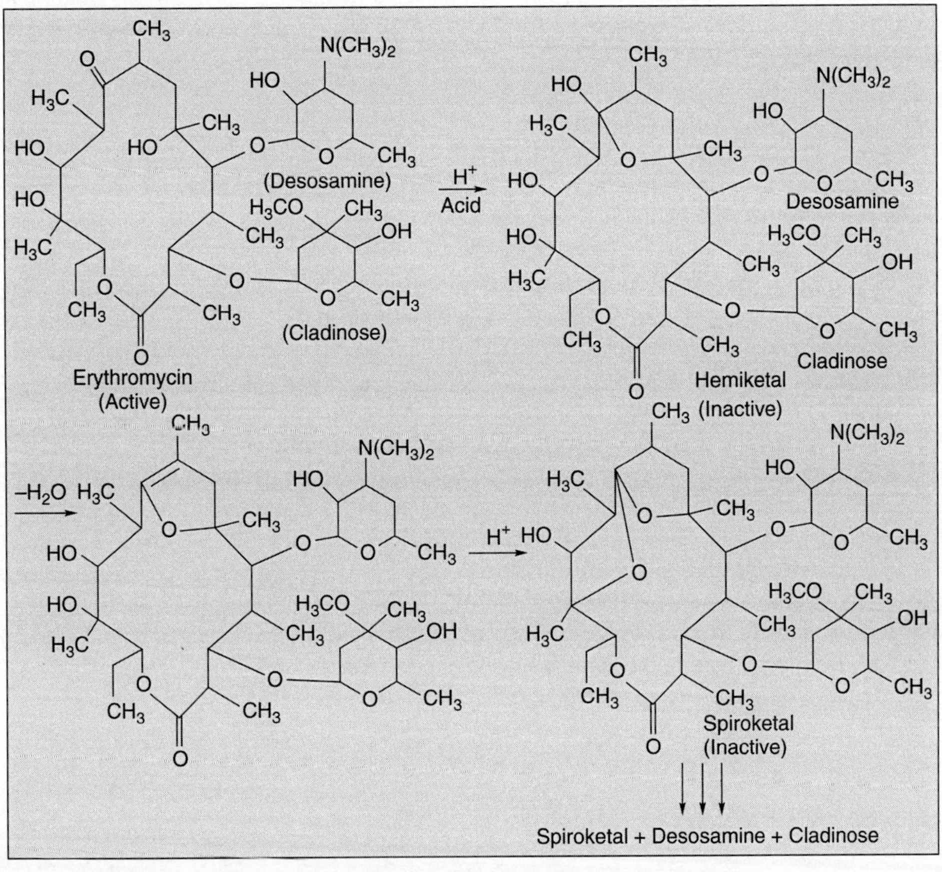

Erythromycin is unstable in acid media. The C-6 hydroxyl group reversibly attacks the C-9 ketone, giving rise to a hemiketal intermediate. Dehydration prevents regeneration of the parent erythromycin, and the C-12 hydroxyl group can subsequently add to produce a spiroketal species. The cladinose group is cleaved from the macrocycle and more harsh conditions lead to the release of desosamine. Useful antibacterial activity lasts upon dehydration of the hemiketal, and the spiroketal is also weakly active.

26.3.1 SAR of Macrolide Antibiotics

1. A number of strategies have been utilized to improve the acid stability of erythromycin.
   (i) The first approach involved the addition of hydroxylamine to the ketone to form oxime—e.g., roxithromycin.
   (ii) The second approach involves an alteration of C-6 hydroxyl group, which is the nucleophilic functionality that initiates erythromycin degradation. Modification that removes the nucleophilic nature of this hydroxyl group can retain antibacterial properties if the size of the group is kept small so as not to affect the ribosomal binding—e.g., clarithromycin.

2. The azalides (e.g., azithromycin) are semi-synthetic 15-membered congeners in which a nitrogen atom has been introduced to expand a 14-membered precursor, and this leads to an extended spectrum of action.

Synthesis
Roxithromycin (3R,4S,5S,6R,7R,9R,11S,12R,13S,14R)-6-[(2S,3R,4S,6R)-4-dimethylamino-3-hydroxy-6-methyloxan-2-yl]oxy-14-ethyl-7,12,13-trihydroxy-4-[(2R,4R,5S,6S)-5-hydroxy-4-methoxy-4,6-dimethyloxan-2-yl]oxy-10-(2-methoxyethoxymethoxyimino)-3,5,7,9,11,13-hexamethyl-1-oxacyclotetradecan-2-one

Roxithromycin is derived from erythromycin, containing the same 14-membered lactone ring. However, an N-oxime side-chain is attached to the lactone ring.

It is prepared by reacting erythromycin with $CH_3OCH_2CH_2OCH_2ONH_2$ (substituted hydroxylamine).
Clarithromycin (2R,3S,4S, 5R,6R,8R,10R,11R,12S,13R)-3-(2,6-dideoxy-3-C-3-O-dimethyl-α-L-*ribo*-hexopyranosyloxy)-6-methoxy-9-oxo-11,12-dihydroxy-2,4,6,8,10,12-hexamethyl-5-(3,4,6-trideoxy-3-dimethylamino-β-D-*xylo*-hexopyranosyloxy)cyclopentadecan-13-olide.

It is a semi-synthetic analogue of erythromycin A, in which the hydroxyl group at C6 is replaced with a methoxyl group. It is prepared by selective methylation of 6-OH group of erythromycin.
Azithromycin [2R-(2R,3S,4R, 5R, 8R,10R,11R,12S,13S,14R)-3-(2,6-dideoxy-3-C-methyl-3-O-methyl-α-L-*ribo*-hexopyranosyloxy)-2-ethyl-3,4,10-trihydroxy-3,5,6,8,10,12,14-heptamethyl-11-[(3,4,6-trideoxy-3-dimethylamino)-β-D-*xylo*-hexopyranosyl]-oxy]-1-oxa-6-azacyclopentadecan-15-one.

It differs chemically from erythromycin in that a methyl-substituted nitrogen atom is incorporated into the lactone ring, thus making the lactone ring 15-membered.

Synthesis

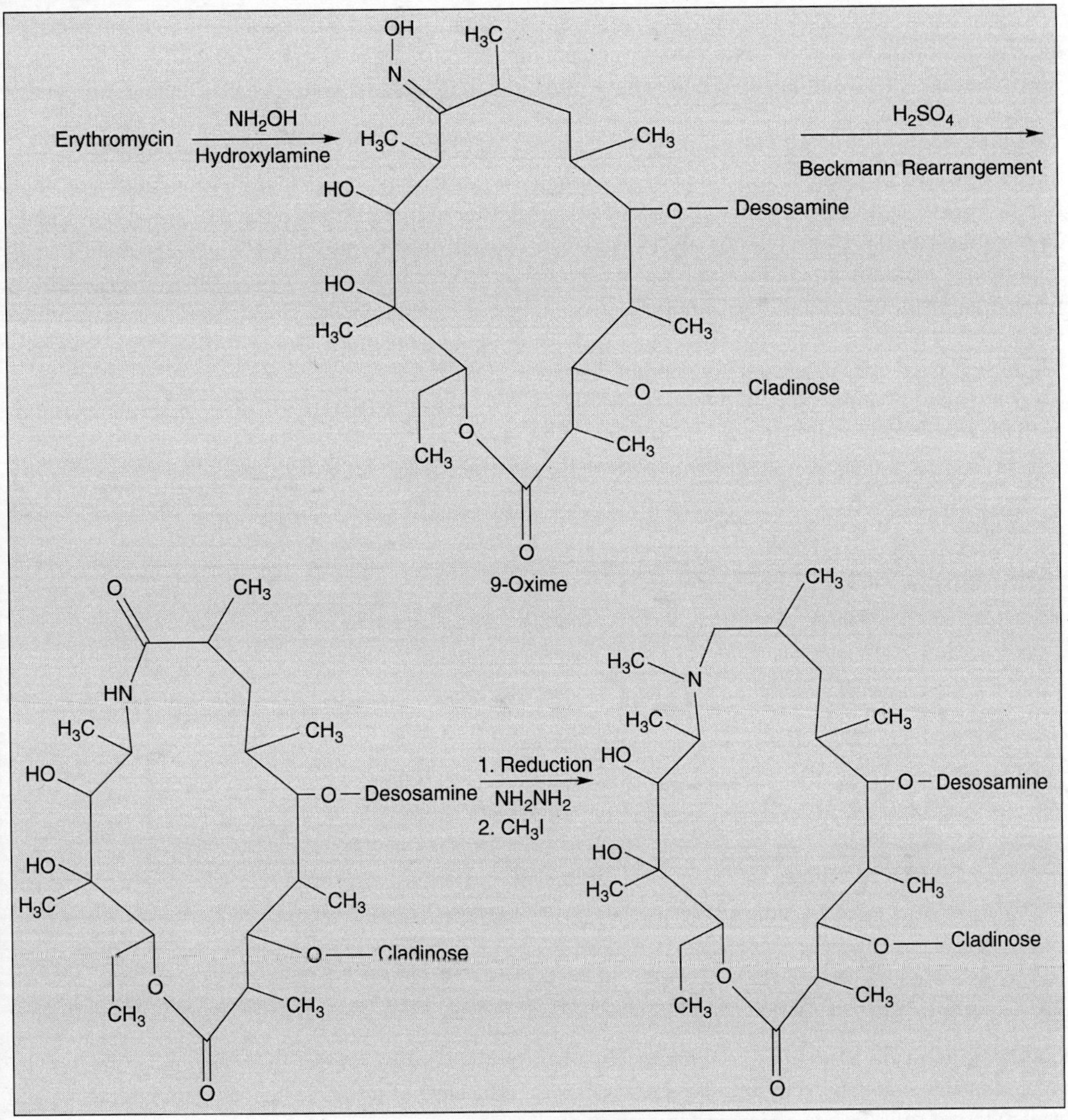

It is prepared from erythromycin by reacting it with hydroxylamine to give oxime derivative, which undergoes Beckmann rearrangement to form ring-enlarged amide derivative. Wolff-Kishner reduction of amide to amine followed by N-methylation with methyl iodide affords azithromycin.

Unlike erythromycin, azithromycin is acid-stable and can, therefore, be taken orally with no need of protection from gastric acids. It is readily absorbed, and its absorption is greater on an empty stomach.

26.4 AMINOGLYCOSIDE ANTIBIOTICS

The aminoglycosides each contains one or more amino sugars, such as glucosamine or neosamine, linked by glycosidic linkages to a basic (amino or guanidine) six-membered carbon ring. These are broad-spectrum antibiotics, and in general they have greater activity against gram-negative than gram-positive bacteria.

Toxicity The toxicity of these agents is dose-related, and therefore every individual can get these side-effects provided the dose is sufficiently high enough. Because of their potential for ototoxicity and nephrotoxicity (kidney toxicity), aminoglycosides are administered in doses based on body weight. Vestibular damage, hearing loss, and tinnitus are irreversible, so care must be taken not to achieve a sufficiently high dose. Concomitant administration of a cephalosporin may lead to increased risk of nephrotoxicity, while administration with a loop diuretic increases the risk of ototoxicity. These properties have limited the use of aminoglycoside chemotherapy to serious systemic indications. Some aminoglycosides can be administered for ophthalmic and topical purposes.

Mechanism of action: Aminoglycosides work by binding to the bacterial 30S ribosomal subunit, inhibiting the translocation of the peptidyl-tRNA from the A-site to the P-site and also causing misreading of mRNA, leaving the bacterium unable to synthesise proteins vital to its growth.

Examples

Name	Source
1. Gentamycin	*Micromonospora purpurea*
2. Neomycin	*Streptomyces fradiae*
3. Streptomycin	*Streptomyces griseus*
4. Tobramycin	*Streptomyces tenebrarius*
5. Framycetin	*Streptomyces decaris*
6. Kanamycin	*Streptomyces kanamyceticus*
7. Amikacin	It is 1-L-(-) 4-amino-2-hydroxybutyryl kanamycin
8. Capreomycin	*Streptomyces capreolus*

26.5 MISCELLANEOUS ANTIBIOTICS

Chloramphenicol D-*threo*-2,2-Dichloro-N-[β-hydroxy-α-(hydroxymethyl)]-N-nitrophenylacetamide

It is also obtained from *Streptomyces venezulae*. Because of its side-effect (blood dyscrasias) and availability of safer agents, the use of this agent declined.

Chloramphenicol inhibits protein synthesis in bacteria, and to a lesser degree in eukaryotic cells. It easily diffuses into the bacterial cell, where it reversibly binds with the 50S ribosomal subunit. This prevents the amino acid ending of tRNA from binding with the binding regions of the 50S ribosome. The binding of aminoacyl tRNA with the 30S subunit is not disturbed. Mammalian cells containing 80S ribosomes are not affected by chloramphenicol. However, this drug inhibits synthesis of mitochondiral proteins in mammalian cells, possibly because of the similarity between mitochondrial and bacterial ribosomes.

Synthesis

The starting material is prepared by aldol condensation of benzaldehyde with 2- nitroethanol to give four enantiomers of nitropropanediol; the total mixture is reduced to the corresponding mixture of aminodiols. The threo isomer is then separated by crystallization and resolved as a diasteromeric salt to give D(−) isomer. Acylation with dichloroacetyl chloride gives triacetate, which on saponification gives the desired amide derivative. The free hydroxyl groups are then acylated with acetic anhydride and the product is nitrated at 4th position. Saponification of ester gives chloramphenicol.

Bacitracin It is a polypeptide antibiotic obtained from *Bacillus subtilis*. Bacitracin interferes with the dephosphorylation of the C_{55}-isoprenyl pyrophosphate, a molecule that carries the building blocks of the peptidoglycan bacterial cell wall outside of the inner membrane. Bacitracin does not work well orally. However, it is very effective topically. Ten individual bacitracins have been isolated: bacitracins A, A1, B, C, D, E, F1, F2, F3, and G. However, the drug itself, named bacitracin, that is used in medicine is a mixture of polypeptide antibiotics.

Polymyxin B sulphate Polymyxines are a group of related polypeptide antibiotics that are produced by spore-forming soil bacteria *Bacillus polymyxa* and *B. circulans*, and they differ in amino acid content. Five different polymyxines have been identified - polymyxines A, B, C, D, and E, which differ in the amino acid content.

Cycloserine 4-Aminoisoxazolidin-3-one

It is obtained from *Streptomyces orchidaceus*. Cycloserine is an antibiotic effective against *Mycobacterium tuberculosis*. The terminal two amino acid residues of the murein precursor lipid II consist of D-alanine, which is produced by the enzyme alanine racemase; the two residues are joined by D-alanine ligase. Both enzymes are competitively inhibited by cycloserine.

Synthesis

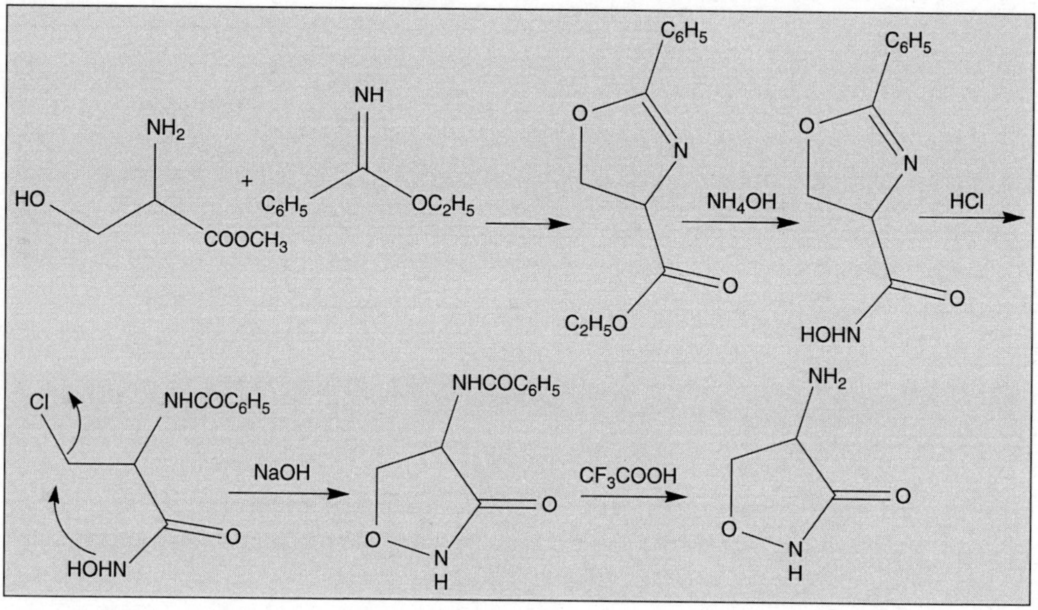

Protection of serine methyl ester as its isoxazoline is attained by reaction with the iminoether obtained from benzonitrile. Ester-amide interchange with hydroxylamine in presence of base gives hydroxamic acid. Reaction of this with hydrogen chloride leads to ring opening with concurrent conversion of hydroxyl group to chloride. Base converts this hydroxamic acid to its anion; the negatively charged oxygen then displaces chlorine to form isoxazoline. Removal of benzoyl group affords cycloserine.

26.6 NEWER DRUGS

Carbapenems Carbapenems are a class of β-lactam antibiotics with a broad spectrum of antibacterial activity, and have a structure that renders them highly resistant to β-lactamases. The carbapenems are structurally very similar to the penicillins, but the sulphur atom in position 1 of the structure has been replaced with a carbon atom, and hence the name of the group, the carbapenems. Examples:

(a) **Imipenem-** ((5R,6S)-3-[2-(aminomethylideneamino)ethylsulphanyl]-6-(1-hydroxyethyl)-7-oxo-1-azabicyclo[3.2.0]hept-2-ene-2-carboxylic acid) and (b) **moropenem** (3-[5-(dimethylcarbamoyl) pyrrolidin-2-yl] sulphanyl-6- (1-hydroxyethyl)-4-methyl-7-oxo- 1-azabicyclo[3.2.0] hept-2-ene-2-carboxylic acid)

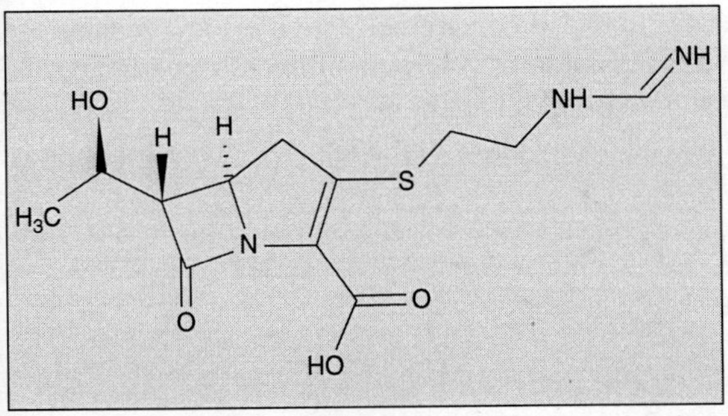

Medicinal Chemistry

[Imipenem structure]
Imipenem

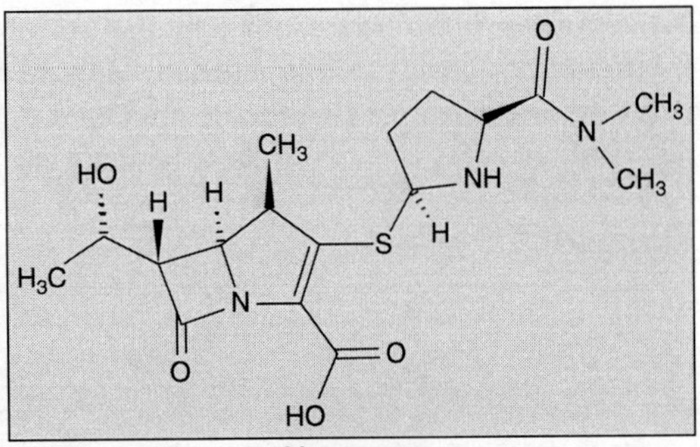

[Moropenem structure]
Moropenem

Fourth-generation cephalosporins Broad spectrum with enhanced activity against gram-positive bacteria and beta-lactamase stability

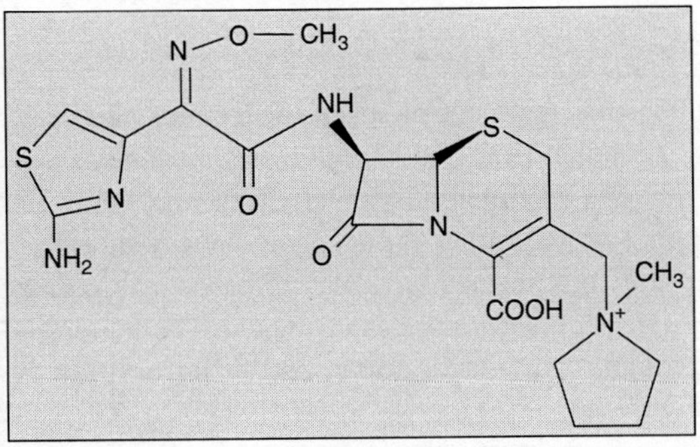

[Cefepime structure]
Cefepime

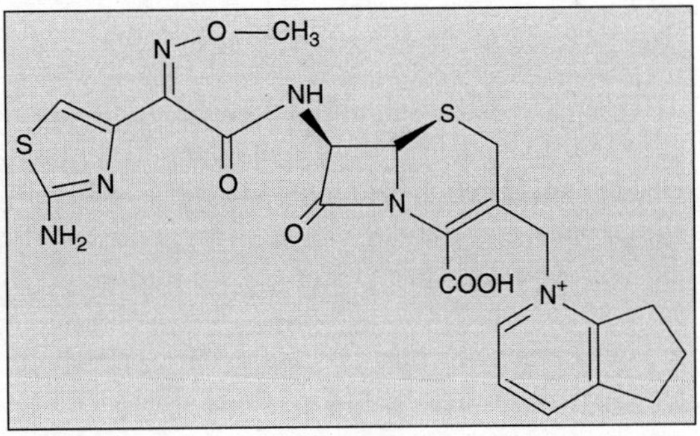

[Cefpirome structure]
Cefpirome

Monobactams: Unlike other beta-lactams, the monobactam contains a nucleus with no fused ring attached. Thus, there is less probability of cross-sensitivity reactions.

Example:
Aztreonam: (3-[2-(2-azaniumyl-1,3-thiazol-4-yl)-2-(1-hydroxy-2-methyl-1-oxo-propan-2-yl)oxyimino-acetyl]amino-2-methyl-4-oxo-azetidine-1-sulphonate)

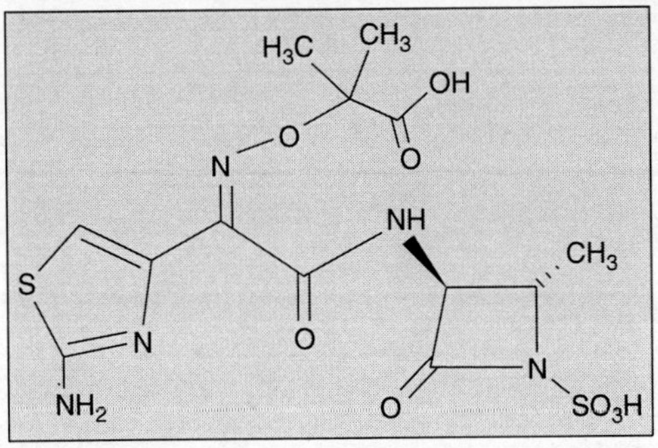

[Aztreonam structure]

β-Lactamase inhibitors: Although they exhibit negligible antimicrobial activity, they contain the β-lactam ring. Their sole purpose is to prevent the inactivation of β-lactam antibiotics by binding the β-lactamases, and, as such, they are co-administered with β-lactam antibiotics. These drugs are irreversible inhibitors of β-lactamase; these bind the enzyme and do not allow it to interact with the antibiotic.

Example:

Clavulanic acid

Tazobactam

Sulbactam

Teicoplanin: It is an antibiotic used in the prophylaxis and treatment of serious infections caused by gram-positive bacteria. It is a glycopeptide antiobiotic extracted from *Actinoplanes teichomyceticus*. Its mechanism of action is to inhibit bacterial cell-wall synthesis. Teicoplanin is actually a mixture of several compounds, five major (named teicoplanin A_2-1 through A_2-5) and four minor (named teicoplanin R_S-1 through R_S.4). All teicoplanins share a same glycopeptide core, termed teicoplanin A_3-1 - a fused ring structure to which two carbohydrates (mannose and *N*-acetylglucosamine) are attached. The major and minor components also contain a third carbohydrate moiety β-D-glucosamine, and differ only by the length and conformation of a side-chain attached to it.

Linezolid: *N*-[[3-(3-Fluoro-4-morpholinophenyl)-aoxooxazolidin-5-yl]methyl]acetamide

It is an antibiotic drug. It was the first commercially available oxazolidinone antibiotic and is usually reserved for the treatment of serious bacterial infections where older antibiotics have failed due to antibiotic resistance. The drug works by inhibiting the initiation of bacterial protein synthesis; it is the only antibiotic to work in this manner. Linezolid is effective against gram-positive pathogens, notably *Enterococcus faecium, Staphylococcus aureus, Streptococcus agalactiae, Streptococcus pneumoniae,* and *Streptococcus pyogenes*. It has almost no effect on gram-negative bacteria and is only bacteriostatic against most *Enterococcus* species.

FURTHER READINGS

1. *The Journal of Antibiotics*, Japan Antibiotic Research Association, Japan.
2. Greene, L.A., 2000. 'New antibiotics put bacteria in bind', *Environmental Health Perspectives*, Vol. 108, 2000.
3. *Macrolide Antibiotics—Chemistry, Biology & Practice*, Satoshi Omura (ed.), Academic Press, 2nd Ed., 2003.
4. *Antibiotics—Antibacterial and Antifungal Agents*, Bryskier, A., A. Bryskier, and J.P. Butzler (ed.), Blackwell Synergy Publishing, 2005.
5. *Tetracyclines in Biology, Chemistry and Medicine*, Nelson, M., W. Hillen, R.A. Greenwald, and Birkhäuser Verlag (ed.), 2001.

MULTIPLE-CHOICE QUESTIONS

1. Agent that acts directly on the cell membrane of the microorganisms affecting permeability is
   a. Penicillin
   b. Nystatin
   c. Tetracycline
   d. Erythromycin

2. The penicillins have a carboxylic acid group placed at
   a. C-3
   b. C-2
   c. C-6
   d. C-7

3. C-12 position is a part of the keto-enol system in
   a. Macrolide antibiotics
   b. Penicillins
   c. Tetracyclines
   d. Aminoglycoside antibiotics

4. The cephalosporin antibiotic with a cyanomethyl side-chain is
   a. Cephalexin
   b. Cefadroxil
   c. Cefamandole
   d. Cephacetrile

5. One of the following statements on the amino function in penicillins is FALSE:
   a. The C-6 amine moiety is necessary for antibacterial activity.
   b. Sulphonation improves antibacterial activity.
   c. Acylation of the amine functionality improves activity.
   d. Carboxamido derivetization is well tolerated.

6. The naturally occurring tetracyclines contain
   a. α-C-4 dimethylamino substituent
   b. α-C-3 dimethylamino substituent
   c. α-C-3-C4 keto-enol group
   d. α-C-3 dihydroxy substituents

7. The antibiotic with an imine functionality is
   a. Ampicillin
   b. Roxithromycin
   c. Doxycycline
   d. Chloramphenicol

8. Chloramphenicol is obtained from
   a. *Streptomyces capreolus*
   b. *Streptomyces venezulae*
   c. *Streptomyces orchidaceus*
   d. *Streptomyces griseus*

9. In cephalosporins, higher resistance to hydrolysis by β-lactamases is shown when
   a. The amino group is acylated
   b. Replacement of sulphur with oxygen
   c. Oxidation of ring sulphur to sulphoxide or sulphone
   d. Introduction of C-7 α-methoxy group

10. The macrolide antibiotics do not have
    a. A large lactone ring
    b. A glycosidically-linked amino sugar

c. A spiroketal group
d. A ketone group
11. Streptomycin is obtained from
    a. *Streptomyces capreolus*
    b. *Streptomyces venezulae*
    c. *Streptomyces orchidaceus*
    d. *Streptomyces griseus*
12. Dimethylamino substituent is present in
    a. Doxycycline
    b. Minocycline
    c. Methacycline
    d. Demeclocycline

QUESTIONS

1. Classify antibiotics based on their mechanism of action.
2. Give an account of antibiotic resistance.
3. Give the structures of various generations of penicillins and write a short note on prodrugs of penicillins.
4. Describe the SAR of tetracyclines.
5. What are the important structural features of β-lactam antibiotics?
6. Write down the complete synthetic protocol for the following:
   a. Doxycycline
   b. Minocycline
   c. Cefuroxime
   d. Cephalexin
   e. Pivampicillin
   f. Dicloxacillin
   g. Roxithromycin

SOLUTION TO MULTIPLE-CHOICE QUESTIONS

1. b;
2. a;
3. c;
4. d;
5. b;
6. a;
7. b;
8. b;
9. d;
10. c;
11. d;
12. b.

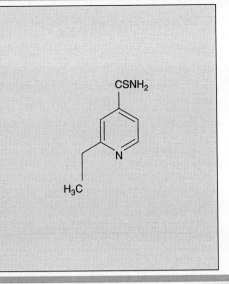

CHAPTER 27

Antitubercular Agents

LEARNING OBJECTIVES

- Define tuberculosis, its causes, and the treatment options
- Describe the chemistry, mode of action, and utility of first-line treatment option for tuberculosis
- Brief the utility of fluoroquinolones as antituberculers

27.1 INTRODUCTION

Tuberculosis is the most prevalent infectious disease worldwide and a leading killer caused by a single infectious agent, i.e., *Mycobacterium tuberculosis*. According to a WHO report, *M. tuberculosis* currently infects over two billion people worldwide, with some 30 million new cases reported every year. This intracellular infection accounts for at least three million deaths annually. Common infection sites of tuberculosis are lungs (primary site), brain, bone, liver, and kidney.

Drugs used in the treatment of tuberculosis can be divided into two major categories:

1. First-line drugs: Isoniazid, Ethambutol, Pyrazinamide, and Rifampicin
2. Second-line drugs: Ethionamide, *p*-Amino salicylic acid, Ofloxacin, Ciprofloxacin, Cycloserine, Amikacin, Kanamycin, and Capreomycin

27.2 MECHANISMS OF ACTION

The large majority of patients with tuberculosis can be treated with first-line drugs. Excellent results can be obtained with a six-month course of treatment; for the first two months, isoniazid, rifampicin or rifampin, and pyrazinamide are given, followed by isoniazid and rifampin for the remaining four months. Second-line drugs are used to mainly treat multidrug-resistant M. *tuberculosis* infections. (*See Fig. 27.1 in the coloured set of pages.*)

Isoniazid Isonicotinic acid hydrazide, INH
Synthesis

4-Picoline on oxidation with potassium permanganate gives isonicotinic acid, which is converted to ethyl ester by reaction with ethanol in presence of acid. Ester-amide interchange takes place by reaction with hydrazine hydrate and affords INH.

Mechanism of action: Isoniazid is a prodrug that is activated on the surface of *M. tuberculosis* by katG enzyme to isonicotinic acid. Isonicotinic acid inhibits the bacterial cell wall mycolic acid and thereby makes *M. tuberculosis* susceptible to reactive oxygen radicals. Isoniazid may be bacteriostatic or bactericidal in action, depending on the concentration of the drug attained at the site of infection and the susceptibility of the infecting organism. The drug is active against susceptible bacteria only during bacterial cell division.

> Isoniazid is bactericidal to rapidly dividing mycobacteria, but is bacteriostatic if the mycobacterium is slow-growing.

Pyrazinamide Pyrazine-2-carboxamide

Pyrazinamide is a prodrug and is activated by M. *tuberculosis* amidase enzyme to pyrazine carboxylic acid, which has bactericidal activity (inhibition of the fatty acid synthetase I).

Synthesis

It is synthesised from quinoxaline by reacting *o*-phenylenediamine with glyoxal. Oxidation of this compound with sodium permanganate gives pyrazin-2,3-dicarboxylic acid. Decarboxylation of the resulting product by heating gives pyrazin-2-carboxylic acid. Esterifying the resulting acid with methanol in the presence of hydrogen chloride and further reaction of this ester with ammonia gives pyrazinamide.

Ethambutol (-)-*N,N*-Ethylenbis-(2-aminobutan-1-ol)

It is a bacteriostatic drug that works by obstructing the formation of cell wall. Mycolic acid attach to the 5'-hydroxyl groups of D-arbinose residue of arabinogalactan and form mycolyl-arabinogalactyl-peptidoglycan (MAP) complex. Ethambutol blocks arabinogalactan synthesis by inhibiting arabinosyl transferase enzyme, thereby prevents the cell wall MAP Complex formation.

Synthesis

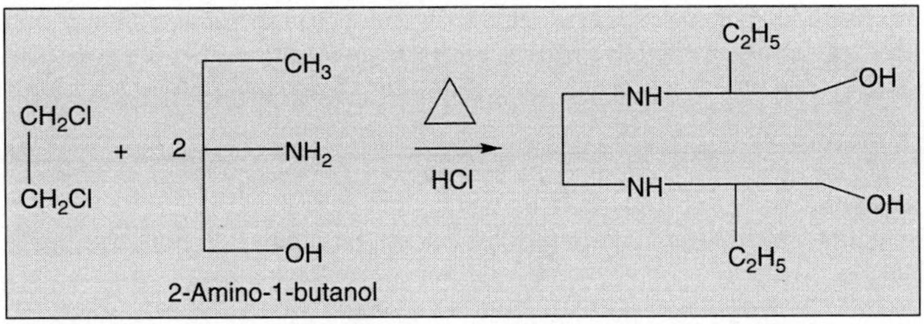

Nitropropane undergoes oxymethylation using formaldehyde, and the nitro group in the resulting 2-nitro-butanol is reduced by hydrogen to an amino group, making racemic (-) 2-aminobutanol. L (-) tartaric acid is used to separate (-) 2-aminobutanol. Reacting this with 1,2-dichloroethane in the presence of sodium hydroxide gives ethambutol.

Rifampin 5,6,9,17,19,21-Hexahydroxy-23-methoxy-2,4,12,16,18,20,22-heptamethyl-8-[*N*-(4-methyl-1-piperazinyl)-formamidoyl]-2,7-(epoxypentadeca-1,11-13-trienimino)-naphtho-[2,1-b] furane-1,11(2H)dion-21 acetate

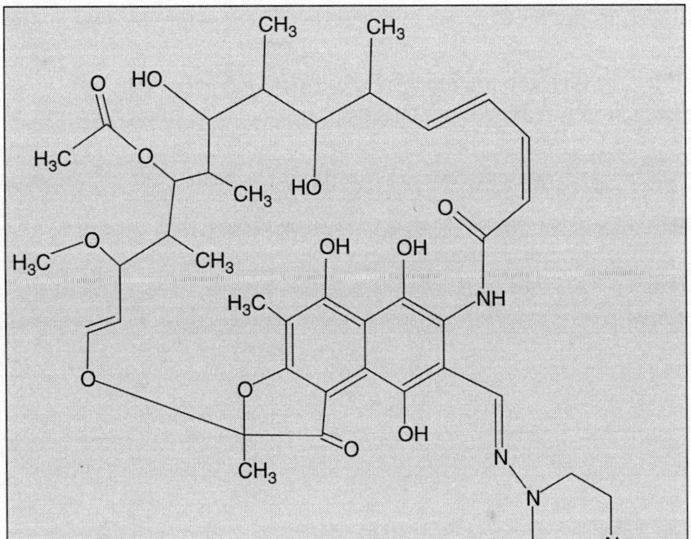

It is an antibiotic obtained from *Streptomyces mediterranei*. Rifampin inhibits DNA-dependent RNA polymerase of mycobacteria by forming a stable drug-enzyme complex, leading to suppression

of initiation of chain formation in RNA synthesis, and acts as a bactericidal drug. Rifampicin is highly active against *Mycobacterium tuberculosis*. Among atypical mycobacteria, it is active against *Mycobacterium kansasii, Mycobacterium marinum,* and most types of *Mycobacterium scrofulaceum* and *Mycobacterium xenopi*. Sensitivity of other mycobacteria varies. Rifampicin also exhibits activity against *Mycobacterium leprae*.

> Nearly one-third of the world's population is infected with latent **Mycobacterium tuberculosis**, and the World Health Organization (WHO) estimates that about 30 million people will be infected within the next 20 years. Currently, available drugs are active against growing bacteria but are ineffective against non-growing bacteria. There is still no specific drug available in the market which could effectively kill this latent/persistent bacillus.

p-**Amino salicylic acid**   PAS, 5-Amino-2-hydroxybenzoic acid
4-Amino salicylic acid is an inhibitor of bacterial folate metabolism in a manner similar to the sulphonamide antibacterials.

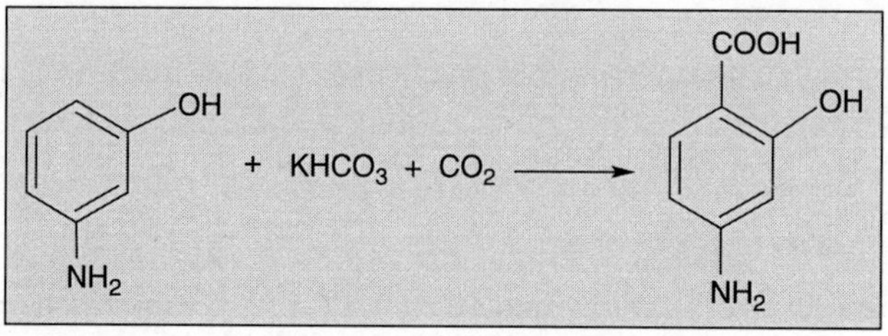

    p-Aminosalicylic acid is synthesised in a Kolbe reaction, which consists of direct interaction of *m*-aminophenol with potassium bicarbonate and carbon dioxide while heating at a moderate pressure of 5–10 atm.

Ethionamide   2-(Ethyl)isonicotinthioamide
The antimycobacterial action of ethionamide seems to be due to an inhibitory effect on the mycolic acid synthesis.

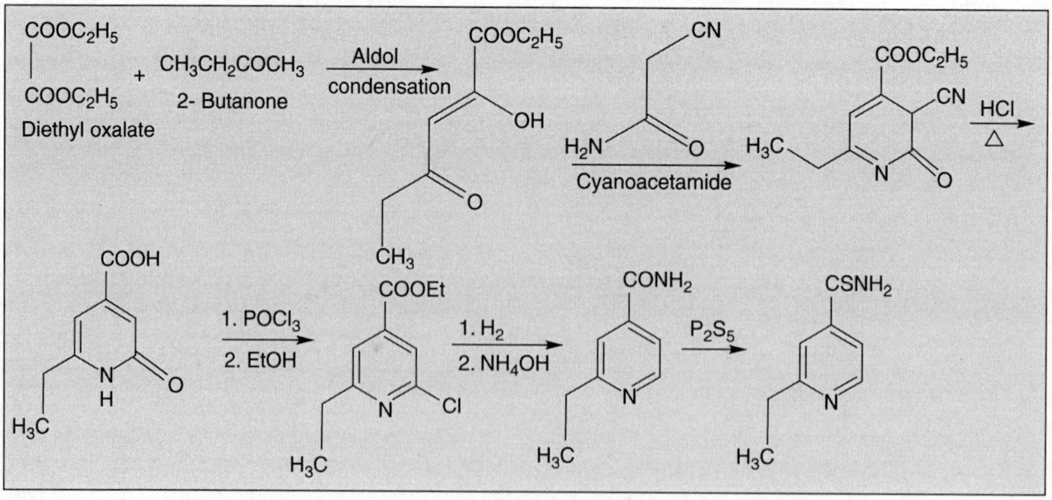

Ethionamide is synthesised by the following scheme. Diethyl oxalate is condensated with methylethylketone in the presence of sodium ethoxide to form the ethyl ester of propionylpyruvic acid. Condensation of this with cyanoacetamide results in heterocyclization, to form 3-cyano-4-carboethoxy-6-ethyl-2-pyridone, which is hydrolysed with hydrochloric acid to give 4-carboxy-6-ethyl-2-pyridone. Reacting this with a mixture of phosphorous oxychloride and pentachloride gives 6-ethyl-2-chloroisonicotinic acid chloride, which is subsequently treated with ethyl alcohol to obtain the ethyl ester of 6-ethyl-2-chloroisonicotinic acid. Reducing this with hydrogen over a palladium catalyst removes the chlorine atom at position 2 of the pyridine ring, giving the ethyl ester of 6-ethylisonicotinic acid. Interacting this with ammonia, followed by phosphorous pentasulfide gives ethionamide.

Fluoroquinolone Recently, fluoroquinolones like ciprofloxacin, ofloxacin, and moxifloxacin were found to be active against *M. tuberculosis*. For the synthesis and mechanisms of action, refer to Chapter 25 on 'Quinolone Antibacterials'.

FURTHER READINGS

Burger's Medicinal Chemistry, Wolff, M.E. (ed.), 4th Ed., Vol. II, John Wiley and Sons, New York, 289–331.

MULTIPLE-CHOICE QUESTIONS

1. One of the following is not a first-line drug for treating tuberculosis:
   a. Isoniazid
   b. Rifampin
   c. Cycloserine
   d. Pyrazinamide

2. The drug useful to treat multi-drug resistant tuberculosis is
   a. Isoniazid
   b. Ethionamide
   c. Rifampin
   d. Pyrazinamide

3. One of the following is not a synthetic drug:
   a. Isoniazid
   b. Rifampin
   c. Pyrazinamide
   d. Ethionamide

4. The mechanism of PAS is
   a. Inhibits mycolic acid synthesis
   b. Inhibits folic acid synthesis
   c. Inhibits DNA-dependent RNA polymerase
   d. Makes the tuberculosis organism susceptible to reactive oxygen

5. Rifampin acts by the following mechanism of:
   a. Inhibits mycolic acid synthesis
   b. Inhibits folic acid synthesis
   c. Inhibits DNA-dependent RNA polymerase
   d. Makes the tuberculosis organism susceptible to reactive oxygen

6. The antitubercular activity of isoniazid is by
   a. Inhibits mycolic acid metabolism
   b. Inhibits folic acid synthesis
   c. Inhibits DNA-dependent RNA polymerase
   d. Makes the tuberculosis organism susceptible to reactive oxygen

7. The conversion of amide to thioamide is achieved with
   a. Thiourea
   b. Phosphorous trisulphide
   c. Phosphorous pentasulphide
   d. Ammonium thiocyanate

QUESTIONS

1. What is TB? How is it treated?
2. Write a short note on antitubercular quinolones.
3. Give the synthetic protocol of isoniazid and ethionamide.
4. How does isoniazid and pyrazinamide act as antitubercular agents?
5. What is multi-drug resistant TB and how are they treated?

SOLUTION TO MULTIPLE-CHOICE QUESTIONS

1. c;
2. b;
3. b;
4. b;
5. c;
6. d;
7. c.

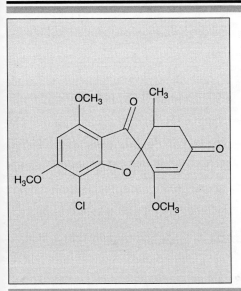

CHAPTER 28

Antifungal Agents

LEARNING OBJECTIVES

- Define fungal infections and the treatment options
- Categorize antifungals
- Describe the chemistry and utility of antifungal antibiotics
- Describe the mode of action and structure of azole antifungals
- Discuss the utility of pyrimidine antifungals
- Discuss the other structural classes of antifungals
- List some new drugs in the antifungal category

28.1 INTRODUCTION

Most fungal infections (mycoses) involve superficial invasion of the skin or mucous membrane of body orifices. These diseases, which can usually be controlled by local application of antifungal agents, are divided into two etiologic groups:

1. The dermatophytes, which are contagious superficial epidermal infections caused by various *Epidermophyton, Microsporum,* and *Trichophyton* species
2. Mycoses caused by pathogenic saprophytic yeasts (*Aspergillus, Blastomyces, Candida, Cryptococcus,* and *Histoplasma*), which are contagious and usually superficial infections involving the skin and mucous membranes. Under certain conditions, these are capable of invading deeper body cavities and causing systemic mycoses. Such infections may become serious and occasionally life-threatening. Moreover, they are difficult to treat

An antifungal drug is medication used to treat fungal infections such as athlete's foot, ringworm, and candidiasis (thrush), and serious systemic infections such as cryptococcal meningitis.

28.2 CLASSIFICATION

1. Antibiotics: Amphotericin B, Nystatin, Griseofulvin
2. Azole (imidazole, triazole) derivatives:
    a. Systemic: Ketoconazole, Fluconazole, and Itraconazole
    b. Locally acting: Clotrimazole, Econazole, Miconazole, Terconazole, and Butoconazole
3. Pyrimidine derivatives: 5-Flucytosine
4. Miscellaneous: Ciclopirox, Tolnaftate, Naftifine, and Terbenafine

28.3 ANTIFUNGAL ANTIBIOTICS

Amphotericin B

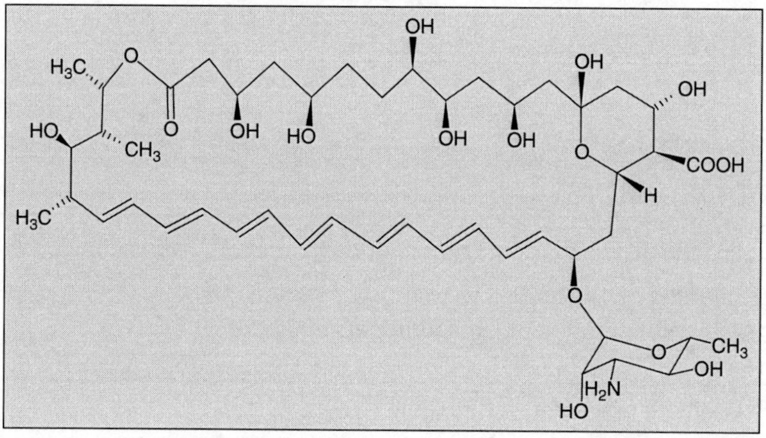

It is a polyene antibiotic obtained from *Streptomyces nodosus*. It is an amphoteric compound composed of a hydrophilic polyhydroxyl chain along one side and a lipophilic polyene hydrocarbon chain on the other.

The antifungal activity of this drug depends at least in part on its binding to a sterol moiety, primarily ergosterol, which is present in the membrane of sensitive fungi. By virtue of their interaction with the sterols of cell membranes, polyenes appear to form pores or channels. The result is an increase in the permeability of the membrane, allowing leakage of a variety of small molecules like intracellular potassium, magnesium, sugars, and metabolites, and then cellular death.

Griseofulvin   7-Chloro-2,4,6-trimethoxy-6-methylspiro[benzofuran-2(3H),1-[2]-cyclohexen]-3,4-dione

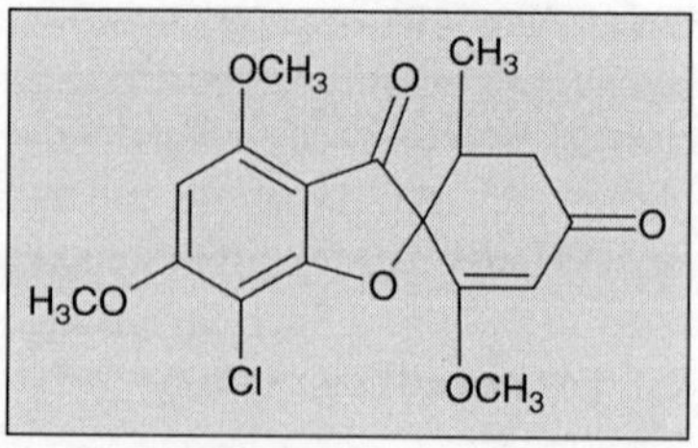

It is obtained from the mould *Penicillium griseofulvum*. Griseofulvin is a fungistatic drug that causes disruption of the mitotic spindle by interacting with polymerised microtubules.

Medicinal Chemistry

Nystatin It is a polyene antibiotic isolated from *Streptomyces noursei*, and it is structurally similar to Amphotericin B and has the same mechanism of action.

28.4 AZOLE ANTIFUNGALS

Mechanism of action: Azole antifungals inhibit sterol-14-α-demethylase, a microsomal cytochrome P450-dependent enzyme system, and thus impair the biosynthesis of ergosterol for the cytoplasmic membrane and lead to the accumulation of 14-α-methyl sterols. These methylsterols may disrupt the close packing of acyl chains of phospholipids, impairing the functions of certain membrane-bound enzyme systems such as ATPase and enzymes of the electron transport system, and thus inhibiting the growth of the fungi.

Ketoconazole (cis-1-Acetyl-4-[4-[2-(2,4-dichlorophenyl)-2-(1H-imidazole-1-ylmethyl)-1,3-dioxolan-4-ylmethyl]phenyl]piperazine)

Ketoconazole is an imidazole antifungal agent. Ketoconazole is a highly lipophilic compound. This property leads to high concentrations of ketoconazole in fatty tissues and purulent exudates. Ketoconazole is active against *Candida* sp. and *Cryptococcus neoformans*.

Ketoconazole is synthesised from 2,4-dichlorophenacyl bromide, the ketalization of which using glycerol gives cis-2-(2,4-dichlorophenyl)-2-bromoethyl-4-hydroxymethyl-1,3-dioxolane. Alkylating the resulting compound with imidazole gives the derivative, which on reaction with methanesulphonyl chloride gives a mesylate. Finally, alkylating with 1-acetyl-4-(4-hydroxyphenyl)piperazine gives ketoconazole.

A study in mice indicated that ketoconazole may have a stimulatory effect on hair growth. Nizoral shampoo has shown to be beneficial in men suffering from androgenic alopecia. One 1998 study showed that Nizoral 2% worked just as well as minoxidil 2% (brand name Rogaine) in men with androgenic alopecia. Both medicines increased hair thickness and increased the number of anagen-phase hair follicles on the scalp

Terconazole   1-[4-[[2-(2,4-Dichlorophenyl)-2-(1H-1,2,4-triazol-1-yl-methyl)-1,3-dioxolan-4-yl]methoxy]phenyl]-4-(1-methylethyl)-piperazine

It is chemically very similar to ketoconazole, the only difference being that instead of an imidazole ring it contains a triazole ring, and the piperazine ring, instead of an acetyl group, is substituted by an isopropyl group. It is synthesised from 2,4-dichloroacetophenone similar to ketoconazole.

Fluconazole   2-(2,4-Difluorophenyl)-1,3-bis(1H-1,2,4-triazol-1-yl)propan-2-ol

Fluconazole is a widely used bis-triazole antifungal agent. Fluconazole is generally considered to be a fungistatic agent. It is principally active against *Candida* sp. and *Cryptococcus* sp. Fluconazole has useful activity against *Coccidioides immitis* and is often used to suppress the meningitis produced by that fungus.

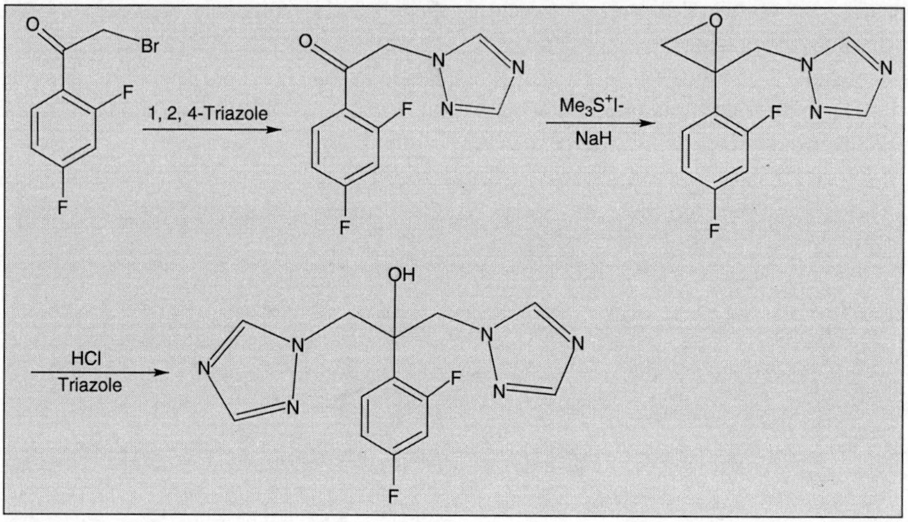

It is prepared from 2,4-difluorophenacyl bromide. Displacement of bromine by triazole affords intermediate; condensation of carbonyl group with the ylide from trimethylsulphonium iodide leads to an addition product. The anion formed on the carbonyl oxygen then internally displaces dimethyl sulphide to give on oxiran (or) epoxide. Reaction of this with triazole leads to epoxide ring opening with consequent incorporation of the second triazole, and affords fluconazole.

Clotrimazole   1-(o-Chloro-α,α-diphenylbenzyl)imidazole

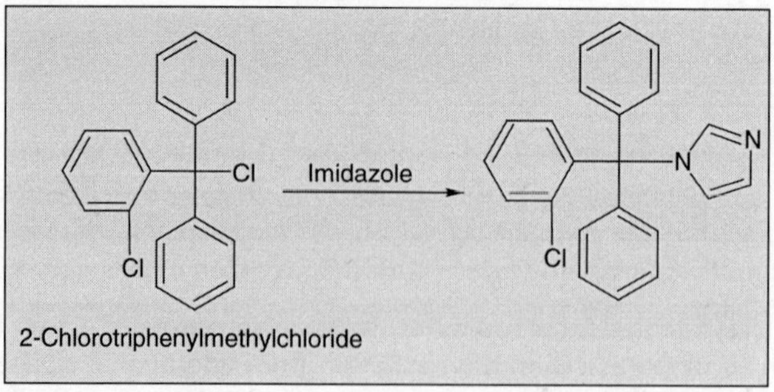

Clotrimazole is synthesised by reacting 2-chlorotriphenylmethylchloride with imidazole in the presence of triethylamine. The starting substance 2-chlorotriphenylmethylchloride is made by the following procedure. Grignard reaction between 2-chlorobenzolphenone and phenylmagnesium bromide is followed by substitution of the hydroxyl group in the resulting 2-chlorotriphenylmethylcarbinol with a chlorine using thionyl chloride.

Miconazole (1-[2,4-Dichloro-β-[(2,4-dichlorobenzyl)oxy]phenethyl]-imidazole) and
Econazole (1-[2,4-Dichloro-β-[(4-chlorobenzyl)oxy]phenethyl]-imidazole)

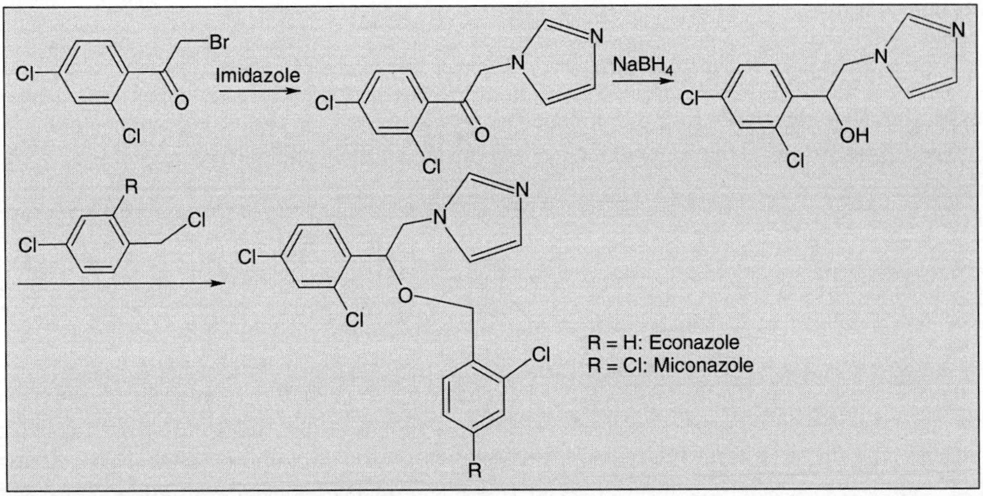

Miconazole, like ketoconazol, is synthesised from 2,4-dichlorophenacylbromide, which is reacted with imidazole to make 1-(2,4-dichlorobenzoylmethyl)-imidazole. Reducing the carbonyl group in this molecule with sodium borohydride gives 1-(2,4-dichlorophenyl)-3-(1-imidazolyl)-ethanol, and the hydroxyl group is alkylated by 2,4-dichlorobenzylchloride using base such as sodium hydride to make miconazole.

Econazole is an analogue of miconazole. It differs in the presence of a single chlorine atom in the benzyl part of the molecule, and is synthesised in the same manner, except that it uses 4-chlorobenzylchloride in the last stage instead of 2,4-dichlorobenzylbromide.

Butoconazol 1-[4-(4-Chlorophenyl)-2-[(2,6-dichlorophenyl)thio]butyl]-1H-imidazole

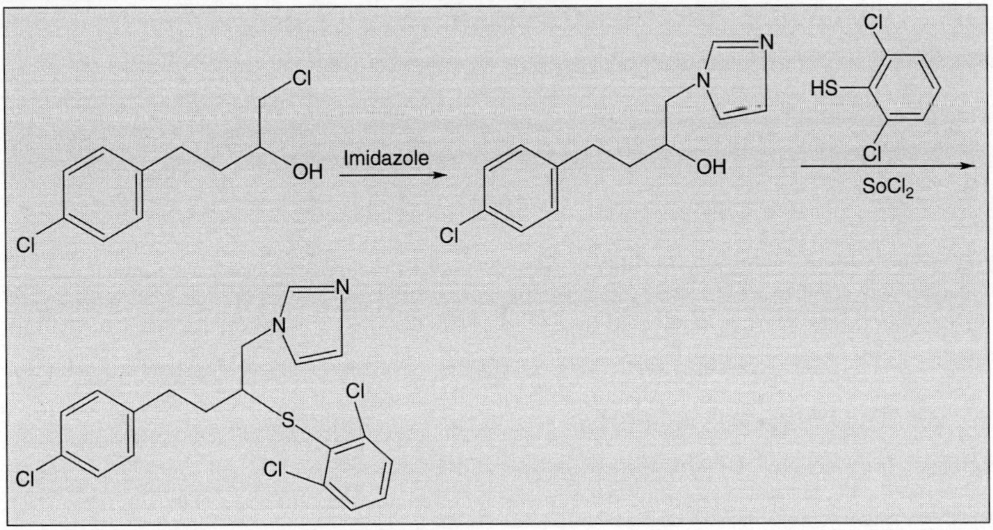

The starting material 4-(4'-chlorophenyl)-1-chlorobutan-2-ol is synthesised by reacting 4-chlorobenzylmagnesium bromide and epichloridrine. This on reaction with imidazole in the presence of sodium makes 4-(4'-chlorophenyl)-1-(1H-imidazolyl)butan-2-ol. The hydroxyl group in the last is replaced with a chlorine atom upon reaction with thionyl chloride, which by the reaction with 2,6-dichlorothiophenol affords butoconazole.

28.5 PYRIMIDINE DERIVATIVES

5-Flucytosine: 4-Amino-5-fluoro-2-pyrimidone
Flucytosine is the only available antimetabolite drug having antifungal activity.
Synthesis

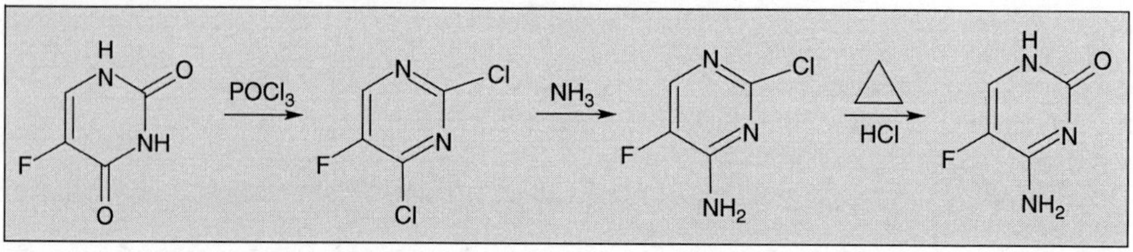

Flucytosine is synthesised from fluorouracil. Fluorouracil is reacted with phosphorous oxychloride in dimethylaniline to make 2,4-dichloro-5-fluoropyrimidine, which is reacted with ammonia to make a product substituted with chlorine at the fourth position of the pyrimidine ring 4-amino-2-chloro-5-fluoropyrimidine. Hydrolysis of the chlorovinyl fragment of this compound in a solution of hydrochloric acid gives the desired flucytosine.

Mechanism of action:

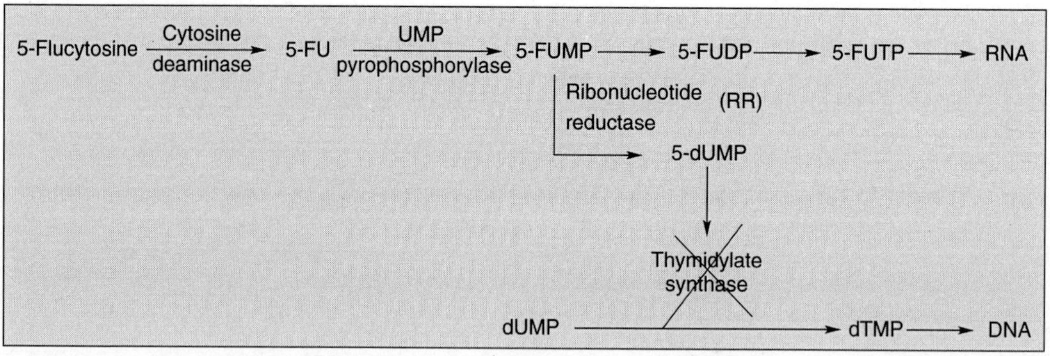

5-Flucytosine is transported into the fungal cell, where it is deaminated to 5-fluorouracil (5-FU). The 5-FU is then converted to 5-FU-ribose monophosphate (5-FUMP) and is then either converted to 5-FUTP and incorporated into RNA or converted by RR to 5dUMP, which is a potent inhibitor of thymidylate synthase.

28.6 MISCELLANEOUS AGENTS

Ciclopirox 6-Cyclohexyl-1-hydroxy-4-methyl-pyridin-2-one
It is available as 1% cream and lotion for the treatment of cutaneous candidiasis and for *Tinea corporis, cruris, pedis,* and *versicolor*.

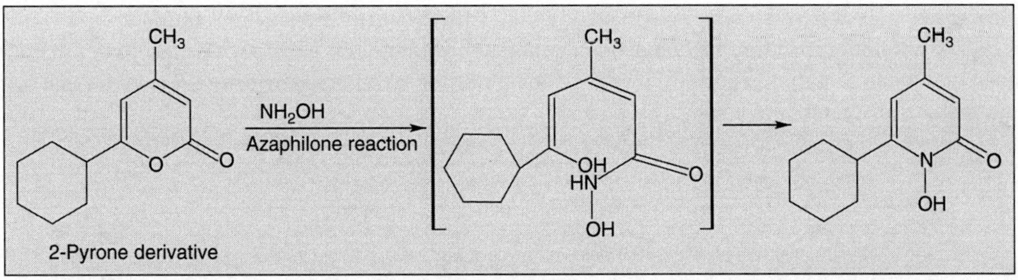

It was formed from 2-pyrone by an azaphilone reaction with hydroxylamine. This may be viewed at least formally as an ester(lactone)-amide exchange to an intermediate oximinoester, which ring closes via an addition-elimination sequence to expel the original lactone ring oxygen in favour of the hydroxylamine nitrogen. Lactone readily converts to lactams by this reaction.

Tolnaftate   o-(2-Naphthyl)-N-methyl-N-(3-tolyl)-thiocarbamate
It is effective in the topical treatment of most cutaneous mycoses caused by *Trichophyton* and *Microsporum*.

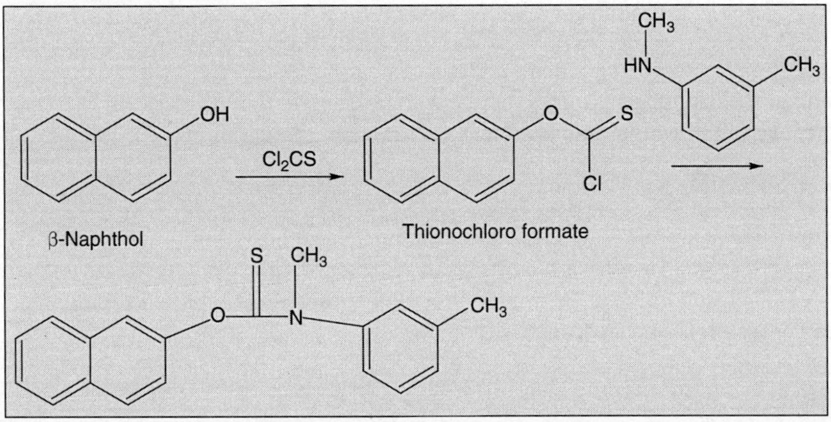

Tolnaftate is synthesised by reacting equimolar amounts of 2-naphthol and thiophosgene to make a monosubstituted product of thiophosgene, which is then reacted with N-methyl-3-toluidine to give the desired tolnaftate.

Naftifine   (E)-N-Methyl N-(3-phenyl-2-propenyl)-1-naphthalinmethanamine The drug has fungicidal activity against *Tinea cruris* and *corporis* species. It is believed that the fungicide activity of this drug is based on its ability to inhibit the fungal enzyme squalene epoxidase, thus lowering the concentration of ergosterol.

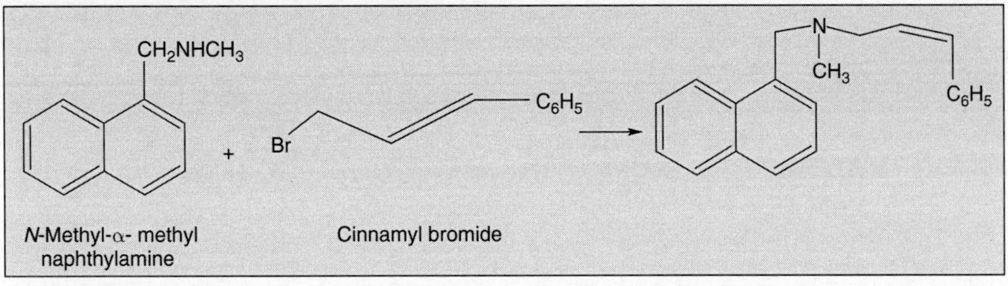

Naftifine is synthesised by alkylating N-methyl-(1-naphthylmethyl)-amine with cinnamyl bromide in the presence of sodium carbonate.

Terbinafine (E)-N,6,6-Trimethyl-N-(naphthalen-1-ylmethyl)hept-2-en-4-yn-1-amine
It is highly lipophilic in nature and tends to accumulate in skin, nails, and fatty tissues. Like other allylamines, terbinafine inhibits ergosterol synthesis by inhibiting squalene epoxidase, an enzyme that is part of the fungal cell-membrane synthesis pathway.

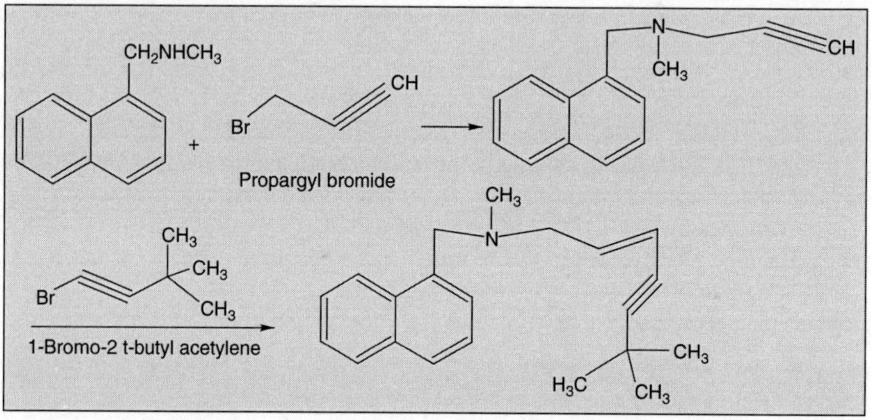

N-Alkylation of naphthalenylmethyl amine with propargyl bromide gives allyl acetylene derivative. This on further alkylation with bromo acetylene derivative in presence of base affords terbinafine.

28.7 NEWER DRUGS

Echinocandins These are antifungal drugs that inhibit the synthesis of glucan in the cell wall, probably via the enzyme 1,3-β-glucan synthase. 1,3-β-glucan synthase is a glucosyltransferase enzyme involved in the generation of β-glucan in fungi. Examples: anidulafungin, caspofungin, micafungin.

Newer imidazole and triazole antifungals

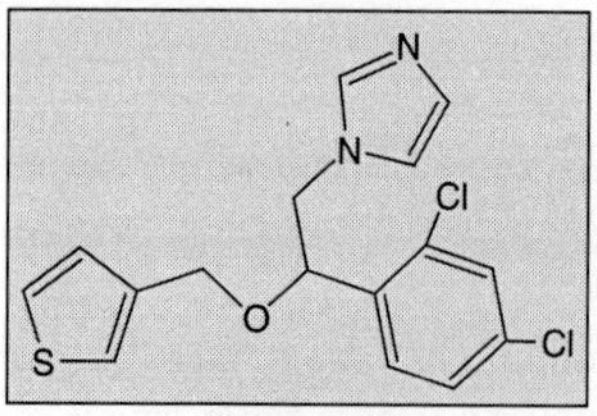

Tioconazole

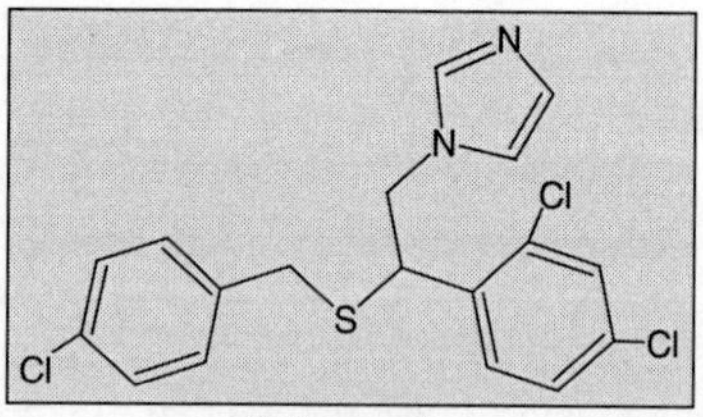

Sulconazole

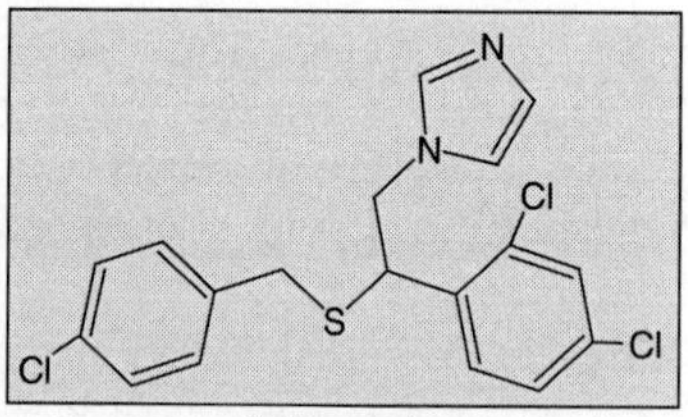

Oxiconazole

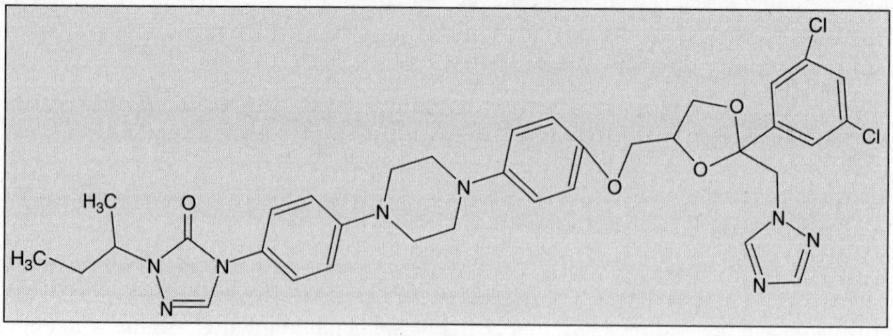

Itraconazole

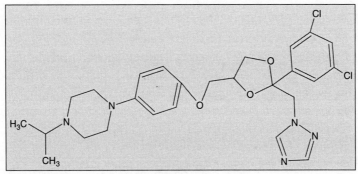
Terconazole

FURTHER READINGS

1. *Fifty Years of Antimicrobials*, by Society for General Microbiology, Cambridge University Press, 1994.
2. *Foye's Principles of Medicinal Chemistry*, by Williams, David A. and and Thomas L. Lemke (ed.), 5th Ed., 2002.

MULTIPLE-CHOICE QUESTIONS

1. One of the following consists of imidazole nucleus in its structure:
   a. Ciclopirox
   b. Butaconazole
   c. Giseofulvin
   d. Co-trimoxazole

2. A potent inhibitor of thymidylate synthase is
   a. Naftifine
   b. 5-Fluocytosine
   c. Ciclopirox
   d. Ketoconazole

3. Inhibitor of sterol-14-α-demethylase is
   a. Naftifine
   b. 5-Fluocytosine
   c. Ciclopirox
   d. Ketoconazole

4. Antifungal antibiotic is
   a. Naftifine
   b. 5-Fluocytosine
   c. Nystatin
   d. Nafimidone

5. Which of the following is not used for the treatment of *tinea pedis*?
   a. Clotrimazole
   b. Ciclopirox
   c. Nystatin
   d. Terbinafine

6. The mechanism of action of naftifine is
   a. Inhibits sterol-14-α-demethylase
   b. Inhibits squalene epoxidase
   c. Inhibits thymidylate synthase
   d. Inhibits microtubule formation

7. The antifungal with bis-triazole nucleus is
   a. Ketoconazole
   b. Butaconazole
   c. Fluconazole
   d. Clotrimazole

QUESTIONS

1. Classify antifungal agents with structural examples for each class.
2. Write a note on the different mechanisms or targets by which antifungals act.
3. Write the synthesis of naftifine, ketoconazole, and fluconazole.
4. Briefly write a note on antimetabolites as antifungals.

SOLUTION TO MULTIPLE-CHOICE QUESTIONS

1. b;
2. b;
3. d;
4. c;
5. c;
6. b;
7. c.

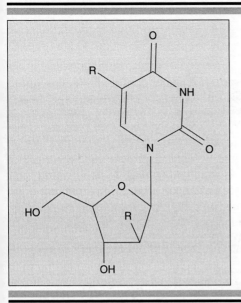

CHAPTER 29

Antiviral Agents

> **LEARNING OBJECTIVES**
>
> - Define HIV and AIDS with a view on the HIV life cycle
> - Describe the potential drug targets in the HIV life cycle
> - Categorize anti-HIV drugs based on the stage of intervention in the replicative cycle
> - Discuss the classes of reverse transcriptase inhibitors and compare nucleoside and non-nucleoside inhibitors
> - Describe the mode of action of each class of anti-HIV drugs
> - Define protease inhibitors and their importance
> - Define anti-HSV drugs and their utility
> - List some miscellaneous classes of antivirals
> - Describe the newer antiviral drugs

In the past few years, the antiviral chemotherapy area has witnessed a remarkable turnover of antiviral compounds, particularly in the domain of nucleoside analogues. Several new compounds have been approved and launched for clinical use of human immunodeficiency virus (HIV), herpes simplex virus (HSV), and other viruses.

29.1 ANTI-HIV AGENTS

Acquired immunodeficiency syndrome (AIDS) is caused by the retrovirus, human immunodeficiency virus. The HIV infection, which targets the lymphocytes, the monocytes, and macrophages expressing surface CD-4 receptors, eventually produces profound defects in cell-mediated immunity. Over time, infection leads to severe depletion of CD-4 T-lymphocytes resulting in opportunistic infection like tuberculosis, fungal, viral, protozoal, and neoplastic diseases, and ultimately death.

29.1.1 HIV—Virus Life Cycle

HIV is the retrovirus of the lentiviridae family, originally referred to as HTLV-III. HIV virus consists of outer lipid bilayer, the surface of which contains gp120 (glycoprotein). Nucleocapsid contains single-stranded RNA. (*See Figs. 29.1 and 29.2 in the coloured set of pages.*)

The gp120 has a greater affinity for CD-4 receptors, which are present on the surface of T-lymphocytes, monocytes, and macrophages. Because of this affinity, HIV **binds** to the host target cell. After binding to the surface of the host cell, the outer membrane of the virus **fuses** with membrane of the host cell. At this point, the virus **uncoats** and unloads the genomic RNA and the enzyme reverse transcriptase (RT). At this point, **RT** performs three important functions:

1. First, using the RNA as template (positive strand), it catalyses an RNA-dependent DNA synthesis to produce single-strand DNA (negative strand)
2. Second, using ribonuclease H (section of RT), the enzyme systemically degrades the genomic RNA strand
3. Third, using the newly synthesised DNA strand as a template, it catalyses a DNA-dependent DNA synthesis of a complementary copy of DNA. This newly formed DNA double helix is also called proviral DNA

In the next stage, the proviral DNA is translocated into the nucleus, where it is integrated into the host genome by the enzyme **integrase.** With integration complete, commonly the virus remains latent for a few months or up to 8–10 years, leaving the person asymptomatic during the entire period. The presence of host factors or gene products of viruses, like the Epstein-Barr virus, herpes simplex virus, and cytomegalovirus, can activate the latent virus. This stimulation leads to **expression** of the viral genes and the production of viral genomic RNA and messenger RNA **(translation)**, followed by synthesis of viral proteins, movement to the surface, and viral budding. At this stage, the **protease** cleaves the viral polyproteins. The budding operation kills the host cell in the process, and the newly formed virus particle becomes matured with the help of **glucosidase** enzyme and leaks out of the target cell to once again begin its life cycle. **Potential targets for anti-HIV agents** The replicative cycle of HIV comprises a number of steps, which could be considered as adequate targets for chemotherapeutic intervention. The important targets are viral adsorption, virus–cell fusion, virion uncoating, reverse transcription (RNA → ds DNA) [RT enzyme], proviral DNA integration, viral transcription (DNA → RNA), viral translation (mRNA → Protein), viral budding (assembly/release), and maturation (protease and glucosidase enzymes). Most of the substances that have been identified as anti-HIV agents can be allocated to one of the ten classes of HIV inhibitors, according to the stage at which they interfere with the HIV replicative cycle (Table 29.1).

Table 29.1
Review of HIV inhibitors according to the stage of intervention in the replicative cycle

1. Adsorption inhibitors

   (a) Polysulphates—Dextran sulphate, Curdlan sulphate, Pentosan polysulphate

   (b) Polysulphonates—Suramin, Evans blue

   (c) Polycarboxylate—Aurin tricarboxylic acid

   (d) Glycyrrhizin

Table 29.1 (*Continued*)

2. **Fusion inhibitors**

    Betulinic acid, Mannose-specific plant lectinoylsulphate

3. **Virus-uncoating inhibitors**

    Bicyclam derivatives

4. **Reverse-transcription inhibitors**

    (a) Nucleoside derivatives—Zidovudine, Stavudine, Lamivudine—Zalcitabine, Didanosine, Abacavir

    (b) Non-Nucleoside derivatives—Nevirapine, Delavirdine, Efavirenz Loviride, Trovirdine, Emivirine

5. **Integration inhibitors**

    Curcumin, L-chicoric acid

6. **DNA-replication inhibitors**

    Antisense constructs

7. **Transcription inhibitors**

    1,4-Benzodiazepine and Fluoroquinolone derivatives

8. **Translation inhibitors**

    Ribozymes, Trichosanthin

9. **Maturation inhibitors**

    (a) Protease inhibitors—Saquinavir, Indinavir, Ritonavir, Nelfinavir, Lopinavir, Amprenavir, Atazanavir, Tipranavir, Darunavir, Amprenavir, Lopinavir

    (b) Glucosidase inhibitors—Castanospermine

10. **Budding inhibitors**

    Interferon, Hypericin

The best-known and most intensively studied drugs active against HIV are the reverse transcriptase and protease inhibitors. There are at present fourteen compounds that have been formally approved for the treatment of HIV infection: six nucleoside RT inhibitors (zidovudine, stavudine, lamivudine, zalcitabine, didanosine, abacavir), three non-nucleoside RT inhibitors (nevirapine, delavirdine, efavirenz), and five protease inhibitors (saquinavir, indinavir, ritonavir, nelfinavir, and amprenavir).

29.1.2 Reverse Transcriptase Inhibitors

Reverse transcriptase (RT) is a key enzyme that plays an essential and multifunctional role in the replication of HIV, and RT is necessary for the catalytic transformation of single-stranded viral RNA into the double-stranded DNA, which is integrated into the host cell chromosome.

$$\text{RNA} \xrightarrow{(a)} \text{RNA-DNA} \xrightarrow{(b)} \text{DNA} \xrightarrow{(c)} \text{DNA-DNA}$$
(single strand)   (complex)   (single strand)   (double strand)

(a) RNA-dependent DNA polymerase
(b) RNase H activity
(c) DNA-dependent DNA polymerase

Classification

1. Nucleoside analogues
    (i) 2', 3'-dideoxy pyrimidine nucleosides: Zidovudine, Stavudine, Lamivudine, Zalcitabine
    (ii) 2', 3'-dideoxy purine nucleosides: Didanosine, Abacavir
    (iii) Acyclic dideoxy purine nucleosides: Adefovir
2. Non-Nucleoside analogues
    (i) Dipyrido diazepine: Nevirapine
    (ii) Bis heteroaryl piperazines (BHAP): Delavirdine
    (iii) Benzoxazinones: Efavirenz
    (iv) α-Anilino phenylacetamide (α-APA): Loviride
    (v) Pyridyl ethyl thiourea (PETT): Trovirdine
    (vi) Hydroxyethoxy phenyl thymine (HEPT): Emivirine

29.1.3 Nucleoside RT Inhibitors

Mechanism of action: All the dideoxynucleoside analogues are phosphorylated to the corresponding mono-, di-, and tri-phosphate by cellular kinases, as illustrated in Fig. 29.3. Nucleoside triphosphate compete with the substrate for RT enzymes and/or, if incorporated, they cause chain termination because they do not have the necessary 3'-hydroxyl group, which forms 3'-5' phosphodiester linkage for further chain elongation.

2', 3'-dideoxythymidine derivatives

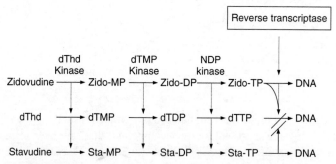

2', 3'-dideoxycytidine derivatives

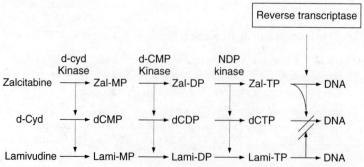

2', 3'-dideoxyadenosine derivatives

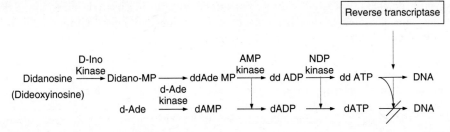

2', 3'-dideoxyguanosidine derivatives

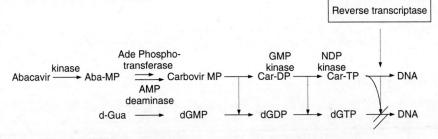

FIGURE 29.3 Mechanism of action of nucleoside RT inhibitors.

Zidovudine 3'-Azido-2',3'-dideoxythymidine
It is an analogue of thymidine.
Synthesis

Primary alcoholic group of 2'-deoxythymidine is protected as trityl derivative by reaction with trityl chloride. This is treated with methansulphonyl chloride in pyridine to make the corresponding mesylate. The mesyl group is replaced with an azide group using lithium azide in dimethylformamaide. Heating this in 55 per cent hydrobromic acid removes the trityl protection, giving zidovudine.

> Zidovudine was the first drug approved for the treatment of AIDS and HIV infection. Jerome Horwitz of Barbara Ann Karmanos Cancer Institute and Wayne State University School of Medicine first synthesised AZT in 1964, under a US National Institutes of Health (NIH) grant. AZT was originally intended to treat cancer.

Stavudine 2',3'-Didehydro-2',3'-dideoxy thymidine

Synthesis

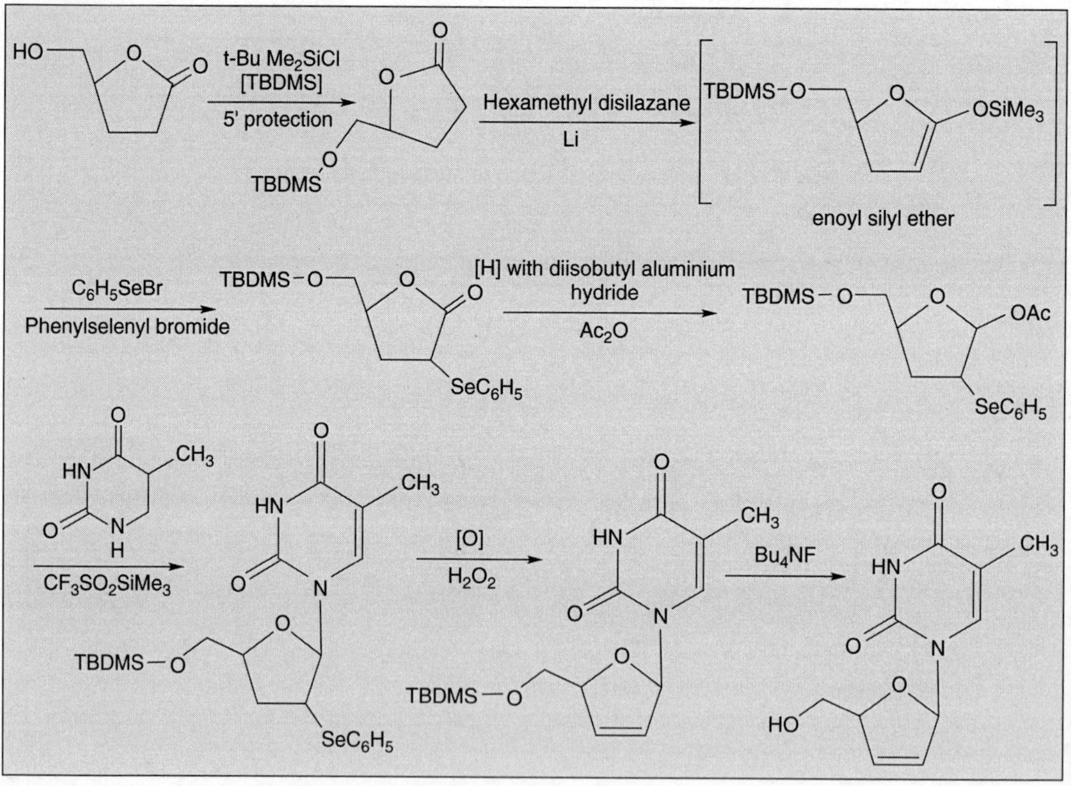

The 5' free hydroxyl group of hydroxybutyrolactone is protected as the silyl derivative. Reaction of this with lithium and hexamethyl disilazane leads to the enol silyl ether. This is then selenated using phenylselenyl bromide. Controlled reduction of the lactone carbonyl group with diisobutylaluminium hydride followed by treatment with acetic anhydride affords acetate derivative. Glycosidation of uracil with acetate intermediate using trimethylsilyl triflate gives nucleoside derivative. Oxidation of selenium with hydrogen peroxide leads to transient selenoxide; this spontaneously splits out phenylselenous acid to form olefin. The silyl protecting group is then removed by treatment with tetrabutylammonium fluoride to afford stavudine.

Zalcitabine   2', 3'-Dideoxy cytidine

Zalcitabine is an analogue of pyrimidine. It is a derivative of the naturally existing deoxycytidine, made by replacing the hydroxyl group in position 3' with hydrogen.

Synthesis

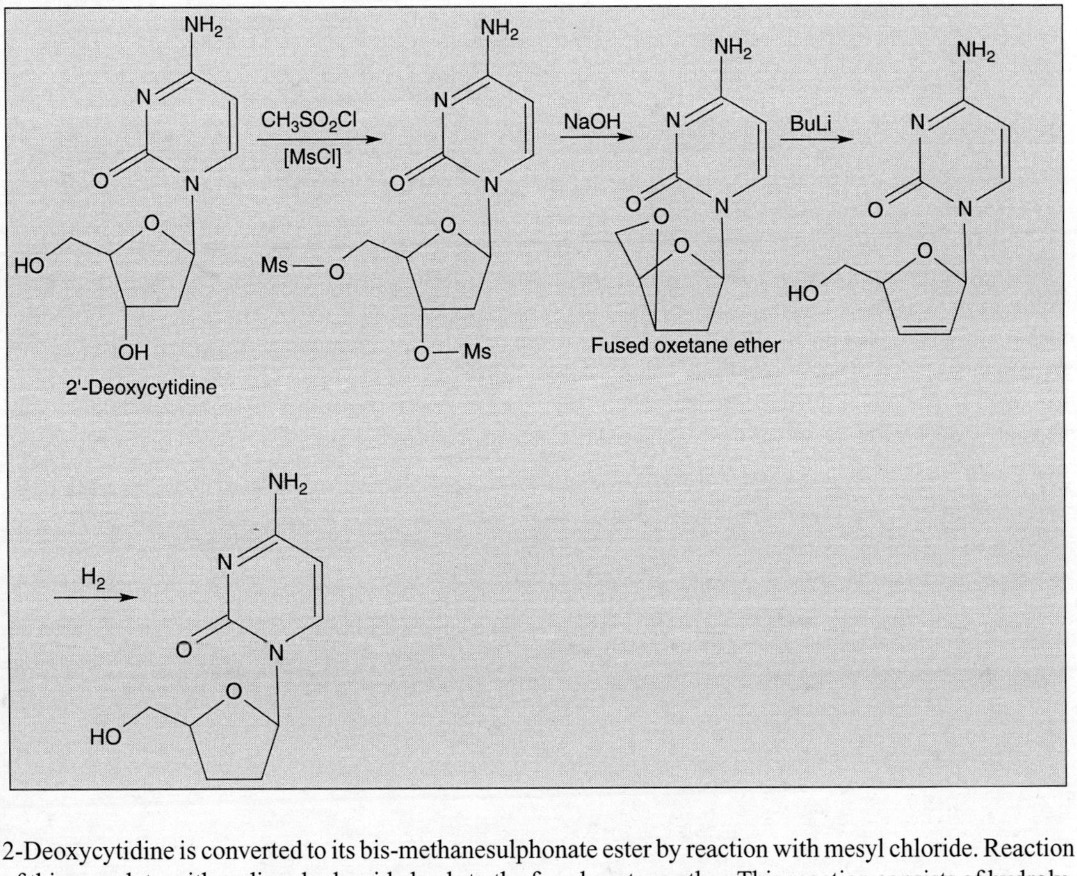

2-Deoxycytidine is converted to its bis-methanesulphonate ester by reaction with mesyl chloride. Reaction of this mesylate with sodium hydroxide leads to the fused oxetane ether. This reaction consists of hydrolysis of the 5'mesylate to an alcohol, followed by backside displacement of the 4'mesylate. Reaction of this oxetane ether with base leads to ring opening and formation of the double bond. Catalytic hydrogenation affords zalcitabine.

Lamivudine   3'-Thia-2',3'- dideoxy cytidine
Synthesis

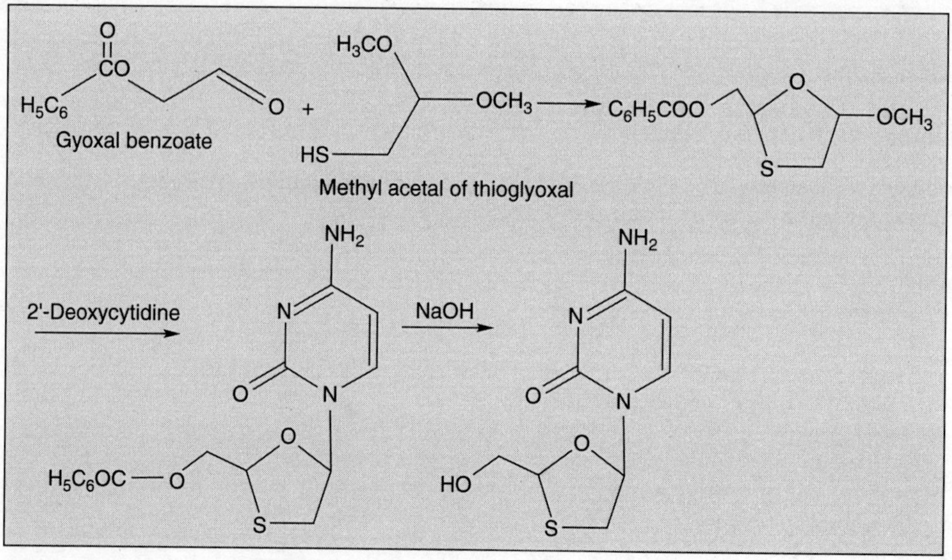

Synthesis starts with the formation of the thioacetal from glyoxal benzoate and the methyl acetal of thioglyoxal. Reaction of this with 2'-deoxycytidine gives nucleoside derivative. The benzoyl group is then removed by reaction with base to afford lamivudine.

Didanosine   2', 3'-Dideoxy inosine

Didanosine (ddI) is a nucleoside analogue of adenosine. It differs from other nucleoside analogues because it does not have any of the regular bases; instead, it has hypoxanthine attached to the sugar ring.

Synthesis

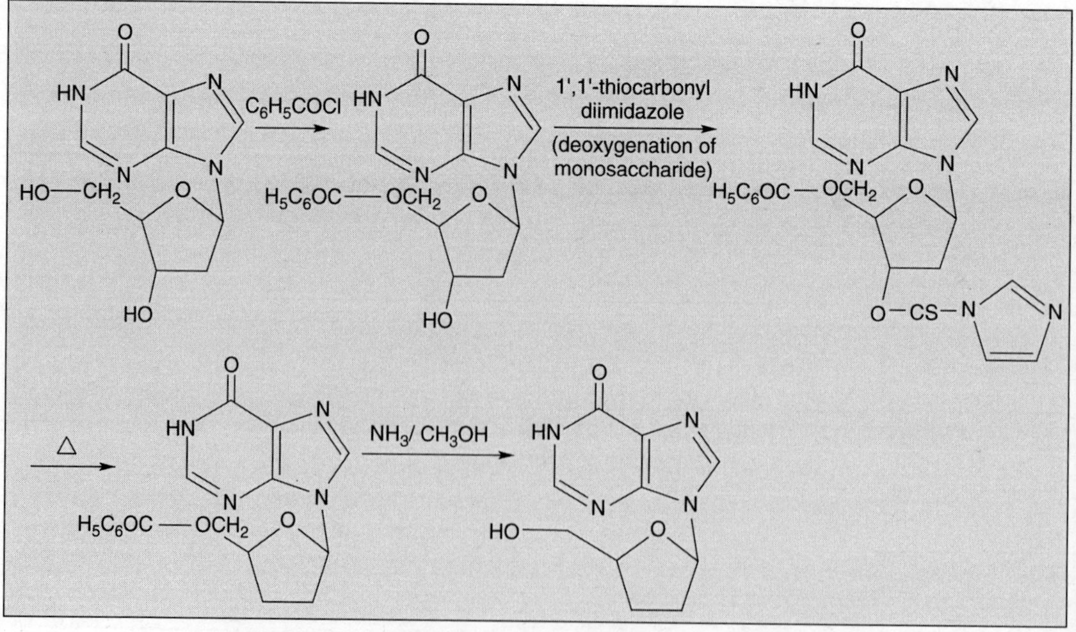

5'-Hydroxyl group of 2'-deoxyinosine is protected as benzoyl ester. 3'-Hydroxyl group on reaction with thiocarbonyl diimidazole gives thiocarbamate intermediate, which on heating undergoes deoxygenation reaction and gives 2',3'-dideoxyinosine derivative. The benzoyl group is then removed by reaction with base to afford didanosine.

Abacavir [(1R)-4-[2-Amino-6-(cyclopropylamino)purin-9-yl]-1-cyclopent-2-enyl]methanol
In this drug, the carbocyclic ring is attached to the base instead of the sugar.

Synthesis

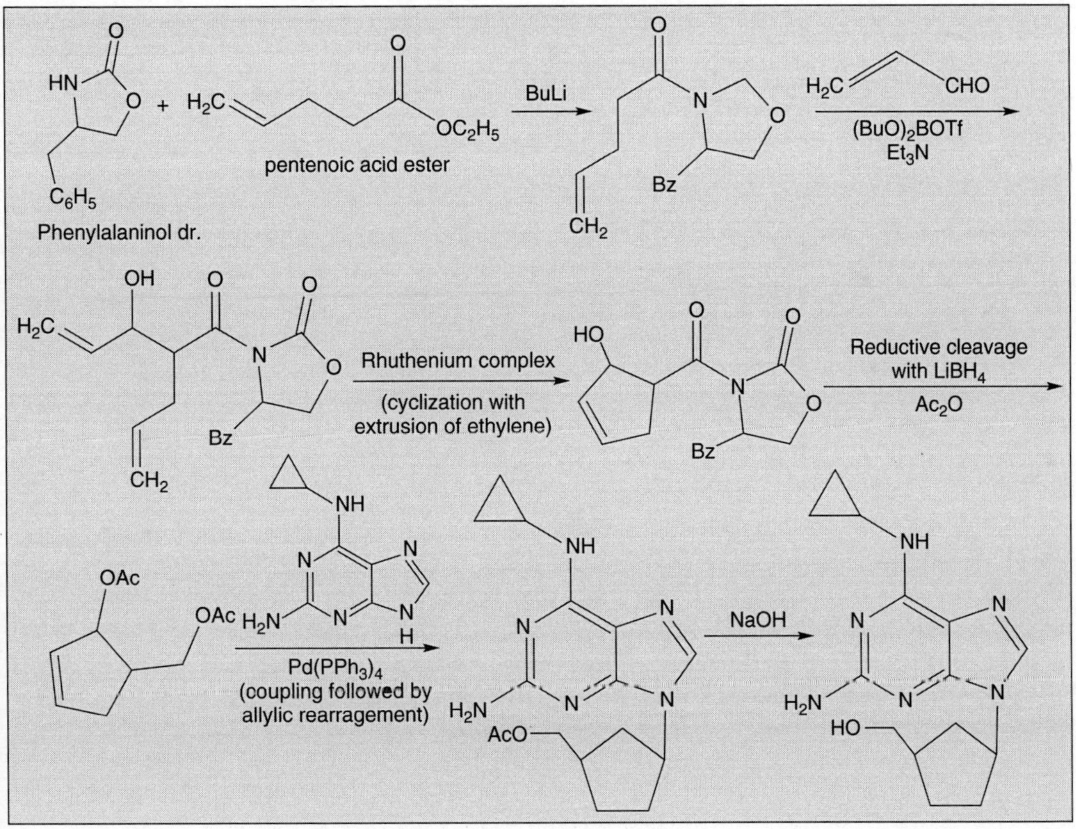

Acylation of phenylalaninol derivative with pentenoic acid ethyl ester gives the imide. This on treatment with triethylamine and the triflate from dibutylboronic acid leads to the transient enol borate. Aldol condensation of the reactive intermediate with acroelin gives adducts, which is treated with ruthenium complex. That catalyses an olefin metathesis reaction involving the bis-terminal diene, which results in cyclization with extrusion of ethylene. Reduction with lithium borohydride leads to initial formation of a carbinolamine, which cleaves and undergoes further reduction to give glycol. Acetylation of the diol then gives diacetate. This on reaction with guanine derivative in presence of palladium leads to coupling, followed by allylic rearrangement to give acetyl derivative. Saponification affords abacavir.

Adefovir 2-(6-Aminopurin-9-yl)ethoxymethylphosphonic acid

Synthesis

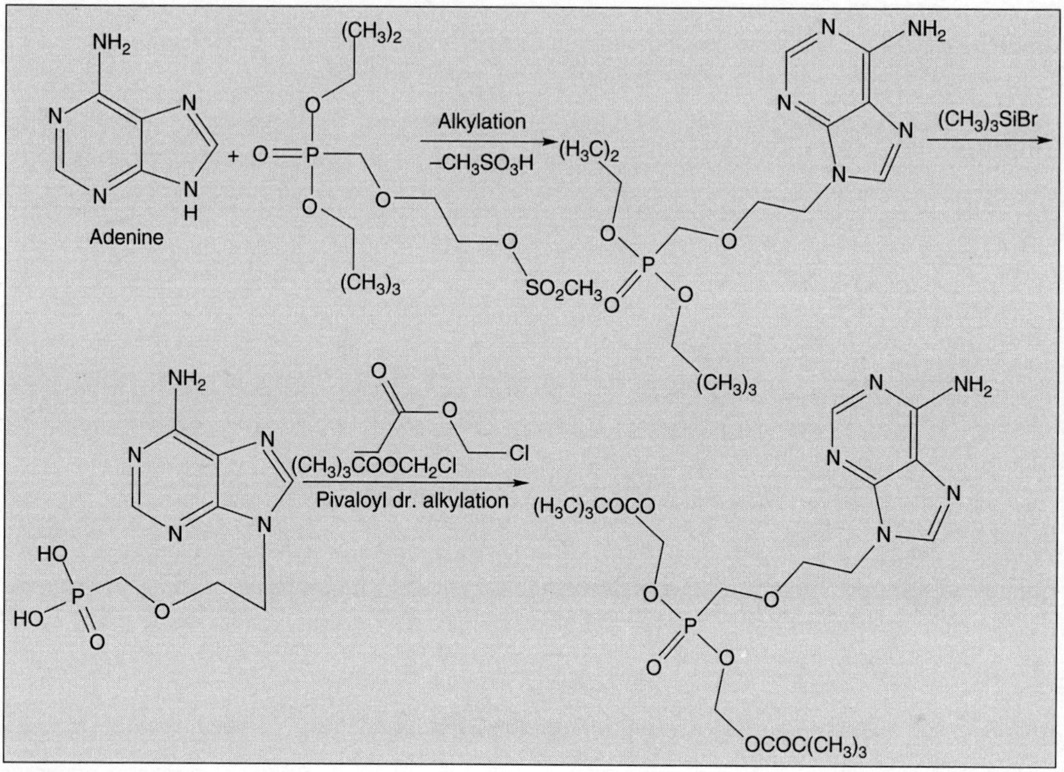

Alkylation of adenine with the side-chain in the presence of base leads to displacement of the mesylate and formation of *N*-alkyl derivative. The phosphate esters are then cleaved by reaction with trimethylsilyl bromide. Alkylation of the hydroxyl groups with pivaloyloxymethyl chloride gives adefovir dipivalate.

29.1.4 Non-nucleoside RT Inhibitors (NNRTI)

Mechanism of action: NNRTIs directly interact with the RT at a non-substrate binding site (allosteric site) and inhibits specifically HIV-1 RT non-competitively. The NNRTI preferentially inhibits the RNA-dependent DNA polymerization step.

Nevirapine 1-Cyclopropyl-5,11-dihydro-4-methyl-6H-dipyrido [3,2-b:2',3'-e][1,4] diazepin-6-one

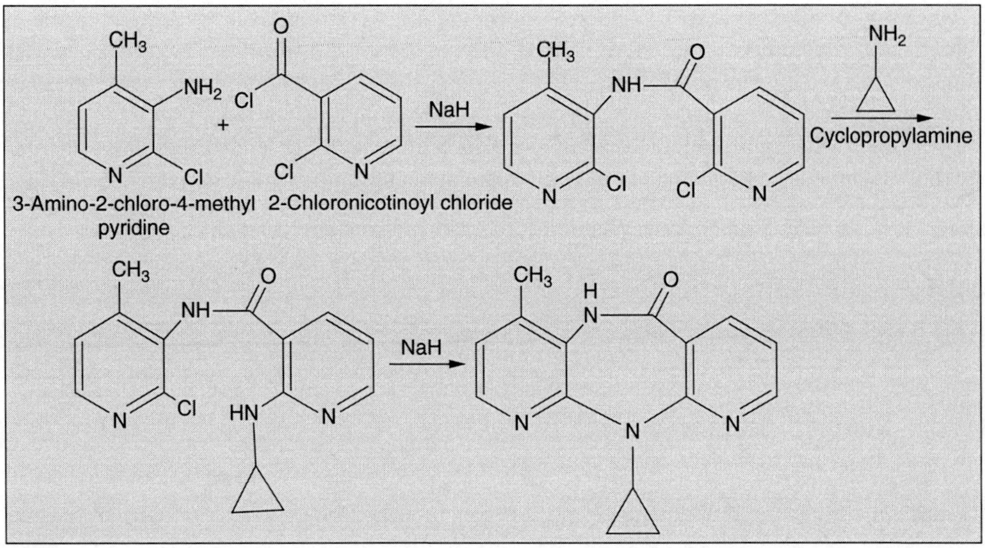

Medicinal Chemistry 515

Synthesis

Acylation of 3-amino-2-chloro-4-methylpyridine with 2-chloronicotinoyl chloride gives dipyridyl amide derivative. One of the chlorine is displaced with cyclopropyl amino group and treatment with strong base leads to cyclization and affords nevirapine.

> Nevirapine was the first NNRTI approved by the Food and Drug Administration (FDA) of the United States. It was approved on 21 June 1996 for adults and on 11 September 1998 for children.

Delavirdine
Synthesis

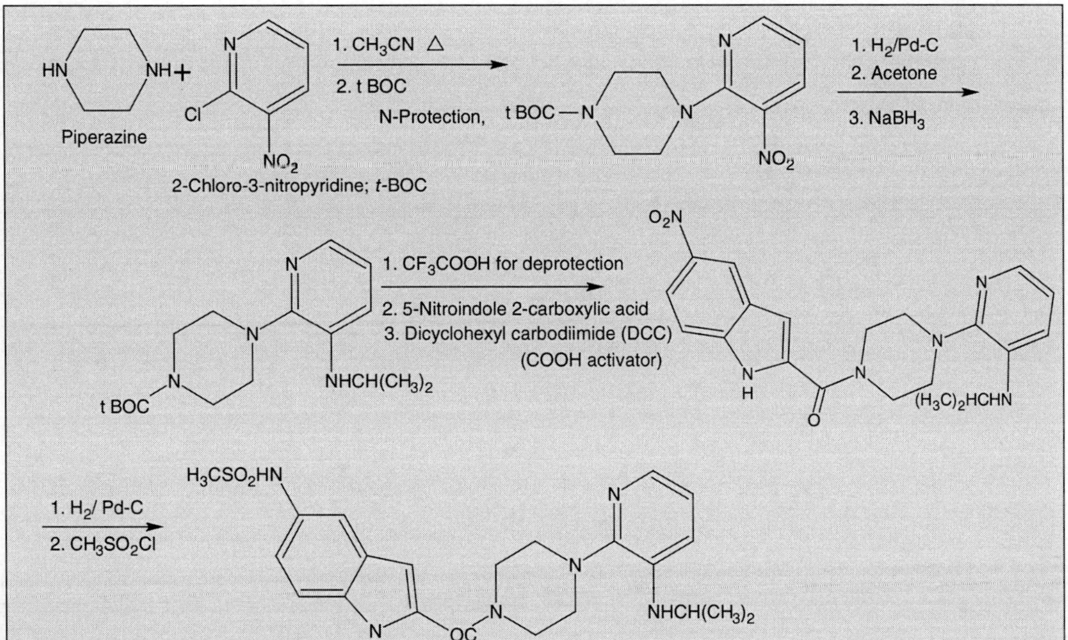

Alkylation of piperazine with 2-chloro-3-nitropyridine gives mono-substituted piperazine derivative. Free amino function is protected as t-BOC and reduction of nitro group gives amine, which on reaction with acetone affords Schiff base, and reduction of this with sodium borohydride gives isopropylamino derivative. t-BOC is deprotected by treatment with trifluoroacetic acid; free amino group forms amide by reaction with 5-nitroindole-2-carboxylic acid in presence of DCC. Nitro group is then reduced and reaction with methanesulphonyl chloride gives the methanesulphonamide derivative delavirdine.

Loviride 2-[(2-Acetyl-5-methylphenyl)amino]-2-(2,6-dichlorophenyl)acetamide

Synthesis

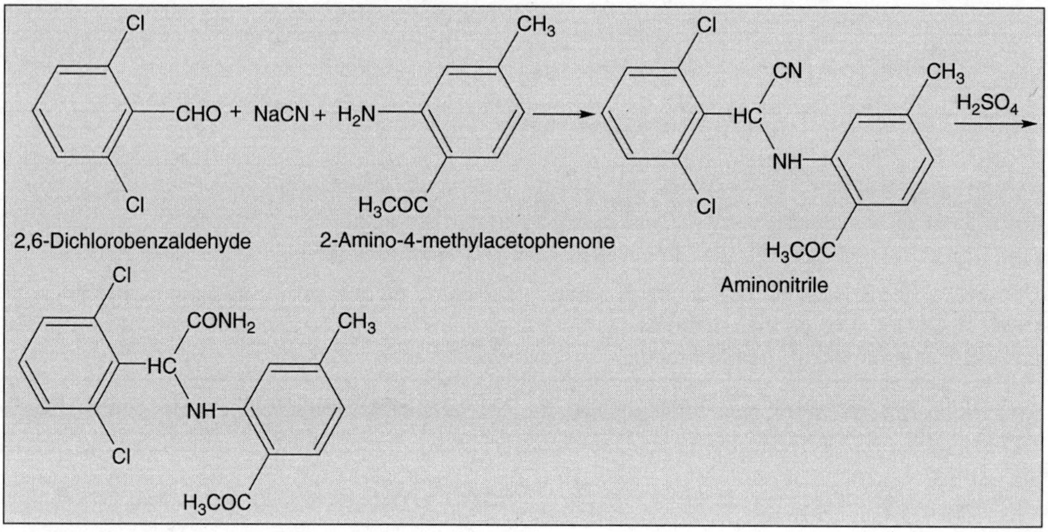

Reaction between 2,6-dichlorobenzaldehyde with sodium cyanide gives cyanohydrin intermediate, which on reaction with 2-amino-4-methylacetophenone affords aminonitrile. Controlled hydrolysis of nitrile affords loviride.

Trovirdine N-(5-Bromo-2-pyridyl)-N'-[2-(2-pyridyl)ethyl]thiourea

Synthesis

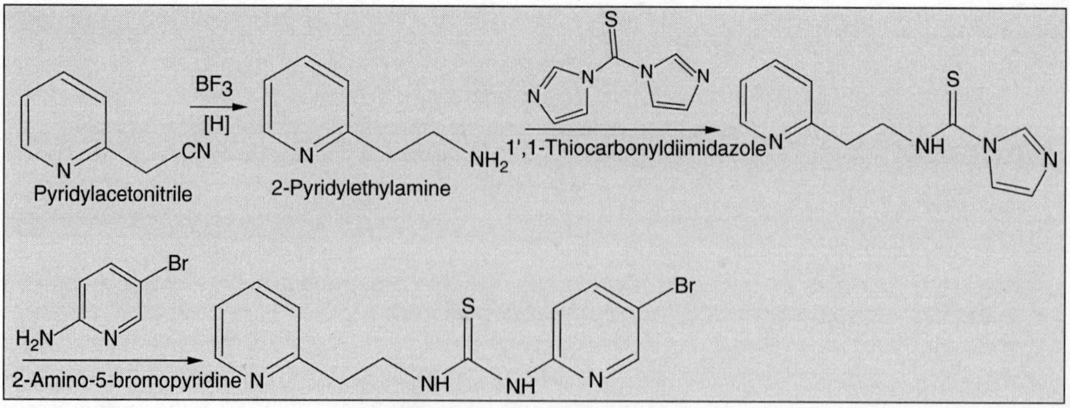

Reduction of pyridylacetonitrile gives pyridyl-2-ethylamine. This on reaction with 1,1'-thiocarbonyldiimidazole gives intermediate thiourea, which on reaction with 2-amino-5-bromopyridine affords pyridylethyl thiourea derivative trovirdine.

Emivirine 6-Benzyl-1-(ethoxymethyl)-5-isopropyl-1,2,3,4-tetrahydro-2,4-pyrimidinedione

Synthesis

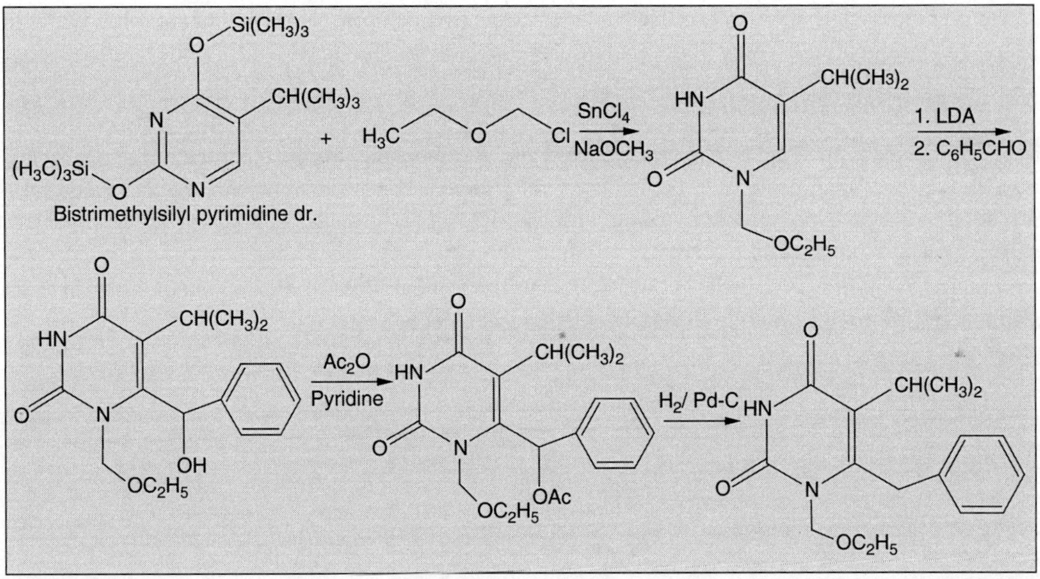

N-Alkylation of bistrimethylsilyl pyrimidine derivative with ethoxymethyl chloride gives 1-(ethoxymethyl)-5-isopropyl-1,2,3,4-tetrahydro-2,4-pyrimidinedione; this on reaction with benzaldehyde in presence of LDA affords 1-(ethoxymethyl)-6-[hydroxy(phenyl)methyl]-5-isopropyl-1,2,3,4-tetrahydro-2,4-pyrimidinedione. Treatment with acetic anhydride gives acetoxy derivative, which on hydrohenolysis with palladium-charcoal affords emivirine.

Efavirenz (S)-6-chloro-4-(cyclopropylethynyl)-4-(trifluoromethyl)-1H-benzo[d][1,3]oxazin-2(4H)-one

Synthesis

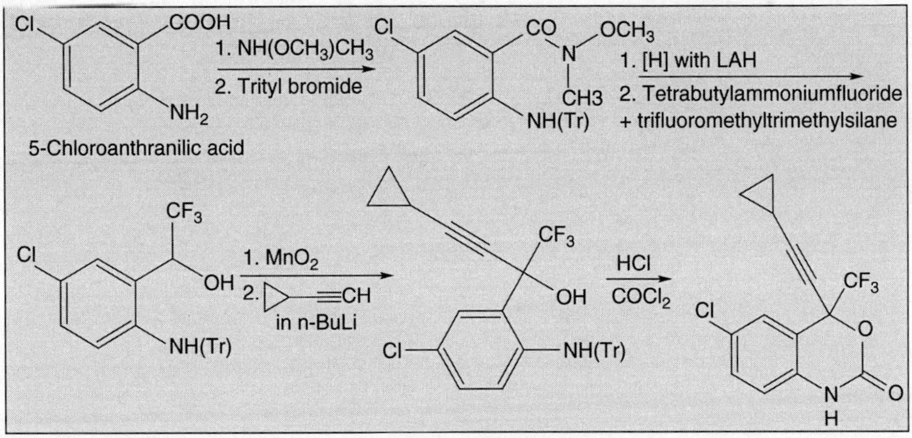

5-Chloroanthranilic acid is converted to Weinreb amide by reaction with methoxymethylamine; this followed by trityl protection of the amine moiety gives intermediate. Weinreb amide with lithium aluminium hydride followed by addition of trifluoromethyl anion (generated *in situ* by the addition of tetrabutylammonium fluoride to a solution of trifluoromethyltrimethylsilane in tetrahydrofuran) results in the formation of the secondary alcohols. Manganese dioxide oxidation of these secondary alcohols to the corresponding amino ketones is followed by treatment with cyclopropyl lithium acetylides in tetrahydrofuran to provide the tertiary alcohols. Detritylation using hydrochloric acid in methanol followed by ring closure with phosgene and Hunig's base (N,N-diisopropylethylamine) in toluene provide benzoxazinones.

29.1.5 Protease Inhibitors

The HIV protease is encoded by the viral genome and is responsible for the cleavage of the *gag-pol* precursor and *pol* precursor proteins to the mature viral proteins. This proteolytic cleavage is needed for the maturation and, hence, the infectivity of the virus particles. HIV protease inhibitors may be expected to suppress virus production and infectivity.

Saquinavir *N*-[1-Benzyl-2-hydroxy-3-[3-(*tert*-butylcarbamoyl)-1,2,3,4,4*a*,5,6,7,8,8*a*- decahydroisoquinolin-2-yl]-propyl]-2-quinolin-2-ylcarbonylamino-butanediamide

Synthesis

Step A Synthesis of decahydroisoquinoline moiety

Phenylalanine on reaction with formaldehyde and hydrochloric acid undergoes chlormethylation, which is followed by dehydrochlorination to form tetrahydroisoquinoline derivative. Reduction with rhodium-charcoal gives perhydroisoquinoline. Amino function is protected with BOC, and carboxyl group is activated with DCC and coupled with *t*-butylamine to give amide. Hydrogenolysis removes BOC and gives intermediate **A**.

Step B

Amino group of phenylalanine is protected as BOC, and reaction with diazomethane gives diazomethyl ketone. Reaction of this with hydrochloric acid gives chloromethyl ketone with the loss of nitrogen, and ketone is further reduced to alcohol with sodium borohydride to give intermediate **B**.

Step C

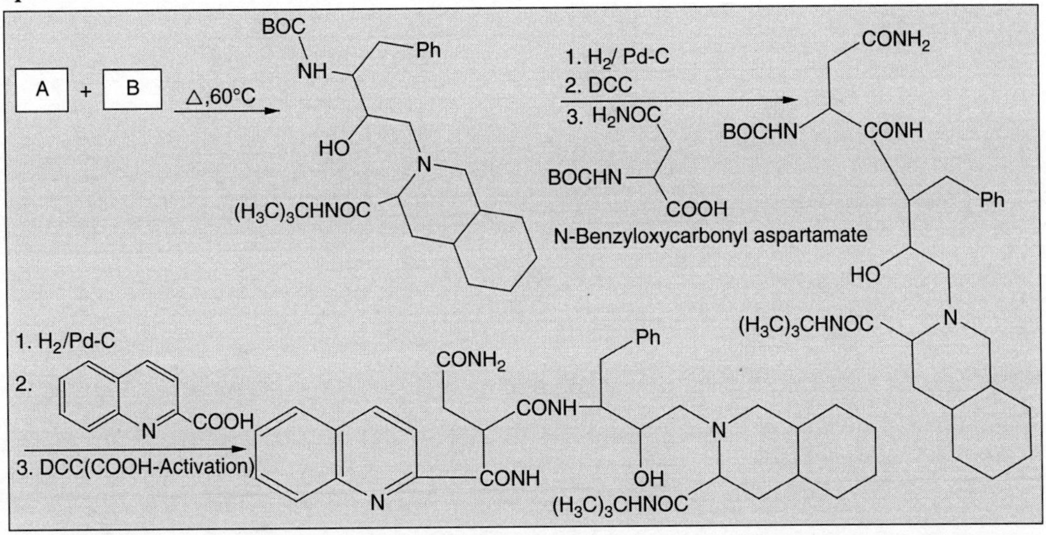

Intermediate **A** is alkylated with **B**, followed by hydrogenolysis to deprotect amine. This reacts with *N*-BOC-protected aspartamate in presence of DCC to give new amide. This on further hydrogenolysis deprotects amine and reacts with quinoline-2-carboxylic acid in presence of DCC to afford saquinavir.

Indinavir 1-[2-Hydroxy-4- [(2-hydroxy-2,3-dihydro- 1*H*-inden-1-yl) carbamoyl]-5 -phenyl-pentyl]-4- (pyridin-3-ylmethyl)- *N-tert*-butyl-piperazine-2-carboxamide

Synthesis

Step A Preparation of indane moiety

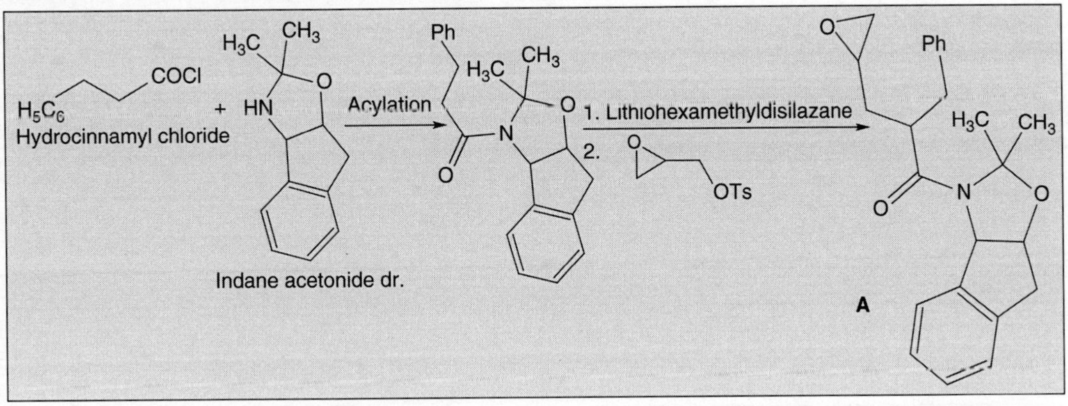

Reaction of 1-amino-2-indanol with acetone gives starting cyclic carbinolamine derivative (indane acetonide). Acylation of this acetonide with hydrocinnamyl chloride gives the amide. Treatment of this amide with lithiohexamethyldisilazane forms anion, where alkylation carried out with tosyl derivative affords glycidol intermediate **A**.

Step B

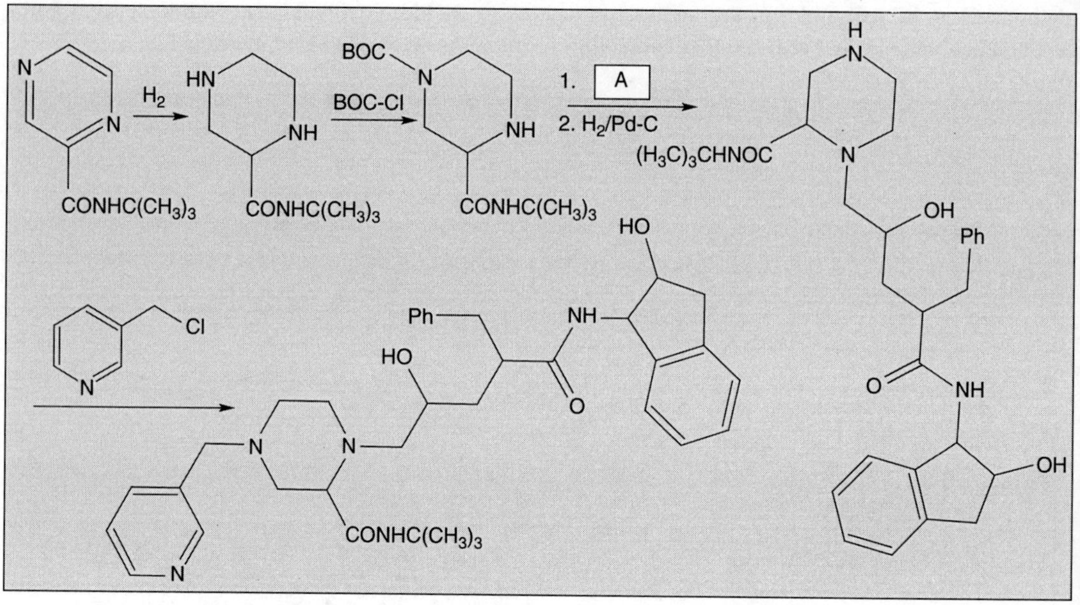

Catalytic reduction of pyrazine carboxamide gives the corresponding piperazine. Treatment of this with BOC-Cl protects selectively at the less steric 4th position. This on reaction with glycidol intermediate **A** leads to attack of the free amino group of piperazine on the epoxide, with consequent ring opening and formation of the alcohol. BOC is then removed and alkylation with 3-chloromethylpyridine affords indinavir.

Ritonavir: 1,3-Thiazol-5-ylmethyl [3-hydroxy-5- [3-methyl-2-[methyl- [(2-propan-2-yl-1,3-thiazol-4-yl)methyl] carbamoyl] amino-butanoyl] amino-1,6-diphenyl-hexan-2-yl] aminoformate

Step A

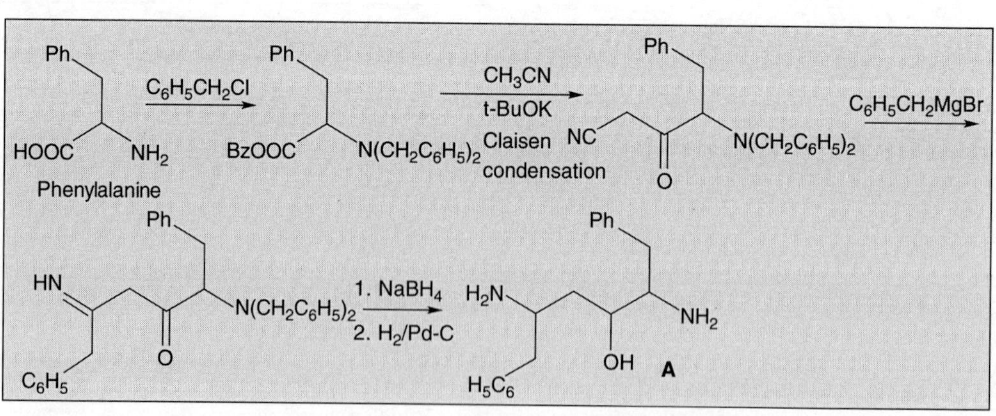

Reaction of phenylalanine with benzyl chloride gives the N,N-bisbenzyl ester. Claisen condensation of the ester with the anion from acetonitrile leads to displacement of benzyloxide and thus forms cyanoketone. Reaction of this with the benzylmagnesium bromide leads to addition to the nitrile to form an imine; the imine is reduced and hydrogenolysis leads to debenzylation and affords diamine intermediate **A**.

Step B

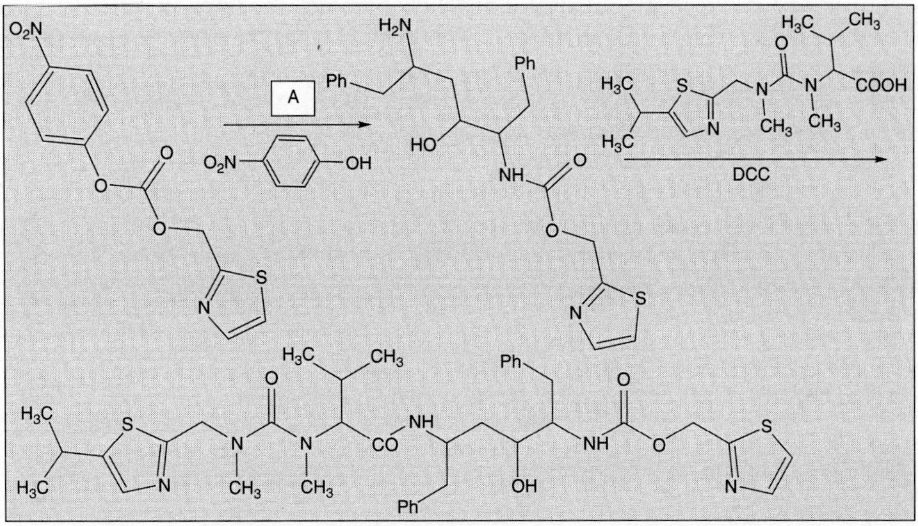

Diamine intermediate A reacts with 4-nitrophenoxyester of the carbonyloxy methyl thiazole, which leads to net replacement of nitrophenol and formation of the urethane. The second fragment thiazolylurea-substituted alanine derivative is condensed with free amine of urethane derivative in presence of DCC to afford ritonavir.

Nelfinavir 2-[2-Hydroxy-3-(3-hydroxy-2-methyl-benzoyl) amino-4-phenylsulphanyl-butyl] -N-tert-butyl- 1,2,3,4,4a,5,6,7,8,8a- decahydroisoquinoline-3-carboxamide
Synthesis

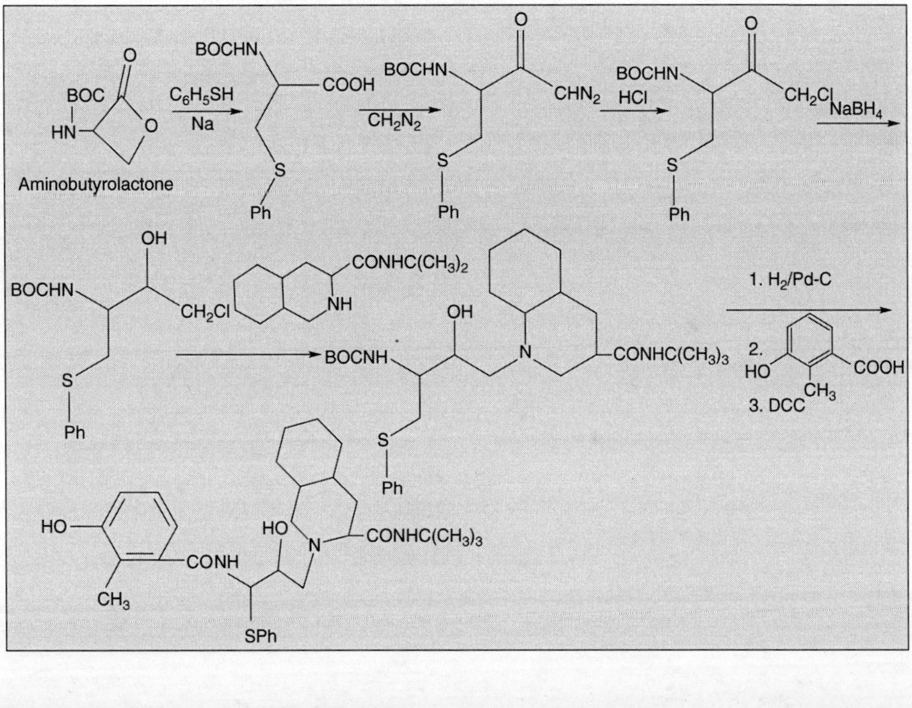

BOC-protected aminobutyrolactone reacts with the anion from thiophenol, and ring opening takes place to form an acid intermediate; reaction of this with diazomethane leads to corresponding diazoketone. This on treatment with hydrogen chloride gives chloro ketone; ketone is reduced to the alcohol with sodium borohydride. Treatment of this product with perhydroisoquinoline derivative (saquinavir synthesis A) leads to N-alkylation. BOC is removed by hydrogenolysis, and amidation with 3-hydroxy-2-methylbenzoic acid in presence of DCC affords nelfinavir.

> Nelfinavir acts broadly against cancer tumours, and is currently under investigation for use as an anti-cancer agent. It inhibits growth and induces apoptosis in prostate cancer and in non-small-cell lung cancer cell lines in the laboratory, and has similar effects in laboratory mice. It also induces protein misfolding in the endoplasmic reticulum of cancer cells, and in a small clinical trial had some success against liposarcoma.

29.2 ANTI-HERPES SIMPLEX VIRUS (HSV) AGENTS

Infection with HSV-1 causes diseases of the mouth, face, skin, hands, oesophagus, or brain, whereas HSV-2 usually causes infections of the genitals, rectum, skin, hands, or meninges.

29.2.1 Classification of Drugs

1. Pyrimidine nucleoside analogues: Brivudine, Sorivudine, Trifluridine, Fialuridine, Idoxuridine, Netivudine
2. Purine nucleosides: Vidarabine
3. Acyclic nucleoside analogues: Acyclovir, Ganciclovir, Valaciclovir, Penciclovir, and Famciclovir

Pyrimidine nucleosides (2-Deoxy thymidine derivatives)

	R	R_1
Brivudine	—CH=CHBr	H
Sorivudine	—CH=CHBr	OH
Netivudine	—CH=CHCH$_3$	OH
Idoxuridine	I	H
Fialuridine	I	F
Trifluridine	CF$_3$	H

Mechanism of Action:

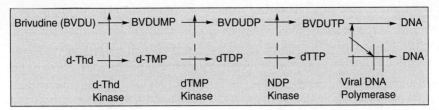

Brivudine is specifically recognized as a substrate by HSV-1-encoded thymidine kinases, which convert the compound subsequently to its 5'-mono-, di-, and tri-phosphate. Brivudine triphosphate can act in a dual fashion with the viral DNA polymerase,

1. as a competitive inhibitor with respect to the natural substrate (dTTP), or
2. as an alternate substrate, which then allows BVDUTP to be incorporated into the DNA chain. This may, in turn, affect both the stability and functioning of DNA.

Brivudine: (5-[(E)-2-Bromoethenyl]-1-[(2R,4S,5R)-4-hydroxy-5-(hydroxymethyl)oxolan-2-yl] pyrimidine-2,4-dione), **and Sorivudine:** (5-[(E)-2-Bromoethenyl]-1-[(2R,3S,4S,5R)-3,4-dihydroxy-5-(hydroxymethyl)oxolan-2-yl]pyrimidine-2,4-dione)

Synthesis

Reaction between uracil arabinoside and mercuric chloride affords the mercurated derivative. Palladium-catalysed coupling of this intermediate with ethyl acrylate adds the side-chain. Saponification of ester gives acid derivative; treatment of this with N-bromosuccinimide results in a Borodin-like reaction, with the net replacement of the carboxyl group by bromine to afford sorivudine.

Idoxuridine 5-Iodo-1-(2-deoxyribofuranosyl)pyrimidin-2,4-(1H.3H)-dione
Synthesis

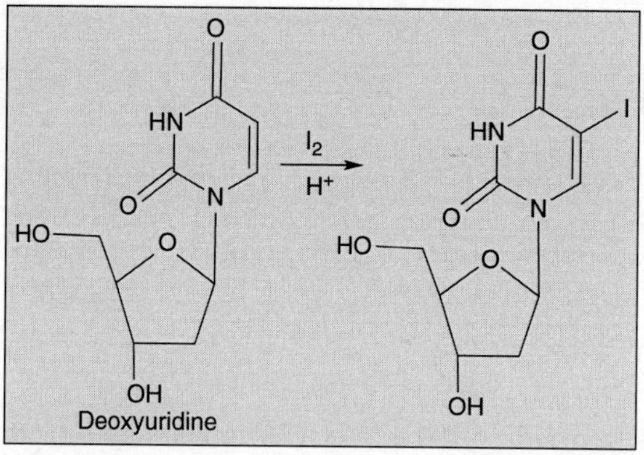

Iodination of 2'-deoxyuridine affords idoxuridine. It is a nucleoside analogue, a modified form of deoxyuridine, similar enough to be incorporated into viral DNA replication, but the iodine atom added to the uracil component blocks base pairing.

Purine nucleosides

Vidarabine   9-β-Arabinofuranosyl-6-amino-9-H-purine

It is an adenosine analogue with an altered sugar (arabinose is the 2'-epimer of ribose). It is active against HSV, poxviruses, rhabdoviruses, hepadenaviruses, and some RNA tumour viruses. The mechanism of action is similar to brivudine, wherein it inhibits viral DNA polymerase activity.

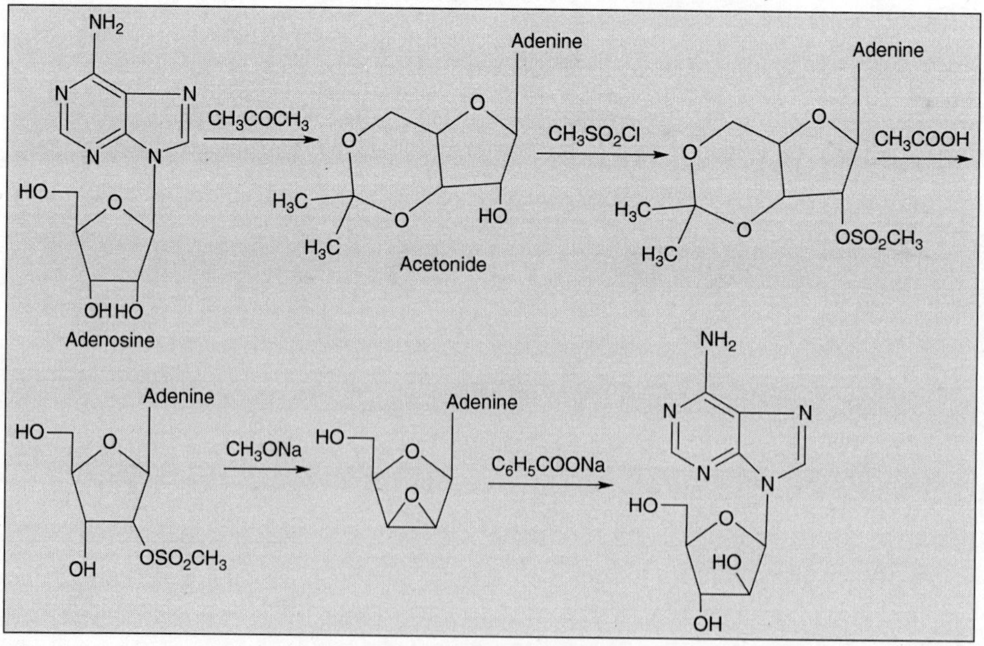

It is synthesised from 9-(3',5'-O-isopropyliden–β-D–xylofuranoside)adenine, which is reacted with methanesulphonyl chloride to make the mesylate 9-(3',5'-O-isopropyliden-2'-O-methansulphonyl-β-D-xylofuranoside)adenine. Prolonged heating in 90 per cent acetic acid removes the acetonyl protective group from the resulting compound, giving the product. Reacting this with sodium methoxide leads to the formation of an epoxide 9-(2',3'-anhydro-β-luxofuranosyl)adenine. Finally, heating this epoxide with sodium opens the epoxide ring in the dimethylformamide–water system to make the corresponding dihydroxy derivative, vidarabine.

Acyclic nucleoside analogues

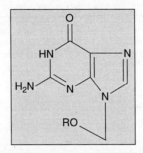

	R
Acyclovir	—$CH_2 CH_2 OH$
Valaciclovir	—$CH_2 CH_2 OCOCH(NH_2)CH(CH_3)_2$
Ganciclovir	—$CH(CH_2OH)_2$
Penciclovir (instead of OR)	—$CH_2 CH_2 CH(CH_2OH)$

Mechanism of Action:

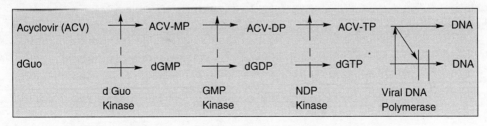

The mechanism of action is similar to brivudine. Acyclovir triphosphate acts as a competitive inhibitor/alternate substrate/chain terminator at the viral DNA polymerase level.

Acyclovir (2-Amino-1,9-dihydro-9-[(2-hydroxyethoxy)methyl]-6H-purin-6-one),

and Valaciclovir (2-[(2-amino-6-oxo-3,9-dihydropurin-9-yl)methoxy] ethyl-2-amino-3-methyl-butanoate)

Synthesis

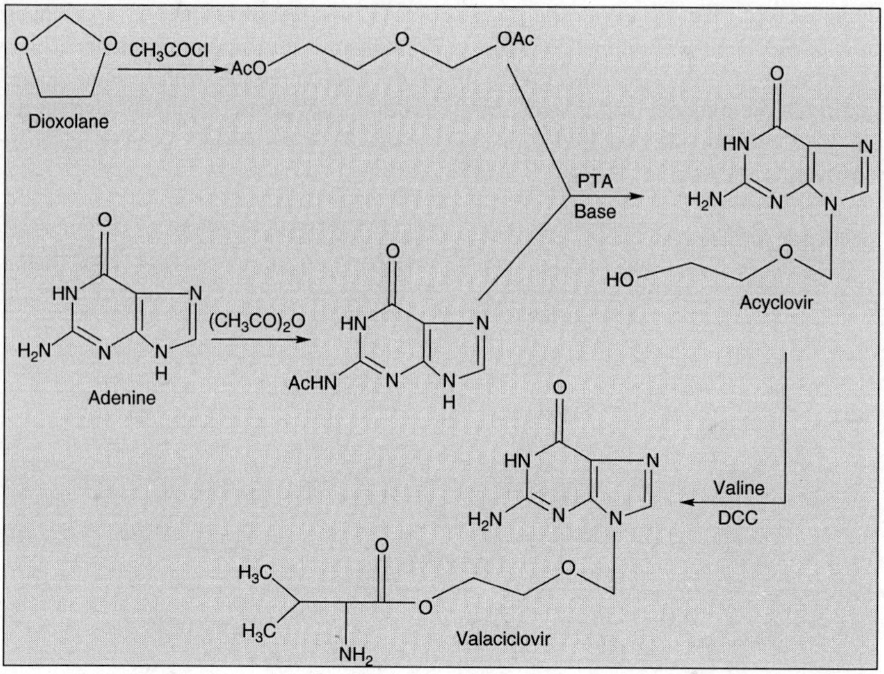

The side-chain is prepared by acylation of dioxolane with acetyl chloride to give the ring-opened diacetate derivative. Reaction of this with acetyl guanine in presence of 4-toluenesulphonic acid leads to open-chain nucleoside derivative. Saponification with base affords acyclovir. Acyclovir is converted to valaciclovir by reaction with valine in presence of DCC.

Acyclovir is active against most species in the herpesvirus family. In descending order of activity: herpes simplex virus type I (HSV-1), herpes simplex virus type II (HSV-2), varicella zoster virus (VZV), Epstein-Barr virus (EBV), and cytomegalovirus (CMV).

Ganciclovir 2-Amino-9-(1,3-dihydroxypropan-2-yloxymethyl)-3H-purin-6-one
Synthesis

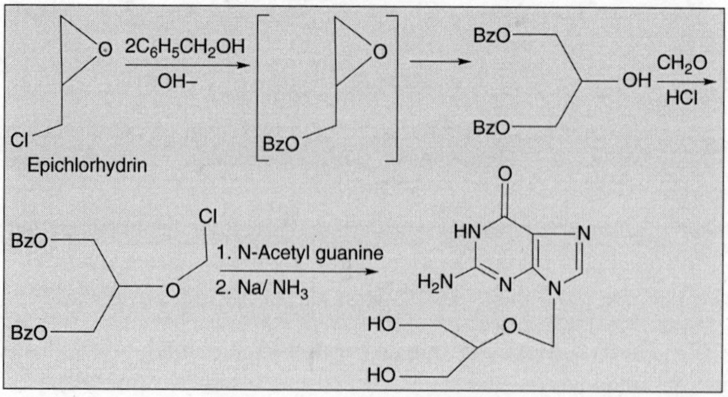

Epichlorhydrin on reaction with two equivalents of benzyl alcohol anion forms first glycidic ether; this opens and reacts with second mole of benzyl alcohol to form triol derivative. Reaction of this with formaldehyde in presence of hydrogen chloride gives chloromethyl ether derivative, which on reaction with N-acetyl guanine followed by reduction with sodium and liquid ammonia results in ganciclovir.

29.3 MISCELLANEOUS AGENTS

Ribavirin 1-β-D-Ribofuranosyl-1H-1,2,4-triazol-3-carboxamide

Ribavirin is active against several DNA and RNA viruses, including herpes, influenza A and B, respiratory syncytial, orthomyxo, paramyxo, arena, bunya, adeno, pox, and retroviruses.

Synthesis

Ribavirin is synthesised by reacting methyl ester of 1,2,4-triazol-3-carboxylic acid with O-2,3,5-triacetyl-β-D-ribofuranose to make methyl ester of 1-O-2,3,5-triacetyl-β-D-ribofuranosyl-1,2,4-triazol-3-carboxylic acid, which is treated with an ammonia solution of methanol to simultaneously deacylate the carbohydrate part and amidation of the carboxyl part of the product to afford ribavirin.

Amantadine Adamantan-1-amine.

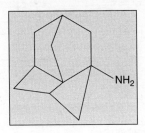

Synthesis: Refer to anti-Alzheimer's agents (Chapter 13).

Mechanism of Action: Amantadine inhibits the replication of influenza A by inhibiting viral uncoating.

29.4 NEWER DRUGS

Emtricitabine 4-Amino-5-fluoro-1- [2-(hydroxymethyl)- 1,3-oxathiolan-5-yl]- pyrimidin-2-one

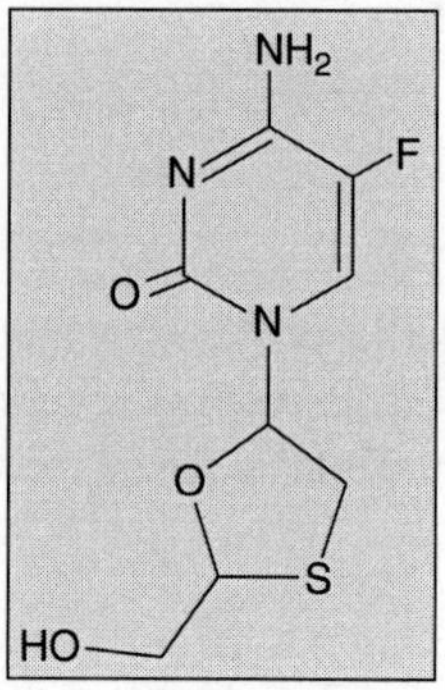

It is an analogue of cytidine. The drug works by inhibiting HIV reverse transcriptase.

Tenofovir 1-(6-Aminopurin-9-yl) propan-2-yloxymethylphosphonic acid

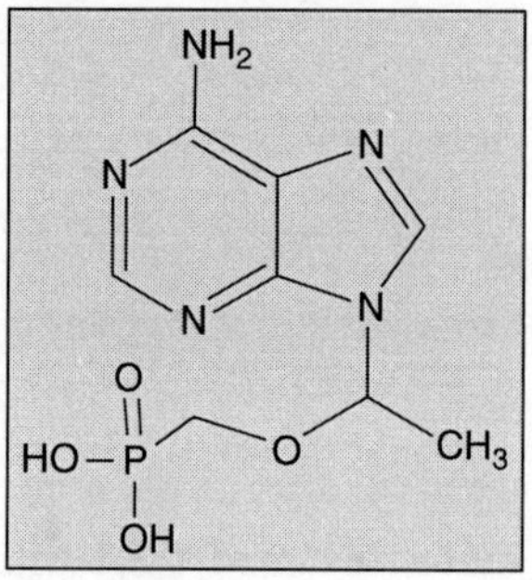

It belongs to a class of antiretroviral drugs known as nucleotide analogue HIV reverse transcriptase inhibitors.

Etravirine 4-[6-Amino-5-bromo-2-[(4-cyanophenyl)amino]pyrimidin-4-yl]oxy-3,5-dimethylbenzonitrile

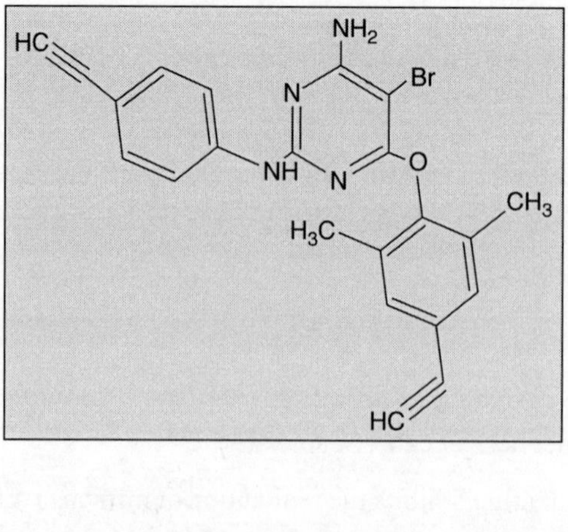

It is a second-generation non-nucleoside HIV reverse transcriptase inhibitor.

Darunavir [(1*R*,5*S*,6*R*)-2,8-Dioxabicyclo[3.3.0]oct-6-yl] *N*-[(2*S*,3*R*)-4- [(4-aminophenyl)sulphonyl-(2-methylpropyl)amino]-3-hydroxy-1-phenyl- butan-2-yl] carbamate

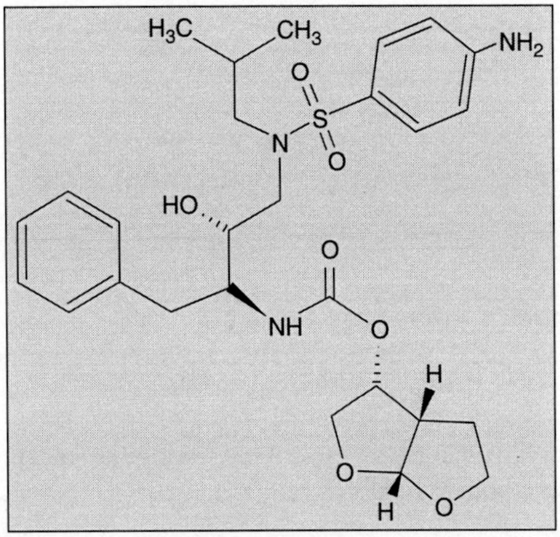

It is a second-generation HIV protease inhibitor (PI), designed specifically to overcome problems with the older agents in this class, such as indinavir. Early PIs often have severe side-effects and drug toxicities, require a high therapeutic dose, are costly to manufacture, and show a disturbing susceptibility to drug-resistant mutations. Such mutations can develop in as little as a year of use, and effectively render the drugs useless.

Raltegravir *N*-(2-(4-(4-Fluorobenzylcarbamoyl) -5-hydroxy-1-methyl-6-oxo-1, 6-dihydropyrimidin-2-yl)propan-2-yl) -5-methyl-1,3,4-oxadiazole-2- carboxamide

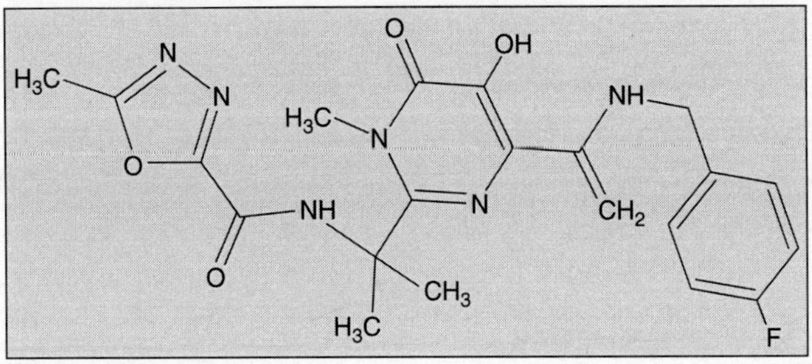

It targets integrase, an HIV enzyme that integrates the viral genetic material into human chromosomes, a critical step in the pathogenesis of HIV.

Maraviroc 4,4-Difluoro-*N*-{(1*S*)-3-[3-(3-isopropyl- 5-methyl-4*H*-1,2,4-triazol-4-yl)-8-azabicyclo[3.2.1]oct-8-yl]-1- phenylpropyl}cyclohexanecarboxamide

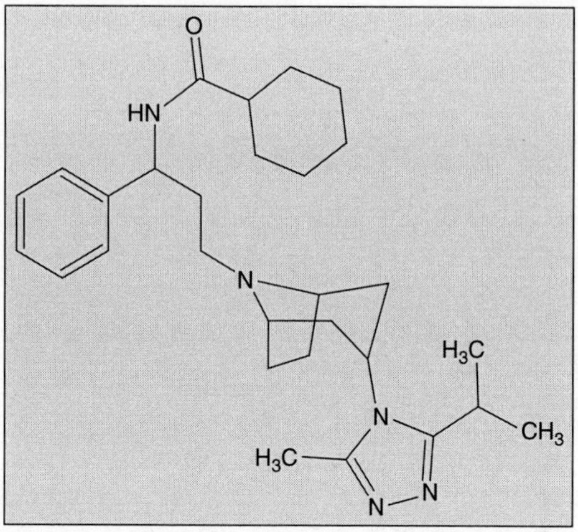

It is an entry inhibitor. Specifically, maraviroc blocks the chemokine receptor CCR5, which HIV uses as a co-receptor to bind and enter a human helper T cell.

FURTHER READINGS

1. Abu-ata et al., 'HIV Therapeutics: Past, Present, and Future', *Advances in Pharmacology*, 49, 1–40, 2000.
2. Challand, R., et al., (ed.), 'Herpes Viruses', Chapter 6, In: Antiviral Chemotherapy, Biochemical & Medicinal Chemistry Series, Spectrum & University Science Books, Sausalito, CA, pp. 54–67, 1996.
3. Clercq, E. De, 'NNRTIs: Past, Present, and Future', *Chemistry and Biodiversity*, 1, 44–64, 2004.
4. Kleymann, G., 'Discovery, SAR and Medicinal Chemistry of Herpesvirus Helicase Primase Inhibitors', *Current Medicinal Chemistry—Anti-Infective Agents*, Volume 3(1), pp. 69–83, March 2004.
5. Sriram, D. and P. Yogeeswari, 'Towards the Design and Development of Agents with Broad Spectrum Chemotherapeutic Properties for the Effective Treatment of HIV/AIDS', *Current Medicinal Chemistry*, 10, 1909–1915, 2003.

MULTIPLE-CHOICE QUESTIONS

1. Mechanism of action of glycyrrhizin is
   a. Uncoating inhibitor
   b. Adsorption inhibitor
   c. Reverse transcriptase inhibitor
   d. Protease inhibitor

2. Mechanism of action of nevirapine is
   a. Uncoating inhibitor
   b. Adsorption inhibitor
   c. Reverse transcriptase inhibitor
   d. Protease inhibitor

3. Famciclovir
   a. Inhibits herpes DNA polymerase
   b. Inhibits viral DNA polymerase
   c. Inhibits HIV reverse transcriptase
   d. Inhibits viral protease
4. In this drug, the carbocyclic ring is attached to the base instead of the sugar:
   a. Efavirenz
   b. Zidovudine
   c. Abacavir
   d. Nevirapine
5. The phosponate drug is
   a. Abacavir
   b. Adefovir
   c. Acyclovir
   d. Ribavirin
6. Ritonovir is synthesised starting from
   a. Epichlorhydrin
   b. Dioxolane
   c. Hydrocinnamyl chloride
   d. Phenylalanine
7. The antiviral drug synthesised starting from 5-chloro anthranilinc acid is
   a. Efavirenz
   b. Emivirine
   c. Loviride
   d. Nevirapine
8. One of the following does not possess purine nucleus:
   a. Ganciclovir
   b. Ribavirin
   c. Adefovir
   d. Didanosine
9. The antiviral drug that is a thiazole analogue is
   a. Nelfinavir
   b. Ritonovir
   c. Saquinavir
   d. Loviride
10. The antiviral drug with no heterocyclic ring system is
    a. Nelfinavir
    b. Loviride
    c. Troviridine
    d. Zidovudine

QUESTIONS

1. Mention the reasons for the inactivity of 2',3'-dideoxy adenine.
2. Match the following with respect to heterocyclic nucleus:

Efavirenz	Purine
Troviridine	Dipyridodiazepine
Nevirapine	Benzoxazinone
Didanosine	Quinoline
Saquinavir	Pyridine

3. Match the following with respect to mechanism of action:

Zidovudine	Glucosidase inhibitor
Nelfinavir	Reverse transcriptase inhibitor
Betulinic acid	Integrase inhibitor
Curcumin	Protease inhibitor
Suramin	Viral budding inhibitor
Interferon	Fusion inhibitor
Castanospermine	Adsorption inhibitor

4. Mention the pharmacokinetic drawbacks of protease inhibitors and explain how these could be overcome.
5. How can zidovudine be converted to a more lipid-soluble compound such that it can cross the blood-brain barrier?
6. What are HSV infections? Classify the drugs useful against these infections.

7. Sketch out the mechanism of action of acyclovir.
8. How are the anti-HSV drugs ribavirin and acyclovir synthesised?

SOLUTION TO MULTIPLE-CHOICE QUESTIONS

1. b;
2. c;
3. b;
4. c;
5. b;
6. d;
7. a;
8. b;
9. b;
10. b.

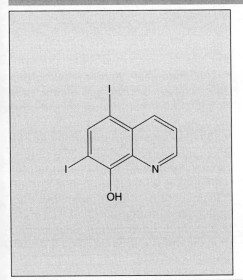

CHAPTER 30

Antiprotozoal Agents

LEARNING OBJECTIVES

- Define antiamoebic agents and their utility
- Categorize amoebicides based on the mode of treatment
- Define malaria and the treatment options
- Classify antimalarials based on the chemical structure
- Describe the life cycle of malaria and the targets for drug development
- Define antihelmintics and their utility
- Categorize antihelmintics
- List other antiprotozoal drugs, their synthesis and utility
- Describe newer drugs in the class of antiprotozoals

30.1   INTRODUCTION

Human beings are host to a wide variety of protozoal parasites, which can be transmitted by insect vector, directly from mammalian reservoirs, or from one person to another. Because protozoa multiply in their hosts and vaccination is not yet an option, chemotherapy has been the only practical way either to treat infected individuals or to reduce transmission in populations. Some of the more common human protozoal infections and the drugs used to treat them are discussed in this chapter.

30.2   ANTIAMOEBIC AGENTS

Worldwide, nearly 480 million people are infected with *Entamoeba histolytica*, of whom 10 per cent develop clinical disease. The infection is transmitted exclusively by the faecal-oral route; human beings are the only known hosts. In most infected individuals, trophozoites exist as commensals in the large intestines—that is, they produce cysts but otherwise cause little harm to the host. In some people, the parasites invade the intestinal mucosa, producing mild to severe colitis (amoebic dysentery). In still other individuals, the parasite invades the extraintestinal tissues, chiefly the liver, producing amoebic abscesses and systemic disease.

30.2.1 Classification of Amoebicides

1. Luminal amoebicides (e.g., diloxanide furoate): It is active only against intestinal forms of amoeba.
2. Systemic amoebicides (e.g., dehydroemetin, chloroquine): These agents have been employed primarily to treat severe amoebic dysentery or hepatic abscesses.
3. Mixed amoebicides (e.g., metronidazole, tinidazole, and ornidazole): These agents are active against both intestinal and systemic forms of amoeba.

Diloxanide furoate   2,2-Dichloro-N-(4-furoyloxyphenyl)-N-methylacetamide
Synthesis

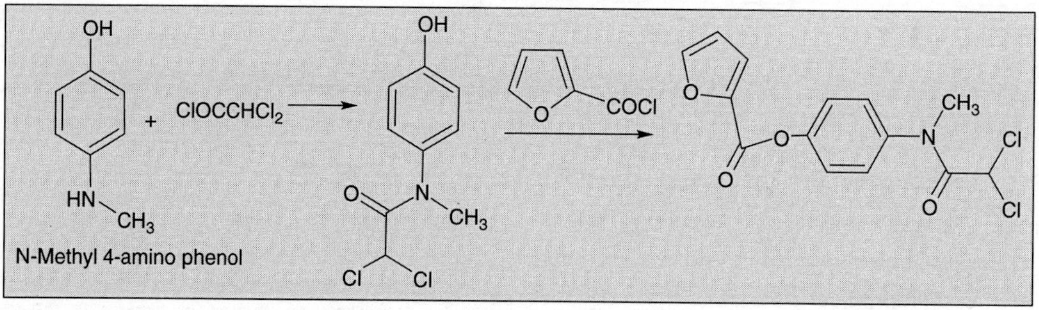

The 2,2-dichloro-N-(4-hydroxyphenyl)-N-methylacetamide is made by N-acylating 4-hydroxy-N-methylaniline with dichloroacetyl chloride. Diloxanide is made by acylating this intermediate with 2-furoyl chloride.

Metronidazole   2-Methyl-5-nitroimidazol-1-ethanol

Synthesis

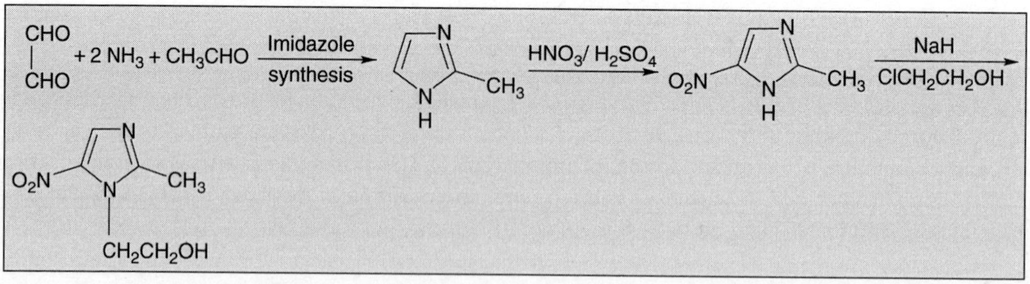

Metronidazole is made by nitrating 2-methylimidazole to make 2-methyl-5-nitroimidazole. The latter is then reacted with 2-chloroethanol, which is easily transformed to the desired metronidazole.

Mechanism of action: The reactive intermediate formed in the parasital reduction of the 5-nitro group of metronidazole covalently binds to the DNA of the parasite triggering the lethal effects. Potential reactive intermediates include the nitroxide, nitroso, hydroxylamine, and amine.

Consuming ethanol (alcohol) while using metronidazole causes a disulfiram-like reaction with effects that can include nausea, vomiting, flushing of the skin, tachycardia (accelerated heart rate), and shortness of breath. Consumption of alcohol should be avoided by patients during systemic metronidazole therapy and for at least 48 hours after completion of treatment.

Tinidazole   1-2-(Ethylsulphonyl)ethyl-2-methyl-5-nitroimidazole

Synthesis

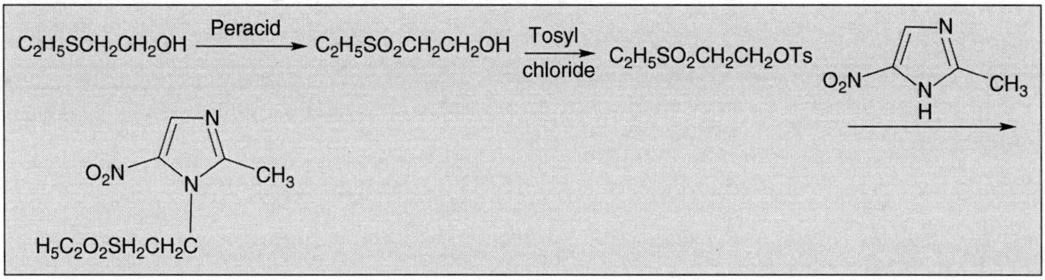

$C_2H_5SCH_2CH_2OH \xrightarrow{\text{Peracid}} C_2H_5SO_2CH_2CH_2OH \xrightarrow[\text{chloride}]{\text{Tosyl}} C_2H_5SO_2CH_2CH_2OTs$

[Structure: O_2N-imidazole-CH_3 with NH]

[Final product: O_2N-imidazole-CH_3 with N-$CH_2CH_2SO_2C_2H_5$ ($H_5C_2O_2SH_2CH_2C$)]

Tinidazole is made from 2-methyl-5-nitroimidazole, which upon being reacted with 2-ethoxysulphonyl-*p*-toluenesulphonate is transformed into the desired tinidazole. The 2-ethoxysulphonyl-*p*-toluenesulphonate necessary for this reaction is, in turn, made by tosylation of 2-ethylsulfonyl ethanol using *p*-toluenesulphonyl chloride.

The mechanism of action is similar to metronidazole.

Ornidazole   1-Chloro-3-(2-methyl-5-nitro-1*H*-1-imidazolyl)-2-propanol

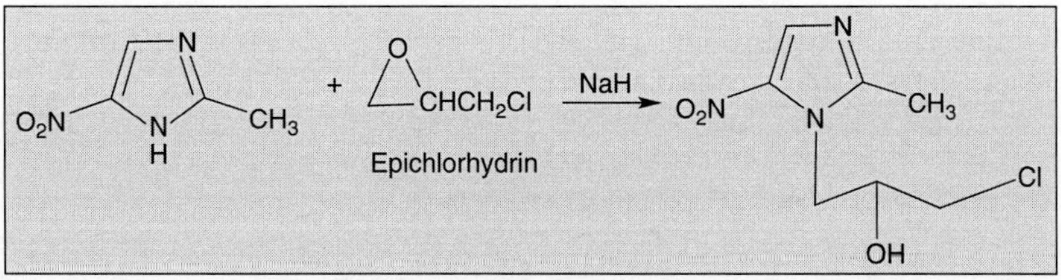

[Structure: O_2N-imidazole-CH_3 with NH + epoxide-$CHCH_2Cl$ (Epichlorhydrin) $\xrightarrow{\text{NaH}}$ O_2N-imidazole-CH_3 with N-$CH_2CH(OH)CH_2Cl$]

Ornidazole is prepared by reaction between 2-methyl-5-nitroimidazole and epichlorhydrin in presence of sodium hydride as base. The mechanism of action is similar to metronidazole. However, it has a longer duration of action than metronidazole.

Nitazoxanide   [2-[(5-Nitro-1,3-thiazol-2-yl)carbamoyl]phenyl]ethanoate

Mechanism of action:

Nitazoxanide is a member of the 5-nitro heterocycles and is a prodrug forming a short-lived redox active intermediate. Nitazoxanide appears to be more selective than metronidazole, with preliminary evidence suggesting interference with pyruvate:ferredoxin oxidoreductase as the site of action. Nitazoxanide, because of its selectivity, does not appear to be mutagenic; nor does it cause DNA fragmentation.

Use: It is used for the treatment of amoebiasis and worm infestations.

Synthesis

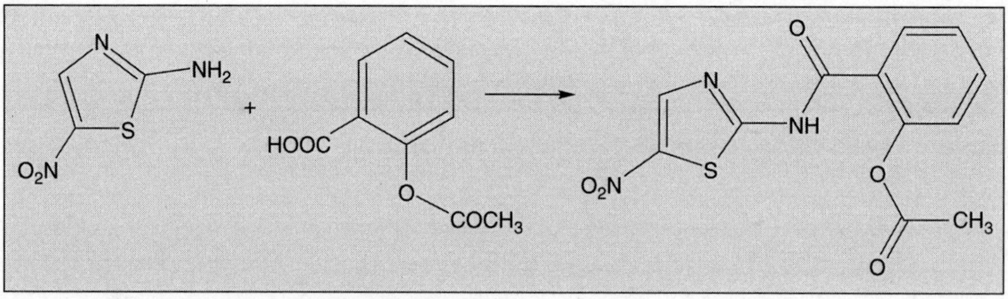

It is a synthetic nitrothiazolyl-salicylamide derivative, prepared by reacting 5-nitro-2-aminothiazole and aspirin in presence of carboxylate activating reagent.

Furazolidone   3-{[(5-Nitro-2-furyl)methylene]amino}-1,3-oxazolidin-2-one

It is indicated in the specific and symptomatic treatment of bacterial or protozoal diarrhoea and enteritis caused by susceptible organisms.

Synthesis

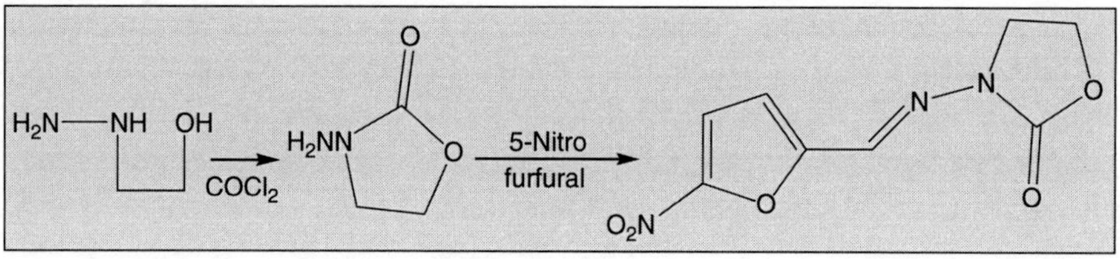

Furazolidone its prepared by condensing 1-amino-1,3-oxazolidin-2-one with 5-nitrofurfural. Oxazolidinone which in turn prepared by reaction between hydroxyethyl hydrazine and phosgene.

Iodoquinol   5,7-Diiodoquinolin-8-ol

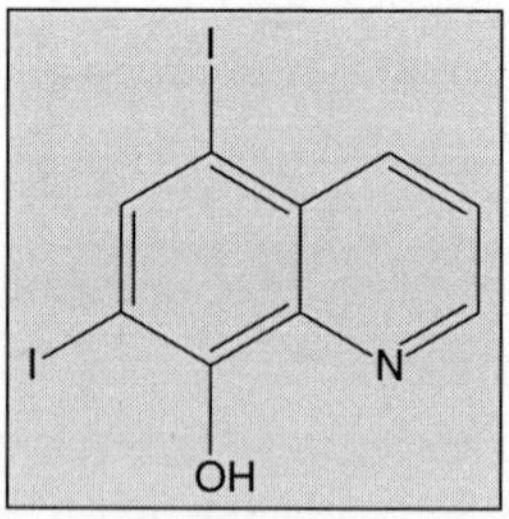

An amoebicide, it treats *Dientamoeba fragilis, Blastocystis hominis, amoebiasis,* and *Balantidium coli.*

30.3   ANTIMALARIAL AGENTS

Malaria remains the world's devastating human infection, with 300 to 500 million clinical cases and nearly 3 million deaths each year. It is caused by several species of the protozoan Plasmodium, of which *P. vivax* and *P. falciparum* are the most common. They all have complex life cycles involving both the Anopheles mosquito and the erythrocyte of the human host. In *vivax*, a persisting tissue phase

Medicinal Chemistry 537

continues to infect the blood at intervals for many years. Thus, an ideal antimalarial should not only eradicate the microzoan from the blood but also from the tissues, to effect radical cure. The several antimalarials differ in their point of interruption of the cycle of the parasite and in the type of malaria affected. Chemotherapy of malaria consists of affecting various stages of the life cycle of the parasite. Antimalarial drugs are subdivided into three corresponding groups: those that have an effect on erythrocyte stage of the life cycle, those that destroy exoerythrocytic (or hepatic stage), and those that affect both stages simultaneously.

30.3.1 Classification of Antimalarials

1. Quinoline derivatives

   (i) 4-Amino quinoline

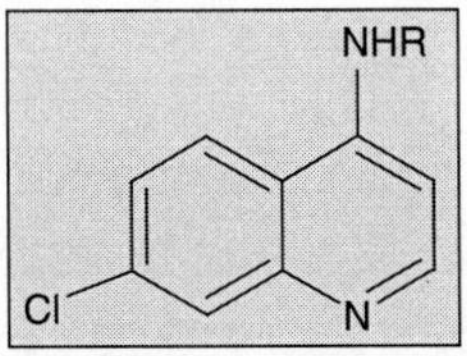

[Structure: 4-amino quinoline with Cl at 7-position and NHR at 4-position]

	R
Chloroquine	—CH(CH$_3$)—(CH$_2$)$_3$—N(C$_2$H$_5$)$_2$
Hydroxy chloroquine	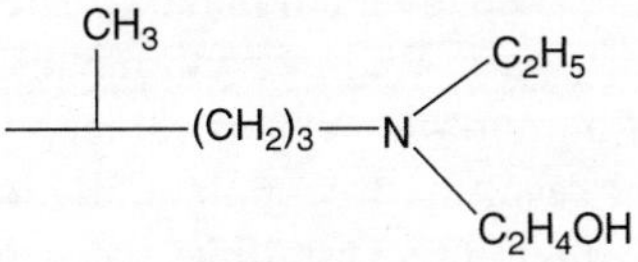 $\underset{\text{CH}_3}{-\text{CH}}-(\text{CH}_2)_3-\text{N}\begin{array}{c}\diagup \text{C}_2\text{H}_5 \\ \diagdown \text{C}_2\text{H}_4\text{OH}\end{array}$
Amodiaquine	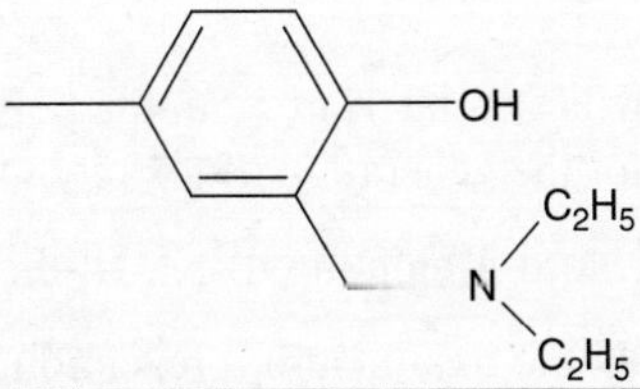 [Structure: phenol ring with —OH and —CH$_2$—N(C$_2$H$_5$)$_2$]

   (ii) 3-Amino quinoline

[Structure: 3-amino quinoline with H$_3$CO at 6-position and NHR at 3-position]

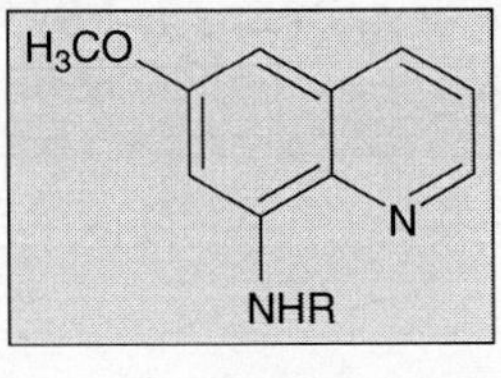

	R
Primaquine	—CH(CH$_3$)—(CH$_2$)$_3$—NH$_2$
Pamaquine	—CH(CH$_3$)—(CH$_2$)$_3$—N(C$_2$H$_5$)$_2$

(iii) Mefloquine

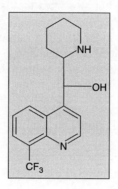

2. 9-Amino acridines: Quinacrine
3. Miscellaneous: Proguanil, Cycloguanil, Halofantrine, Artemether, Artether, Artesunate, Pyrimethamine, and Sulphadoxine

Quinoline antimalarials

Mechanism of action: Following human infection caused by the bite of an infected female Anopheles, Plasmodium parasites accumulate in hepatocytes and then invade the erythrocytes for the next stage of their maturation. After a few days, the infected red cells burst open and the merozoites are released causing periodic fevers of malaria. These merozoites infect new erythrocytes and the intra-erythrocyte cycles start again. Within erythrocytes, the parasite degrades haemoglobin of the host and digests 30 per cent or more of the protein moiety, using it as a source of amino acids for the synthesis of its own protein. The resulting free, potentially toxic haeme(ferriprotoporphyrin IX) left after digestion are polymerized as a microcrystalline, redox inactive iron (III) haeme pigment called haemozoin (non-toxic).

Quinoline antimalarials interfere with the polymerization of toxic haeme to non-toxic haemozoin, probably accounting for the generation of toxicity to the parasite.

Chloroquine N^4-(7-Chloro-4-quinolinyl)-N^1, N^1-diethyl-1, 4-pentane diamine

It has long been used in the treatment or prevention of malaria. As it also mildly suppresses the immune system, it is used in some autoimmune disorders, such as rheumatoid arthritis and lupus erythematosus. Chloroquine is in clinical trials as an investigational antiretroviral in humans with HIV-1/AIDS, and as a potential antiviral agent against chikungunya fever and SARS. Moreover, the radio-sensitizing and chemo-sensitizing properties of chloroquine are beginning to be exploited in anticancer strategies in humans.

Against rheumatoid arthritis, it operates by inhibiting lymphocyte proliferation, phospholipase A, release of enzymes from lysosomes, release of reactive oxygen species from macrophages, and production of IL-1.

As an antiviral agent, it impedes the completion of the viral life cycle by inhibiting some processes occurring within intracellular organelles and requiring a low pH. As for HIV-1, chloroquine inhibits the glycosylation of the viral envelope glycoprotein gp120, which occurs within the Golgi apparatus.

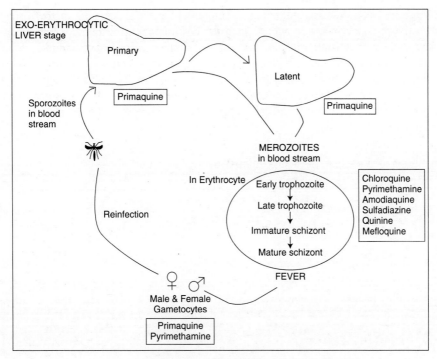

FIGURE 30.1 Life cycle of malaria.

The mechanisms behind the effects of chloroquine on cancer are currently being investigated. The best known effects (investigated in clinical and pre-clinical studies) include radio-sensitizing effects through lysosome permeabilization, and chemo-sensitizing effects through inhibition of drug efflux pumps (ATP-binding cassette transporters) or other mechanisms.

Synthesis

***Step I*:** Synthesis of quinoline moiety (4,7-dichloroquinoline)

For preparing of 4,7-dichloroquinoline, 3-chloroaniline reacts with the diethyl ester of oxaloacetic acid in the presence of acetic acid to give the corresponding enamine, which when heated to 250°C undergoes heterocyclization to the ethyl ester of 7-chloro-4-hydrozyquinolin-2-carboxylic acid accompanied with a small amount of 5-chloro-4-hydroxyquinolin-2-carboxylic acid, which is separated from the main product by crystallization from acetic acid. Alkaline hydrolysis of the ethylester of the 7-chloro-4-hydroxyquinolin-2-carboxylic acid and subsequent high-temperature decarboxylation of the resulting acid gives 7-chloro-4-hydroxyquinolin. Reacting this with phosphorus oxychloride gives 4,7-dichloroquinoline.

Step II: Synthesis of side-chain (4-diethylamino-1-methylbutylamine)

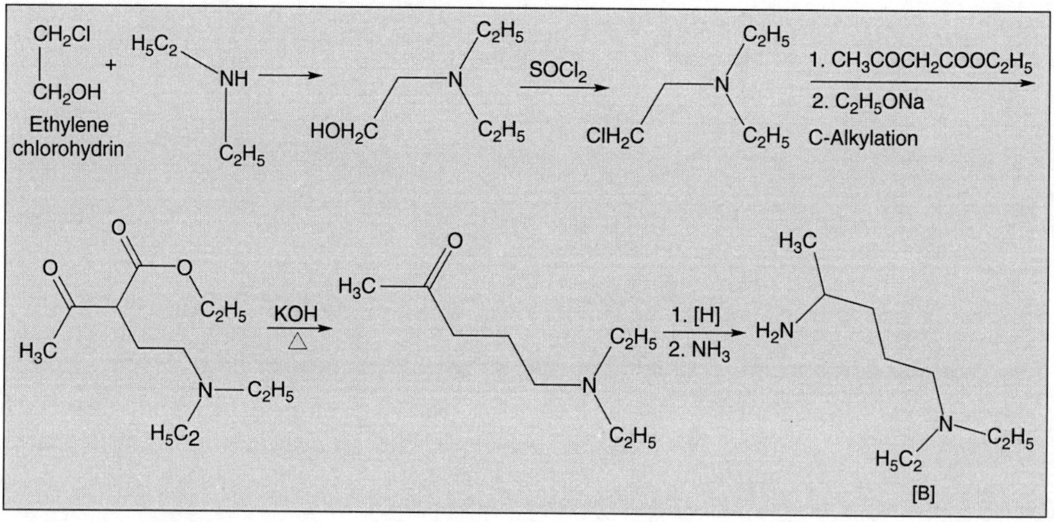

Ethylene chlorohydrine on reaction with diethylamine gives diethylamino ethanol, which on reaction with thionyl chloride gives diethylamino ethyl chloride. Alkylating acetoacetic ester with 2-diethylaminoethylchloride gives 2-diethylaminoethylacetoacetic acid ester, which upon basic hydrolysis and simultaneous decarboxylation makes 1-diethylamino-4-pentanone. Reductive amination of this compound with hydrogen and ammonia using Raney nickel as a catalyst gives 4-diethylamino-1-methylbutylamine.

Step III:

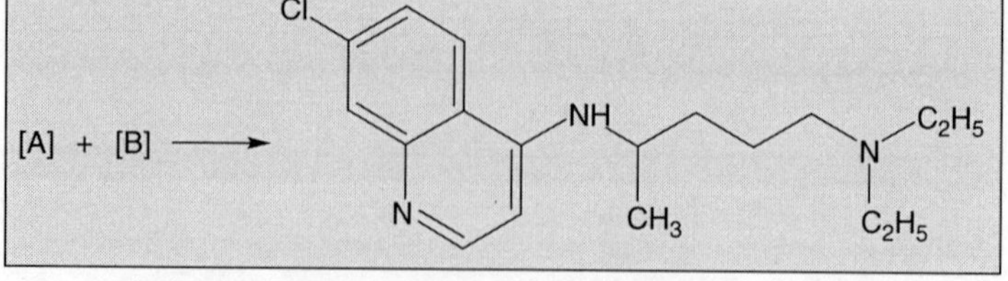

Chloroquine is made by reacting 4,7-dichloroquinoline [A] with 4-diethylamino-1-methylbutylamine [B] at 180°C.

Amodiaquine 4-[(7-Chloro-4-quinolinyl) amino]-2-[(diethylamino) methyl] phenol
Amodiaquine has been shown to be more effective than chloroquine in treating CRPF (chloroquine-resistant *Plasmodium falciparum*) malaria infections and may afford more protection than chloroquine when used as weekly prophylaxis.
Synthesis

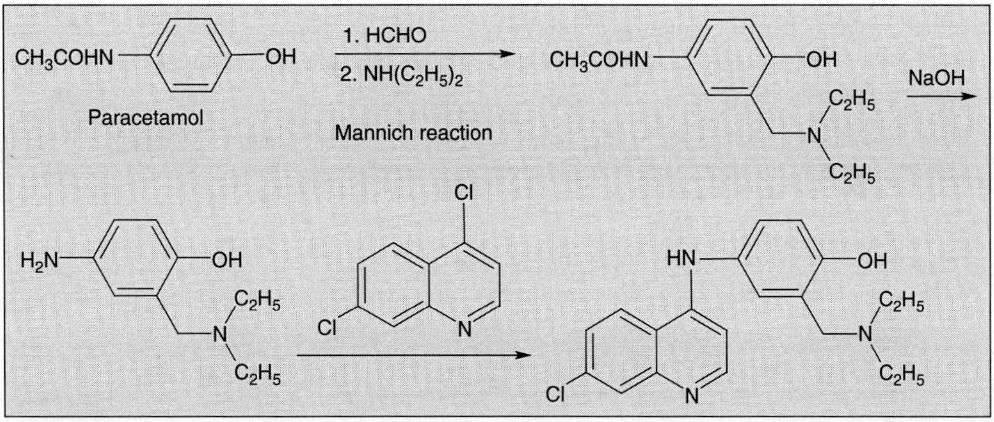

Paracetamol undergoes aminomethylation reaction (Mannich reaction) using formaldehyde and diethylamine. Saponification of this gives amino derivative, which in turn reacts with 4,7-dichloroquinoline to afford amodiaquine.

Pamaquine 8-[(4-Diethylamino-1-methylbutyl) amino]-6-methoxy quinoline
Pamaquine is an 8-aminoquinoline drug used for the treatment of malaria. It is closely related to primaquine (de-diethylamino derivative).

Moving the side-chain from the fourth position of the quinoline ring to the eighth position completely changes the compound's spectrum of activity. Unlike the 4-substituted aminoquinolines, primaquine and pamaquine has practically no effect on erythrocyte forms of the malaria parasite. Its activity is limited to tissue forms of the parasite in mammals and in the mosquitoes themselves. This makes primaquine and pamaquine an especially valuable drug, allowing radical recovery and simultaneous prevention, which is usually not achieved by using erythrocyte drugs.

Synthesis

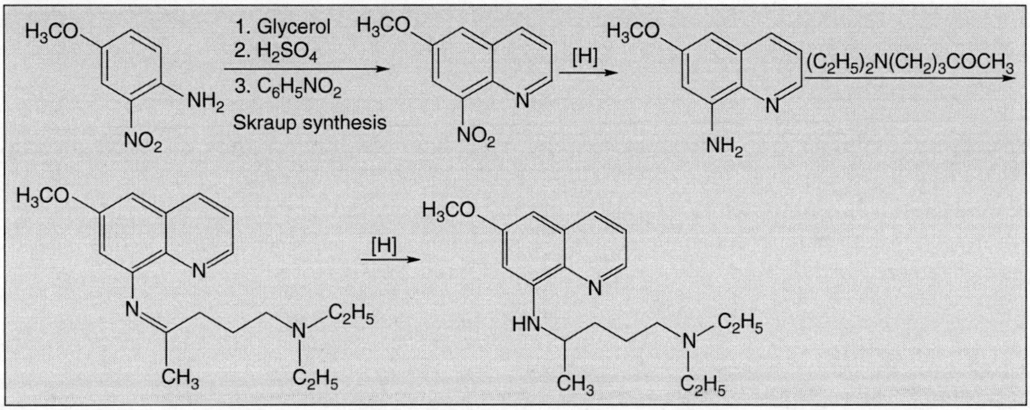

It is made from 6-methoxy-8-nitroquinoline, which is synthesised in a Skraup reaction from 4-methoxy-2-nitroaniline and glycerol in the presence of sulphuric acid. The nitro group in this compound is reduced to make 6-methoxy-8-aminoquinoline. Reductive alkylation of this with 5-(diethylamino)-2-pentanone affords pamaquine.

Mefloquine D,L-*erythro*-α-2-Piperidyl-2,8-*bis*-(trifluoromethyl)-4-quinolinmethanol
Mefloquine is an analogue of quinine, and it differs from it in that the side-chain at C_4 of the quinoline ring contains a piperidine fragment instead of a quinuclidine fragment, and positions C_2 and C_8 are substituted with trifluoromethyl groups.

Mefloquine has an unexpectedly long half-life in serum, 385±150 hours (16 days), and permits a dosage regime that consists of a single oral tablet taken at weekly intervals.

Synthesis

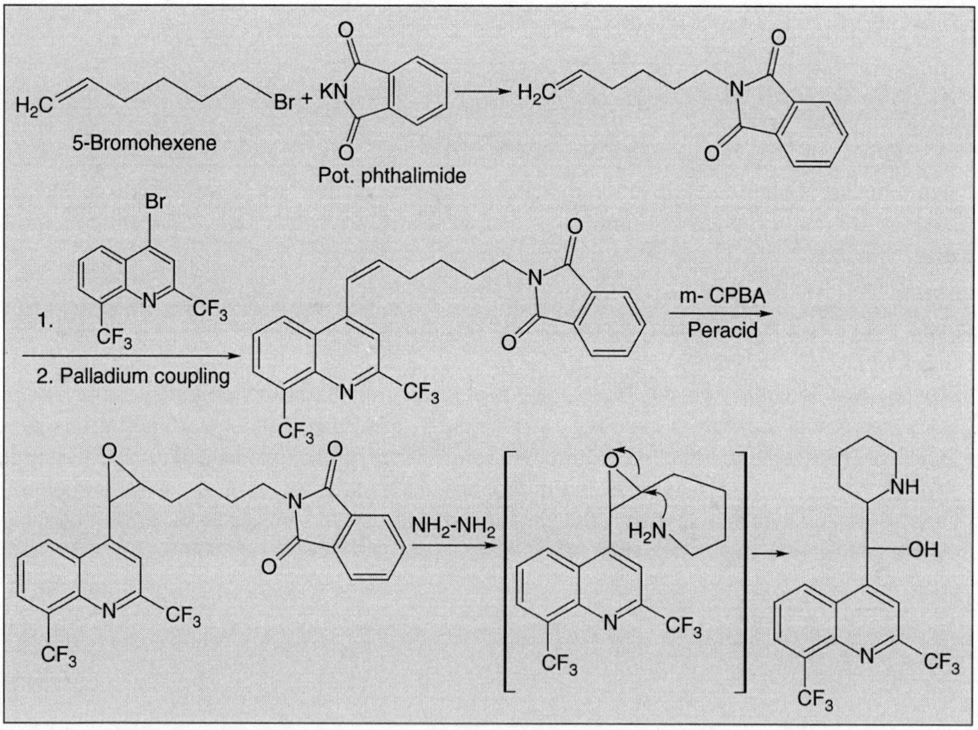

Displacement of bromine in 5-bromohexene by potassium phthalimide gives alkylated derivative. Palladium-mediated coupling of this olefin with 4-bromo-2,8-*bis* trifluoromethyl quinoline affords vinylated product, which is converted to epoxide by treatment with peracid. Phthalimide group is removed by reaction with hydrazine to afford amine; this opens the oxiran with consequent formation of a piperidine ring, mefloquine.

9-Amino acridines

Quinacrine 6-Chloro-9-[[4-(diethylamino)-1-methylbutyl]amino]-2-methoxy acridine

Quinacrine is a derivative of acridine that is chemically and clinically very similar to 4-aminoquinolines used as an antimalarial drug. It has also been used as an antibiotic in the treatment of giardiasis (an intestinal parasite), and in research as an inhibitor of phospholipase A_2. It has also been proposed for use in systemic lupus erythematosus.

Synthesis

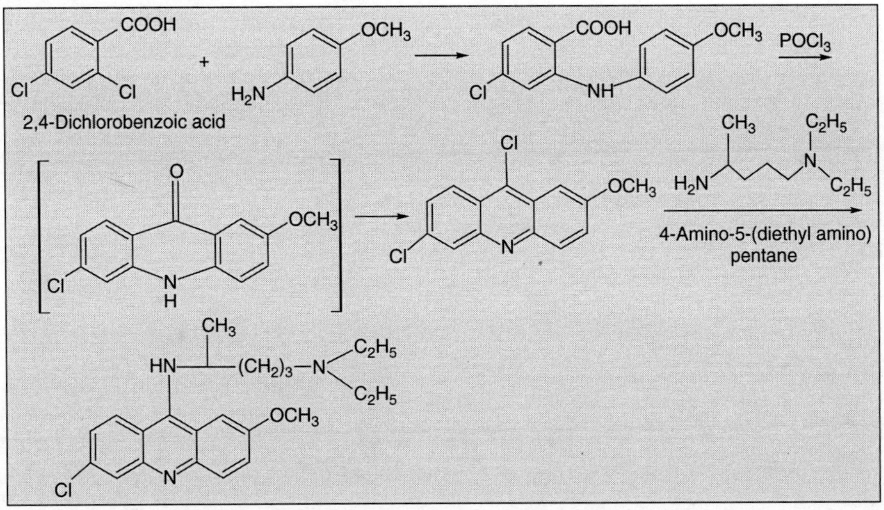

It is synthesised from 6,9-dichloro-2-methoxyacridine and 4-diethylamino-1-methylbutylamine. The 6,9-dichloro-2-methoxyacridine necessary for the synthesis is made in two stages. The initial reaction of 2,4-dichlorobenzoic acid and 4-anisidine in the presence of copper dust and potassium carbonate gives 2-(4-methoxyanilino)-4-chlorobenzoic acid), which upon reaction with phosphorus oxychloride turns into the necessary 6,9-dichloro-2-methoxyacridine.

Miscellaneous Drugs

Proguanil 1-(4-Chlorophenyl)-2-(N′-propan-2-ylcarbamimidoyl) guanidine

Proguanil is a prodrug that is metabolised in the liver to a diaminotriazine (cycloguanil), which acts as a dihydrofolate reductase inhibitor of Plasmodium species and inhibits DNA synthesis.

Synthesis

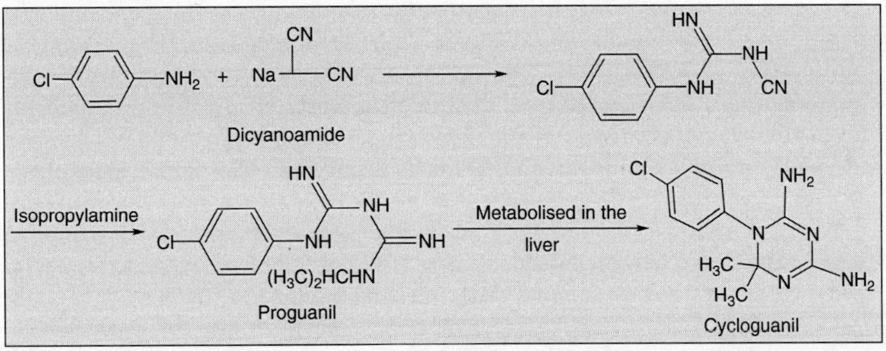

It is made from 4-chloroaniline and sodium dicyanoamide, the interaction of which results in the formation of (4-chlorophenyl)dicyanodiamide. Reacting this with isopropylamine gives the desired proguanil.

Halofantrine 3-Dibutylamino-1-[1,3-dichloro-6-(trifluoromethyl)phenanthren-9-yl]-propan-1-ol

Synthesis

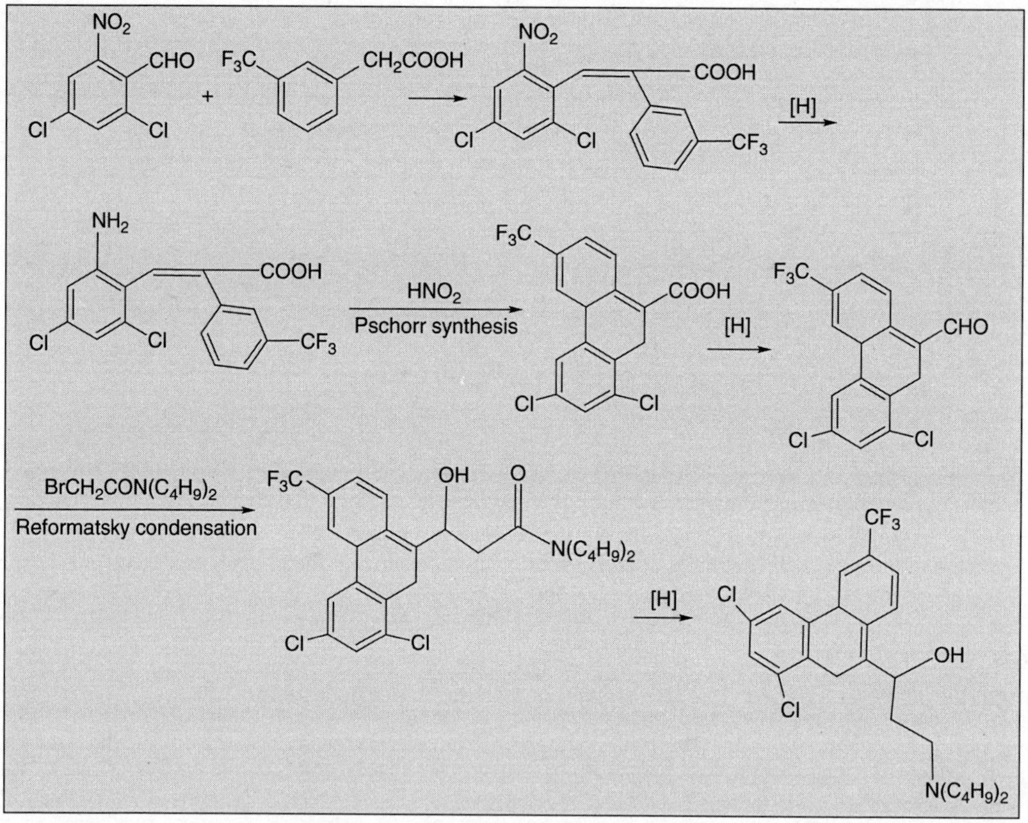

Aldol condensation of substituted benzaldehyde with 3-trifluoromethylphenylacetic acid gives cinnamic acid derivative. Reduction of nitro group gives amino derivative; this is then cyclized to the phenanthrene by Pschorr synthesis. Reduction of carboxylic acid to alcohol followed by oxidation gives aldehyde derivative. Reformatsky condensation of this with N,N-di-N-butyl bromoacetamide and zinc gives amidoalcohol. This is reduced to the amino alcohol and affords halofantrine.

Artemether and artether These are oil-soluble derivatives of natural product artemisinin (Octahydro-3,6,9-trimethyl-3,12-epoxy-12 H pyrano-(4,3-di)-1,2-benzodioxepin-10-(3H)-one).

Artemisinin interacts strongly with haemin, which leads to adduct formation and presumably interferes with the normal conversion of haemin to haemozoin.

Synthesis

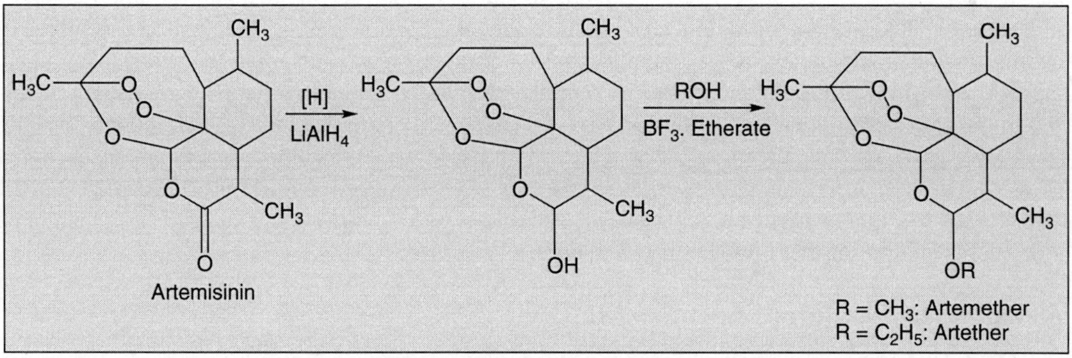

Artemisinin is reduced to dihydroartemisinin, which on reaction with methanol/ethanol in boron-trifluoride etherate medium affords artemether and artether, respectively.

Artesunate This is a water-soluble derivative of artemisinin.

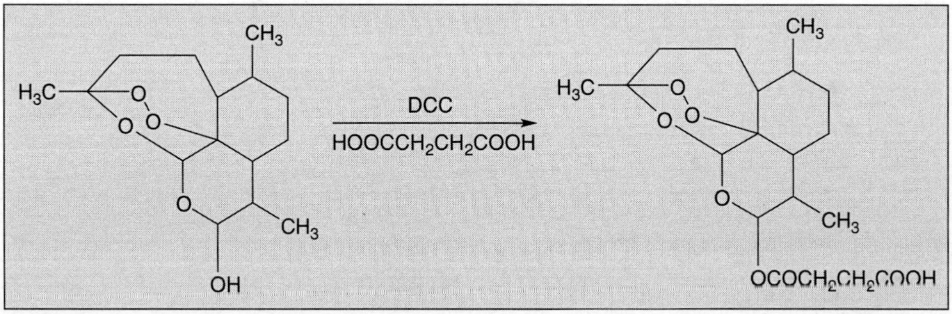

It is prepared by esterification of dihydroartemisinin with succinic acid in presence of DCC.

Artemisinin is under early research and testing for treatment of cancer, primarily by researchers at the University of Washington. Artemisinin has a peroxide lactone group in its structure. It is thought that when the peroxide comes into contact with high iron concentrations (common in cancerous cells), the molecule becomes unstable and releases reactive oxygen species. It has been shown to reduce angiogenesis and the expression of vascular endothelial growth factor in some tissue cultures.

Pyrimethamine and Sulfadoxine: Discussed in Chapter 24.

30.4 ANTHELMINTICS

The geographic distribution of helminthiasis or infection with parasitic worms is cosmopolitan with over two billion people affected. In tropical regions, where the prevalence is greatest, simultaneous infestations with more than one type of helminth are common. Worms that are pathogenic to human beings—namely metazoa—are conventionally classified into round worms (nematodes) and two types of flat worms: flukes (trematodes) and tapeworms (cestodes). Most nematode infections are localized in the intestinal tract, although a few of them can pass into other organs, including the heart, liver, lungs, muscles, and so on, from which removal is significantly harder. Cestode infections are usually localized in the gastrointestinal tract, but there have been cases of them passing into the circulatory system. Trematodes cause chronic infection, called schistosomiasis, in which the blood vessels are attacked and various organ structures (liver, intestines, and urinary tract) are damaged. Anthelmintics are drugs that act either locally to expel worms from the gastrointestinal tract or systemically to eradicate adult helminths or developmental forms that invade organs and tissues.

30.4.1 Classification of Anthelmintics

1. Benzimidazoles

	R_1	R_2
Mebendazole	—$NHCO_2CH_3$	—COC_6H_5
Albendazole	—$NHCO_2CH_3$	—$SCH_2CH_2CH_3$
Thiabendazole	(thiazole ring)	H

2. Quinolines and isoquinolines: Oxamniquine and Praziquentel
3. Piperazines: Piperazine citrate, and Diethyl carbamazine
4. Vinyl pyrimidines: Pyrantel pamoate
5. Amides: Niclosamide
6. Imidazothiazoles: Levamisole
7. Organophosphanes: Metrifonate

Benzimidazoles

These are versatile anthelmintic agents, particularly effective against gastrointestinal nematodes. These are highly effective in ascariasis, intestinal capillariasis, enterobiasis, trichuriasis, and hookworm infections as single or mixed infections.

Mechanism of action: The primary action of these drugs is to inhibit microtubule polymerization by binding to β-tubulin, and thereby arrest nematodes' cell division in metaphase. These also produce many biochemical changes in susceptible nematodes, e.g., inhibition of mitochondrial fumarate reductase, reduced glucose transport, and uncoupling of oxidative phosphorylation.

Mebendazole   Methyl-5-benzoyl-2-benzimidazole carbamate

Synthesis

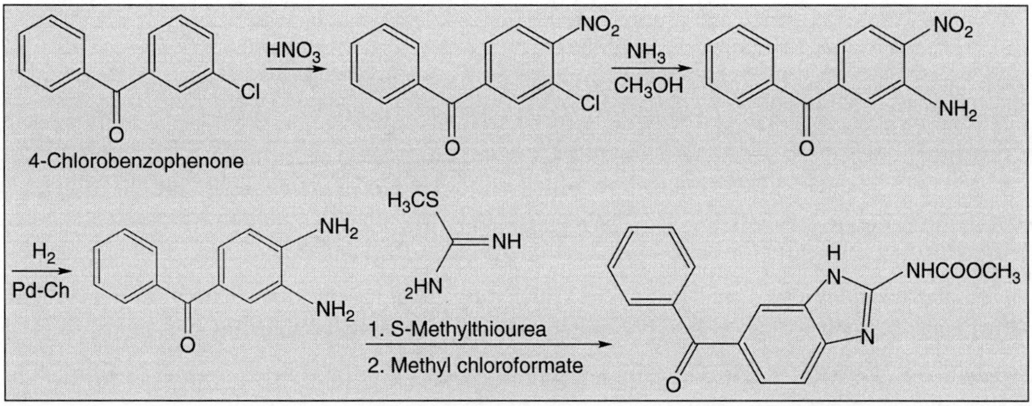

Nitration of 4-chlorobenzophenone with nitric acid at a temperature lower than 5°C gives 4-chloro-3-nitrobenzophenone, in which the chlorine atom is replaced with an amino group by heating it to 125°C in a solution of ammonia in methanol to make 4-amino-3-nitrobenzophenone. Reducing the nitro groups in this compound with hydrogen using a palladium on carbon catalyst gives 3,4-diaminobenzophenone. This on reaction with N-methoxycarbonyl-S-methylthiourea (prepared by reacting methyl chloroformate with S-methylthiourea) affords mebendazole.

Albendazole   Methyl-[5-(propylthio)-1H-benzoimidazol-2-yl]carbamate

Synthesis

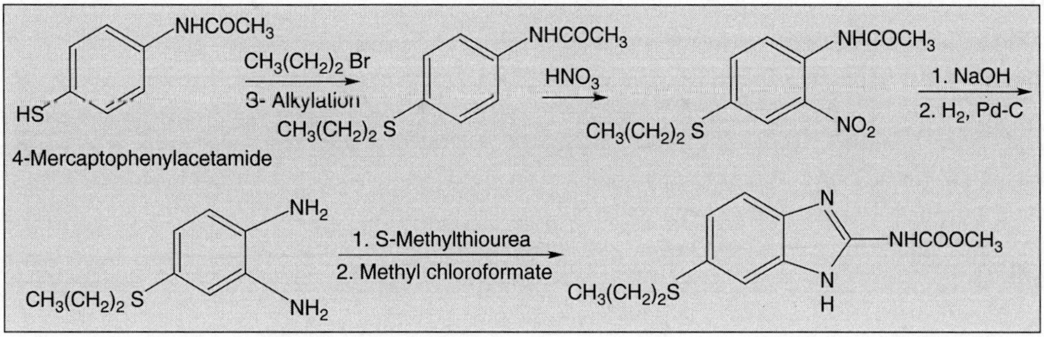

It is also made by the heterocyclization of a derivative of phenylenediamine to a derivative of benzimidazole. Alkylation of 4-mercaptophenylacetamide with N-propyl bromide gives S-alkylated product, and nitration followed by saponification and reduction gives o-phenylenediamine derivative. This, in turn, cyclized with N-methoxycarbonyl-S-methylthiourea affords albendazole.

Thiabendazole 2-(4'-Thiazolyl)benzimidazole

Synthesis

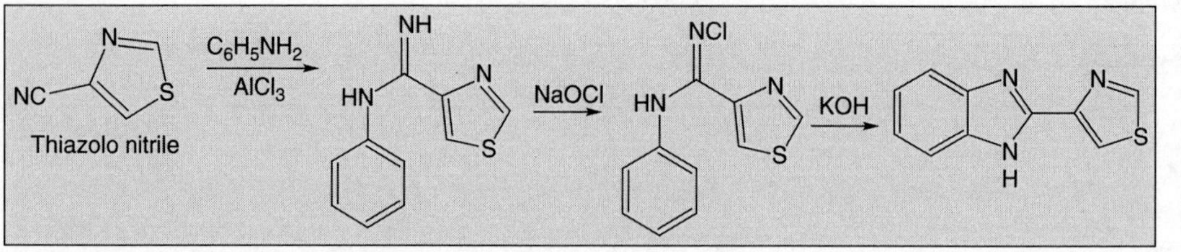

Condensation of thiazolo nitrile with aniline, which is catalysed by aluminium chloride, affords amidine addition product. This is converted to its reactive *N*-chloro derivative by reaction with sodium hypochlorite. Treatment of this with base leads to cyclization and, thus, thiabendazole.

Thiabendazole is also a chelating agent, which means that it is used medicinally to bind metals in cases of metal poisoning, such as lead poisoning, mercury poisoning, or antimony poisoning.

Niridazole 1-(5-Nitro-1,3-thiazol-2-yl)imidazolidin-2-one

Synthesis

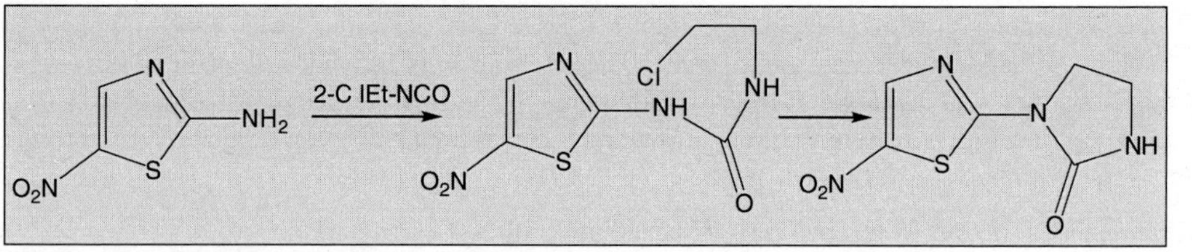

2-Amino-5-nitrothiazole on reaction with 2-chloroethylisocyanate leads to urea intermediate, which in presence of base undergoes cyclization to form niridazole.

Piperazine derivatives

Piperazine citrate It is highly effective against both *Ascaris lumbricoides* (roundworm) and *Enterobius vermicularis* (pinworm). Piperazine blocks the response of worm muscle to acetylcholine, apparently by altering the permeability of the cell membrane to ions that are responsible for the maintenance of the resting potential. The drug causes hyperpolarization and suppression of spontaneous spike potentials with accompanying paralysis that result in the expulsion of the worm by peristalsis.

Synthesis

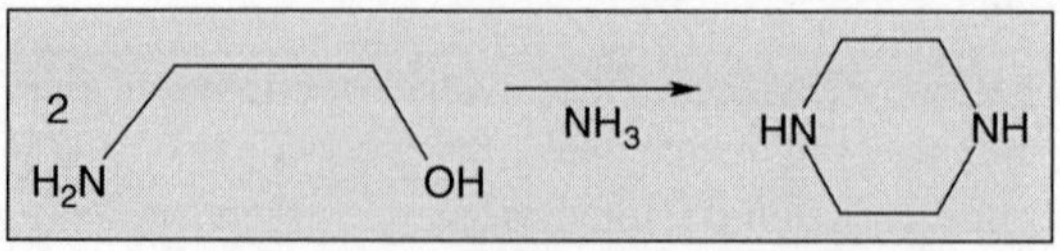

It is made from ethanolamine by heating it in ammonia at a temperature of 150°C–220°C and a pressure of 100 atm–250 atm. It is used as a drug in the form of a salt, and as a rule, in the form of adipate orcitrate.

Diethyl carbamazine   N, N-Diethyl-4-methyl-1-piperazincarboxamide

It is active against the filarial parasite *Wuchereria bancrofti*. It decreases the muscular activity and eventually immobilizes the organism; this may result from a hyperpolarizing effect of the piperazine moiety. It also produces alteration in the microfilarial surface membranes, thereby rendering them more susceptible to host defence mechanisms.

Synthesis

$H_3C-N\bigcirc NH \xrightarrow[\text{Phosgene}]{COCl_2} H_3C-N\bigcirc N-COCl \xrightarrow{HN(C_2H_5)_2} H_3C-N\bigcirc N-CON(C_2H_5)_2$

N-Methyl piperazine

It is prepared by reacting N-methylpiperazine with phosgene and diethylamine.

Quinolines and isoquines

Oxamniquine   1,2,3,4-Tetrahydro-2-isopropylaminomethyl-7-nitro-6-quinolylmethanol

Schistosoma mansoni is highly susceptible to oxamniquine. ATP-dependent enzymatic activation of the drug in susceptible schistosomes results in an unstable phosphate ester that dissociates to yield a chemically reactive carbocation. This intermediate alkylates the DNA.

Synthesis

[Reaction scheme: 2,6-dimethylquinoline → (Chlorination) → 2-chloromethyl-6-methylquinoline → ($H_2NCH(CH_3)_2$) → 6-methyl-2-(isopropylaminomethyl)quinoline → ([H] Cat.) → tetrahydro derivative → (HNO_3/H_2SO_4) → 7-nitro tetrahydroquinoline → (*Aspergillus sclerotium* [O]) → final hydroxymethyl (oxamniquine) product]

Chlorination of 2-methyl group of 2,6-dimethylquinoloine, followed by displacement of chlorine with isopropyl amine, gives quinoline derivative. Catalytic hydrogenation of this gives terahydroquinoline derivative, which is nitrated and followed by oxidation of remaining methyl group to carbinol by fermentation with *Aspergillus sclerotium*.

Praziquantel   2-(Cyclohexylcarbonyl)-1,2,3,6,7,11b-hexahydro-4*H*pyrazino[2,1a]isoquinolin-4-one

It is used for the treatment of schistosomiasis, and liver fluke infections. It causes tegumental damage to the worms, which activates host defence mechanisms and results in the destruction of the worms.

Synthesis

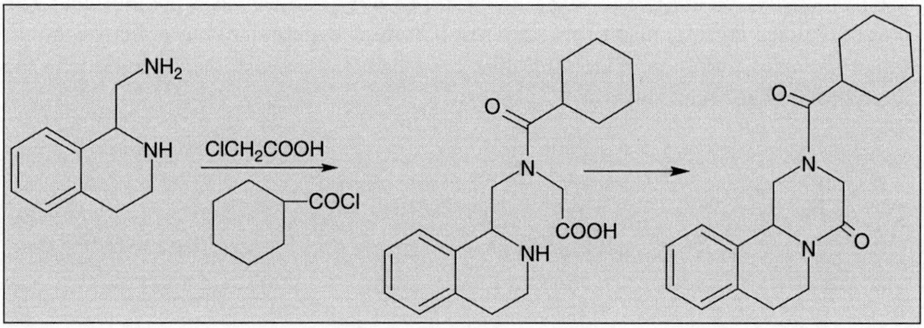

Praziquantel is a derivative of pyrazinoisoquinoline that is made by alkylating 1-aminomethyl-1,2,3,4-tetrahydroisoquinoline with chloroacetic acid. The resulting amine is acylated with cyclohexanecarbonyl chloride to make 1-(N-carboxymethyl-N-cyclohexylcarbonylaminomethyl)-1,2,3,4-tetra-hydroisoquinoline, which is heated at 150°C to give the desired praziquantel.

Pyrantel 1,4,5,6-Tetrahydro-1-methyl-2-[trans-2-(2-thienyl)vinyl]pyrimidine

Pyrantel is a depolarizing neuromuscular blocking agent. It induces marked persistent activation of nicotinic receptors, which results in spastic paralysis of the worm. It is an alternative to mebendazole in the treatment of ascariasis and enterobiasis.

Synthesis

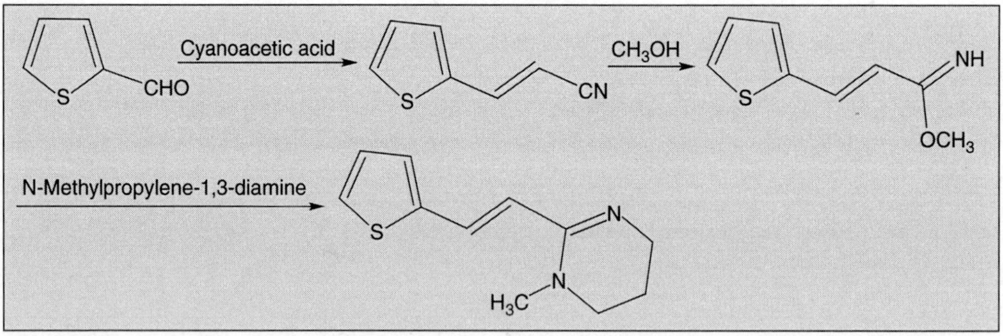

Pyrantel, a derivative of tetrahydropyrimidine, is made from 3-(2-thienyl)acrylonitrile, which is made in a Knoevangel condensation of 2-formylthiophen with cyanoacetic acid. It is then methanolyzed in the presence of acid to afford imino ether. Condensation with N-methylpropylene-1,3-diamine proceeds probably by addition-elimination of each amino group in turn with imino ether and affords pyrantel.

Niclosamide 2',5-Dichloro-4-nitrosaicylanilide

The principal action of the drug may be to inhibit anaerobic phosphorylation of ADP by the mitochondria of the parasite, an energy-producing process.

Medicinal Chemistry

Synthesis

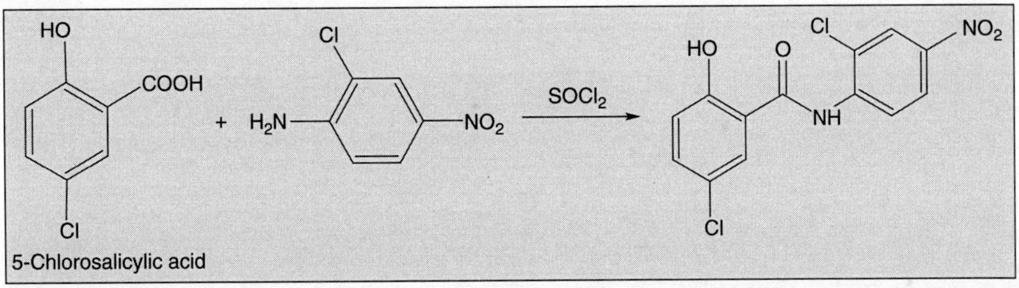

Niclosamide is made by reacting 5-chlorosalicylic acid with 2-chloro-4-nitroaniline in the presence of phosphorus trichloride/thionyl chloride.

Levamisole (-)-2,3,5,6-Tetrahydro-6-phenylimidazo[2,1-b]thiazol

It is a *levo* rotatory isomer and a potent stereospecific inhibitor of fumarate reductase in many nematodes.

Synthesis

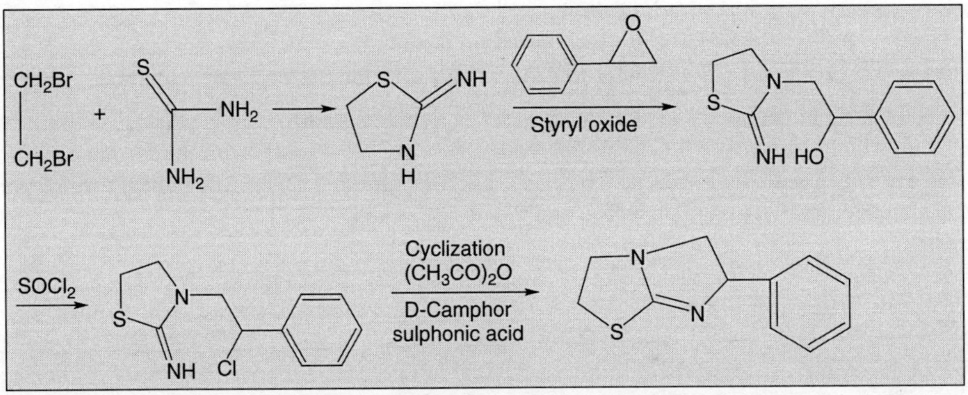

Reaction of styrene oxide and 2-imino-1,3-thiazolidine and subsequent treatment of the resulting product with thionyl chloride and then with acetic anhydride lead to the formation of tetramizole. Treating this with D-10-camphorsulphonic acid isolates the desired L-isomer levamisole.

Metrifonate 2,2,2-Trichloro-1-dimethoxyphosphorylethanol

Metrifonate is metabolised and rearranged *in vivo* to dichlorvos, which inhibits acetylcholinesterase.

Synthesis

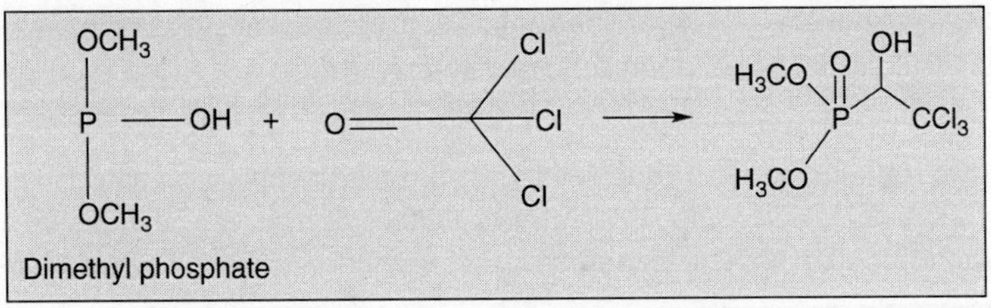

Reaction between 2,2,2-trichloroacetaldehyde and dimethyl phosphate affords metrifonate.

30.5 MISCELLANEOUS ANTIPROTOZOAL DRUGS

Eflornithine   2,5-Diamino-2-(difluoromethyl)pentanoic acid

It is used for the treatment of African trypanosomiasis (sleeping sickness) due to *Trypanosoma brucei*. It is an irreversible catalytic inhibitor of ornithine decarboxylase, the enzyme that catalyses the first and the rate-limiting step in the biosynthesis of polyamines. The polyamines putrescine and spermine are required for cell division and for the normal cell differentiation.
Synthesis

The amino groups of ornithine are protected as imine by reaction with two moles of benzaldehyde; this is alkylated with chlorodifluoromethane in presence of lithium diisopropylamide. Treatment with acid deprotects amino groups and affords eflornithine.

Nifurtimox   *N*-(3-Methyl-1,1-dioxo-1,4-thiazinan-4-yl)-1-(5-nitro-2-furyl)methanimine

It is the drug of choice for the treatment of trypanosomiasis caused by *T. cruzei*. The trypanocidal action of nifurtimox appears to be due to the reduction of the nitro group with the formation of chemically reactive superoxide, hydrogen peroxide, and hydroxy radicals. These free radicals are responsible for DNA damage, inactivation of enzymes, and membrane injury.
Synthesis

It is prepared by reacting 5-nitrofurfural and 4-amino-3-methylhexahydro-1λ^6,4-thiazine-1,1-dione.

Pentamidine   4-[5-(4-Carbamimidoylphenoxy) pentoxy]benzenecarboximidamide

It is effective against trypanosomiasis, leishmaniasis, and pneumocystosis. It interferes with polyamine synthesis in trypanosomes by reversible inhibition of S-adenosyl-L-methionine decarboxylase. The drug also inhibits ATP-dependent topoisomerase in *P. carnii*.
Synthesis

4-Aminophenol on diazotization followed by reaction with cuprous cyanide gives 4-cyanophenol. *O*-Alkylation of this with 1,5-dibromopentane gives diether derivative. Ethanolysis and reaction with ammonia gives diamidino derivative pentamidine.

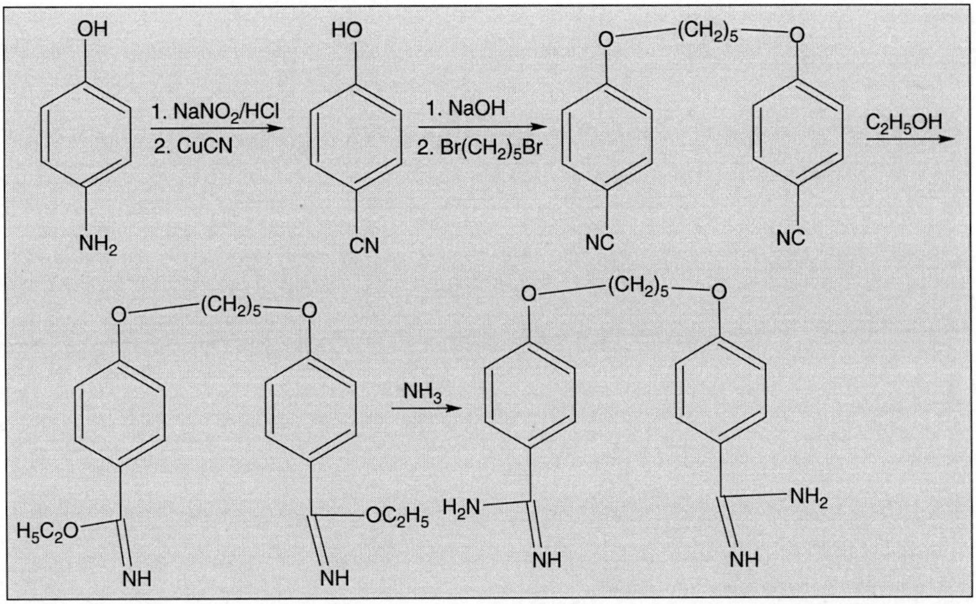

Sodium stibogluconate 2,4:2',4'-*O*-(Oxydistibylidyne)*bis*[D-gluconic acid] Sb,Sb'dioxide trisodium salt nonahydrate

Synthesis

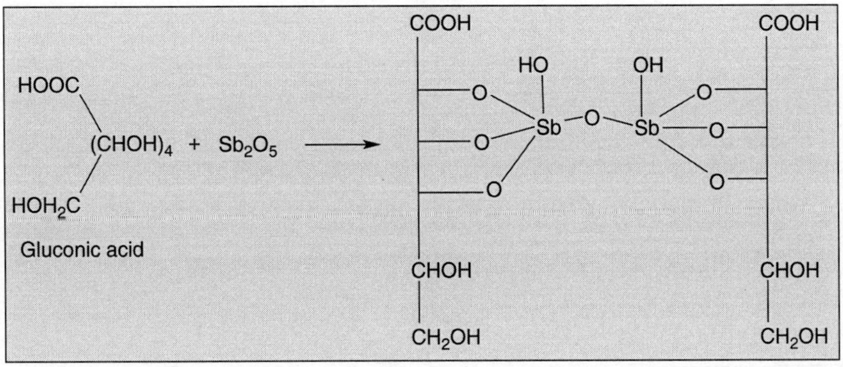

It is the drug of choice for all forms of leishmaniasis, and prepared by reacting two moles of gluconic acid with antimony pentaoxide.

Suramin sodium 8-[(4-methyl-3-{[3-({[3-({2-methyl-5-[(4,6,8-trisulpho-1-naphthyl)carbamoyl]phenyl}carbamoyl)phenyl]carbamoyl}amino)benzoyl]amino}benzoyl)amino]naphthalene-1,3,5-trisulphonic acid

Synthesis

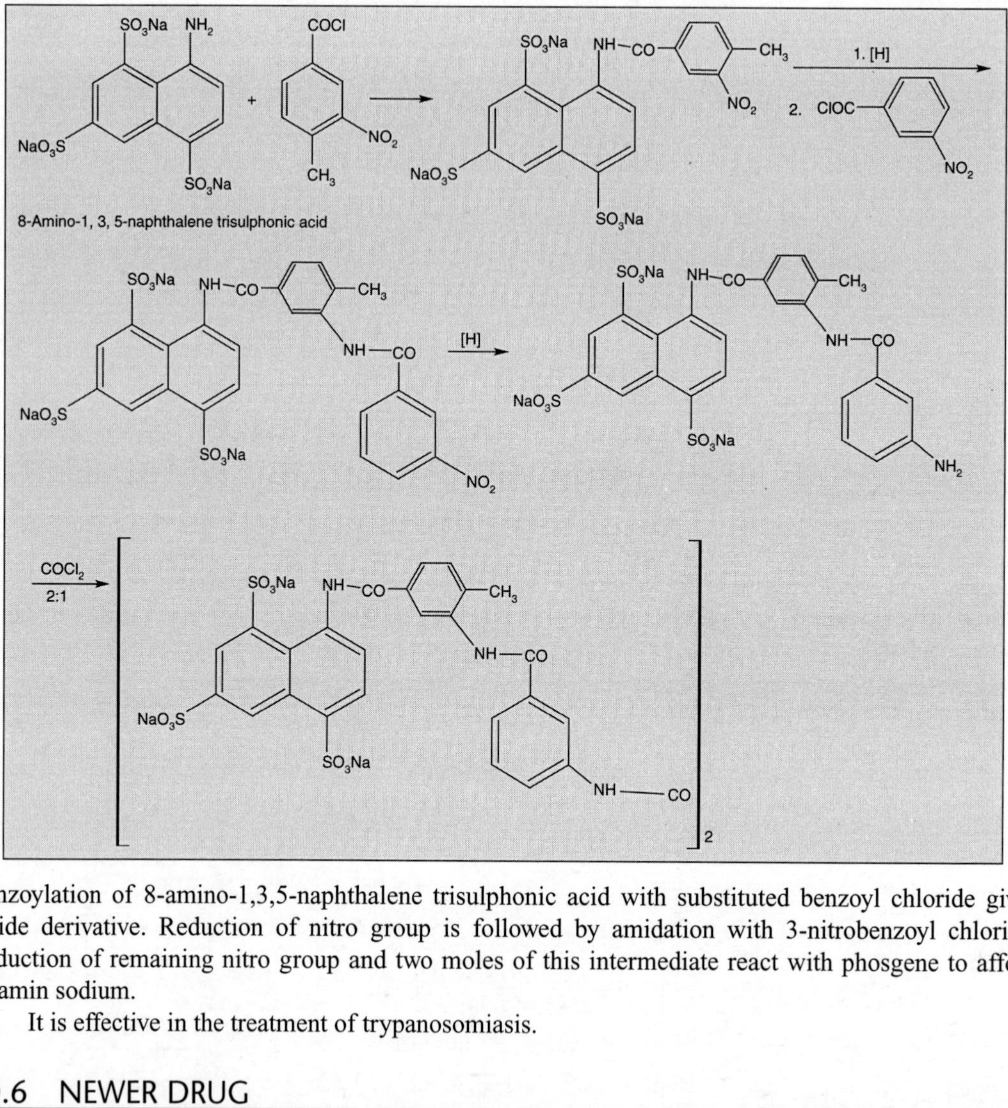

Benzoylation of 8-amino-1,3,5-naphthalene trisulphonic acid with substituted benzoyl chloride gives amide derivative. Reduction of nitro group is followed by amidation with 3-nitrobenzoyl chloride. Reduction of remaining nitro group and two moles of this intermediate react with phosgene to afford suramin sodium.

It is effective in the treatment of trypanosomiasis.

30.6 NEWER DRUG

Ivermectin It is an anti-parasite medication derived from the bacterium *Streptomyces avermitilis* and kills by interfering with the target animal's nervous system. Ivermectin is a mixture of two compounds, 22,23-dihydroavermectin B1a and 22,23-dihydroavermectin B1b, which differ by a single ethyl or methyl side-group. Ivermectin is an antihelmintic used mainly in the treatment of onchocerciasis in humans, and also for strongyloidiasis, ascariasis, trichuriasis, and enterobiasis.

Ivermectin inactivates parasitic nematodes, arachnids, and insects. It binds to glutamate gated chloride channels (GABA-mediated, present in nerves and muscle cells). This binding promotes increased membrane permeability to chloride ions, which causes hyperpolarization of the nerve or muscle cell. This results in neuro-muscular paralysis, which may lead to death.

FURTHER READINGS

1. *Approaches to design and synthesis of antiparasitic drugs*, Anand, N. and S. Sharma (ed.), Elsevier, 1997.
2. *Burger's Medicinal Chemistry and Drug Discovery*, Wolf, M.E., 6th Ed., Wiley, 2003.

MULTIPLE-CHOICE QUESTIONS

1. Chloroquine is a
   a. Luminal amoebicide
   b. Systemic amoebicide
   c. Mixed amoebicide
   d. Oral amoebicide

2. Amodiaquine is a derivative of
   a. 3-Amino quinoline
   b. 4-Amino quinoline
   c. 2-Amino quinoline
   d. 5-Amino quinoline

3. The mechanism of action of levamisole is
   a. Blocks the response to acetylcholine
   b. Reversible inhibition of S-adenosyl-L-methionine decarboxylase
   c. Inhibits acetylcholinesterase
   d. Stereospecific inhibitor of fumarate reductase

4. The mechanism of action of pentamidine is
   a. Blocks the response to acetylcholine
   b. Reversible inhibition of S-adenosyl-L-methionine decarboxylase
   c. Inhibits acetylcholinesterase
   d. Stereospecific inhibitor of fumarate reductase

5. Match the following drugs with respect to the heterocyclic ring system present in the structure:

Drug	Heterocyclic System	Drug	Heterocyclic System
1. Metrifonate	A. Imidazole	6. Thiabendazole	F. Thiazole
2. Nitazoxanide	B. Indole	7. Mefloquine	G. Triazine
3. Nifurtimox	C. Acridine	8. Cycloguanil	H. Furan
4. Pyrantel	D. Benzimidazole	9. Quinacrine	I. Piperazine
5. Diethylcarbazine	E. None	10. Ornidazole	J. Thiophene

6. Match the following drugs with their mechanism of action:

Drug	Mechanism of Action	Drug	Mechanism of Action
1. Pentamidine	A. Interferes with the polymerization of haeme to haemozoin	6. Pyrantel	F. Inhibits microtubule polymerization
2. Nifurtimox	B. Covalently binds to the DNA, triggering lethal effects	7. Mebendazole	G. Free radical generation and DNA damage
3. Eflornithine	C. Persistent activation of nicotinc receptor	8. Proguanil	H. Irreversible catalytic inhibitor of ornithine decarboxylase
4. Metrifonate	D. Inhibition of dihydrofolate reductase	9. Chloroquine	I. Stereospecific inhibitor of fumarate reductase
5. Levamisole	E. Reversible inhibition of S-adenosyl-L-methionine decarboxylase	10. Metronidazole	J. Inhibits AchE

QUESTIONS

1. Classify antimalarial drugs with structural examples for each.
2. Give the synthetic protocol of chloroquine.
3. With the help of the life cycle of malarial parasite, explain the various targets for therapy.
4. Classify antiamoebic agents and sketch down the synthesis and mechanism of action of metronidazole and ornidazole.
5. Describe the importance of anthelmintics, and classify and write down the synthesis of one drug from each class.

SOLUTION TO MULTIPLE-CHOICE QUESTIONS

1. b;
2. b;
3. d;
4. b;
5. [1. E; 2. F; 3. H; 4. J; 5. I; 6. D; 7. B; 8. G; 9. C; 10. A];
6. [1. E; 2. G; 3. H; 4. J; 5. I; 6. C; 7. F; 8. D; 9. A; 10. B].

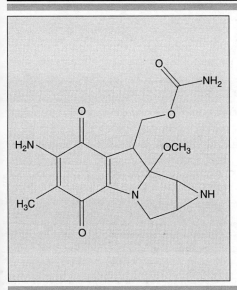

CHAPTER 31

Anticancer Agents

LEARNING OBJECTIVES

- Define cancer and their types
- Describe the treatment options for cancer and their limitations
- Categorize the anticancer agents based on the mode of action or based on the source
- Describe the mode of action of cyclophosphamide
- Describe the antibiotics useful in the treatment of cancer
- Describe drugs obtained from natural source for the treatment of cancer

31.1 INTRODUCTION

Antineoplastic agents are used for the treatment of cancer. Neoplasm (In Greek, 'neo' means 'new' and 'plasm' means 'formation') refers to a group of diseases caused by several agents—namely, chemical compounds and radiant energy. Cancer is characterized by an abnormal and uncontrolled division of cells, which produce tumours and invade adjacent normal tissues. Often, cancer cells separate themselves from the primary tumour and are carried by the lymphatic system, reach distant sites of the organs, where they divide and form secondary tumours (metastasis).

Cancer is classified according to the tissues and type of cells in which new growth occurs:

- **Carcinoma:** Malignant tumours derived from epithelial cells. This group represents the most common cancers, including the common forms of breast, prostate, lung and colon cancer
- **Sarcoma:** Malignant tumours derived from connective tissue
- **Lymphoma** and **leukaemia:** Malignancies derived from haematopoietic (blood-forming) cells
- **Germ cell tumour:** Tumours derived from totipotent cells. In adults, most often found in the testicle and ovary; in foetuses, babies, and young children, most often found on the body midline, particularly at the tip of the tailbone
- **Blastic tumour:** A tumour (usually malignant) that resembles an immature or embryonic tissue. Many of these tumours are most common in children

31.2 LIMITATIONS OF THERAPY

1. Biochemical and morphological differences between normal and neoplastic cells are slight; therefore, antineoplastic agents are devoid of selective toxicity to tumour cells.
2. Some tumours are poorly irrigated by blood, and this hampers the easy access of drugs to the cancer cells.
3. Antineoplastic agents kill cells by first-order kinetics—that is, they kill a constant fraction of cells. Some of the cancer cells elude killing, and this one cell can restabilize the tumour. It is extremely difficult to kill all the malignant cells.
4. Cancer cells very rapidly develop resistance to antineoplastic drugs.
5. Most antineoplastic drugs are highly toxic to the patient.

Treatment of cancer includes surgical intervention, radiation, immunotherapy, and chemotherapy using neoplastic drugs. Chemotherapy is currently used in addition to surgical intervention in order to remove possible metastatic cells that still remain. Moreover, some types of tumours are currently treated first with chemotherapeutic agents.

Cancer chemotherapy is generally non-specific. This means that drugs kill not only cancerous cells, but also normal cells. Because of the fact that it is non-specific, special strategies are developed to increase the potential of destroying cancerous cells and lessening toxic effects on normal tissue.

31.3 CLASSIFICATION

1. Alkylating agents
   (i) Nitrogen mustard—Mechlorethamine, Melphalan, Chlorambucil, Estramustine, Uracil mustard, Cyclophosphamide, and Ifosphamide
   (ii) Nitrosourea—Carmustine, Lomustine, Semustine, Chlorozotocin
   (iii) Aziridines—Thiotepa, Benzotepa, Altretamine
   (iv) Aryl sulphonates—Busulphan
   (v) Miscellaneous—Dacarbazine, Streptozocin

2. Antimetabolites
   (i) Folic acid antagonists—Methotrexate and Trimetrexate
   (ii) Purine antagonists—6-Mercapto purine, 6-Thioguanine, and Fludarabine
   (iii) Pyrimidine antagonists—5- Fluoro uracil and Cytarabine

3. Antibiotics
   (i) Anthracyclines—Daunorubicin, Doxorubicin, Carminomycin, Idarubicin, and Epirubicin
   (ii) Miscellaneous—Actinomycin D, Mithramycin, Bleomycin, Mitomycin C

4. Plant products: Vincristine, Vinblastine, Paclitaxel, Docetaxel, Camptothecin, Topotecan, Irinotecan, Colchicine, Etoposide, Teniposide

5. Hormones and their antagonists: Ethinyl estradiol, Testolactone, Megestrol, Tamoxifen, Flutamide, Aminoglutethimide

6. Miscellaneous: Procarbazine, Hydroxy urea, and Cisplatin

31.4 ALKYLATING AGENTS

The chemotherapeutic alkylating agents have the common property of becoming strong electrophiles through the formation of carbonium ion intermediates, which in turn react with nucleophilic moieties of the target molecule (DNA). The 7- nitrogen atom of guanine is particularly susceptible to the formation of a covalent bond with alkylating agents. Other atoms in the purine and pyrimidine bases of DNA were the 1, 3- nitrogen of the adenine, the 3-nitrogen of cytosine, the 6-oxygen of guanine, and, to a lesser extent, the phosphate atom of the DNA.

Mechanism of action of mechlorethamine

Step 1 Activation of mechlorethamine

The 2-chloro ethyl side-chain undergoes first-order intramolecular cyclization, with the release of Cl⁻ and formation of a highly reactive ethylene iminium intermediate. By this reaction, the tertiary amine is converted into quaternary ammonium compound, which can react strongly with a variety of sites that possess high electron density.

Step 2 Nucleophilic attack of unstable aziridine ring by electron donor

The alkylation of 7-nitrogen may exert several effects. Normally, guanine residue in DNA exists predominantly as the keto tautomer and readily makes Watson-Crick base pair with cytosine by hydrogen bonding. However, the 7-nitrogen of guanine is alkylated, the guanine residue is more acidic, and enol tautomer is favoured. The modified guanine can mispair with thymine residue during DNA synthesis, leading to the substitution of an adenine–thymine base pair for guanine-cytosine base pair. Secondly, alkylation of 7-nitrogen labilizes the imidazole ring, making possible the opening of the imidazole ring of guanine or depurination. Either of these seriously damages the DNA molecule.

Thirdly, with bifunctional alkylating agents such as nitrogen mustard, the second CH_2CH_2Cl side-chain undergoes a similar cyclization reaction and alkylates a 2-guanine residue, resulting in the cross-linking of two nucleic acid chains and leading to a major disruption of nucleic acid function. The ultimate cause of cell death related to DNA damage is not known.

31.4.1 Nitrogen Mustards

Several nations stock-piled large amounts of munitions containing nitrogen mustard gas during the Second World War, but none were used in combat. As with all types of mustard gas, nitrogen mustard is a powerful and persistent radulititc compound blister agent.

Mechlorethamine 2, 2-Dichloro-N-methyldiethylamine
Synthesis

$$H_3C-NH_2 + 2\ \text{(ethylene oxide)} \xrightarrow{H^+} H_3C-N(CH_2CH_2OH)_2 \xrightarrow{SOCl_2} H_3C-N(CH_2CH_2Cl)_2$$

Ethylene oxide

It is effective in Hodgkin's disease.

It is made by reacting methylamine with ethylene oxide, forming *bis*-(2-hydroxyethyl)methylamine, which upon reaction with thionyl chloride turns into the desired mechlorethamine.

It is a vigorously active agent used only by I.V. route and not for oral use. The most serious toxicity of this drug is bone-marrow depression. Various structural modifications were carried out so that greater selectivity, stability, and therefore less toxicity are achieved.

SAR: Aryl-substituted nitrogen mustards, i.e., *bis* (2-chloroethyl) groups, have been linked to phenylalanine (melphalan), substituted phenyl groups (aminophenyl butyric acid in chlorambucil), pyrimidine bases (uracil mustard), and several entities in an effort to make more stable and orally available forms.

Electron-withdrawing property of aromatic ring decreases the nucleophilicity of the nitrogen atom and so reduces the rate of cyclization and carbonium ion formation. These compounds, therefore, reach distant sites in the body before reacting with components of blood and other tissues.

Melphalan 4-[*Bis*(2-chloroethyl)amino]-L-phenylalanine
Synthesis

Phenylalanine $\xrightarrow[\text{EtOH/HCl}]{\text{HNO}_3, \text{H}_2\text{SO}_4}$ (O$_2$N-phenyl-CH$_2$-CH(NH$_2$)-COOC$_2$H$_5$) $\xrightarrow[\text{2. [H]}]{\text{1. Phthalic anhydride}}$ (phthalimide intermediate) $\xrightarrow[\text{2. SOCl}_2, \text{3. HCl/NH}_2\text{NH}_2]{\text{1. 2 ethylene oxide}}$ (ClH$_2$CH$_2$C)$_2$N-phenyl-CH$_2$-CH(NH$_2$)-COOH

Medicinal Chemistry 561

It is synthesised from L-phenylalanine, the nitration of which with nitric acid gives 4-nitro-L-phenylalanine. Reacting this with an ethanol in the presence of hydrogen chloride gives the hydrochloride of 4-nitro-L-phenylalanine ethyl ester, the amino group of which is protected by changing it to phthalimide by a reaction with phthalic anhydride. The nitro group in this molecule is reduced to an amino group using palladium on calcium carbonate as a catalyst. The resulting aromatic amine is then reacted with ethylene oxide, which forms a *bis*-(2-hydroxyethyl)-amino derivative. The hydroxyl groups in this molecule are replaced with chlorine atoms upon reaction with thionyl chloride, after which treatment with hydrochloric acid and hydrazine removes the ethyl ester and phthalamide protection, respectively, giving melphalan.

It is active against multiple myeloma.

Chlorambucil 4-[*p*-[*Bis*(2-chloroethyl)amino]phenyl] butyric acid

Synthesis

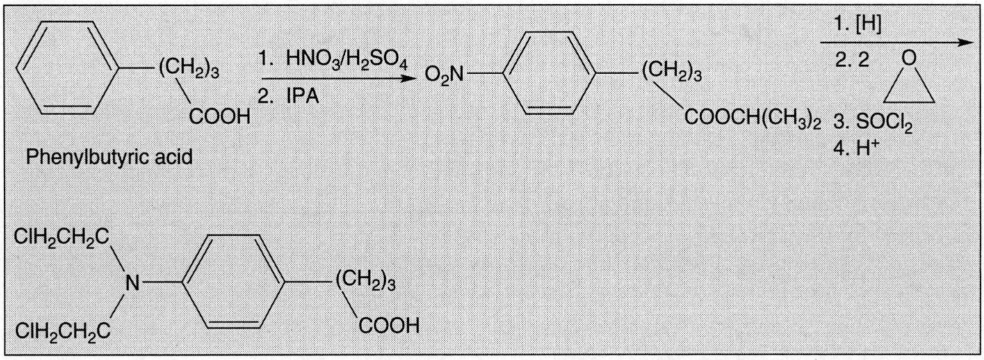

Phenylbutyric acid on nitration followed by reaction with isopropyl alcohol gives isopropyl ester derivative. Reduction of nitro group is followed by reaction with ethylene oxide, which forms a *bis*-(2-hydroxyethyl)-amino derivative. The hydroxyl groups in this molecule are replaced with chlorine atoms upon reaction with thionyl chloride, after which treatment with hydrochloric acid gives chlorambucil.

It is indicated especially in the treatment of chronic lymphocytic leukaemia and primary macroglobulinemia. Other indications are lymphosarcoma and Hodgkin's disease.

Estramustine Estradiol-3-*bis* (2-chloroethyl) carbamate

Synthesis

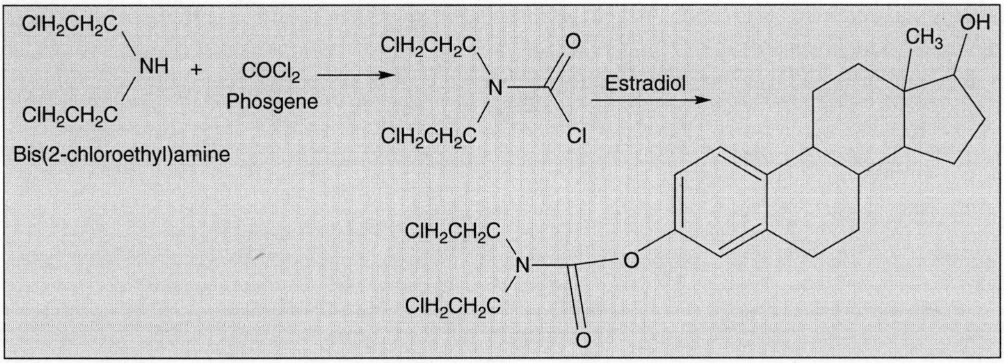

It is prepared by reacting *bis*(2-chloroethyl)amine with phosgene to give carbamoyl chloride intermediate, which in turn reacts with estradiol to form carbamate derivative estramustine.

It is used for the treatment of prostate cancer.

Uracil mustard   5-[*bis*(2-Chloroethyl)amino] uracil

Synthesis

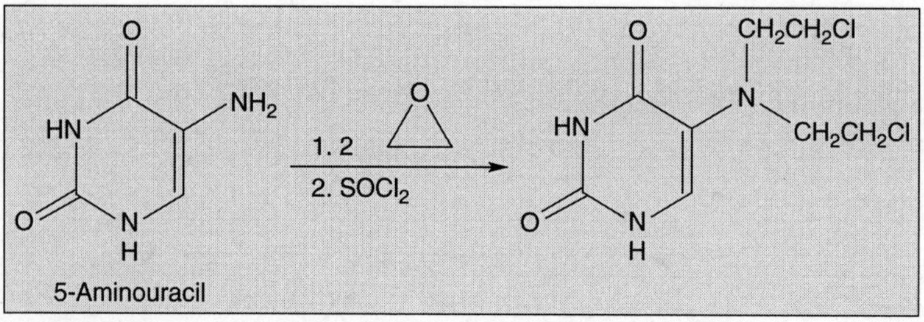

[Reaction scheme: 5-Aminouracil + 1. 2 ethylene oxide, 2. SOCl$_2$ → 5-[bis(2-chloroethyl)amino]uracil with CH$_2$CH$_2$Cl groups]

It is prepared by reacting 5-aminouracil with two moles of ethylene oxide, which forms a *bis*-(2-hydroxyethyl)-amino derivative. The hydroxyl groups in this molecule are replaced with chlorine atoms upon reaction with thionyl chloride, and affords uracil mustard.

It is used for the treatment of primary thrombocytosis.

None of the above modifications has produced an agent highly selective for malignant cells. It was hoped that neoplastic tissues might possess high phosphatase or phosphamidase activity. Then, the replacement of methyl group of mechlorethamine with phosphate, i.e., phosphamides, was prepared which cleave selectively in the neoplastic cells.

Cyclophosphamide   N,N-*bis*(2-Chloroethyl)tetrahydro1,3,2-oxazaphosphorin-2-amide

Synthesis

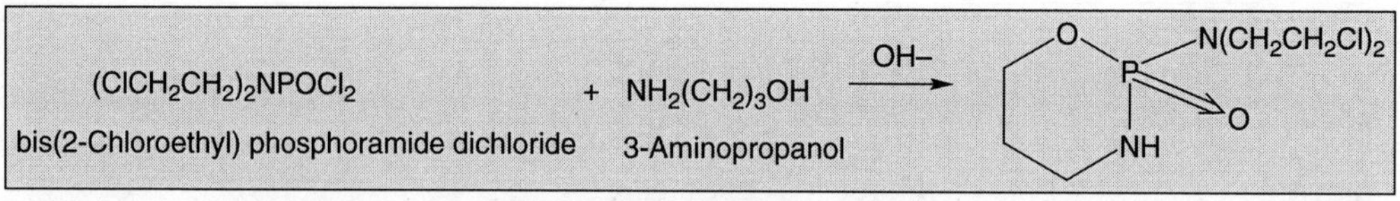

[Reaction scheme: (ClCH$_2$CH$_2$)$_2$NPOCl$_2$ (bis(2-Chloroethyl) phosphoramide dichloride) + NH$_2$(CH$_2$)$_3$OH (3-Aminopropanol) →(OH−) cyclophosphamide ring with N(CH$_2$CH$_2$Cl)$_2$]

It is made by reacting *bis*(2-chloroethyl)amine with phosphorous oxychloride, giving N,N-*bis*-(2-chloroethyl)dichlorophosphoramide, which upon subsequent reaction with 3-aminopropanol is transformed into cyclophosphamide.

It is a prodrug and is activated by hepatic cytochrome P450 system.

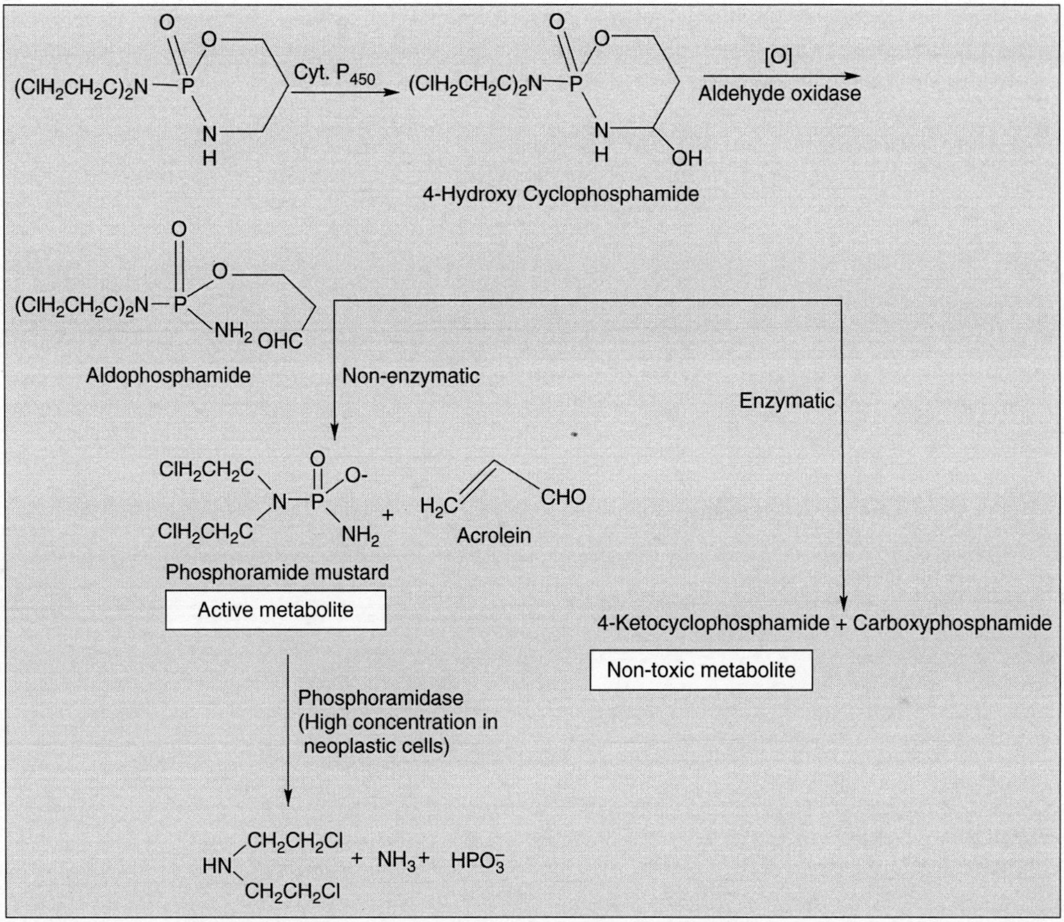

In tumour cells, aldophosphamide may cleave spontaneously to phosphoramide mustard, which is cyclized to aziridinium ion, the principal cross-linking alkylator formed from cyclophosphamide.

It is active against multiple myeloma, chronic lymphocytic leukaemia, and acute leukaemia of children.

Ifosfamide 3-(2-Chloroethyl)-2-[(2-chloroethyl)amino]tetrahydro-2H-1,3,2-oxazaphosphorin-2-oxide
It is an isomer of cyclophosphamide in which one of the 2-chloroethyl substituents is on the ring nitrogen, and also has potent antitumour activity.

Synthesis

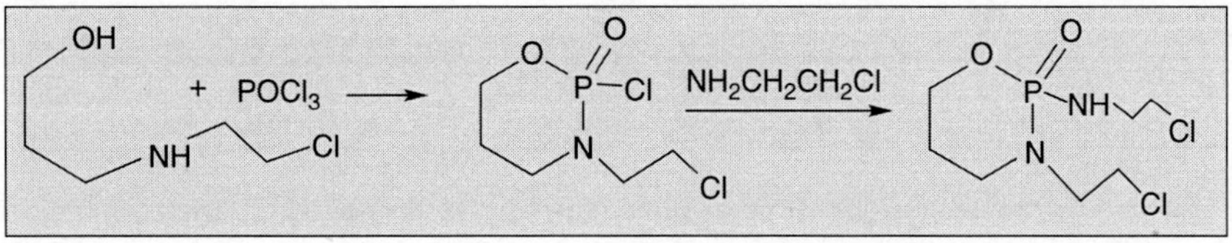

It is made by reacting *N*-(2-chloroethyl)-*N*-(3-hydroxypropyl)amine with phosphorous oxychloride, giving 3-(2-chloroethyl)-2-chlorotetrahydro-2*H*-1,3,2-oxazaphosphorin-2-oxide, which is reacted with *N*-(2-chloroethyl)amine, forming the desired ifosfamide.

31.4.2 Nitrosoureas

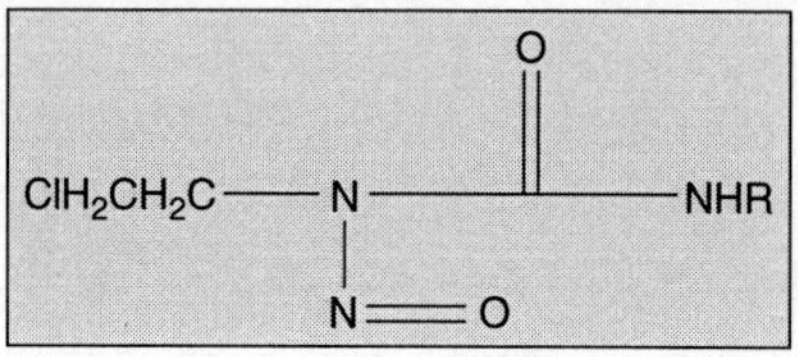

	R
Carmustine	—CH$_2$CH$_2$Cl
Lomustine	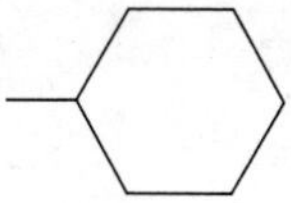 —cyclohexyl
Semustine	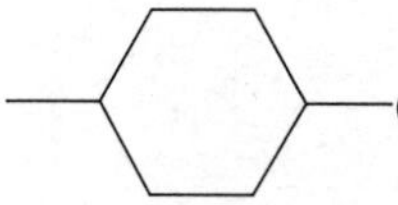 —4-methylcyclohexyl
Chlorozotocin	—Glucose

Carmustine 1,3-*bis*(2-Chloroethyl)-1-nitrosourea
Synthesis

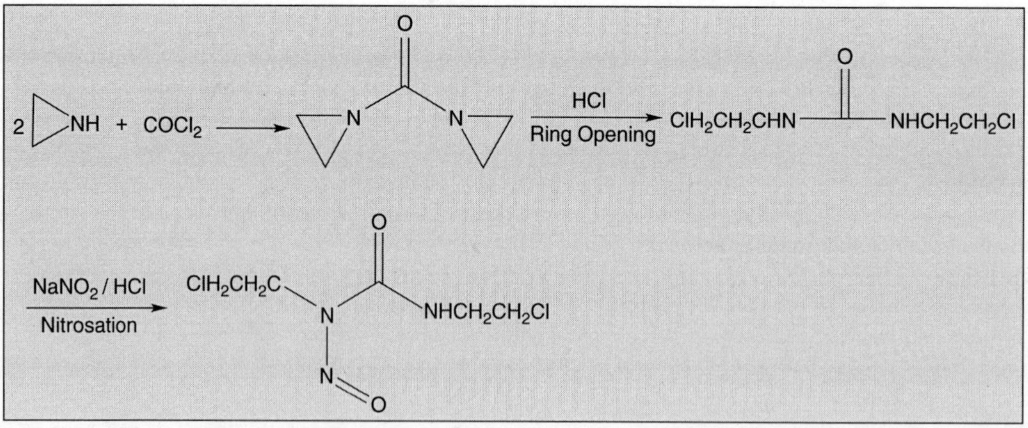

Two moles of aziridine react with one mole of phosgene to give di-1-aziridinylmethanone, on treatment of which with hydrochloric acid ring opening takes place to form *bis*-2-chloroethyl urea derivative. This on nitrosation with sodium nitrite and acid affords carmustine.

It is used against brain tumours, Hodgkin's disease, and other lymphomas.

Mechanism of action: The non-nitrosated nitrogen supplies electron for an intramolecular displacement of Cl⁻ to give an intermediate iminoether, which collapses to isocyanate and highly reactive vinyldiazohydroxide, which in turn results in vinyl carbonium ion.

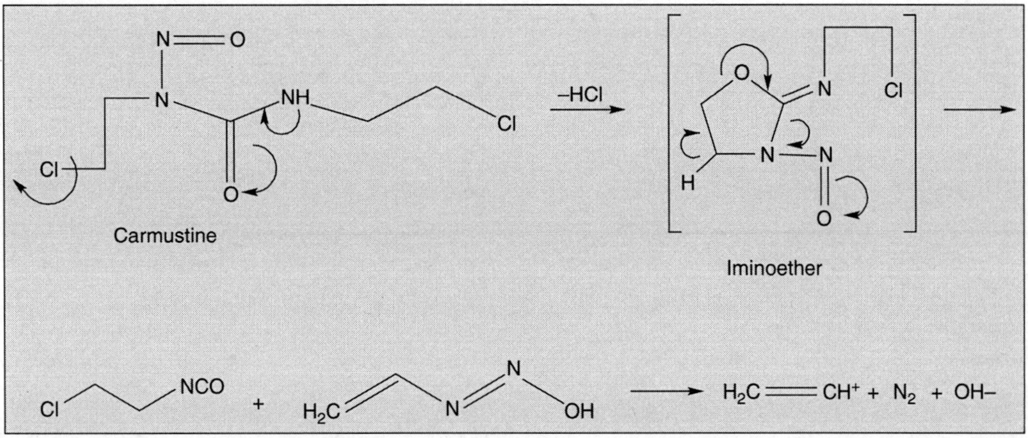

Vinyl carbonium ion is an alkylating agent, which inhibits DNA synthesis. The intermediate 2-chloroethyl isocyanate reacts with -NH_2 group of lysine of DNA-repairing enzyme guanine-O^6-alkyl transferase (GOAT) to form urea and inactivate the repairing enzyme.

Lomustine and Semustine   1-(2-Chloroethyl)-3-(sub) cyclohexyl-1-nitrosourea
Synthesis

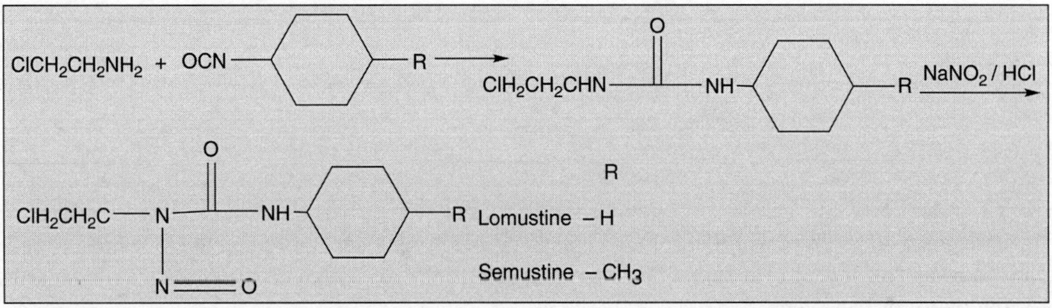

Lomustine is made by reacting 2-chloroethylamine with cyclohexylisocyanate, which forms 1-(2-chloroethyl)-3-cyclohexylurea. This is nitrosated with acid and sodium nitrite to give lomustine.

Mechanism of action:

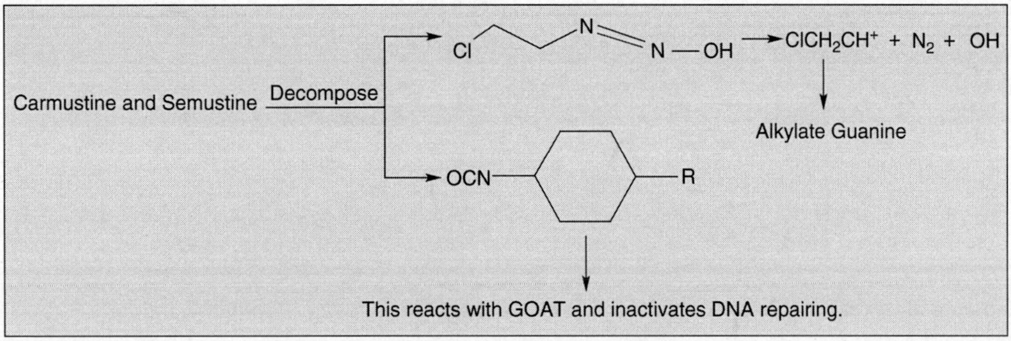

AZIRIDINES:

Thiotepa Tris (1-aziridinyl) phosphine sulphate
Synthesis

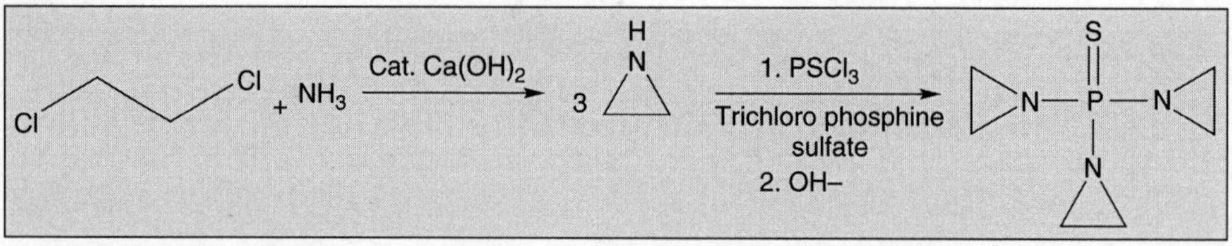

It is made by reacting ethylenimine (aziridine) with phosphorous sulphochloride.

The aziridine rings open after protonation of the ring nitrogen, leading to a reactive molecule. It is active against breast, ovarian, and bronchogenic carcinomas and malignant lymphomas.

Benzotepa
Synthesis

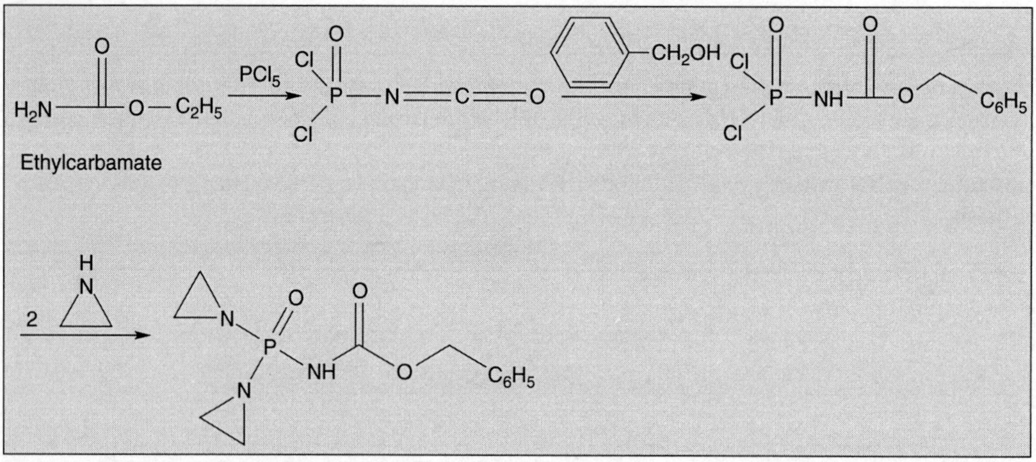

Ethylcarbamate on reaction with phosphorous pentachloride gives isocyanate derivative, which on reaction with benzyl alcohol affords carbamate intermediate. This on reaction with two moles of ethylenimine (aziridine) affords benzotepa.

Altretamine N_2,N_2,N_4,N_4,N_6,N_6-Hexamethyl-1,3,5-triazine-2,4,6-triamine
Synthesis

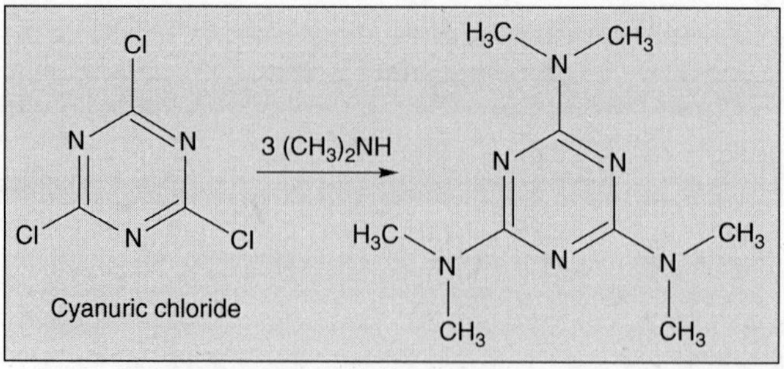

It is prepared by reacting 2,4,6-trichloro-1,3,5-triazine with three moles of dimethylamine.

Alkyl sulphonates

Busulphan 1,4-Di (methane sulphonyloxy) butane
Synthesis

$$2\ CH_3SO_2Cl + HO-(CH_2)_4-OH \longrightarrow CH_3SO_2O(CH_2)_4OSO_2CH_3$$
Methanesulphonyl chloride

It is made by reacting 1,4-butandiol with methanesulphonyl chloride.

It is mainly used for the treatment of chronic granulocytic leukaemia. The exact mechanism is not well understood; cross-linked guanine residues have been identified in DNA incubated *in vitro* with busulphan.

31.4.3 Miscellaneous

Dacarbazine 5-(3, 3-Dimethyl-1-triazenyl) imidazole-4-carboxamide
Synthesis

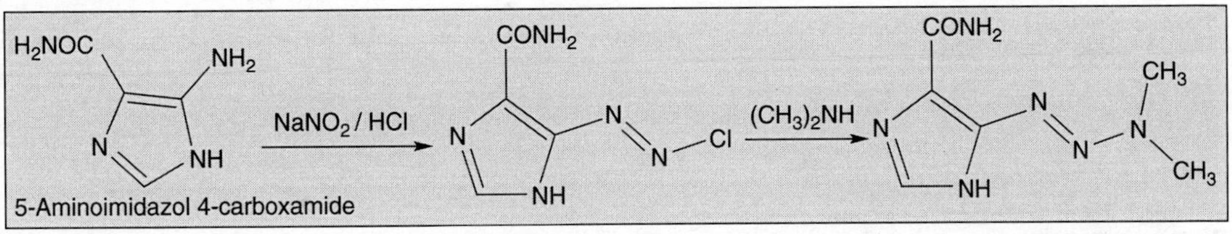

[Reaction scheme: 5-Aminoimidazol 4-carboxamide → (NaNO₂/HCl) → diazo intermediate → ((CH₃)₂NH) → Dacarbazine]

It is made by diazotization of 5-aminoimidazol-4-carboxamide with nitrous acid, which results in the formation of 5-diazoimidazol-4-carboxamide. Reacting this with dimethylamine gives the desired dacarbazine.

Dacarbazine is indicated for the treatment of metastatic malignant melanoma.

Mechanism of action:

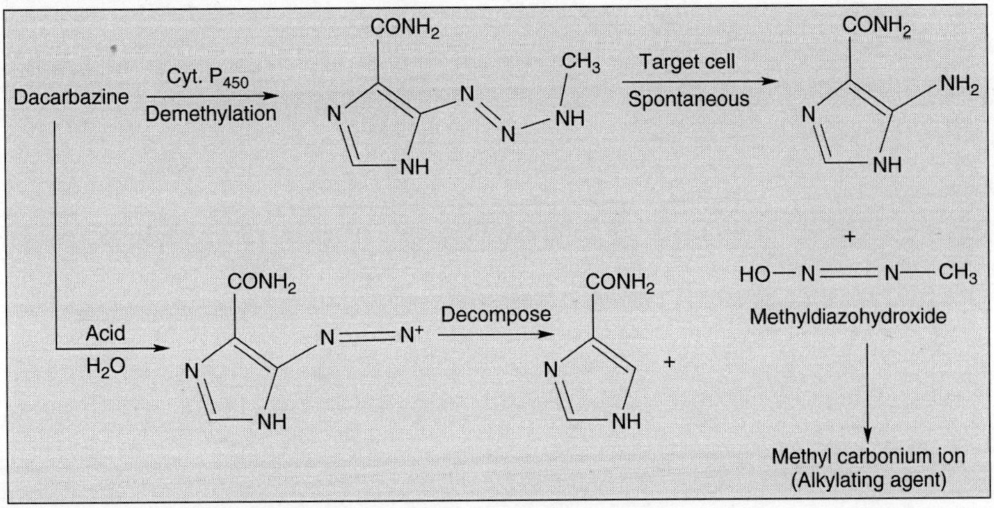

[Scheme: Dacarbazine → (Cyt. P₄₅₀ Demethylation) → monomethyltriazene → (Target cell, Spontaneous) → 5-aminoimidazole-4-carboxamide + Methyldiazohydroxide (HO—N=N—CH₃) → Methyl carbonium ion (Alkylating agent); alternative pathway: Dacarbazine → (Acid, H₂O) → diazonium → Decompose → products]

31.5 ANTIMETABOLITES

Antimetabolites are compounds that prevent the biosynthesis or utilization of normal cellular metabolites. They usually are closely related in structure to the metabolite.

31.5.1 Folic Acid Antagonists

Folic acid is an essential dietary factor from which is derived a series of tetrahydro folate cofactors that provide single carbon group for the synthesis of precursors of DNA (thymidylate and purines) and RNA (purines).
Synthesis of folate cofactor (1 C donor)

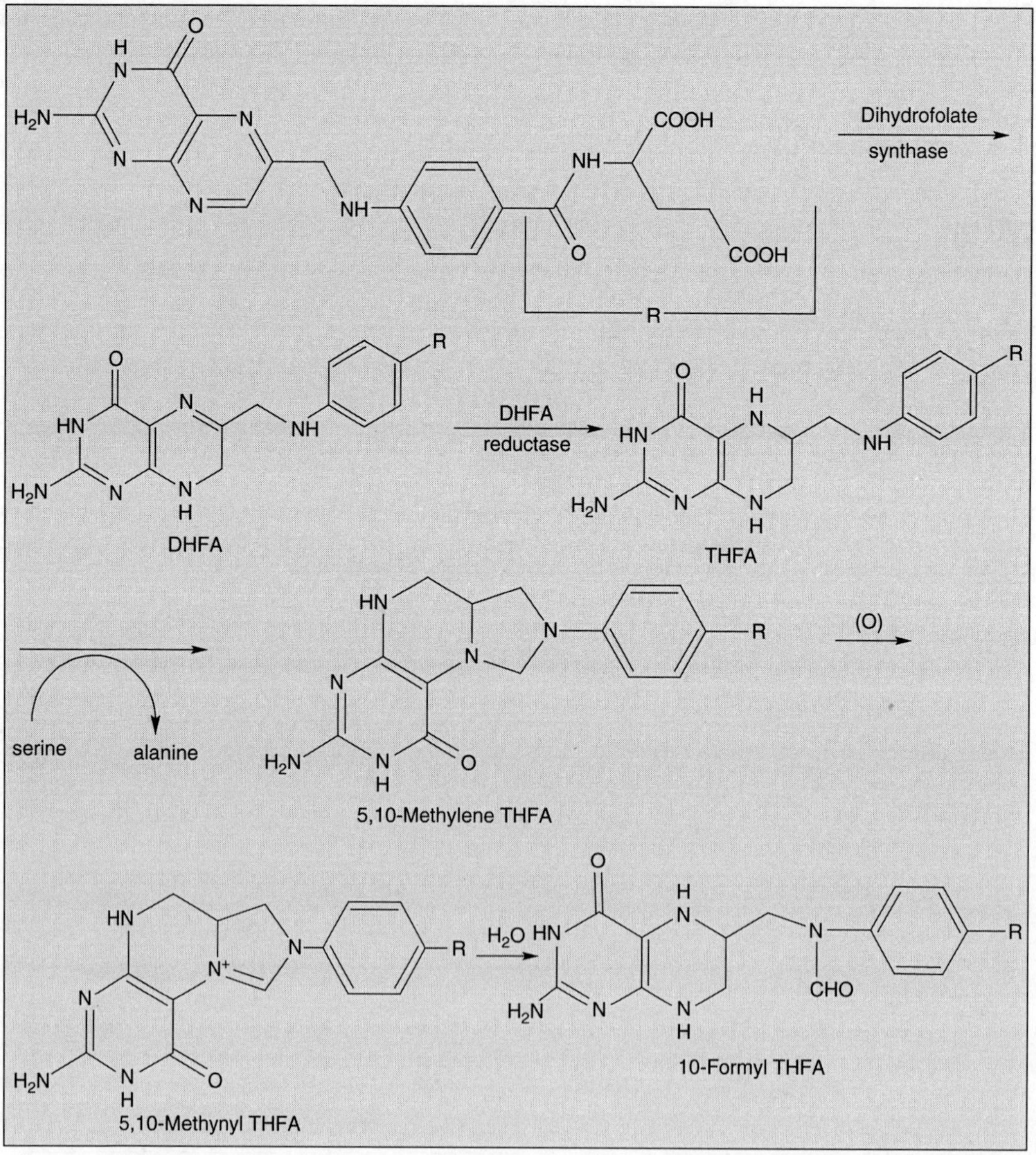

5, 10-Methylene THFA is the 1 C donor for the synthesis of thymidylate.

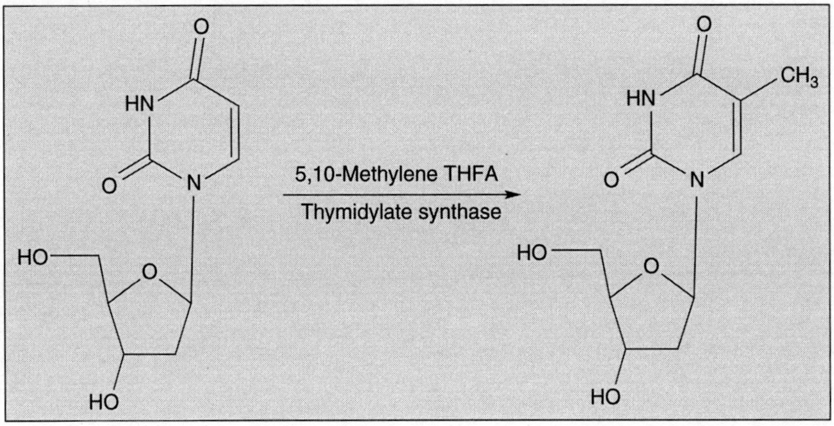

10-formyl THFA is the 1 C donor for the synthesis of purines.

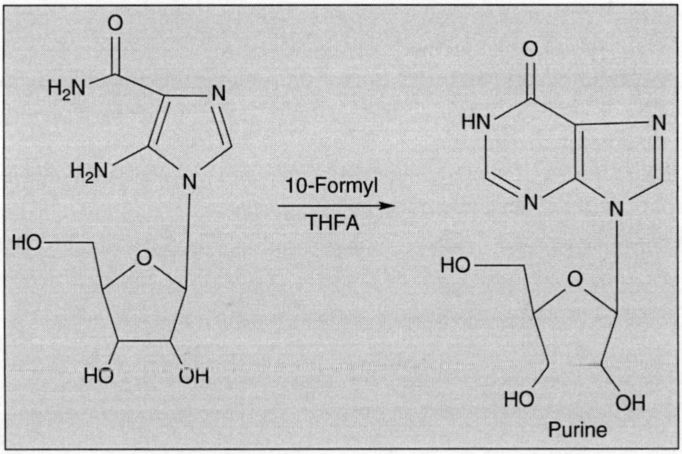

Mechanism of action:
Folic acid antagonists, methotrexate and trimetrexate inhibit DHFR, which in turn leads to the depletion of folate cofactors and thereby prevent the biosynthesis of DNA and RNA in cancerous cells.

Methotrexate 4-Amino-N¹⁰-methyl pteroylglutamic acid
Synthesis

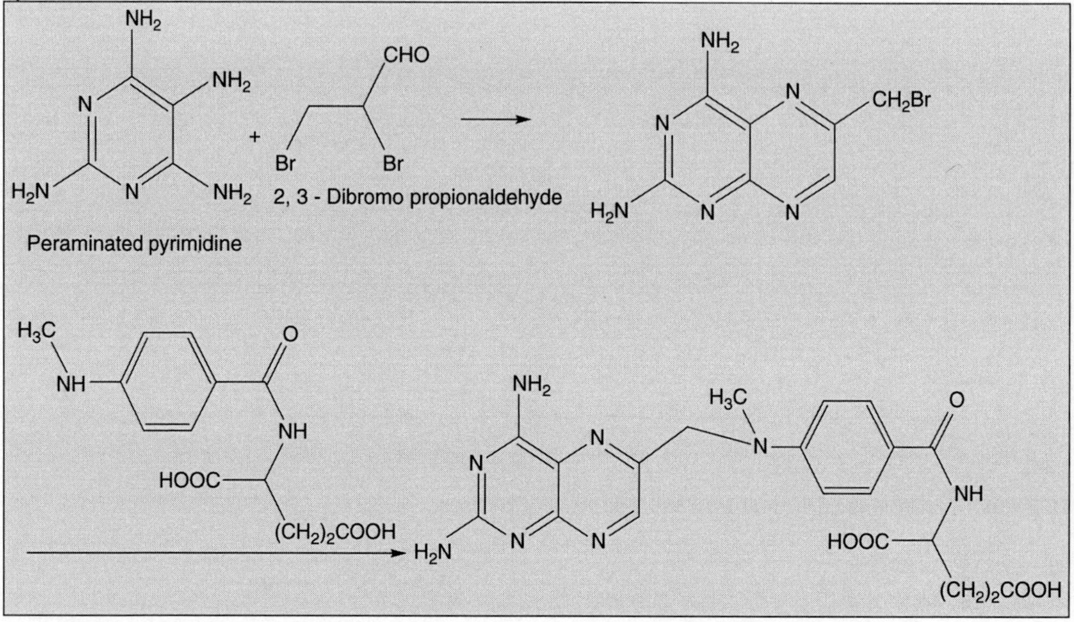

2,4,5,6-Tetraaminopyrimidine upon reacting with 1,2-dibromopropionic aldehyde gives 2,4-diamino-6-bromomethylpteridine. Alkylating the amine nitrogen atom of N-(4-methylaminbenzoyl)glutamic acid with resulting bromide gives methotrexate.

It is used for the treatment of acute lymphocytic leukaemia, acute lymphoblastic leukaemia, breast cancer, epidermoid cancer of the head and neck, and lung cancer.

Trimetrexate 5-Methyl-6-[(3,4,5-trimethoxyphenyl) aminomethyl] quinazoline-2,4-diamine
Synthesis

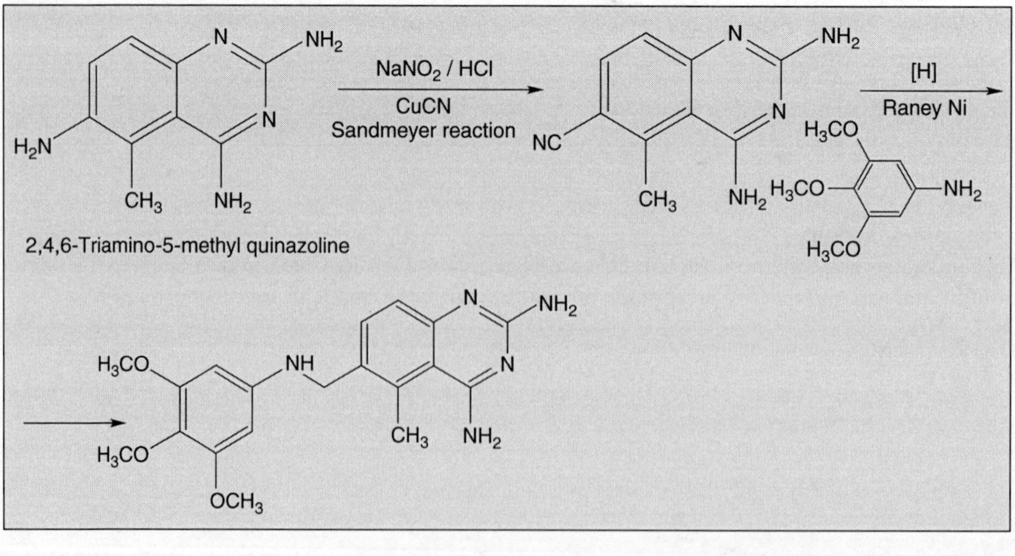

Treatment of 2,4,6-triamino-5-methylquinazoline with nitrous acid gives unstable diazonium salt, which on treatment with cuprous cyanide gives 6-cyano-2,4-diamino-5-methylquinazoline. Hydrogenation of nitrile in the presence of 3,4,5-trimethoxyaniline over Raney nickel affords trimetrexate.

31.5.2 Purine Antagonists

6-Mercaptopurine 3,7-Dihydropurine-6-thione
Synthesis

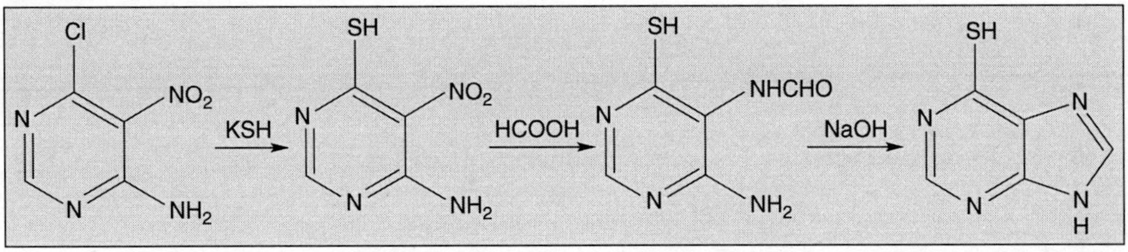

4-Amino-6-chloro-5-nitropyrimidine on reaction with potassium hydrogensulphide gives 6-mercapto derivatives, which on reaction with formic acid affords 6-mercaptopurine.

Mechanism of action:

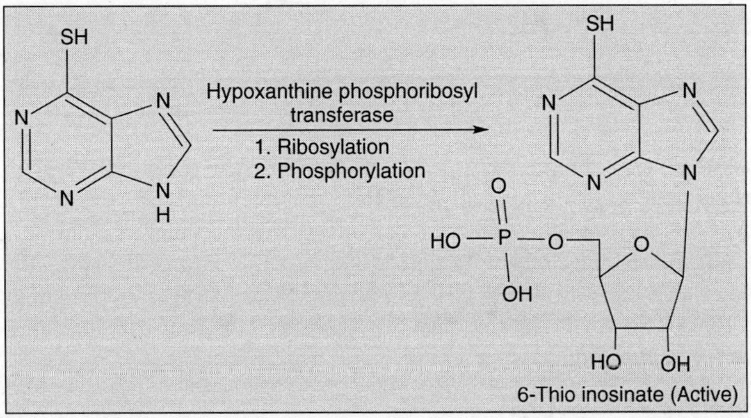

6-Thio inosinate inhibits the conversion of inosinic acid to adenylic acid and xanthylic acid, and thereby prevents the purine biosynthesis. It is used primarily for treating acute leukaemia.

6-Thioguanine 2-Aminopurin-6-thiol
Synthesis

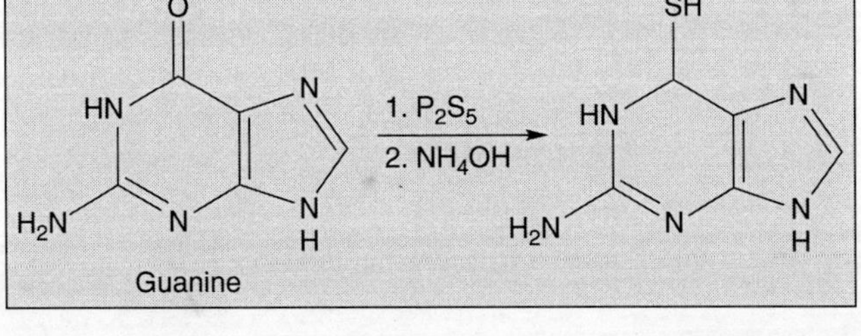

It is prepared by treating guanine with phosphorous penta sulphide in presence of base.

6-Thio guanine is converted to 6-thio guanine ribonucleotide, which acts as 6-mercapto purine.

Fludarabine 2-(6-Amino-2-fluoro-9H-9-purinyl)-5-(hydroxymethyl)tetrahydro-3,4-furandiol

Synthesis

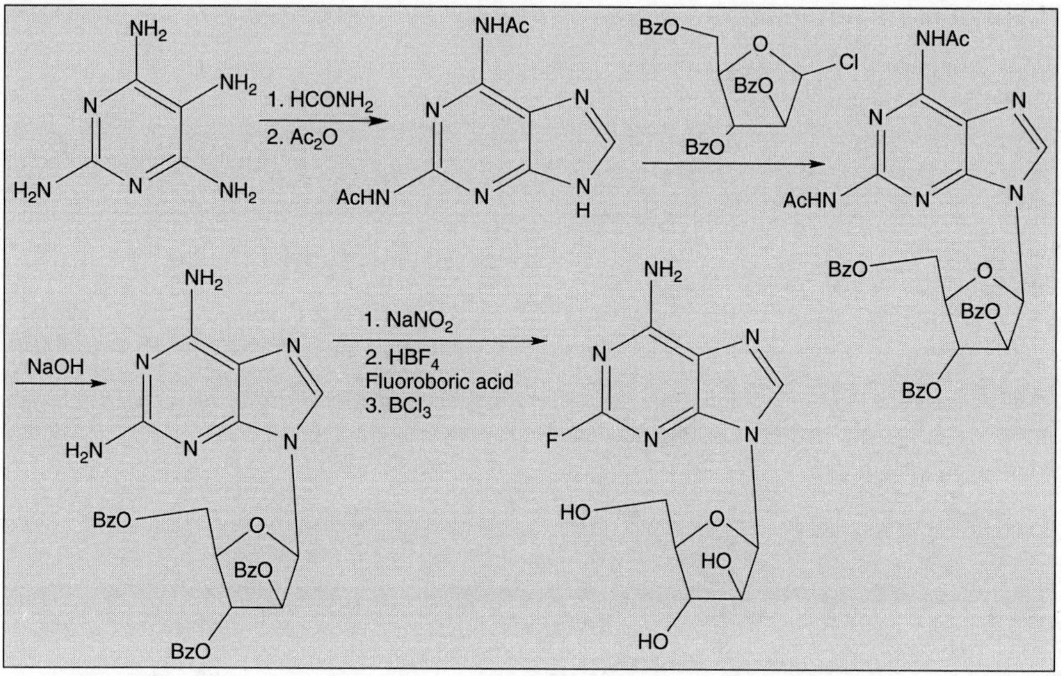

Peraminated pyrimidine reacts with formamide to form diaminopurine, and treatment of this with acetic anhydride gives the corresponding diacetyl derivative. Displacement of chlorine in the fully benzylated arabinioside chloro sugar results in glycosylation of the purine and formation of nucleoside. Saponification followed by treatment with nitrous acid in presence of fluoroboric acid leads to diazotization, followed by replacement with fluorine. The benzyl-protecting groups are then cleaved by treatment with boron trifluoride to afford fludarabine.

31.5.3 Pyrimidine Antagonists

5-Fluorouracil: 5-Fluoro-1H-pyrimidine-2,4-dione

Synthesis

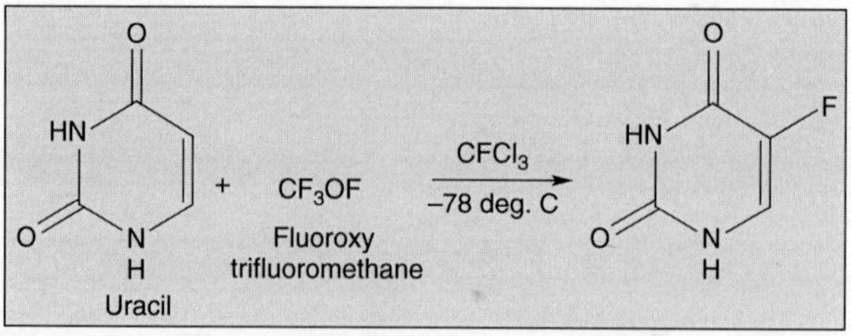

It is prepared by treating uracil with fluoroxy trifluoromethane at −78°C.

It is used in the management of the breast, colon, pancreas, and stomach carcinomas.
Mechanism of action:

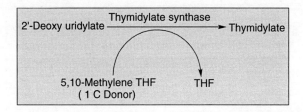

Thymidylate is biosynthesised from uridylic acid.

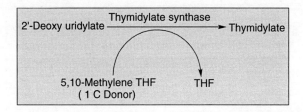

The active metabolite of 5-FU competes with 2'-deoxy uridylate, and it reacts with enzyme and 5, 10-methylene THF to form a stable complex as a terminal product and inhibits the thymidylate biosynthesis.

Cytarabine   4-Amino-1-β-arabinofuranosyl-2(1H)pyrimidone

Synthesis

Fully benzylated arabinoside chloro sugar on treatment with 2,4-dimethoxypyrimidine displaces chlorine to form nucleoside intermediate, which on treatment with ammonium hydroxide leads to 4-amino derivative. Benzyl deprotection affords cytarabine.

Mechanism of action: It is a cytidine arabinoside. Arabinose moiety is epimeric at the 2'-position with ribose. This modification after anabolism to cytarabine-triphosphate inhibits the ribonucleotide reductase, which converts cytidylic acid to 2'-deoxy cytidylic acid during DNA synthesis. It is used for the treatment of acute leukaemia of adults and children.

31.6 ANTICANCER ANTIBIOTICS

Anthracycline antibiotics: Anthracyclines occur as glycosides of the anthracyclinone. The glycosidic linkage usually involves the 7-hydroxyl group of the anthracyclinone and the β-anomer of a sugar with L-configuration. Anthracyclinone refers to an aglycone containing the anthraquinone chromophore within a linear hydrocarbon skeleton. The anthracyclines differ from each other in the number and location of the phenolic hydroxylgroups, the degree of oxidation of the two carbon side-chains at position 9, and the presence of carboxylic acid ester at position 10.

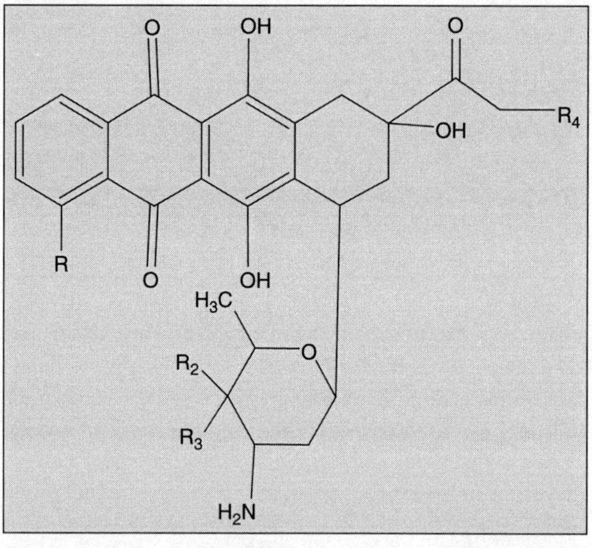

	R_1	R_2	R_3	R_4
Daunorubicin	—OCH$_3$	H	OH	H
Doxorubicin	—OCH$_3$	H	OH	OH
Carminomycin	OH	H	OH	OH
Idrabucin	H	H	OH	H
Epirubicin	—OCH$_3$	OH	H	OH

Daunorubicin and doxorubicin were commonly used and were obtained from *Streptomyces peucetium*. These drugs bind to the DNA and inhibit nucleic acid synthesis, inhibit mitosis, and promote chromosomal aberration. These drugs are used for the treatment of acute myelocytic leukaemia, primary hepatocellular carcinoma, ovarian endometrial, pancreatic, bladder, and lung carcinoma, gastric adenocarcinoma, and cervical and testicular carcinoma.

31.7 MISCELLANEOUS DRUGS

Actinomycin D

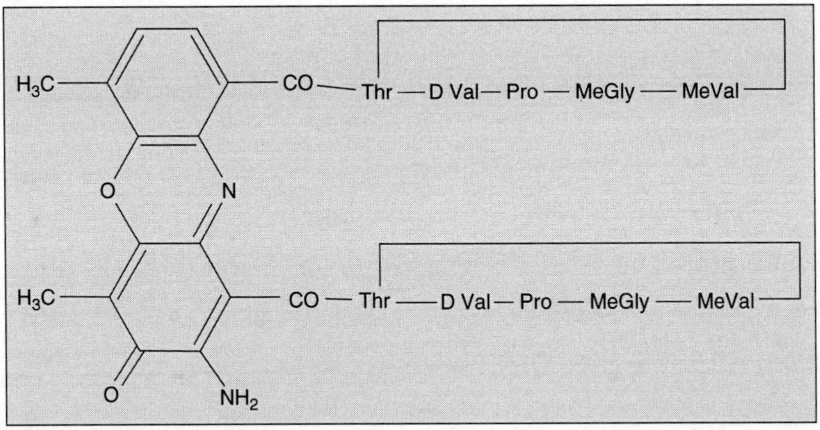

It is obtained from cultures of *Streptomyces antibioticus*. It intercalates into the double-helical DNA. The main biochemical consequence of the intercalation of actinomycin into DNA is the inhibition of DNA and RNA synthesis, which in turn causes depletion of protein and leads to cell death.

Mithramycin

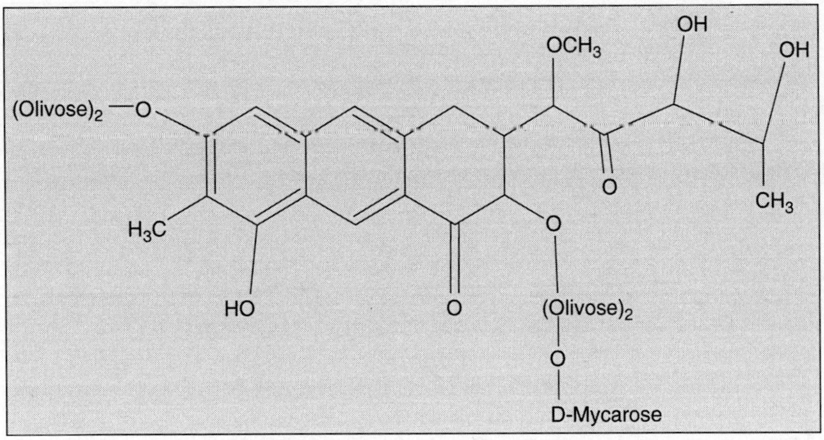

It is an aureolic acid derivative obtained from *Streptomyces plicatus*. It forms a complex with divalent metals such as magnesium and calcium. Such complex formation is required before binding with DNA. It inhibits DNA-dependent RNA polymerase that leads to cell death.

Bleomycin sulphate

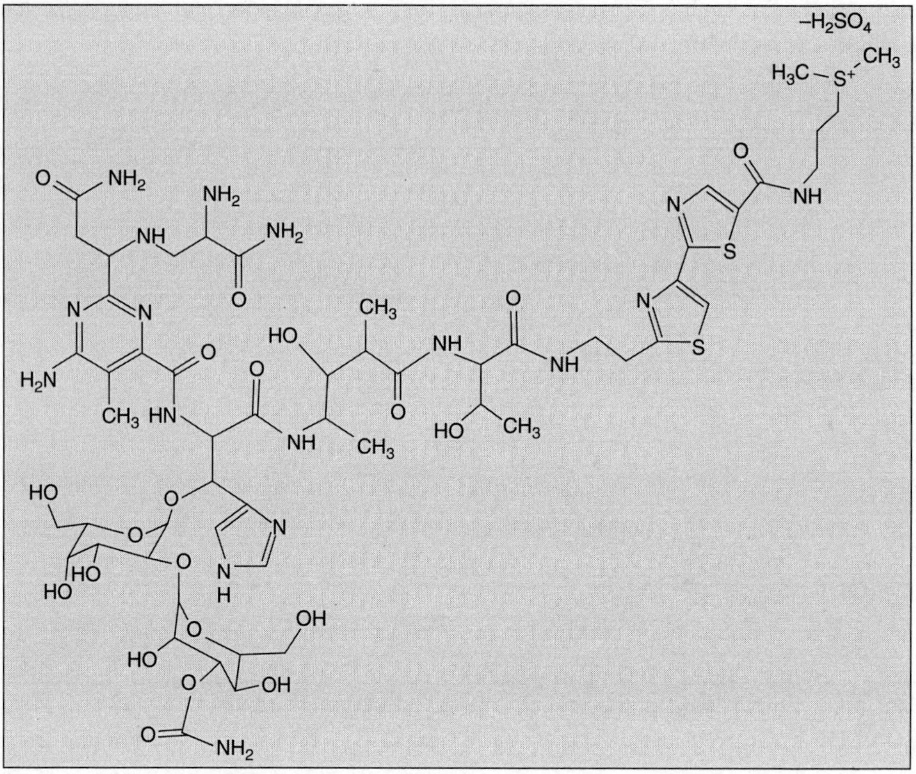

[Structure of Bleomycin sulphate shown, with –H$_2$SO$_4$ annotation]

It is a glycopeptide antibiotic obtained from *Streptomyces verticillus*. Inside the cell, bleomycin forms a complex with Fe (II) that gives rise to hydroxyl radical and superoxide radicals. These radicals cleave the phosphodiester bond of DNA. This degradation of DNA strands is thought to be a lethal event in cells.

Mitomycin C

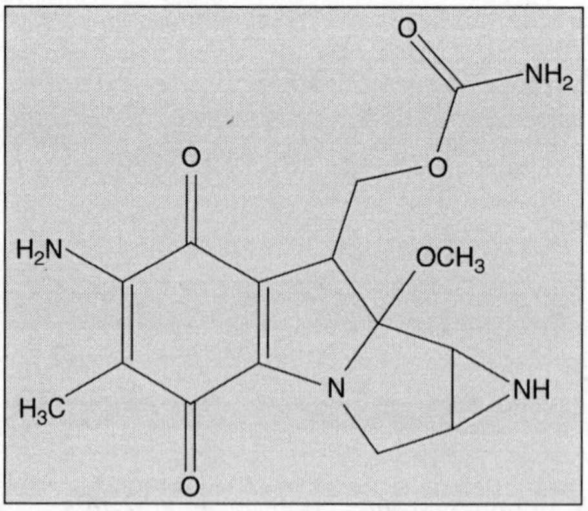

[Structure of Mitomycin C shown]

It is obtained from *Streptomyces caespitosus*. It contains three different carcinostatic functions: quinone, carbamate, and aziridine. Chemical and enzymatic reduction of quinone to the corresponding hydroquinone is followed by the loss of methanol (from -CH₃O group), and the resulting indolo hydroquinone becomes a bifunctional alkylating agent capable of cross-linking double-helical DNA.

31.8 ANTICANCER PLANT PRODUCTS

Vinca alkaloids

Vinca alkaloids are isolated from *Catharanthus roseus*. They have complex structures composed of an indole-containing moiety named catharanthine and an indoline-containing moiety named vindoline. Four closely related compounds have antitumour activity: vincristine, vinblastine, vinrosidine, and vinleurosine.

Vinca alkaloids cause mitotic arrest by promoting the dissolution of microtubules in cells. They are used for the treatment of acute leukaemia, Hodgkin's disease, testicular cell tumour, lymphocytic lymphoma, histicytic lymphoma, and carcinoma of the breasts.

Taxol derivatives

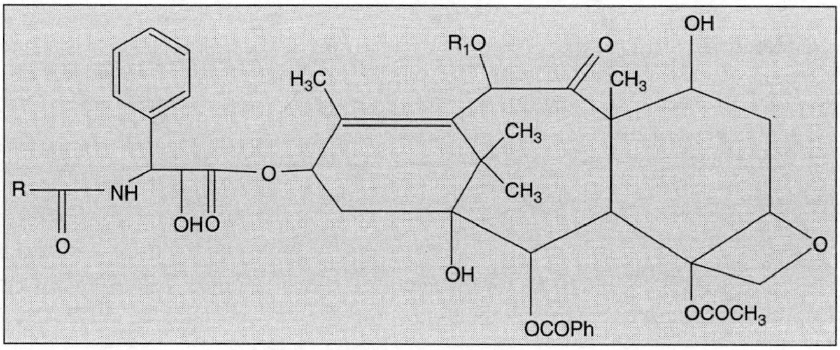

	R	R_1
Paclitaxel	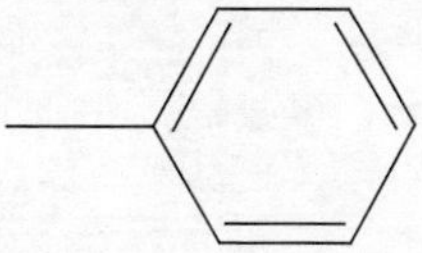 C₆H₅	COCH₃
Docetaxel	(CH₃)₃CO—	H

The taxol derivative paclitaxel is isolated from the western yew tree *Taxus brevifolia*. These drugs bind specifically to the β-tubulin subunit of the microtubules and appear to antagonise the disassembly of the key cytoskeletal proteins, and arrest in mitosis follows.

Camptothecin derivatives

Camptothecin (CPT) is a pentacyclic alkaloid originally isolated from *Camptotheca acuminata*.

CPT has a planar pentacyclic ring structure that includes a pyrrolo[3,4-β]-quinoline moiety (rings A, B, and C), conjugated pyridone moiety (ring D), and one chiral centre at position 20 within the α-hydroxy lactone ring with (S) configuration (the E-ring). Its planar structure is thought to be one of the most important factors in topoisomerase inhibition.

These drugs are used for the treatment of colorectal and ovarian cancers.

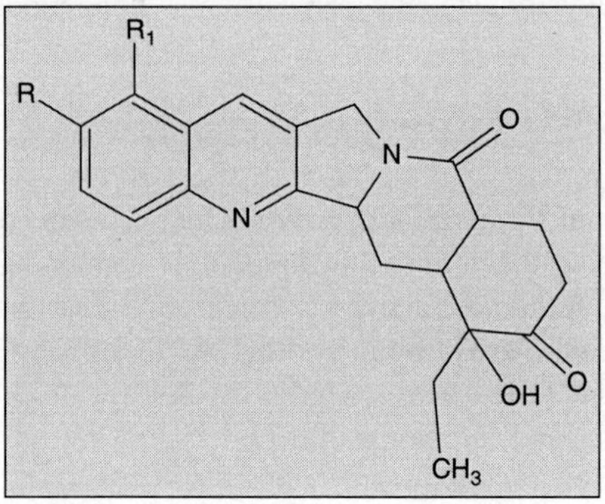

	R	R_1
Camptothecin	H	H
Irinotecan	—C_2H_5	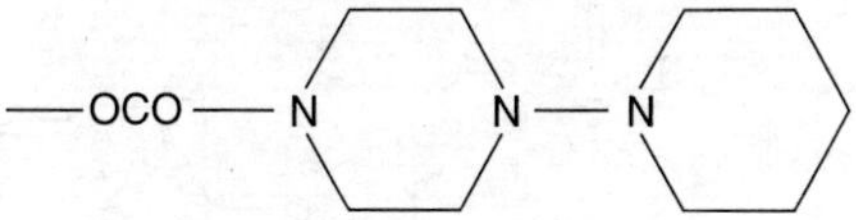 —OCO—N⌐⌐N—N⌐ (piperazine-piperidine)
Topotecan	—$CH_2N(CH_3)_2$	OH

Mechanism of action: DNA normally exists as a super-coiled double helix. During replication, it unwinds with the single strands serving as templates for the synthesis of new strands. Topoisomerase facilitates this process, and camptothecin derivatives inhibit the topoisomerase enzyme and lead to DNA synthesis arrest.

Podophyllotoxins

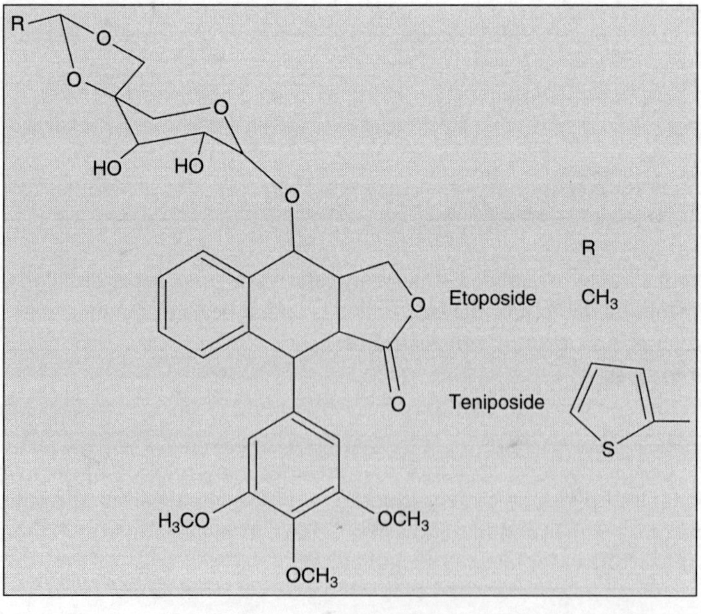

Etoposide: R = CH_3

Teniposide: R = thiophene

These are semi-synthetic derivatives of podophyllotoxin used for the treatment of paediatric leukaemia, small cell lung carcinoma, testicular tumours, and Hodgkin's disease. These drugs inhibit mitosis by destroying the structural organization of the mitotic apparatus.

31.9 HORMONES AND THEIR ANTAGONISTS

Tamoxifen   (Z)-2-[p-(1,2-Diphenyl-1-butenyl)phenoxy]N,N-dimethylethylamine

Tamoxifen is a nonsteroidal agent that has shown potent antioestrogenic properties. It is useful in the treatment of advanced breast cancer in postmenopausal women.

Synthesis

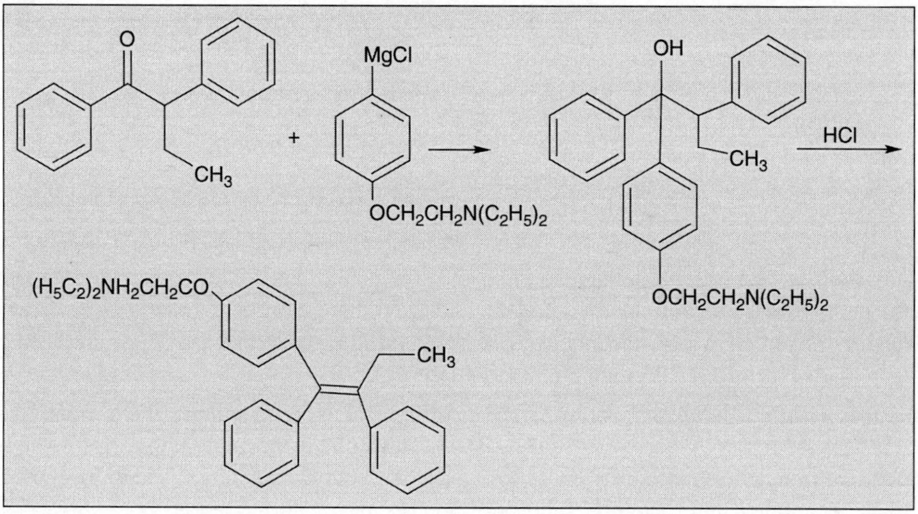

Reaction of α-ethyldeoxybenzoin and the 4-(2-dimethylaminoethoxy)phenylmagnesium bromide, and further dehydration of the resulting carbinol are followed by subsequent separation of the mixture of E and Z isomers.

Mechanism of action: Under normal physiological conditions, oestrogen stimulation increases tumour cell production of transforming growth factor-β (TGF-β), an autocrine inhibitor of tumour cell growth. Tamoxifen is a competitive inhibitor of estradiol binding to the oestrogen receptor. By blocking the TGF-β pathway, it is used to decrease the autocrine stimulation of breast cancer growth.

Flutamide   4'-Nitro-3'-trifluoromethylisobutyranilide

Synthesis

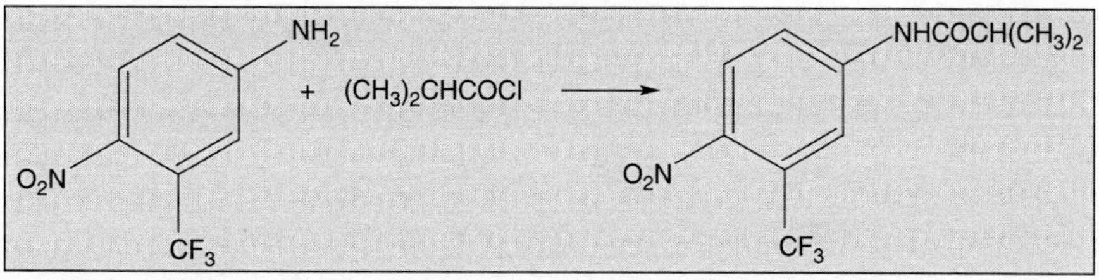

Flutamide, is made by acylating 4-nitro-3-trifluoromethylaniline with isobutyric acid chloride.

Mechanism of action: It is a nonsteroidal antiandrogen drug, which inhibits the translocation of the androgen receptor to the nucleus from the cytoplasm in the hypothalamus and prostate. It is used for the treatment of prostate cancer.

Aminoglutethimide   (-)-2-(4-Aminophenyl)-2-ethylglutarimide

Synthesis

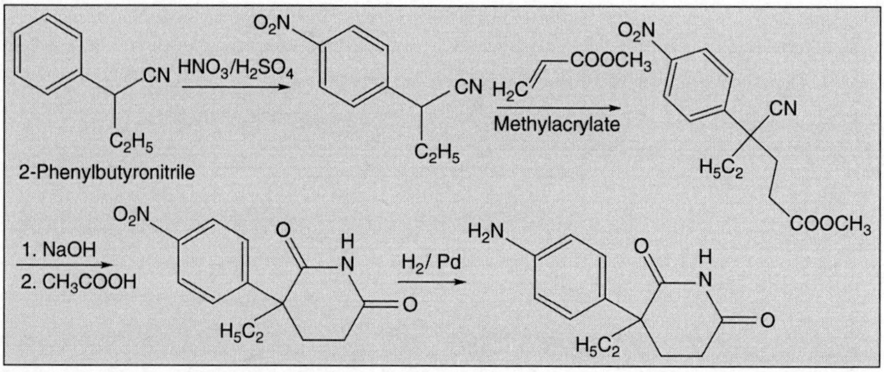

Synthesis starts with 2-phenylbutyronitrile, which is nitrated under cold conditions, forming 2-(4-nitrophenyl)butyronitrile. The last, in Michael addition-reaction conditions, in the presence of benzyltrimethylammonia hydroxide is added to methylacrylate, and the obtained product undergoes basic hydrolysis, during which a cyclization to 2-(4-nitrophenyl)-2-ethylglutarimide takes place, and this product is reduced by hydrogen to give the desired product aminoglutethimide.

Mechanism of action: It inhibits the conversion of cholesterol to pregnenolone, the first step in the synthesis of cortisol. Inhibition of cortisol synthesis, however, results in a compensatory rise in the secretion of adrenocorticotrophic hormone sufficient to overcome adrenal blockade. It is used to treat patients with adrenocortical carcinoma and Cushing's syndrome, and metastatic hormone-dependent breast cancer.

31.10   MISCELLANEOUS DRUGS

Procarbazine   1-Methyl-2-(n-isopropylcarbamoylbenzyl)hydrazine

Synthesis

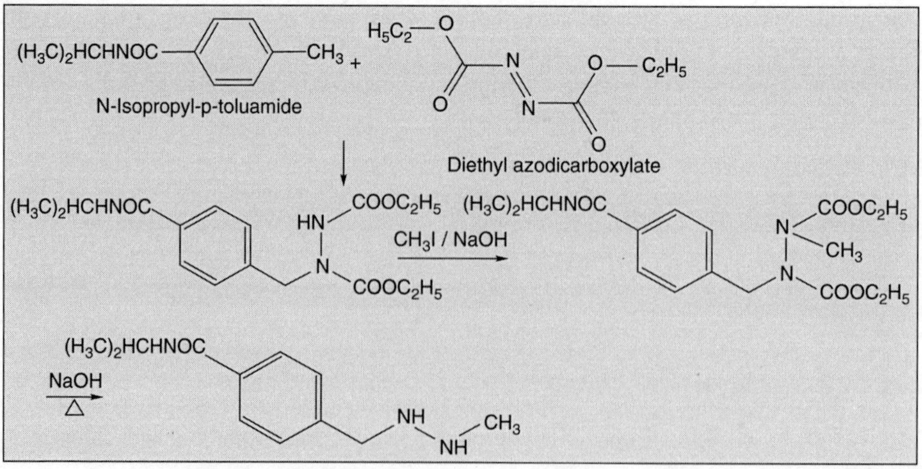

Reaction of *N*-isopropyl-4-toluamide with ethyl azodicarboxylate leads to the substituted hydrazine. Methylation with methyl iodide gives *N*-methyl derivative. Saponification of carbethoxy group affords procarbazine.

Mechanism of action:

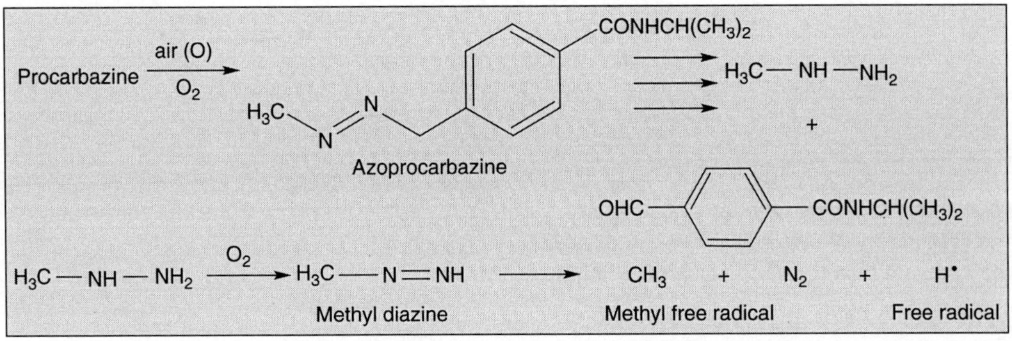

Procarbazine $\xrightarrow{\text{air (O)}, O_2}$ Azoprocarbazine (H$_3$C—N=N—CH$_2$—C$_6$H$_4$—CONHCH(CH$_3$)$_2$) $\rightarrow$ H$_3$C—NH—NH$_2$ + OHC—C$_6$H$_4$—CONHCH(CH$_3$)$_2$

H$_3$C—NH—NH$_2$ $\xrightarrow{O_2}$ H$_3$C—N=NH (Methyl diazine) $\rightarrow$ CH$_3$ (Methyl free radical) + N$_2$ + H• (Free radical)

Procarbazine is activated to methyl free radical, which acts as an alkylating agent for DNA.

Hydroxyurea

Synthesis

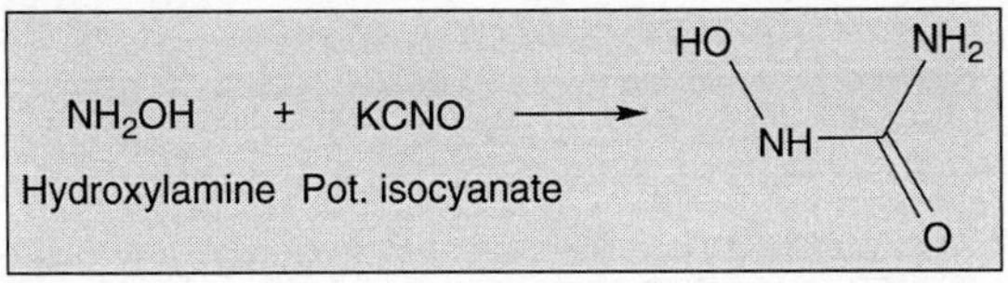

NH$_2$OH + KCNO $\rightarrow$ HO—NH—C(=O)—NH$_2$

Hydroxylamine Pot. isocyanate

It is made by reacting potassium cyanate with hydroxylamine.

It inhibits ribonucleoside diphosphate reductase. This enzyme catalyses the reductive conversion of ribonucleotide to deoxy ribonucleotide during the biosynthesis of DNA. It is used for the treatment of melanoma, ovarian cancer, and myelocytic leukaemia.

Cisplatin *cis*-Diaminodichloroplatinum

Synthesis

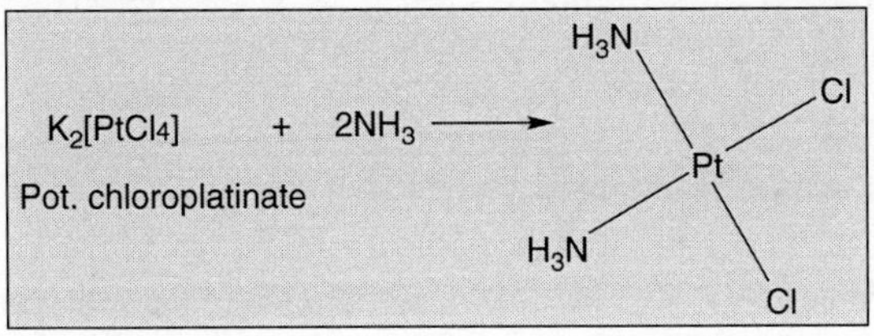

K$_2$[PtCl$_4$] + 2NH$_3$ $\rightarrow$ *cis*-[Pt(NH$_3$)$_2$Cl$_2$]

Pot. chloroplatinate

Cisplatin is made by reacting potassium tetrachloroplatinate with ammonia.

It is a potent inhibitor of DNA polymerase. Its activity resembles those of the alkylating agents. In DNA cross-linking by the platinum complex, the two chlorides are displaced by nitrogen or oxygen atom of purines. It is used for the treatment of ovarian and testicular cancers.

31.11 NEWER DRUGS

Temozolomide 4-Methyl-5-oxo-2,3,4,6,8-pentazabicyclo [4.3.0] nona-2,7,9-triene- 9-carboxamide

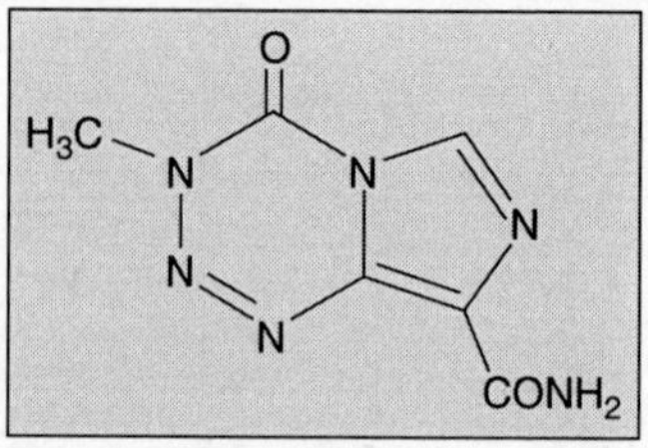

It is an imidazotetrazine derivative of the alkylating agent dacarbazine. It is a prodrug and undergoes rapid chemical conversion in the systemic circulation at physiological pH to the active compound, MTIC (3-methyl-(triazen-1-yl)imidazole-4-carboxamide). It is used for the treatment of refractory anaplastic astrocytoma - a type of cancerous brain tumour.

Pemetrexed 2-[4-[2-(4-Amino-2-oxo-3,5,7-triazabicyclo[4.3.0] nona-3,8,10-trien-9-yl)ethyl] benzoyl] aminopentanedioic acid

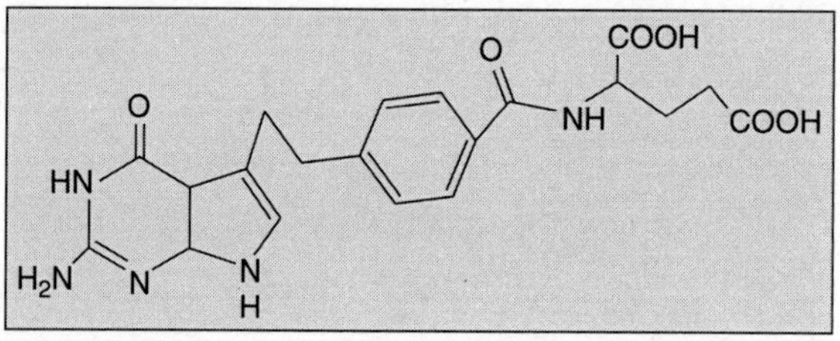

It is chemically similar to folic acid and is in the class of chemotherapy drugs called folate antimetabolites. It is approved for treatment of malignant pleural mesothelioma, a type of tumour of the lining of the lung, in combination with cisplatin. It is also used as a second-line agent for the treatment of non-small cell lung cancer.

FURTHER READINGS

1. *Wilson and Giswold's Textbook of Organic Medicinal & Pharmaceutical Chemistry*, Block, J.H. and Beale, J.M., Jr (ed.), 11th Ed., 2004.
2. *Foye's Principles of Medicinal Chemistry*, Williams, D.A. and Lemke, T.L. (ed.), 5th Ed., 2002.

MULTIPLE-CHOICE QUESTIONS

1. Cancer in glands is called
   a. Sarcoma
   b. Tumorous
   c. Carcinoma
   d. Leukaemia

2. One of the following statements is TRUE on antineoplastics:
   a. Selective toxicity to cancer cells
   b. Kill cells by zero-order kinetics
   c. Easy access to cancer cells of any region
   d. Easily develop resistance

3. One of the following is not a nitrosourea:
   a. Carmustine
   b. Semustine
   c. Estramustine
   d. Chlorozotocin

4. The residue in DNA that exists predominantly as the keto tautomer:
   a. Cytosine
   b. Guanine
   c. Adenine
   d. Thymidine

5. The principal active alkylator formed from ayclophosphamide is
   a. Aldophosphamide
   b. 4-Ketocycophosphamide
   c. 4-Hydroxycyclophosphamide
   d. Phosphoramide mustard

6. The drug effective in Hodgkin's disease is
   a. Melphalan
   b. Paclitaxol
   c. Estramustine
   d. Mechlormethamine

7. The 1 C donor for the synthesis of purine is
   a. Folic acid
   b. 10-Formyl THFA
   c. Tetrahydro folic acid (THFA)
   d. 5, 10-Methylene THFA

8. Nonsteroidal antiandrogen useful as anticancer agent is
   a. Tamoxifen
   b. Flutamide
   c. Etoposide
   d. Aminoglutethimide

9. A free radical alkylating drug is
   a. Carmustine
   b. Thiotepa
   c. Procarbazine
   d. Altretamine

10. The drug that prevents conversion of cholesterol to pregnolone is
    a. Tamoxifen
    b. Topotecan
    c. Anthracyclines
    d. Aminoglutethimide

11. One of the following is a glycopeptide antibiotic:
    a. Bleomycin
    b. Actinomycin D
    c. Mithramycin
    d. Mitomycin

12. One of the following is a topoisomerase inhibitor:
    a. Paclitaxol
    b. Vincristine
    c. Topotecan
    d. Etoposide

QUESTIONS

1. Classify anticancer agents based on mechanism of action and give structural examples for each class.

2. What is the rationale behind the design of antimetabolites? Explain in detail with structural examples.

3. Why are phosphoramides more effective than other class of compounds as anticancer agents?

4. Write a brief note on hormones and their antagonists as anticancer agents.

5. Describe in detail some plant-derived anticancer compounds and their mechanism of action.

6. Derive the synthetic protocol of methotrexate, carmustine, cytarbine, procarbazine, 6-mercaptopurine, dacarbazine, benzotepa, and mechlorethamine.

7. Explain the mechanism of action of cyclophosphamide with structures.

SOLUTION TO MULTIPLE-CHOICE QUESTIONS

1. c;
2. d;
3. c;
4. b;
5. d;
6. d;
7. b;
8. b;
9. c;
10. d;
11. a;
12. c.

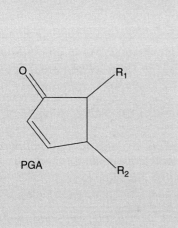

CHAPTER 32

Prostaglandins

LEARNING OBJECTIVES

- Define prostaglandins and their important functions
- Describe the biosynthetic pathway of various prostaglandins
- Describe the naming pattern used for prostaglandins
- Describe various synthetic methods available for prostaglandins
- List out the therapeutically useful prostaglandins
- Derive the relationship between structure and activity of prostaglandins

32.1 INTRODUCTION

Prostaglandins were first discovered and isolated from human semen in the 1930s by Ulf von Euler of Sweden. Thinking they had come from the prostate gland, he named them prostaglandins. It has since been determined that they exist and are synthesised in virtually every cell of the body. Prostaglandins are like hormones in that they act as chemical messengers, but they do not move to other sites—they work right within the cells where they are synthesised.

> The name 'prostaglandin' derives from the prostate gland. When prostaglandin was first isolated from seminal fluid in 1935 by the Swedish physiologist Ulf von Euler, and independently by M.W. Goldblatt, it was believed to be part of the prostatic secretions (in actuality, prostaglandins are produced by the seminal vesicles); it was later shown that many other tissues secrete prostaglandins for various functions.

Prostaglandins are unsaturated carboxylic acids, consisting of a 20-carbon skeleton that also contains a five-membered ring, and are based upon the fatty acid, arachidonic acid. There are a variety of structures—one, two, or three double bonds. On the five-membered ring there may also be double bonds, a ketone, or alcohol groups.

32.2 FUNCTIONS OF PROSTAGLANDINS

There are a variety of physiological effects:

1. There may be activation of the inflammatory response, production of pain, and fever. When tissues are damaged, white blood cells flood to the site to try to minimize tissue destruction. Prostaglandins are produced as a result.
2. Blood clots form when a blood vessel is damaged. A type of prostaglandin called thromboxane stimulates constriction and clotting of platelets. Conversely, PGI_2 is produced to have the opposite effect on the walls of blood vessels where clots should not be forming.
3. Certain prostaglandins are involved with the induction of labour and other reproductive processes. $PGF_{2\alpha}$ causes uterine contractions and has been used to induce labour.
4. Prostaglandins are involved in several other organs such as the gastrointestinal tract (inhibit acid synthesis and increase secretion of protective mucus); they increase blood flow in kidneys; and leukotrienes promote constriction of bronchi associated with asthma.

32.3 BIOSYNTHESIS OF PROSTAGLANDINS

Prostaglandins are found in virtually all tissues and organs. They are autocrine and paracrine lipid mediators that act upon platelet, endothelium, uterine, and mast cells, among others. They are synthesised in the cell from arachidonic acid produced by phospholipase A_2. The intermediate is then passed into either the cyclooxygenase pathway or the lipoxygenase pathway to form either prostaglandin and thromboxane or leukotriene. The cyclooxygenase pathway produces thromboxane, prostacyclin, and prostaglandin D, E, and F. The lipoxygenase pathway is active in leukocytes and in macrophages, and synthesises leukotrienes. Prostaglandins are released through the prostaglandin transporter on the cell's plasma membrane.

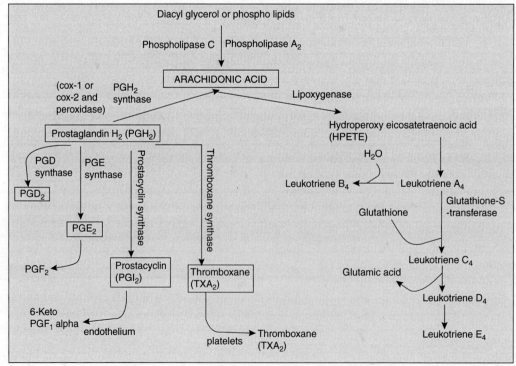

32.4 NOMENCLATURE

Naturally occurring prostaglandins are classified according to their ring substituents.

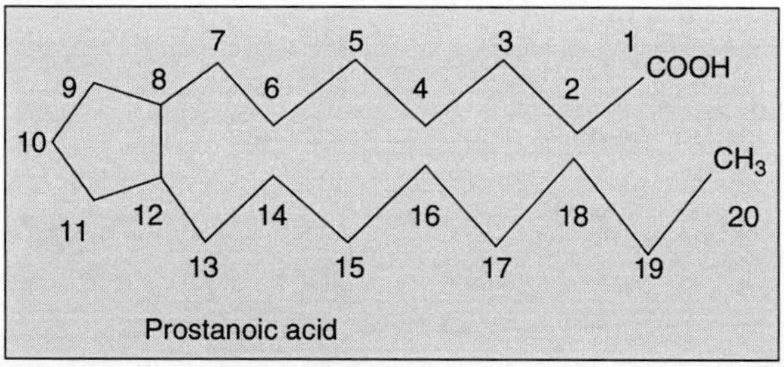

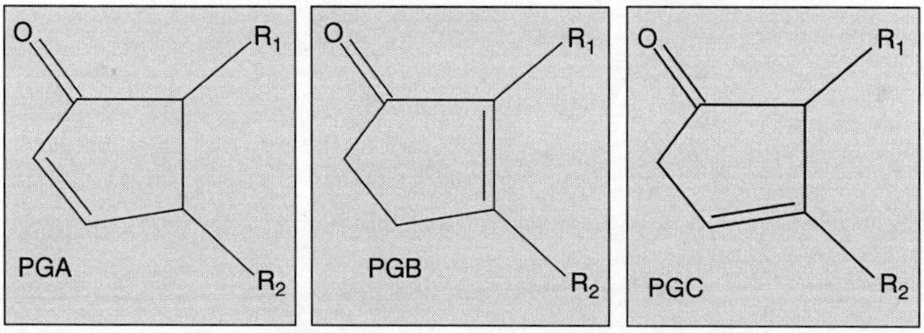

Prostaglandins A, B, and C are α- or β-unsaturated ketones, D and E are β-hydroxy ketones, and F are 1, 3-diols. The main classes are further subdivided in accordance with the number of double bonds present in the side-chain. This is indicated by subscripts 1 or 2—e.g., PGE_1 and $PGF_2 \alpha$.

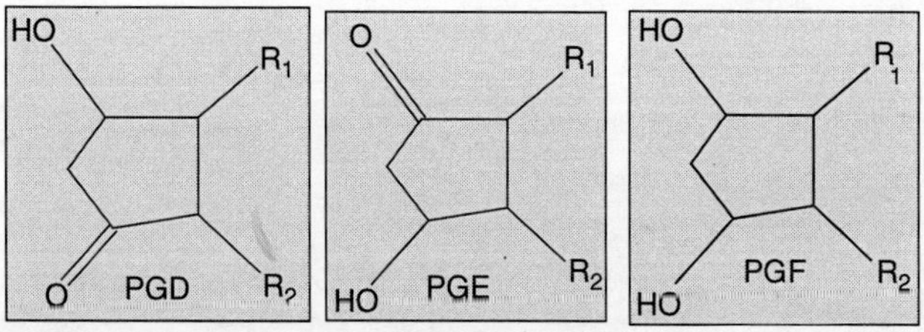

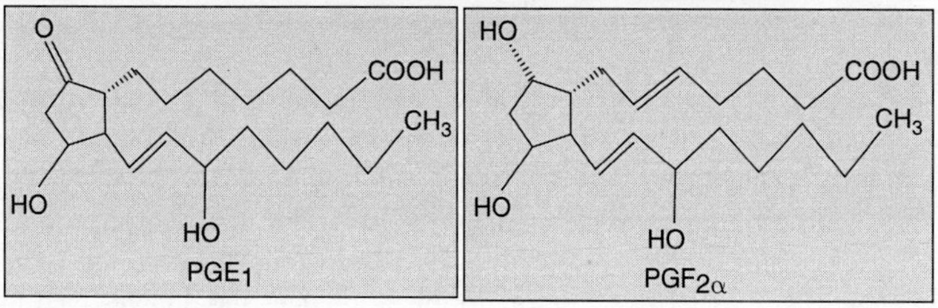

An α-substituent is one projecting down from the plane of the molecule and a β-substituent is one projecting above the plane of the molecule. In $PGF_2\alpha$, the 9-hydroxyl group projects below the plane. The two arms (side-chains) are trans to each other in natural PGs. The carboxy-bearing upper arm is attached in the α-configuration.

32.4.1 Synthesis of Prostaglandins

Prostaglandin E_1 Miyano's synthesis from acyclic precursor β

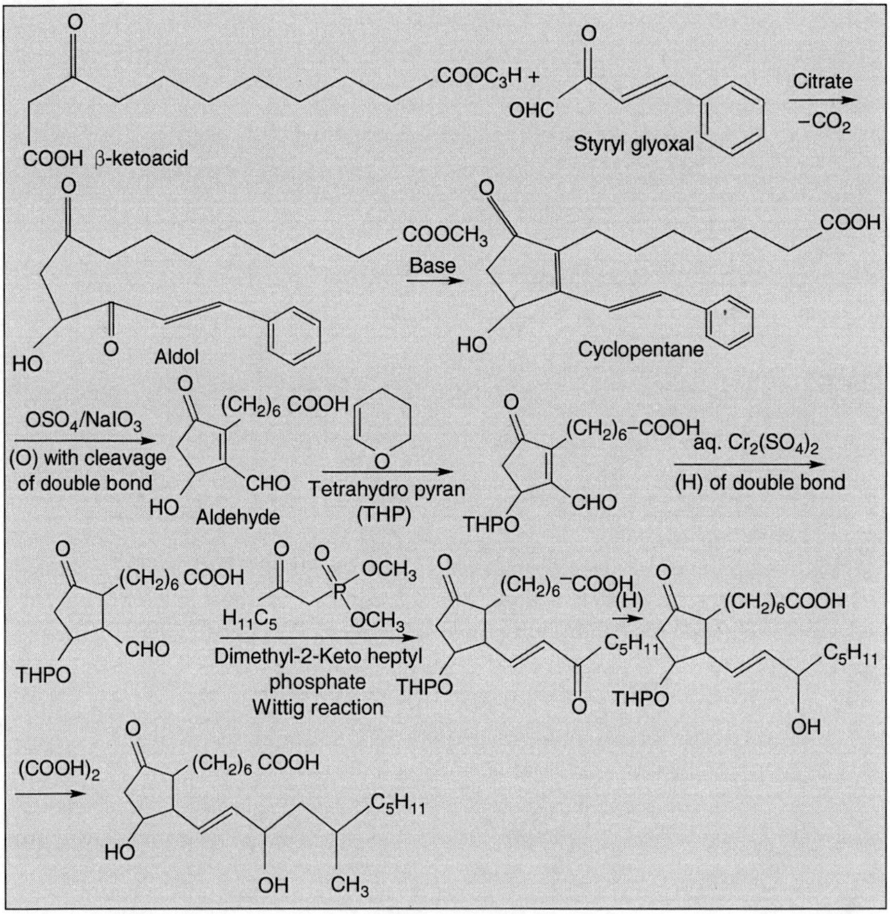

Reaction between β-keto acid 11-methoxy-3,11-dioxoundecanoic acid and styryl glyoxal gives aldol, namely methyl 11-hydroxy-9,12-dioxo-14-phenyl-13-tetradecenoate. Aldol undergoes cyclization in presence of base to give cyclopentane, namely 7-(2-styryl-3-hydroxy-5-oxo-1-cyclopentenyl)heptanoic acid. Reaction of this with osmium tetroxide oxidizes the lower arm with cleavage to form 7-(2-formyl-3-hydroxy-5-oxo-1-cyclopentenyl)heptanoic acid. 3-Hydroxy group is protected as tetrahydropyranyl derivative, and treatment with aqueous chromous sulphate reduces the double bond to give cyclopentane dertivative. This on reaction with dimethyl-2-ketoheptyl phosphate gives olefin (Wittig reaction). Reduction of lower chain keto function leads to alcohol, and deprotection of 3-hydroxyl group done with treatment with oxalic acid affords PGE_1.

Prostaglandin F$_{2\alpha}$ — Corey's synthesis from cyclopentane precursor

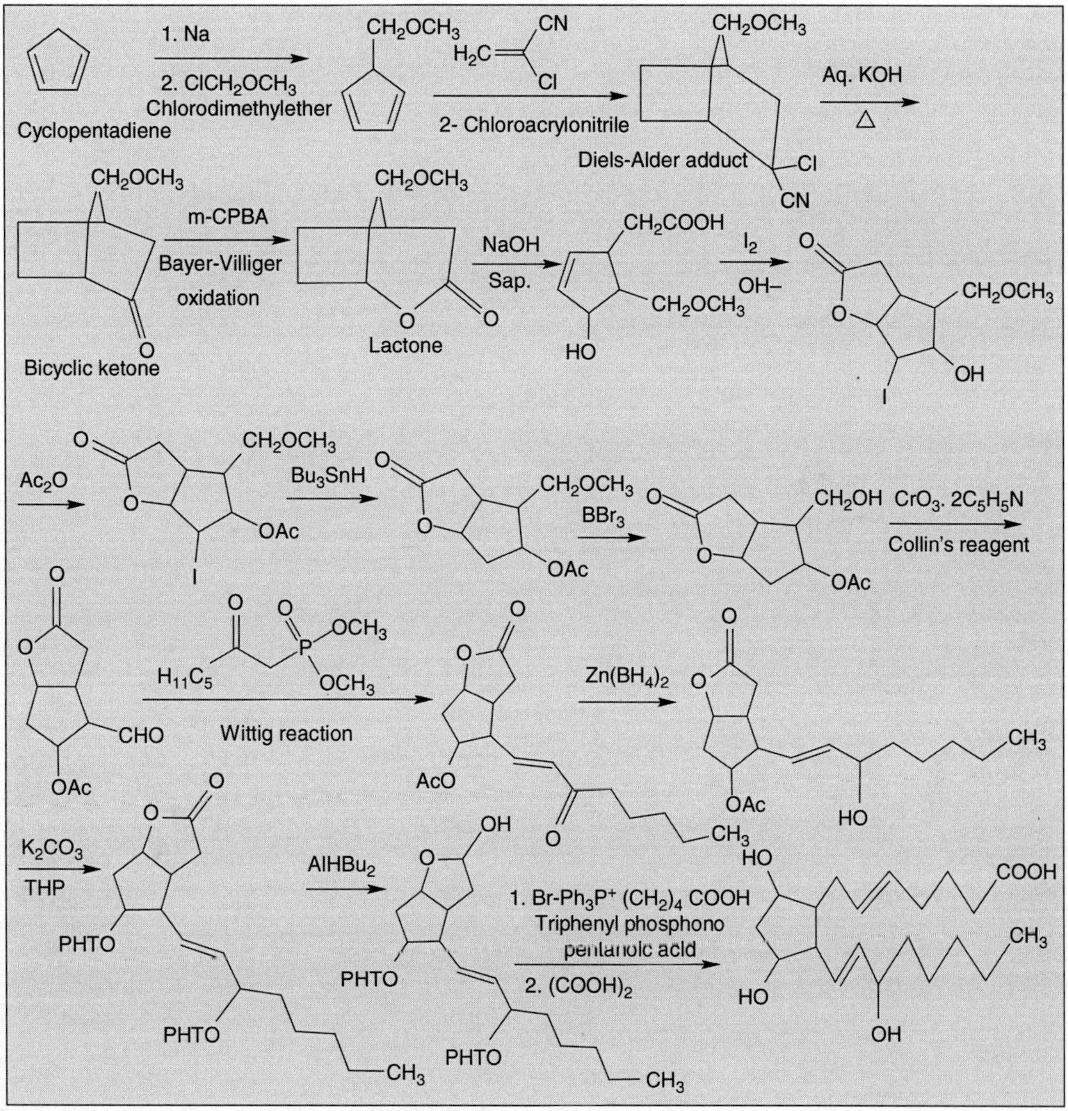

Cyclopentadiene on reaction with chlorodimethylether in presence of sodium gives methoxymethyl ether derivative, which on reaction with 2-chloroacrylonitrile gives Diels-Alder adduct. This on treatment with base followed by heating gives bicyclic ketone, which on treatment with peracid undergoes Bayer-Villiger oxidation to give cyclic ester. Saponification of cyclic ester gives cyclopentane acetic acid derivative. Iodination gives diiodo derivative, in presence of base, one of halide converted to alcohol and cyclic ester formed by reaction between acetic acid moiety and alcoholic group. The free hydroxyl group is protected as acetoxy derivative; iodo group is removed by treatment with tributyl tin hydride. Reaction with boron tribromide deprotects ether; treatment with Collin's reagent oxidizes primary alcohol to aldehyde. Reaction of this with phosphorous ylide undergoes Wittig reaction to form alkene. Reduction of lower chain keto function leads to alcohol, and cyclopentyl hydroxy group is protected as tetrahydropyranyl derivative. Reduction of lactone leads to lactol, and is further reacted with triphenyl phosphono pentanoic acid, and treatment with oxalic acid affords $PGF_{2\alpha}$.

32.5 STRUCTURES OF THERAPEUTICALLY USEFUL PROSTAGLANDINS

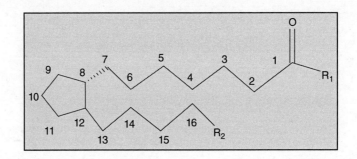

Name	OH	Double Bond	R_1	R_2	Other
1. Antiulcer agents					
Arbaprostil	11α, 15β	5–6, 13–14	—OH	—C_4H_9	9-oxo, 15- αCH$_3$
Enisoprost	11α, 16β	5–6, 13–14	—OCH$_3$	—C_4H_9	9-oxo, 15- βCH$_3$
Enprostil	11α, 15α	4–5–6 (allene), 13–14	—OCH$_3$	—OC$_6$H$_5$	9-oxo, 16- βCH$_3$
Misoprostil	11α, 16β	13–14	—OCH$_3$	—C_4H_9	9-oxo, 16- βCH$_3$
Rioprostil	11α, 16β (1—OH)	13–14	= CH$_2$OH	—C_4H_9	9-oxo
2. Abortifacients					
Carboprost	9α, 11α, 15α	5–6, 13–14	—OH	—C_4H_9	15- βCH$_3$
Dinoprostone	9α, 15α	5–6, 13–14	—OH	—C_4H_9	9-oxo
Fluprostenol	9α, 11α, 15α	5–6, 13–14	—OCH$_3$	—C_4H_9	9-oxo, 16-DiCH$_3$

Name	OH	Double Bond	R_1	R_2	Other
Metenprost	11α, 15α	5–6, 13–14	—OH	—C_4H_9	9 CH_2=
Sulprostone	11α, 15α	5–6, 13–14	—$NHSO_2CH_3$	—OC_6H_5	9-oxo
3. Impotency					
Fenprostalene	9α, 11α, 15α	4–5–6 (allene), 13–14	—OCH_3	—OC_6H_5	—
4. Bronchodilators					
Alprostidil	11α, 15α	13–14	—OH	—C_4H_9	9-oxo
5. Hypotensives					
Viprostol	11α, 16β	5–6, 13–14	—OCH_3	—C_4H_9	9-oxo, 16-βvinyl

32.6 SAR OF PROSTAGLANDINS

1. In the upper chain: Methyl esters (misoprostal), sulphonamide(sulprostone), and hydroxyl group (rioprost) possess greater activity than natural prostaglandins.
2. In the cyclopentane ring: Variation in the cyclopentane ring has led to reduction in PG activity. Enlargement of the ring or reduction of ring leads to inactive compounds. Replacement of carbon atom of cyclopentane ring by O, S, and N leads to inactive compounds. Replacement of 9 keto group with =CH_2 group gives active (metenprost) PG.
3. In the lower chain: The 15-hydroxyl group has been protected (from metabolism) by the introduction of methyl group at C-15 and gem dimethyl group at C-16. The shifting of C-15 hydroxyl to C-16 position increases metabolic stability. Alkoxy and phenoxy (enprostil, sulprostone) analogues were more active than natural PGs. Introduction of acetylinic group at C13-14 increases luteolytic activity.

FURTHER READINGS

1. *Prostaglandins*, Bergstrom, Sune and Samuelsson, Bengt (ed.), Interscience Publishers, NY, London and Sydney, 1967.
2. Euler, Ulf S. von (Ulf Svante) and Eliasson, Rune, 1967. *Prostaglandins*, Academic Press, New York.
3. Euler, U.S. von. *Prostaglandins (Medicinal Chemistry*, Vol. 8), Academic Press.

MULTIPLE-CHOICE QUESTIONS

1. The following statements describe the chemistry of prostaglandins except:
   a. 5-membered ring system
   b. 21-carbon skeleton
   c. 1-3 double bonds
   d. Ketone group

2. The following statements describe the biology of prostaglandins except:
   a. Produce pain
   b. Stimulates clotting
   c. Induction of labour
   d. Decreases blood flow to kidneys
3. PGI$_2$ produces
   a. Produce fever
   b. Stimulates clotting
   c. Induction of labour
   d. Inhibits clotting
4. The cyclooxygenase pathway does not produce
   a. PGH$_2$
   b. PGI$_2$
   c. Leukotriene E$_4$
   d. TXA$_2$
5. The hydroxyl groups in metenprost is in the following configuration:
   a. 11α, 15α
   b. 9α, 11α, 15α
   c. 11α, 16β
   d. 11α, 15β

QUESTIONS

1. Write a note on the chemistry and nomenclature of prostaglandins.
2. Discuss the SAR of prostaglandins.
3. Derive the biosynthetic pathway involving cyclooxygenases and lipoxygenases.
4. What are the synthetic approaches for prostaglandins? Explain briefly.

SOLUTION TO MULTIPLE-CHOICE QUESTIONS

1. b;
2. d;
3. d;
4. c;
5. a.

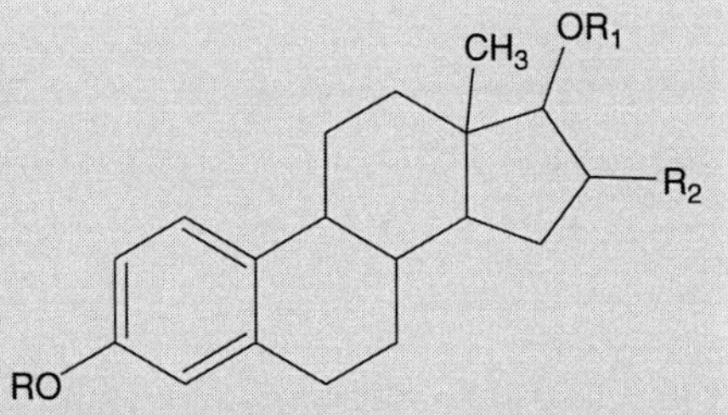

CHAPTER 33

Steroids

LEARNING OBJECTIVES

- Define steroids and the biosynthetic pathway of steroidal hormones
- Describe the mode of action of steroidal hormones
- Define oestrogens and describe their functions
- Describe the therapeutically useful oestrogens and progestins
- Define androgens and their major functions
- Describe the glucocorticoids and their mode of action
- Describe the importance of structure and activity profile of all steroidal hormones

33.1 INTRODUCTION

Steroids are structurally related compounds that have a common cyclopentano perhydro phenanthrene nucleus. They are widely distributed in the plant and animal kingdom. The fused tetracyclic steroid nucleus provides the carbon framework for at least four large groups of mammalian hormones; these hormones comprise the oestrogens, androgens, progestins, and corticosteroids. Many of these are of great medicinal value, and since animal sources yield them only at considerable cost, methods have been developed for manufacturing some of them from **sapogenins.** (Saponin is a type of glycoside widely distributed in plants. It consists of a sapogenin as the aglycon moiety, and a sugar. The sapogenin may be a steroid or a triterpene and the sugar may be glucose, galactose, a pentose, or a methylpentose.) Steroids are mainly obtained from a glycoside dioscine, and the Mexican wild yam root *Dioscoria fluribanda* constituted the first plant source for steroid drugs.

Other compounds can be mentioned as steroids, example of which are vitamin D, cardiac glycosides etc.

33.2.1 Nomenclature and Stereochemistry of Steroid Hormones

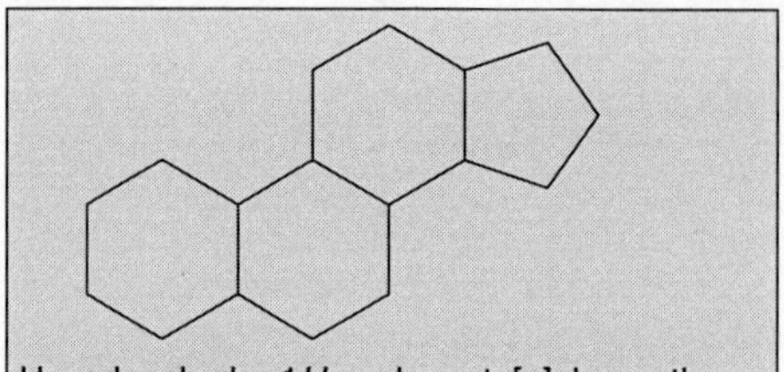

Hexadecahydro-1H-cyclopenta[a]phenanthrene

There are four steroid skeletons present in the steroidal drugs; these are the **gonane** skeleton, the **estrane** skeleton, the **androstane** skeleton, and the **pregnane** skeleton.

- The **gonane** skeleton is the simplest one. The C19 methyl group at C10 and the C18 methyl group at C13 are absent in this skeleton.
- In the **estrane** skeleton, only the methyl group at C10 is missing. This skeleton is also indicated as a 19-nor androstane skeleton, but that is not an official name. The prefix **nor** means that one C-atom is missing—in this case, C19. One of the best-known 19-nor steroids is nandrolone, which is also known as 19-nor-testosterone.
- In the **androstane** skeleton, both methyl groups at C10 and C13 are present.
- The **pregnane** skeleton has in addition an ethyl group in the β-position at C17.

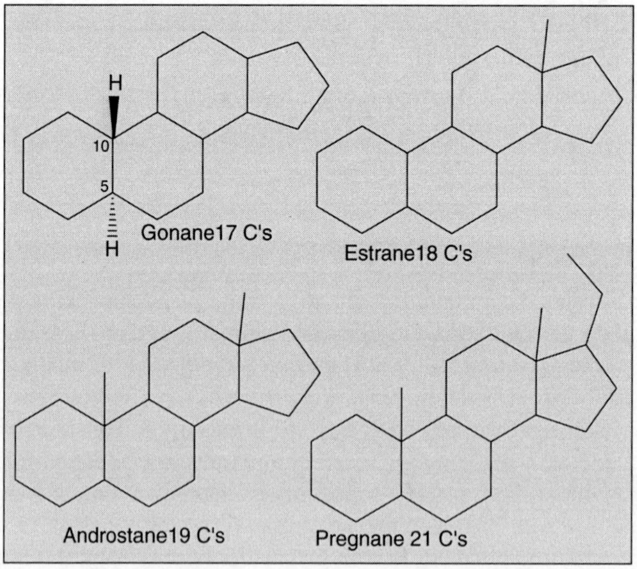

Gonane 17 C's

Estrane 18 C's

Androstane 19 C's

Pregnane 21 C's

In all the four basic skeletons, the connection between the A- and B-rings, **5α** or **5β**, has to be indicated. If both positions are possible for the H-atom at C5, or when its position is unknown, this is shown with a waving line in the formula or with the indication 5ζ in the name. When the H-atom at C5 is pointing to the bottom side, this is indicated as **5α** in the name and it will be drawn with a hatched bond in the formula. When this H-atom is pointing to the top side, it is indicated as **5β** in the name and drawn with a thick wedge-shaped bond in the structural formula.

Stereochemistry of a common steroid nucleus is depicted in the accompanying figure.

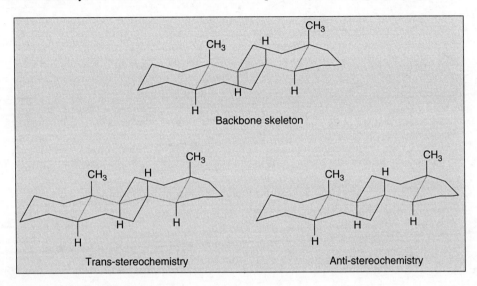

Classification

Steroidal drugs are classified as follows:

- Sex hormones: Androsterone, Androstendion, Oestrone, Progesterone
- Glucocorticoids (GCs): Hydrocortisone, 11-Dehydrocorticosterone, Corticosterone
- Mineralocorticoids (MCs): Aldeosterone, 11-Deoxycorticosterone, 11-Deoxy-17-oxycorticosterone
- Cardiac glycosides: Digitoxin, Digoxin

33.2 BIOSYNTHESIS OF STEROID HORMONES

Numerous steroids are secreted by endocrine glands (ovaries, testes, adrenal glands) and are deposited directly into the bloodstream. These steroids are called steroidal hormones. Steroid hormones possess numerous activities. In fact, they are so important that absence of these hormones may be lethal. The major classes of steroid hormones are:

- Female sex hormones, the major classes being oestrogens and progesterones
- Male sex hormones, the major class being androgens
- Adrenocorticoids that include glucocorticoids and mineralocorticoids

Let us look at a simplified picture of how these steroids are biosynthesised in our body.

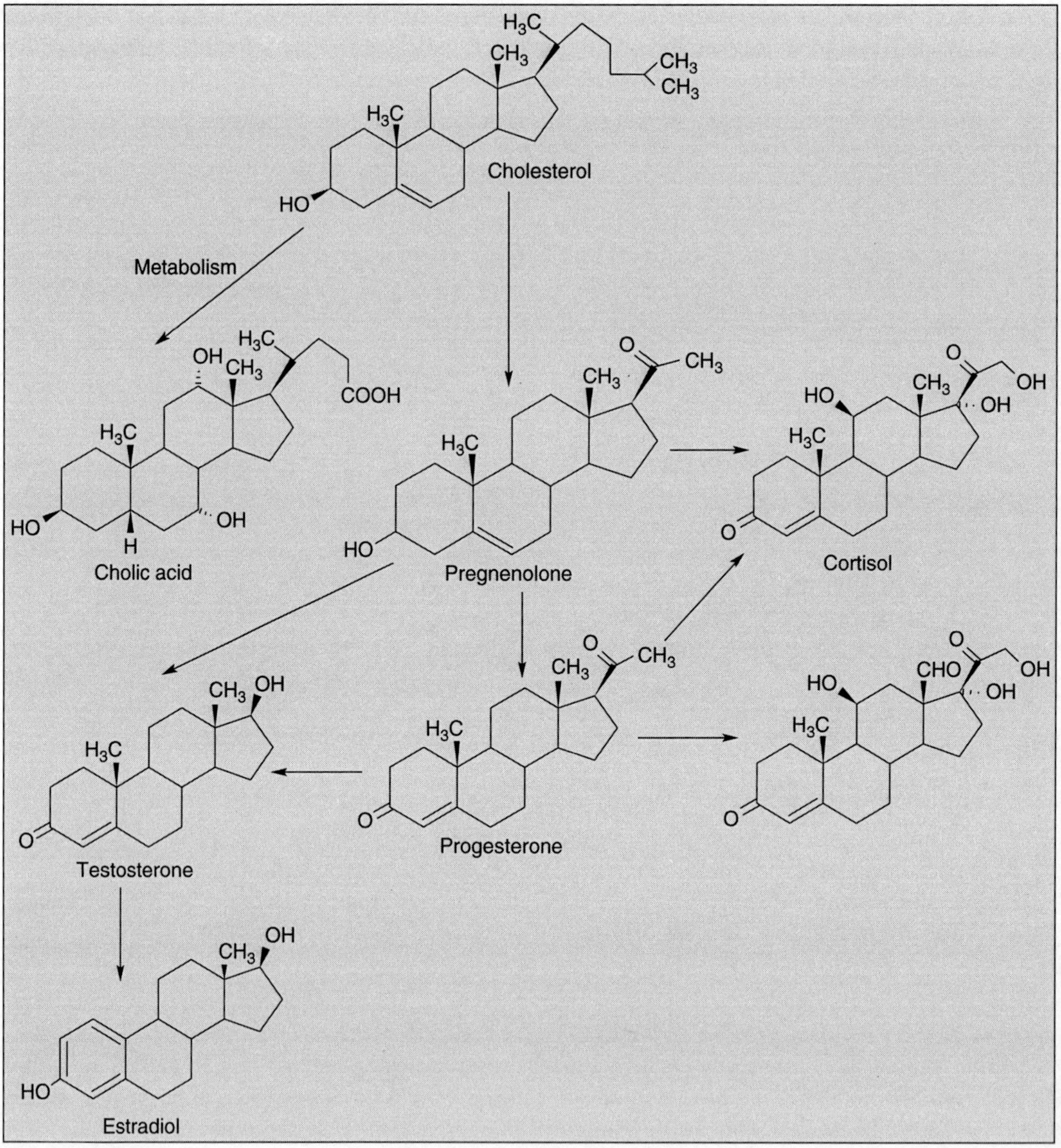

As might be apparent from this figure, the inter-conversion occurs between male and female sex hormones. Thus, the process occurs in males as well as females ensuring that each one of us has all these steroids. It turns out that females have more oestrogens and progesterones than males, while androgens are present to a considerably greater extent in males than in females. A typical example—during menstruation, females secrete ~20 mg progesterone per day. Males usually have ~2–5 mg progesterones.

33.3 OVERALL MECHANISM OF ACTION OF STEROID HORMONES

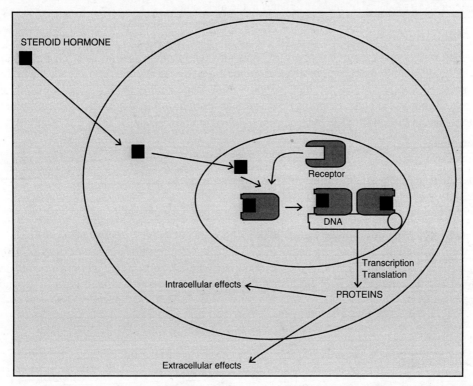

The overall mechanism of steroid hormone action is the regulation of gene expression. The lipophilic steroid hormones are carried into the bloodstream with the majority of hormone reversibly bound to carrier proteins and a small amount of free steroids. The free steroid diffuses through the cell membrane and enters the cells. The sensitive cells contain a high-affinity steroid hormone receptor either in the cytosol or in the nucleus. The steroid receptor complex enters the nucleus and initiates a conformational change that involves dimerization to activate the complex to interact with specific regions on cellular DNA, referred to as hormone-responsive elements (HRE). This initiates the process of transcription to produce mRNA and translations to produce proteins. These proteins regulate cell function, growth differentiation, etc. So, it is the process of expression of proteins that these hormones regulate. The proteins perform their regular tasks.

Hormones: Of the numerous hormones that regulate body function, two steroid hormones are extremely important: oestradiol, a female sex hormone, and testosterone, a male sex hormone present in the body in insignificant amounts. They regulate sexual differentiation and reproduction as well as affect the performance of many other physiological systems.

33.3.1 Classification

- The two primary classes of female sexual hormones are **oestrogens** and **progestins,** which are formed in the ovaries, to a lesser degree in the adrenal cortex, and in the placenta during pregnancy.

- **Androgens,** commonly referred to as male sex hormones or anabolic steroids, and in particular, testosterone, are produced by male sex glands in the male body.

33.4 OESTROGENS

Oestrogens (or estrogens) are a group of steroid compounds that function as the primary female sex hormone.

The three naturally occurring oestrogens are oestradiol, oestriol, and oestrone. In the body, these are all produced from androgens through enzyme action. Oestradiol is produced from testosterone and oestrone from androstenedione. Oestrone is weaker than oestradiol, and in post-menopausal women more oestrone is present than oestradiol.

Classification

They are of two types:

- **Steroidal oestrogens:** They essentially possess a steroidal nucleus and have oestrogenic activity—oestrone, oestriol, and oestradiol
- **Non-steroidal oestrogens:** Diethylstilbestrol, Hexestrol, Dienoestrol

Chemistry Synthesis of oestrogens starts in theca interna cells in the ovary, by the synthesis of androstenedione from cholesterol. Androstenedione is a substance of moderate androgenic activity. This compound crosses the basal membrane into the surrounding granulosa cells, where it is converted to oestrone or oestradiol, either immediately or through testosterone. Presence of both cell types is essential for oestrogen synthesis.

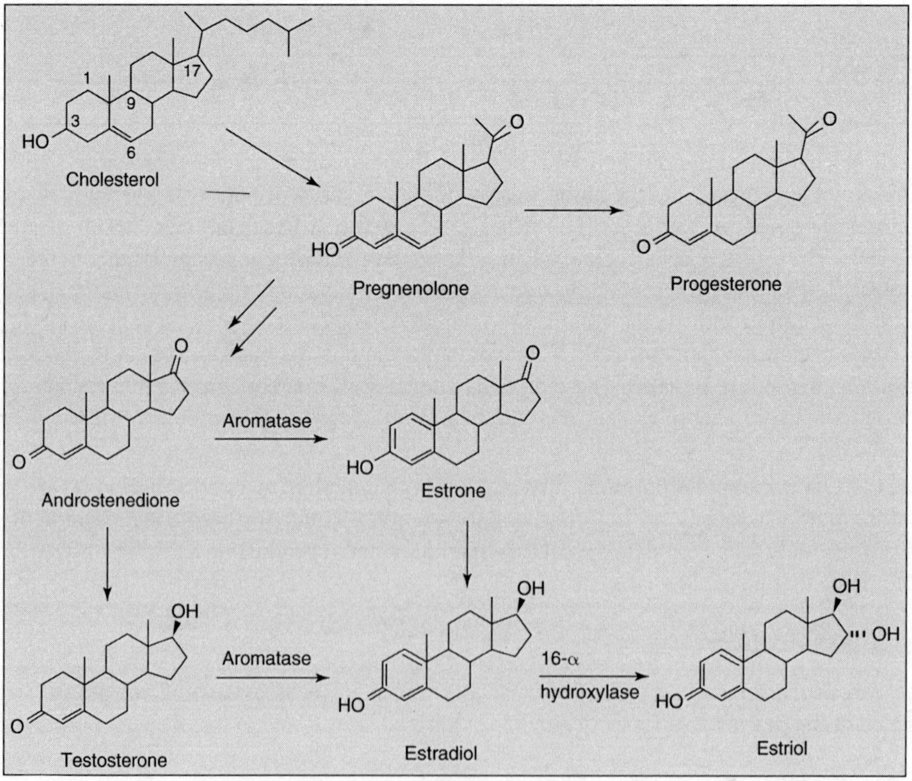

Sources Oestrogens are produced primarily by developing follicles in the ovaries, the corpus luteum, and the placenta. Some oestrogens are also produced in smaller amounts by other tissues such as liver, adrenal glands, and breasts. These secondary sources of oestrogen are especially important in post-menopausal women.

Effects While oestrogens are present in both men and women, they are found in women in significantly higher quantities. They promote the development of female secondary sexual characteristics, such as breasts, and are also involved in the thickening of the endometrium and other aspects of regulating the menstrual cycle, which is why many oral contraceptives contain oestrogens.

Therapeutic uses The use of oestrogen, especially together with a progestin, is a treatment for the symptoms of menopause. This is controversially alleged to do as much harm as good many professional bodies have suggested that the problems may be due only to the particular product(s) (premarin alone and with provera as *Prempro*) and fixed doses used in all the recent studies, and extensive research is underway to absolutely confirm the precise processes of such variations.

> About 80 per cent of breast cancers, once established, rely on supplies of the hormone oestrogen to grow: they are known as hormone-sensitive or hormone-receptor-positive cancers. Suppression of production in the body of oestrogen is a treatment for these cancers.

Medicinally important oestrogens and their derivatives

	R	R_1	R_2
Oestradiol	—H	—H	—H
Oestradiol valerate	—H	—H	—CO(CH$_2$)$_3$CH$_3$
Oestradiol cypionate	—H	—H	—OC(H$_2$C)$_2$—(cyclopentyl)
Ethinyl oestradiol	—H	—C≡CH	—H
Mestranol	—CH$_3$	—C≡CH	—H
Quinestrol	—(cyclopentyl)	—C≡CH	—H
Oestrone	—H	—	=O(OR$_1$)
Oestrone sulphate	—SO$_3$H	—	=O(OR$_1$)
Equilin	—H	—	=O(OR$_1$)

Also contains 7, 8 double bond

SAR The most potent naturally occurring oestrogens in humans are 17-β-oestradiol, oestrone, and oestriol. Each of these molecules is an 18-carbon steroid, containing a phenolic-A ring (an aromatic ring with hydroxyl group at C-3), and a β-hydroxyl group or ketone in position 17 of ring-D. The phenolic-A ring is the principal structural feature for selective, high-affinity binding to oestrogen receptors. Most alkyl substitutions on the phenolic-A ring impair such binding, but substitution on rings C or D may be tolerated. Ethinyl substitution at the C-17 position greatly increases oral potency by inhibiting the first-pass hepatic metabolism. Esters of oestradiol are absorbed more slowly than oestradiol itself—e.g., oestradiol valerate and cypionate.

Oestradiol   Estra-1,3,5(10)-trien-3,17β-diol

Synthesis

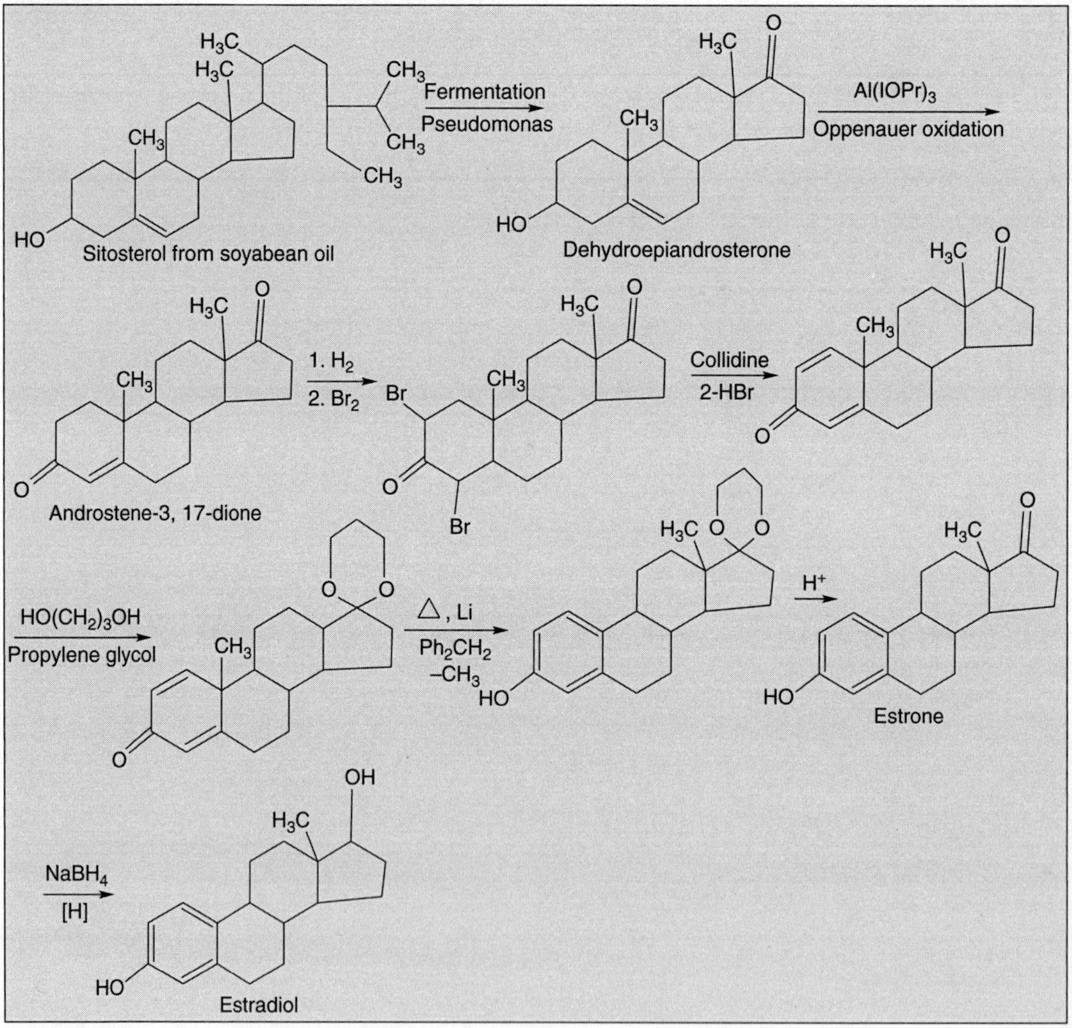

Sitosterol on fermentation with Pseudomonas species gives dehydroepiandrosterone. Oppenauer oxidation of this with aluminium isopropoxide in 2-propanol gives androstene-3,17-dione. Reduction of double bond followed by treatment with bromine gives dibromo derivative. Treatment with collidine leads to dehydrobromination, to give dieneone. 17-Ketone group is protected with propylene glycol, and upon heating this compound at a temperature of about 530°C with lithium and diphenyl methane, methyl molecule is detached from position 10 to form aromatic ring-A and the desired oestrone. Oestradiol is most easily made by reducing the keto-group of oestrone by various reducing agents, in particular sodium borohydride.

Oestradiol esters

Synthesis

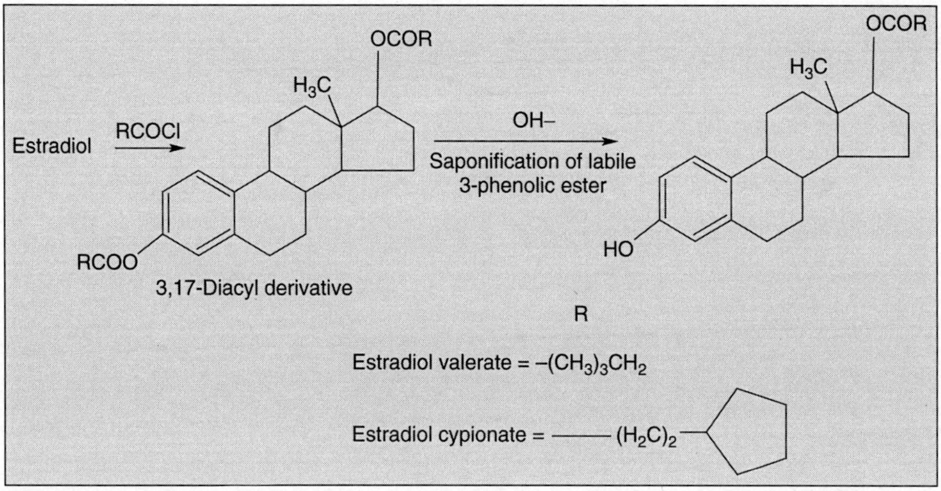

Oestradiol esters are prepared by treating oestradiol with two moles of required acid chloride, followed by saponification of labile phenolic ester.

Ethinyl oestradiol (17α-Ethynyl-1,3,5(10)-estratrien-3-17β-diol)

and Mestranol (17-Ethynyl-3-methoxy-13-methyl-7,8,9,11,12,14,15,16-octahydro-6H-cyclopenta[a]phenanthren-17-ol)

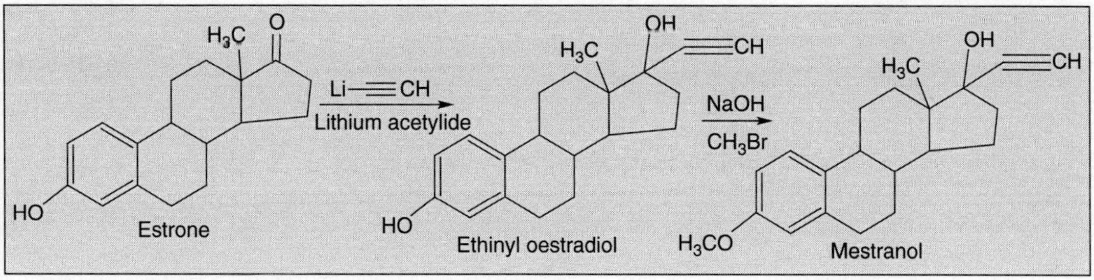

Ethinyl oestradiol is made either by condensing oestrone with acetylene in the presence of potassium hydroxide (Favorskii reaction), or by reacting lithium acetylenide in liquid ammonia with oestrone.

Methyl ether of ethinyl oestradiol (mestranol) is prepared by treatment with methyl bromide in presence of sodium hydroxide. It is a biologically inactive prodrug of ethiny oestradiol, to which it is demethylated in the liver.

Diethylstilbestrol *trans*-α,β-Diethyl-4,4'-stilbendiol

Synthesis

1,2-*bis*(4-Methoxyphenyl)ethanone is alkylated by ethyl iodide in the presence of sodium ethoxide, and the resulting ketone is reacted in a Grignard reaction with ethylmagnesium bromide, which forms the carbinol. Dehydration of this compound by distillation in the presence of *p*-toluenesulphonic acid gives dimethyl ether of stilbestrol, methyl groups of which are removed by heating it at high temperatures with potassium hydroxide, thus forming diethylstilbestrol.

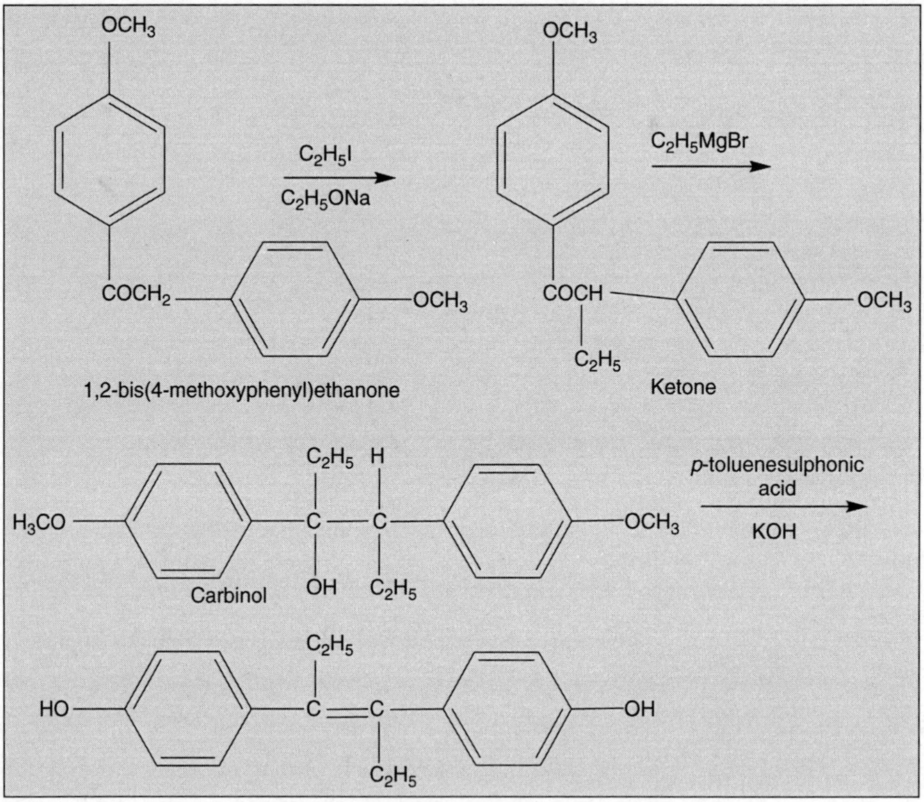

Dienoestrol It is a dehydro derivative of diethylstilbestrol.

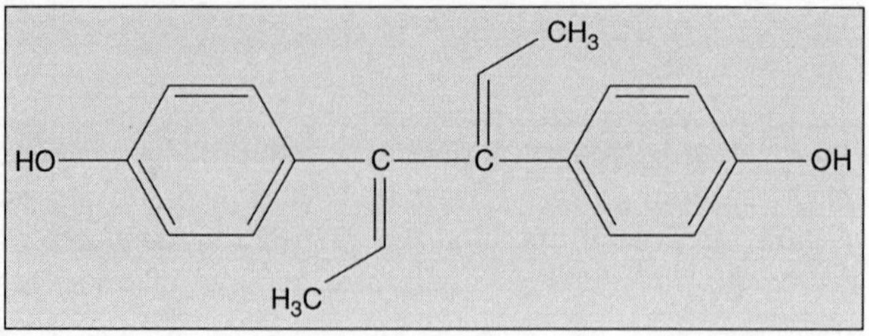

Hexestrol 4,4'-(1,2-Diethylethylene)diphenol
Synthesis

Hexestrol is made in a Wurtz dimerization reaction of 1-bromo-1-(4-methoxyphenyl)propane in the presence of sodium. The initial 1-bromo-1-(4-methoxyphenyl)propane is made in turn by addition reaction of hydrogen bromide to 1-methoxy-4-(prop-1-enyl)benzene. Subsequent removal of the methoxy protective groups from the resulting dimerization product using hydroiodic acid gives hexestrol.

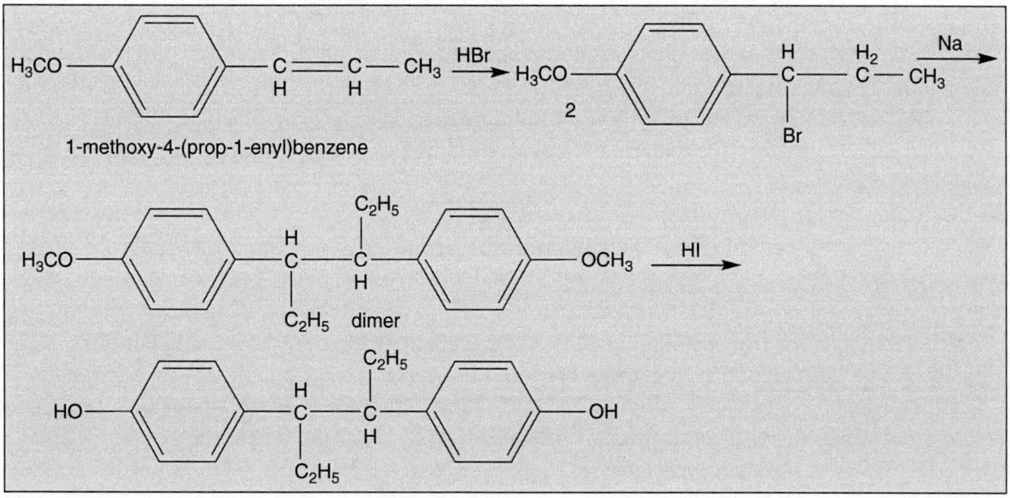

Hexoestrol is 2.8 times, dienoestrol 2.3 times, and diethylstilbestrol twice as active as oestrone.

33.5 PROGESTINS

Progesterone is a C-21 steroid hormone involved in the female menstrual cycle, pregnancy (supports *gestation*), and embryogenesis of humans and other species. Progesterone belongs to a class of hormones called progestagens, and is the only naturally occurring human progestagen.

Chemistry Like other steroids, progesterone consists of four interconnected cyclic hydrocarbons. Progesterone contains ketone and aldehyde functional groups, as well as two methyl branches. Like all steroid hormones, it is hydrophobic. This is mostly due to its lack of very polar functional groups.

Biosynthesis Progesterone, like all other steroid hormones, is synthesised from pregnenolone, a derivative of cholesterol. This conversion takes place in two steps. The 3-hydroxyl group is converted to a keto group and the double bond is moved to C-4, from C-5.

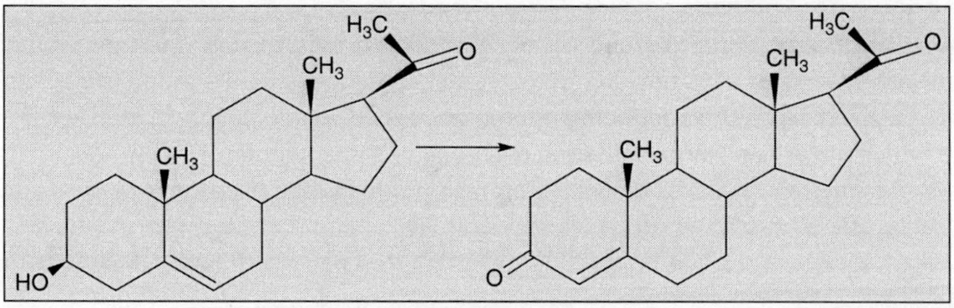

Progesterone is the precursor of the aldosterone, and after conversion to 17-hydroxyprogesterone of cortisol, and after further conversion to androstenedione of testosterone, oestrone, and oestradiol.

Sources Progesterone is produced in the adrenal glands, the gonads—specifically after ovulation in the corpus luteum, the brain, and during pregnancy—in the placenta. In humans, increasing amounts of progesterone are produced during pregnancy; initially, the source is the corpus luteum that has been 'rescued' by the presence of human chorionic gonadotropins (hCG) from the conceptus, but after the 8th week production of progesterone shifts over to the placenta. The placenta utilizes maternal cholesterol as the initial substrate, and most of the produced progesterone enters the maternal circulation, but some is picked up by the foetal circulation and is used as substrate for foetal corticosteroids. At term, the placenta produces about 250 mg progesterone/day.

Effects Progesterone exerts its action via the intracellular progesterone receptor. It has a number of physiological effects, usually to counteract effects caused by oestrogen. Oestrogen is required to induce a progesterone receptor.

Reproduction: Progesterone's reproductive function serves to convert the endometrium to its secretory stage to prepare the uterus for implantation. If pregnancy does not occur, progesterone levels will decrease leading to menstruation in the human. Normal menstrual bleeding is a progesterone-withdrawal bleeding.

During implantation and gestation, progesterone appears to decrease the maternal immune response to allow for the acceptance of the pregnancy. Progesterone decreases contractility of the uterine musculature. The foetus metabolises placental progesterone in the production of adrenal mineralo- and glucosteroids. A drop in progesterone levels is possibly one step that facilitates the onset of labour. In addition, progesterone inhibits lactation during pregnancy. The fall in progesterone levels following delivery is one of the triggers for milk production. Progesterone has an effect upon vaginal epithelium and cervical mucus.

Neurosteroids: Progesterone, like pregnenolone and dehydroepiandrosterone, belongs to the group of neurosteroids that are found in high concentrations in certain areas in the brain and are synthesised there. Neurosteroids affect synaptic functioning, are neuroprotective, and affect myelinization. They are investigated for their potential to improve memory and cognitive ability. Progesterone as a neuroprotectant affects regulations of apoptotic genes. Its effect as a neurosteroid works predominately through the GSK-3 beta pathway, as an inhibitor. Other GSK-3 beta inhibitors include bipolar mood stabilizers, lithium, and valproic acid.

Other systems: Progesterone has multiple effects outside of the reproductive system. Progesterone is thermogenic, raising the core temperature. It reduces spasm and relaxes smooth muscle tone. Gall bladder activity is reduced. Bronchi are widened. Progesterone acts as an anti-inflammatory agent and reduces the immune response. Other effects include normalizing blood clotting and vascular tone, zinc and copper levels, cell oxygen levels, and use of fat stores for energy. Progesterone also assists in thyroid function and bone building by osteoblasts. As an oestrogen antagonist, progesterone appears to prevent endometrial cancer (involving the uterine lining) and breast cancer.

Therapeutic applications Progesterone is poorly absorbed by oral ingestion unless micronized and in oil, or with fatty foods; it does not dissolve in water. Products such as prometrium, utrogestan, and microgest are, therefore, capsules containing micronized progesterone in oil - in all three mentioned, that is peanut oil, which may cause serious allergic reactions in some people, but compounding pharmacies, which have the facilities and licenses to make their own products, can use alternatives. Vaginal and rectal application is also effective, with products such as cyclogest, which is progesterone in cocoa butter in the form of pessaries. Progesterone can be given by injection, but because it has a short half-life, they need to be daily.

Implants for a longer period are also available. Marketing of progesterone pharmaceutical products, country to country, varies considerably, with many countries having no oral progesterone products marketed, but they can usually be specially imported by pharmacies through international wholesalers.

In some countries, 'natural progesterone' products are heavily marketed, often said to contain extract of yams, with extensive claims and without need of prescription. If they contain any actual progesterone, the strength is always considerable lower value than the pharmaceutical products.

Progesterone is used to control anovulatory bleeding. It is also used to prepare uterine lining in infertility therapy and to support early pregnancy. Patients with recurrent pregnancy loss due to inadequate progesterone production may receive progesterone.

Progesterone is being investigated as potentially beneficial in multiple sclerosis, since the characteristic deterioration of the myelin insulation of nerves halts during pregnancy, when progesterone levels are raised, but commences again when the levels drop.

Since most progesterone in males is produced in the process of testicular production of testosterone, and most in females by the ovaries, the shutting down (whether by natural or chemical means), or removal, of those inevitably causes a considerable reduction in progesterone levels. Previous concentration upon the role of progestogens in female reproduction, when progesterone was simply considered a 'female hormone', obscured the significance of progesterone elsewhere in both sexes. The tendency for progesterone to have a regulatory effect, and the presence of progesterone receptors in many types of body tissue, and the pattern of deterioration (or tumour formation) in many of those increasing in later years when progesterone levels have dropped, is prompting widespread research into the potential value of maintaining progesterone levels in both males and females.

Progesterone is used in hormone therapy for transsexual women, and some intersex women—especially when synthetic progestins have been ineffective or caused side-effects—since normal breast tissue cannot develop except in the presence of both progesterone and oestrogen. Mammary glandular tissue is otherwise fibrotic, the breast shape conical, and the areola immature. Progesterone can correct those even after years of inadequate hormonal treatment. Research usually cited against such value was conducted using provera, a synthetic progestin. Progesterone also has a role in skin elasticity and bone strength, in respiration, in nerve tissue, and in female sexuality, and the presence of progesterone receptors in certain muscle and fat tissue may hint at a role in sexually dimorphic proportions of those.

These roles of progesterone may not be fulfilled by synthetic progestins, which were designed solely to mimic progesterone's uterine effects. It is necessary to be suspicious of research that claims that progesterone is ineffective when only a synthetic progestin was tested.

Progesterone receptor antagonists, such as RU-486 (mifepristone), can also be used to prevent conception or induce medical abortions. Oral birth control pills do not contain progesterone but a progestin.

Classification

1. One group of progestins contains the 21-carbon skeleton of progesterone. These agents are highly selective and have a spectrum of activity very similar to that of endogenous hormones. These compounds are most frequently used in conjunction with oestrogen for hormone replacement therapy in post-menopausal women—e.g., hydroxy progesterone caproate, medroxy progesterone acetate, megesterol acetate, chlormadinone acetate, and delmadinone acetate.
2. A second major class of compounds is derived from 19-nor testosterone. These so called 19-nor compounds lack the C-19, -20, and -21 carbons found in progesterone and resemble testosterone in the vicinity of the D-ring. Compounds in this 19-nor category historically have been the progestin component of combination oral contraceptives. These 19-nor compounds have potent progestational

activity, but they also have an androgenic and other activities that are thought to contribute to their side-effects—e.g., norethindrone, norethynodrel, norgestrel, tigestol, and ethynerone.

Structural features and SAR of progestins
Progesterones:

1. Unlike the oestrogen receptors, which require the phenolic-A-ring for high-affinity binding, the progesterone receptors favour a Δ^4-3-one A-ring structure in an inverted 1-β, 2-α conformation.
2. 17- α-Hydroxy progesterone itself is inactive but some of its ester derivatives have progestational activity. Compounds such as hydroxyprogesterone caproate have progestational activity but must be used parenterally due to first-pass hepatic metabolism.
3. However, substitution of such 17-esters at the C-6 position of the B-ring yields orally active compounds such as medroxy progesterone acetate, megesterol acetate, chlormadinone acetate, and delmadinone acetate.

19-Nor compounds

1. The spectrum of activity of these compounds is highly dependent upon specific substituent groups, especially the nature of the C-17 substituents in the D-ring, the presence of a C-19 methyl group, and the presence of an ethyl group at C-13.
2. An ethinyl substituent at C-17 decreases hepatic metabolism and yields 19-nor testosterone analogues such as norethindrone and norethynodrel, which are orally active.
3. Addition of a 13-ethyl group to norethindrone yields norgestrel, which is more potent than parent compounds.
4. These compounds are less selective than progesterone derivatives and have varying degree of androgenic, oestrogenic, and anti-oestrogenic activities.

Synthesis of pregnane derivatives
Progesterone: Pregn-4-en-3,20-dione

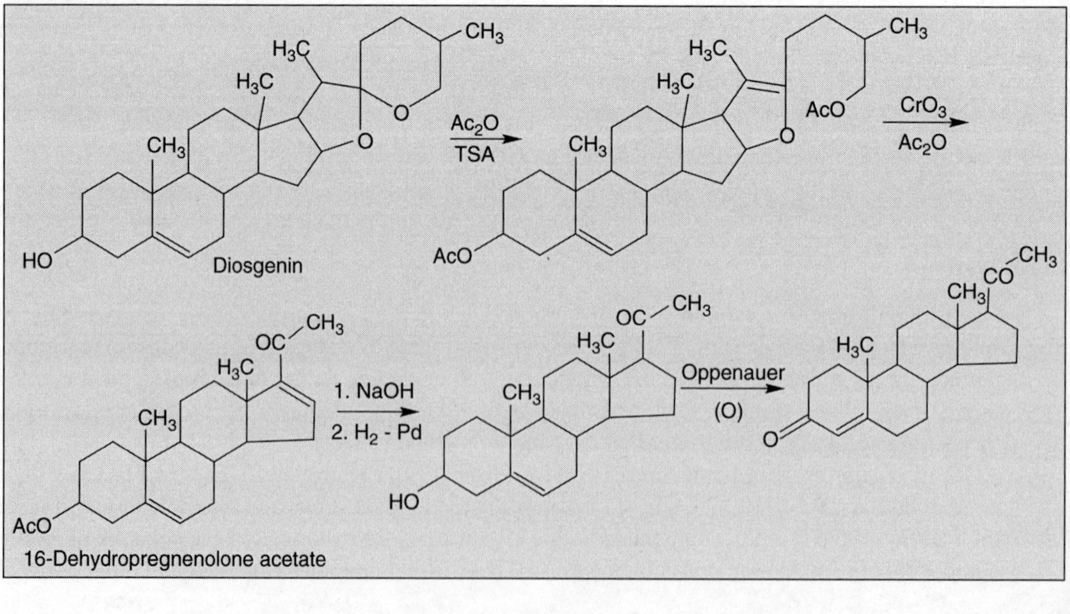

Diosgenin on treatment with acetic anhydride gives diosan, which on oxidation followed by dehydration gives 16-dehydropregnenolone acetate. Saponification of this, reduction of C_{16-17} double bond, and Oppenauer oxidation gives progesterone.

Hydroxyprogesterone caproate: 17α-Hydroxypregn-4-en-3,20-dione

The potency and oral activity of progestins in the pregnane class are enhanced by introducing 17-α-acyloxy group.

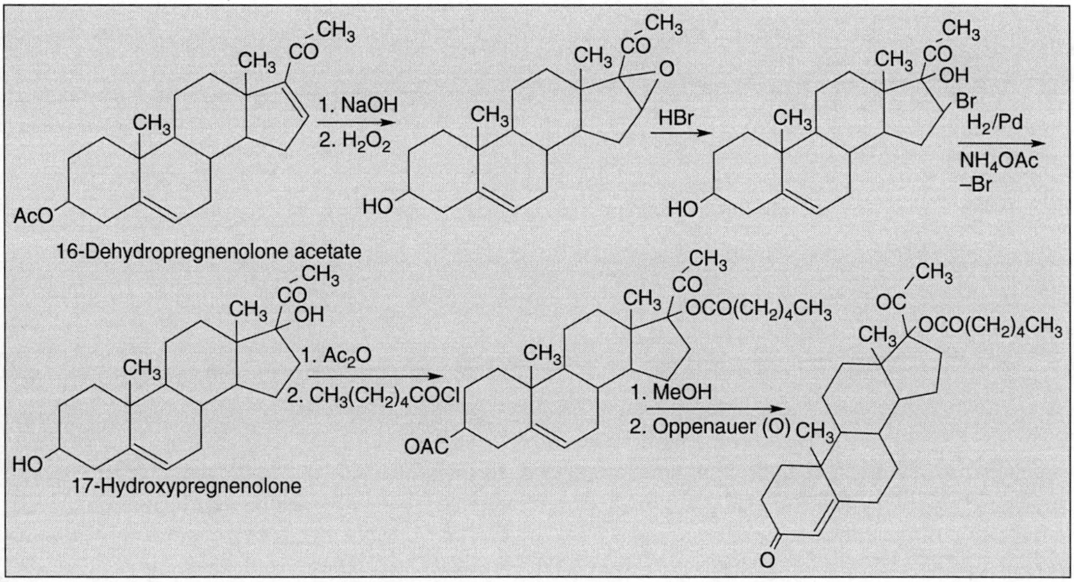

Hydroxyprogesterone is synthesised from dehydropregnenolone. The double bond at C_{16}–C_{17} of dehydropregnenolone is oxidized by hydrogen peroxide in the presence of a base to give an epoxide. Interaction of the resulting epoxide with hydrogen bromide in acetic acid forms a bromohydrin. Reduction by hydrogen over a palladium catalyst removes the bromine atom at C_{16}, forming 17-hydroxypregnenolone. Treatment with acetic anhydride protects C_3 hydroxy group, and acylation of C_{17} hydroxy group is done with hexanoyl chloride. Deprotection of C_3 hydroxyl group and Oppenauer oxidation of this gives 17-hydroxyprogesterone caproate.

> It has been observed in animal models that females have reduced susceptibility to traumatic brain injury and this protective effect has been hypothesised to be caused by increased circulating levels of oestrogen and progesterone in females. A number of additional animal studies have confirmed that progesterone has neuroprotective effects when administered shortly after traumatic brain injury.

Medroxyprogesterone acetate: 17α-Hydroxy-6α-methylpregn-4-en-3,20-dione

Medroxyprogesterone differs from hydroxyprogesterone in the presence of an additional methyl group at C_6.

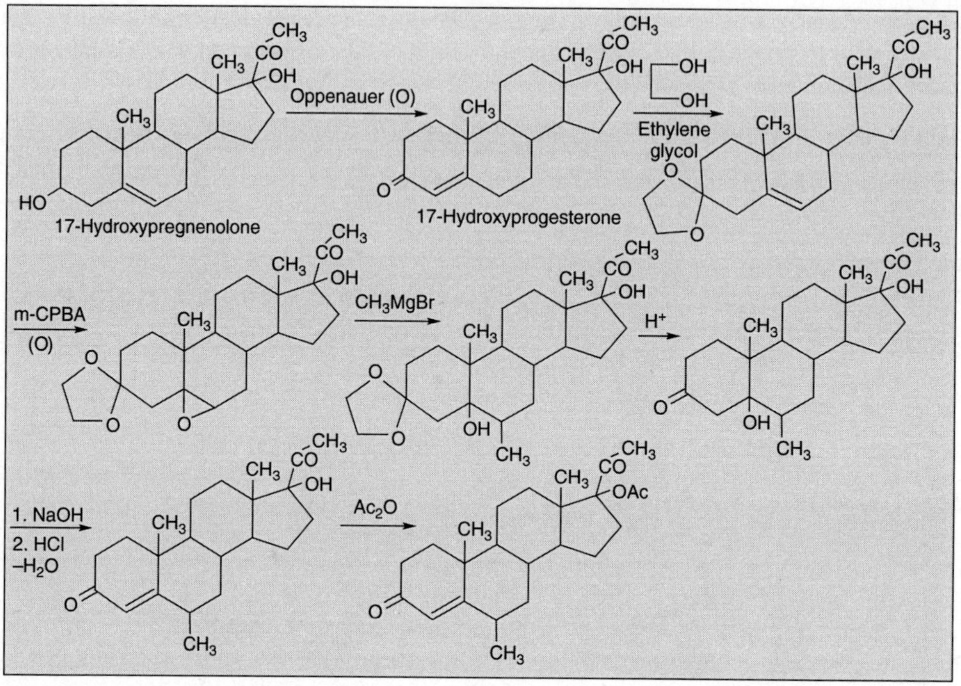

Oppenauer oxidation of 17-hydroxypregnenolone leads to 17-hydroxyprogesterone. Reaction of this with ethylene glycol in presence of acid gives bis-acetal. Reaction of this with *m*-chloro per benzoic acid leads to formation of the epoxide. Condensation of this with Grignard reagent leads to 6-methyl derivative via diaxial opening of the oxirane. The acetal group is then removed by treatment with acid to give β-hydroxy ketone. This undergoes reverse Michael addition of water on treatment with base to give conjugated ketone. Acetylation with acetic anhydride gives medroxyprogesterone acetate.

Megestrol acetate: 17-Acetyl-17-hydroxy-6,10,13-trimethyl-2,8,9,11,12,14,15,16-octahydro-1H-cyclopenta[*a*] phenanthren-3-one

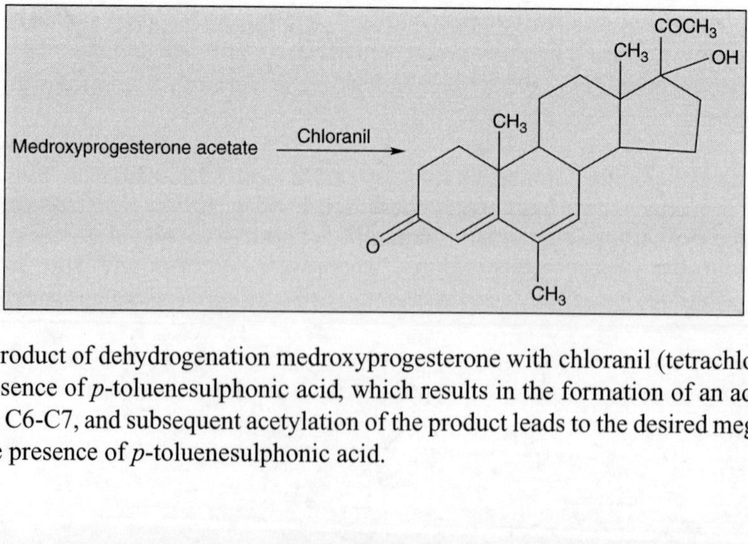

Megestrol is a product of dehydrogenation medroxyprogesterone with chloranil (tetrachloro-*p*-benzoquinone) in the presence of *p*-toluenesulphonic acid, which results in the formation of an additional double bond at position C6-C7, and subsequent acetylation of the product leads to the desired megestrol by acetic anhydride in the presence of *p*-toluenesulphonic acid.

Chlormadinone acetate: (17-Acetyl-6-chloro-10,13-dimethyl-3-oxo-2,3,8,9,10,11,12,13,14,15,16,17-dodecahydro-1*H*-cyclopenta[α]phenanthren-17-yl acetate) and **Delmadinone acetate** (17-Acetyl-6-chloro-10,13-dimethyl-3-oxo-8,9,10,11,12,13,14,15,16,17-decahydro-3*H*-cyclopenta[*a*]phenanthren-17-yl acetate)

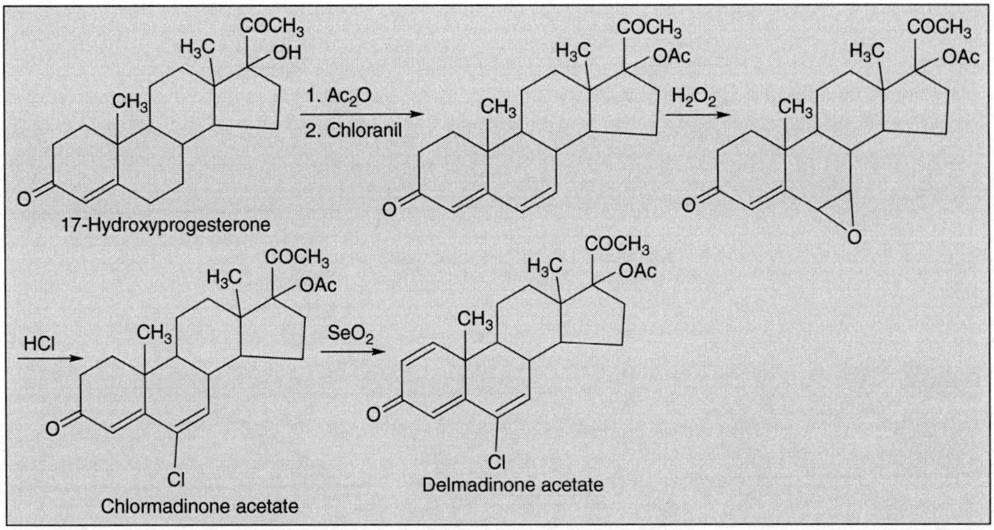

Acetylation of 17-hydroxyprogesterone and treatment with chloranil introduces 6,7 double bond by dehydrogenation. Reaction of this dienone with hydrogen peroxide gives epoxide; this in turn with anhydrous hydrogen chloride leads to opening of oxiran ring and dehydration to form chlormadinone acetate.

Dehydrogenation of chlormadinone acetate with selenium dioxide leads to introduction of the double bond at the 1-position to afford delmadinone acetate.

Synthesis of 19-nor derivatives

Norethindrone (17α-Ethynyl-17β-hydroxyestra-4-en-3-one) and **Norethynodrel**

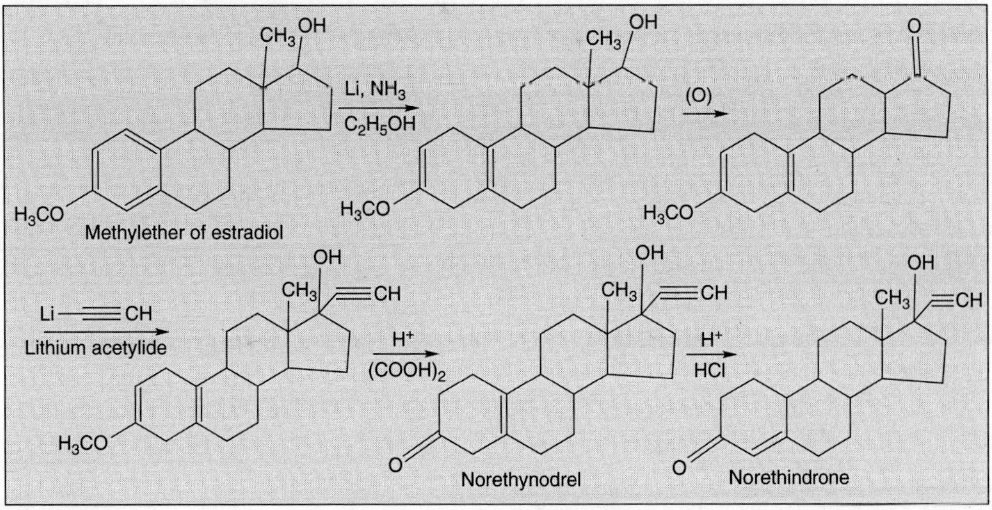

Norethindrone is made from methylether of oestradiol, which on treatment with lithium in liquid ammonia leads to partial reduction of aromatic ring, and this is then oxidized to a keto-group by chromium (VI) oxide in acetic acid. Reaction of 17 keto group with lithium acetylide gives alchol, which is further hydrogenated and oxidized to form norethynodrel. Treatment with acid shifts the double bond and affords norethindrone.

Norgestrel: (8R,9S,10R,13S,14S,17S)- 13-Ethyl-17-ethynyl-17-hydroxy- 1,2,6,7,8,9,10,11,12,14,15, 16- dodecahydrocyclopenta[a]phenanthren-3-one

It differs from norethindrone in the presence of an ethyl group at C13 of the steroid system. The presence of an ethyl group at C13 of the steroid system makes it less active than progesterone; however, it retains activity when taken orally, which provides effective contraception.

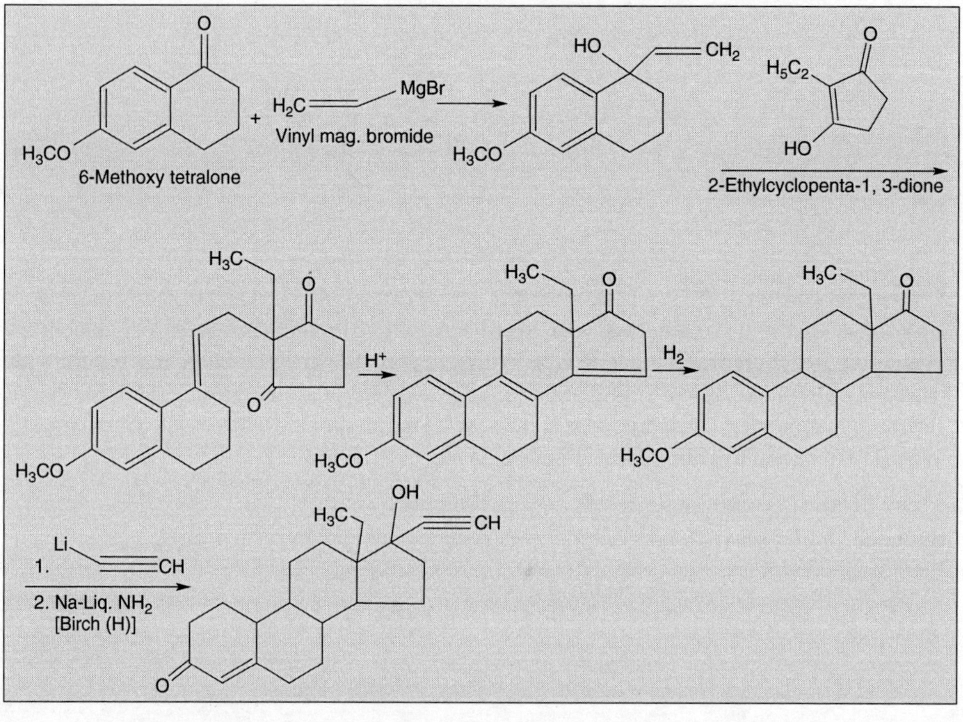

Vinyl magnesium bromide reacts with 6-methoxy tetralone to give alcohol intermediate. This on reaction with 2-ethylcyclopenta-1,3-dione gives cyclopenta dione intermediate, which undergoes cyclization and reduction to give oestradiol-3-methyl ether derivative. This on various steps like Birch reduction, oxidation, ethynylation, and hydrolysis affords norgestrel.

Tigestol: 17-(1-Ethynyl)-13-methyl-2,3,4,6,7,8,9,11,12,13,14,15,16,17-tetradecahydro-1H-cyclopenta[α]phenanthren-17-ol

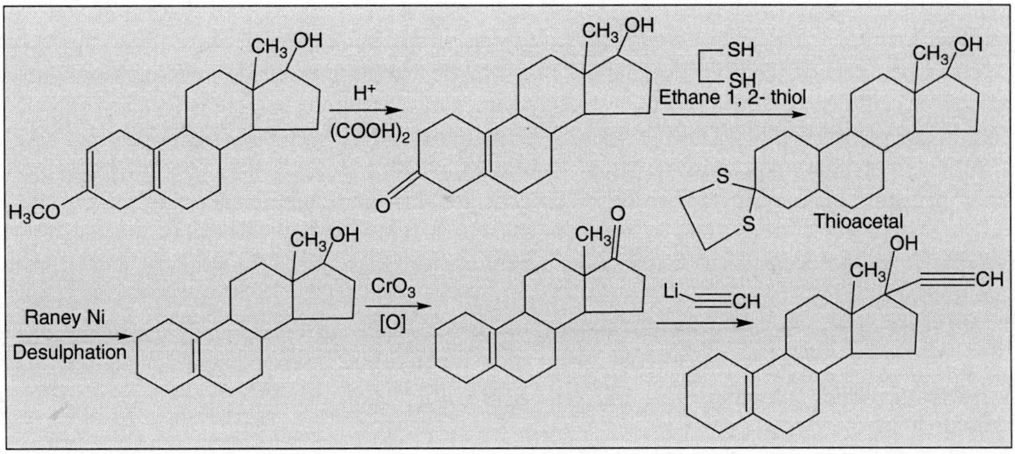

The starting material 3-methoxy-13-methyl-4,6,7,8,9,11,12,13,14,15,16,17-dodecahydro-1H-cyclopenta[a] phenanthren-17-ol (obtained from norethindrone synthesis) is treated with acid to deprotect the 3-hydroxyl group, and oxidize to ketone. This unconjugated 5,9 olefin is converted to a thioacetal, desulphurized, oxidized, and ethynylated to afford tigestol.

Ethynerone: 17-(2-Chloro-1-ethynyl)-17-hydroxy-13-methyl-2,3,6,7,8,11,12,13,14,15,16,17-dodecahydro-1H-cyclopenta[a]phenanthren-3-one

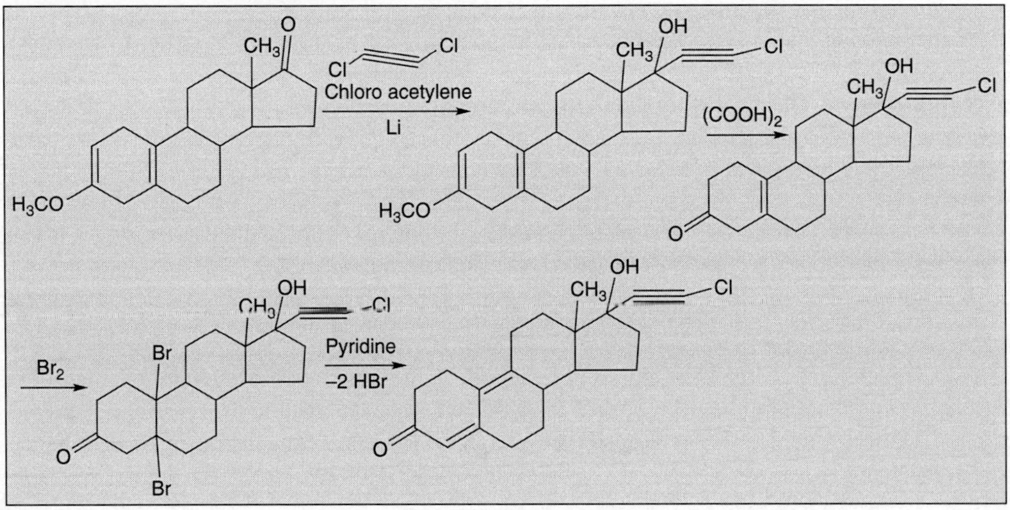

The starting material 3-methoxy-13-methyl-4,6,7,8,9,11,12,13,14,15,16,17-dodecahydro-1H-cyclopenta[α] phenanthren-17-ol (obtained from norethindrone synthesis) is treated with 1,2-dichloacetylene in presence of lithium and introduces 2-chloroethynyl derivative. Methoxy group at 3rd position is hydrolysed followed by oxidation and gives unconjugated ketone. This on treatment with bromine gives dibromo derivative; this loses hydrogen bromide in the presence of pyridine to give the diene ethynerone.

33.6 ANDROGENS

Androgen is the generic term for any natural or synthetic compound, usually a steroid hormone, that stimulates or controls the development and maintenance of masculine characteristics in vertebrates by binding to androgen receptors. This includes the activity of the accessory male sex organs and development of male secondary sex characteristics. Androgens, which were first discovered in 1936, are also called **androgenic hormones or testoids**. Androgens are also the original anabolic steroids. All natural androgens are steroid derivatives of androstane (19-carbon tetracyclic hydrocarbon nucleus, $C_{19}H_{32}$). They are also the precursor of all oestrogens, the female sex hormones. The primary, and most well-known, androgen is testosterone.

A subset of androgens, **adrenal androgens**, includes any of the 19-carbon steroids synthesised by the adrenal cortex, an adrenal gland, that functions as weak steroids or steroid precursors, including dehydroepiandrosterone (DHEA), dehydroepiandrosterone sulphate (DHEA-S), and androstenedione.

Besides testosterone, other androgens include:

- Dehydroepiandrosterone (DHEA): A steroid hormone produced from cholesterol in the adrenal cortex, which is the primary precursor of natural oestrogens. DHEA is also called dehydroisoandrosterone or dehydroandrosterone
- Androstenedione: An androgenic steroid, which is produced by the testes, adrenal cortex, and ovaries. While androstenediones are converted metabolically to testosterone and other androgens, they are also the parent structure of oestrone. Use of androstenedione as an athletic or body-building supplement has been banned by the International Olympic Committee
- Androstanediol: The steroid metabolite that is thought to act as the main regulator of gonadotropin secretion
- Androsterone: A chemical byproduct created during the breakdown of androgens, or derived from progesterone, that also exerts minor masculinising effects, but with one-seventh the intensity of testosterone. It is found in approximately equal amounts in the plasma and the urine of both males and females
- Androstenolone: An androgenic steroid secreted by the adrenal cortex and the testes, which is a major precursor of testosterone, but is even weaker than androsterone
- Dihydrotestosterone (DHT): A metabolite of testosterone that is actually a more potent androgen in that it binds more strongly to androgen receptors

SAR of androgens Oral administration of testosterone is followed by absorption into the portal blood and prompts degradation by the liver, so that insignificant amounts of hormone reach the systemic circulation. Chemical modification of testosterone was done to retard the catabolism or to enhance the androgenic potency.

1. Esterification of the 17-β hydroxyl group with carboxylic acids decreases the polarity of the molecule, makes it more soluble in the lipid vehicles used for injection, and, hence, slows the release of the injected steroid into the circulation. The longer the carbon chain in the ester, the more lipid-soluble it is, and the steroid becomes more prolonged in action.
2. Alkylation at the 17-α position allows androgens to be effective orally, because the alkylated derivatives are slowly catabolised by the liver.

Pathways of testosterone action

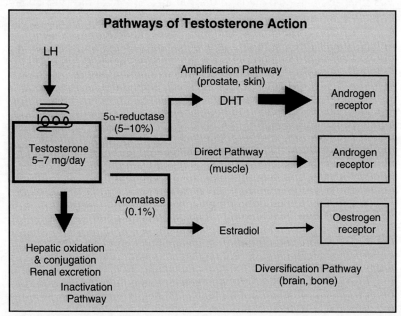

In men, most (>95%) testosterone is produced under LH stimulation through its specific receptor, a heptahelical G-protein coupled receptor located on the surface membrane of the steroidogenic Leydig cells. The daily production of testosterone (5–7 mg) is disposed along one of four major pathways. The direct pathway of testosterone action is characteristic of skeletal muscle in which testosterone itself binds to and activates the androgen receptor. In such tissues, there is little metabolism of testosterone to biologically active metabolites. The amplification pathway is characteristic of the prostate and hair follicle in which testosterone is converted by the type 2 5α reductase enzyme into the more potent androgen, dihydrotestosterone. This pathway produces local tissue-based enhancement of androgen action in specific tissues according to where this pathway is operative. The local amplification mechanism was the basis for the development of prostate-selective inhibitors of androgen action via 5a reductase inhibition, the forerunner being finasteride. The diversification pathway of testosterone action allows testosterone to modulate its biological effects via oestrogenic effects that often differ from androgen receptor-mediated effects. The diversification pathway, characteristic of bone and brain, involves the conversion of testosterone to oestradiol by the enzyme aromatase, which then interacts with the ERs a and/or β. Finally, the inactivation pathway occurs mainly in the liver with oxidation and conjugation to biologically inactive metabolites that are excreted by the liver into the bile and by the kidney into the urine.

Classification

1. Parenteral androgens: Testosterone, Testosterone propionate, Testosterone enanthate, Testosterone cypionate
2. Oral and buccal androgens: Danazol, Fluoxy mestrone, Methyl testosterone, Oxandrolone, Nandrolone, Stanazolol

> Ben Johnson was stripped of his gold medal in the 100-metre sprint at the 1988 Summer Olympics, when he tested positive for stanozolol after winning the final.

Testosterone   17β-Hydroxyandrost-4-ene-3-one

Synthesis

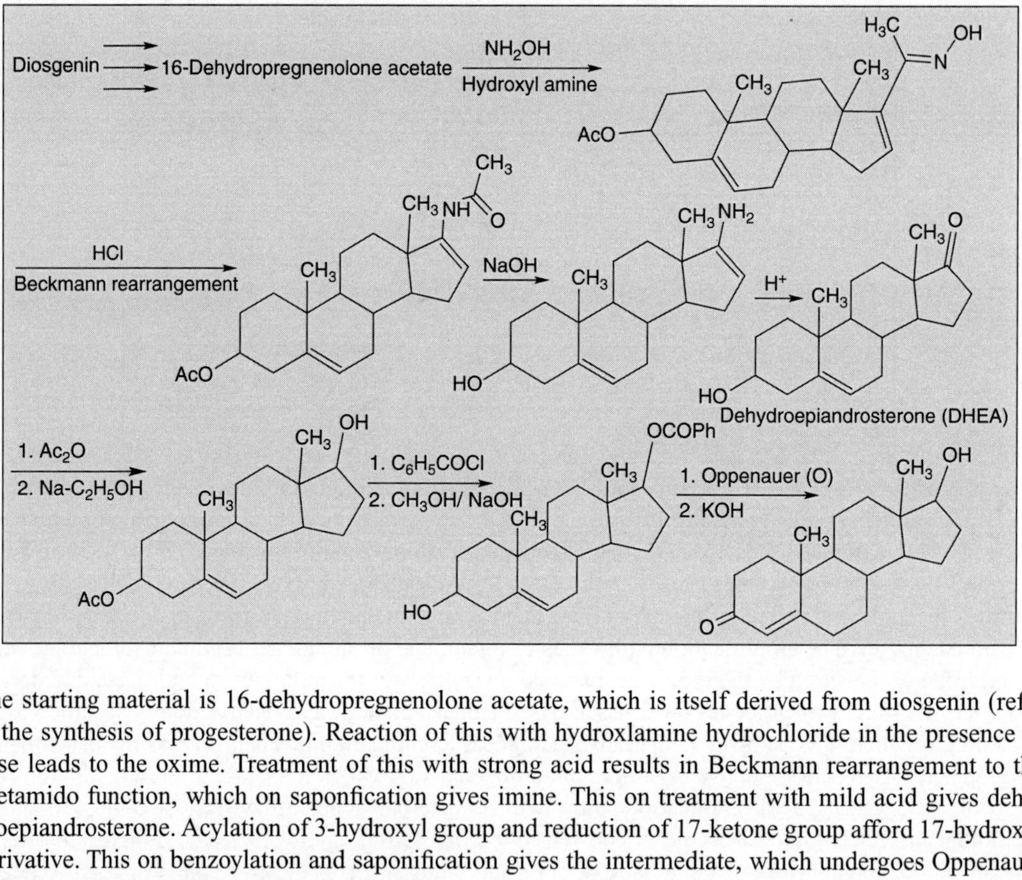

The starting material is 16-dehydropregnenolone acetate, which is itself derived from diosgenin (refer to the synthesis of progesterone). Reaction of this with hydroxlamine hydrochloride in the presence of base leads to the oxime. Treatment of this with strong acid results in Beckmann rearrangement to the acetamido function, which on saponfication gives imine. This on treatment with mild acid gives dehydroepiandrosterone. Acylation of 3-hydroxyl group and reduction of 17-ketone group afford 17-hydroxyl derivative. This on benzoylation and saponification gives the intermediate, which undergoes Oppenauer oxidation and hydrolysis to afford testosterone.

> Appropriate testosterone therapy can prevent or reduce the likelihood of osteoporosis, type 2 diabetes, cardiovascular disease (CVD), obesity, depression and anxiety, and the statistical risk of early mortality. Low testosterone also brings with it an increased risk for the development of Alzheimer's disease.

Testosterone enanthate, propionate, and cypionate

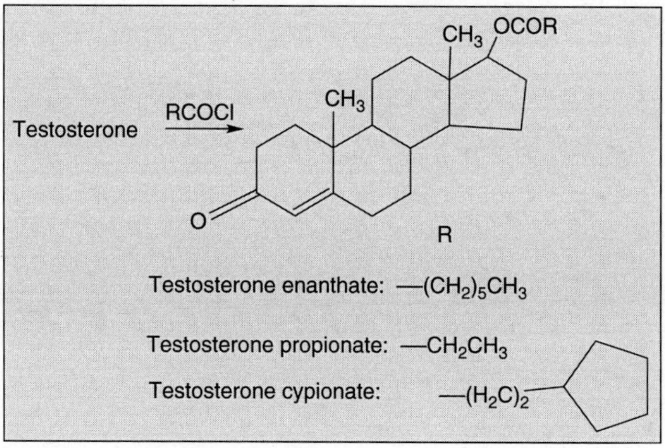

Testosterone $\xrightarrow{\text{RCOCl}}$ [steroid structure with CH$_3$, CH$_3$, OCOR groups]

Testosterone enanthate: —(CH$_2$)$_5$CH$_3$

Testosterone propionate: —CH$_2$CH$_3$

Testosterone cypionate: —(H$_2$C)$_2$—[cyclopentyl]

Esterification of testosterone with appropriate acid chloride gives ester derivatives. It markedly prolongs the activity.

Danazol 17β-Hydroxy-2,4,17α-pregnadien-20-yno[2,3-D]isoxazole

Synthesis

Heterocyclic ring fused to the ring-A was much more active than testosterone.

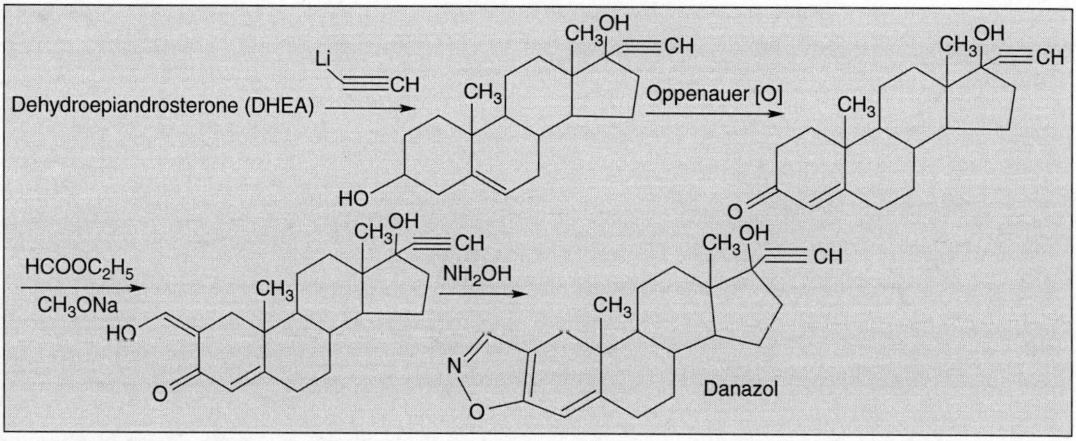

Dehydroepiandrosterone (DHEA) $\xrightarrow{\text{Li}\equiv\text{CH}}$ [17-ethynyl DHEA intermediate] $\xrightarrow{\text{Oppenauer [O]}}$ [17-ethynyl testosterone]

$\xrightarrow[\text{CH}_3\text{ONa}]{\text{HCOOC}_2\text{H}_5}$ [2-hydroxymethylene derivative] $\xrightarrow{\text{NH}_2\text{OH}}$ Danazol

Dehydroepiandrosterone on reaction with lithium acetylide followed by Oppenauer oxidation gives 17-ethynyl testosterone. Condensation of this with ethyl formate in the presence of base gives 2-hydroxymethylene derivative, and reaction of this with hydroxylamine leads to the danazol.

Stanazolol   17α-Methyl-5α-androstano[3,2-c]pyrazol-17β-ol

Synthesis

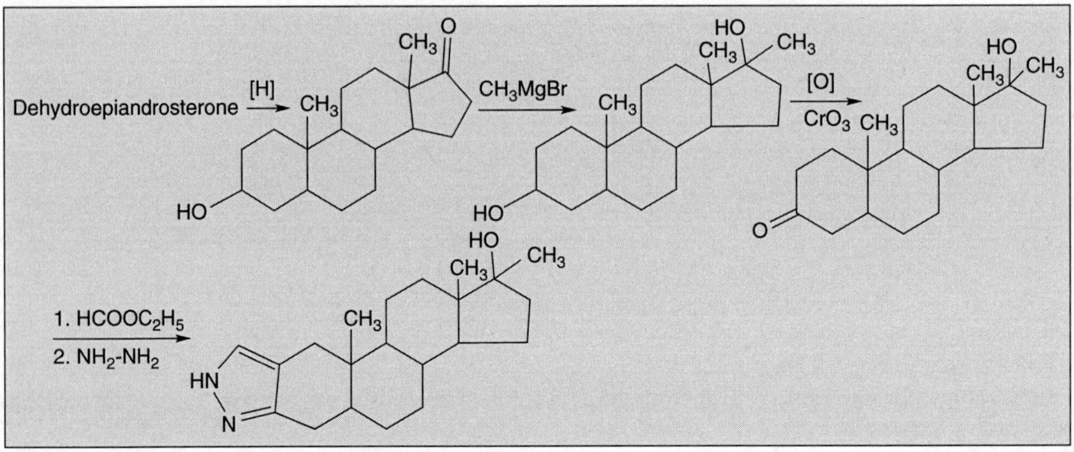

Dehydroepiandrosterone on reduction at C_{17} followed by reaction with methyl magnesium bromide gives 17-methyl testosterone. Oxidation of C_3 hydroxy group gives 3-keto derivative. This on reaction with ethyl formate and hydrazine gives stanazolol.

Nandrolone   17β-Hydroxyester-4-en-3-one
Synthesis

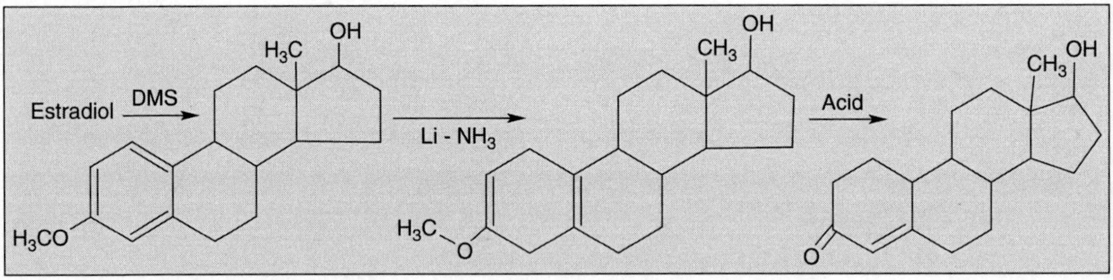

Nandrolone is made from oestradiol. The phenol hydroxyl group undergoes methylation by dimethylsulphate in the presence of sodium hydroxide, forming the corresponding methyl ether, and then the aromatic ring is reduced by lithium in liquid ammonia, which forms enol ether. Hydrolysing this compound with a mixture of hydrochloric and acetic acids leads to the formation of a keto group, and simultaneous isomerization of the double bond from C_5—C_{10} to position C_4—C_5 gives the desired nandrolone.

Methyl testosterone   17β-Hydroxy-17α-methylandrost-4-ene-3-one

Synthesis
Introduction of C17-α- methyl group confers oral activity on testosterone.

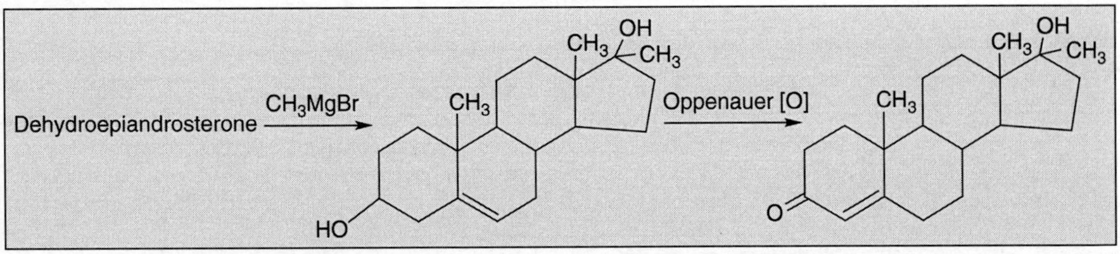

Dehydroepiandrosterone on reaction with methyl magnesium bromide followed by Oppenauer oxidation gives 17-methyl testosterone.

Oxandrolone   17β-Hydroxy-17α-methyl-2-oxa-5-androstan-3-one

Synthesis
Introduction of an oxygen atom instead of methylene group in ring-A resulted in oral anabolic activity.

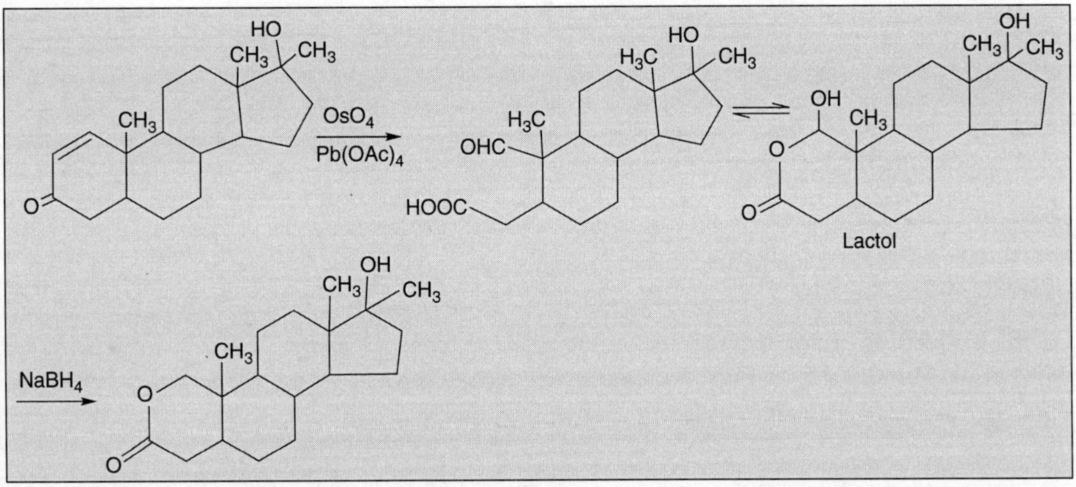

Oxandrolone is made by oxidation of the C_1—C_2 double bond of 17β-hydroxy-17α-methyl-1-androsten-3-one by a mixture of lead tetraacetate and osmium tetroxide with an opening of the A-ring of the steroid system, which forms an aldehyde acid. Upon reducing the aldehyde group with sodium borohydride, intramolecular cyclization takes place, directly forming a lactone, which is the desired oxandrolone.

Fluoxymestrone   9-Fluoro-11,17-dihydroxy-13,17-dimethyl-2,3,6,7,8,9,10,11,12,13,14,15,16,17-tetradecahydro-1*H*-cyclopenta[α]phenanthren-3-one

Synthesis

Introduction of 9-fluoro and 11-hydroxy groups makes orally active androgen, about 5–10 times more potent than testosterone.

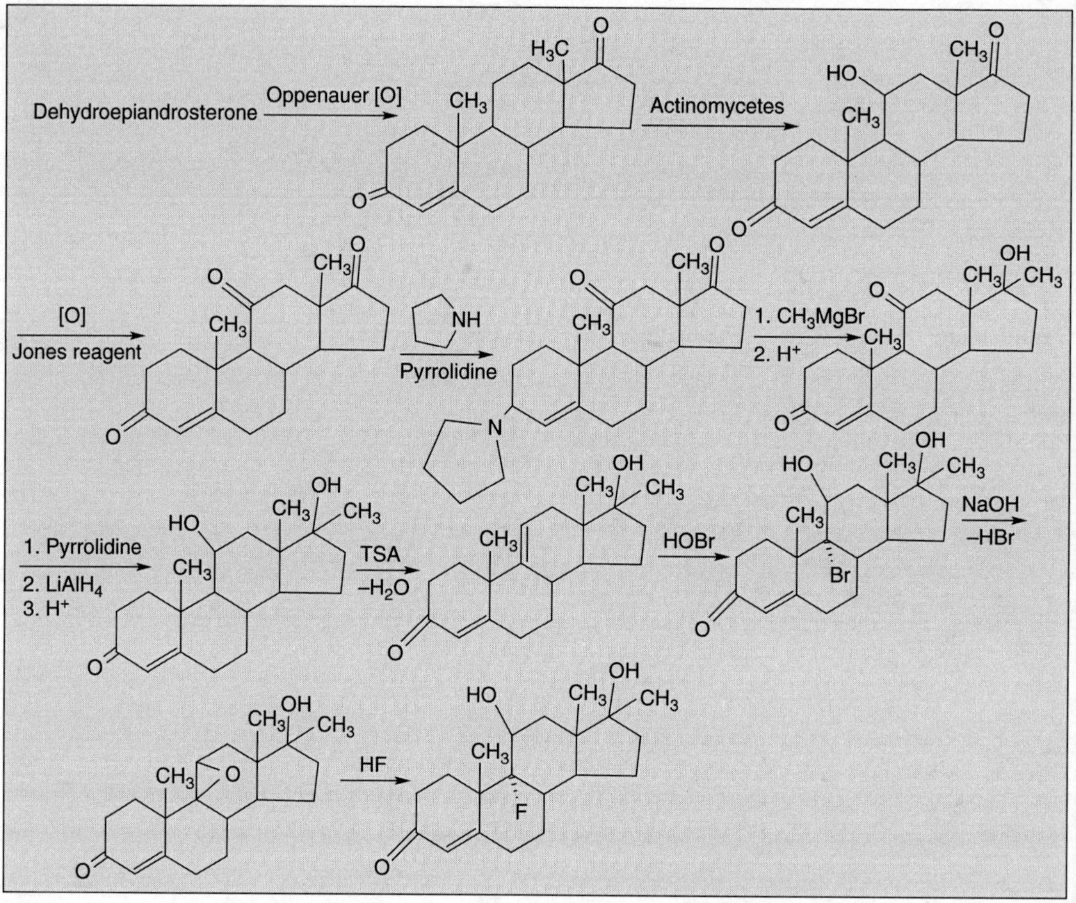

Oppenauer oxidation of dehydroepiandrosterone gives androstenedione. Microbial oxidation of this gives 11α-hydroxyl derivative, which on oxidation with chromium trioxide gives triketone derivatve. Treatment of this with pyrrolidine under controlled condition leads to selective eneamine of least hindered ketone at 3. Reaction with methyl magnesium bromide followed by removal eneamine yields 17β-hydroxy-17α-methyl derivative. The ketone at 3 is then again converted to eneamine and treatment with lithium aluminium hydride followed by hydrolysis of the eneamine gives dihydroxyketone. The 11 hydroxyl group on dehydration gives olefin, which on treatment with hypobromous acid gives 9-bromo-11-hydroxy derivative. Treatment with base leads to displacement of bromine and forms epoxide. Ring opening of oxiran with hydrogen fluoride affords fluoxymestrone.

33.7 GLUCOCORTICOIDS

Glucocorticoids are endogenous compounds that have an effect on carbohydrate, lipid, and protein metabolism, and exhibit anti-inflammatory, desensitizing, and anti-allergy action. They have an effect on the cardiovascular system, the gastrointestinal tract, the skeletal musculature, the skin, the connective tissue, the blood, and the endocrine system. The direct indication for using them is severe and chronic adrenal insufficiency. They are used for collagenosis, rheumatoid arthritis, rheumatism, eczema, neurodermatitis and other skin diseases, allergies, tissue transplants, bronchial asthma, and many other diseases.

Effects The name 'glucocorticoid' derived from the observation that these hormones were involved in glucose metabolism. In the fasted state, cortisol or hydrocortisone stimulates several processes that collectively serve to increase and maintain normal concentrations of glucose in blood. These effects include:

- It stimulates gluconeogenesis: This effect results in the synthesis of glucose from non-sugar substrates such as amino acids and lipids.
- It mobilizes the amino acids from extra-hepatic tissues: These also serve as substrates for gluconeogenesis.
- It inhibits the glucose uptake in muscle and adipose tissues: This mechanism also conserves glucose.
- It stimulates fat breakdown in adipose tissues: The fatty acids released during the lipid breakdown are used for production of energy in tissues, and the released glycerol provide another substrate for gluconeogenesis.
- Glucocorticoids have potent anti-inflammatory and immunosuppressive properties.

Mechanism of action: Glucocorticoid mainly binds to the cytosolicglucocorticoid receptor. After a hormone binds to the corresponding receptor, the receptor-ligand complex translocates into the cell nucleus, where it binds to glucocorticoid response elements (GRE) in the promoter region of the target genes. The DNA-bound receptor then interacts with basic transcription factors, causing the increase in expression of specific target genes. This process is called transactivation. The opposite mechanism is called transrepression. The activated hormone receptor interacts with specific transcription factors and prevents the transcription of targeted genes.

The ordinary glucocorticoids do not distinguish among transactivation and transrepression, and influence both the 'wanted' immune and the 'unwanted' genes regulating the metabolic and cardiovascular functions.

Pharmacological properties A variety of synthetic glucocorticoids, some far more potent than cortisol, have been created for therapeutic use. They differ in the pharmacokinetics and in pharmacodynamics properties. Because they absorb well through the intestines, they are primarily administered by oral route, and also by other ways, like topically on skin. More than 90 per cent of them bind different plasma proteins, though with a different binding specificity. In the liver, they quickly metabolise by conjugatiion with a sulphate or glucuronic acid and are secreted in the urine. Glucocorticoid potency, duration of effect, and overlapping mineralocorticoid potency vary (see accompanying table).

Comparative potencies of steroids

Name	Glucocorticoid Potency	Mineralocorticoid Potency	Duration of Action ($t_{1/2}$ in hours)
Hydrocortisone	1	1	8
Cortisone acetate	0.8	0.8	By oral 8, intramuscular >18
Prednisone	3.5–5	0.8	16–36
Prednisolone	4	0.8	16–36
Methylprednisolone	5–7.5	0.5	18–40
Dexamethasone	25–80	0	36–54
Betamethasone	25–30	0	36–54
Triamcinolone	5	0	12–36

SAR of corticosteroids

1. The 4,5 double bond and the 3-keto group on ring-A are essential for both glucocorticoid (GC) and mineralocorticoid (MC) activity.
2. An 11-β hydroxyl group on ring-C is required for GC activity but not MC activity.
3. A hydroxyl group at C-21 on ring-D is present in all natural corticosteroids and most of the active synthetic analogues, and seems to be an absolute requirement for MC activity but not GC activity.
4. The 17-α-hydroxyl group on ring-D is a substituent on cortisol and all of the currently used synthetic glucocorticoids. While steroids without 17-α-hydroxyl group (corticosteroids) have appreciable GC activity, the 17-α-hydroxyl group gives optimal potency.
5. Introduction of an additional double bond in the 1,2 position of ring-A, as in prednisolone or prednisone, selectively increases GC activity (approximately fourfold compared to hydrocortisone), resulting in an enhanced GC to MC potency ratio. This modification also results in compounds that are metabolised more slowly than hydrocortisone.
6. Fluorination at the 9- α position on ring-B enhances both GC and MC activity, and possibly is related to an electron-withdrawing effect on the nearby 11- β hydroxyl group. Fludrocortisone has enhanced activity at the GC receptor (10-fold relative to cortisol), but even greater activity at the MC receptor (125-fold relative to cortisol).
7. When combined with the 1, 2 double bond in ring-A and other substitutions at C-16 on ring-D, the 9- α fluoro derivatives (triamcinolone, betamethasone, and dexamethasone) have marked GC activity. These substitutions at C-16 virtually eliminate MC activity.
8. 6- α substitution on ring-B has somewhat unpredictable effects. 6- α-methyl cortisol has increased GC and MC activity, whereas 6- α-methylprednisolone has somewhat greater GC activity and less MC activity than prednisolone.
9. A number of modifications convert GC to a more lipophilic molecule with enhanced topical to systemic potency ratios. Examples include the introduction of an acetonide between hydroxyl groups at C-16 and C-17, esterification of the hydroxyl group with valerate at C-17, esterification of the hydroxyl groups with propionate at C-17 and C-21, and substitution of the hydroxyl group at C-21 with chlorine.
10. Other approaches to achieve local GC activity while minimizing systemic effects involve the formation of analogues that are rapidly inactivated following absorption. Examples of this later group include C-21 carboxylate or carbothiolate glucocorticoid esters, which are rapidly metabolised to inactive C-21 carboxylic acids.

Classification based on the time taken to act
   a. Short- to medium-acting glucocorticoids: Hydrocortisone, Prednisone, Methylprednisolone, Prednisolone
   b. Intermediate-acting glucocorticoids: Triamcinolone, Fluprednisolone
   c. Long-acting glucocorticoids: Betamethasone, Dexamethasone

Hydrocortisone (11β,17α,21-Trihydroxypregn-4-en-3,20-dione)
and Cortisone acetate (17α,21-Dihydroxypregn-4-en-3,11,20-trione)

Synthesis

In the first stage of the synthesis, progesterone undergoes microbiological oxidation, which forms 11α-hydroxyprogesterone. The resulting hydroxyl group is oxidized by chromium trioxide in acetic acid, giving 11-ketoprogesterone. This is reacted with diethyloxalate in the presence of sodium ethoxide, forming the corresponding α-ketoester in the form of the sodium enolate, which undergoes bromination by two equivalents of bromine, giving the dibromoketone. The resulting dibromoketone undergoes Favorskii rearrangement and is further hydrolysed, giving an unsaturated acid. Then, the carbonyl group at position C_3 is ketalized using ethylenglycol in the presence of 4-toluenesulphonic acid, during which a migration of the double bond between carbon atoms C_5 and C_6 takes place, forming a ketal. The resulting product is reduced by lithium aluminium hydride. During this, the carboxyl and ketogroups at C_{11} are reduced to alcohol groups, forming a diol. The ketal-protecting group is subsequently removed in acidic conditions, during which the double bond again migrates back to the initial position between C_4 and C_5, and the primary hydroxyl group is acylated by acetic anhydride in pyridine forming the product. The double bond in this compound is oxidized using hydrogen peroxide in the presence of osmium tetroxide in N-methylmorpholine, forming hydrocortisone acetate.

Oxidation of 11-hydroxyl group with chromium trioxide results in cortisone acetate.
Prednisone acetate   17α,21-Dihydroxypregna-1,4-dien-3,11-20-trione
Synthesis
Prednisone differs from cortisone in the presence of an additional double bond between C_1 and C_2.

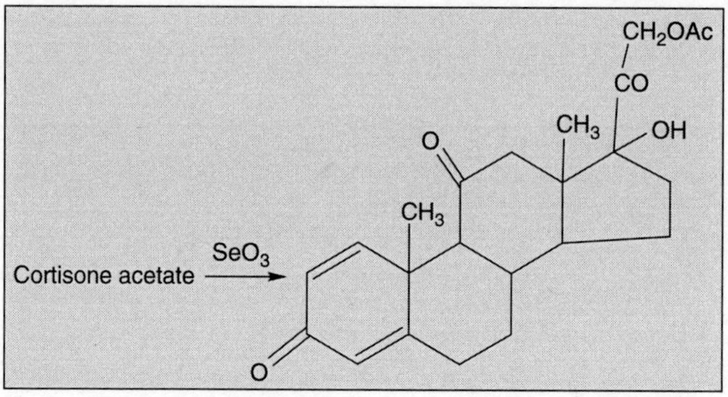

Cortisone acetate $\xrightarrow{SeO_3}$ [structure of prednisone acetate]

The double bond between C_1 and C_2 is introduced in the cortisone acetate by treatment with selenium trioxide.
Prednisolone acetate   11β,17α,21-Trihydroxypregna-1,4-dien-3,20-dione
Synthesis
Structurally, prednisolone differs from prednisone in that the keto group at C_{11} of prednisone is replaced by a hydroxyl group.

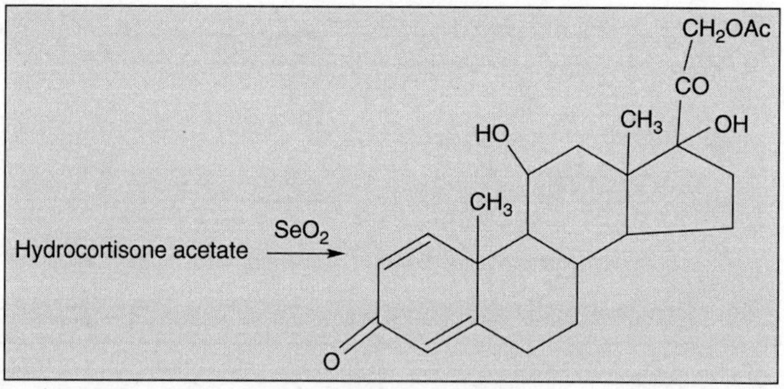

Hydrocortisone acetate $\xrightarrow{SeO_2}$ [structure of prednisolone acetate]

The double bond between C_1 and C_2 is introduced in the hydrocortisone acetate by treatment with selenium trioxide.
Fludrocortisone acetate   9α-Fluoro-11β,17α,21-trihydroxypregn-4-en-3,20-dione

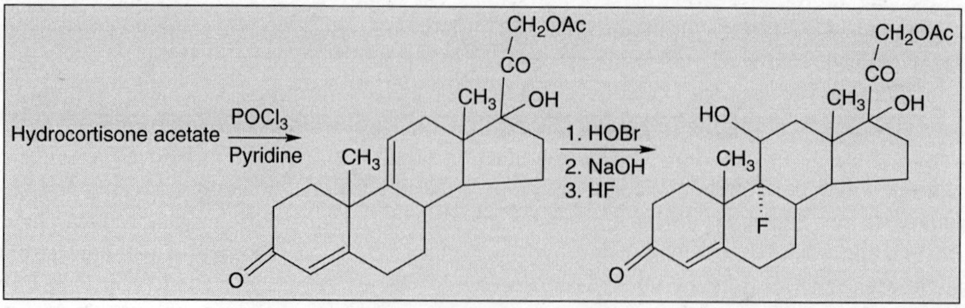

Hydrocortisone acetate $\xrightarrow[\text{Pyridine}]{POCl_3}$ [intermediate structure] $\xrightarrow[\substack{\text{1. HOBr}\\\text{2. NaOH}\\\text{3. HF}}]{}$ [structure of fludrocortisone acetate]

Synthesis

Fludrocortisone is synthesised from hydrocortisone acetate. In the first stage of synthesis, dehydration of the hydrocortisone molecules is accomplished using phosphorous chloride in pyridine, which forms a product with a double bond at C_9—C_{11}. The resulting double bond is synthesised into an epoxide by an initial transformation to a bromohydrine using hypobromous acid and subsequent dehydrobromination using sodium hydroxide, which forms 21-O-acetoxy-9d-11β-epoxy-17α-hydroxy-4-pregnen-3,20-dione. The epoxide ring is opened by hydrofluoric acid, which results in the formation of the 21-O-acetate of fludrocortisone.

Methylprednisolone 11β,17α,21-Trihydroxy-6a-methylpregna-1,4-dien-3,20-dione

Methylprednisolone differs from prednisolone in the presence of a methyl group at position C_6 of the steroid skeleton of the molecule.

Synthesis

It is synthesised from hydrocortisone, the carbonyl group of which initially undergoes ketalization by ethylene glycol in the presence of traces of acid, during which the double bond at position C_4—C_5 is shifted to position C_5—C_6, giving the diethyleneketal. The product is oxidized to an epoxide using perbenzoic acid. Next, the resulting epoxide is reacted with methylmagnesium bromide, and subsequent removal of the ketal protection by hydrogen reduction gives the 5-hydroxy-6-methyl derivative of dihydrocortisone. The resulting β-hydroxyketone is dehydrated using an alkaline, and then the resulting 6α-methylcortisone undergoes microbiological dehydrogenation at position C_1—C_2, giving the desired methylprednisolone.

Dexamethasone (9α-Fluoro-16α-methyl-11β,17,21-trihydroxypregna-1,4-dien-3,20-dione); **Betamethasone and Beclomethasone**

Synthesis

In dexamethasone, C—16 methyl group is in α-configuration, whereas in betamethasone it is 16-β-methyl group. Beclomethasone synthesis is similar to dexamethasone but in the last step the treatment is with HCl, so that 9-α-Cl derivatives result instead of 9-α-F compounds.

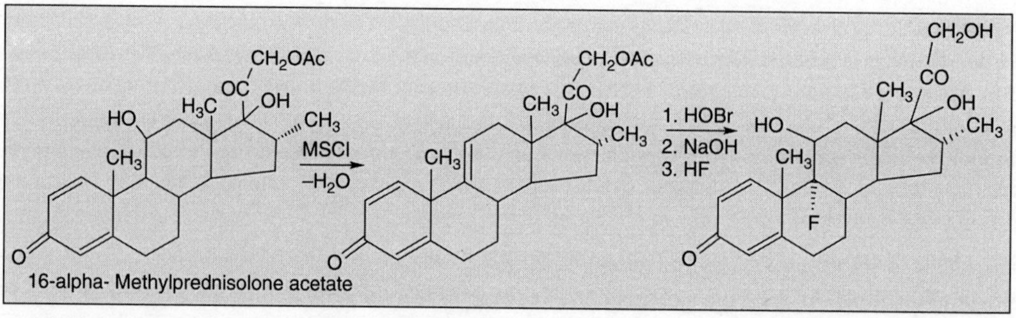

The 11 hydroxyl group of 16α-methylprednisolone on dehydration gives olefin, which on treatment with hypobromous acid gives 9-bromo-11-hydroxy derivative. Treatment with base leads to displacement of bromine and forms epoxide. Ring opening of oxiran with hydrogen fluoride affords dexamethasone.

Triamcinolone acetonide 9α-Fluoro-11b,16a,17,21-tetrahydroxypregna-1,4-dien-3,20-dione

Synthesis

Triamcinolone differs from dexamethasone in terms of chemical structure, in that the methyl group at C_{16} is replaced with a hydroxyl group.

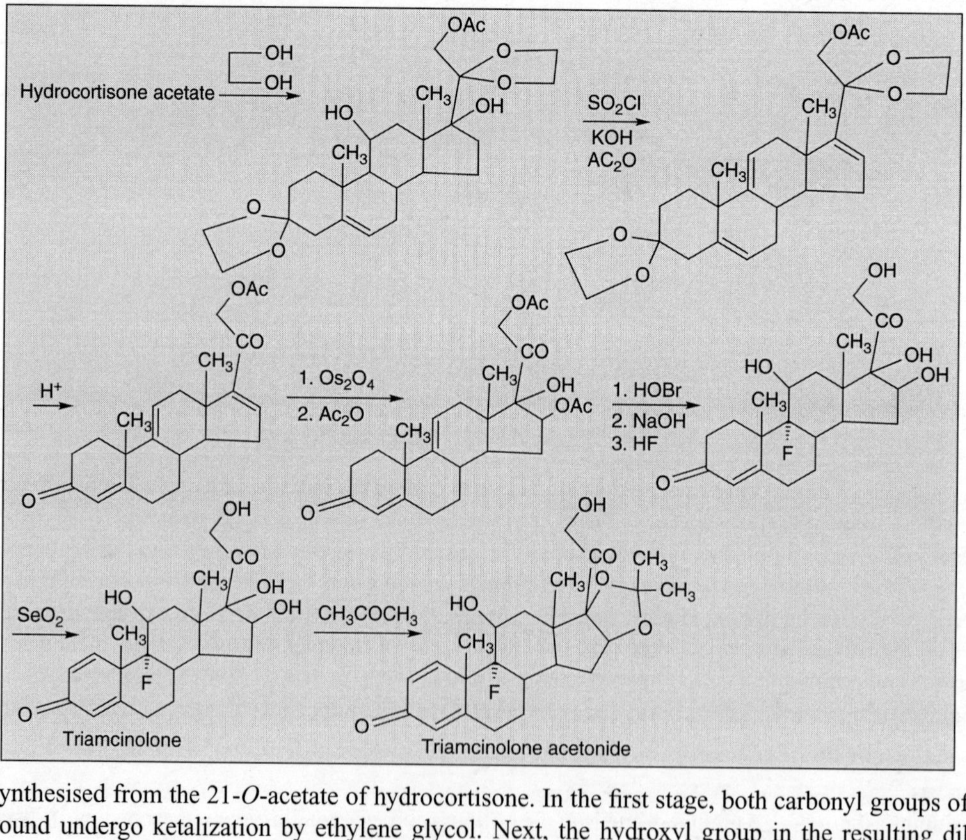

It is synthesised from the 21-O-acetate of hydrocortisone. In the first stage, both carbonyl groups of this compound undergo ketalization by ethylene glycol. Next, the hydroxyl group in the resulting diketal is replaced with chlorine using thionyl chloride, and the product undergoes dehydrochlorination using an alkali, during which the 21-O-acetyl group also is hydrolysed. Acetylating the hydroxyl group once again with acetic anhydride gives a triene. Reacting this with osmium tetroxide gives the vicinal diol. The secondary hydroxyl group at C_{16} of this product undergoes acetylation by acetic anhydride in pyridine, which forms the diacetate. Treating the product with hypobromous acid gives a bromohydrin, which upon reaction with alkali is transformed to an epoxide. Opening of the epoxide ring, using hydrofluoric acid, gives the corresponding 9-fluoro-11-hydroxy derivative. Upon selenium dioxide dehydrogenation, the C_1—C_2 bond is oxidized to a double bond, forming triamcinolone acetate, the acetyl group of which is hydrolysed, forming the desired triamcinolone. Reaction with acetone forms acetonide derivative.

Mineralocorticoids

Mineralocorticoids play a major role in regulating the balance of electrolytes and water, especially in the kidneys, where they facilitate reabsorption of sodium ions and water from urine. The main endogenic mineralocorticoid is **aldosterone**. Aldosterone is produced in the cortex of the adrenal gland and its secretion is mediated principally by angiotensin II, but also by adrenocorticotrophic hormone (ACTH) and local potassium levels.

Aldosterone acts on the kidneys to provide active reabsorption of sodium and an associated passive reabsorption of water, as well as the active secretion of potassium in the principal cells of the cortical collecting tubule and active secretion of protons via proton ATPases in the lumenal membrane of the intercalated cells of the collecting tubule. This in turn results in an increase of blood pressure and blood volume.

An example of synthetic mineralocorticoids is fludrocortisone.

Aldosterone: 11β,21-Dihydroxypregn-4-en-2,18,20-trione
Synthesis
It is synthesised from 21-*O*-acetylcorticosterone, which when reacted with nitrosyl chloride in pyridine gives the nitrite. When photochemically irradiated, this compound is transformed to the oxime, which is hydrolysed by nitrous acid and forms the semiacetal, which is an acetate of the desired aldosterone. Alkaline hydrolysis of the acetyl group of this compound leads to the desired aldosterone.

Fludrocortisone: 9α-Fluoro-11β,17α,21-trihydroxypregn-4-en-3,20-dione
Synthesis
It is synthesised from hydrocortisone acetate. In the first stage of synthesis, dehydration of the hydrocortisone molecules is accomplished using phosphorous chloride in pyridine, which forms a product with a double bond at C_9—C_{11}. The resulting double bond is synthesised into an epoxide by an initial transformation to a bromohydrine using N-bromoacetamide and subsequent dehydrobromination using sodium acetate, which forms 21-O-acetoxy-9d-11β-epoxy-17α-hydroxy-4-pregnen-3,20-dione. As described above, the epoxide ring is opened by hydrofluoric acid, which results in the formation of the 21-O-acetate of fludrocortisone. Hydrolysis of the acetyl group of this compound using potassium acetate gives fludrocortisone.

Cardiac glycosides

Glycosides are isolated from leaves of various types of foxglove *Digitalis lanta, Digitalis purpurea,* and strophanthus *Strophantus kombe,* and also a number of other plants, which exhibit a direct effect on the myocardium and which strengthen its contractions. The most simple cardiac genins, digitoxigenin, gitoxigenin, and strophanthidin are aglycons of the most important glycosides *Digitalis lanta, Digitalis purpurea,* and *Strophantus kombe.*

It acts on the contractibility of the heart by affecting the process of calcium ion transfer through the membrane of myocardiocytes. The effect of cellular membranes in electric conductivity is mediated by transport of sodium, calcium, and potassium ions, which is a result of indirect inhibitor action on the (Na^+ –K^+) ATPase of cell membranes.

Cardiac glycosides are used for treating severe, chronic cardiac insufficiency, for certain forms of cardiac arrhythmia, and for cardiac shock. The following unique structural features are characteristic of glycons of cardiac glycosides: a coupling of rings A and B-*cis*, rings B and C-*trans*, rings C and D-*cis*, and a butenolide region located at position 17β.

Physiologically active compounds must contain a hydroxyl group at 14β. OH and CO groups at positions 11, 12, 16, and 19 evidently have less of an effect on activity.

The main purpose of sugar residues, which esterify hydroxyl groups at C_3, evidently lies in facilitating the solubility of genins.

Digitoxin: 3β,14β-Dihydroxy-5β-card-20(22)enolide-3-tridigitoxide

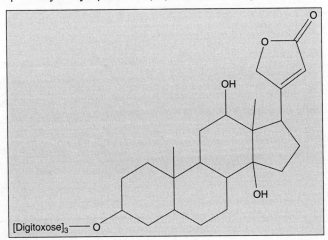

Digitoxin is used for chronic cardiac insufficiency, tachyarrhythmia form of atrial fibrillation, paroxysmal ciliary arrhythmia, and paroxysmal supraventricular tachycaria.

Digoxin: 3β,14β,16β -Triihydroxy-5β-card-20(22)-enolide-3-tridigitoxide

It differs from digitoxin in that it has an additional hydroxyl group at C_{16} of the steroid skeleton.

FURTHER READINGS

1. *Advances in Medicinal Chemistry*, Maryanoff, B.E. and C.A. Maryanoff (ed.), Volume 2, Elsevier.
2. *Burger's Medicinal Chemistry & Drug Discovery*, Volume 5: Therapeutic Agents, Wolf, Manfred E. (ed.), Wiley Interscience.

MULTIPLE-CHOICE QUESTIONS

1. The following are naturally occurring female hormones except:
   a. Oestradiol
   b. Oestriol
   c. Mestranol
   d. Oestrone
2. Oestrogens are synthesised from
   a. Pregnenolone
   b. Androstenedione
   c. Dehydroepiandrosterone
   d. Aldosterone
3. Progesterone has the following effects except:
   a. Used in hormone therapy for transsexual women
   b. Thermogenic
   c. Bronchodilator
   d. Increase immune response
4. The inactive derivative of progesterone is
   a. 17-α-Hydroxyprogesterone
   b. Hydroxyprogesterone caproate
   c. Medroxyprogesterone acetate
   d. Chlormadinone acetate
5. The steroid metabolite that acts as the main regulator of gonadotropin secretion is
   a. Testosterone
   b. Androstenolone
   c. Androstanediol
   d. Androstenedione
6. The steroid having a role in promoting maturation of the lung in foetus is
   a. Androgens
   b. Oestrogens
   c. Progestins
   d. Glucocorticoids
7. The structural feature essential for both glucocorticoid (GC) and mineralocorticoid (MC) activity is (are)
   a. 6-α Substitution on ring-B
   b. 3-Keto group on ring-A
   c. Double bond in the 1,2 position of ring-A
   d. Without 17-α-hydroxyl group
8. One of the following is not a parenteral androgen:
   a. Testosterone
   b. Testosterone propionate
   c. Testosterone enanthate
   d. Methyltestosterone
9. One of the following is an orally active progestin:
   a. Progesterone
   b. Norethindrone
   c. Chlormadinone acetate
   d. Medroxyprogesterone acetate
10. The reagent used for the conversion of hydrocortisone acetate to cortisone acetate is
    a. CrO_3
    b. SeO_3
    c. HOBr
    d. m-CPBA

QUESTIONS

1. Discuss the chemistry of steroids.
2. Discuss the structural features and therapeutic applications of progestins.
3. Derive the complete synthetic protocol for mestranol, hydroxyprogesterone acetate, norgesterol, testosterone cypionate, stanzolol, fluoxymestrone, prednisolone acetate, and beclomethasone.
4. Write in detail on the SAR of glucocorticoids and their biologic effects.

SOLUTION TO MULTIPLE-CHOICE QUESTIONS

1. c;
2. b;
3. d;
4. a;
5. c;
6. d;
7. b;
8. d;
9. b;
10. a.

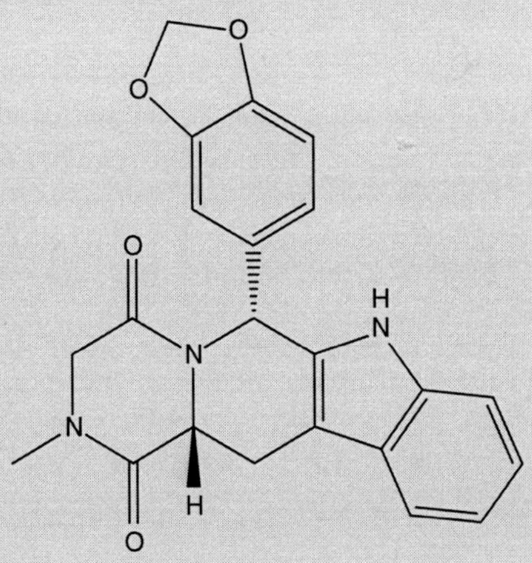

CHAPTER 34

Miscellaneous Agents

LEARNING OBJECTIVES

- Describe obesity, risks of obesity and currently available treatments
- Male erectile dyfunction (MED), mechanism of action, and synthesis of drugs used for treatment of MED
- Migraine and its treatment
- Brief study on anti-asthma drug, anti-thyroid drugs antiplatelet, anti-cough drugs
- To study about diagnostic agents, antiseptics and pharmaceutical aid

34.1 ANTI-OBESITY DRUGS

Obesity is the most common metabolic disease in developed nations. The World Health Organization has estimated that worldwide over one billion adults are overweight, with at least 300 million of them being obese. The increasing prevalence of obesity among children and adolescents is of great concern and suggests a likelihood of worsening obesity trends in future adults.

Obesity leads to, or significantly increases the risk of, co-morbidities involving various body systems including: (1) *cardiovascular* (hypertension, congestive cardiomyopathy, varicosities, pulmonary embolism, coronary heart disease), (2) *neurological* (stroke, idiopathic intracranial hypertension, meralgia parethetica), (3) *respiratory* (dyspnea, obstructive sleep apnea, hypoventilation syndrome, Pickwickian syndrome, asthma), (4) *musculoskeletal* (immobility, degenerative osteoarthritis, low back pain), (5) *skin* (striae distensae or 'stretch marks', venous stasis of the lower extremities, lymphoedema, cellulitis, intertrigo, carbuncles, acanthosisnigricans, skin tags), (6) *gastrointestinal* (GI; gastro-oesophageal reflux disorder, nonalcoholic fatty liver/steatohepatitis, cholelithiasis, hernias, colon cancer), (7) **genitourinary** (stress incontinence, obesity-related glomerulopathy, breast and uterine cancer),(8) **psychological** (depression and low self-esteem, impaired quality of life), and (9) **endocrine** (metabolic syndrome, type 2 diabetes, dyslipidemia, hyperandrogenemia in women, polycystic ovarian syndrome, dysmenorrhoea, infertility, pregnancy complications, male hypogonadism).

Anti-obesity medication or weight loss drugs refer to all pharmacological agents that reduce or control weight. These drugs alter one of the fundamental processes of the human body, weight regulation, by altering appetite, metabolism, or absorption of calories.

34.1.1 Current Therapies

- **Amphetamines** (dextroamphetamine) have been used as anti-obesity drugs, but can cause unacceptable tachycardia and hypertension. They also have a high rate of abuse potential and do not have a US Food and Drug Administration indication for the treatment of obesity. Other sympathomimetic adrenergic agents, including **phentermine**, **benzphetamine**, **phendimetrazine**, **mazindol**, and **diethylpropion**, have less abuse potential than amphetamines; but these agents may have adverse cardiovascular side-effects, and their indicated use is only short-term (~12 weeks) for the treatment of what is commonly a chronic metabolic disease. In 2000, the appetite suppressant **phenylpropanolamine** was removed from the over-the-counter market in the United States because of unacceptable risks of stroke, especially in adult women.
- **Dexfenfluramine** and **fenfluramine** were dual 5-HT reuptake inhibitors and serotonin-releasing agents that were not indicated for treatment of depression, but had previously been used for suppression of appetite as anti-obesity drugs. They were subsequently withdrawn from the market because of the onset of heart valve abnormalities thought to be related to the stimulation of peripheral (heart) 5-hydroxytryptamine (5-HT) 2b receptors. Investigational 'selective' 5-HT 2c receptor agonists under development may induce satiety by selective effects on the hypothalamus while avoiding toxicities to the heart.

34.1.2 FDA-approved Drugs

Orlistat (S)-((S)-1-((2S,3S)-3-Hexyl-4-oxooxetan-2-yl)tridecan-2-yl)2-formamido-4-methylpentanoate

Orlistat acts by inhibiting pancreatic lipase, an enzyme that breaks down triglycerides in the intestine. Without this enzyme, triglycerides from the diet are prevented from being hydrolysed into absorbable free fatty acids and are excreted undigested. Only trace amounts of orlistat are absorbed systemically; the primary effect is local lipase inhibition within the GI tract after an oral dose. The primary route of elimination is through the faeces.

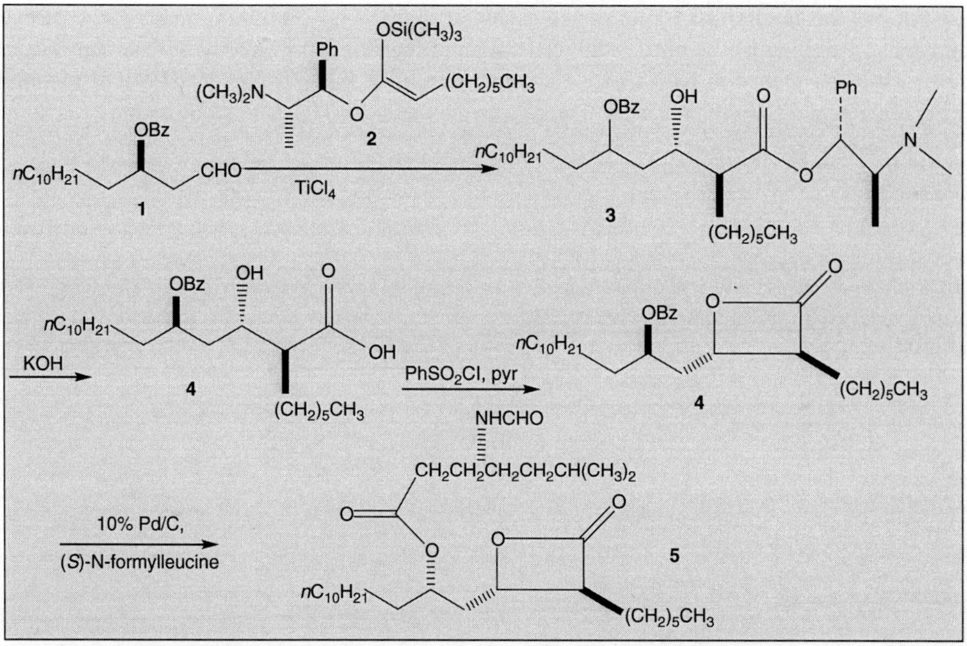

One of the early syntheses of orlistat utilized the Mukaiyama aldol reaction as the key convergent. Therefore, in the presence of TiCl₄, aldehyde **1** was condensed with ketene silyl acetal **2** containing a chiral auxiliary to assemble ester **3** as the major diastereomer in a 3:1 ratio. After removal of the amino alcohol chiral auxiliary via hydrolysis, the α-hydroxyl acid **4** was converted to β-lactone **5** through the intermediacy of the mixed anhydride. The benzyl ether on **5** was unmasked via hydrogenation and the (S)-N-formylleucine side-chain was installed using the Mitsunobu conditions to fashion orlistat.

Sibutramine   1-[1-(4- Chlorophenyl) cyclobutyl]- N,N,3- trimethylbutan- 1-amine

Sibutramine is a neurotransmitter reuptake inhibitor that reduces the reuptake of serotonin (by 53 per cent), nor-epinephrine (by 54 per cent), and dopamine (by 16 per cent), thereby increasing the levels of these substances in synaptic clefts and helping enhance satiety (of being full, of being at maximum capacity); the serotonergic action, in particular, is thought to influence appetite.

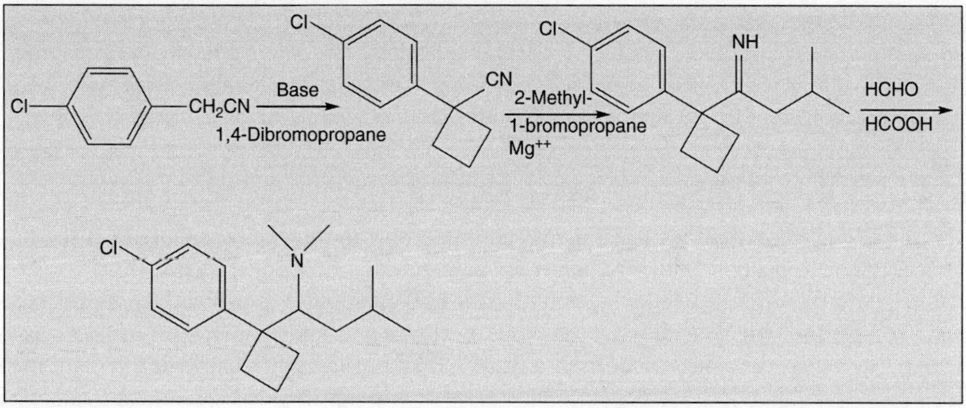

Bis-alkylation of the anion from 4-chlorophenylacetonitrile with 1,4-dibromopropane gives the cyclobutane derivative. Reaction of the nitrile with the Grignard reagent from 2-methyl-1-bromopropane gives the imine, which on reduction with LiAlH$_4$ affords primary amine. *Bis N*-methylation of the amine by means of formaldehyde and formic acid (Clark-Eshweiler reaction) leads to sibutramine.

Rimonabant   5-(4-Chlorophenyl)-1-(2,4-dichloro-phenyl)-4-methyl-*N*-(piperidin-1-yl)-1*H*-pyrazole-3-carboxamide

It is an inverse agonist for the cannabinoid receptor CB1. Its main avenue of effect is reduction in appetite.

> Rimonabant may also be found to be effective in assisting some smokers to quit smoking. Sanofi-Aventis is currently conducting studies to determine the possible value of rimonabant in smoking-cessation therapy.

The ketone reacted with diethyl oxalate in the presence of lithium *bis*(trimethylsilyl)amide to provide the appropriate lithium salt. This was allowed to react with 2,4-dichlorophenylhydrazine hydrochloride to yield substituted 1*H*-pyrazole-3-carboxylic acid ethyl esters. The esters obtained were hydrolysed to carboxylic acid and then converted to the corresponding acid chlorides, which reacted with 1-aminopiperidine in the presence of triethylamine to give rimonabant.

34.2 PDE-5 INHIBITORS FOR MALE ERECTILE DYSFUNCTION (ED)

When a man becomes sexually aroused, increased blood flow to the genital area readies the body for intercourse. The penis becomes enlarged and erect. In men with erectile dysfunction (ED), however, this physical response does not happen as it should. With ED, intercourse is difficult or impossible. Once thought to be a psychological condition, most cases of ED are now known to have a physical cause, such as a disease, an injury, or a side-effect from a drug. Certain medications can interfere with the nerve signals that cause an erection. Hardening of the arteries and high blood pressure can damage blood vessels and interfere with blood flow to the penis. Smoking is a major risk factor for these conditions as well as for ED. Diabetes can damage nerves and interfere with erection. Surgery for prostate cancer may cause ED. Other possible physical causes include alcoholism, liver failure, hormonal abnormalities (such as low testosterone), and neurological disorders. In most cases of ED, even when there is also a definite physical cause, men may feel anxious, guilty, or depressed, which can make the problem worse.

Drug therapy with PDE-5 inhibitors Viagra® **(sildenafil)** was approved by the US Food and Drug Administration in 1998. Men who have heart problems and take medications that help widen the coronary arteries are not good candidates because the drug combination can lower blood pressure dangerously. In 2003 the FDA approved Levitra® **(vardenafil)** for the treatment of ED. The way it works is similar to sildenafil. The latest medication approved for ED is Cialis® **(tadalafil)**. It differs from the other two drugs only because its effects persist for 36 hours, rather than just a few hours.

> Sildenafil was synthesised by a group of pharmaceutical chemists working at Pfizer's Sandwich, Kent, research facility in England. It was initially studied for use in hypertension and angina pectoris. Phase I clinical trials suggested that the drug had little effect on angina, but that it could induce marked penile erections. Pfizer, therefore, decided to market it for erectile dysfunction, rather than for angina. The drug was patented in 1996, and approved for use in erectile dysfunction by the US Food and Drug Administration on 27 March, 1998.

34.2.1 Mechanism of Action of PDE-5 Inhibitors

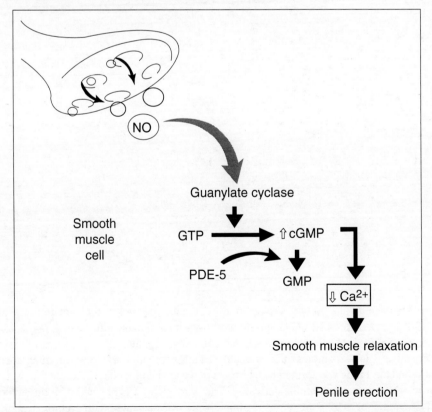

FIGURE 34.1

Sexual stimulation leads to the release of nitric oxide (NO) within the blood vessels of the penis, where it stimulates guanylate cyclase to increase cGMP levels in the corpus cavernosum. Elevated levels of peripheral cGMP, in turn, promote the efflux of Ca^{2+} ions from the cavernosa smooth muscle cells. This induces muscle relaxation, facilitates blood flow into the corpora cavernosa, and thereby helps to obtain and maintain penile erection (Figure 34.1) .It turns out there are high levels of PDE-5 in the corpus cavernosum muscle of the penis, which is able to degrade cGMP and cause termination of the erection. Nitric oxide production may be impaired in patients suffering from erectile dysfunction, leading to low levels of cGMP, which can be quickly degraded by PDE-5. Inhibition of PDE-5 by sildenafil, vardenafil, and tadalafil slows the breakdown of cGMP, allowing for higher concentrations to build up in the corpus cavernosum leading to an erection.

Sildenafil 1-[4-Ethoxy-3-(6,7-dihydro-1-methyl-7-oxo-3-propyl-1*H*-pyrazolo[4,3-*d*]pyrimidin-5-yl) phenylsulphonyl]-4-methylpiperazine
In clinical trials, the most common adverse effects of sildenafil use included headache, flushing, dyspepsia, nasal congestion, and impaired vision, including photophobia and blurred vision. Some sildenafil users have complained of seeing everything tinted blue (cyanopsia). Some complained of blurriness and loss of peripheral vision.

The synthesis of sildenafil started with the reaction of 2-pentanone and diethyloxalate in the presence of sodium ethoxide to give diketoester. Condensation of this with hydrazine afforded the pyrazole derivative. Selective methylation of the 1-pyrazole nitrogen with dimethylsulphate followed by hydrolysis of the ester gave acid derivative. The 4-position of the pyrazole was nitrated with a mixture of fuming nitric acid and sulphuric acid, and the amide was formed by treatment of the acid with thionyl chloride followed by ammonium hydroxide to furnish carboxamide derivative. The nitro group was reduced with tin dichloride dihydrate in refluxing ethanol to give the requisite amine, which was acylated with 2-ethoxybenzoyl chloride using triethylamine and dimethylaminopyridine as base in dichloromethane. The resulting benzamido carboxamide was cyclized to the pyrazolo[4,3-d]pyrimidin-'1-one using aqueous sodium hydroxide and hydrogen peroxide in ethanol underreflux. Subsequently, it was selectively sulphonated para to the ethoxy group with chlorosulphonic acid to yield the sulphonyl chloride, which was reacted with 1-methyl piperazine in ethanol to provide sildenafil.

Vardenafil 4-[2-Ethoxy-5-(4-ethylpiperazin-1-yl)sulphonyl-phenyl]-9-methyl-7-propyl-3,5,6,8-tetrazabicyclo[4.3.0] nona-3,7,9-trien-2-one

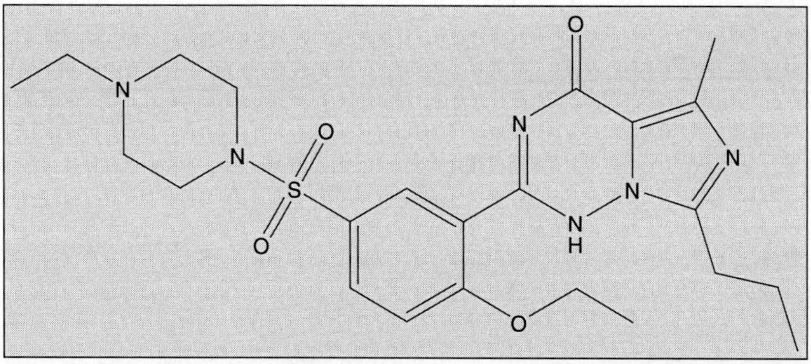

The structure of vardenafil looks very similar to sildenafil, except that vardenafil contains a slightly different purine-isosteric imidazo[5,1-f][1,2,4]triazin-4(3H)-one heterocyclic core.

Tadalafil (6R-trans)-6-(1,3-Benzodioxol-5-yl)- 2,3,6,7,12,12a-hexahydro-2-methyl-pyrazino [1',2':1,6] pyrido[3,4-b]indole-1,4-dione

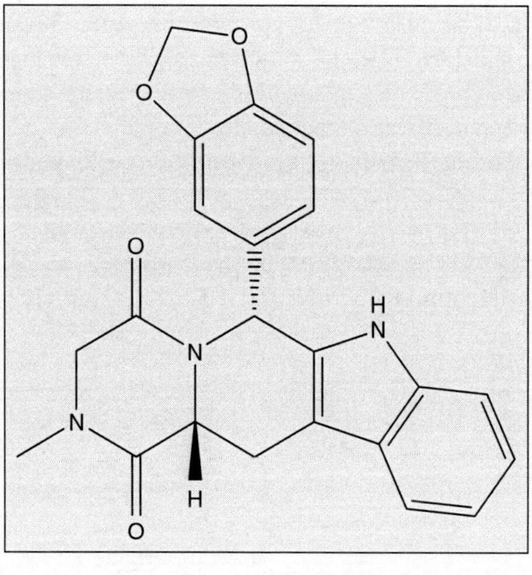

Its pharmacologic distinction is its longer half-life (17.50 hours) compared to sildenafil (4.0–5.0 hours) and vardenafil (4.0–5.0 hours), resulting in longer duration of action.

34.3 ANTI-MIGRAINE DRUGS

Migraine is a neurological syndrome characterized by altered bodily perceptions, headaches, and nausea. The pain of a migraine headache is often described as an intense pulsing or throbbing pain in one area of the head. It is often accompanied by extreme sensitivity to light and sound, nausea, and vomiting. Migraine is three times more common in women than in men. Some individuals can predict the onset of a migraine because it is preceded by an 'aura', visual disturbances that appear as flashing lights, zigzag lines, or a temporary loss of vision. People with migraine tend to have recurring attacks triggered by a lack of food or sleep, exposure to light, or hormonal irregularities (only in women). Anxiety, stress, or relaxation after stress can also be triggers. For many years, scientists believed that migraines were linked to the dilation and constriction of blood vessels in the head. Investigators now believe that migraine is caused by inherited abnormalities in genes that control the activities of certain cell populations in the brain.

Non-steroidal anti-inflammatory drugs (NSAIDs) such as aspirin, paracetamol, ibuprofen, and naproxen may be effective in relieving mild migraine pain, but do not work in the majority of migraine patients. Ergot alkaloids have been used for over a century for the treatment of migraines. Ergotamine is a powerful vasoconstrictor, but its effects are long-lasting and are not specific to the cranial vessels, which leads to side-effects.

Long after the discovery of the ergots, it was recognized that their beneficial effects resulted from activation of 5-HT1-like receptors, specifically 5-$HT_{1B/1D}$ receptors. This led to the development of **sumatriptan**, a selective 5-$HT_{1B/1D}$ agonist, as the first specific anti-migraine medication. It is believed that 5-$HT_{1B/1D}$ agonists (*triptans*) elicit their anti-migraine action by selective vasoconstriction of excessively dilated intracranial, extracerebral arteries and/or inhibiting the release of inflammatory neuropeptides from perivascular trigeminal sensory neurons. It has been suggested that 5-HT_{1B} receptor activation results in vasoconstriction of intracranial vessels, while inhibition of neuropeptide release is mediated via the 5-HT_{1D} receptor. Selective 5-HT_{1D} agonists have recently been identified and are being studied to determine the relative importance of these receptor-mediated events on the anti-migraine activity.

Sumatriptan has proved to be an effective anti-migraine drug; however, it has several limitations including low bioavailability, short half-life, and a high headache recurrence rate. Since the introduction of sumatriptan, several second-generation triptans have entered the marketplace, including zolmitriptan, naratriptan, rizatriptan, almotriptan, frovatriptan, and eletriptan. The second-generation triptans generally have improved pharmacokinetics, higher oral bioavailability, and a longer plasma half-life.

> 'Triptan' is a word commonly used for a class of anti-migraine drugs that are selective 5- hydroxytryptamine1B/1D (5-HT1B/1D) agonists.

Sumatriptan: 1-[3-(2-Dimethylaminoethyl)-1*H*-indol-5-yl]- N-methyl-methanesulphonamide

The synthesis began with hydrogenation of *N*-methyl-4-nitrobenzenemethanesulphonamide to give amine, which was treated with sodium nitrite to give a diazonium salt, which was reduced with tin chloride to give the hydrazine. Condensation of the hydrazine with 4,4-dimethoxy-N,N-dimethylbutylamine aqueous HCl provided the hydrazone. Hydrazone was treated with polyphosphateester (PPE) in refluxing chloroform to affect the Fischer indole reaction, resulting in the formation of sumatriptan.

34.4 ANTI-ASTHMATICS

Asthma is a very common chronic disease involving the respiratory system in which the airways constrict, become inflamed, and are lined with excessive amounts of mucus, often in response to one or more triggers. These episodes may be triggered by such things as exposure to an environmental stimulant such as an allergen, environmental tobacco smoke, cold or warm air, perfume, pet dander, moist air, exercise or exertion, or emotional stress. This airway narrowing causes symptoms such as wheezing, shortness of breath, chest tightness, and coughing.

The pharmacological control of asthma can be achieved in most patients with chromones and inhaled glucocorticoids. Corticosteroids are considered the mainstay of asthma therapy. The clinical efficacy of these agents is probably the result of their inhibitory effect on leukocyte recruitment into the airways—e.g., **beclomethasone** (refer to Chapter 33 on 'Steroids') and **fluticasone**. (*See Fig. 34.2 in the coloured set of pages*).

The β_2-adrenergic agonists are the most prescribed class of drugs for the treatment of asthma. These drugs produce their effects through stimulation of specific β_2-adrenergic receptors located in the plasma membrane, resulting in alterations in adenylyl cyclase and elevations in intracellular AMP. Cyclic AMP is responsible for the relaxation of smooth muscle with bronchodilation in the bronchi and a reduction in mucus viscosity—e.g., **salmeterol**.

The newest therapy available for the treatment of asthma arises from the recognition of the role of the leukotrienes (LTs) in the initiation and propagation of airway inflammation. The evidence to support the role of leukotrienes in bronchial asthma includes: a) cells known to be involved in asthma produce LTs; b) cysteinyl LTs (LTC4, LTD4, and LTE4) cause airway abnormalities and mimic those seen in asthma; and c) the production of LTs is increased in the airways of people with asthma. The leukotrienes exert their effects through G-protein coupled receptors regulating a signal transduction pathway that ultimately causes calcium release from the cells. There are two classes of leukotriene receptors, BLT1 receptors and CysLT receptors 1 and 2. It is these latter receptors that mediate the actions of the cysteinyl leukotrienes in asthma. **Montelukast** is a prime example of this class and is typified by high intrinsic potency, good oral bioavailability, and long duration of action.

Fluticasone S-(Fluoromethyl) (6S,8S,9R,10S,11S,13S,14S,16R,17R)-6,9-difluoro-11,17-dihydroxy-10,13,16-trimethyl-3-oxo-6,7,8,11,12,14,15,16-octahydrocyclopenta[a]phenanthrene-17-carbothioate

The synthesis of fluticasone utilizes commercially available flumethasone. Oxidation of flumethasone with periodic acid gave the etianic acid, whose imidazolide when treated with hydrogen sulphide gas gave the carbothioic acid. Treatment with excess propionyl chloride followed by aminolysis of the mixed anhydride with diethylamine gave an intermediate. This on treatment with bromofluoromethane affords fluticasone.

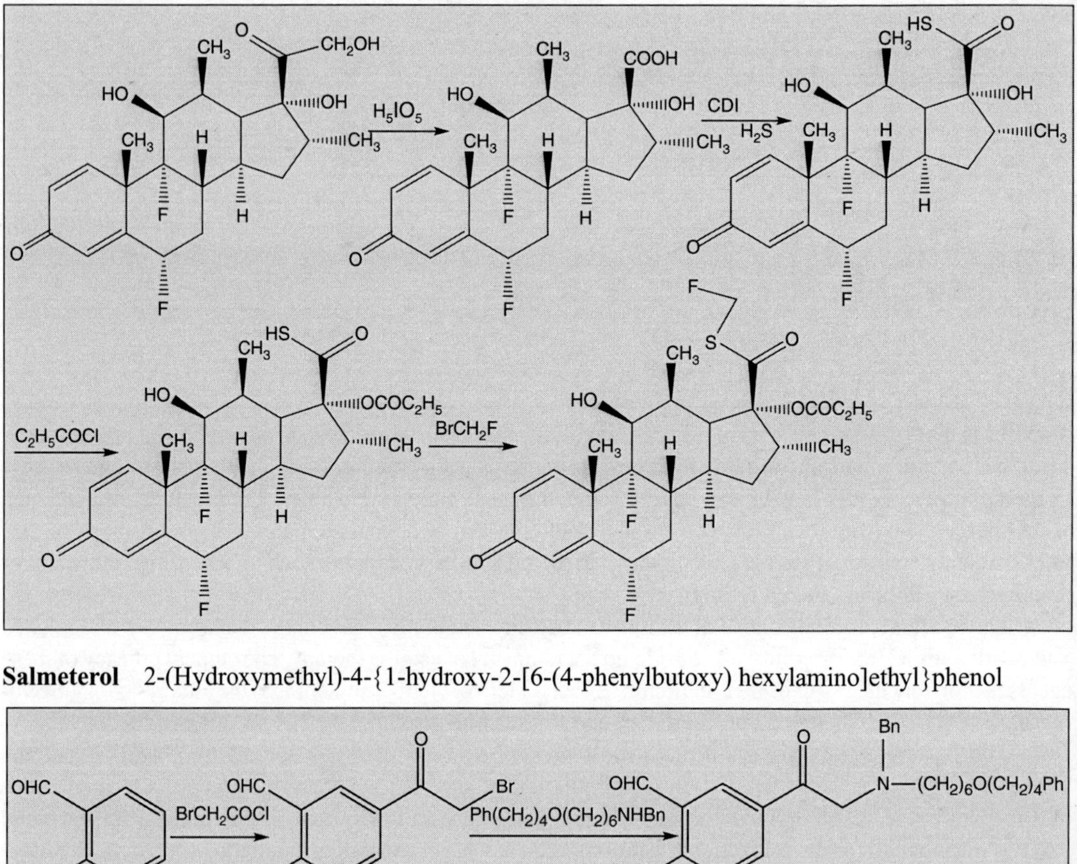

Salmeterol 2-(Hydroxymethyl)-4-{1-hydroxy-2-[6-(4-phenylbutoxy) hexylamino]ethyl}phenol

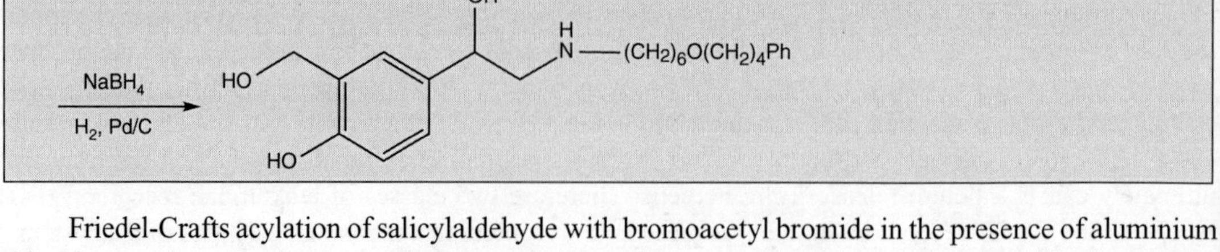

Friedel-Crafts acylation of salicylaldehyde with bromoacetyl bromide in the presence of aluminium chloride gave the acetophenone. Alkylation of amine with bromoacetyl in refluxing acetonitrile gave the

ketone. Reduction of ketone with sodium borohydride in methanol followed by catalytic hydrogenolysis of the benzyl group over 10 per cent Pd/C gave salmeterol.

Montelukast 2-[1-[[(1R)-1-[3-[2-(7-Chloroquinolin-2-yl)ethenyl]phenyl]-3-[2-(2-hydroxypropan-2-yl)phenyl]propyl]sulphanylmethyl]cyclopropyl]acetic acid

Montelukast is a CysLT$_1$ antagonist; that is, it blocks the action of leukotriene D4 on the cysteinyl leukotriene receptor CysLT$_1$ in the lungs and bronchial tubes by binding to it.

34.5 THYROID AND ANTI-THYROID DRUGS

Thyroxine, or 3,5,3',5'-Tetraiodothyronine (often abbreviated as T$_4$), a form of thyroid hormones, is the major hormone secreted by the follicular cells of the thyroid gland. T$_4$ is involved in controlling the rate of metabolic processes in the body and influencing physical development.

Thyroid hormones are used clinically primarily to treat hypothyroidism. This disease is characterized by a decrease or lack of endogenic thyroid hormone secretion. When originating in childhood, it can be clinically described as cretinism (infantile hypothyroidism), and in adults as myxedema (adult hypothyroidism), which is expressed in a loss of mental or physical ability to work, suppression of metabolic processes in the body, and oedema. Currently, a very small number of various drugs such as drugs of animal thyroid glands and synthetic drugs is used to treat hypothyroidism. They are: thyroidin-dried thyroid, which is made from cow, sheep, and pig thyroid glands; thyroglobulin, a purified extract of pig thyroid glands; synthetic drugs levothyroxine and lyothyronin, and also lotrix, a mixture of synthetic levothyroxine and lyothyronin in a 4:1 ratio.

Levothyroxine L-3-[4-(4-Hydroxy-3,5-diiodophenoxy)-3,5-diiodophenyl] alanine
It is synthesised from 4-hydroxy-3-iodo-5-nitrobenzaldehyde. Reacting this with benzenesulphonylchloride in pyridine gives the corresponding benzenesulphonate, the benzenesulphonyl group of which is easily replaced with a 4-methoxyphenyloxy group upon reaction with 4-methoxyphenol. The resulting 3-iodo-4-(4-methoxyphenoxy)benzaldehyde is reacted further with *N*-acetylglycine in the presence of sodium acetate in a Knoevenagel reaction, in which the resulting ylidene compound cyclizes to an oxazolone derivative. The oxazolone ring of this compound is opened upon reaction with sodium methoxide,

forming the desired cinnamic acid derivative. The nitro group of this product is reduced to an amino group by hydrogen in the presence of a Raney nickel catalyst, forming the corresponding amine, and subsequent diazotization and replacement of the diazo group of this with iodine gives the methyl ester of a-acetamido-3,5-diiodo-4-(4-methoxyphenoxy)crotonic acid. The resulting compound undergoes simultaneous reaction with hydrogen iodide and phosphorous in acetic acid, in which the double bond in the crotonic acid is reduced, and the methoxy protection is removed from the phenol ring. During this, a simultaneous hydrolysis of the acetyl group on the nitrogen atom also takes place, forming D,L-3,5-diiodothyronine. The amino group in this product is once again protected by the reaction with formic acid in the presence of acetic anhydride, which gives D,L-N-formyl-3,5-diiodothyronine. Separation of isomers in the resulting racemic mixture is accomplished using brucine, giving D-(-)-N-formyl-3, 5-diiodothyronine L-(-)-N-formyl-3,5-diiodothyronine. The protecting formyl group is hydrolysed using hydrobromic acid,giving L-(-)-3,5-diiodothyronine, which undergoes direct iodination using iodine in the presence of potassium iodide in aqueous methylamine, to give the desired levothyroxine.

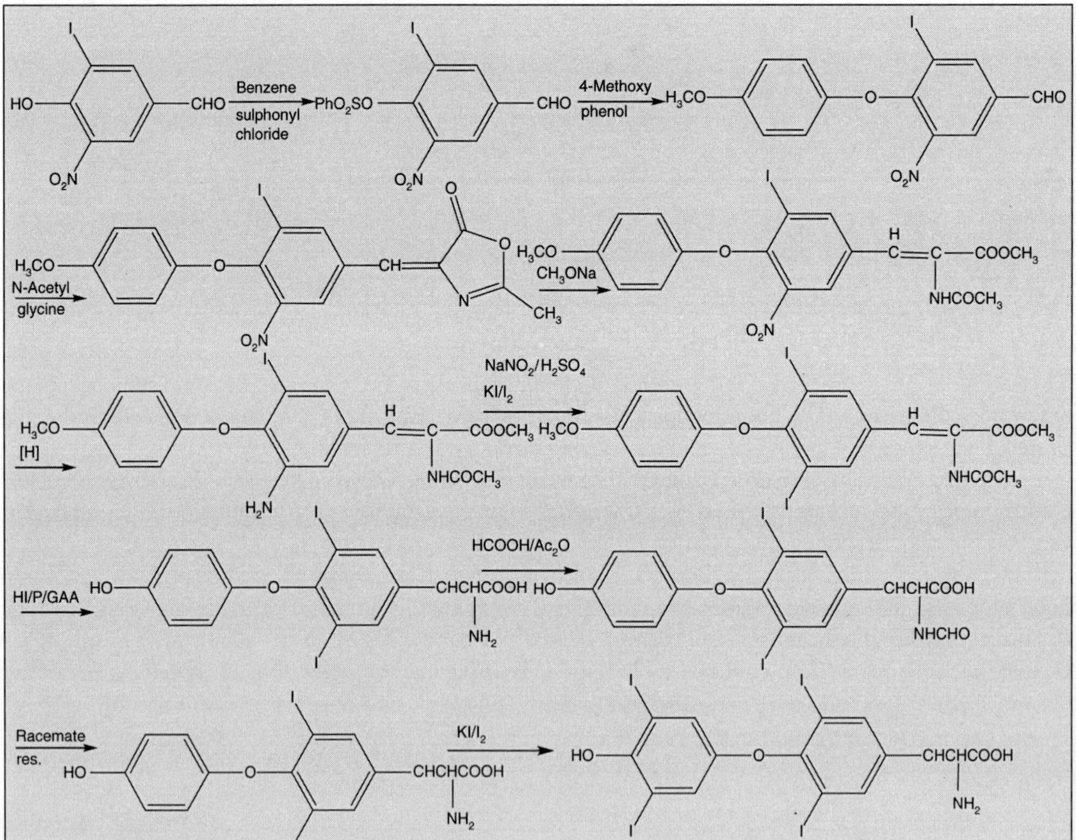

Levothyronine L-3-[4-(4-Hydroxy-3-iodophenoxy)-3,5-diiodophenyl]alanine

It is synthesised in the exact same manner using one equivalent of iodine during iodination of L-(−)-3,5-diiodothyronine. It has properties of levothyroxine; however, it acts faster and binds less with blood proteins.

In a hyperfunctioning of the thyroid gland, secretion of an excess quantity of thyroid hormones leads to a hyperthyroid condition (Basedow's disease, goitre). In this condition, drugs (**anti-thyroid**) are used that suppress production of thyrotropic hormones in the anterior lobe of the hypophysis (diiodotyrosine) and in the thyroid gland (propylthiouracil, methylthiouracil, methimazole, and carbimazole), as well as drugs that destroy thyroid gland follicles (radioactive iodine).

Diiodotyrosine 3,5-Diiodotyrosine

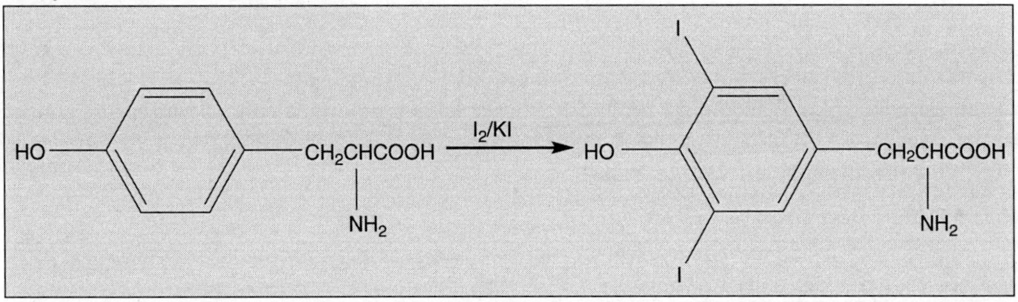

It is synthesised by directly iodinating tyrosine with iodine in the presence of sodium iodide in aqueous ethylamine.

Methylthiouracil 6-Methyl-2-thio-2,4-(1H,3H)-pyrimidindione

It is synthesised by condensing ethyl acetoacetate with thiourea in the presence of sodium ethoxide.

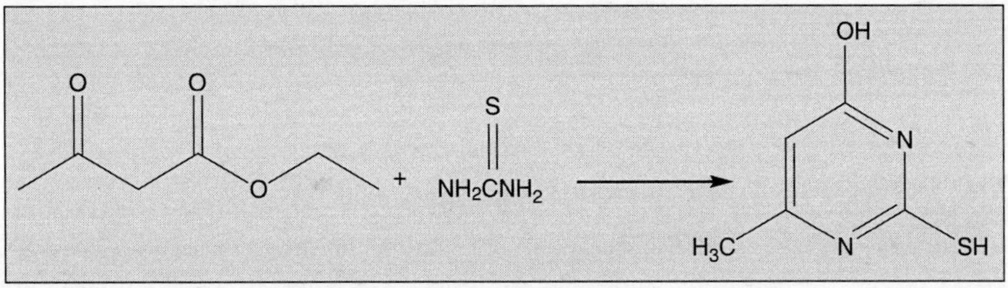

Propylthiouracil 6-Propyl-2-thio-2,4-(1H,3H)-pyrimidindione

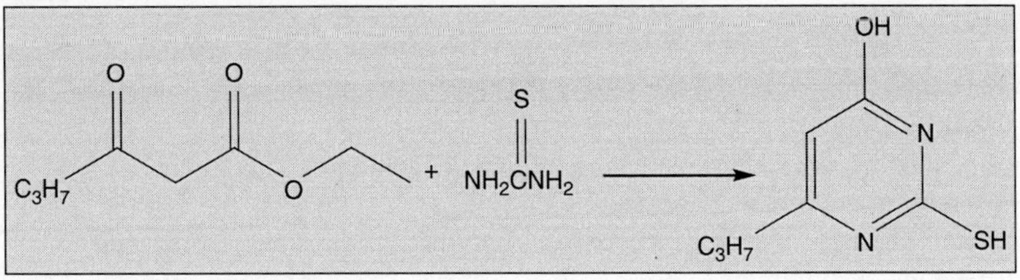

It is synthesised by condensating ethyl butyroacetate with thiourea in the presence of sodium ethoxide.

Methimazole 1-Methyl-2-imidazolthiol

It is synthesised by reacting aminoacetic aldehyde diethylacetal with methylisothiocyanate and subsequent hydrolysis of the acetal group of the resulting disubstituted urea derivative by a solution of sulphuric acid, during which a simultaneous cyclization reaction takes place, forming the imidazole ring of the desired methimazole.

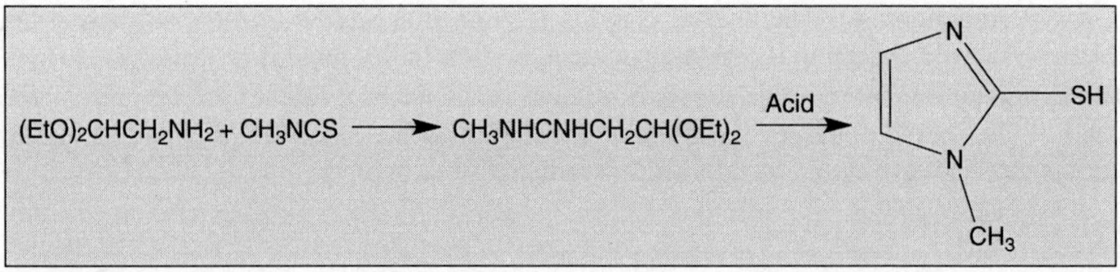

Carbimazole Ethyl ester of 3-methyl-2-thioimidazolin-1-carboxylic acid
It is synthesised by a simultaneous reaction of ethylenacetal of bromoacetaldehyde with methylamine and potassium isocyanate, forming 3-methyl-2-imidazolthione, which is further acylated at the nitrogen atom by ethyl chloroformate, giving carbimazole.

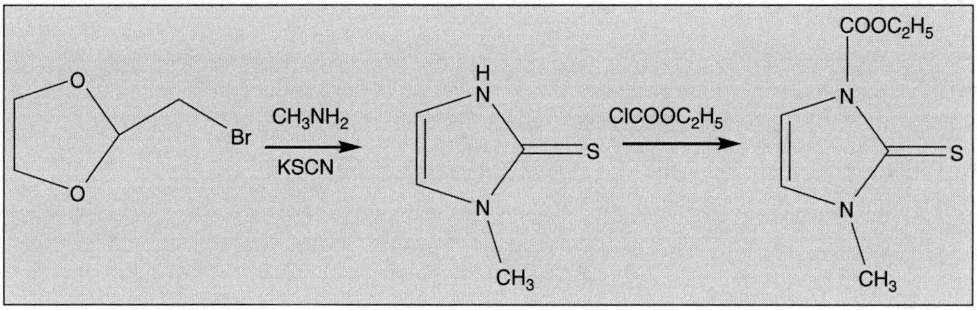

34.6 ANTI-PLATELET DRUGS (ANTITHROMBOTICS)

Anti-platelet drugs decrease platelet aggregation and inhibit thrombus formation. They are effective in the arterial circulation, where anticoagulants have little effect. They are widely used in primary and secondary prevention of thrombotic cerebrovascular or cardiovascular disease. The most important anti-platelet drugs are:

1. Cyclooxygenase inhibitors: Aspirin (Refer to Chapter 12 on 'Antipyretics and Non-Steroidal Anti-Inflammatory Drugs')
2. Adenosine diphosphate (ADP) receptor inhibitors: Clopidogrel, Ticlopidine
3. Phosphodiesterase inhibitors: Cilostazol
4. Glycoprotein IIB/IIIA inhibitors: Abciximab, Eptifibatide, Tirofiban, Defibrotide
5. Adenosine reuptake inhibitors: Dipyridamole (Refer to Chapter 19 on 'Antianginal Drugs')

Adenosine diphosphate (ADP) receptor inhibitors Both ticlopidine and clopidogrel inhibit platelet aggregation induced by adenosine diphosphate (ADP), a platelet activator that is released from red blood cells, activated platelets, and damaged endothelial cells. The mechanisms of action for both ticlopidine and clopidogrel are the same through the antagonism of the P2Y12 purinergic receptor and prevention of binding of ADP to the P2Y12 receptor.

Ticlopidine: 5-(*o*-Chlorobenzyl)-4,5,6,7-tetrahydrothieno[3,2-c]pyridine

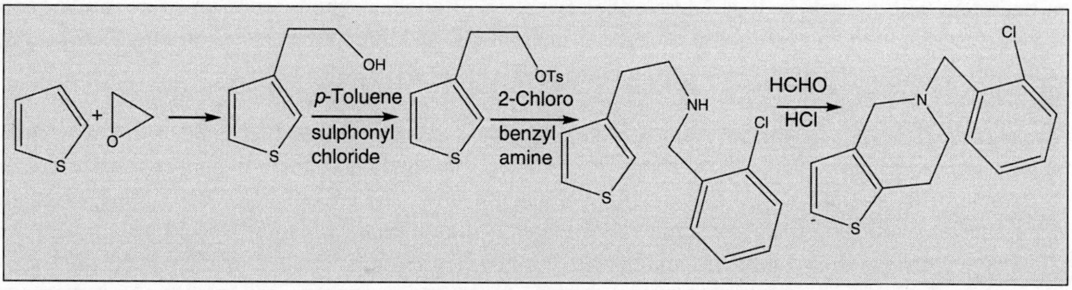

Alkylating thiophene with ethylene oxide forms 2-(2'-hydroxy)ethylthiophene, which reacts with *p*-toluenesulphonic acid chloride to give the corresponding tosylate. Substitution of the tosyl group using 2-chlorobenzylamine gives an amine, which under reaction conditions for chloromethylation cyclizes to the desired ticlopidine.

Clopidogrel: (+)-(*S*)-Methyl 2-(2-chlorophenyl)- 2-(6,7-dihydrothieno[3,2-c]pyridin-5(4*H*)-yl)acetate

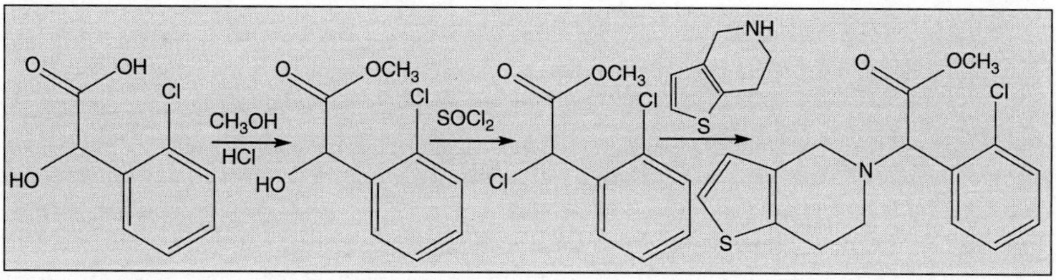

Methyl mandelate was prepared by refluxing chlorinated mandelic acid with methanol in the presence of concentrated HCl. Chlorination of this ester with thionyl chloride gave methyl α-chloro-(2-chlorophenyl) acetate. S_N2 displacement of acetate intermediate with thieno[3,2-c]pyridine affords (+)-clopidogrel.

Phosphodiesterase inhibitor
Cilostazol: 6-[4-(1-Cyclohexyl-1*H*-tetrazol-5-yl)butoxy]-3,4-dihydro-2(1*H*)-quinolinone
Cilostazol is a selective cAMP phosphodiesterase inhibitor. It inhibits platelet aggregation and is a direct arterial vasodilator.

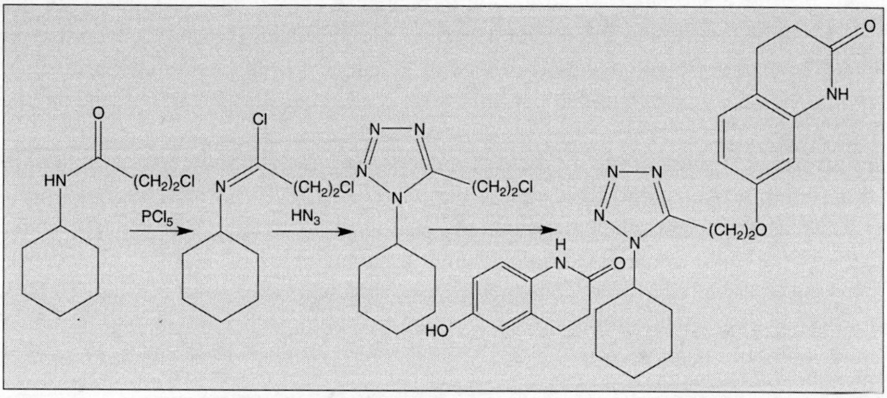

Reaction of the cyclohexyl amide derivative with PCl_5 leads to the formation of the imino chloride, which on treatment with hydrazoic acid leads to the formation of terazole; the reaction involves initial addition–elimination to the imino chloride function. Internal cycloaddition then closes the ring. This is used to alkylate the phenoxate of 3,4-dihydroquinolin-2(1H)-one to afford cilostazol.

Glycoprotein IIB/IIIA inhibitors Glycoprotein IIb/IIIa is an integrin found on platelets. It is a receptor for fibrinogen and aids in platelet activation. The complex is formed via calcium-dependent association of gpIIb and gpIIIa, a required step in normal platelet aggregation and endothelial adherence. **Abciximab** is a monoclonal antibody made from the Fab fragments of an immunoglobulin. **Eptifibatide** is a cyclic heptapeptide derived from a protein found in the venom of the southeastern pygmy rattlesnake (*Sistrurus miliarius barbouri*). It belongs to the class of the so-called arginin-glycin-aspartate-mimetics and reversibly binds to platelets. **Tirofiban** [(S)-2-(butylsulphonamino)-3-(4-[4-(piperidin-4-yl)butoxy]phenyl) propanoic acid] is a synthetic non-peptide inhibitor. **Defibrotide** is a deoxyribonucleic acid derivative (single-stranded) derived from cow lung or porcine mucosa.

34.7 IMMUNOPHARMACOLOGICAL DRUGS

The ability of the body to independently protect itself from certain diseases is called immunity. In a medical sense, immunity is a state of having sufficient biological defences to avoid infection, disease, or other unwanted biological invasion. Various drugs are capable of affecting specific immune reactions. They can both increase the general resistivity of the body or its nonspecific immunity, as well as suppress the body's immune reactions. It is evident that immunopharmacological drugs are of great significance in diseases of the immune system, organ transplants, viral infections, and, in particular, in the treatment of AIDS.

Immunostimulants Enhancing the overall resistivity of the body is observed upon treatment with a number of known drugs: immunostimulants (caffeine, phenamine, methyluracil), vitamins (retinol, ascorbic acid, vitamins of group B), nucleic acid derivatives, proteins such as lyphokines, in particular, interleukin-2 and 3, and the glycoprotein interferon. Practically, the only purely synthetic immunostimulant drug that is used is levamisole, which was initially proposed as an anthelminthic agent.

Levamisole (Synthesis in Chapter 30 on 'Antiprotozoal Agents') has immunomodulating activity. It is believed that it regulates cellular mechanisms of the immune system, and the mechanism of its action may be associated with activation and proliferative growth of T-lymphocytes, increased numbers of monocytes and activation of macrophages, and also with increased activity and haemotaxis of neutrophylic granulocytes. Levamisole exhibits anthelmint action. It also increases the body's overall resistivity and restores altered T-lymphocyte and phagocyte function. It can also fulfil an immunomodulatory function by strengthening the weak reaction of cellular immunity, weakening strong reaction, and having no effect on normal reaction.

Immunosuppressants These drugs suppress immunogenesis and antibody production (which is especially important in transplantation of various tissues and organs, during which the body produces antibodies that cause death of transplanted tissue), and also used for treating a few autoimmune diseases. Substances of various pharmacological groups exhibit immunodepressive activity: glucocorticoids (cortisone, prednisone, methylprednisolone, betamethasone, dexamethasone, triamcinolone [refer to Chapter 33 on 'Steroids']), cytotoxics (azathioprine and cyclophosphamide [refer to Chapter 31 on 'Anticancer Agents']), and antibiotics (cyclosporine).

Azathioprine: 6-[(1-Methyl-4-nitroimidazol-5-yl)thio]purine

It is synthesised by heteroarylation of the sulphhydrile group of 6-mercaptopurine with 5-chloro-1-methyl-4-nitroimidazol in the presence of sodium acetate as a weak base.

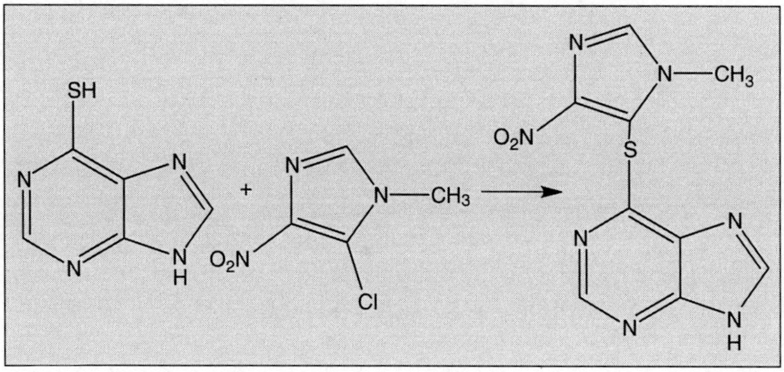

Cyclosporine:

Cyclosporine, the main form of the drug, is a cyclic nonribosomal peptide of 11 amino acids (an undecapeptide) produced by the fungus Tolypocladium inflatum Gams, and contains D-amino acids, which are rarely encountered in nature.

It was the first immunosuppressive drug that allowed selective immunoregulation of T cells without excessive toxicity. Cyclosporin was isolated from the fungus *Tolypocladium inflatum*. Cyclosporin is approved for use in organ transplantation to prevent graft rejection in kidney, liver, heart, lung, and combined heart–lung transplants. It is used to prevent rejection following bone marrow transplantation and in the prophylaxis of host-versus-graft disease. It is also used in the treatment of psoriasis, atopic dermatitis, rheumatoid arthritis, and nephrotic syndrome.

34.8 ANTI-COUGH AND EXPECTORANTS

Cough is a sudden and often repetitively occurring defence reflex that helps to clear the large breathing passages from excess secretions, irritants, foreign particles, and microbes. The cough reflex consists of three phases: an inhalation, a forced exhalation against a closed glottis (the complex of the vocal folds), and a violent release of air from the lungs following opening of the glottis, usually accompanied by a distinctive sound. Coughing can happen voluntarily as well as involuntarily, though for the most part, involuntarily. Anti-cough drugs can have an effect at the 'cough centre' level in the medulla, as well as an effect on various regions of the tracheobronchial tree. Drugs that exhibit anti-cough effects are divided into two groups. They are **centrally acting drugs:** narcotic anti-cough drugs or opiates such as codeine and hydrocodone (refer to chapter 11 on 'Narcotic Analgesics'), as well as various groups of drugs displaying **both central and peripheral effects** that suppress coughing, and the so-called non-narcotic anti-cough drugs (dextromethorphan, benzonatate).

Dextromethorphan (9α,13α,14α)-3-Methoxy-17-methylmorphinane

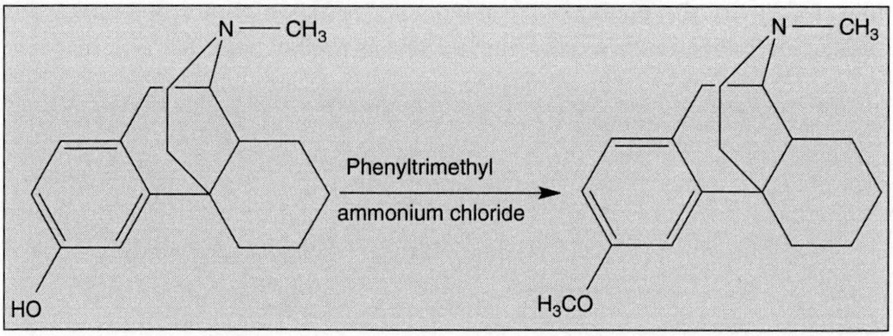

It is synthesised from (−)-3-hydroxy-*N*-methylmorphinane by methylating the phenolic hydroxyl group using phenyltrimethylammonium chloride and sodium methoxide in methanol. The resulting racemic product (−)-3-methoxy-*N*-methylmorphinane is separated into isomers using D-tartaric acid, which produces dextromethorphan.

Benzonatate *p*-Butylaminobenzoate 2,5,8,11,14,17,20,23,26-nonaoctacozan-28-ol

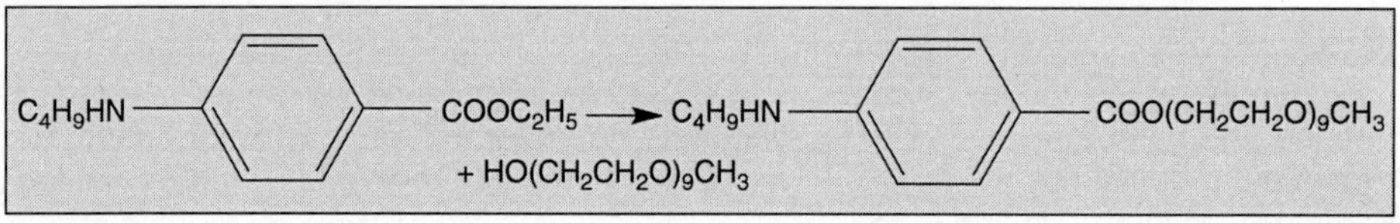

It is synthesised by re-esterifying the ethyl ester of 4-butylaminobenzoic acid with the monomethyl ether nonaethyleneglycol.

34.9 DIAGNOSTIC AGENTS

Medical imaging refers to the techniques and processes used to create images of the human body (or parts thereof) for clinical purposes (medical procedures seeking to reveal, diagnose, or examine disease) or medical science (including the study of normal anatomy and physiology). The practice of clinical diagnostic radiology has been made possible by advances not only in diagnostic equipment and investigative techniques, but also in the contrast media that permit visualization of the details of the internal structure or organs that would not otherwise be demonstrable.

Bentiromide (S)-4-((2-(Benzoylamino)-3-(4-hydroxyphenyl) -1-oxopropyl)amino)benzoic acid

It is synthesised by reaction between ethyl-4-aminobenzoate and N-benzoyl tyrosine to give amide, which on hydrolysis affords bentiromide.

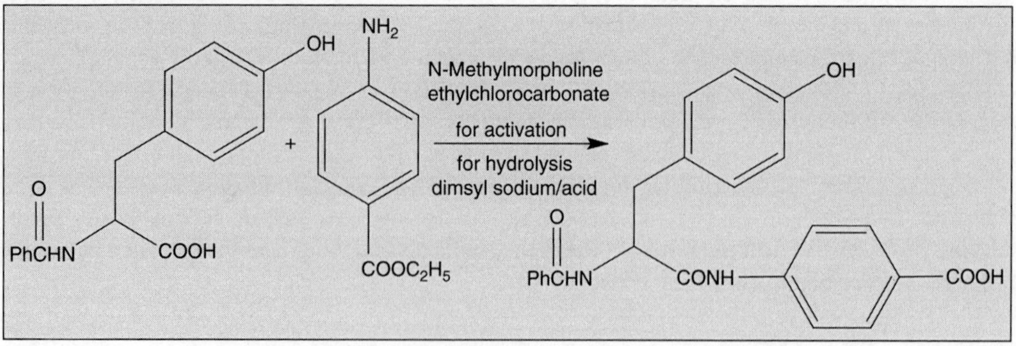

It is a peptide used as a screening test for exocrine pancreatic insufficiency and to monitor the adequacy of supplemental pancreatic therapy. It is given by mouth as a non-invasive test. The amount of 4-aminobenzoic acid and its metabolites excreted in the urine is taken as a measure of the chymotrypsin-secreting activity of the pancreas.

Congo red Sodium salt of benzidinediazo-*bis*-1-naphtylamine-4-sulphonic acid
It is synthesised by diazitizing amino groups of benzidine followed by coupling with 4-amino-1-naphthalene sulphonic acid.

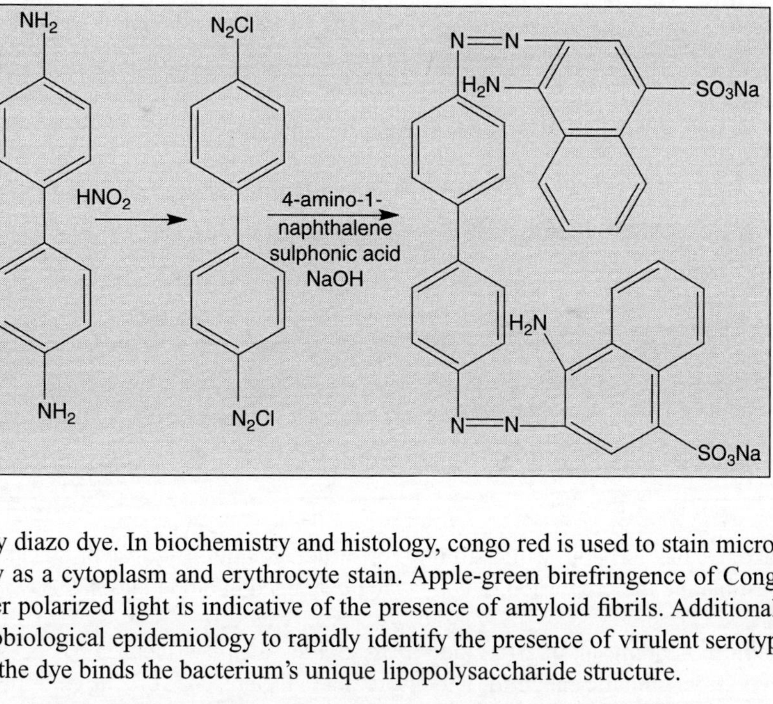

It is a secondary diazo dye. In biochemistry and histology, congo red is used to stain microscopic preparates, especially as a cytoplasm and erythrocyte stain. Apple-green birefringence of Congo red-stained preparates under polarized light is indicative of the presence of amyloid fibrils. Additionally, congo red is used in microbiological epidemiology to rapidly identify the presence of virulent serotype 2a *Shigella flexneri*, where the dye binds the bacterium's unique lipopolysaccharide structure.

Diatrizoic acid (Amidotrizoic acid) 3,5-Diacetamido-2,4,6-triiodobenzoic acid
Reduction of 3,5-dinitro benzoic acid gives 3,5-diamino benzoic acid, which on iodination with iodine monochloride gives triiodo derivative. Acetylation of diamino groups gives diatrizoic acid.

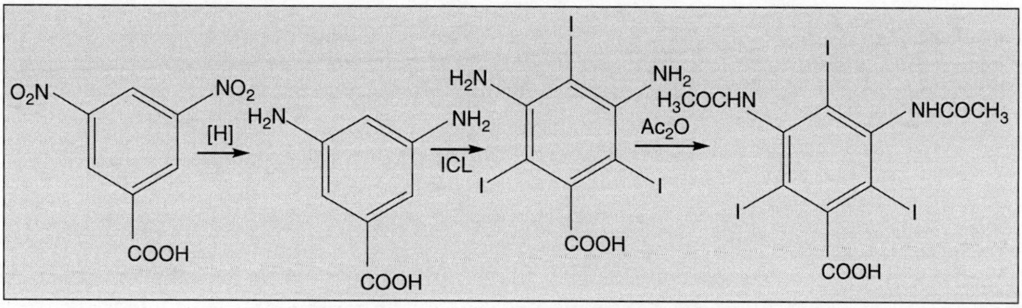

It is a radiocontrast agent containing iodine. Diatrizoic acid may be used as an alternative to barium sulphate for medical imaging of the gastrointestinal tract.

Evans Blue Sodium salt of 6,6'-(3,3'-dimethylbiphenyl-4,4'-diyl)bis(diazene-2,1-diyl)bis(4-amino-5-hydroxynaphthalene-1,3-disulphonate)

It is synthesised by diazotizing amino groups of 3,3'-dimethylbiphenyl-4,4'-diamine, followed by coupling with 1-amino-8-naphthol-2,4-disulphonicacid.

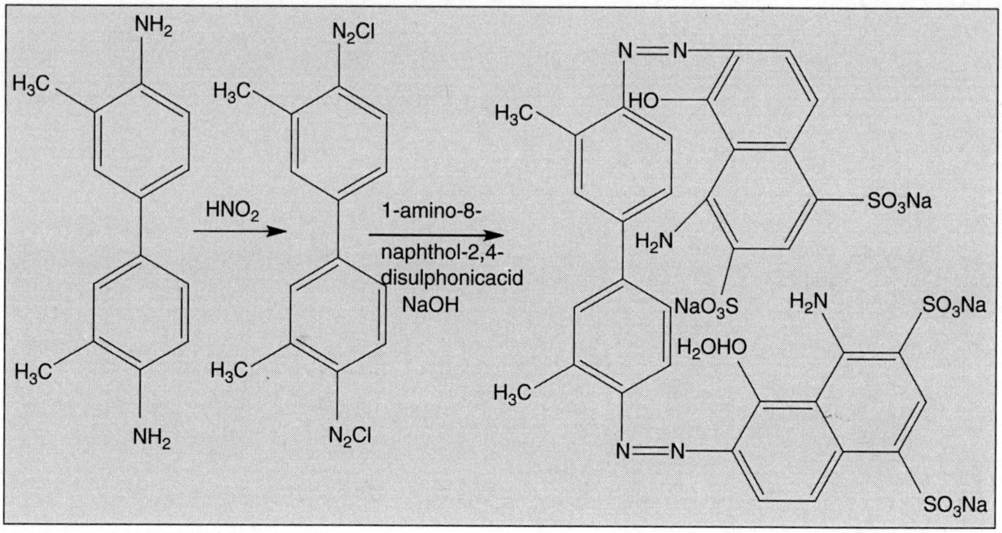

It is an azo dye that has a very high affinity for serum albumin. Because of this, it can be useful in physiology in estimating the proportion of body water contained in blood plasma.

Fluorescein sodium Sodium salt of 2-(6-oxido-3-oxo-3H-xanthen-9-yl)benzoate
It can be prepared from phthalic anhydride and resorcinol in the presence of zinc chloride via the Friedel-Crafts reaction.

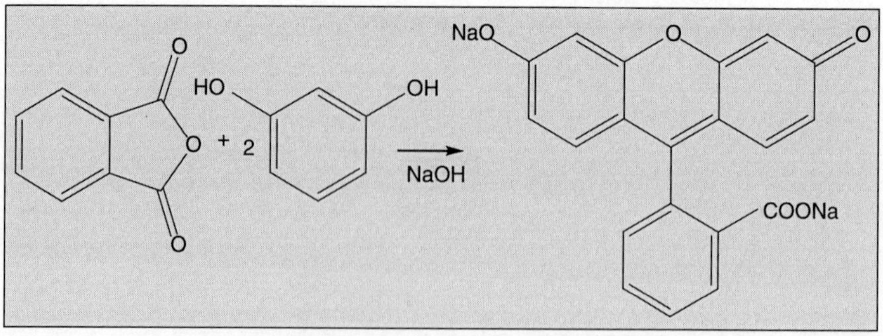

It is used extensively as a diagnostic tool in the field of ophthalmology, where topical fluorescein is used in the diagnosis of corneal abrasions, corneal ulcers, and herpetic corneal infections. Intravenous or oral fluorescein is used in fluorescein angiography in research and to diagnose and categorize vascular disorders in legs, including retinal disease macular degeneration, diabetic retinopathy, inflammatory intraocular conditions, and intraocular tumours, and increasingly during surgery for brain tumours.

Erythrosine Disodium salt of 2,4,5,7-tetraiodofluorescein

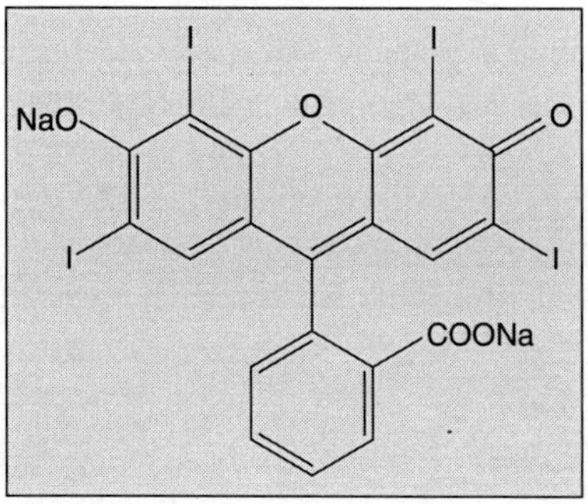

It is a cherry-pink, coal-based fluorone food dye. It is used as a food dye, in printing inks, as a biological stain, a dental plaque-disclosing agent, and a radiopaque medium.

Metyrapone (Metopirone) 2-Methyl-1,2-dipyridin-3-yl-propan-1-one
Two molecules of 3-acetyl pyridine undergo self-condensation to give 2-(pyridin-3-yl)-3-(pyridin-4-yl) butane-2,3-diol, which on treatment with sulphuric acid undergoes rearrangement to form metyrapone.

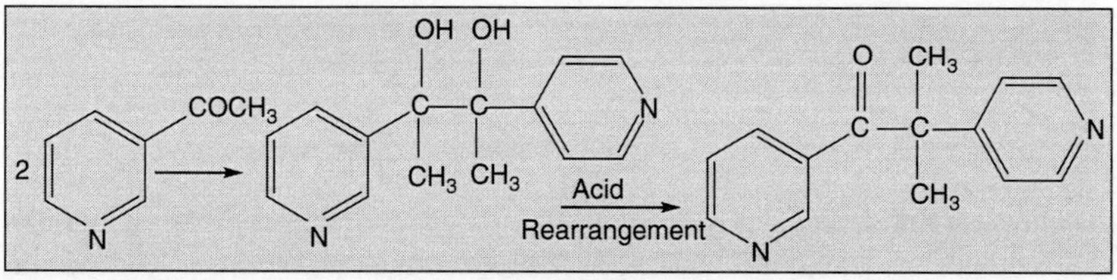

It is used in the diagnosis of adrenal insufficiency. Metyrapone 30 mg/kg, maximum dose 3000 mg, is administered at midnight, usually with a snack. The plasma cortisol and 11-deoxycortisol are measured the next morning between 8:00 a.m. and 9:00 a.m. A plasma cortisol less than 220 nmol/l indicates adequate inhibition of 11β-hydroxylase. In patients with intact hypothamalmo-pituitary-adrenal (HPA) axis, CRH and ACTH levels rise as a response to the falling cortisol levels. This results in an increase of the steroid precursors in the pathway. Therefore, if 11-deoxycortisol levels do not rise and remain less than 7 mcg/dl, then it is highly suggestive of impaired HPA axis.

Methacholine chloride 2-Acetoxy-N,N,N-trimethylpropan-1-aminium
It is prepared by acylating 2-hydroxy-N,N,N-trimethylpropan-1-aminium salt.

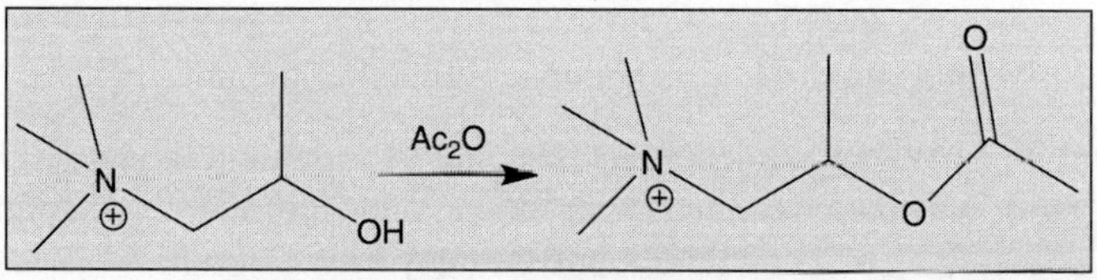

It is a synthetic choline ester. The primary clinical use of methacholine is to diagnose bronchial hyperreactivity, which is the hallmark of asthma and also occurs in chronic obstructive pulmonary disease. This is accomplished through the methacholine challenge test.

Pentagastrin   L-Phenylalaninamide,   N-((1,1-Dimethylethoxy)carbonyl)-β-alanyl-L-tryptophyl-L-methionyl-L-α-aspartyl

It is a synthetic polypeptide used as a diagnostic aid as the pentagastrin-stimulated calcitonin test. The pentagastrin-stimulated calcitonin test is a diagnostic test for medullary carcinoma of the thyroid (MTC).

Phenol red   3,3-*bis*(4-Hydroxyphenyl)-2,11^6-benzoxathiole-1,1(3*H*)-dione

Reaction between 2-sulphobenzenecarboxylic acid and phosphorous pentoxide gives 2,11^6-benzoxathiole-1,1(3*H*)-dione, which on fusion with two moles of phenol gives phenol red.

It is a pH indicator that is frequently used in cell biology laboratories. Most living tissues prosper at a near-neutral pH – that is, a pH close to 7. The pH of blood ranges from 7.35 to 7.45, for instance. When cells are grown in tissue culture, the medium in which they grow is held close to this physiological pH. A small amount of phenol red added to this growth medium will have a pink-red colour under normal conditions. Typically, 15 mg/1L is used for cell culture.

In the event of problems, waste products produced by dying cells or overgrowth of contaminants will cause a change in pH, leading to a change in indicator colour. For example, a culture of relatively slowly-dividing mammalian cells can be quickly overgrown by bacterial contamination. This usually results in an acidification of the medium, turning it yellow. Many biologists find this a convenient way to rapidly check on the health of tissue cultures. In addition, the waste products produced by the mammalian cells themselves will slowly decrease the pH, gradually turning the solution orange and then yellow. This colour change is an indication that even in the absence of contamination, the medium needs to be replaced (generally, this should be done before the medium has turned completely orange).

Tyropanoic acid   2-[(3-Butanamido-2,4,6-triiodophenyl)methyl]butanoic acid

It is synthesised from 2-(3-nitrobenzylidene)butanoic acid by reduction to give 2-(3-aminobenzyl)butanoic acid (iopanoic acid), which on treatment with iodine monochloride gives triiodo derivative. Amide (tyropanoic acid) is prepared by reacting it with butyric anhydride.

It is radiopaque contrast media used in cholecystography (X-ray diagnosis of gallstones).

34.10 ANTISEPTICS AND DISINFECTANTS

Both of these substances are agents that kill, or at least control the growth of, microbes. Antiseptics are agents that are used on living tissue. Antiseptics are antimicrobial substances that are applied to living tissue/skin to reduce the possibility of infection, sepsis, or putrefaction. They should generally be distinguished from antibiotics that destroy bacteria within the body, and from disinfectants, which destroy microorganisms found on non-living objects. Disinfectants are used on non-living things such as floors, countertops, and dishes. They are usually stronger and are too toxic to be used on living tissue.

Antiseptics and disinfectants are used extensively in hospitals and other healthcare settings for a variety of topical and hard-surface applications. In particular, they are an essential part of infection control practices and aid in the prevention of nosocomial infections.

Mechanism of action: Antiseptics and disinfectants are acting by any one of the following processes: lysis and leakage of intracellular constituents, perturbation of cell homeostasis, effects on model membranes, inhibition of enzymes, electron transport, and oxidative phosphorylation, interaction with macromolecules, and effects on macromolecular biosynthetic processes.

Classification, examples, and uses

1. **Alcohols: Ethyl alcohol** (ethanol, alcohol), **isopropyl alcohol** (isopropanol, propan-2-ol) and *n*-propanol are the most widely used. Alcohols exhibit rapid broad-spectrum antimicrobial activity against vegetative bacteria (including mycobacteria), viruses, and fungi. Generally, the antimicrobial activity of alcohols is significantly lower at concentrations below 50 per cent and is optimal in the 60 per cent to 90 per cent range. Little is known about the specific mode of action of alcohols, but based on the increased efficacy in the presence of water, it is generally believed that they cause membrane damage and rapid denaturation of proteins, with subsequent interference with metabolism and cell lysis.

2. **Aldehydes: Glutaraldehyde** has a broad spectrum of activity against bacteria and their spores, fungi, and viruses. Glutaraldehyde is more active at alkaline than at acidic pHs. The mechanism of action of glutaraldehyde involves a strong association with the outer layers of bacterial cells, specifically with unprotonated amines on the cell surface, possibly representing the reactive sites. Such an effect could explain its inhibitory action on transport and on enzyme systems, where access of

substrate to enzyme is prohibited. **Formaldehyde** (methanal, CH_2O) is a monoaldehyde that exists as a freely water-soluble gas. Formaldehyde solution (formalin) is an aqueous solution containing *ca*. 34 per cent to 38 per cent (wt/wt) CH_2O with methanol to delay polymerization. Its clinical use is generally as a disinfectant. It is difficult to pinpoint accurately the mechanism(s) responsible for formaldehyde-induced microbial inactivation. Clearly, its interactive and cross-linking properties must play a considerable role in this activity.

3. **Anilides: Triclocarban** is particularly active against gram-positive bacteria but significantly less active against gram-negative bacteria and fungi. The anilides are thought to act by adsorbing to and destroying the semi-permeable character of the cytoplasmic membrane, leading to cell death.
4. **Biguanides: Chlorhexidine** is probably the most widely used biocide in antiseptic products, in particular in hand washing and oral products but also as a disinfectant and preservative. This is due in particular to its broad-spectrum efficacy, substantively for the skin, and low irritation. The mechanism of action is membrane disruption.
5. **Halogen-releasing agents**: **Chlorine-releasing agents** (CRA), the most important types of CRAs, are sodium hypochlorite, chlorine dioxide, and the N-chloro compounds such as sodium dichloroisocyanurate, with chloramine-T being used to some extent. Sodium hypochlorite solutions are widely used for hard-surface disinfection. In water, sodium hypochlorite ionizes to produce Na1 and the hypochlorite ion, OCl_2, which establishes an equilibrium with hypochlorous acid, HOCl. Between pH4 and pH7, chlorine exists predominantly as HClO, the active moiety, whereas above pH9, OCl2 predominates. CRAs are highly active oxidizing agents and thereby destroy the cellular activity of proteins. **Iodine** and **iodophors** are less reactive than chlorine; iodine is rapidly bactericidal, fungicidal, tuberculocidal, virucidal, and sporicidal. Although aqueous or alcoholic (tincture) solutions of iodine have been used for about 150 years as antiseptics, they are associated with irritation and excessive staining. In addition, aqueous solutions are generally unstable; in solution, these problems were overcome by the development of iodophors ('iodine carriers' or 'iodine-releasing agents'); the most widely used are **povidone-iodine** and **poloxamer-iodine** in both antiseptics and disinfectants. Iodophors are complexes of iodine and a solubilizing agent or carrier, which acts as a reservoir of the active 'free' iodine. Iodine rapidly penetrates into microorganisms and attacks key groups of proteins (in particular, the free sulphur amino acids cysteine and methionine), nucleotides, and fatty acids, which culminates in cell death.
6. **Silver compounds: Silver sulphadiazine** a combination of two antibacterial agents, Ag^+ and sulphadiazine, and has a broad spectrum of activity and, unlike silver nitrate, produces surface and membrane blebs in susceptible (but not resistant) bacteria. It binds to cell components, including DNA. Bacterial inhibition would then presumably be achieved when silver binds to sufficient base pairs in the DNA helix, thereby inhibiting transcription.
7. **Peroxygens: Hydrogen peroxide** (H_2O_2) is a widely used biocide for disinfection, sterilization, and antisepsis. H_2O_2 demonstrates broad-spectrum efficacy against viruses, bacteria, yeasts, and bacterial spores. H_2O_2 acts as an oxidant by producing hydroxyl free radicals ($\cdot OH$) that attack essential cell components, including lipids, proteins, and DNA. It has been proposed that exposed sulphhydryl groups and double bonds are particularly targeted.
8. **Phenols:** They have often been referred to as 'general protoplasmic poisons'; they have membrane-active properties that also contribute to their overall activity. Phenol induces progressive leakage of intracellular constituents, including the release of K^+. The *bis*-**phenols** are hydroxy-halogenated derivatives of two phenolic groups connected by various bridges. **Triclosan** (2,4,4'-trichloro-2'-

hydroxydiphenyl ether) and hexachlorophene (2,2'-dihydroxy-3,5,6,3',5',6'-hexachlorodiphenyl-methane) are the most widely used biocides in this group, especially in antiseptic soaps and hand rinses. The specific mode of action of triclosan is unknown, but it has been suggested that the primary effects are on the cytoplasmic membrane. The primary action of hexachlorophene is to inhibit the membrane-bound part of the electron transport chain, and the other effects noted above are secondary ones that occur only at high concentrations. It induces leakage, causes protoplast lysis, and inhibits respiration.

Chloroxylenol (4-chloro-3,5-dimethylphenol; *p*-chloro-*m*-xylenol) is the key **halophenol** used in antiseptic or disinfectant

9. **Quaternary ammonium compounds:** Surface-active agents (surfactants) have two regions in their molecular structures, one a hydrocarbon, water-repellent (hydrophobic) group and the other a water-attracting (hydrophilic or polar) group. Depending on the basis of the charge or the absence of ionization of the hydrophilic group, surfactants are classified into cationic, anionic, nonionic, and ampholytic (amphoteric) compounds. Of these, the cationic agents, as exemplified by quaternary ammonium compounds (QACs) **cetylpyridium chloride** and **benzalkonium chloride**, are the most useful antiseptics and disinfectants. The following sequence of events (mechanism of action) take place with microorganisms exposed to cationic agents: (i) adsorption and penetration of the agent into the cell wall; (ii) reaction with the cytoplasmic membrane(lipid or protein) followed by membrane disorganization; (iii) leakage of intracellular low-molecular-weight material; (iv) degradation of proteins and nucleic acids; and (v) wall lysis caused by autolytic enzymes. There is, thus, a loss of structural organization and integrity of the cytoplasmic membrane in bacteria, together with other damaging effects to the bacterial cell.

34.11 SOME IMPORTANT ANTISEPTICS AND DISINFECTANTS

Povidone-iodine 2-Pyrrolidinone, 1-ethenyl-, homopolymer, compd with iodine

It is a stable chemical complex of polyvinylpyrrolidone (povidone) and elemental iodine. It contains from 9.0 per cent to 12.0 per cent available iodine, calculated on a dry basis.

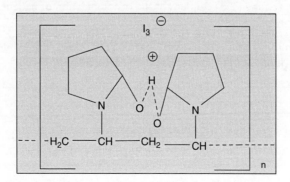

Cetrimide (Cetrimonium bromide)

The cetrimonium (or hexadecyltrimethylammonium) cation is an effective antiseptic agent against bacteria and fungi.

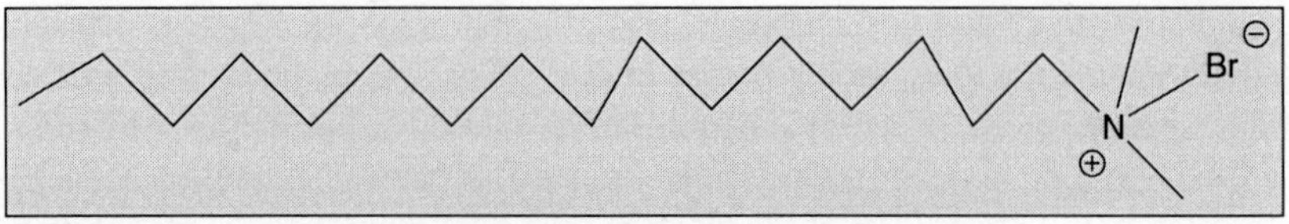

Benzalkonium chloride (Alkyldimethylbenzylammonium chloride)
It is a mixture of alkylbenzyldimethylammonium chlorides of various even-numbered alkyl chain lengths. The greatest biocidal activity is associated with the C12-C14 alkyl derivatives.

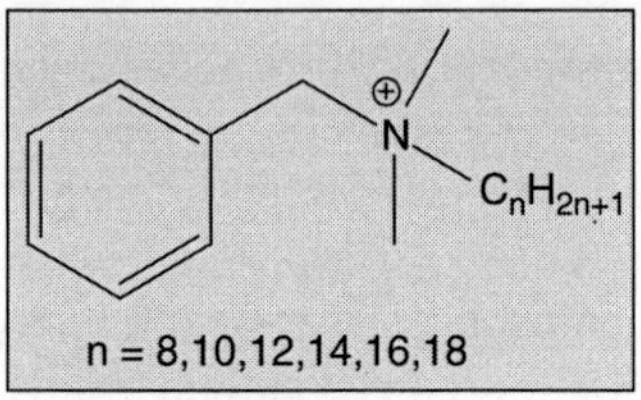

$n = 8, 10, 12, 14, 16, 18$

Triclosan   5-Chloro-2-(2,4-dichlorophenoxy)phenol)

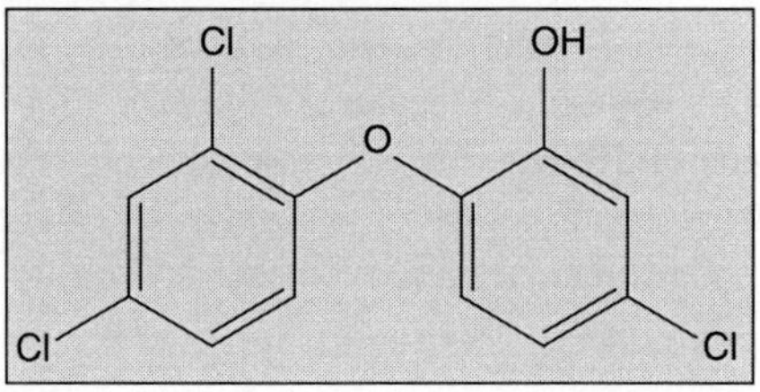

It is a potent, wide-spectrum antibacterial and antifungal agent. Triclosan is found in soaps (0.15 per cent–0.30 per cent), deodorants, toothpastes, shaving creams, mouthwashes, and cleaning supplies, and is infused in an increasing number of consumer products, such as kitchen utensils, toys, bedding, socks, and trash bags. Triclosan has been shown to be effective in reducing and controlling bacterial contamination on the hands and on treated products. More recently, showering or bathing with two per cent triclosan has become a recommended regimen for the decolonization of patients whose skin is carrying methicillin resistant *Staphylococcus aureus*.

Chlorhexidine   1-[Amino-[6-[amino-[amino-(4-chlorophenyl)amino-methylidene]amino-methylidene] aminohexylimino]methyl]imino-N-(4- chlorophenyl)-methanediamine

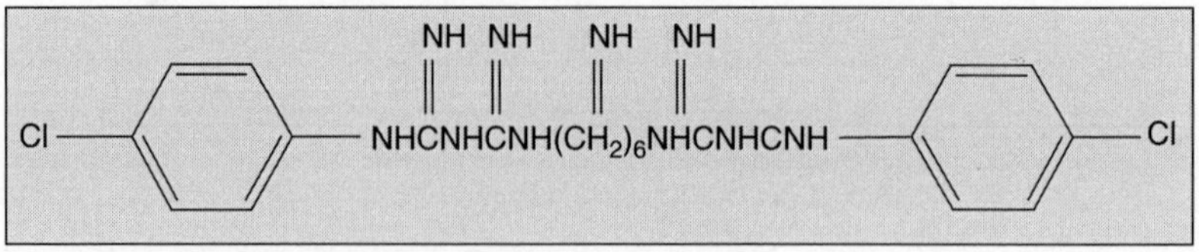

It is a chemical antiseptic. It kills (is bactericidal to) both gram-positive and gram-negative microbes, although it is less effective with some gram-negative microbes.

34.12 PHARMACEUTICAL AID

A **preservative** is a natural or synthetic chemical substance that prevents or inhibits microbial growth, which may be added to pharmaceuticals products for this purpose to avoid consequent spoilage of the product by microorganisms.

Parabens Methyl/ethyl/butyl ester of 4-hydroxybenzoic acid
Parabens are synthesised by esterifying 4-hydroxybenzoic acid with corresponding alcohol in presence of acid.

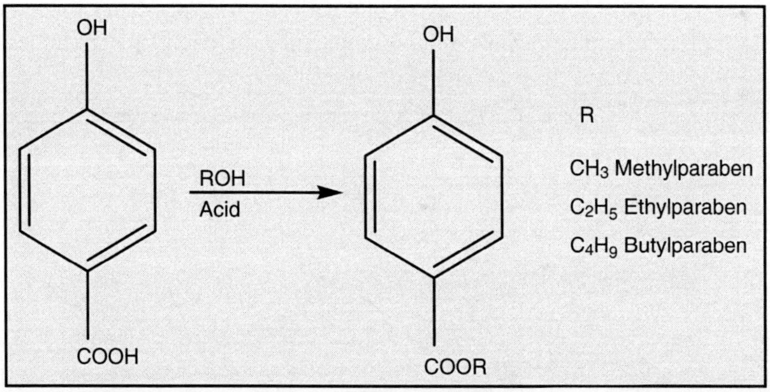

[Reaction scheme: 4-hydroxybenzoic acid + ROH/Acid → 4-hydroxybenzoate ester]

R = CH_3 Methylparaben
R = C_2H_5 Ethylparaben
R = C_4H_9 Butylparaben

These compounds are mainly used as ophthalmic preservative at the 0.1 per cent concentration.

Organic mercurials: Phenyl mercuric nitrate and thiomersal

[Structures of Phenyl mercuric nitrate and Thiomersal]

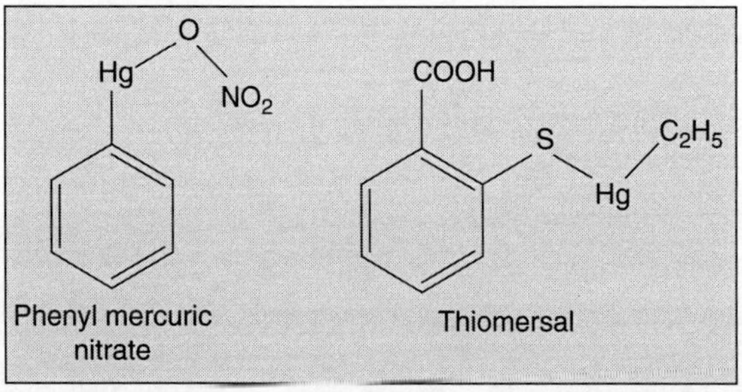

Phenyl mercuric nitrate and thiomersal at the concentrations of 0.001 per cent–0.01 per cent are used as a preservative in vaccines, immunoglobulin preparations, skin test antigens, antivenoms, and ophthalmic and nasal products.

Sodium benzoate C_6H_5COONa

It is the sodium salt of benzoic acid and exists in this form when dissolved in water. It can be produced by reacting sodium hydroxide with benzoic acid. Sodium benzoate is a preservative. It is bacteriostatic and fungistatic under acidic conditions.

Benzalkonium chloride

It is a mixture of alkylbenzyldimethylammonium chlorides of various even-numbered alkyl chain lengths. The greatest biocidal activity is associated with the C12–C14 alkyl derivatives.

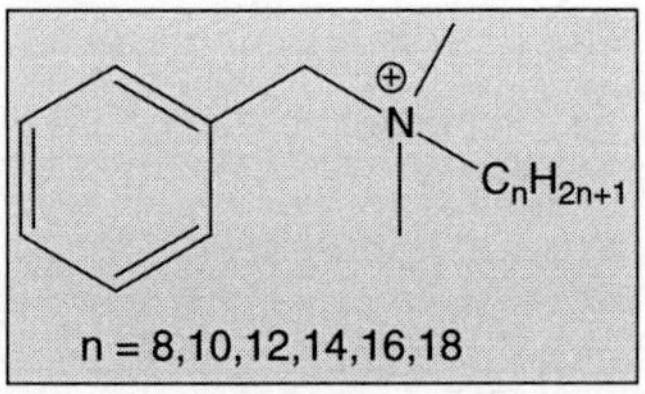

It is mainly used as an ophthalmic preservative at 0.004 per cent–0.02 per cent concentrations.

An **antioxidant** is a substance capable of inhibiting oxidation, which may be added for this purpose to pharmaceutical products subject to decomposition by oxidative processes.

Butylated hydroxyanisole   2-*tert*-Butyl-4-methoxyphenol

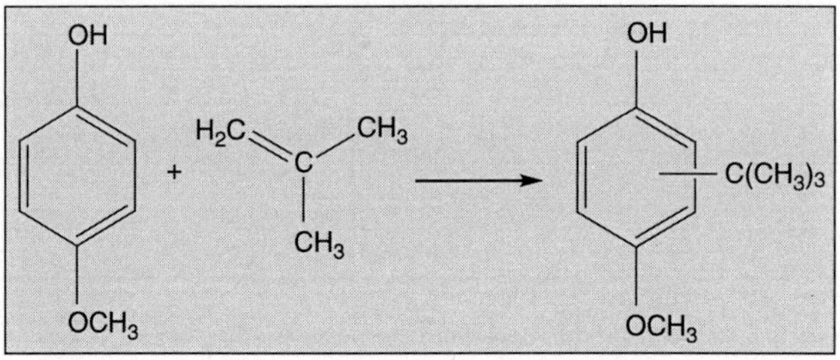

It is a mixture of two isomeric organic compounds, 2-*tert*-butyl-4-hydroxyanisole and 3-*tert*-butyl-4-hydroxyanisole. It is prepared from 4-methoxyphenol and isobutylene. It is a waxy solid that exhibits antioxidant properties.

Butylated hydroxytoluene   2,6-Di-*tert*-butyl-4-methylphenol

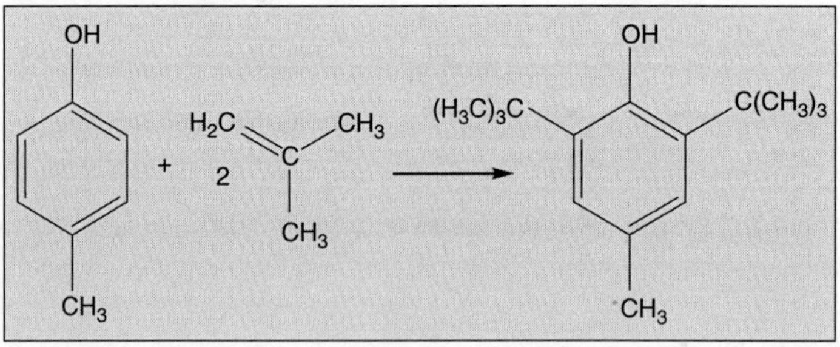

It is a lipophilic (fat-soluble) organic compound that is primarily used as an antioxidant in pharmaceuticals. It is prepared by the reaction of *p*-cresol (4-methylphenol) with isobutylene (2-methylpropene) catalysed by sulphuric acid.

Inorganic substances: Sodium/potassium metabisulphite (Na/$K_2S_2O_5$) and sodium bisulphate (NaHSO$_3$) are used as antioxidants in pharmaceutical formulations.

A **sweetening agent** is an additive that duplicates the effect of sugar in taste, but usually has less food energy. Some sugar substitutes are natural and some are synthetic. Those that are not natural are, in general, referred to as artificial sweeteners.

Saccharin Benzoic sulphinide
It is synthesised from anthranilic acid by successively reacting with nitrous acid, sulphur dioxide, chlorine, and then ammonia to yield saccharin.

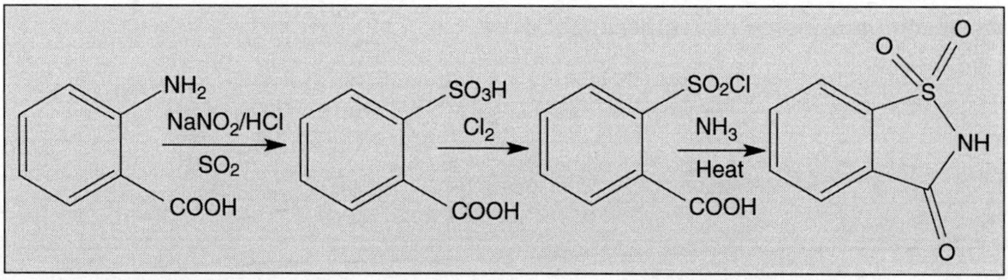

It is an artificial sweetener. The basic substance, benzoic sulfinide, has effectively no food energy and is much sweeter than sucrose, but has an unpleasant bitter or metallic after taste, especially at high concentrations.

Aspartame Aspartyl-phenylalanine-1-methyl ester

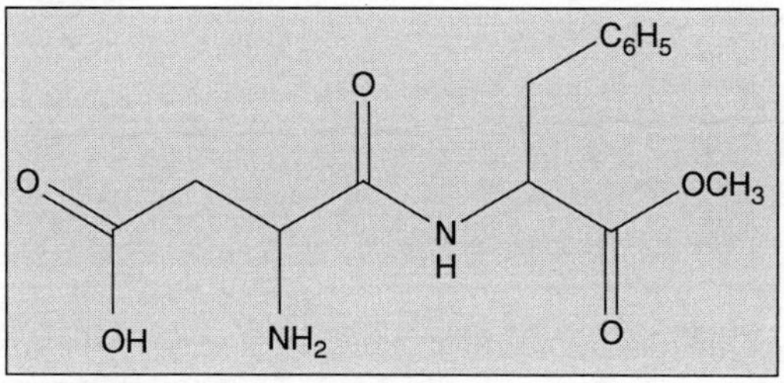

It is an artificial, non-saccharide sweetener. It is a methyl ester of the dipeptide of the amino acids aspartic acid and phenylalanine. It is 180 times sweeter than sugar in typical concentrations, without the high energy value of sugar.

Sucralose (2R,3R,4R,5R,6R)-2-((2R,3S,4S,5S)-2,5-bis(Chloromethyl)-3,4-dihydroxytetrahydrofuran-2-yloxy)-5-chloro-6-(hydroxymethyl)tetrahydro-2H-pyran-3,4-diol

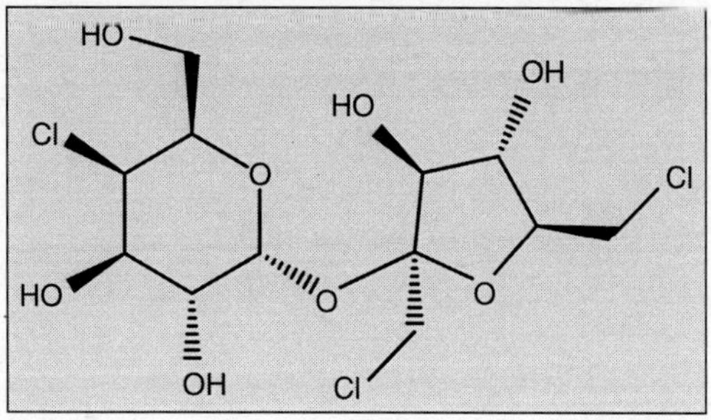

It is approximately 600 times as sweet as sucrose, twice as sweet as saccharin, and three-and-a-third times as sweet as aspartame.

Colouring agents may be defined as compounds employed in pharmaceutical formulation solely for the purpose of imparting colour. Colourants are classified into two types: (a) natural: obtained from minerals and plant and animal sources—e.g., red ferric oxide, titanium dioxide, chlorophyll, cochineal, rutin, hesperidine, tyrian purple, and carbon black; and (b) synthetic: coal tar dyes include more than a dozen groups among which are nitroso dyes, nitro dyes, azo dyes, oxazines, acridines, thiazines, pyrazolones, and quinolines.

Structures of commonly used pharmaceutical dyes

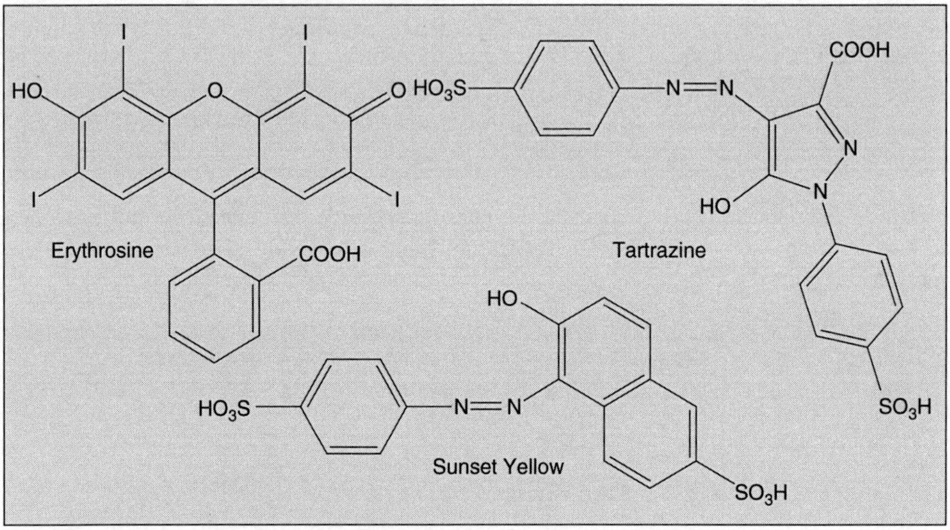

The compounds employed as **flavouring agent** considerably vary in their chemical structures, ranging from esters (methyl salicylate), alcohol (glycerin), and aldehyde (vanillin), to carbohydrates (honey) and the complex volatile oil (anise oil).

Structures of commonly used pharmaceutical flavours

34.13 VITAMINS

Vitamins are complex organic compounds which are needed in the diet in very small quantities. They do not serve as a source of energy, but instead participate in various essential chemical processes in the body. The term 'vitamin' first became popular in the early 1800's as a contraction to the words 'vital' and 'mineral', though the actual meaning of the word has developed somewhat since that time.

Vitamins can be divided into two types, namely, water-soluble and fat-soluble compounds. The fat-soluble vitamins have structures which are mostly complex hydrocarbons with only a few functional groups.

Vitamins A, D, E, and K are fat-soluble. They are stored in the fatty tissues of the body. The water-soluble vitamins, on the other hand, have lots of functional groups, which makes them more likely to dissolve in water (since many of the functional groups are attracted to water). The body has a limited capacity to store these vitamins, so they are needed regularly in the diet. Vitamin C and all of the vitamins B are water-soluble.

Vitamins have diverse biochemical functions acting as hormones (e.g. vitamin D), antioxidants (e.g. vitamin E), and mediators of cell signaling and regulators of cell and tissue growth and differentiation (e.g. vitamin A). The vitamins of B complex function as precursors for enzyme cofactor biomolecules (coenzymes) that help act as catalysts and substrates in metabolism. When acting as part of a catalyst, vitamins are bound to enzymes and are called prosthetic groups. For example, biotin is part of enzymes involved in making fatty acids. Vitamins also act as coenzymes to carry chemical groups between enzymes. For example, folic acid carries various forms of carbon groups—methyl, formyl and methylene—in the cell. Although these roles in assisting enzyme reactions are best-known function of vitamins, the other vitamin functions are also equally important.

Table 34.1
List of vitamins and their details

Vitamin	Chemical Name	Solubility	RDA Adult	Deficiency Disease	Upper Intake Level Adult	Overdose Disease
Vitamin A	Retinol, retinal, various retinoids,	Fat	900 µg	Night-blindness and Keratomalacia	3,000 µg	Hypervitaminosis A
Vitamin B_1	Thiamine	Water	1.2 mg	Beriberi, Wernicke-Korsakoff syndrome	N/D	Drowsiness or muscle relaxation with large doses
Vitamin B_2	Riboflavin	Water	1.3 mg	Ariboflavinosis	N/D	None
Vitamin B_3	Niacin, niacinamide	Water	16.0 mg	Pellagra	35.0 mg	Liver damage (dose > 2g/day)
Vitamin B_5	Pantothenic acid	Water	5.0 mg	Paresthesia	N/D	Diarrhea, nausea, and heartburn
Vitamin B_6	Pyridoxine, pyridoxamine, pyridoxal	Water	1.3–1.7 mg	Anemia, peripheral neuropathy	100 mg	Impairment of proprioception, nerve damage (dose > 100 mg/day)
Vitamin B_7	Biotin	Water	30.0 µg	Dermatitis, enteritis	N/D	None
Vitamin B_9	Folic acid, folinic acid	Water	400 µg	Deficiency during pregnancy is associated with birth defects, such as neural tube defects	1,000 µg	May mask symptoms of vitamin B_{12} deficiency

(Continued)

Table 34.1 (Continued)

Vitamin	Chemical Name	Solubility	RDA Adult	Deficiency Disease	Upper Intake Level Adult	Overdose Disease
Vitamin B$_{12}$	Cyanocobalamin	Water	2.4 µg	Megaloblastic anemia	N/D	None
Vitamin C	Ascorbic acid	Water	90.0 mg	Scurvy	2,000 mg	Nil
Vitamin D	Ergocalciferol, cholecalciferol	Fat	5.0 µg–10 µg	Rickets and osteomalacia	50 µg	Hypervitaminosis D
Vitamin E	Tocopherols, tocotrienols	Fat	15.0 mg	Deficiency is very rare; mild hemolytic anemia in newborn infants	1,000 mg	Increased congestive heart failure
Vitamin K	Phylloquinone, menaquinones	Fat	120 µg	Bleeding diathesis	N/D	Increases coagulation in patients taking warfarin

Vitamin A Retinol, 3,7-Dimethyl-9-(2,6,6-trimethyl-1-cyclohexenyl)nona-2,4,6,8-tetraen-1-ol
Vitamin A is the immediate precursor to two important active metabolites: retinal, which plays a critical role in vision, and retinoic acid, which serves as an intracellular messenger that affects transcription of a number of genes. Vitamin A does not occur in plants, but many plants contain carotenoids such as β-carotene that can be converted to vitamin A within the intestine and other tissues. All forms of vitamin A have a β-ionone ring to which an isoprenoid chain is attached, called a *retinyl group*. This structure is essential for vitamin activity.

Retinol: CH$_2$OH
Retinal: CHO
Retinoic acid: COOH

Physiologic effects of vitamin A:
Vitamin A and its metabolites retinal and retinoic acid appear to serve a number of critical roles in physiology. Some of the well-characterized effects of vitamin A include:

- **Vision:** Retinal is a necessary structural component of rhodopsin or visual purple, the light sensitive pigment within rod and cone cells of the retina. If inadequate quantities of vitamin A are present, vision is impaired.
- **Resistance to infectious disease:** In almost every infectious disease studied, vitamin A deficiency has been shown to increase the frequency and severity of disease. Several large trials with malnourished

children have demonstrated dramatic reductions in mortality from diseases such as measles by the simple and inexpensive procedure of providing vitamin A supplementation. This "anti-infective" effect is undoubtedly complex, but is due, in part, to the necessity for vitamin A in normal immune responses. Additionally, many infections are associated with inflammatory reactions that lead to reduced synthesis of retinol-binding protein and thus, reduced circulating levels of retinol.
- **Epithelial cell integrity:** Many epithelial cells appear to require vitamin A for proper differentiation and maintenance. Lack of vitamin A leads to dysfunction of many epithelia—the skin becomes keratinized and scaly, and mucus secretion is suppressed. It seems likely that many of these effects are due to impaired transcriptional regulation due to deficits in retinoic acid signalling.
- **Bone remodeling:** Normal functioning of osteoblasts and osteoclasts is dependent upon vitamin A.
- **Reproduction:** Normal levels of vitamin A are required for sperm production, reflecting a requirement for vitamin A by spermatogenic epithelial (Sertoli) cells. Similarly, normal reproductive cycles in females require adequate availability of vitamin A.

Synthesis of tretinoin:

It is prepared from β-ionone, condensation of which with carbanion from acetonitrile followed by dehydration of the initially formed carbinol gives cyano intermediate. Reduction of cyano group by DIBAL leads to the corresponding imine; this hydrolyses to aldehyde. Base catalysed aldol condensation of this aldehyde with β-methylglutaconic anhydride involves condensation with the active methylene group of the anhydride leading to the formation of condensation product. The anhydride is then hydrolysed to the vinylgous β-dicarboxylic acid; the superfluous carboxyl group is removed by heating the compound with quinoline in presence of copper to give acid derivative. The terminal double bond is then isomerized by treatment with iodine to afford tretinoin

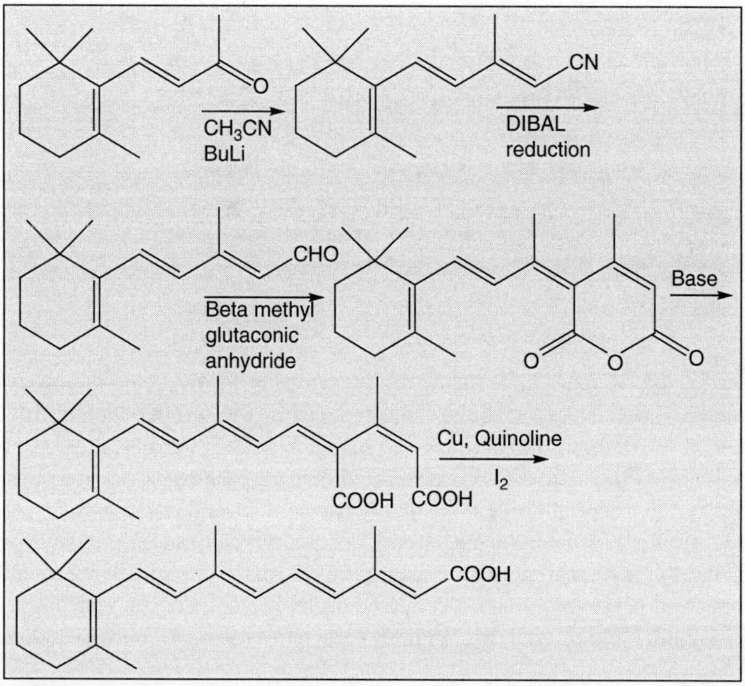

Tretinoin is the acid form of vitamin A and so also known as all-*trans* retinoic acid. It is a drug commonly used to treat acne vulgaris and keratosis pilaris. It is available as a cream or gel. It is also used to treat acute promyelocytic leukemia.

Vitamin D http://vitamind.ucr.edu/Image/chem1.gif

The structures of vitamin D_2 (ergocalciferol) and vitamin D_3 (cholecalciferol) are important in this class of vitamins. Vitamin D is a generic term and indicates a molecule of the general structure shown for rings A, B, C, and D with differing side chain structures. The A, B, C, and D ring structure is derived from the cyclopentanoperhydrophenanthrene ring structure for steroids. Technically, vitamin D is classified as a *seco*-steroid. Seco-steroids are those in which one of the rings has been broken; in vitamin D, the 9,10 carbon-carbon bond of ring B is broken and it is indicated by the inclusion of "9,10-seco" in the official nomenclature.

Vitamin D (calciferol) is named according to the revised rules of the International Union of Pure and Applied Chemists (IUPAC). Because vitamin D is derived from a steroid, the structure retains its numbering from the parent compound cholesterol. Asymmetric centers are designated by using the R, S notation; the configuration of the double bonds are notated E for "entgegen" or *trans*, and Z for "zuzammen" or *cis*. Thus, the official name of vitamin D_3 is 9,10-seco(5Z,7E)-5,7,10(19)cholestatriene-3b-ol, and the official name of vitamin D_2 is 9,10-seco(5Z,7E)-5,7,10(19), 22-ergostatetraene-3b-ol.

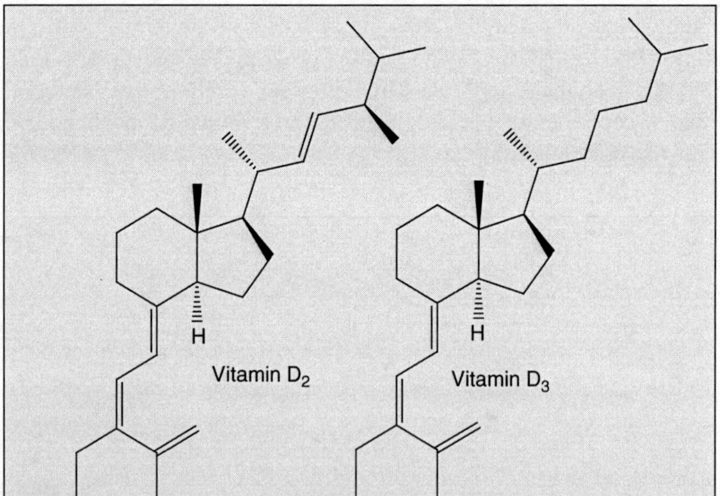

It is to be emphasized that Vitamin D_3 is the naturally occurring form of the vitamin; it is produced from 7-dehydrocholesterol, which is present in the skin produced by the action of sunlight. Vitamin D_2 (which is equivalently potent to vitamin D_3 in humans and many mammals, but not birds) is produced commercially by the irradiation of the plant sterol ergosterol with ultraviolet light.

Vitamin D plays an important role in the maintenance of several organ systems. However, its major role is to increase the flow of calcium into the bloodstream by promoting absorption of calcium and phosphorus from food in the intestines and reabsorption of calcium in the kidneys enabling normal mineralization of bones and preventing hypocalcemic tetany. It is also necessary for bone growth and bone remodeling by osteoblasts and osteoclasts. Without sufficient vitamin D, bones can become thin, brittle, or their shapes may get distorted. Deficiency can arise from inadequate intake coupled with inadequate sunlight exposure; disorders that limit its absorption; conditions that impair conversion of vitamin D into active metabolites, such as liver or kidney disorders; or, rarely, by a number of hereditary disorders. Vitamin D deficiency results in impaired bone mineralization and leads to bone softening diseases like rickets in children and osteomalacia in adults and possibly contributes to osteoporosis. Vitamin D also plays a role in the inhibition of calcitonin release from the thyroid gland. Calcitonin acts directly on osteoclasts resulting in the inhibition of bone resorption and cartilage degradation. Vitamin D can also inhibit parathyroid hormone secretion from the parathyroid gland, modulate neuromuscular and immune function and reduce inflammation.

Ergosterol (ergosta-5,7,22-trien-3β-ol), a sterol, is a biological precursor (a provitamin) to vitamin D_2. It is turned into viosterol by ultraviolet light and is then converted into ergocalciferol, which is a form of vitamin D.

Vitamin E:

Vitamin E is a generic term for tocopherols and tocotrienols. It is a family of α-, β-, γ-, and δ-tocopherols and corresponding four tocotrienols and is a fat-soluble antioxidant that stops the production of reactive oxygen species formed when fat undergoes oxidation. Of these, α-tocopherol (also written as a-tocopherol) has been most studied as it has the highest bioavailability.

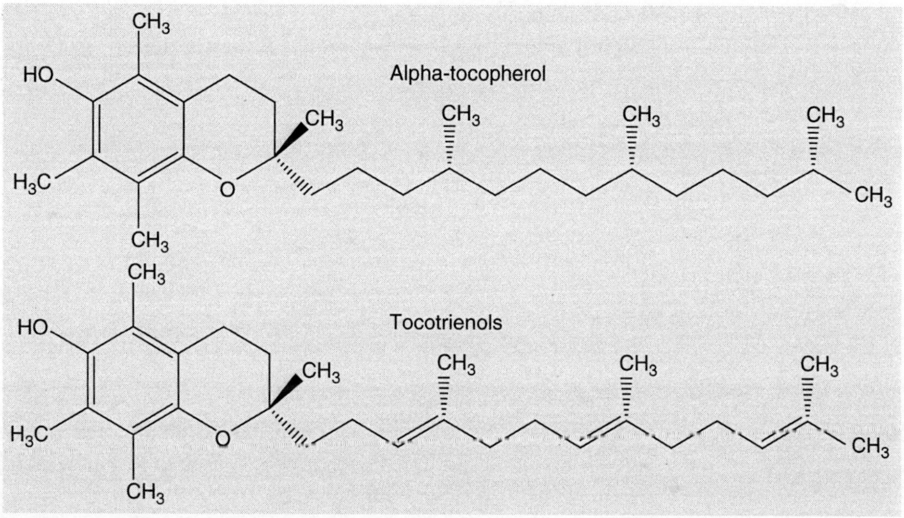

Vitamin E has two types of structures, namely, tocopherol and tocotrienol structures. Both structures are similar except that the tocotrienol structure has double bonds on the isoprenoid units. There are many derivatives of these structures as different substituents are possible on the aromatic ring at positions 5, 6, 7, and 8. Notice that there are three chiral centers, at positions 2', 4', and 8', in the phytyl tail. Thus, there is a possibility of eight stereoisomers. The most abundant of the naturally-occurring forms is the R, R, R form. The tocotrienols share the same ring structure, but have an unsaturated tail.

Tocopherols are equipped to perform a unique function. They can interrupt free radical chain reactions by capturing the free radical hence are considered to possess antioxidant properties. The free hydroxyl group on the aromatic ring is responsible for the antioxidant properties. The hydrogen from this group is donated to the free radical resulting in a relatively stable free radical form of the vitamin.

It has been claimed that α-tocopherol is the most important lipid-soluble antioxidant and that it protects cell membranes from oxidation by reacting with lipid radicals produced in the lipid peroxidation chain reaction. This would remove the free radical intermediates and prevent the oxidation reaction from continuing. The oxidized α-tocopheroxyl radicals produced in this process may be recycled back to the active reduced form through reduction by other antioxidants, such as ascorbate, retinol or ubiquinol.

Vitamin K:
Vitamin K is a necessary participant in synthesis of several proteins that mediate both coagulation and anticoagulation. Its deficiency is manifested as a tendency to bleed excessively. At least two naturally-occuring forms of vitamin K have been identified and these are designated as vitamin K_1 [also known as phylloquinone or phytomenadione] and vitamin K_2 [(menaquinone, menatetrenone)]. Both are quinone derivatives. The structure of vitamin K_1 is here.

All members of vitamin K group share a methylated naphthoquinone ring structure, and vary in the aliphatic side chain attached at the 3-position. Phylloquinone (also known as vitamin K_1) invariably contains in its side chain four isoprenoid residues one of which is unsaturated.

Physiologic Effects of Vitamin K: Vitamin K serves as an essential cofactor for a carboxylase that catalyzes carboxylation of glutamic acid residues on vitamin K-dependent proteins. The key vitamin K-dependent proteins include:

- Coagulation proteins: factors II (prothrombin), VII, IX and X
- Anticoagulation proteins: proteins C, S and Z
- Others: bone proteins osteocalcin and matrix-Gla protein and certain ribosomal proteins

These proteins have in common the requirement to be post-translationally modified by carboxylation of glutamic acid residues (forming gamma-carboxyglutamic acid) in order to become biologically active. For example, Prothrombin has 10 glutamic acids in the amino-terminal region of the protein which are carboxylated. Without vitamin K, the carboxylation does not occur and the proteins that are synthesized are biologically inactive.

Vitamin B_1 2-[3-[(4-Amino-2-methyl-pyrimidin-5-yl)methyl]-4-methyl-thiazol-5-yl] ethanol

Thiamine or thiamin, sometimes called aneurin is a water-soluble vitamin of the B complex (vitamin B_1), whose phosphate derivatives are involved in many cellular processes. The best characterized form is thiamine diphosphate, a coenzyme in the catabolism of sugars and amino acids. Its structure contains a pyrimidine ring and a thiazole ring linked by a methylene bridge.

Thiamine is synthesized in bacteria, fungi and plants. The insufficient intake of this vitamin results in a disease called beriberi affecting the peripheral nervous system (polyneuritis) and/or the cardiovascular system with fatal outcome if not cured by thiamine administration.

Vitamin B_2 7,8-Dimethyl-10-((2S,3S,4R)-2,3,4,5-tetrahydroxypentyl)benzo[g]pteridine-2, 4(3H,10H)-dione

Riboflavin also known as vitamin B_2, is an easily absorbed micronutrient which plays a key role in maintaining health in human beings and animals. It is the central component of the cofactors FAD and FMN and is therefore required by all flavoproteins. Vitamin B_2 is also required for a wide variety of cellular processes. Like the other B vitamins, it plays a key role in energy metabolism and is required for the metabolism of fats, ketone bodies, carbohydrates and proteins.

Riboflavin has been used in several clinical and therapeutic situations. For over 30 years, riboflavin supplements have been used as part of the phototherapy treatment of neonatal jaundice. The light used to irradiate the infants breaks down not only the toxin causing the jaundice, but also the naturally occurring riboflavin within the blood of the infants as well.

Of late, there has been growing evidence of the fact that supplemental riboflavin may be a useful additive along with beta-blockers in the prevention of migraine headaches.

Recently, riboflavin has also been used to slow or stop the progression of the corneal disorder known as keratoconus. This is called *corneal collagen crosslinking*. In corneal crosslinking, riboflavin drops are applied to the patient's corneal surface. Once the riboflavin has penetrated through the cornea, ultraviolet A light therapy is applied. This induces collagen crosslinking, which increases the tensile strength of the cornea. This treatment has been sccusseful in several studies where the disease keratoconus has stabilized.

Vitamin B₃ Pyridine-3-carboxylate

Niacin, also known as vitamin B_3 or nicotinic acid is an organic water-soluble vitamin. It is a derivative of pyridine with a carboxyl group at the 3-position. Other forms of vitamin B_3 include the corresponding amide, nicotinamide ("niacinamide"), where the carboxyl group gets replaced by a carboxamide group as well as the more complex amides and a variety of esters.

Niacin is a precursor to NADH, NAD^+, $NADP^+$ and NADPH, which plays an essential metabolic role in living cells. Niacin is involved in both DNA repair and the production of steroid hormones in the adrenal gland.

Vitamin B₅ 3-[(2,4-Dihydroxy-3,3-dimethylbutanoyl)amino] propanoic acid

Pantothenic acid, also called vitamin B_5 is a water-soluble vitamin required to sustain life (essential nutrient). Only the dextrorotatory (D) isomer of pantothenic acid possesses biological activity. The levorotatory (L) form may antagonize the effects of the dextrorotatory isomer.

Pantothenic acid is used in the synthesis of coenzyme A (CoA). Coenzyme A may act as an acyl group carrier to form acetyl-CoA and other related compounds; this is a way to transport carbon atoms within the cell. CoA is important in energy metabolism for pyruvate to enter the tricarboxylic acid cycle (TCA cycle) as acetyl-CoA and for α-ketoglutarate to be transformed to succinyl-CoA in the cycle. CoA is also important in the biosynthesis of many important compounds such as fatty acids, cholesterol and acetylcholine. CoA is incidentally also required in the formation of ACP, which is also required for fatty acid synthesis in addition to CoA. Pantothenic acid in the form of CoA is also required for acylation and acetylation, which, for example, are involved in signal transduction and enzyme activation and deactivation, respectively.

Vitamin B₆ 4,5-bis(Hydroxymethyl)-2-methyl-pyridin-3-ol

Vitamin B_6 is a water-soluble vitamin and is part of the vitamin B complex group. Pyridoxal phosphate (PLP) is the active form.

Pyridoxal phosphate is involved in many aspects of macronutrient metabolism, neurotransmitter synthesis, histamine synthesis, hemoglobin synthesis and function and gene expression. Pyridoxal phosphate generally serves as a coenzyme for many reactions and can help facilitate decarboxylation, transamination, racemization, elimination, and replacement and β-group inter conversion reactions.

Vitamin B$_7$ 5-((1S,3aS,6aR)-4,6-Dioxohexahydro-1H-thieno[3,4-c]pyrrol-1-yl) pentanoic acid
Biotin, also known as vitamin H or B$_7$, is a water-soluble B-complex vitamin which is composed of an ureido (tetrahydroimidizalone) ring fused with a tetrahydrothiophene ring. A valeric acid substituent is attached to one of the carbon atoms of the tetrahydrothiophene ring.

Biotin is a coenzyme in the metabolism of fatty acids and leucine and it plays a role in gluconeogenesis.

Vitamin B$_9$ 2-(4-((2-Amino-4-hydroxypteridin-6-yl)methylamino)benzamido)pentanedioic acid
Folic acid (also known as vitamin B$_9$ or folacin) and folate (the naturally occurring form), as well as pteroyl-L-glutamic acid and pteroyl-L-glutamate, are forms of the water-soluble vitamin B$_9$. Folic acid is itself not biologically active with its biological importance due to tetrahydrofolate and other derivatives after its conversion to dihydrofolic acid in the liver.

Vitamin B$_9$ (folic acid and folate inclusive) is essential to numerous bodily functions ranging from nucleotide biosynthesis to the remethylation of homocysteine. The human body needs folate to synthesize, repair and methylate DNA as well as to act as a cofactor in biological reactions involving folate. It is especially important during periods of rapid cell division and growth. Both children and adults require folic acid to produce healthy red blood cells and prevent anemia. A lack of dietary folic acid leads to folate deficiency. This can result in many health problems, most notably neural tube defects in developing embryos.

Vitamin B₁₂

Vitamin B_{12}, also called cobalamin is a water soluble vitamin with a key role in the normal functioning of the brain and nervous system and also ensures formation of blood. Vitamin B_{12} is normally involved in the metabolism of every cell of the body, especially affecting the DNA synthesis and regulation but also fatty acid synthesis and energy production.

Vitamin C: (R)-3,4-Dihydroxy-5-((S)-1,2-dihydroxyethyl)furan-2(5H)-one

Vitamin C is purely the L-enantiomer of ascorbate; the opposite D-enantiomer has no physiological significance. Both forms are mirror images of the same molecular structure.

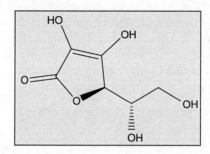

When L-ascorbate which is a strong reducing agent carries out its reducing function it is converted to its oxidized form L-dehydroascorbate. L-dehydroascorbate can then be reduced back to the active L-ascorbate form in the body by enzymes and glutathione. During this process semidehydroascorbic acid radical is formed. Ascorbate free radical reacts poorly with oxygen, and a superoxide is not created. Instead, two semidehydroascorbate radicals will react and form one ascorbate and one dehydroascorbate. With the help of glutathione, dehydroxyascorbate is converted back to ascorbate. The presence of glutathione is crucial since it spares ascorbate and improves antioxidant capacity of blood. Without it dehydroxyascorbate could not convert back to ascorbate.

FURTHER READINGS

Burger's Medicinal Chemistry, Wolff, M.E. (ed.), 4th edition, Vol.II, JohnWiley and Sons, New York. pp. 289–331.

Remington: The Science and Practice of Pharmacy, 21st edition.

MULTIPLE-CHOICE QUESTIONS

1. The heterocyclic nucleuses present in anti-obesity drug rimonabant
   a. Piperidine
   b. Pyrrolidine
   c. Pyrazole and piperidine
   d. Pyrazole and pyrrolidine

2. Give an example of PDE-5 inhibitors:
   a. Tadalafil
   b. Sumatriptan
   c. Nicorandil
   d. Albuterol

3. Anti-obesity drug orlistat acts by
   a. Blocking the neurotransmitter release
   b. Inhibiting pancreatic lipase
   c. Blocking cholesterol synthesis
   d. Acting on cannabinoid receptor
4. Give an example of 5-$HT_{1B/1D}$ agonist:
   a. Sildenafil
   b. Droperidol
   c. Loxapine
   d. Sumatriptan
5. Montelukast is used for the treatment of
   a. Asthma
   b. Migraine
   c. Cough
   d. Male erectile dysfunction
6. Mechanism of action of clopidogrel
   a. Cysteine M synthase inhibitors
   b. Adenosine diphosphate (ADP) receptor inhibitors
   c. Glycoprotein IIB/IIIA inhibitors
   d. PDE-4 inhibitors

QUESTIONS

1. Write the synthesis and mechanism of action of the following drugs:
   a. Sildenafil
   b. Montelukast
   c. Sibutramine
   d. Sumatriptan
2. Classify antiseptics and disinfectants. Give the synthesis of triclosan.
3. Give the synthesis and uses of tyropanoic acid.
4. Classify antiplatelet drugs with examples. Give synthesis of any two.

SOLUTION TO MULTIPLE-CHOICE QUESTIONS

1. c;
2. a;
3. b;
4. d;
5. a;
6. b.

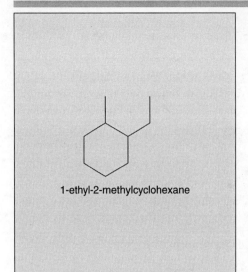

1-ethyl-2-methylcyclohexane

CHAPTER 35

Nomenclature of Medicinal Compounds

LEARNING OBJECTIVES

- To learn IUPAC naming of chemical structure
- Common heterocyclic nucleus occured in medicinal chemistry
- Some examples of drug molecule

35.1 INTRODUCTION

'**Nomenclature**', in chemistry, is a system of naming compounds using various nomenclatural operations in accordance with a set of principles, rules, and conventions. International Union of Pure and Applied Chemistry (IUPAC) have developed systematic nomenclature, a set of rules that allows any compound to be given a unique name that can be deduced directly from its chemical structure. Systematic names can be divided into three parts: one describes the **basic framework/skeleton;** one describes the **functional groups;** and one indicates **where the functional groups are attached** to the skeleton.

35.2 GENERAL RULES

- The names of the first 10 *simple alkanes* are methane, ethane, propane, butane, pentane, hexane, heptane, octane, nonane, and decane.
- *Branched alkanes* have alkyl substituents branching off from the main chain. When naming a branched alkane, identify the longest chain and number it from the end nearest the branch point. Identify the substituent and its position on the longest chain.

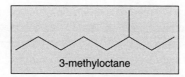
3-methyloctane

- If there is more than one alkyl substituent present in the structure (***multi-branched alkanes***), then the substituents are named in alphabetical order, numbering again from the end of the chain nearest the substituents. If a structure has identical substituents, then the prefixes di-, tri-, tetra-, *etc.*, are used to represent the number of substituents.

3-ethyl-4-methylhexane   3,3,4-trimethylhexane

- ***Cycloalkanes*** are simply named by identifying the number of carbons in the ring and prefixing the alkane name with 'cyclo'. Examples: cyclopropane, cycolobutane cyclopentane, cyclohexane.
- Cycloalkanes consisting of a cycloalkane moiety linked to an alkane moiety (***branched-cycloalkanes***) are usually named such that the cycloalkane is the parent system and the alkane moiety is considered to be an alkyl substituent. Note that there is no need to number the cycloalkane ring when only one substituent is present. If the alkane moiety contains more carbon atoms than the ring, the alkane moiety becomes the parent system and the cycloalkane group becomes the substituent.

Mothylcyclohexane   1-cyclohexylbutane

- Branched cycloalkanes having different substituents (***multi-branched cycloalkanes***) are numbered such that the alkyl substituent having alphabetical priority is at position 1. The numbering of the rest of the ring is then carried out such that the substituent positions add up to a minimum.

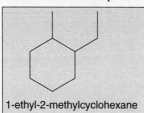
1-ethyl-2-methylcyclohexane

A functional group is a portion of an organic molecule which consists of atoms other than carbon and hydrogen, or which contains bonds other than C—C and C—H bonds. Many of the nomenclature rules for alkanes hold true for molecules containing a functional group, but extra rules are needed in order to define the type of functional group present and its position within the molecule. The main rules are as follows:

- The main chain must include the functional group, and so may not necessarily be the longest chain.

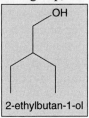

2-ethylbutan-1-ol

- The presence of some functional groups is indicated by replacing -ane for the parent alkane chain with the following suffixes: **functional group (*suffix*)** alkene (*-ene*), alkyne (*-yne*), alcohol (*-anol*), aldehyde (*-anal*), ketone (*-anone*), carboxylic acid (*-anoic acid*), acid chloride (*-anoyl chloride*), and amine (*-ylamine*).
- Numbering must start from the end of the main chain nearest the functional group.
- The position of the functional group must be defined in the name.
- Other substituents are named and ordered in the same way as for alkanes.

The best-known aromatic structure is benzene. If an alkane chain is linked to a benzene molecule, then the alkane chain is usually considered to be an alkyl substituent of the benzene ring. However, if the alkane chain contains more than six carbons, then the benzene molecule is considered to be a **phenyl** substituent of the alkane chain.

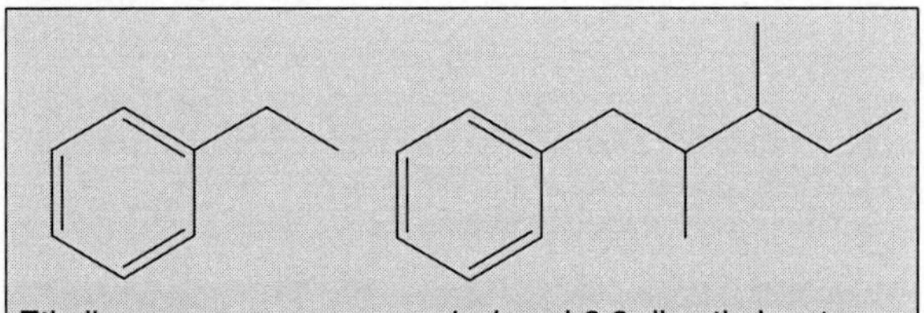

Ethylbenzene   1-phenyl-2,3-dimethylpentane

- With disubstituted aromatic rings, the position of substituents must be defined by numbering around the ring such that the substituents are positioned at the lowest numbers possible. The following structure is 1,3-dichlorobenzene and not 1,5-dichlorobenzene.

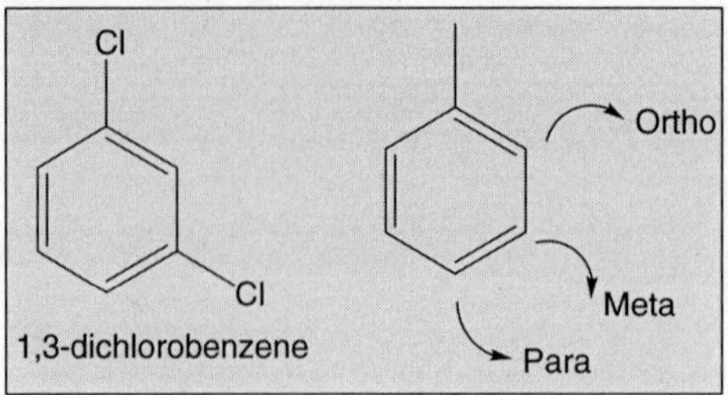

1,3-dichlorobenzene

Ortho
Meta
Para

Alternatively, the terms *ortho*, *meta*, and *para* can be used. These terms define the relative position of one substituent to another. Thus, 1,3-dichlorobenzene can also be called *meta*-dichlorobenzene. This can be shortened to *m*-dichlorobenzene.

How Should you Name Compounds?

Draw a structure first and worry about the name afterwards
- Learn the names of the functional groups (ester, nitrile, etc.)
- Learn and use the names of a few simple compounds used by all chemists
- In speech,, refer to compounds as 'that acid' (or whatever) while pointing to a diagram
- Grasp the principles of systematic (IUPAC) nomenclature and use it for compounds of medium size
- Keep a notebook to record acronyms, trivial names, structures, etc. that you might need later

---

35.3 HETEROCYCLIC COMPOUNDS

A cyclic organic compound containing all carbon atoms in ring formation is referred to as a ***carbocyclic compound.*** If at least one atom other than carbon forms a part of the ring system, then it is designated as a ***heterocyclic*** compound. Nitrogen, oxygen, and sulphur are the most common heteroatoms, but heterocyclic rings containing other heteroatoms are also widely known.

In heterocyclic chemistry, there is a special name for each individual ring system and a trivial name for each compound. Trivial names convey little or no structural information but they are still widely used. The systematic name, in contrast, is designed so that one may deduce from it the structure of the compound. They tend to be long. However, a systematic nomenclature is still indispensable. In recent years, the IUPAC has made efforts to systematize the nomenclature of heterocyclic compounds. According to this system, single three-to-ten-membered rings are named by combining the appropriate prefix or prefixes (listed in Table 35.1) with a stem from Table 35.2.

Table 35.1
Prefix for heteroatoms

Hetero Atom	Prefix
O	Oxa
N	Aza
S	Thia
P	Phospha
As	Arsa

Table 35.2
Common name endings for heterocyclic compounds

Ring Size	Suffixes for Fully Saturated Compounds		Suffixes for Fully Unsaturated Compounds	
	With N	Without N	With N	Without N
3	-irine	-irene	-iridine	-irane
4	-ete	-ete	-etidine	-etane
5	-ole	-ole	-olidine	-olane
6	-ine	-in		-ane
7	-epine	-epin		-epane
8	-ocine		-ocin	-ocane

Accordingly, some examples of compounds named on the basis of tables 1 and 2 are cited here:

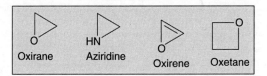

Saturated or hydrogenated ring systems are named by varying the ending or by placing prefixes such as '*dihydro-*' and '*tetrahydro-*'. The ending of the name will depend on the presence or absence of nitrogen. Two or more similar atoms contained in a ring are indicated by the prefixes '*di-*', '*tri*', etc.

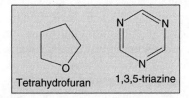

If two or more different heteroatoms occur in the ring, then it is named by combining the prefixes in Table 1 with the ending in Table 2 in order of their preference, i.e., O, S, and N. This is illustrated by the examples given here:

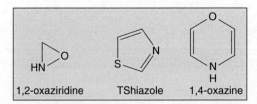

Numbering of the heterocyclic rings becomes essential when substituents are placed on the ring. Conventionally, the heteroatom is assigned position 1 and the substituents are then counted around the ring in a manner so as to give them the lowest possible numbers. While writing the name of the compound, the substituents are placed in an alphabetical order. In case the heterocyclic ring contains more than one heteroatom, the order of preference for numbering is O, S, and N. The ring is numbered from the atom of preference in such a way so as to give the smallest possible number to the other heteroatoms in the ring. As a result, the position of the substituent plays no part in determining how the ring is numbered in such compounds. The examples here illustrate this rule:

1,2–isoxazole   4–methylthiazole   Thiirene1,1–dioxide

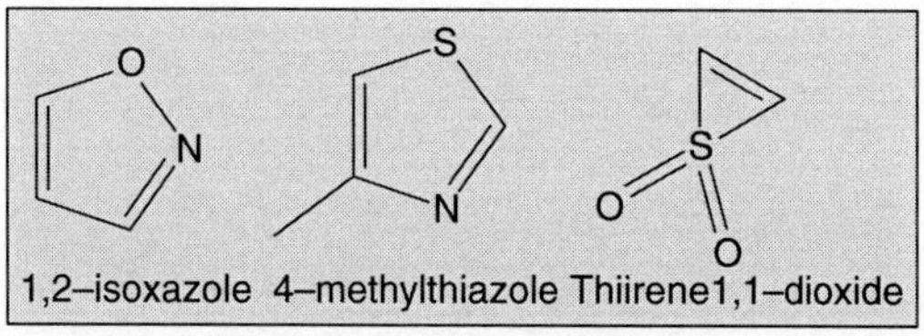

There are a large number of important ring systems that do not possess any systematic names; rather, non-systematic or common names are used for them. Examples include the following:

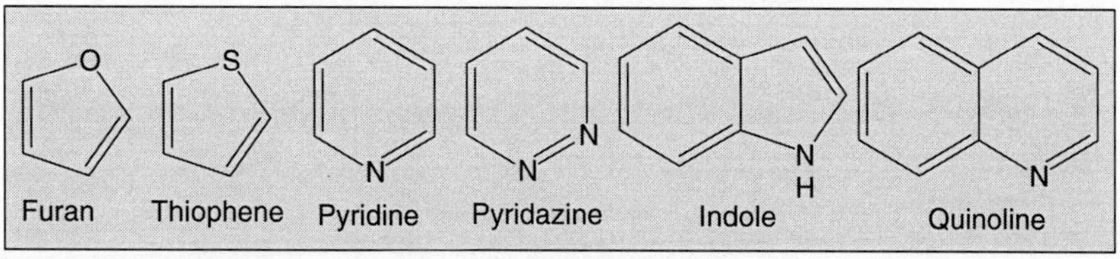

Furan   Thiophene   Pyridine   Pyridazine   Indole   Quinoline

In a heterocyclic ring with maximum unsaturation, if the double bonds can be arranged in more than one way, then their positions are specified by numbering those nitrogen or carbon atoms that are not multiply-bonded, i.e., bear an *'extra'* hydrogen atom, by italic capital '1*H*', '2*H*', '3*H*', etc. The numerals indicate the position of these atoms having the extra hydrogen atom. The examples here illustrate this rule:

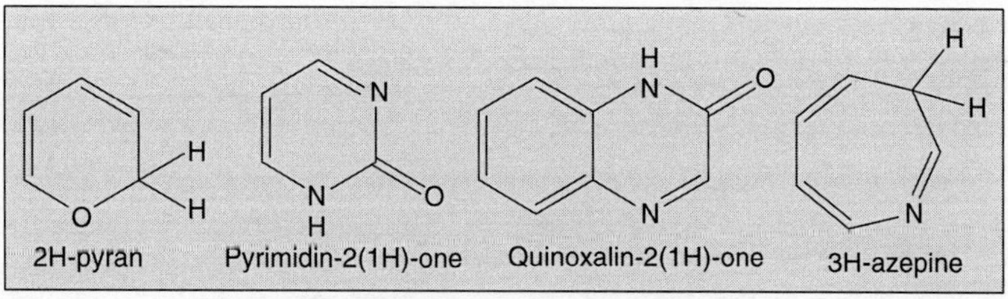

2H-pyran   Pyrimidin-2(1H)-one   Quinoxalin-2(1H)-one   3H-azepine

35.4 EXAMPLES

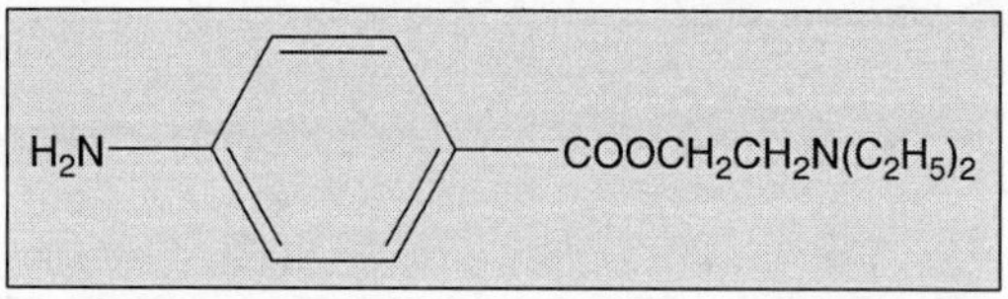

H$_2$N—⟨⟩—COOCH$_2$CH$_2$N(C$_2$H$_5$)$_2$

Select the basic molecule; here it is *p*-aminobenzoate. Give the numbering for other portion; here it is 2-diethylaminoethyl chain. Hence, the name is ***2-(diethylamino)ethyl 4-aminobenzoate***.

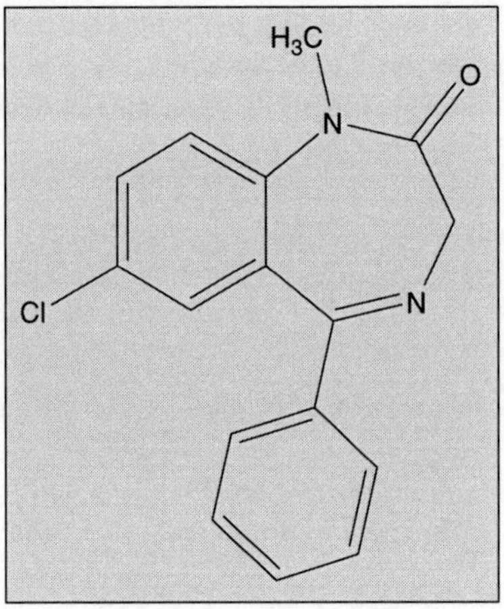

Here, the basic skeleton is 1,4-benzodiazepine-2-one derivative. Substituents are there in 1, 5, and 7 positions. Applying common rules, the name is *7-chloro-1-methyl-5-phenyl-1H-benzo[e][1,4]diazepin-2(3H)-one*.

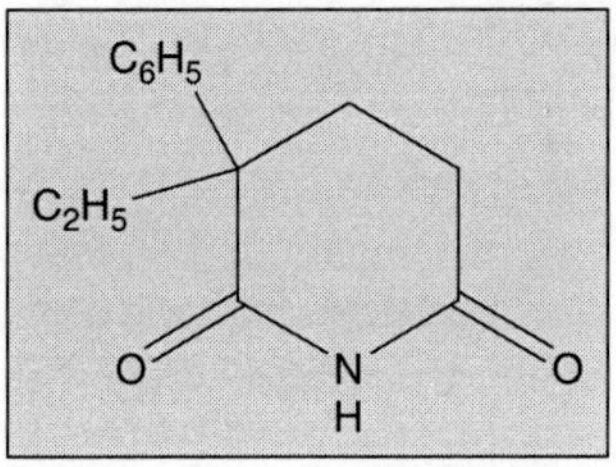

In this molecule the basic nucleus is piperidine-2,6-dione, and in third position ethyl and phenyl groups are attached. Hence, it is *3-ethyl-3-phenylpiperidine-2,6-dione*.

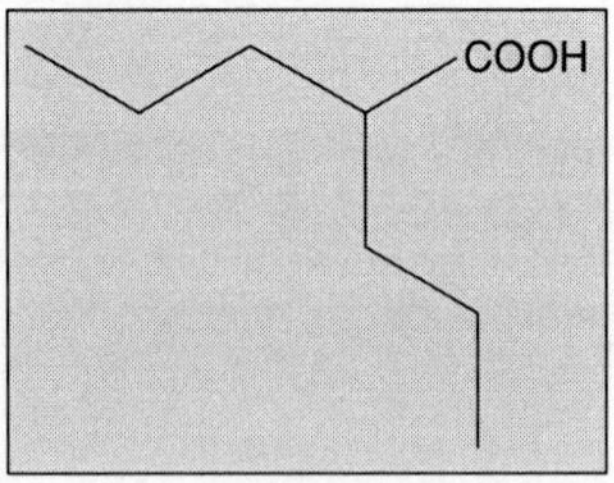

Select the longest chain with functional group. Give the functional group lowest number; here the name of the compound is *2-propylpentanoic acid*.

FURTHER READINGS

1. *Advances in Heterocyclic Chemistry*, Vols 1 to 27, Katritzky, A.R. and J.A. Boulton (ed.), Academic Press, New York (1963–1980).
2. *Organic Chemistry* by Clayden, Greeves, Warren and Wotheres, Oxford Press, 2002.

QUESTIONS

Draw the structures for the following IUPAC names:
a. 1-Ethyl-7-methyl-4-oxo-[1,8]naphthyridine-3-carboxylic acid
b. 1-[1-(4- Chlorophenyl) cyclobutyl]- *N,N*,3-trimethylbutan- 1-amine
c. 2,2-Dichloro-*N*-(4-furoyloxyphenyl)-*N*-methylacetamide
d. 4-[(7-Chloro-4-quinolinyl) amino]-2-[(diethylamino) methyl] phenol
e. 3'-Azido-2',3'-dideoxythymidine
f. 2-(Ethyl)isonicotinthioamide
g. 2,4-Diamino-5-(3',4',5'-trimethoxy benzyl) pyrimidine
h. 1-(4-Amino-6,7-dimethoxy-2-quinazolinyl)-4-(2-furoyl) piperazine
i. 4,5-Dichloro-1,3-benzenedisulphonamide
j. Ethyl 1-[2-(4-aminophenyl)ethyl]-4-phenyl-piperidine-4-carboxylate
k. 4-Butyl-1-(4-hydroxyphenyl)-2-phenylpyrazolidine-3,5-dione
l. 4-[4-(*p*-Chlorophenyl)-4-hydroxypiperidino]-4'-fluorobutyrophenone
m. 5-Ethyl-5-phenyldihydropyrimidine-4,6(1*H*,5*H*)-dione
n. 2-Chloro-1,1,2-trifluoroethyldifluoromethyl ether

Index

2-(p-aminobenzenesulphonamido)-5-isopropyl-thiadiazole (IPTD), 386
4-hydroxycoumarin, 7
5-fluorouracil, 72, 572–573
6-mercaptopurine, 571
6-thioguanine, 571–572
9-amino acridines, 542–543
9-fluorenylmethoxycarbonyl, 66
19-nor derivatives, 609–611
α-adrenergic antagonists, 409
α-adrenoreceptor agonists, 402
$β_1$-adrenergic agonists, 402, 409
$β_2$-selective adrenergic agonists, 402–403
β-adrenergic blockers, 336–339, 377
β-adrenoreceptors, 336
β-Keto ester, 444
β-lactam antibiotics, 448–472
β-lactamase inhibitors, 486

A

abacavir, 513
abciximab, 644
acarbose, 392
acecainide, 357
aceclofenac, 250 251
acenocoumarin, 398
acetaminophen, 10
acetycholine chloride, 416
acetylcholine esterase inhibitors, 268–270
acetylation, 55
acetylcholinesterase (AChE), 418–422
acid-base chemistry, 29–34
acidity, 32–34
    factors affecting, 32–34
acquired immunodeficiency syndrome (AIDS). See HIV
acrivastine, 292–293

actinomycin D, 575
activation-aggravation theory, 41–42
acyclic nucleoside analogues, 525–527
acyclovir, 73, 525
acyl cyanide, 167
adefovir, 514
adenosine diphosphate (ADP) receptor inhibitors, 642–643
adrenal androgens, 612
adrenaline, 404–405
adrenergic blocking agents, 408–411
adrenergic drugs, 401–411
adrenergic receptors, 401
agonist, 43–44
albendazole, 547
alcohols, 23, 651
aldehydes, 24, 651–652
alkylating agents, 558–567
alkynes, 20, 22
allopurinol, 248
almitrine, 273
aloxidone, 159
altretamine, 577
Alzheimer's disease (AD), 266
Alzheimer's disease treatment, 266–270
amantadine, 258, 527
ambenonium chloride, 419
amdinocillin, 458
amides, 26, 35, 63
amidines, 35
amileridine, 217–218
amines, 25, 63
amino alkyl ethers, 278–279
aminoalkyl ether derivatives, 282–284
aminoglutethimide, 580
aminoglycoside antibiotics, 482

amiodarone, 356–357
amisulpride, 191–192
amodiaquine, 541
amoebicides, classification of, 534
amoxapine, 200
amoxicillin, 456
amphetamines, 271, 630
amphotericin B, 496
ampicillin, 456, 64–65
aneurin, 665
angina pectoris, 373
angiotensin II antagonists, 345–347
angiotensin-converting enzyme (ACE) inhibitors, 342–345
anilides, 121–122, 652
aniline, 34
anilodes, 119–122
anisindione, 397
anisotropine methyl bromide, 423–424
antagonist, 45
antazoline, 289
anthelmintics, 546–551
anthracycline antibiotics, 574–575
anthranilic acid derivatives, 235–236
antiamoebic agents, 533–536
antianginal drugs, 373–392
anti-arrhythmic drugs, 351–371
    class I agents, 352–356
    class II agents, 356
    class III agents, 356–357
anti-asthmatics, 637–639
antibacterials, 439–446
antibiotics, 447–486
anticancer agents, 557–582
anticancer antibiotics, 574–575
anticancer plant products, 577–579
anticoagulants, 394–400
anticonvulsants, 169. See also epilepsy

anticonvulsants, 164–170
anti-cough drugs, 645–646
antidepressants, 195–205
 classification, 196
 mechanism of action, 196
 side effects, 196
 use of, 196
antifungal antibiotics, 496–497
antifungal drugs, 495–503
anti-gout drugs, 247–248
anti-herpes simplex virus (HSV) agents, 522–527
antihistamines, 278
anti-HIV agents, 505–522
antihyperlipidemic agents, 359–371
antihypertensive drugs, 328–348
anti-inflammatory agents, 228
antimalarial agents, 536–545
antimuscarinic drugs, 422–426
antineoplastic agents, 557–582
anti-obesity drugs, 629–632
antioxidant, 656
anti-parkinsonism agents, 255–260
anti-platelet drugs, 642–644
anti-pseudomonal penicillins, 451
antipsychotic drugs, 173–192
antipyretics, 227–228
antiseptics, 651–654
antithrombotics, 642–644
antitubercular agents, 489–493
anxiolytic agents, 148–150
aripiprazole, 192
aromatic amines, 34
aromatic hydrocarbons, 22
arrhythmia, 351–352
artemisinin, 64
artesunate, 545
aryl alkanoic acid derivatives, 236
aryl-and heteroaryl acetic/propionic acid derivatives, 240–244
arylcyclohexylamine, 101–102
aryloxyisobutyric acid derivatives, 365–367
ascorbate, 668
aspartame, 657
aspirin, 230, 378
astemizole, 293
asthma, 637
atenolol, 337

atherosclerosis, 359–360
atorvastatin, 369–370
atypical antidepressants, 200–202
atypical antipsychotic agents, 185–191
ATZ antagonists, 345–347
azatadine, 289–290
azathioprine, 645
azelastine, 302
aziridines, 566
azithromycin, 480–481
azlocillin, 457
azo reduction, 71
azole antifungals, 497–500
aztreonam, 485

B

bacitracin, 483
baclofen, 265
barbiturates, 130–131, 135–136
basicity, 32–34
 factors affecting, 32–34
beclomethasone, 623–624
benoxinate, 118
bentiromide, 646–647
benxonatate, 646
benzalkonium chloride, 653
benzazocin (benzomorphan) derivatives, 220–221
benzodiazepines, 102–104, 131–132, 136–143, 160, 264
benzodiazepine antagonists, 143–144
benzoic acid derivatives, 113–116, 118–119, 388–389
benzotepa, 566
benztropine, 259
besipirdine, 266–267
betamethasone, 623–624
bile acid sequestrants, 367
bioisosteres, 39
bioisosteric replacement, 39
bioprecursor prodrug, 69–73
biotin, 667
biperidine, 259
bis-phenols, 652
bis-thiophenyl ketone, 162
blastic tumour, 557

bleomycin sulphate, 576
blood clotting mechanism, 395–396
blood, coagulation of, 394–396
bretylium, 357
brivudine, 523
broad-spectrum penicillins, 450–451
broad-spectrum ureido penicillins, 451
brodimoprim, 437
bromazepam, 141
bromindione, 397
Bronsted-Lowry theory, 29
brotizolam, 142
buccal androgens, 613
buclizine, 288
buformin, 384
bupivacaine hydrochloride, 120–121
buprenorphine, 215
bupropion, 201
buspirone, 149–150
butylated hydroxyanisole, 656
butylated hydroxytoluene, 656
butyrophenone derivatives, 181–183

C

caffeine, 271
calciferol, 662–663
calcium channel blockers, 339–342, 357, 377
camptothecin (CPT), 577–578
cancer, 557–582
candesartan, 345–346
caprofen, 242
captopril, 343
carbachol, 417
carbamate, 27
carbinoxamine, 283
carbocyclic compound, 673
carbon, 20
carbonate, 27
carbonic anhydrase (CA) inhibitors, 309–311
carboxylic acids, 26
carbutamide, 383
carcinoma, 557
cardiac glycosides, 626–627
carisoprodol, 262–263

Index

carmustine, 564–565
carrier-linked bipartite prodrugs, 64–67
carrier-linked prodrugs, 58–69
cefaclor, 460, 467–468
cefadroxil, 459, 465–466
cefamandole, 460, 468
cefazolin, 460
cefepime, 485
cefoperazone, 461, 471–472
cefotaxime, 460, 469–470
cefpirome, 485
ceftazidime, 461, 470–471
ceftizoxime, 460, 470
ceftriaxone, 461
cefuroxime, 460, 468–469
celecoxib, 249
centrally acting antihypertensive drugs, 332–334
cephacetrile, 460
cephalexin, 459, 465
cephalosporins, 459–472, 485
cephalothin, 466
cephradine, 459, 466
cetirizine, 10, 291–292
cetrimide, 653
cetylpyridium chloride, 653
charge-transfer complexes, 48–49
chemokine receptors, 50
chemotherapeutic alkylating agents, 558
chemotheraphy, 558
chirality, 27–28
chlorambucil, 561
chloramphenicol, 482–483
chloramphenicol acetate, 67
chlordiazepoxide, 141
chlorhexidine, 652, 654
chlorine-releasing agents (CRA), 652
chlormadinoneacetate, 609
chloroprocaine, 117
chloroquine, 8, 538–540
chloroxylenol, 653
chlorozoxanzone, 263
chlorphenesin carbamate, 261–262
chlorpheniramine, 14
chlorprothixene, 180
chlorthalidone, 323

cholestyramine, 368
cholestyramine resin, 368
cholinergic agonist, 266–268
cholinergic blocking agents, 422–426
cholinergic drugs, 413–427
cholinergic nerves, 413
cholinergic receptors, 414–415
cholinomimetics, 415–418
cholonoblockers, 329
cisapride, 427
cisplatin, 581
clarithromycin, 480
classical antipsychotic agents, 174–185
clemastine, 284
clenbuterol, 411
clidinium bromide, 424
climdamycin HCl injection, 67
clinical studies, 6
 phase I, 5
 phase II, 5–6
 phase III, 6
clobazam, 160
clofibrate, 366
clomethiazole, 150
clonazepam, 137
clonidine, 332–333
clopamide, 321
clopidogrel, 643
clorazepate, 138–139
clorexolone, 323–324
clotrimazole, 498–499
cloxacillin, 455
clozapine, 185–186
CNS stimulants, 270–274
coagulation of blood, 394–396
cobalamin, 668
cocaine, 8, 113–114
colestipol, 368
colourants, 657–658
colouring agents, 657–658
compounds, 670
 nomenclature of, 670–676
computer-aided drug design (CADD), 75–90
congo red, 647
conjugation, 55
corneal crosslinking, 666

coronary artery disease (CAD), 359
corticosteroids, 620
cortisone acetate, 621
coumarin derivatives, 397–399
covalent bonding, 45–46
cox-2 inhibitors, 248–250
cyclandelate, 378
cyclazocin, 221
cyclization, 73
cycloalkanes, 671
cyclobenzaprine, 265
cyclomethcaine sulphate, 115
cyclopentolate, 424–425
cyclophosphamide, 562–563
cyclopropane, 98
cycloserine, 483–484
cyclosporine, 645
cyproheptadine, 288–289
cytarabine, 573

D

D_2 antagonist, 185–186
D_3 receptors, 176
D_4 receptor antagonist, 185
dacarbazine, 567
dalvastatin, 364–365
danazol, 615
dantrolene, 265
darunavir, 529
database searching method, 87–88
daunorubicin, 574–575
defibrotide, 644
dehydroepiandrosterone (DHEA), 612
delavirdine, 515–516
demecarium bromide, 419–420
denzimol, 165
depressive syndrome, 195
dermatophytes, 495
desflurane, 107
dextromoramide, 220
dezinamide, 165
diabetes, 381–382
dibenzocycloheptanes, 199–200
dibenzoxazepine, 184
dibucaine, 124–125
dichlorphenamide, 311
diclofenac, 240
dicloxacillin, 455

dicoumarol, 398
dicyclomine, 425
didanosine, 512–513
dienoestrol, 602
diethyl carbamazine, 549
diethylpropion, 274
diethylstilbestrol, 602
diflunisal, 231
digitoxin, 627
dihydrofloate reductase (DHFR) inhibitors, 435–437
dihydroindole, 185–186
dihydrotestosterone (DHT), 612
dihydroxyheptanoic acid, 361
diiodotyrosine, 641
diloxanide furoate, 534
diltiazem, 341–342, 357
dimethisoquin hydrochloride, 125–126
diperodon, 123
diphenadione, 397
diphenhydramine, 14, 282–283
diphenoxylate, 219
diphenyl heptanones, 219–220
diphenylbutyl piperidine, 183
dipole-dipole bonding, 47
dipyridamole, 377–378
direct vasodilators, 334–336
disinfectants, 651–654
disopyramide, 353–354
distomer, 38
disulphide reduction, 72
diuretics, 305–325
  classification, 306–307
  uses, 305–306
donepenzil, 269–270, 426
DOPA decarboxylase inhibitor, 258
dopamine, 256–257
dopamine agonist, 256–257
dopamine receptors, 176
dopamine-releasing agent, 258
dorzolamide, 13, 324–325
doxapram, 272
doxazosin, 331–332
doxylamine, 283
droperidol, 182
drug action, 36–38
  physicochemical properties and, 36–38

drug derivatives, 56
drug design, procedures followed in, 6–16
drug metabolism, 54–55
drug-receptor bonding, 40–49
  agonists and antagonists, 43–45
  covalent and non-covalent bonding, 45–49
  interaction of drug with receptor, 42
  receptor theories, 40–42
drug-receptor interaction, 41
drugs, 4, 195
  adrenergic drugs, 401–411
  antianginal drugs, 373
  antiarrhythmic, 351–371
  anti-asthmatics, 637–639
  antibiotics, 447–486
  anticancer agents, 557–582
  anticoagulants, 394–400
  anti-cough and expectorants, 645–646
  antidepressants, 195–205
  antifungal agents, 495–503
  antihistamines and anti-ulcer drugs, 277–302
  antihyperlipidemic agents, 359–371
  antihypertensive drugs, 328–348
  anti-migraine drugs, 636–637
  anti-obesity drugs, 629–632
  antiprotozoal agents, 533–554
  antipsychotic drugs, 173–192
  antipyretics and non-steroidal anti-inflammatory drugs, 227–252
  antiseptics and disinfectants, 651–654
  antitubercular agents, 489–493
  antiviral drugs, 505–530
  cholinergic drugs, 413–427
  diuretics, 305–325
  exploitation of side effects of, 11
  immunophrmacological drugs, 644–645
  insulin and oral hypoglycaemic agents, 381–392

interaction with receptor, 42
preclinical development, 4
prostaglandins, 585–591
steroids, 593–627
sulphones, 435
suplhonamide drugs, 429–434
thyroid and anti-thyroid drugs, 639–642
duloxetine, 204–205
dyclonine, 123–124

E
early penicillins, 449
echinocandins, 502
echothiophate iodide, 420
econazole, 499
elcosanoids, 228
  biosynthesis of, 228
electronegativity, 32–33
electronic parameters, molecule, 77–78
emivirine, 517
emtricitabine, 528
enalapril, 343–344
enantiomers, 27
encainide, 354–355
enflurane, 96
enfuvirtide, 14
ephedrine, 406–407
epidural anaesthesia, 111
epilepsy, 153–154, 169. *See also* anticonvulsants
epileptic disorders, 157–158
  agents used for treating, 157–158
epileptic seizures, 154
eplerenone, 324
epoxidation, 71
eprosartan, 347–348
eptifibatide, 644
erectile dysfunction (ED), 632–633
ergosterol, 663
erythrityl tetranitrate, 376
erythromycin, 59, 479–480
ethambutol, 491
ethchlorvynol, 144
ether, 24, 98
ethinamate, 144–145
ethinyl oestradiol, 601
ethionamide, 492–493

Index

ethopropazine, 260
ethosuximide, 160
ethotoin, 158
ethyl piperidyl side-chain, 178
ethylenediamine derivatives, 279, 284
ethynerone, 611
etidocaine hydrochloride, 121
etodolac, 244
etomidate, 105–106
etoricoxib, 251
etorphine, 215–216
etravirine, 528–529
eutomer, 38
evans blue, 648
ezetimibe, 370–371

F

famotidine, 297
fat soluble vitamins, 658–659
fenamates, 235–236
fenfluramine, 630
fenofibrate, 367
fenoprofen, 241
fentanyl, 218
fentanyl citrate, 104
Ferguson principle, 37
fibrates, 365–367
field block anaesthesia, 111
fludrocortisone, 626
fludrocortisones acetate, 622–623
flufenamic acid, 236
flumazenil, 143–144
fluorescein sodium, 648–649
fluoroquinolone, 439–446, 493
folic acid antagonists, 568–571
fondaparinux, 399
formaldehyde (HCHO), 67, 652
formoterol, 411
Friedel-Crafts acylation, 137
functional group, 20–27, 672
furazolidone, 536
furosemide, 320

G

G protein-coupled receptors (GPCR), 49–50
G protein-linked receptors (GPLR), 49

gabapentin, 162
ganciclovir, 526–527
gaseous anaesthetics, 98–99
gastric acid, 294
gemfibrosil, 369
general anaesthetics, 95–107
  inhalation anaesthetics, 96–99
  intravenous anaesthetics, 99–106
  mechanism of action, 95
  newer drugs, 107
generalized seizures, 154
germ cell tumour, 557
gestational diabetes, 382
glibenclamide, 384
glipizide, 384
glucocorticoids, 619–627
glucosidase enzymes, 65
glutaraldehyde, 651
glutethimide, 145
glycerin, 308
glycerol monoether, 261–262
glyceryl trinitrate, 375–376
glycoprotein IIB/IIIA inhibitors, 644
glycopyrrolate, 425–426
glycosides, 626–627
gout, 247–248
griseofulvin, 496
guanabenz, 334

H

H_1 receptor antagonist, 278–294
H_2 receptor antagonist, 294–298
H_3-receptor agonists, 298
H_3-receptor antagonists, 298
H_4 receptors antagonists, 299
halofantrine, 544
halogen-releasing agents, 652
haloperidol, 181–182
halothane, 97–98
halozepam, 137
Hammett coefficient, 82
Hammett substituent coefficient, 37–38
Henderson–Hasselbach equation, 30
heteroatoms, 23–27
heterocyclic compounds, 673–675
hetramine, 14

hexobarbital, 134–135
hexylcaine hydrochloride, 114
high-performance liquid chromatography (HPLC), 78–79
histamine, 277–278
histamine H_3 receptors, 298,
histamine H_4 receptors, 299
HIV, 505, 506–507
HIV protease, 12–13
HMG-CoA reductase inhibitors (statins), 361–365
hormones, 579–580
hydantoins, 158
hydralazine, 335
hydrocarbons, 20–27
  bonded to heteroatoms, 23–27
  simple hydrocarbons, 20–22
hydrocortisone, 621
hydrogen bonding, 48
hydrogen peroxide, 652
hydrolysis, 55
hydrophobic effect, 48
hydrophobicity descriptors, 78–79
hydroxyprogesterone caproate, 607
hyperlipidemia, 359
hypertension, 328–329
hypnotics, 130–148
hypolipidemic agents, 360

I

ibuprofen, 240
icopezil, 270
idoxuridine, 524
ifosfamide, 563–564
imidazobenzodiazepines, 140
imidazole, 35
imidazoline derivatives, 407–408
imipenem, 484
immunopharmacological drugs, 644–645
immunostimulants, 644
immunosuppressants, 644
indanedione derivatives, 399
indapamide, 320–321
indene acetic acid derivative, 238
inhalation anaesthetics, 96–99
inhibitors, 248–250
insomnia, 130

insulin, 9–10, 382–383
intravenous anaesthetics (synthesis, uses, and dose), 99–106
intravenous regional anaesthesia, 111
inverse agonist, 44
investigational drug, 4–6
  testing of, 5–6
investigational new drug (IND), 5
iodophors, 652
iodoquinol, 536
ion-dipole bonding, 47
isobucaine hydrochloride, 114–115
isocarboxazide, 197
isoflurane, 96–97
isoflurophate, 421
isomeric potency, 38
isoniazid, 490
isoprenaline, 405
isoprophylhydrazide, 11
isoquines, 549–551
isosorbide, 308
isosorbide dinitrate, 376

K

kappa (k) receptors, 208
katamine, 101–102
ketoconazole, 497
ketoemidone, 218
ketones, 24
ketoprofen, 241–242
ketorolac, 243

L

labetalol, 338
lamivudine, 512
lamotrigine, 167
lanzoprazole, 301
lasix, 320
latentiated drugs, 56
lead identification, 4
lead optimization, 4
lead compounds, 2
  molecular modification of, 14–16
  search for, 7–13
leukaemia, 557
leukotrienes, 637
levallorphan, 214
levamisole, 551, 644
levermectin, 554

levetiracetam, 170
levobupivacaine, 126
levocetirizine, 292
levodopa, 256–257
levofloxacin, 445
levorphanol, 38, 215
levothyronine, 640–641
lidocaine hydrochloride, 119–120
ligand building method, 88–89
linear free energy relationship, 81–82
linezolid, 486
linogliride, 390–391
linopirdine, 267
lipids, 360
liquid anaesthetics, 96–98
lisinopril, 344
local anaesthetics, 110–126
  classification of, 111–113
  clinical uses of, 111
  mechanism of action, 110
  properties of, 110
lofentanil, 218–219
lomustine, 565
loop diuretics, 318–320
losartan, 345
lovastatin, 363
loviride, 516
loxapine, 184
lumiracoxib, 251–252
lvabradine, 378–379
lymphoma, 557

M

macrolide antibiotics, 477–481
macromolecular perturbation theory, 41
macromolecule-based drug design, 86–89
major tranquillizers, 174
marketing authorization application (MAA), 6
mebendazole, 547
mechlorethamine, 559–560
mecillinam, 458
meclizine, 288
meclofenamate, 236
medroxyprogesterone, 15

medroxyprogesterone acetate, 607–608
medrylamine, 283
mefenemic acid, 236
mefenide, 434
mefloquine, 542
megestrol acetate, 608
meglitinides, 388–389
melphalan, 560–561
meperidine, 8, 217
mephenesin, 261
metabolite, 10–11
metformin, 384
methacholine chloride, 649–650
methacycline, 476
methocarbamol, 262
methohexital sodium, 100–101
methotrexate, 570
methoxyflurane, 97
methrarpone, 649
methsuximide, 160
methyl prednisolone, 64
methyl testosterone, 616–617
methylation, 55
methyldopa, 333
methylphenidate, 271–272
methylprednisolone, 623
methylthiouracil, 641
mevastatin, 362
mexiletine, 354
miconazole, 499
midazolam, 102–104
miglitol, 392
migraine, 636
milameline, 267–268
minaxolone, 107
mineralocorticoids, 625
minocycline, 477
minoxidil, 11, 335–336
mithramycin, 575
mitomycin C, 576
mixed D_2 and $5HT_{2A}$ antagonist, 185–186
molar refractivity (MR), 79–80
molecular connectivity index, 81
molecular descriptors, 76–77
molecular modification, 14–16
molindone, 185–186

Index

monoamine oxidase inhibitor (MAOI), 10
monoamino oxidase (MAO) inhibitors, 196–198
monobactams, 485
montelukast, 639
moropenem, 484
morphine, 8, 26, 212–216
mu (μ) receptors, 208
muscarinic receptors, 414–415
muscle relaxants, 260–266
mutual prodrugs, 67–69
mycoses, 495

N
nafimidone, 167
naftifine, 501
naleglinide, 388
nalidixic acid, 445
naphazoline, 408
naproxen, 241
narcotic analgesics, 104–105, 207–225
    classification, 209–212
    mechanism of action, 208
    pharmacophore requirement, 208–209
    use of, 209
narcotic antagonists, 221–223
nateglinide, 391
N-dealkylation, 69
nelfinavir, 521–522
neostigmine, 422
nerve block anaesthesia, 111
neuroleptics. *See* antipsychotic drugs
neurosteroids, 604
nevirapine, 514–515
new drug application (NDA), 6
newer anticonvulsants, 164–170
newer drugs, 107, 126, 150, 274
    analgesics, 223–224
    antidepressants, 203–205
    antipsychotic drugs, 191–192
    general anaesthetics, 107
    local anaesthetics, 126
    Parkinson's disease, 274

niacin, 369, 666
niclosamide, 550–551
nicorandil, 376
nicotinic receptors, 414
nitazoxanide, 535–536
nitrogen mustards, 560–564
nitrogen, 20–21
nitrogen-containing compounds, 34–36
nitrous oxide, 98–99
nizatidine, 297
noitrosoureas, 564–565
nomenclature, 670–676
    general rules, 670–673
    heterocyclic compounds, 673–675
non- steroidal oestrogens, 598
non-covalent bonding, 46–49
non-nucleoside RT inhibitors (NNRTI), 514–518
non-sedative anti-histamines, 290–294
non-selective α-antagonists, 409
non-selective β-antagonists, 409
non-specific agents, 37
non-steroidal anti-inflammatory (NSAIDS), 228–229
nonsteroidal anti-inflammatory or fever-reducing drugs–NSAID, 207
norethynodrel, 609–611
N-oxidation, 71
NSAIDS, 229
    side-effects, 229
nucleophilic substitution, 25
nucleoside RT inhibitors, 508–514
nucleotide activation, 72–73
nystatin, 497

O
obesity, 629–630
occupancy theory, 40
octanol, 78
O-dealkylation, 70
oestradiol esters, 601
oestradiol, 600–601
oestrogens, 598–603
ofloxacin, 444–445
omeprazole, 73, 300

open door concept, 43
oral androgens, 613
oral anticoagulants, 394–400
oral hypoglycaemic agents, 383–385
organic nitrates, 374–376
orlistat, 630–631
ornidazole, 535
orphenadrine, 264–265
osmotic diuretics, 307–308
Overton–Meyer hypothesis, 37
oxacillin, 455
oxandrolone, 617
oxazepam, 137
oxazolidinediones, 159
oxygen, 21
oxymetazoline, 408
oxypertine, 185
oxyphenbutazone, 234

P
paliperidone, 192
pamaquine, 541–542
p-aminobenzoic acid derivatives, 116–119
p-aminophenol derivatives, 231–233
pantoprazole, 300–301
pantothenic acid, 666
parabens, 655
paracetamol, 232–233
paramethadione, 159
parenteral androgens, 613
pargyline, 198
Parkinsonism, 255–256
partial seizures, 154
PDE-5 inhibitors , 632–635
pemetrexed, 582
pemoline, 272
penams. *See* penicillin
penicillin, 8–9, 449–454
penicillinase-resistant penicillins, 450
pentagastrin, 650
pentamidine, 552
pentazocine, 220–221
pentobarbitone, 16
pergolide, 257
perhexiline, 379

peripheral anti-adrenergic drugs, 330–332
peroxygens, 652
p-ethylphenol, 23
pharmacology/toxicology, 4
pharmacophore, 83
pharmacophore-based drug design, 83–85
phase I clinical studies, 5
phase I reactions, 54
phase II clinical studies, 5–6
phase II reactions, 55
phase III clinical studies, 6
phase IIIb/IV studies, 6
phenacaine hydrochloride, 122
phenacemide, 164
phenformin, 384, 387
phenindione, 397
pheniramine, 285
phenobarbital, 134
phenol, 23, 652
phenol red, 650
phenothiazine derivatives, 280, 286–287
phenothiazine ring, 39
phenothiazines, 176–179
phenoxybenzamine, 409–410
phenprocoumon, 398–399
phenserine, 269
phensuximide, 160
phentermine, 273
phentolamine, 410
phenyl (ETHYL) piperidines, 216–219
phenyl mercuric nitrate, 655
phenylbutazone, 234
phenylephrine, 406
phenylethanolamine derivatives, 404–407
phenylethyl hydantoin, 158
phenytoin, 158
phosphodiesterase inhibitor, 643–644
phosphorus, 21
phosphorylation activation, 73
phototoxicity, 442
physostigmine, 421–422
pilocarpine, 417–418
pimozide, 183

pioglitazone, 385, 391–392
piperacillin, 450
piperazine citrate, 548
piperazine derivatives, 281, 287–288, 548–549
piperocaine hydrochloride, 115–116
piroxicam, 245
pivampicillin, 456–457
p-nitrophenol, 23
podophyllotoxins, 578–579
polymyxin B sulphate, 483
polymyxines, 483
post-approval studies, 6
potassium-sparing diuretics, 314–317
povidone-iodine, 653
pralidoxime, 71
pramipexole, 274
pramoxine hydrochloride, 123
pregabalin, 170
pregnane derivatives, 606–609
procainamide, 353
procaine hydrochloride, 117
procarbazine, 580–581
prochlorperazine, 179
procyclidine, 259
prodrugs, 55–73
  characteristics of, 57–58
  classification of, 58–73
  concept, 56–57
  for improved absorption and distribution, 64–65
  for increased water solubility, 64
  for prolonged release, 66
  for site specificity, 65
  for stability, 65–66
  purpose of, 57
  to minimize toxicity, 66
progabide, 161
progesterone, 15
progesterone, 603–611
progestins, 603–611
proguanil, 543–544
proliferators-activated receptors (PPARα), 365
promethazine, 286
propafenone, 356
proparacaine hydrochloride, 118
propofol, 106

propoxycaine, 118
propranolol, 65, 337
propyl dialkylamino side-chain, 177
propyl piperazine side-chain, 178
propylamine derivatives, 279–280, 285–286
propylthiouracil, 641
prostaglandin (PGE$_2$), 228
prostaglandins, 227, 585–591
  biosynthesis of, 586
  functions of, 586
  nomenclature, 587–590
  structures of therapeutically useful, 590–591
  synthesis of, 588–590
prostaglandin E$_1$, 588
prostaglandin E$_{2\alpha}$, 589
protease inhibitors, 518–522
proteases, 12
protein databank, 88
proton pump inhibitors, 299–301
pryimidine antagonist, 572–574
psychomotor stimulants, 270–271
psychotropics. *See* antipsychotic drugs
purine antagonist, 571–572
purine nucleosides, 524–525
pyrantel, 550
pyrazinamide, 490
pyrazole, 35
pyrazolidinedione derivatives, 233–235
pyridine, 34
pyridostigmine bromide, 422
pyridoxal phosphate (PLP), 667
pyrimidine nucleosides, 522–524
pyrrobutamine, 286
pyrrole acetic acid derivatives, 238–239
pyrrole nitrogen, 35

Q

quantitative structure-activity relationships (QSAR), 37, 76–82
quaternary amines, 423

quaternary ammonium compounds, 653
quazepam, 137
quetiapine, 186–187
quilostigmine, 268
quinacrine, 542–543
quinethazone, 321
quinine, 8
quinolines, 439–446 549–551
quinoline antimalarials, 538–542
quinolone antibacterials, 439–446

R

rabeprazole, 301
ralitoline, 168
raltegravir, 529
randic branching index, 81
random screening, 11
ranitidine, 296
ranolazine, 379
rate theory, 40
rational drug design, 3, 12–13
receptor theories, 40–42
   activation-aggravation theory, 41–42
   induced fit theory, 41
   macromolecular perturbation theory, 41
   occupancy theory, 40
   rate theory, 40
receptor-based drug design, 12–13
reduction, 54
remifentanil, 223
renin, 13
renin angiotensin system, 342–345
repaglinide, 388–389
resins, 368
resonance, 33
respiratory stimulants, 271
retinol, 660
reverse transcriptase (RT), 507
reverse transcriptase inhibitors, 507–508
ribavirin, 527
riboflavin, 665–666
rifampin, 490–492
rimonanbant, 632
risperidone, 187–188
ritonavir, 520–521

rivastigmine, 426
rosiglitazone, 390
rous sarcoma virus protease, 12
roxatidine, 298
roxithromycin, 480

S

saccharin, 657
salbutamol, 405
salicylic acid derivatives, 230–231
salmeterol, 411, 638–639
salsalate, 230–231
saquinavir, 518
structure-activity relationship (SAR), 188–119, 230
   anilides, 121–122
   antranilic acid derivatives, 235–236
   aryl alkanoic acid derivatives, 236
   barbiturates, 135–136
   benzodiazepines, 142–143
   butyrophenones, 182–183
   cephalosporins, 463–464
   cholinomimetics, 415–416
   diphenyl heplanones, 219
   glycerol derivatives, 262
   H_2 receptor antagonists, 295
   indole acetic acid derivatives, 236
   macrolide antibiotics, 480
   mehchlorethamine, 560
   morphine, 213
   oxicams, 244–245
   p-aminophenol derivatives, 232
   penicillins, 453–454
   phenothiazines, 179
   phenyl (ethyl) piperidines, 216
   pyrazolidinedione derivatives, 233–234
   quinolones, 441–442
   sulphonamides, 431
   tetracyclines, 472–473
   tmazide diuretics, 312
sarcoma, 557
sarcoma virus protease, 12
saturated analogues, 279–280
schizophrenia, 173
s-configuration, 28
secondary alcohols, 23

second-generation H_1-antihistamines, 290–291
sedation, 130
sedatives, 130–148
seizures, 153–154
selective serotonin reuptake inhibitor (SSRI), 204
selective serotonin reuptake inhibitors ssRIs, 13
selective α-antagonists, 409
selective $β_1$-antagonists, 409
selegiline, 274
semustine, 565
serine proteases, 12
serotonin/dopamine antagonist (SDA), 187
sertindole, 189–190
sertraline, 202–203
seven transmembrane domaine receptors. *See* G protein-coupled receptors (GPCR),
sevoflurane, 107
sibutramine, 631–632
sigma receptors (σ), 208
sildenafil, 9, 11, 633, 634–635
sliver sulphadiazine, 433
small molecule-based drug design, 76–86
sodium benzoate, 655
sodium channel blockers, 110
sodium channel inhibitors, 315–317
sodium stibogluconate, 553
sodium valproate, 163–164
specific agents, 37
spinal anaesthesia, 111
spironolactone, 317
stanazolol, 616
statins, 361–365
steroid hormone, 594–595
   biosynthesis of, 595–596
   mechanism of action of, 597–598
   nomenclature and stereochemistry of, 595–595
steroidal oestrogens, 598
stiripentol, 169
substituted alkane diols, 262–264
substituted benzoic acid derivatives (meglitinides), 384

substituted benzoic acid derivatives, 388–389
succinimides, 159–160
sucralose, 657
suldinac, 71–72
sulfinpyrazone, 235
sulindac, 238
sulphadiazine, 432
sulphadoxine, 434
sulphaguanidin, 433
sulphamethoxazole, 432
sulphasalazine, 69, 71, 231, 433
sulphation, 55
sulphcetamide, 433
suramin sodium, 553–554
surfactants, 653
sweetening agent, 656
sympathetic (adrenergic) blocking agents, 408–411
sympatholytics, 408–411
sympathomimetic drugs, 402–411
synthetic anti-cholinergics, 259–260

T

tacrine, 268–269
tadalafil, 635
taludipine, 339–340
tamoxifen, 579
tapentadol, 224
taxol derivatives, 577
tazomeline, 268
teicoplanin, 486
temozolomide, 582
tenofovir, 528
tertiary alcohols, 23
tertiary amines, 423
testoids, 612–618
testosterone, 612–614
tetracaine, 117
tetracycline antibiotics, 472–477
thiabendazole, 548
thiamin, 72
thiamine, 665
thiamyal sodium, 101
thiazide diuretics, 311–314
thiazolidinediones (glitazones), 385, 389–390
thiomersal, 655
thiopentone sodium, 99–100

thiopentone, 16
thiotepa, 566
thiothixene derivatives, 175
thiothixene, 180–181
thioxanthenes, 179–181
third-generation H_1-antihistamines, 291
thonzylamine, 284
three-dimensional database searching, 87
three-dimensional QSAR (3D QSAR), 85–86
thyroid hormones, 639
thyroxine, 639
tiagabin, 162
ticarcillin, 457
tolbutamide, 383
tolcapone, 274
tolmetin, 238–239
tolnaftate, 501
tolserine, 269
topical anaesthesia, 111
topiramate, 168
topless approach, 82
tretinoin, 661
triamcinolone acetonide, 624–625
triamterene, 316–317
triazolobenzodiazepines, 139–140
triclocarban, 652
triclosan, 652–653
tricyclic antidepressants, 199–200
trihexyphenidyl, 259
trimazosin, 331
trimeprazine, 286
tripelennamine, 15, 284
triprolidine, 285–286
tropicamide, 426
trovirdine, 516–517
tuberculosis, 489
type-I diabetes, 381
type-II diabetes, 381–382
typical antipsychotic agents, 174–185
tyropanoic acid, 650–651

U

ultrashort acting barbiturates, 99–101
undetermined seizures, 154

unsaturated analogues, 280
uracil mustard, 562
urea, 27

V

valaciclovir, 525–526
valdecoxib, 250
valsartan, 346–347
Van der waals forces, 49
vardenafil, 635
vardenfil, 633
venlafaxine, 204
verapamil, 341, 357
viagra, 9, 633
vidarabine, 524–525
vigabatrin, 161–162
vinca alkaloids, 577
vitamin A, 660
vitamin B_1, 665
vitamin B_{12}, 668
vitamin B_3, 666
vitamin B_4, 666
vitamin B_6, 667–667
vitamin B_7, 667
vitamin B_9, 667
vitamin C, 668
vitamin D, 662–663
vitamin E, 663–664
vitamin K, 664–665
vitamine B_2, 665–666
vitamins, 658–668

W

warfarin, 398
water-soluble vitamins, 659

X

ximelagatran, 399–400
xipamide, 321
xylometazoline, 407–408

Z

zalcitabine, 511
zaleplon, 146–147
zanamivin, 14
zidovudine (AZT), 11, 509–51
ziprasidone, 188–189
zolpidem, 147–148
zomepirac, 239
zonisamide, 169
zopiclone, 145–146

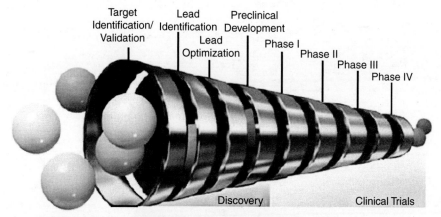

FIGURE 1.2 Drug discovery process.

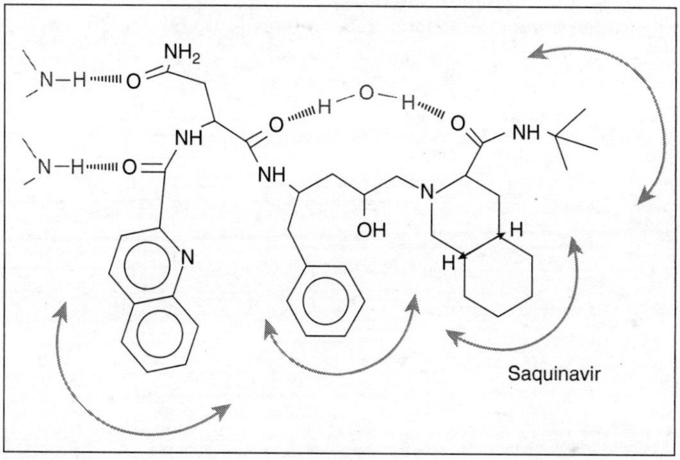

FIGURE 1.6a Saquinavir interactions with HIV protease active site.
Blue = important hydrogen bond interactions, Red = interactions with specificity sites of the protease.

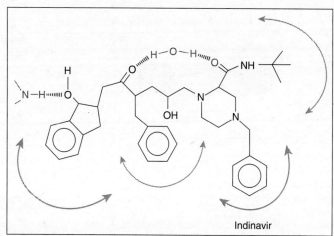

FIGURE 1.6b Indinavir interactions with HIV protease active site.
Blue = important hydrogen bond interactions, Red = interactions with specificity sites of the protease.

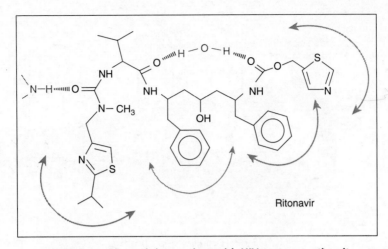

FIGURE 1.6c Ritonavir interactions with HIV protease active site. Blue = important hydrogen bond interactions, Red = interactions with specificity sites of the protease.

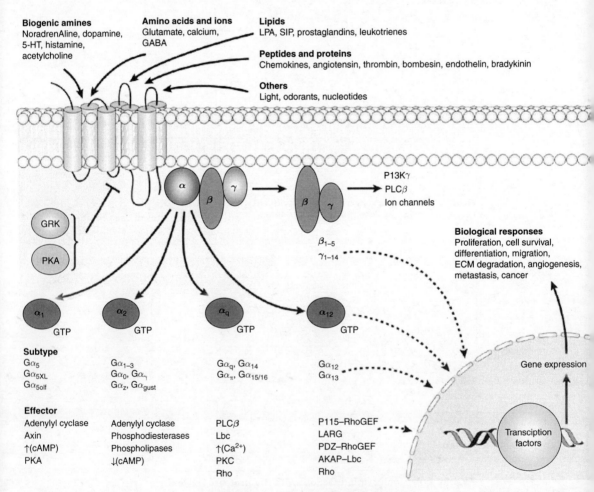

FIGURE 2.6 G-Protein coupled receptors.

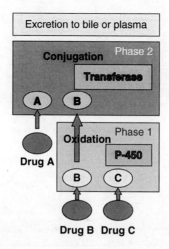

FIGURE 3.1 Stages of drug metabolism.

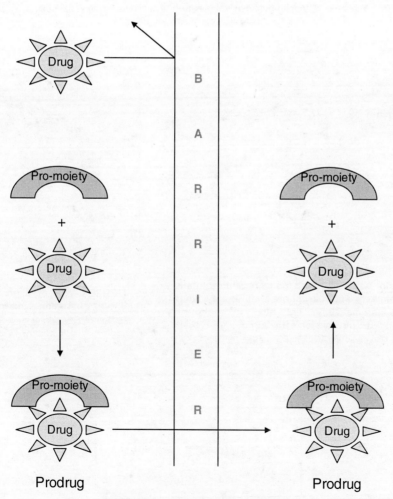

FIGURE 3.5 Targeting with prodrugs.

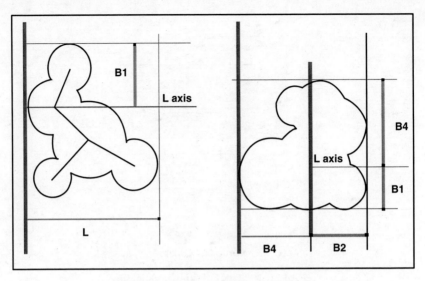

FIGURE 4.1 Verloop's sterimol parameters.
Source: Modified from Verloop et al. 'Application of new steric substituent parameters in drug design' in *Drug Design*, Vol. VII, Academic Press, New York, 1976.

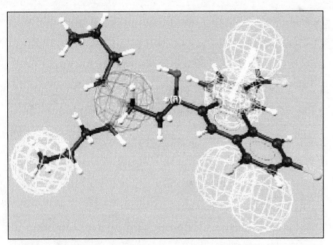

FIGURE 4.4(b) 4-Point pharmacophore model for antiarrythmic potassium channel blockers. Pharmacophore features are colour-coded with: orange = aromatic rings; light-blue = hydrophobic groups; red = positive ionizable atom
Source: You et al., *Bioorganic and Medicinal Chemistry*, 14, 2004.

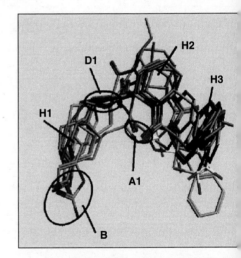

FIGURE 4.4(d) 3D alignment of thrombin inhibitors based on a pharmacophore model
Source: Schneider et al., *Journal of Medicinal Chemistry*, 47, 2004.

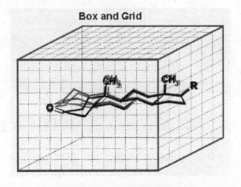

FIGURE 4.5 3D Grid structure.

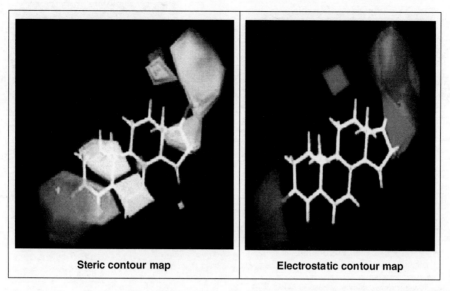

FIGURE 4.6 Steric bulkiness is favourable for activity in the green regions and unfavourable in the yellow regions. High-electron density (negative charge) is expected at red regions and low-electron density (positive charge) at blue regions favour activity.

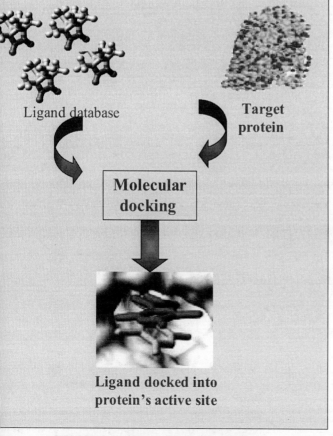

FIGURE 4.7 Crystal structure of MurF enzyme in complex with an inhibitor.
Source: Stamper et al., *Chemical Biology & Drug Design*, 67, 2006.

FIGURE 4.8 Structure-based drug design.

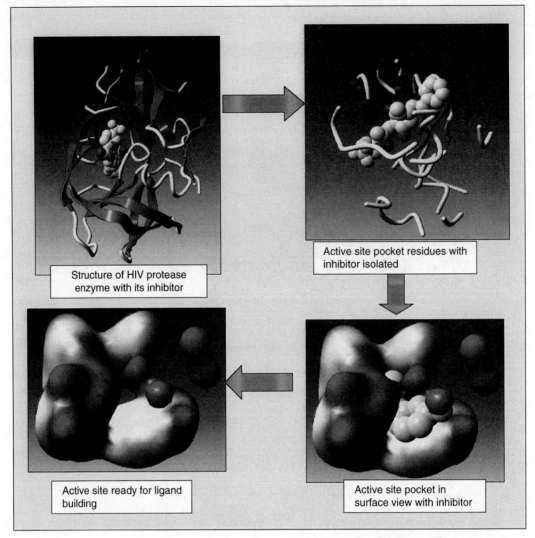

FIGURE 4.9 HIV-Protease inhibitor development.

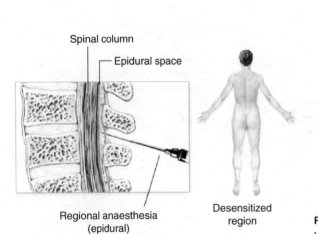

FIGURE 6.1 Epidural anaesthesia.

FIGURE 6.2 For over a thousand years South America indigenous peoples have chewed the coca leaf (*Erythro coca*), a plant that contains vital nutrients as well as nu ous alkaloids, including cocaine. The cocaine alkaloid w first isolated by the German chemist Friedrich.

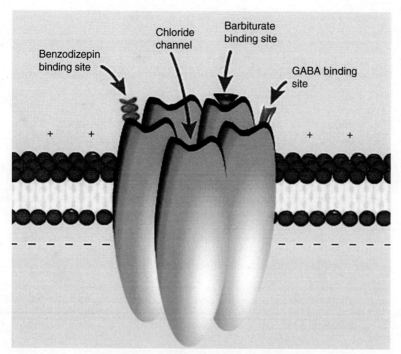

FIGURE 7.1 Barbiturate action on the CNS.

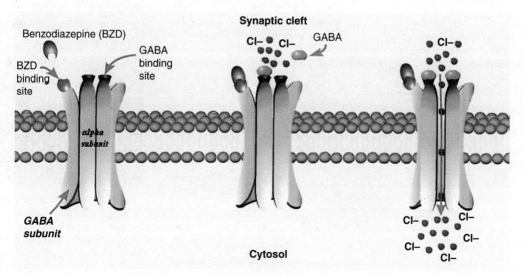

FIGURE 7.2 Mechanism of action by Benzodiazepines.

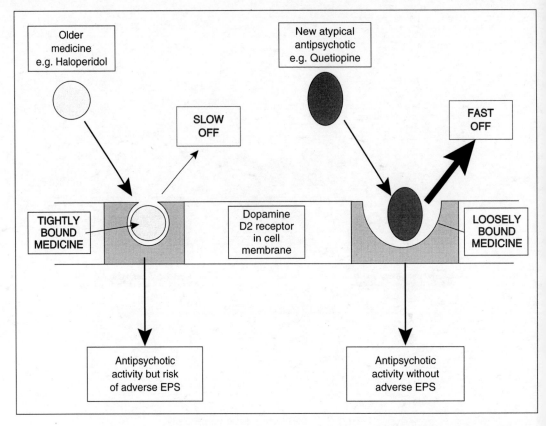

FIGURE 9.1 Action of antipsychotic drugs.

TYPES OF CEREBRAL PALSY

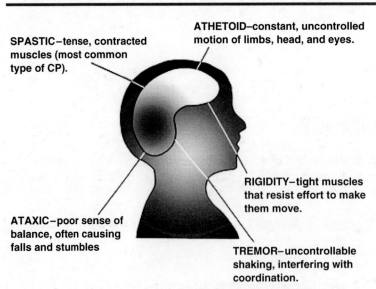

SPASTIC—tense, contracted muscles (most common type of CP).

ATHETOID—constant, uncontrolled motion of limbs, head, and eyes.

RIGIDITY—tight muscles that resist effort to make them move.

ATAXIC—poor sense of balance, often causing falls and stumbles

TREMOR—uncontrollable shaking, interfering with coordination.

FIGURE 9.2 Parts of brain which control movement.

FIGURE 11.3 Opium is a narcotic formed from the latex (i.e., sap) released by lacerating (or "scoring") the immature seed pods of opium poppies (*Papaver somniferum*). It contains up to 12% morphine, an opiate alkaloid.

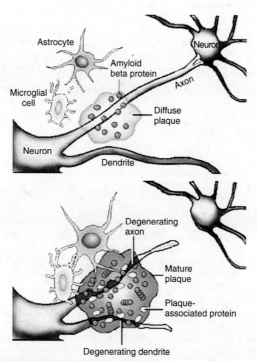

FIGURE 13.1 Alzheimer's disease has been identified as a protein misfolding disease (proteopathy), caused by accumulation of abnormally folded A-beta and tau proteins in the brain. Plaques are made up of small peptides, 39–43 amino acids in length, called β-amyloid.

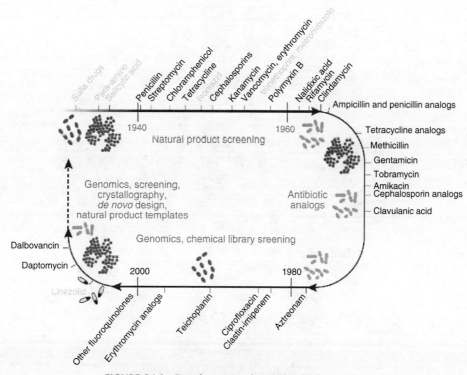

FIGURE 26.1 Development of antimicrobials.

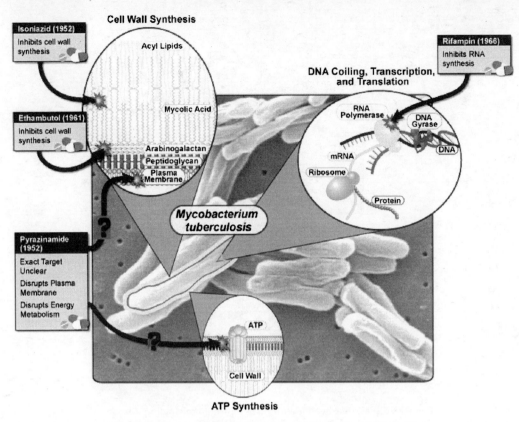

FIGURE 27.1 Mechanism of action of first-line anti-TB drugs.

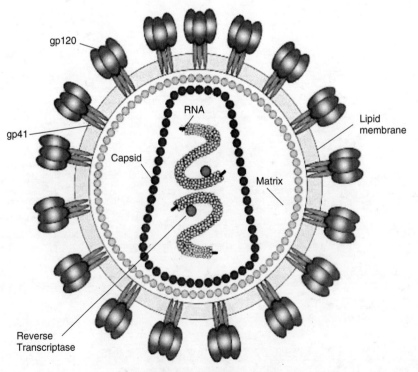

FIGURE 29.1 HIV virus structure.

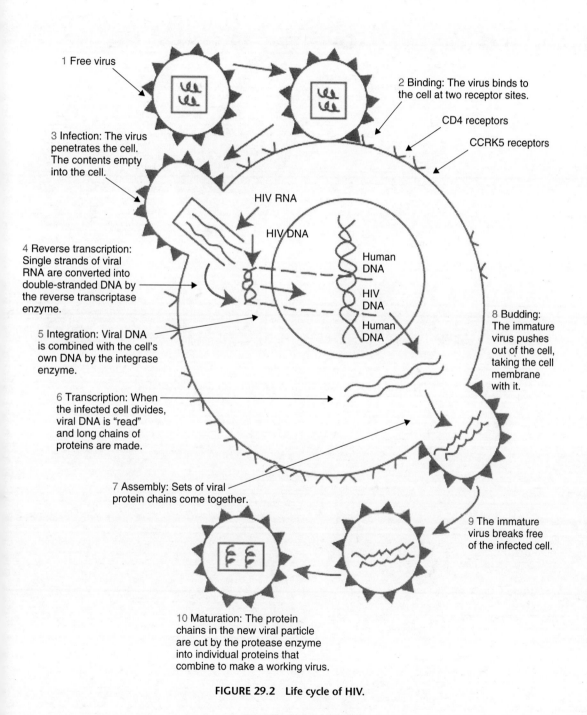

FIGURE 29.2 Life cycle of HIV.

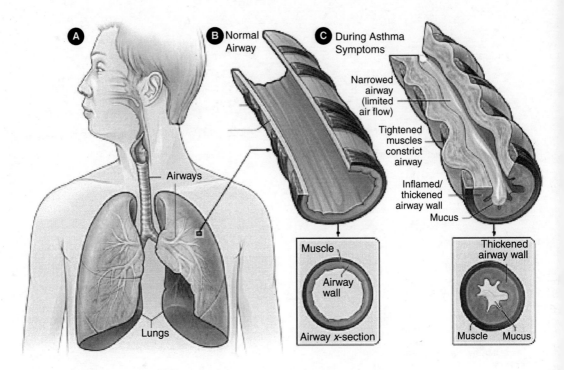

FIGURE 34.2 Airways of normal and asthmatic conditions.